APPLICATIONS

Introduction to the Practice of Statistics presents a wide variety of applications from diverse disciplines. The following list indicates the number of Examples and Exercises related to different fields. Note that some items appear in more than one category.

Examples by Application

Agriculture:
3.18, 14.8, 14.9, 15.1, 15.2, 15.3, 15.4, 15.5, 15.13, 15.14

Business and consumers:
1.1, 1.2, 1.6, 1.7, 1.8, 1.9, 1.14, 1.15, 1.21, 1.22, 1.23, 1.24, 1.25, 1.26, 1.27, 1.28, 1.42, 1.43, 1.45, 2.2, 2.7, 2.30, 2.40, 2.41, 3.6, 3.8, 3.9, 3.10, 3.11, 3.12, 3.13, 3.15, 3.16, 3.21, 3.22, 3.23, 3.24, 3.28, 4.32, 4.33, 4.37, 5.2, 5.3, 5.4, 5.5, 5.6, 5.14, 5.15, 5.16, 5.19, 5.20, 5.21, 5.22, 6.2, 6.4, 6.5, 6.6, 6.7, 6.8, 6.9, 6.10, 6.11, 6.12, 6.13, 6.14, 6.20, 8.4, 10.10, 10.11, 10.12, 11.2, 12.1, 12.2, 12.25, 13.1, 13.2, 14.6, 14.7, 14.10, 14.11, 15.6, 15.7, 16.1, 16.2, 16.3, 16.4, 16.5, 16.6, 16.8, 16.9, 16.10, 16.13, 16.54, 16.55

Demographics and characteristics of people:
1.17, 1.32, 1.34, 1.35, 3.2, 4.30, 4.44, 5.1, 5.17, 5.23, 5.24, 6.4, 6.5, 6.6, 6.7, 6.8, 6.9, 6.10, 6.11, 6.12, 6.13, 6.14, 6.20, 7.1, 7.2, 7.3, 7.13, 13.7

Economics and finance:
2.12, 2.13, 2.14, 2.15, 2.16, 3.1, 6.31, 7.4, 7.5, 7.6

Education and child development:
1.3, 1.4, 1.5, 1.13, 1.16, 1.17, 1.18, 1.31, 1.32, 1.33, 1.36, 1.37, 1.39, 1.40, 1.41, 2.11, 2.31, 3.2, 3.7, 3.19, 3.20, 3.36, 4.18, 4.22, 4.44, 6.3, 6.16, 6.19, 7.14, 7.15, 7.18, 11.1, 12.26, 12.27, 15.8, 15.9, 15.10, 16.11, 16.12, 17.19

Environment:
1.32, 1.44, 6.1, 6.17, 6.18, 6.30, 14.8, 14.9

Ethics:
3.32, 3.33, 3.34, 3.35, 3.36, 3.37, 3.38, 3.39, 6.24, 6.25, 6.26

Health and nutrition:
1.10, 1.11, 1.12, 1.19, 1.20, 1.29, 2.1, 2.3, 2.4, 2.5, 2.6, 2.18, 2.19, 2.20, 2.21, 2.22, 2.23, 2.25, 2.27, 2.28, 2.29, 2.32, 2.33, 2.34, 2.35, 2.36, 2.37, 2.38, 2.39, 2.43, 2.44, 3.14, 3.17, 3.32, 3.35, 3.37, 4.20, 4.21, 4.38, 4.40, 5.1, 5.17, 5.18, 6.2, 6.15, 6.24, 6.29, 7.16, 7.17, 7.19, 7.20, 7.21, 7.22, 7.23, 8.3, 8.4, 8.9, 9.7, 9.8, 9.9, 9.10, 9.11, 9.14, 9.15, 10.1, 10.2, 10.2, 10.4, 10.5, 10.6, 10.7, 10.8, 10.9, 10.13, 10.14, 10.15, 10.16, 10.17, 10.18, 10.19, 10.20, 10.21, 10.22, 10.23, 10.24, 10.25, 13.3, 13.4, 13.5, 13.6, 13.7, 13.8, 13.10

Humanities and social sciences:
1.13, 1.16, 2.1, 3.3, 3.4, 3.5, 3.15, 3.25, 3.26, 3.27, 3.33, 3.38, 3.39, 4.9, 4.10, 4.11, 4.26, 4.45, 5.10, 5.11, 5.12, 5.13, 6.25, 6.27, 7.7, 7.8, 7.9, 7.12, 8.8, 8.10, 8.11, 9.1, 9.2, 9.3, 9.4, 9.5, 9.6, 9.12, 9.13, 11.2, 12.3, 12.4, 12.5, 12.6, 12.7, 12.8, 12.9, 12.10, 12.11, 12.12, 12.13, 12.14, 12.15, 12.16, 12.17, 12.18, 12.19, 12.20, 12.21, 12.22, 12.23, 12.24, 13.8, 13.9, 15.6, 15.7, 16.14, 17.20, 17.21

International:
1.21, 1.22, 1.23, 1.24, 1.25, 1.143, 2.11, 2.12, 2.13, 2.14, 2.15, 2.16, 3.1, 11.2

Law and government data:
1.6, 3.1, 3.2, 3.3, 3.24, 3.25, 3.34

Manufacturing, products, and processes:
3.21, 5.13, 5.25, 6.17, 6.18, 6.30, 6.32, 10.10, 10.11, 10.12, 17.1, 17.2, 17.3, 17.4, 17.5, 17.6, 17.7, 17.8, 17.9, 17.11, 17.12, 17.13, 17.14, 17.15, 17.17, 17.18, 17.19, 17.20, 17.21

Science:
1.30, 3.31, 4.19, 6.28, 15.24

Sports and leisure:
1.1, 1.2, 1.36, 1.37, 1.39, 1.40, 1.45, 2.17, 4.17, 4.27, 4.31, 4.34, 4.35, 4.36, 4.41, 4.43, 4.45, 7.1, 7.2, 7.3, 7.10, 7.11, 8.1, 8.2, 8.6, 8.7, 14.1, 14.2, 14.3, 14.4, 14.5, 15.11, 15.12, 16.7, 16.10

Students:
1.3, 1.4, 1.5, 1.17, 1.18, 1.31, 1.36, 1.37, 1.39, 1.40, 2.3, 2.5, 2.7, 2.32, 2.33, 2.34, 2.35, 2.36, 2.37, 2.38, 2.39, 3.9, 3.22, 3.23, 4.44, 5.1, 5.9, 5.17, 6.3, 6.4, 6.5, 6.6, 6.7, 6.8, 6.9, 6.10, 6.11, 6.12, 6.13, 6.14, 6.15, 6.16, 6.19, 6.20, 8.5, 8.6, 8.7, 9.8, 9.9, 9.10, 9.11, 10.1, 10.2, 10.3, 10.4, 10.5, 10.6, 10.7, 10.8, 10.9, 10.14, 10.24, 10.25, 11.1, 12.3, 12.4, 12.5, 12.6, 12.7, 12.8, 12.9, 12,10, 12.11, 12.12, 12.13, 12.14, 12.15, 12.16, 12.17, 12.18, 12.19, 12.20, 12.21, 12.22, 12.23, 12.24

Technology and the Internet:
1.1, 1.2, 2.1, 2.3, 2.8, 2.9, 2.10, 2.24, 2.26, 3.6, 3.10, 3.11, 3.12, 3.16, 4.42, 4.46, 4.47, 5.7, 5.8, 7.1, 7.2, 7.3, 7.10, 7.11, 12.3, 12.4, 12.5, 12.6, 12.7, 12.8, 12.9, 12.10, 12.11, 12.12, 12.13, 12.14, 12.15, 12.16, 12.17, 12.18, 12.19, 12.20, 12.21, 12.22, 12.23, 12.24

Exercises by Application

Agriculture:
1.67, 1.123, 1.168, 2.163, 3.27, 6.122, 12.8, 12.13, 12.15, 12.17, 13.39, 13.40, 13.41, 13.42, 13.43, 13.44, 13.45, 13.46, 14.19, 14.20, 15.10, 15.30, 15.36, 15.37, 17.71, 17.72

Business and consumers:
1.1, 1.5, 1.6, 1.14, 1.15, 1.19, 1.20, 1.21, 1.23, 1.26, 1.31, 1.47, 1.49, 1.50, 1.53, 1.159, 1.61, 1.62, 1.63, 1.64, 1.75, 1.77, 1.99, 1.159, 1.169, 2.3, 2.9, 2.15, 2.16, 2.17, 2.18, 2.19, 2.28, 2.34, 2.43, 2.44, 2.51, 2.71, 2.72, 2.88, 2.91, 2.100, 2.104, 2.113, 2.130, 2.145, 2.148, 2.149, 2.150, 2.151,

2.152, 2.153, 2.154, 2.158, 2.176, 3.3, 3.4, 3.5, 3.9, 3.11, 3.15, 3.16, 3.17, 3.21, 3.25, 3.32, 3.34, 3.36, 3.40, 3.41, 3.42, 3.55, 3.56, 3.59, 3.72, 3.85, 3.119, 3.123, 3.126, 4.34, 4.79. 4.85, 4.91, 4.92, 4.93, 4.94, 4.115, 4.116, 4.117, 4.141, 4.149, 5.25, 5.26, 5.28, 5.64, 5.69, 5.77, 5.82, 6.7, 6.8, 6.9, 6.20, 6.22, 6.23, 6.27, 6.36, 6.53, 6.54, 6.55, 6.72, 6.84, 6.85, 6.98, 6.104, 6.113, 6.114, 6.116, 6.117, 6.123, 6.124, 6.125, 6.132, 7.1, 7.3, 7.6, 7.8, 7.9, 7.14, 7.34, 7.38, 7.42, 7.45, 7.54, 7.55, 7.65, 7.66, 7.67, 7.73, 7.74, 7.77, 7.78, 7.79, 7.80, 7.83, 7.86, 7.87, 7.88, 7.93, 7.103, 7.104, 7.110, 7.118, 7.119, 7.122, 7.123, 7.129, 7.134, 7.135, 7.136, 7.141, 7.142, 8.1, 8.3, 8.5, 8.6, 8.7, 8.8, 8.9, 8.11, 8.22, 8.23, 8.24, 8.32, 8.40, 8.41, 8.44, 8.48, 8.49, 8.50, 8.51, 8.67, 8.72, 8.73, 8.74, 8.75, 8.76, 8.77, 8.78, 8.79, 8.86, 8.88, 8.89, 8.101, 9.10, 9.22, 9.25, 9.36, 9.37, 9.46, 10.14, 10.15, 10.16, 10.17, 10.18, 10.35, 10.36, 10.50, 11.13, 11.18, 11.19, 11.20, 11.21, 11.22, 11.23, 11.24, 11.25, 11.26, 12.14, 12.16, 12.18, 12.19, 12.23, 12.24, 12.28, 12.33, 12.34, 12.36, 12.40, 12.41, 12.45, 12.69, 13.7, 13.8, 13.9, 13.10, 13.13, 13.14, 13.15, 13.25, 13.31, 14.3, 14.4, 14.5, 14.6, 14.7, 14.8, 14.11, 14.12, 14.14, 14.16, 14.17, 14.18, 14.22, 14.23, 14.24, 14.25, 14.26, 14.28, 14.29, 14.30, 14.37, 14.40, 15.1, 15.2, 15.3, 15.4, 15.5, 15.6, 15.7, 15.17, 15.18, 15.19, 15.20, 15.21, 15.22, 15.23, 15.41, 15.42, 16.1, 16.10, 16.11, 16.12, 16.17, 16.18, 16.19, 16.22, 16.26, 16.28, 16.31, 16.35, 16.37, 16.39, 16.40, 16.41, 16.42, 16.43, 16.47, 16.58, 16.67, 16.70, 16.71, 16.73, 16.76, 16.79, 16.86, 16.87, 17.5, 17.6, 17.14, 17.26, 17.47, 17.65, 17.66, 17.67, 17.68, 17.69, 17.79, 17.80

Demographics and characteristics of people:

1.21, 1.30, 1.73, 1.170, 2.15, 2.16, 2.17, 2.84, 2.106, 2.113, 2.125, 2.126, 2.127, 2.136, 2.146, 2.147, 4.37, 4.52, 4.60, 4.88, 4.99, 4.100, 4.109, 4.110, 4.111, 4.112, 4.113, 4.116, 4.122, 4.123, 4.124, 4.125, 4.126, 4.127, 4.145, 4.146, 5.34, 5.53, 5.65, 5.74,

6.21, 7.24, 7.140, 8.35, 8.36, 9.27, 9.28, 11.31, 11.32, 11.33, 16.25, 16.62

Economics and finance:

1.160, 1.161, 2.10, 2.11, 2.12, 2.40, 2.41, 2.58, 2.88, 2.89, 2.90, 3.78, 4.39, 5.30, 6.64, 10.5, 10.12, 10.33, 10.43, 10.45, 10.46, 10.47, 14.28, 14.29, 14.30, 16.45, 16.49, 16.61

Education and child development:

1.2, 1.3, 1.4, 1.7, 1.10, 1.11, 1.12, 1.13, 1.30, 1.35, 1.41, 1.43, 1.44, 1.45, 1.48, 1.51, 1.52, 1.53, 1.55, 1.56, 157, 1.60, 1.80, 1.82, 1.100, 1.101, 1.102, 1.103, 1.104, 1.105, 1.106, 1.107, 1.108, 1.114, 1.115, 1.131, 1.132, 1.133, 1.134, 1.135, 1.136, 1.137, 1.138, 1,139. 1.140. 1.141, 1.165, 1.176, 2.1, 2.24, 2.25, 2.26, 2.29, 2.30, 2.47, 2.48, 2.49, 2.50, 2.59, 2.61, 2.78, 2.87, 2.105, 2.125, 2.126, 2.127, 2.131, 2.135, 2.138, 2.139, 2.142, 2.155, 2.168, 2.170, 2.173, 2.175, 3.12, 3.13, 3.14, 3.19, 3.48. 3.93, 3.94, 3.105, 3.132, 4.2, 4.11, 4.29, 4.46, 4.75, 4.78, 4.99, 4.100, 4.109, 4.110, 4.111, 4.122, 4.123, 4.142, 4.145, 4.146, 5.18, 5.20, 5.58, 5.80, 6.54, 6.64, 6.69, 6.84, 6.111, 7.30, 7.31, 7.41, 7.43, 7.48, 7.111, 7.131, 7.132, 7.139, 8.87, 8.92, 8.94, 9.17, 9.18, 9.19, 9.20, 9.21, 9.27, 9.28, 9.45, 9.49, 10.10, 10.11, 10.28, 10.30, 10.34, 10.51, 10.52, 10.53, 11.1, 11.3, 11.5, 11.6, 11.27, 11.28, 11.29, 11.30, 12.13, 12.14, 12.15, 12.16, 12.17, 12.18, 12.30, 12.42, 12.43, 12.44, 12.61, 12.62, 12.64, 12.66, 13.7, 13.8, 13.26, 13.28, 13.29, 13.30, 13.48, 13.49, 13.50, 13.51, 14.43, 14.44, 14.45, 14.46, 15.11, 15.12, 15.13, 15.14, 15.16, 15.28, 15.40, 16.6, 16.14, 16.23, 16.51, 16.53, 16.56, 16.59, 16.80, 16.81, 16.82, 17.75

Environment:

1.24, 1.25, 1.36, 1.38, 1.68, 1.74, 1.125, 1.150, 1.156, 1.164, 2.31, 2.66, 2.67, 2.68, 2.69, 2.70, 2.110, 2.140, 2.161, 2.166, 3.33, 3.44, 3.49, 3.60, 3.69, 5.19, 5.71, 5.73, 6.33, 6.66, 6.67, 6.108, 6.109, 7.10, 7.27, 7.44, 7.73, 7.74, 7.81, 7.82, 7.89, 7.91, 7.92, 7.93, 7.103, 7.104, 7.105, 7.106, 7.108, 7.109, 7.124, 7.125, 7.126, 7.127,

7.137, 8.64, 8.85, 8.98, 9.41, 9.55, 10.19, 10.22, 10.23, 10.24, 10.25, 10.26, 10.27, 10.54, 10.55, 10.56, 10.57, 11.16, 11.40, 11.41, 11.42, 11.43, 11.44, 11.45, 11.46, 11.47, 11.48, 11.49, 11.50, 12.8, 12.14, 12.16, 12.18, 12.35, 12.37, 13.38, 13.39, 13.40, 13.41, 13.42, 13.43, 13.44, 13.45, 13.46, 13.47, 14.19, 14.20, 15.15, 15.24, 15.29, 15.38, 16.5, 16.20, 16.24, 16.60, 16.63, 16.74

Ethics:

3.96, 3.97, 3.98, 3.99, 3.100, 3.101, 3.102, 3.103, 3.104, 3.105, 3.106, 3.107, 3.108, 3.109, 3.110, 3.111, 3.112, 3.113, 3.114, 3.115, 3.116, 3.117, 3.119, 3.120, 3.132, 3.134, 3.135, 3.136, 3.137, 3.138, 4.105, 4.106, 6.86, 6.88, 6.90, 7.34, 7.45, 8.28, 8.68, 8.84, 8.98, 8.99, 9.15, 9.16, 9.24, 9.25, 9.36, 9.37, 9.47, 16.86, 16.87

Health and nutrition:

1.8, 1.9, 1.33, 1.34, 1.37, 1.69, 1.70, 1.71, 1.78, 1.97, 1.120, 1.142, 1.143, 1.144, 1.145, 1.157, 1.158, 2.2, 2.35, 2.62, 2.63, 2.64, 2.82, 2.98, 2.99, 2.102, 2.103, 2.107, 2.115, 2.116, 2.117, 2.118, 2.119, 2.120, 2.121, 2.122, 2.123, 2.124, 2.132, 2.133, 2.134, 2.137, 2.141, 2.143, 2.144, 2.156, 2.157, 2.165, 2.178, 3.7, 3.10, 3.20, 3.22, 3.23, 3.24, 3.28, 3.30, 3.31, 3.35, 3.39, 3.45, 3.47, 3.88, 3.100, 3.101, 3.103, 3.104, 3.110, 3.112, 3.114, 3.115, 3.124, 3.130, 3.131, 3.136, 4.7, 4.25, 4.26, 4.31, 4.35, 4.42, 4.43, 4.44, 4.45, 4.56, 4.73, 4.81, 4.88, 4.103, 4.104, 4.107, 4.108, 4.129, 4.130, 4.131, 4.148, 5.10, 5.11, 5.17, 5.21, 5.23, 5.50, 6.17, 6.18, 6.24, 6.31, 6.37, 6.60, 6.65, 6.90, 6.110, 6.116, 6.121, 7.28, 7.32, 7.33, 7.39, 7.40, 7.47, 7.51, 7.53, 7.59, 7.60, 7.68, 7.69, 7.70, 7.71, 7.72, 7.75, 7.76, 7.85, 7.90, 7.94, 7.96, 7.97, 7.99, 7.101, 7.102, 7.107, 7.119, 7.120, 7.121, 7.128, 7.130, 7.134, 7.135, 7.136, 8.5, 8.6, 8.7, 8.8, 8.9, 8.11, 8.15, 8.33, 8.34, 8.37, 8.38, 8.42, 8.43, 8.70, 9.3, 9.4, 9.7, 9.8, 9.9, 9.23, 9.29, 9.35, 9.48, 10.2, 10.3, 10.4, 10.13, 10.21, 10.40, 10.41, 10.48, 10.58, 10.60, 10.61, 11.17, 11.34, 11.35, 11.36, 11.37,

11.38, 11.39, 12.25, 12.26, 12.39, 12.47, 12.48, 12.49, 12.50, 12.68, 13.11, 13.12, 13.16, 13.17, 13.18, 13.19, 13.20, 13.27, 13.32, 13.33, 13.34, 14.13, 14.15, 14.31, 14.33, 14.35, 14.39, 14.41, 14.42, 14.47, 15.24, 15.26, 15.30, 15.31, 15.32, 15.33, 15.34, 15.35, 15.39, 15.44, 15.45, 15.46, 15.48, 16.64, 16.65, 16.66, 16.72, 17.10, 17.32, 17.34, 17.35, 17.50, 17.60, 17.74

Humanities and social sciences:

1.10, 1.32, 1.37, 1.44, 1.45, 1.72, 1.82, 1.89, 1.112, 1.113, 1.130, 2.20, 2.21, 2.22, 2.53, 2.79, 2.83, 2.92, 2.93, 2.97, 2.137, 2.164, 2.171, 2.172, 3.8, 3.20, 3.20, 3.34, 3.57, 3.71, 3.74, 3.75, 3.76, 3.79. 3.87, 3.96, 3.97, 3.98, 3.99, 3.113, 3.116, 3.117, 3.137, 3.138, 4.10, 4.12, 4.13, 4.28, 4.38, 4.65, 4.66, 4.76, 4.82, 4.105, 4.106, 4.114, 4.128, 4.152, 5.16, 5.28, 5.50, 5.54, 5.56, 5.59, 5.60, 5.61, 5.63, 6.53, 6.61, 6.62, 6.68, 6.69, 6.132, 7.12, 7.13, 7.25, 7.29, 7.30, 7.31, 7.41, 7.48, 7.65, 7.66, 7.80, 7.86, 7.87, 7.110, 7.131, 7.132, 7.133, 8.16, 8.26, 8.27, 8.28, 8.29, 8.35, 8.63, 8.68, 8.69, 8.71, 8.82, 8.83, 8.90, 8.91, 9.6, 9.24, 9.25, 9.26, 9.33, 9.34, 9.36, 9.37, 9.43, 9.44, 10.44, 10.48, 10.58, 10.59, 11.13, 11.15, 11.30, 11.31, 11.32, 11.33, 12.22, 12.31, 12.32, 12.36, 12.38, 12.46, 12.63, 12.67, 13.13, 13.14, 13.15, 13.21, 13.22, 13.25, 13.31, 13.37, 14.32, 14.34, 14.36, 15.7, 15.8, 15.9, 15.14, 15.32, 16.4, 16.32, 16.50, 16.70, 16.71, 16.77, 16.78

International:

1.28, 1.29, 1.36, 1.37, 1.47, 1.49, 1.69, 1.70, 1.71, 1.74, 1.113, 1.150, 1.160, 1.161, 1.162, 1.163, 2.10, 2.11, 2.12, 2.20, 2.21, 2.22, 2.33, 2.40, 2.45, 2.46, 2.88, 2.89, 2.90, 2.165, 2.170, 3.32, 3.48, 3.78, 3.86. 3.88, 3.110, 3.115, 4.26, 4.28, 4.140, 5.46, 5.48, 5.70, 6.70, 6.93, 7.28, 8.15, 8.16, 8.64, 9.19, 9.20, 9.21, 10.12, 10.37, 10.38, 10.39, 10.45, 10.46, 10.47, 11.31, 11.32, 11.33, 12.22, 12.31, 12.32, 12.67, 13.16, 13.17, 13.18, 13.19, 13.20, 14.23, 14.38, 15.27, 16.5, 16.45, 16.49, 16.61, 17.22

Law and government data:

1.6, 1.76, 3.63, 3.65, 3.66, 3.76, 3.111, 3.120, 4.124, 4.125, 4.126, 4.127, 6.42, 7.111, 8.98, 8.99, 9.15, 9.16, 9.47, 10.5

Manufacturing, products, and processes:

2.130, 2.163, 2.166, 4.85, 5.13, 5.15, 5.24, 5.27, 5.31, 5.71, 5.77, 5.78, 5.79, 6.28, 6.29, 6.34, 6.37, 6.71, 6.73, 7.26, 7.35, 7.36, 7.37, 7.39, 7.40, 7.47, 7.48, 7.49, 7.50, 7.51, 8.86, 9.10, 9.46, 10.20, 10.31, 10.32, 10.62, 11.51, 11.52, 11.53, 11.54, 11.55, 11.56, 11.57, 11.58, 11.59, 12.54, 12.55, 12.56, 12.57, 13.7, 13.8, 13.35, 13.36, 17.9, 17.10, 17.12, 17.13, 17.15, 17.16, 17.17, 17.18, 17.19, 17.20, 17.26, 17.27, 17.28, 17.31, 17.32, 17.33, 17.34, 17.35, 17.36, 17.37, 17.38, 17.39, 17.40, 17.41, 17.42, 17.43, 17.44, 17.45, 17.46, 17.47, 17.49, 17.53, 17.54, 17.55, 17.56, 17.58, 17.59, 17.60, 17.61, 17.62, 17.65, 17.66, 17.67, 17.68, 17.69, 17.70, 17.71, 17.72, 17.73, 17.74, 17.75, 17.76, 17.77, 17.79, 17.80, 17.83, 17.84, 17.85, 17.86, 17.87

Science:

1.42, 1.79, 1.81, 1.86, 1.87, 1.95, 1.96, 1.152, 1.153, 1.167, 2.31, 2.32, 2.36, 2.52, 2.80, 2.81, 2.86, 2.101, 2.108, 2.162, 2.167, 5.76, 9.41, 10.29, 10.37, 10.38, 10.39, 11.15, 12.29, 12.51, 12.52, 12.53, 12.58, 12.59, 13.38, 15.24, 15.29, 15.43, 15.47

Sports and leisure:

1.1, 1.22, 1.23, 1.46, 1.66, 1.72, 1.166, 1.174, 1,175, 2.23, 2.34, 2.37, 2.51, 2.92, 2.109, 2.159, 2.160, 2.177, 3.1, 3.26, 3.43, 3.47, 3.127, 4.3, 4.4, 4.5, 4.6, 4.9, 4.16, 4.18, 4.23, 4.24, 4.30, 4.32, 4.33, 4.54, 4.57, 4.59, 4.67, 4.74, 4.77, 4.84, 4.95, 4.96, 4.97, 4.98, 4.101, 4.136, 4.137, 4.139, 4.140, 4.143, 4.144, 4.147, 5.1, 5.9, 5.12, 5.14, 5.22, 5.36, 5.46, 5.48, 5.50, 5.55, 5.57, 6.14, 6.15, 6.16, 6.70, 6.93, 6.123, 6.128, 7.11, 7.24, 7.28, 7.122, 7.123, 7.128, 8.2, 8.4, 8.17, 8.18, 8.30, 8.55, 8.56, 8.75, 8.58, 8.59, 8.65, 8.66, 8.72, 8.73, 8.86, 8.88, 8.89, 8.100,

9.39, 10.15, 10.16, 10.21, 12.27, 15.1, 15.2, 15.3, 15.4, 15.5, 15.6, 15.18, 15.19, 15.20, 15.21, 15.22, 15.23, 16.6, 16.8, 16.13, 16.16, 16.37, 16.38, 16.47, 16.62, 16.76, 16.83, 16.84, 16.85, 17.46, 17.49, 17.61, 17.62, 17.70

Students:

1.2, 1.3, 1.4, 1.5, 1.7, 1.11, 1.12, 1.13, 1.16, 1.17, 1.18, 1.39, 1.41, 1.48, 1.51, 1.52, 1.53, 1.55, 1.56, 1.57, 1.60, 1.80, 1.100, 1.114, 1.115, 2.1, 2.24, 2.25, 2.26, 2.28, 2.47, 2.48, 2.49, 2.50, 2.59, 2.115, 2.116,. 2.117, 2.118, 2.119, 2.120, 2.125, 2.126, 2.127, 2.131, 2.134, 2.135, 2.168, 2.170, 2.177, 3.50, 3.51, 3.55, 3.59, 3.70, 3.79, 3.129, 4.99, 4.100, 4.109, 4.110, 4.111, 4.118, 4.119, 4.120, 4.121, 4.122, 4.123, 5.10, 5.11, 5.18, 5.20, 5.32, 5.62, 5.81, 6.1, 6.2, 6.3, 6.5, 6.6, 6.9, 6.13, 6.15, 6.16, 6.36, 6.52, 6.55, 6.61, 6.63, 6.69, 6.84, 6.91, 6.110, 6.113, 6.115, 7.7, 7.24, 7.43, 7.69, 7.70, 7.111, 7.117, 7.128, 7.138, 8.10, 8.20, 8.21, 8.22, 8.23, 8.24, 8.25, 8.28, 8.33, 8.42, 8.43, 8.77, 8.78, 8.79, 8.80, 8.81, 8.92, 8.101, 9.3, 9.4, 9.17, 9.18, 9.19, 9.20, 9.21, 9.35, 9.39, 9.45, 9.49, 10.2, 10.3, 10.4, 10.10, 10.11, 10.34, 10.40, 10.41, 10.51, 10.52, 10.53, 11.1, 11.3, 12.13, 12.15, 12.17, 12.22, 12.26, 12.27, 12.45, 12.46, 12.63, 14.41, 14.42, 16.80, 16.81, 16.82, 17.4, 17.7, 17.8, 17.11

Technology and the Internet:

1.1, 1.22, 1.23, 1.25, 1.27, 1.28, 1.29, 1.162, 1.163, 1.166, 1.169, 1.172, 2.5, 2.6, 2.7, 2.8, 2.20, 2.21, 2.22, 2.38, 2.39, 2.45, 2.46, 2.96, 3.9, 3.12, 3.13, 3.14, 3.19, 3.42, 3.50, 3.51, 3.60, 3.80, 3.133, 4.21, 4.22, 4.23, 4.24, 4.36, 4.50, 4.51, 4.55, 5.1, 5.6, 5.16, 5.32, 5.33, 5.45, 5.46, 5.47, 5.48, 5.49, 5.58, 5.70, 6.25, 6.26, 6.113, 7.7, 7.24, 7.54, 7.55, 7.71, 7.77, 7.78, 7.79, 7.101, 7.120, 8.25, 8.26, 8.52, 8.53, 8.54, 8.65, 8.66, 8.74, 8.75, 8.76, 8.82, 8.83, 8.84, 9.11, 9.12, 9.13, 9.14, 9.38, 9.42, 10.42, 12.14, 12.16, 12.18, 12.46, 12.63, 14.38, 14.40, 16.6, 16.7, 16.13, 16.16, 16.38, 16.44, 16.48, 16.68, 16.69, 17.33

Introduction to the Practice of
STATISTICS

SEVENTH
EDITION

David S. Moore
George P. McCabe
Bruce A. Craig
Purdue University

W. H. FREEMAN AND
COMPANY *New York*

Publisher: Ruth Baruth

Acquisitions Editor: Karen Carson

Executive Marketing Manager: Jennifer Somerville

Developmental Editor: Anne Scanlan-Rohrer and Andrew Sylvester

Media Editor: Laura Capuano

Associate Editors: Katrina Wilhelm

Assistant Media Editor: Catriona Kaplan

Editorial Assistant: Lauren Kimmich

Photo Editor: Cecilia Varas

Photo Researcher: Julie Tesser

Cover and Text Designer: Blake Logan

Project Management, Illustrations, and Composition: MPS Limited, a Macmillan Company

Production Manager: Julia DeRosa

Printing and Binding: Worldcolor Versailles

Library of Congress Control Number: 2010928450

Student Edition:
ISBN-13: 978-1-4292-4020-8
ISBN-10: 1-4292-4020-2 (hardcover)

ISBN-13: 978-1-4292-7407-4
ISBN-10: 1-4292-7407-7 (paper)

ISBN-13: 978-1-4292-6173-9
ISBN-10: 1-4292-6173-0 (extended version)

ISBN-13: 978-1-4292-4032-1
ISBN-10: 1-4292-4032-6 (package with CD)

Instructor Complementary Copy:
ISBN-13: 978-1-4292-7406-7
ISBN-10: 1-4292-7406-9

Printed in the United States of America

Second printing

W. H. Freeman and Company
41 Madison Avenue
New York, NY 10010
Houndmills, Basingstoke RG21 6XS, England
www.whfreeman.com

BRIEF CONTENTS

CONTENTS

PART I: Looking at Data

CHAPTER 1   Looking at Data— Distributions   1

CHAPTER 2   Looking at Data— Relationships   79

CHAPTER 3   Producing Data   163

PART II: Probability and Inference

CHAPTER 4   Probability: The Study
of Randomness   227

CHAPTER 5 Sampling Distributions **297**

CHAPTER 6 Introduction to
       Inference **341**

CHAPTER 7 Inference for
              Distributions   403

CHAPTER 8 Inference for
              Proportions   473

PART III: Topics in Inference

CHAPTER 9 Analysis of Two-Way Tables   511

CHAPTER 10 Inference for Regression   545

CHAPTER 11 Multiple Regression   591

CHAPTER 12 One-Way Analysis of Variance   623

CHAPTER 13 Two-Way Analysis of Variance 669

Companion Chapters (on the IPS Web site www.whfreeman.com/ips7e and CD-ROM)

CHAPTER 14 Logistic Regression 14-1

CHAPTER 15 Nonparametric Tests 15-1

CHAPTER 16 Bootstrap Methods and Permutation Tests* 16-1

CHAPTER 17 Statistics for Quality:
                Control and Capability 17-1

Statistics is the science of data. *Introduction to the Practice of Statistics* (*IPS*) is an introductory text based on this principle. We present methods of basic statistics in a way that emphasizes working with data and mastering statistical reasoning. *IPS* is elementary in mathematical level, but conceptually rich in statistical ideas. After completing a course based on our text, we would like students to be able to think objectively about conclusions drawn from data and use statistical methods in their own work.

In *IPS* we combine attention to basic statistical concepts with a comprehensive presentation of the elementary statistical methods that students will find useful in their work. *IPS* has been successful for several reasons:

1. *IPS* examines the nature of modern statistical practice at a level suitable for beginners. We focus on the production and analysis of data as well as the traditional topics of probability and inference.

2. *IPS* has a logical overall progression, so data production and data analysis are a major focus, while inference is treated as a tool that helps us draw conclusions from data in an appropriate way.

3. *IPS* presents data analysis as more than a collection of techniques for exploring data. We emphasize systematic ways of thinking about data. Simple principles guide the analysis: always plot your data; look for overall patterns and deviations from them; when looking at the overall pattern of a distribution for one variable, consider shape, center, and spread; for relations between two variables, consider form, direction, and strength; always ask whether a relationship between variables is influenced by other variables lurking in the background. We warn students about pitfalls in clear cautionary discussions.

4. *IPS* uses real examples to drive the exposition. Students learn the technique of least-squares regression and how to interpret the regression slope. But they also learn the conceptual ties between regression and correlation, the importance of looking for influential observations.

5. *IPS* is aware of current developments both in statistical science and in teaching statistics. Brief optional "Beyond the Basics" sections give quick overviews of topics such as density estimation, scatterplot smoothers, data mining, nonlinear regression, and meta-analysis. Chapter 16 gives an elementary introduction to the bootstrap and other computer-intensive statistical methods.

The title of the book expresses our intent to introduce readers to statistics as it is used in practice. Statistics in practice is concerned with drawing conclusions from data. We focus on problem solving rather than on methods that may be useful in specific settings.

Statistical Thinking As the continuing revolution in computing relieves us of many of the burdens associated with doing calculations and making graphs, we are able to focus more attention on the big picture. Where do data come from?

How do observational studies and experiments differ? Why are randomized comparative experiments so important? We have seen many mistakes made in drawing conclusions from data, but few have involved doing the arithmetic incorrectly.

Data Students develop statistical thinking by analyzing real data. Data are numbers with a context, as we say in "To Students: What Is Statistics?" A newborn who weighs 10.3 pounds is a big baby, and the birth weight could not plausibly be 10.3 ounces or 10.3 kilograms. Because context makes numbers meaningful, our examples and exercises use real data with real contexts that we briefly describe. Calculating the mean of five numbers is arithmetic, not statistics. The presence of background information, even in exercises intended for routine drill, encourages students to always consider the meaning of their calculations as well as the calculations themselves. Note in this connection that a calculation or a graph or "reject H_0" is rarely a full answer to a statistical problem. We strongly encourage requiring students to always state a brief conclusion in the context of the problem. A course based on *IPS* is an ideal setting for students to develop the kinds of communication skills that they will need in the future.

Mathematics The foundation of applied statistics lies in the mathematical theories based on probability theory and mathematical statistics. Applied statistics is not, however, a branch of mathematics and it should not be taught as if it were. There is a difference between understanding the mathematics that underlies a particular statistical method and how it should be used in practice. In *IPS* we require only the ability to read and use equations without having each step parsed. No algebraic derivations are presented and calculus is not used. Because this is a *statistics* text, it is richer in ideas and requires more thought than the low mathematical level suggests.

Calculators and Computers Software to perform statistical calculations and to make graphs is widely available today. We encourage instructors to use software in their course. By removing the drudgery associated with calculating a correct number, students are empowered and can concentrate on why the number is being computed and how to use it once the calculation is complete. We provide displays from several statistical software packages. A student who knows the basics can interpret almost any output.

Judgment Statistics in practice requires judgment. It is easy to list the mathematical assumptions that justify use of a particular procedure, but not so easy to decide when the procedure can be safely used in practice. Because judgment develops through experience, an introductory course should present clear guidelines and not make unreasonable demands on the judgment of students. We have given guidelines that we follow ourselves. Similarly, many exercises require students to use some judgment and to explain their choices in words. Many students and instructors would prefer that work be evaluated solely by using the correct method to obtain the correct numerical answer. We are convinced that the proper use of statistics requires more than this and we believe that requiring students to think is a rewarding exercise for all involved.

Teaching Recommendations We have used *IPS* in courses taught to a variety of student audiences. For general undergraduates from mixed disciplines, we recommend covering Chapters 1 to 8 and Chapter 9, 10, or 12, omitting all optional material. For a quantitatively strong audience—sophomores planning to major in actuarial science or statistics—we recommend moving more quickly. Add Chapters 10 and 11 to the core material in Chapters 1 to 8 and include most of the optional content. In general, we recommend de-emphasizing the material on probability because these students will take a probability course later in their program. For beginning graduate students in such fields as education, family studies, and retailing, we recommend that the students read the entire text (Chapters 11 and 13 lightly), again with reduced emphasis on Chapter 4 and some parts of Chapter 5. In all cases, beginning with data analysis and data production (Part I) helps students overcome their fear of statistics and builds a sound base for studying inference. We believe that *IPS* can easily be adapted to wide variety of audiences.

The Seventh Edition: *What's New?*

- **Text Organization** The introduction to Chapter 1 has been rewritten with additional information about organizing data in a suitable format for analysis. The log transformation has been moved from a "Beyond the Basics" section to an optional section within Chapter 2. Other parts of this chapter have been streamlined. Chapter 5 has been reorganized so the sampling distribution of the sample mean and the central limit theorem come before the sampling distribution of the sample proportion and binomial distribution, to allow more initial focus on understanding the sampling distribution of the mean.

- **Exercises and Examples** Over 30% of the exercises are new and many are revised or updated. To maintain the attractiveness of the examples to students, we have replaced or updated a large number of these. A list of exercises and examples categorized by application area is provided.

- **Use Your Knowledge Exercises** We have found the Use Your Knowledge Exercises to be a very useful learning tool. Therefore, we have increased the number and variety of these exercises. They are listed, with page numbers, at the end of each section.

- **Design** A new design incorporates colorful, revised figures throughout to aid the students' understanding of text material. Photographs related to chapter examples and exercises make connections to real-life applications and provide a visual context for topics.

In addition to the new seventh edition enhancements, *IPS* has retained the successful pedagogical features from previous editions:

◄ LOOK BACK
- **Look Back** At key points in the text, Look Back margin notes direct the reader to the first explanation of a topic, providing page numbers for easy reference.

- **Caution** Warnings in the text, signaled by a caution icon, help students avoid common errors and misconceptions.

- **Challenge Exercises** More challenging exercises are signaled with an icon. Challenge exercises are varied: some are mathematical, some require open-ended investigation, and others require deeper thought about the basic concepts.

- **Applets** Applet icons are used throughout the text to signal where related interactive statistical applets can be found on the *IPS* text Web site and CD.

- **Statistics in Practice** Formerly found at the end of each chapter, these accounts by professionals who use statistics on the job are now located on the *IPS* Web site and CD.

Acknowledgements

We are pleased that the first six editions of *Introduction to the Practice of Statistics* have helped to move the teaching of introductory statistics in a direction supported by most statisticians. We are grateful to the many colleagues and students who have provided helpful comments and we hope that they will find this new edition another step forward. In particular, we would like to thank the following colleagues who offered specific comments on the new edition:

Leyla Batakci
Elizabethtown College
Dan Brick
University of St. Thomas
Pinyuen Chen
Syracuse University
Smiley W. Cheng
University of Manitoba
Scott D. Crawford
Texas A&M University
Yuehua Cui
Michigan State University
Gerarda Darlington
University of Guelph
A. R. de Leon
University of Calgary
Steven T. Garren
James Madison University
Suzan Gazioglu
Montana Tech
Mary W. Gray
American University
Susan Herring
Sonoma State University
David Himes
Cascadia Community College
Don Holbert
East Carolina University

Monica C. Jackson
American University
Stephen Sauchi Lee
University of Idaho
David Loewen
University of Manitoba
Xuewen Lu
University of Calgary
Henry Mesa
Portland Community College
Carrie Paquette
University of Manitoba
Jamis J. Perrett
Texas A & M University
Kathryn Prewitt
Arizona State University
Enrico Rogora
University of Utah
Nancy Roper
Portland Community College
Laura J. Shick
Clemson University
Angela P. Stabley
Portland Community College
Cove Sturtevant
San Diego State University
Yolande Tra
Rochester Institute of Technology

Brian Travers
Salem State University
D. Alexander Varakin
Knox College
Paul W. Vos
East Carolina University
Howard Wainer
University of Pennsylvania
Lan Wang
University of Minnesota

Michelle M. Wiest
University of Idaho
Todd Will
University of Wisconsin La Crosse
Lang Wu
University of British Columbia
Yuehua Wu
York University
Hongling Yang
University of Texas El Paso

The professionals at W. H. Freeman and Company, in particular Mary Louise Byrd, Ruth Baruth, Jamina Ward, Laura Capuano, and Lauren Kimmich have contributed greatly to the success of *IPS*. In addition, we would like to thank Anne Scanlan-Rohrer, Pam Bruton, Leo Kelly, Jackie Miller, Darryl Nester, and Julie Tesser for their valuable contributions to the seventh edition. Most of all, we are grateful to the many friends and collaborators whose data and research questions have enabled us to gain a deeper understanding of the science of data. Finally, we would like to acknowledge the contributions of John W. Tukey whose contributions to data analysis have had such a great influence on us as well as a whole generation of applied statisticians.

MEDIA AND SUPPLEMENTS

NEW! STATS P⬤RTAL

www.yourstatsportal.com (Access code or online purchase required.) StatsPortal is the digital gateway to *Introduction to the Practice of Statistics,* seventh edition, designed to enrich the course and enhance students' study skills through a collection of Web-based tools. StatsPortal integrates a rich suite of diagnostic, assessment, tutorial, and enrichment features, enabling students to master statistics at their own pace. It is organized around three main teaching and learning components:

1. **Interactive eBook** offers a complete and customizable online version of the text, fully integrated with all the media resources available with *IPS* 7e. The eBook allows students to quickly search the text, highlight key areas, and add notes about what they're reading. Instructors can customize the eBook to add, hide, and reorder content, add their own material, and highlight key text for students.

2. **Resources** organizes all the resources for *IPS* 7e into one location for ease of use. The resources include

   - **NEW! The Statistical Video Series** consists of StatClips, StatClips Examples, and Statistically Speaking "Snapshots." View animated lecture videos, whiteboard lessons, and documentary-style footage that illustrate key statistical concepts and help students visualize statistics in real-world scenarios.

   - **StatTutor Tutorials** offer audio-multimedia tutorials tied directly to the textbook, containing videos, applets, and animations.

   - **Statistical Applets** offer 17 interactive applets to help students master key statistical concepts and work exercises from the text.

   - **CrunchIt! 2.0®️ Statistical Software** allows users to analyze data from any Internet location. Designed with the novice user in mind, the software is not only easily accessible, but also easy to use. CrunchIt! 2.0®️ offers all the basic statistical routines covered in introductory statistics courses and more.

   - **Stats@Work Simulations** put students in the role of the statistical consultant, helping them better understand statistics interactively within the context of real-life scenarios.

   - **EESEE Case Studies,** developed by the Ohio State University Statistics Department, teach students to apply their statistical skills by exploring actual case studies using real data.

   - **Data Sets** are available in ASCII, Excel, TI, Minitab, SPSS, an IBM Company,[1] S-PLUS, and JMP formats.

---

[1] SPSS was acquired by IBM in October 2009.

- **Student Solutions Manual** provides detailed solutions to the odd-numbered exercises.

- **Statistical Software Manuals** for TI-83/84, Minitab, Excel, JMP, and SPSS provide instruction, examples, and exercises using specific statistical software packages.

- **Interactive Table Reader** allows students to use statistical tables interactively to seek the information they need.

Resources (instructors only)

- **Instructor's Guide with Full Solutions** includes teaching suggestions, chapter comments, and detailed solutions to all exercises.

- **The Test Bank** offers hundreds of multiple-choice questions.

- **Lecture PowerPoint slides** offer a detailed lecture presentation of statistical concepts covered in each chapter of *IPS* 7e.

- **NEW! SolutionMaster** is a Web-based version of the solutions in the Instructor's Guide with Full Solutions. This easy-to-use tool allows instructors to generate a solution file for any set of homework exercises. Solutions can be downloaded in PDF format for convenient printing and posting. For more information or a demonstration, contact your local W. H. Freeman sales representative.

3. **Assignments** organizes assignments and guides instructors through an easy-to-create assignment process providing access to questions from the Test Bank and Exercises from the text, including many algorithmic problems. The Assignment Center enables instructors to create their own assignments from a variety of question types for machine-gradable assignments. This powerful assignment manager allows instructors to select their preferred policies in regard to scheduling, maximum attempts, time limitations, feedback, and more!

Online Study Center 2.0 www.whfreeman.com/osc/ips7e **(Access code or online purchase required.)** The Online Study Center offers all the resources available in *StatsPortal*, except the eBook and Assignment Center.

Companion Web site www.whfreeman.com/ips7e This open-access Web site includes statistical applets, data sets, supplementary exercises, statistical profiles, self-quizzes. The Web site also offers four optional companion chapters covering bootstrap methods and permutation tests and statistics for quality control and capability.

Interactive Student CD-ROM is included with every new copy of *IPS* 7e. The CD contains access to all the content available on the Companion Web site. CrunchIt! 2.0® statistical software and EESEE case studies are available via an access-code-protected Web site. (Access code is included with every new text.)

Special Software Packages with student versions of JMP, Minitab, S-PLUS, and SPSS are available on a CD-ROM packaged with the textbook. This software is not sold separately and must be packaged with a text or a manual.

Contact your W.H. Freeman representative for information or visit www.whfreeman.com.

NEW! Video Tool Kit **(Access code or online purchase required.)** This new Statistical Video Series consists of two types of videos aimed to illustrate key statistical concepts and help students visualize statistics in real-world scenarios:

- **StatClips** lecture videos, created and presented by Alan Dabney, PhD, Texas A&M University, are innovative visual tutorials that illustrate key statistical concepts. In 3–5 minutes, each StatClips video combines dynamic animation, data sets, and interesting scenarios to help students understand the concepts in an introductory statistics course.

- **StatClips Examples** which are linked to the StatClips videos, are also created and presented by Alan Dabney. Each example walks students through step-by-step examples related to the StatClips lecture videos to reinforce the concepts through problem solving.

- **SnapShots** videos, abbreviated, student-friendly versions of the **Statistically Speaking** video series, bring the world of statistics into the classroom. Based on the successful PBS series *Against All Odds Statistics*, **Statistically Speaking** uses new and updated documentary footage and interviews that show real people using data analysis to make important decisions in their careers and in their daily lives. From business to medicine, from the environment to understanding the Census, **SnapShots** focus on why statistics is important for students' careers, and how statistics can be a powerful tool to understand their world.

Printed Student Solutions Manual by Darryl Nester, Bluffton University. This printed manual provides stepped-through solutions for all odd-numbered exercises in the text. ISBN: 1-4292-7371-2

Software Manuals Software manuals covering Minitab, Excel, SPSS, TI-83/84, and JMP are offered within StatsPortal and the Online Study Center. These manuals are also available in printed versions through custom publishing. They serve as basic introductions to popular statistical software options and guides to their use with *IPS* 7e.

FOR INSTRUCTORS

The **Instructor's Web site** (www.whfreeman.com/ips7e) requires user registration as an instructor and features all the student Web materials plus

- Instructor version of **EESEE** (Electronic Encyclopedia of Statistical Examples and Exercises), with solutions to the exercises in the student version.

- **PowerPoint slides** containing all textbook figures and tables.

- **Lecture PowerPoint slides** offering a detailed lecture presentation of statistical concepts covered in each chapter of *IPS* 7e.

- **Full answers** to the **Supplementary Exercises** supplement on the student Web site.

Printed Instructor's Guide with Full Solutions by Darryl Nester, Bluffton University. This printed guide includes full solutions to all exercises and provides additional examples and data sets for class use, Internet resources, and sample examinations. It also contains brief discussions of the *IPS* approach for each chapter. ISBN: 1-4292-7353-4

Test Bank   The test bank contains hundreds of multiple-choice questions to generate quizzes and tests for each chapter of the text. It is available in print as well as electronically on CD-ROM (for Windows and Mac), where questions can be downloaded, edited, and resequenced to suit each instructor's needs.

Printed Version, ISBN: 1-4292-7370-4

Computerized (CD) Version, ISBN: 1-4292-7405-0

Enhanced Instructor's Resource CD-ROM   This CD-ROM allows instructors to **search** and **export** (by key term or chapter) all the material from the student CD, plus

- All text images and tables
- Instructor's Guide with full solutions
- PowerPoint files and lecture slides
- Test bank files

ISBN: 1-4292-7373-9

Course Management Systems   W. H. Freeman and Company provides courses for Blackboard, WebCT (Campus Edition and Vista), Angel, Desire2 Learn, Moodle, and Sakai course management systems. These are completely integrated solutions that you can easily customize and adapt to meet your teaching goals and course objectives. Visit www.bfwpub.com/lms for more information.

i·clicker

i-clicker   is a two-way radio-frequency classroom response solution developed by educators for educators. University of Illinois physicists Tim Stelzer, Gary Gladding, Mats Selen, and Benny Brown created the i>clicker system after using competing classroom response solutions and discovering they were neither classroom appropriate nor student friendly. Each step of i>clicker's development has been informed by teaching and learning. i>clicker is superior to other systems from both a pedagogical and technical standpoint. To learn more about packaging i>clicker with this textbook, please contact your local sales rep or visit www.iclicker.com.

Statistics is the science of collecting, organizing, and interpreting numerical facts, which we call *data*. We are bombarded by data in our everyday lives. The news mentions movie box-office sales, the latest poll of the president's popularity, and the average high temperature for today's date. Advertisements claim that data show the superiority of the advertiser's product. All sides in public debates about economics, education, and social policy argue from data. A knowledge of statistics helps separate sense from nonsense in this flood of data.

The study and collection of data are also important in the work of many professions, so training in the science of statistics is valuable preparation for a variety of careers. Each month, for example, government statistical offices release the latest numerical information on unemployment and inflation. Economists and financial advisors, as well as policy makers in government and business, study these data in order to make informed decisions. Doctors must understand the origin and trustworthiness of the data that appear in medical journals. Politicians rely on data from polls of public opinion. Business decisions are based on market research data that reveal consumer tastes and preferences. Engineers gather data on the quality and reliability of manufactured products. Most areas of academic study make use of numbers, and therefore also make use of the methods of statistics. This means it is extremely likely that your undergraduate research projects will involve, at some level, the use of statistics.

Learning from Data

The goal of statistics is learn from data. To learn, we often perform calculations or make graphs based on a set of numbers. But to learn from data, we must do more than calculate and plot, because data are not just numbers; they are numbers that have some context that helps us learn from them.

Two-thirds of Americans are overweight or obese according to the Center for Disease Control and Prevention (CDC) Web site (www.cdc.gov/nchs/nhanes. htm). What does it mean to be obese or to be overweight? To answer this question we need to talk about body mass index (BMI). Your weight in kilograms divided by the square of your height in meters is your BMI. A person who is 6 feet tall (1.83 meters) and weighs 180 pounds (81.65 kilograms) will have a BMI of $81.65/(1.83)^2 = 24.4$ kg/m^2. How do we interpret this number? According to the CDC, a person is classified as overweight or obese if their BMI is 25 kg/m^2 or greater, and as obese if their BMI is 30 kg/m^2 or more. Therefore, two-thirds of Americans have a BMI of 25 kg/m^{2}or more. The person who weighs 180 pounds and is 6 feet tall is not overweight or obese, but if he gains 5 pounds, his BMI would increase to 25.1 and he would be classified as overweight.

When you do statistical problems, even straightforward textbook problems, don't just graph or calculate. Think about the context and state your conclusions in the specific setting of the problem. As you are learning how to do statistical calculations and graphs, remember that the goal of statistics is not calculation for its own sake, but gaining understanding from numbers. The calculations and graphs can be automated by a calculator or software, but you must supply the understanding. This book presents only the most common specific procedures for statistical analysis. A thorough grasp of the principles of statistics will enable you to quickly learn more advanced methods as needed. On the other hand, a fancy computer analysis carried out without attention to basic principles will often produce elaborate nonsense. As you read, seek to understand the principles as well as the necessary details of methods and recipes.

The Rise of Statistics

Historically, the ideas and methods of statistics developed gradually as society grew interested in collecting and using data for a variety of applications. The earliest origins of statistics lie in the desire of rulers to count the number of inhabitants or measure the value of taxable land in their domains. As the physical sciences developed in the seventeenth and eighteenth centuries, the importance of careful measurements of weights, distances, and other physical quantities grew. Astronomers and surveyors striving for exactness had to deal with variation in their measurements. Many measurements should be better than a single measurement, even though they vary among themselves. How can we best combine many varying observations? Statistical methods that are still important were invented in order to analyze scientific measurements.

By the nineteenth century, the agricultural, life, and behavioral sciences also began to rely on data to answer fundamental questions. How are the heights of parents and children related? Does a new variety of wheat produce higher yields than the old, and under what conditions of rainfall and fertilizer? Can a person's mental ability and behavior be measured just as we measure height and reaction time? Effective methods for dealing with such questions developed slowly and with much debate.

As methods for producing and understanding data grew in number and sophistication, the new discipline of statistics took shape in the twentieth century. Ideas and techniques that originated in the collection of government data, in the study of astronomical or biological measurements, and in the attempt to understand heredity or intelligence came together to form a unified "science of data." That science of data—statistics—is the topic of this text.

The Organization of This Book

Part I of this book, called simply "Data," concerns data analysis and data production. The first two chapters deal with statistical methods for organizing and describing data. These chapters progress from simpler to more complex data. Chapter 1 examines data on a single variable, Chapter 2 is devoted to relationships among two or more variables. You will learn both how to examine data produced by others and how to organize and summarize your own data. These summaries will first be graphical, then numerical, and then, when appropriate, in the form of a mathematical model that gives a compact description of the overall pattern of the data. Chapter 3 outlines arrangements (called designs) for producing data that answer specific questions. The principles presented in this chapter will help you to design proper samples and experiments for your research projects, and to evaluate other such investigations in your field of study.

Part II, consisting of Chapters 4 to 8, introduces statistical inference—formal methods for drawing conclusions from properly produced data. Statistical inference uses the language of probability to describe how reliable its conclusions are, so some basic facts about probability are needed to understand inference. Probability is the subject of Chapters 4 and 5. Chapter 6, perhaps the most important chapter in the text, introduces the reasoning of statistical inference. Effective inference is based on good procedures for producing data (Chapter 3), careful examination of the data (Chapters 1 and 2), and an understanding of the nature of statistical inference as discussed in Chapter 6. Chapters 7 and 8 describe some of the most common specific methods of inference, for drawing conclusions about means and proportions from one and two samples.

The five shorter chapters in Part III introduce somewhat more advanced methods of inference, dealing with relations in categorical data, regression and correlation, and analysis of variance. Supplementary chapters, available from the text Web site, bound separately, present additional statistical topics.

What Lies Ahead

Introduction to the Practice of Statistics is full of data from many different areas of life and study. Many exercises ask you to express briefly some understanding gained from the data. In practice, you would know much more about the background of the data you work with and about the questions you hope the data will answer. No textbook can be fully realistic. But it is important to form the habit of asking "What do the data tell me?" rather than just concentrating on making graphs and doing calculations.

You should have some help in automating many of the graphs and calculations. You should certainly have a calculator with basic statistical functions. Look for keywords such as, "two-variable statistics" or "regression" when you shop for a calculator. More advanced (and more expensive) calculators will do much more, including some statistical graphs. You may be asked to use software as well.

There are many kinds of statistical software, from spreadsheets to large programs for advanced users of statistics. The kind of computing available to learners varies a great deal from place to place—but the big ideas of statistics don't depend on any particular level of access to computing.

Because graphing and calculating are automated in statistical practice, the most important assets you can gain from the study of statistics are an understanding of the big ideas and the beginnings of good judgment in working with data. Ideas and judgment can't (at least yet) be automated. They guide you in telling the computer what to do and in interpreting its output. This book tries to explain the most important ideas of statistics, not just teach methods. Some examples of big ideas that you will meet are "always plot your data," "randomized comparative experiments," and "statistical significance."

You learn statistics by doing statistical problems. "Practice, practice, practice." Be prepared to work problems. The basic principle of learning is persistence. Being organized and persistent is more helpful in reading this book than knowing lots of math. The main ideas of statistics, like the main ideas of any important subject, took a long time to discover and take some time to master. The gain will be worth the pain.

David S. Moore is Shanti S. Gupta Distinguished Professor of Statistics, Emeritus, at Purdue University and was 1998 president of the American Statistical Association. He received his A.B. from Princeton and his Ph.D. from Cornell, both in mathematics. He has written many research papers in statistical theory and served on the editorial boards of several major journals. Professor Moore is an elected fellow of the American Statistical Association and of the Institute of Mathematical Statistics and an elected member of the International Statistical Institute. He has served as program director for statistics and probability at the National Science Foundation.

In recent years, Professor Moore has devoted his attention to the teaching of statistics. He was the content developer for the Annenberg/Corporation for Public Broadcasting college-level telecourse *Against All Odds: Inside Statistics* and for the series of video modules *Statistics: Decisions through Data*, intended to aid the teaching of statistics in schools. He is the author of influential articles on statistics education and of several leading texts. Professor Moore has served as president of the International Association for Statistical Education and has received the Mathematical Association of America's national award for distinguished college or university teaching of mathematics.

George P. McCabe is the Associate Dean for Academic Affairs in the College of Science and a Professor of Statistics at Purdue University. In 1966 he received a B.S. degree in mathematics from Providence College and in 1970 a Ph.D. in mathematical statistics from Columbia University. His entire professional career has been spent at Purdue with sabbaticals at Princeton, the Commonwealth Scientific and Industrial Research Organization (CSIRO in Melbourne (Australia), the University of Berne (Switzerland), the National Institute of Standards and Technology (NIST) in Boulder, Colorado, and the National University of Ireland in Galway. Professor McCabe is an elected fellow of the American Association for the Advancement of Science and of the American Statistical Association; he was 1998 Chair of its section on Statistical Consulting. In 2008–2010, he served on the Institute of Medicine Committee on Nutrition Standards for the National School Lunch and Breakfast Programs. He has served on the editorial boards of several statistics journals. He has consulted with many major corporations and has testified as an expert witness on the use of statistics in several cases.

Professor McCabe's research interests have focused on applications of statistics. Much of his recent work has focused on problems in nutrition including nutrient requirements, calcium metabolism, and bone health. He is author or coauthor of over 160 publications in many different journals.

Bruce A. Craig is Professor of Statistics and Director of the Statistical Consulting Service at Purdue University. He received his B.S. in mathematics and economics from Washington University in St. Louis and his Ph.D. in statistics from the University of Wisconsin-Madison. He is an active member of the American Statistical Association and was chair of its section on Statistical Consulting in 2009. He is also an active member of the Eastern North American Region of the International Biometrics Society and was elected by the voting membership to the Regional Committee between 2003 and 2006. Professor Craig has served on the editorial board of several statistical journals and has been a member of several data and safety monitoring boards, including Purdue's IRB.

Professor Craig's research interest focuses on the development of novel statistical methodology to address research questions in the life sciences. Areas of current interest are protein structure determination, diagnostic testing, and animal abundance estimation. In 2005, he was named Purdue University Faculty Scholar.

DATA TABLE INDEX

BEYOND THE BASICS INDEX

Looking at Data—Distributions

Introduction

Statistics is the science of learning from data. Data are numerical facts. In this chapter, we will master the art of examining data.

A statistical analysis starts with a set of data. We construct a set of data by first deciding what *cases* or units we want to study. For each case, we record information about characteristics that we call *variables*.

CASES, LABELS, VARIABLES, AND VALUES

Cases are the objects described by a set of data. Cases may be customers, companies, subjects in a study, or other objects.

A **label** is a special variable used in some data sets to distinguish the different cases.

A **variable** is a characteristic of a case.

Different cases can have different **values** for the variables.

1

EXAMPLE

1.1 Over 5 billion sold. Apple's music-related products and services generated $1.05 billion in the first quarter of 2008 and accounted for 13% of the company's revenue. Since Apple started marketing iTunes in 2003, they have sold over 5 billion songs. Lets take a look at this remarkable product. Figure 1.1 is part of an iTunes playlist named IPS. The four songs shown are cases. They are numbered from 1 to 4 in the first column. These numbers are the labels that distinguish the four songs. The following five columns give name (of the song), time (the length of time it takes to play the song), artist, album, and genre.

FIGURE 1.1 Part of an iTunes playlist, for Example 1.1.

Some variables, like the name of a song and the artist simply place cases into categories. Others, like the length of a song, take numerical values for which we can do arithmetic. It makes sense to give an average length of time for a collection of songs, but it does not make sense to give an "average" album. We can, however, count the numbers of songs for different albums and we can do arithmetic with these counts.

CATEGORICAL AND QUANTITATIVE VARIABLES

A **categorical variable** places a case into one of several groups or categories.

A **quantitative variable** takes numerical values for which arithmetic operations such as adding and averaging make sense.

The **distribution** of a variable tells us what values it takes and how often it takes these values.

EXAMPLE

1.2 Categorical and quantitative variables in iTunes playlist. The IPS iTunes playlist contains five variables. These are the name, time, artist, album, and genre. The time is a quantitative variable. Name, artist, album, and genre are categorical variables.

An appropriate label for your cases should be chosen carefully. In our iTunes example, a natural choice of a label would be the name of the song. However, if you have more than one artist performing the same song, or the same artist performing the same song on different albums, then the name of the song would not uniquely label each of the songs in your playlist.

A quantitative variable such as the time in the iTunes playlist requires some special attention before we can do arithmetic with its values. The first song in the playlist has time equal to 3:29, that is 3 minutes and 29 seconds. To do arithmetic with this variable, we should first convert all of the values so that they have a single unit. We could convert to seconds; 3 minutes is 180 seconds, so the total time is $180 + 29$ or 209 seconds. An alternative would be to convert to minutes; 29 seconds is .483 minutes, so time in this way is 3.483 minutes.

USE YOUR KNOWLEDGE

1.1 Time in the iTunes playlist. In the iTunes playlist, do you prefer to convert the time to seconds or minutes? Give a reason for your answer.

In practice, any set of data is accompanied by background information that helps us understand the data. When you plan a statistical study or explore data from someone else's work, ask yourself the following questions:

1. **Who?** What **cases** do the data describe? **How many** cases appear in the data?

2. **What?** How many **variables** do the data contain? What are the **exact definitions** of these variables? In what **unit of measurement** is each variable recorded?

3. **Why? What purpose** do the data have? Do we hope to answer some specific questions? Do we want to draw conclusions about cases other than the ones we actually have data for? Are the variables that are recorded suitable for the intended purpose?

EXAMPLE

1.3 Data for students in a statistics class. Figure 1.2 shows part of a data set for students enrolled in an introductory statistics class. Each row gives the data on one student. The values for the different variables are in the columns. This data set has eight variables. ID is an identifier for each student. Exam1, Exam2, Homework, Final, and Project give the points earned, out of a total of 100 possible, for each of these course requirements. Final grades are based on a possible 200 points for each exam and the final, 300 points for Homework, and 100 points for Project. TotalPoints is the variable that gives the composite score. It is computed by adding 2 times Exam1, Exam2, and Final, 3 times Homework plus 1 times Project. Grade is the grade earned in the course. This instructor used cut-offs of 900, 800, 700, etc. for the letter grades.

Microsoft Excel

	A	B	C	D	E	F	G	H
1	ID	Exam1	Exam2	Homework	Final	Project	TotalPoints	Grade
2	101	89	94	88	87	95	899	B
3	102	78	84	90	89	94	866	B
4	103	71	80	75	79	95	780	C
5	104	95	98	97	96	93	962	A
6	105	79	88	85	88	96	861	B

FIGURE 1.2 Spreadsheet for Example 1.3.

USE YOUR KNOWLEDGE

1.2 Who, what, and why for the statistics class data. Answer the who, what, and why questions for the statistics class data set.

1.3 Read the spreadsheet. Refer to Figure 1.2. Give the values of the variables Exam1, Exam2, and Final for the student with ID equal to 105.

1.4 Calculate the grade. A student whose data do not appear on the spreadsheet scored 86 on Exam1, 82 on Exam2, 77 for Homework, 90 on the Final, and 80 on the Project. Find TotalPoints for this student and give the grade earned.

spreadsheet

The display in Figure 1.2 is from an Excel **spreadsheet.** Spreadsheets are very useful for doing the kind of simple computations that you did in Exercise 1.4. You can type in a formula and have the same computation performed for each row.

Note that the names we have chosen for the variables in our spreadsheet do not have spaces. For example, we could have used the name "Exam 1" for the first exam score rather than Exam1. In some statistical software packages, however, spaces are not allowed in variable names. For this reason, when creating spreadsheets for eventual use with statistical software, it is best to avoid spaces in variable names. Another convention is to use an underscore (_) where you would normally use a space. For our data set, we could use Exam_1, Exam_2, and Final_Exam.

EXAMPLE

1.4 Cases and variables for the statistics class data. The data set in Figure 1.2 was constructed to keep track of the grades for students in an introductory statistics course. The cases are the students in the class. There are 8 variables in this data set. These include an identifier for each student and scores for the various course requirements. There are no units for ID and grade. The other variables all have "points" as the unit.

EXAMPLE

1.5 Statistics class data for a different purpose. Suppose the data for the students in the introductory statistics class were also to be used to study relationships between student characteristics and success in the course. For

FIGURE 1.3 Spreadsheet for Example 1.5.

Microsoft Excel

	A	B	C	D	E	F
1	ID	TotalPoints	Grade	Gender	PrevStat	Year
2	101	899	A	F	Yes	4
3	102	866	B	M	Yes	3
4	103	780	C	M	No	3
5	104	962	A	M	No	1
6	105	861	B	F	No	4

this purpose, we might want to use a data set like the spreadsheet in Figure 1.3. Here, we have decided to focus on the TotalPoints and Grade as the outcomes of interest. Other variables of interest have been included: Gender, PrevStat (whether or not the student has taken a statistics course previously), and Year (student classification as first, second, third, or fourth year). ID is a categorical variable, total points is a quantitative variable, and the remaining variables are all categorical.

In our example, the possible values for the grade variable are A, B, C, D, and F. When computing grade point averages, many colleges and universities translate these letter grades into numbers using A = 4, B = 3, C = 2, D = 1, and F = 0. The transformed variable with numeric values is considered to be quantitative because we can average the numerical values across different courses to obtain a grade point average.

Sometimes, experts argue about numerical scales such as this. They ask whether or not the difference between an A and a B is the same as the difference between a D and an F. Similarly, many questionnaires ask people to respond on a 1 to 5 scale with 1 representing strongly agree, 2 representing agree, etc. Again we could ask about whether or not the five possible values for this scale are equally spaced in some sense. From a practical point of view, the averages that can be computed when we convert categorical scales such as these to numerical values frequently provide a very useful way to summarize data.

USE YOUR KNOWLEDGE

1.5 Apartment rentals. A data set lists apartments available for students to rent. Information provided includes the monthly rent, whether or not cable is included free of charge, whether or not pets are allowed, the number of bedrooms, and the distance to the campus. Describe the cases in the data set, give the number of variables, and specify whether each variable is categorical or quantitative.

The context of data includes an understanding of the variables that are recorded. Often the variables in a statistical study are easy to understand: height in centimeters, study time in minutes, and so on. But each area of work also has its own special variables. A psychologist uses the Minnesota Multiphasic Personality Inventory (MMPI), and a physical fitness expert measures "VO2 max," the volume of oxygen consumed per minute while exercising at your maximum

instrument

capacity. Both of these variables are measured with special **instruments.** VO2 max is measured by exercising while breathing into a mouthpiece connected to an apparatus that measures oxygen consumed. Scores on the MMPI are based on a long questionnaire, which is also an instrument. Part of mastering your field of work is learning what variables are important and how they are best measured. Because details of particular measurements usually require knowledge of the particular field of study, we will say little about them.

Be sure that each variable really does measure what you want it to. A poor choice of variables can lead to misleading conclusions. Often, for example, the

rate

rate at which something occurs is a more meaningful measure than a simple count of occurrences.

EXAMPLE

1.6 Injuries in the workplace. The Bureau of Labor Statistics keeps track of occupational injuries in various categories. In a recent year, there were 4613 fatal injuries among wage and salary workers but only 1044 fatal injuries for self-employed workers.[1] Does this mean that the risk of an injury causing death is greater for wage and salary workers than for the self-employed? Not necessarily! A total of 136,670,000 persons were wage and salary workers whereas only 10,544,000 persons were self-employed. Let's compute the fatal injury rates for the two groups. Since the actual numbers of fatal injuries are small while the numbers of workers are large, rates such as these are expressed as fatal injuries per 100,000 workers. To calculate the rate, we take the ratio of fatal injuries to workers and multiply by 100,000. For wage and salary workers the rate is

$$\frac{4613}{136,670,000} 100,000 = 3.4 \text{ per } 100,000 \text{ workers}$$

For self-employed workers the rate is

$$\frac{1044}{10,544,000} 100,000 = 9.9 \text{ per } 100,000 \text{ workers}$$

When we compare the rates, we see that the self-employed are almost three times more likely than wage and salary workers to have a fatal injury at work.

USE YOUR KNOWLEDGE

1.6 Compare using a different type of rate. Refer to the previous example on fatal workplace injuries.

(a) Find the rates per worker for the two groups.

(b) Find the rates per 10,000 workers for the two groups.

(c) Compare the rates that you calculated in (a) and (b) with the rates given in the example. Which do you prefer for effectively communicating the results to a general audience?

The preceding exercise illustrates an important point about presenting the results of your statistical calculations. *Always consider how to best communicate*

your results to a general audience. The numbers produced by your calculator or by statistical software may need some work before they are ready for public consumption.

1.1 Displaying Distributions with Graphs

exploratory data analysis

Statistical tools and ideas help us examine data in order to describe their main features. This examination is called **exploratory data analysis.** Like an explorer crossing unknown lands, we want first to simply describe what we see. Here are two basic strategies that help us organize our exploration of a set of data:

- Begin by examining each variable by itself. Then move on to study the relationships among the variables.

- Begin with a graph or graphs. Then add numerical summaries of specific aspects of the data.

We will follow these principles in organizing our learning. This chapter presents methods for describing a single variable. We will study relationships among several variables in Chapter 2. Within each chapter, we will begin with graphical displays, then add numerical summaries for more complete description.

Categorical variables: bar graphs and pie charts

distribution of a categorical variable

The values of a categorical variable are labels for the categories, such as "yes" and "no." The **distribution of a categorical variable** lists the categories and gives either the **count** or the **percent** of cases who fall in each category.

 GPS

EXAMPLE

1.7 GPS market share. The Global Positioning System (GPS) uses satellites to transmit microwave signals that enable GPS receivers to determine the exact location of the receiver. Here are the market shares for the major GPS receiver brands sold in the United States.[2]

Company	Percent (%)
Garmin	47
TomTom	19
Magellan	17
Mio	7
Other	10

Company is the categorical variable in this example and the values are the names of the companies that provide GPS receivers in this market.

Note that the last value of the variable Company is "Other" which includes all receivers sold by companies other than the four listed by name. For data sets that have a large number of values for a categorical variable, we often create a category such as this that includes categories that have relatively small counts or percents. Careful judgment is needed when doing this. You don't want to cover up some important piece of information contained in the data by combining data in this way.

When we look at the GPS market share data set, we see that Garmin dominates the market with almost half of the sales. By using graphical methods, we can easily see this information and other characteristics of the data easily. We now examine two graphical ways to do this.

GPS

bar graph

EXAMPLE

1.8 Bar chart for the GPS market share data. Figure 1.4 displays the GPS market share data using a **bar graph.** The heights of the five bars show the market shares for the four companies and the "Other" category.

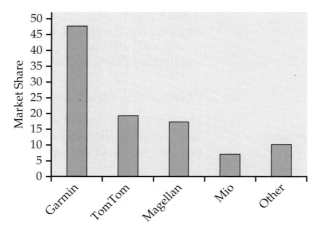

FIGURE 1.4 Bar chart for the GPS data in Example 1.8.

The categories in a bar graph can be put in any order. In Figure 1.4, we ordered the companies based on their market share with the "Other" category coming last. For other examples, an alphabetical ordering or some other arrangement might be used to produce a more useful graphical display. You should always consider the best way to order the values of the categorical variable in a bar chart. Choose an ordering that will be useful to you. If you have difficulty, ask a friend if your choice communicates what you expect.

GPS

pie chart

EXAMPLE

1.9 Pie chart for the GPS market share data. The **pie chart** in Figure 1.5 helps us see what part of the whole each group forms. Here it is very easy to see Garmin has about half of the market.

To make a pie chart, you must include all the categories that make up a whole. A category such as "Other" in this example can be used, but the sum of the percents for all of the categories should be 100%.

Market Share

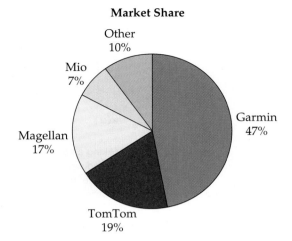

FIGURE 1.5 Pie chart for the GPS data in Example 1.9.

Bar graphs are more flexible. For example, you can use a bar graph to compare the numbers of students at your college majoring in biology, business, and political science. A pie chart cannot make this comparison because not all students fall into one of these three majors.

Quantitative variables: stemplots

A *stemplot* (also called a stem-and-leaf plot) gives a quick picture of the shape of a distribution while including the actual numerical values in the graph. Stemplots work best for small numbers of observations that are all greater than 0.

STEMPLOT

To make a **stemplot,**

1. Separate each observation into a **stem** consisting of all but the final (rightmost) digit and a **leaf,** the final digit. Stems may have as many digits as needed, but each leaf contains only a single digit.

2. Write the stems in a vertical column with the smallest at the top, and draw a vertical line at the right of this column.

3. Write each leaf in the row to the right of its stem, in increasing order out from the stem.

EXAMPLE

1.10 Vitamin D. Your body needs vitamin D to use calcium when building bones. It is particularly important that young adolescents have adequate supplies of this vitamin because their bodies are growing rapidly. Vitamin D in the form 25 hydroxy vitamin D is measured in the blood and represents the stores available for the body to use. The units are nanograms per

1		1	6		1	6
2		2	3 5 8 6		2	3 5 6 8
3		3	8 6 5 7 4 3 5		3	3 4 5 5 6 7 8
4		4	3 8 2 3 0 1 2		4	0 1 2 2 3 3 8
5		5	1		5	1
(a)		**(b)**			**(c)**	

FIGURE 1.6   Making a stemplot of the data in Example 1.10. (a) Write the stems. (b) Go through the data and write each leaf on the proper stem. For example, the values on the 2 stem are 23, 25, 28, and 26 in the order given in the display for the example. (c) Arrange the leaves on each stem in order out from the stem. The 2 stem now has leaves 3, 5, 6, and 8.

milliliter (ng/ml) of blood. Here are some values measured on a sample of 20 adolescent girls aged 11 to 14 years:[3]

16	43	38	48	42	23	36	35	37	34
25	28	26	43	51	33	40	35	41	42

To make a stemplot of these data, use the first digits as stems and the second digits as leaves. Figure 1.6 shows the steps in making the plot. The girl with a measured value of 16 ng/ml for vitamin D appears on the first stem with a leaf of 1 while the girl with a measured value of 43 ng/ml for appears on the stem labeled 4 with a leaf of 5.

The overall pattern of the data is fairly regular. The lowest value of 16 ng/ml is somewhat lower than the next highest value of 23 but it is not particularly extreme.

USE YOUR KNOWLEDGE

STATCOURSE

1.7   Make a stemplot.   Here are the scores on the first exam in an introductory statistics course for 30 students in one section of the course:

80	73	92	85	75	98	93	55	80	90	92	80	87	90	72
65	70	85	83	60	70	90	75	75	58	68	85	78	80	93

Use these data to make a stemplot. Then use the stemplot to describe the distribution of the first-exam scores for this course.

back-to-back stemplot   When you wish to compare two related distributions, a **back-to-back stemplot** with common stems is useful. The leaves on each side are ordered out from the common stem.

EXAMPLE

VITDBOYS

1.11   Vitamin D for boys.   Here are the 25 hydroxy vitamin D values for a sample of adolescent boys aged 11 to 14 years:

18	28	28	28	37	31	24	29	8	27
24	12	21	32	27	24	23	33	31	29

```
        Girls        Boys
                  0 | 8
             6    1 | 28
          8653    2 | 134447788899
       8765543    3 | 11237
       8332210    4 |
             1    5 |
```

FIGURE 1.7 A back-to-back stemplot to compare the distributions of vitamin D for samples of adolescent girls and boys, for Example 1.11.

Figure 1.7 gives the back-to-back stemplot for the girls and the boys. The values on the left give the vitamin D measures for the girls while the values on the right give the measures for the boys. The values of the boys tend to be lower than those for the girls.

splitting stems

trimming

There are two modifications of the basic stemplot that can be helpful in different situations. You can double the number of stems in a plot by **splitting each stem** into two: one with leaves 0 to 4 and the other with leaves 5 through 9. When the observed values have many digits, it is often best to **trim** the numbers by removing the last digit or digits before making a stemplot. You must use your judgment in deciding whether to split stems and whether to trim, though statistical software will often make these choices for you. Remember that the purpose of a stemplot is to display the shape of a distribution. If there are many stems with no leaves or only one leaf, trimming will reduce the number of stems. Let's take a look at the effect of splitting the stems for our vitamin D data.

EXAMPLE

1.12 Stemplot for vitamin D with split stems. Figure 1.8 gives the data from Examples 1.10 and 1.11 with split stems. Notice that we only needed one stem for 0 because there are no values between 0 and 4.

FIGURE 1.8 A back-to-back stemplot with split stems to compare the distributions of vitamin D for samples of adolescent girls and boys, for Example 1.12.

```
        Girls        Boys
                  0 | 8
                  1 | 2
             6    1 | 8
             3    2 | 13444
           865    2 | 7788899
            34    3 | 1123
         87655    3 | 7
        332210    4 |
             8    4 |
             1    5 |
```

> ## USE YOUR KNOWLEDGE
>
> **1.8 Which stemplot do you prefer?** Look carefully at the stemplots for the vitamin D data in Figures 1.7 and 1.8. Which do you prefer? Give reasons for your answer.
>
> **1.9 Why should you keep the space?** Suppose that you had a data set for girls similar to the one given in Example 1.10, but that the observation of 23 ng/ml was changed to 25 ng/ml.
>
> (a) Make a stemplot of these data for girls only using split stems.
>
> (b) Should you use one stem or two stems for the 20s? Give a reason for your answer. (*Hint:* How would your choice reveal or conceal a potentially important characteristic of the data?)

Histograms

Stemplots display the actual values of the observations. This feature makes stemplots awkward for large data sets. Moreover, the picture presented by a stemplot divides the observations into groups (stems) determined by the number system rather than by judgment. Histograms do not have these limitations.

histogram   A **histogram** breaks the range of values of a variable into classes and displays only the count or percent of the observations that fall into each class. You can choose any convenient number of classes, but you should always choose classes of equal width. Histograms are slower to construct by hand than stemplots and do not display the actual values observed. For these reasons we prefer stemplots for small data sets. The construction of a histogram is best shown by example. Most statistical software packages will make a histogram for you.

> ### EXAMPLE
>
> **1.13 Distribution of IQ scores.** You have probably heard that the distribution of scores on IQ tests is supposed to be roughly "bell-shaped." Let's look at some actual IQ scores. Table 1.1 displays the IQ scores of 60 fifth-grade students chosen at random from one school.
>
> 1. Divide the range of the data into classes of equal width. The scores in Table 1.1 range from 81 to 145, so we choose as our classes
>
> $$75 \leq \text{IQ score} < 85$$
> $$85 \leq \text{IQ score} < 95$$
> $$\vdots$$
> $$145 \leq \text{IQ score} < 155$$
>
> Be sure to specify the classes precisely so that each individual falls into exactly one class. A student with IQ 84 would fall into the first class, but IQ 85 falls into the second.
>
> 2. Count the number of individuals in each class. These counts are called **frequencies,** and a table of frequencies for all classes is a **frequency table.**

frequency
frequency table

TABLE 1.1

IQ test scores for 60 randomly chosen fifth-grade students

145	139	126	122	125	130	96	110	118	118
101	142	134	124	112	109	134	113	81	113
123	94	100	136	109	131	117	110	127	124
106	124	115	133	116	102	127	117	109	137
117	90	103	114	139	101	122	105	97	89
102	108	110	128	114	112	114	102	82	101

Class	Count	Class	Count
75 to 84	2	115 to 124	13
85 to 94	3	125 to 134	10
95 to 104	10	135 to 144	5
105 to 114	16	145 to 154	1

3. Draw the histogram. First, on the horizontal axis mark the scale for the variable whose distribution you are displaying. That's the IQ score. The scale runs from 75 to 155 because that is the span of the classes we chose. The vertical axis contains the scale of counts. Each bar represents a class. The base of the bar covers the class, and the bar height is the class count. There is no horizontal space between the bars unless a class is empty, so that its bar has height zero. Figure 1.9 is our histogram. It does look roughly "bell-shaped."

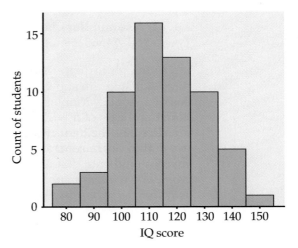

FIGURE 1.9 Histogram of the IQ scores of 60 fifth-grade students, for Example 1.13.

Large sets of data are often reported in the form of frequency tables when it is not practical to publish the individual observations. In addition to the frequency (count) for each class, we may be interested in the fraction or percent of the observations that fall in each class. A histogram of percents looks just like a frequency histogram such as Figure 1.9. Simply relabel the vertical scale to read in percents. Use histograms of percents for comparing several distributions that have different numbers of observations.

1.10 **Make a histogram.** Refer to the first-exam scores from Exercise 1.7. Use these data to make a histogram using classes 50–59, 60–69, etc. Compare the histogram with the stemplot as a way of describing this distribution. Which do you prefer for these data?

Our eyes respond to the *area* of the bars in a histogram. Because the classes are all the same width, area is determined by height and all classes are fairly represented. There is no one right choice of the classes in a histogram. Too few classes will give a "skyscraper" graph, with all values in a few classes with tall bars. Too many will produce a "pancake" graph, with most classes having one or no observations. Neither choice will give a good picture of the shape of the distribution. You must use your judgment in choosing classes to display the shape. Statistical software will choose the classes for you. The software's choice is often a good one, but you can change it if you want.

You should be aware that the appearance of a histogram can change when you change the classes. The histogram function in the *One-Variable Statistical Calculator* applet on the text CD and Web site allows you to change the number of classes by dragging with the mouse, so that it is easy to see how the choice of classes affects the histogram.

1.11 **Change the classes in the histogram.** Refer to the first-exam scores from Exercise 1.7 and the histogram you produced in Exercise 1.10. Now make a histogram for these data using classes 40–59, 60–79, and 80–100. Compare this histogram with the one that you produced in Exercise 1.7. Which do you prefer? Give a reason for your answer.

1.12 **Use smaller classes.** Repeat the previous exercise using classes 55–59, 60–64, 65–69, etc.

Although histograms resemble bar graphs, their details and uses are distinct. A histogram shows the distribution of counts or percents among the values of a single variable. A bar graph compares the size of different items. The horizontal axis of a bar graph need not have any measurement scale but simply identifies the items being compared. Draw bar graphs with blank space between the bars to separate the items being compared. Draw histograms with no space, to indicate that all values of the variable are covered. *Some spreadsheet programs, which are not primarily intended for statistics, will draw histograms as if they were bar graphs, with space between the bars. Often, you can tell the software to eliminate the space to produce a proper histogram.*

Data analysis in action: don't hang up on me

Many businesses operate call centers to serve customers who want to place an order or make an inquiry. Customers want their requests handled thoroughly. Businesses want to treat customers well, but they also want to avoid wasted time on the phone. They therefore monitor the length of calls and encourage their representatives to keep calls short.

CALLCENTER80

EXAMPLE

1.14 Calls to a customer service center. We have data on the length of all 31,492 calls made to the customer service center of a small bank in a month. Table 1.2 displays the lengths of the first 80 calls.[4]

Take a look at the data in Table 1.2. In this data set the *cases* are calls made to the bank's call center. The *variable* recorded is the length of each call. The *units* are seconds. We see that the call lengths vary a great deal. The longest call lasted 2631 seconds, almost 44 minutes. More striking is that 8 of these 80 calls lasted less than 10 seconds. What's going on?

TABLE 1.2

Service times (seconds) for calls to a customer service center

77	289	128	59	19	148	157	203
126	118	104	141	290	48	3	2
372	140	438	56	44	274	479	211
179	1	68	386	2631	90	30	57
89	116	225	700	40	73	75	51
148	9	115	19	76	138	178	76
67	102	35	80	143	951	106	55
4	54	137	367	277	201	52	9
700	182	73	199	325	75	103	64
121	11	9	88	1148	2	465	25

We started our study of the customer service center data by examining a few cases, the ones displayed in Table 1.2. It would be very difficult to examine all 31,492 cases in this way. How can we do this? Let's try a histogram.

CALLCENTER

EXAMPLE

1.15 Histogram for customer service center call lengths. Figure 1.10 is a histogram of the lengths of all 31,492 calls. We did not plot the few lengths greater than 1200 seconds (20 minutes). As expected, the graph shows that most calls last between about a minute and 5 minutes, with some lasting much longer when customers have complicated problems. More striking is the fact that 7.6% of all calls are no more than 10 seconds long. It turned out that the bank penalized representatives whose average call length was too long—so some representatives just hung up on customers in order to bring their average length down. Neither the customers nor the bank were happy about this. The bank changed its policy, and later data showed that calls under 10 seconds had almost disappeared.

tails   The extreme values of a distribution are in the **tails** of the distribution. The high values are in the upper, or right, tail and the low values are in the lower, or left, tail. The overall pattern in Figure 1.10 is made up of the many moderate call lengths and the long right tail of more lengthy calls. The striking departure

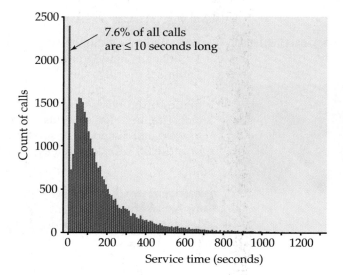

FIGURE 1.10 The distribution of call lengths for 31,492 calls to a bank's customer service center, for Example 1.15. The data show a surprising number of very short calls. These are mostly due to representatives deliberately hanging up in order to bring down their average call length.

from the overall pattern is the surprising number of very short calls in the left tail.

Our examination of the call center data illustrates some important principles:

- After you understand the background of your data (cases, variables, units of measurement), the first thing to do is **plot** your data.

- When you look at a plot, look for an **overall pattern** and also for any **striking departures** from the pattern.

We now turn to the kinds of graphs that are used to describe the distribution of a quantitative variable. We will explain how to make the graphs by hand, because knowing this helps you understand what the graphs show. However, making graphs by hand is so tedious that software is almost essential for effective data analysis unless you have just a few observations.

Examining distributions

Making a statistical graph is not an end in itself. The purpose of the graph is to help us understand the data. After you make a graph, always ask, "What do I see?" Once you have displayed a distribution, you can see its important features as follows.

EXAMINING A DISTRIBUTION

In any graph of data, look for the **overall pattern** and for striking **deviations** from that pattern.

You can describe the overall pattern of a distribution by its **shape, center,** and **spread.**

An important kind of deviation is an **outlier,** an individual value that falls outside the overall pattern.

In Section 1.2, we will learn how to describe center and spread numerically. For now, we can describe the center of a distribution by its *midpoint*, the value with roughly half the observations taking smaller values and half taking larger values. We can describe the spread of a distribution by giving the *smallest and largest values*. Stemplots and histograms display the shape of a distribution in the same way. Just imagine a stemplot turned on its side so that the larger values lie to the right. Some things to look for in describing shape are

modes
unimodal
- Does the distribution have one or several major peaks, called **modes**? A distribution with one major peak is called **unimodal.**

symmetric
skewed
- Is it approximately symmetric or is it skewed in one direction? A distribution is **symmetric** if the values smaller and larger than its midpoint are mirror images of each other. It is **skewed to the right** if the right tail (larger values) is much longer than the left tail (smaller values).

Some variables commonly have distributions with predictable shapes. Many biological measurements on specimens from the same species and sex—lengths of bird bills, heights of young women—have symmetric distributions. Money amounts, on the other hand, usually have right-skewed distributions. There are many moderately priced houses, for example, but the few very expensive mansions give the distribution of house prices a strong right-skew.

EXAMPLE

1.16 Examine the histogram. What does the histogram of IQ scores (Figure 1.9) tell us? **Shape:** The distribution is *roughly symmetric* with a *single peak* in the center. We don't expect real data to be perfectly symmetric, so we are satisfied if the two sides of the histogram are roughly similar in shape and extent. **Center:** You can see from the histogram that the midpoint is not far from 110. Looking at the actual data shows that the midpoint is 114. **Spread:** The spread is from 81 to 145. There are no outliers or other strong deviations from the symmetric, unimodal pattern.

The distribution of call lengths in Figure 1.10, on the other hand, is strongly *skewed to the right*. The midpoint, the length of a typical call, is about 115 seconds, or just under 2 minutes. The spread is very large, from 1 second to 28,739 seconds.

The longest few calls are *outliers*. They stand apart from the long right tail of the distribution, though we can't see this from Figure 1.10, which omits the largest observations. The longest call lasted almost 8 hours—that may well be due to equipment failure rather than an actual customer call.

USE YOUR KNOWLEDGE

STATCOURSE

1.13 Describe the first-exam scores. Refer to the first-exam scores from Exercise 1.7. Use your favorite graphical display to describe the shape, the center, and the spread of these data. Are there any outliers?

Dealing with outliers

In data sets smaller than the service call data, you can spot outliers by looking for observations that stand apart (either high or low) from the overall pattern of a histogram or stemplot. *Identifying outliers is a matter for judgment. Look for points that are clearly apart from the body of the data, not just the most extreme observations in a distribution.* You should search for an explanation for any outlier. Sometimes outliers point to errors made in recording the data. In other cases, the outlying observation may be caused by equipment failure or other unusual circumstances.

COLLEGEBYSTATE

EXAMPLE

1.17 College students. How does the number of undergraduate college students vary by state? Figure 1.11 is a histogram of the numbers of undergraduate students in each of the United States.[5] Notice that about 50% of the states are included in the first bar of the histogram. These states have less than 200,000 undergraduates. The next bar includes another 40% of the states. These have between 200,000 and 600,000 students. The bar at the far right of the histogram corresponds to the state of California which has over 2,000,000 undergraduates. California certainly stands apart from the other states for this variable. It is an outlier.

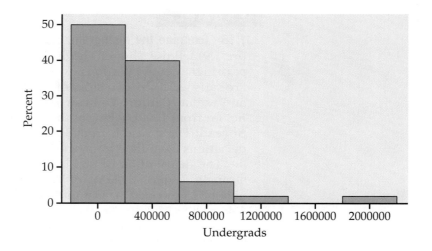

FIGURE 1.11 The distribution of the numbers of undergraduate college students for the 50 United States, for Example 1.17.

The state of California is an outlier in the previous example because it has a very large number of undergraduate students. California has the largest population of all the states so we might expect it to have a large number of undergraduate college students. Let's look at these data in a different way.

COLLEGEBYSTATE

EXAMPLE

1.18 College students per 1000. To account for the fact that there is large variation in the populations of the states, we divide the number of undergraduate students by the population and then multiply by 1000. This gives the undergraduate college enrollment expressed as the number per 1000 people in each state. Here is a stemplot of the distribution:

```
3 | 7
4 | 1 1 1 1 1 3 3
4 | 5 5 5 5 6 6 6 7 7 7 7 7 8 8 8 8 9
5 | 0 0 1 1 1 2 2 2 4 4 4 4
5 | 5 6 6
6 | 0 0 0 0 1
6 | 7 9
6 | 1 2
7 | 7
```

California has 60 undergraduate students per 1000 people. This is one of the higher values in the distribution but it is clearly not an outlier.

In Example 1.17 we looked at the distribution of the number of undergraduate students while in Example 1.18 we adjusted these data by expressing the counts as number per 1000 people in each state. Which way is correct? The answer depends upon why you are examining the data. If you are interested in marketing a product to undergraduate students, the unadjusted numbers would be of interest. On the other hand, if you are interested in comparing states with respect to how well they provide opportunities for higher education to their residents, the population adjusted values would be more suitable. *When presenting the results of any statistical analysis, always think about why you are doing the analysis and this will guide you in choosing an appropriate analytic strategy.*

Here is an example with a different kind of outlier.

 PTH

EXAMPLE

1.19 Healthy bones and PTH. Bones are constantly being built up (bone formation) and torn down (bone resorption). Young people who are growing have more formation than resorption. When we age, resorption increases to the point where it exceeds formation. (The same phenomenon occurs when astronauts travel in space.) The result is osteoporosis, a disease associated with fragile bones that are more likely to break. The underlying mechanisms that control these processes are complex and involve a variety of substances. One of these is parathyroid hormone (PTH). Here are the values of PTH measured on a sample of 30 boys and girls aged 12 to 15 years:[6]

| 39 | 59 | 30 | 48 | 71 | 31 | 25 | 31 | 71 | 50 | 38 | 63 | 49 | 45 | 31 |
| 33 | 28 | 40 | 127 | 49 | 59 | 50 | 64 | 28 | 46 | 35 | 28 | 19 | 29 | |

The data are measured in picograms per milliliter of blood (pg/ml). The original data were recorded with one digit after the decimal point. They have been rounded to simplify our presentation here. Here is a stemplot of the data:

```
 1 | 9
 2 | 5 8 8 8 9
 3 | 0 1 1 1 3 5 8 9
 4 | 0 5 6 8 9 9
 5 | 0 9 9
 6 | 3 4
 7 | 1 1
 8 |
 9 |
10 |
11 |
12 | 7
```

The observation 127 clearly stands out from the rest of the distribution. A PTH measurement on this individual taken on a different day was similar to the rest of the values in the data set. We conclude that this outlier was caused by a laboratory error or a recording error and we are confident in discarding it for any additional analysis.

Time plots

Whenever data are collected over time, it is a good idea to plot the observations in time order. *Displays of the distribution of a variable that ignore time order, such as stemplots and histograms, can be misleading when there is systematic change over time.*

TIME PLOT

A **time plot** of a variable plots each observation against the time at which it was measured. Always put time on the horizontal scale of your plot and the variable you are measuring on the vertical scale.

VITAMIND

EXAMPLE

1.20 Seasonal variation in vitamin D. Although we get some of our vitamin D from food, most of us get about 75% of what we need from the sun. Cells in the skin make vitamin D in response to sunlight. If people do not get enough exposure to the sun, they can become deficient in vitamin D, resulting in weakened bones and other health problems. The elderly, who need more vitamin D than younger people, and people who live in northern areas where there is relatively little sunlight in the winter, are particularly vulnerable to these problems. Figure 1.12 is a plot of the serum levels of vitamin D versus time of year for samples of subjects from Switzerland.[7] Units for these measures are nanomoles per liter (nmol/l). The observations are grouped into periods of two months for the plot. Means are marked by dark circles and are connected in the plot. The effect of the lack of sunlight in the winter months on vitamin D levels is clearly evident in the plot.

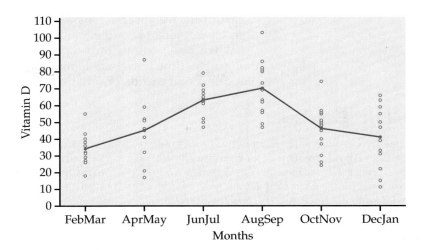

FIGURE 1.12 Plot of vitamin D versus months of the year, for Example 1.20.

The data described in the previous exercise is based on a subset of the subjects in a study of 248 subjects. They were particularly concerned about subjects whose levels were deficient, defined as a serum vitamin D level of less than 50 nmol/l. They concluded that there was a 3.8 fold higher deficiency rate in February–March versus August–September, 91.2% versus 24.3%. To ensure that individuals from this population have adequate levels of vitamin D, some form of supplementation is needed, particularly during certain times of the year.

SECTION 1.1 Summary

A data set contains information on a number of **cases.** Cases may be people, animals, or things. For each case, the data give values for one or more **variables.** A variable describes some characteristic of a case, such as a person's height, gender, or salary. Variables can have different **values** for different cases.

Some variables are **categorical** and others are **quantitative.** A categorical variable places each individual into a category, such as male or female. A quantitative variable has numerical values that measure some characteristic of each case, such as height in centimeters or annual salary in dollars.

Exploratory data analysis uses graphs and numerical summaries to describe the variables in a data set and the relations among them.

The **distribution** of a variable tells us what values it takes and how often it takes these values.

Bar graphs and **pie charts** display the distributions of categorical variables. These graphs use the counts or percents of the categories.

Stemplots and **histograms** display the distributions of quantitative variables. Stemplots separate each observation into a **stem** and a one-digit **leaf.** Histograms plot the **frequencies** (counts) or the percents of equal-width classes of values.

When examining a distribution, look for **shape, center,** and **spread** and for clear **deviations** from the overall shape.

Some distributions have simple shapes, such as **symmetric** or **skewed.** The number of **modes** (major peaks) is another aspect of overall shape. Not all distributions have a simple overall shape, especially when there are few observations.

Outliers are observations that lie outside the overall pattern of a distribution. Always look for outliers and try to explain them.

When observations on a variable are taken over time, make a **time plot** that graphs time horizontally and the values of the variable vertically. A time plot can reveal **trends** or other changes over time.

SECTION 1.1 Exercises

For Exercise 1.1, see page 3; for Exercises 1.2 to 1.4, see page 4; for Exercise 1.5, see page 5; for Exercise 1.6, see page 6; for Exercise 1.7, see page 10; for Exercises 1.8 and 1.9, see page 12; for Exercise 1.10, see page 14; for Exercises 1.11 and 1.12, see page 14; and for Exercise 1.13, see page 17.

1.14 Employee application data. The personnel department keeps records on all employees in a company. Here is the information that they keep in one of their data files: employee identification number, last name, first name, middle initial, department, number of years with the company, salary, education (coded as high school, some college, or college degree), and age.

(a) What are the cases for this data set?

(b) Describe each of these items as label, a quantitative variable, or a categorical variable.

(c) Set up a spreadsheet which could be used to record the data. Give appropriate column headings and five sample cases.

1.15 City rankings. Various organizations rank cities and produce lists of the top 10 or the top 100 best based on various measures. Create a list of criteria that you would use to rank cities. Include at least eight variables and give reasons for your choices. Explain whether each variable is quantitative or categorical.

1.16 Survey of students. A survey of students in an introductory statistics class asked the following questions: (a) age; (b) do you like to dance? (yes, no); (c) can you play a musical instrument (not at all, a little, pretty well); (d) how much did you spend on food last week? (e) height; (f) do you like broccoli? (yes, no). Classify each of these variables as categorical or quantitative and give reasons for your answers.

1.17 What questions would you ask? Refer to the previous exercise. Make up your own survey questions with at least six questions. Include at least two categorical variables and at least two quantitative variables. Tell which variables are categorical and which are quantitative. Give reasons for your answers.

1.18 Choosing a college or university. Popular magazines rank colleges and universities on their "academic quality" in serving undergraduate students. Describe five variables that you would like to see measured for each college if you were choosing where to study. Give reasons for each of your choices.

1.19 Favorite colors. What is your favorite color? One survey produced the following summary of responses to that question: blue, 42%; green, 14%; purple, 14%; red, 8%; black, 7%; orange, 5%; yellow, 3%; brown, 3%; gray, 2%; and white, 2%.[8] Make a bar graph of the percents and write a short summary of the major features of your graph. 🖲 FAVORITECOLORS

1.20 Least-favorite colors. Refer to the previous exercise. The same study also asked people about their least-favorite color. Here are the results: orange, 30%; brown, 23%; purple, 13%; yellow, 13%; gray, 12%; green, 4%; white, 4%; red, 1%; black, 0%; and blue, 0%. Make a bar graph of these percents and write a summary of the results. 🖲 LEASTFAVCOLORS

1.21 Ages of survey respondents. The survey about color preferences reported the age distribution of the people who responded. Here are the results:

Age group (years)	1–18	19–24	25–35	36–50	51–69	70 and over
Count	10	97	70	36	14	5

(a) Add the counts and compute the percents for each age group.

(b) Make a bar graph of the percents.

(c) Describe the distribution.

(d) Explain why your bar graph is not a histogram.

1.22 Mobile browsing and iPhones. Users of iPhones were asked to respond to the statement, "I do a lot more browsing on the iPhone than I did on my previous mobile phone" and responded as follows:[9] 🖲 BROWSING

Response	Percent (%)
Strongly agree	54
Mildly agree	22
Mildly disagree	16
Strongly disagree	8

(a) Make a bar graph to display the distribution of the responses.

(b) Display the distribution with a pie chart.

(c) Summarize the information in these charts.

(d) Do you prefer the bar chart or the pie chart? Give a reason for your answer.

1.23 What did the iPhone replace? The survey in the previous exercise also asked iPhone users what phone, if any, did the iPhone replace. Here are the responses:

Response	Percent (%)	Response	Percent (%)
Motorola Razr	23.8	BlackBerry	13.0
Symbian	3.9	Windows Mobile	13.9
Sidekick	4.1	Replaced nothing	10.0
Palm	6.7	Other phone	24.5

Make a bar graph for these data. Carefully consider how you will order the responses. Explain why you made the ordering that you chose. 🖳 PHONEREPLACE

1.24 Garbage. The formal name for garbage is "municipal solid waste." Here is a breakdown of the materials that made up American municipal solid waste:[10] 🖳 GARBAGE

Material	Weight (million tons)	Percent of total (%)
Food scraps	31.7	12.5
Glass	13.6	5.3
Metals	20.8	8.2
Paper, paperboard	83.0	32.7
Plastics	30.7	12.1
Rubber, leather, textiles	19.4	7.6
Wood	14.2	5.6
Yard trimmings	32.6	12.8
Other	8.2	3.2
Total	254.1	100.0

(a) Add the weights and the percents for the nine materials given, including "Other." Each entry, including the total, is separately rounded to the nearest tenth. So the sum and the total may differ slightly because of **roundoff error.**

(b) Make a bar graph of the percents. The graph gives a clearer picture of the main contributors to garbage if you order the bars from tallest to shortest.

(c) If you use software, also make a pie chart of the percents. Comparing the two graphs, notice that it is easier to see the small differences among "Food scraps," "Plastics," and "Yard trimmings" in the bar graph.

1.25 Recycled garbage. Refer to the previous exercise. The following table gives the percentage of the weight that was recycled for each of the categories. 🖳 GARBAGE

Material	Weight (million tons)	Percent recycled (%)
Food scraps	31.7	2.6
Glass	13.6	23.7
Metals	20.8	34.8
Paper, paperboard	83.0	54.5
Plastics	30.7	6.8
Rubber, leather, textiles	19.4	14.7
Wood	14.2	9.3
Yard trimmings	32.6	64.1
Other	8.2	0.0
Total	254.1	

(a) Use a bar chart to display the percent recycled for these materials. Use the order of the materials given in the table above.

(b) Make another bar chart where the materials are ordered by the percent recycled, largest percent to smallest percent.

(c) Which bar chart, (a) or (b), do you prefer? Give a reason for your answer.

(d) Explain why it is inappropriate to use a pie chart to display these data. 🖳 GARBAGE

1.26 Market share for search engines. The following table gives the market share for the major search engines.[11] 🖳 SEARCHENGINES

Search engine	Market share (%)	Search engine	Market share (%)
Google—Global	79.9	Microsoft Live Search	1.6
Yahoo—Global	11.3	Ask—Global	1.2
MSN—Global	3.4	Other	0.2
AOL—Global	2.4		

(a) Use a bar graph to display the market shares.

(b) Summarize what the graph tells you about market shares for search engines.

1.27 Spam. Email spam is the curse of the Internet. Here is a compilation of the most common types of spam:[12]

Type of spam	Percent (%)
Adult	14.5
Financial	16.2
Health	7.3
Leisure	7.8
Products	21.0
Scams	14.2

Make two bar graphs of these percents, one with bars ordered as in the table (alphabetical) and the other with bars in order from tallest to shortest. Comparisons are easier if you order the bars by height. A bar graph ordered from tallest to shortest bar is sometimes called a **Pareto chart,** after the Italian economist who recommended this procedure.

1.28 Facebook users by country. The following table gives the numbers of Facebook users by country for the top 20 countries (excluding the United States) as of September 29, 2008:[13] FACEBOOKCOUNTRY

Country	Facebook users (in millions)	Country	Facebook users (in millions)
United Kingdom	11.39	Venezuela	1.01
Canada	9.51	South Africa	0.97
Turkey	3.50	Hong Kong	0.91
Australia	3.36	Egypt	0.80
Colombia	2.69	Denmark	0.79
Chile	2.46	Spain	0.77
France	2.45	India	0.77
Norway	1.14	Germany	0.70
Sweden	1.14	Israel	0.61
Mexico	1.01	Italy	0.57

(a) Use a bar chart to describe these data.

(b) Describe the major features of your chart in a short paragraph.

1.29 Facebook use increases by country. Facebook use has been increasing rapidly. Data are available on the increases between February 8, 2008 and September 29, 2008.[14] The following table gives the percent increase in the numbers of Facebook users for the same 20 countries that we studied in the previous exercise. Note that there is no entry for Hong Kong because the number of users as of February 8, 2008 is not reported.
 FACEBOOKINCR

Country	Percent increase in Facebook users (%)	Country	Percent increase in Facebook users (%)
United Kingdom	31	Venezuela	683
Canada	9	South Africa	33
Turkey	23	Hong Kong	
Australia	43	Egypt	31
Colombia	246	Denmark	92
Chile	2197	Spain	132
France	92	India	42
Norway	7	Germany	44
Sweden	4	Israel	42
Mexico	69	Italy	139

(a) Summarize the data by carefully examining the table. Are there any extreme outliers? Which ones would you classify in this way?

(b) Use a stemplot to describe these data. You can list any extreme outliers separately from the plot.

(c) Describe the major features of these data using your plot and your list of outliers.

(d) How effective is the stemplot for summarizing these data? Give reasons for your answer.

1.30 Women seeking graduate and professional degrees. Here are the percents of women among students seeking various graduate and professional degrees:[15] GRADDEGREES

Degree	Percent female (%)
Master's in business administration	39.8
Master's in education	76.2
Other master of arts	59.6
Other master of science	53.0
Doctorate in education	70.8
Other PhD degree	54.2
Medicine (MD)	44.0
Law	50.2
Theology	20.2

(a) Explain clearly why we cannot use a pie chart to display these data.

(b) Make a bar graph of the data. (Comparisons are easier if you order the bars by height.)

1.31 Vehicle colors. Vehicle colors differ among types of vehicle. Here are data on the most popular colors for luxury cars and for intermediate cars in North America.[16]
 VEHICLECOLORS

Color	Luxury car percent (%)	Intermediate car percent (%)
Black	22	10
Silver	16	25
White Pearl	14	4
Gray	12	12
White	11	8
Blue	7	13
Red	7	10
Yellow/Gold	6	4
Other	5	14

(a) Make a bar graph for the luxury car percents.

(b) Make a bar graph for the intermediate percents.

(c) Now, be creative: make *one* bar graph that compares the two vehicle types as well as comparing colors. Arrange your graph so that it is easy to compare the two types of vehicle.

1.32 Shakespeare's plays. Figure 1.13 is a histogram of the lengths of words used in Shakespeare's plays. Because there are so many words in the plays, we use a histogram of percents. What is the overall shape of this distribution? What does this shape say about word lengths in Shakespeare? Do you expect other authors to have word length distributions of the same general shape? Why?

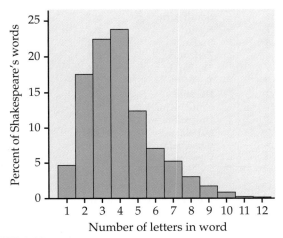

FIGURE 1.13 Histogram of the lengths of words used in Shakespeare's plays, for Exercise 1.32.

1.33 Diabetes and glucose. People with diabetes must monitor and control their blood glucose level. The goal is to maintain "fasting plasma glucose" between about 90 and 130 milligrams per deciliter (mg/dl). Here are the fasting plasma glucose levels for 18 diabetics enrolled in a diabetes control class, five months after the end of the class:[17]

141	158	112	153	134	95	96	78	148
172	200	271	103	172	359	145	147	255

Make a stemplot of these data and describe the main features of the distribution. (You will want to trim and also split stems.) Are there outliers? How well is the group as a whole achieving the goal for controlling glucose levels? GLUCOSE

1.34 Compare glucose of instruction and control groups. The study described in the previous exercise also measured the fasting plasma glucose of 16 diabetics who were given individual instruction on diabetes control. Here are the data:

128	195	188	158	227	198	163	164
159	128	283	226	223	221	220	160

Make a back-to-back stemplot to compare the class and individual instruction groups. How do the distribution shapes and success in achieving the glucose control goal compare? GLUCOSE

1.35 Vocabulary scores of seventh-grade students. Figure 1.14 displays the scores of all 947 seventh-grade students in the public schools of Gary, Indiana, on the vocabulary part of the Iowa Test of Basic Skills.[18] Give a brief description of the overall pattern (shape, center, spread) of this distribution.

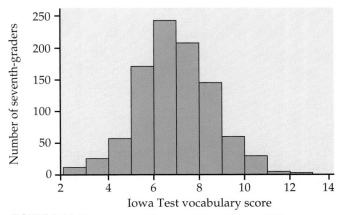

FIGURE 1.14 Histogram of the Iowa Test of Basic Skills vocabulary scores of seventh-grade students in Gary, Indiana, for Exercise 1.35.

1.36 Carbon dioxide from burning fuels. Burning fuels in power plants or motor vehicles emits carbon dioxide (CO_2), which contributes to global warming. Table 1.3 displays CO_2 emissions per person from countries with population at least 20 million.[19] CO_2

(a) Why do you think we choose to measure emissions per person rather than total CO_2 emissions for each country?

(b) Display the data of Table 1.3 in a graph. Describe the shape, center, and spread of the distribution. Which countries are outliers?

1.37 Thinness in Asia. Asian culture does not emphasize thinness, but young Asians are often influenced by Western culture. In a study of concerns about weight among young Korean women, researchers administered the Drive for Thinness scale (a questionnaire) to 264 female college students in Seoul, South Korea.[20] Drive for Thinness measures excessive concern with weight and dieting and fear of weight gain. Roughly speaking, a score of 15 is typical of Western women with eating disorders but is unusually high (90th percentile) for other Western women. Graph

TABLE 1.3

Carbon dioxide emissions (metric tons per person)

Country	CO_2	Country	CO_2	Country	CO_2
Algeria	2.3	Iran	3.8	Poland	8.0
Argentina	3.9	Iraq	3.6	Romania	3.9
Australia	17.0	Italy	7.3	Russia	10.2
Bangladesh	0.2	Japan	9.1	Saudi Arabia	11.0
Brazil	1.8	Kenya	0.3	South Africa	8.1
Canada	16.0	Korea, North	9.7	Spain	6.8
China	2.5	Korea, South	8.8	Sudan	0.2
Columbia	1.4	Malaysia	4.6	Tanzania	0.1
Congo	0.0	Mexico	3.7	Thailand	2.5
Egypt	1.7	Morocco	1.0	Turkey	2.8
Ethiopia	0.0	Myanmar	0.2	Ukraine	7.6
France	6.1	Nepal	0.1	United Kingdom	9.0
Germany	10.0	Nigeria	0.3	United States	19.9
Ghana	0.2	Pakistan	0.7	Uzbekistan	4.8
India	0.9	Peru	0.8	Venezuela	5.1
Indonesia	1.2	Philippines	0.9	Vietnam	0.5

the data and describe the shape, center, and spread of the distribution of Drive for Thinness scores for these Korean students. Are there any outliers? THINNESS

1.38 Acidity of rainwater. Changing the choice of classes can change the appearance of a histogram. Here is an example in which a small shift in the classes, with no change in the number of classes, has an important effect on the histogram. The data are the acidity levels (measured by pH) in 105 samples of rainwater. Distilled water has pH 7.00. As the water becomes more acidic, the pH goes down. The pH of rainwater is important to environmentalists because of the problem of acid rain.[21] ACIDRAIN

4.33	4.38	4.48	4.48	4.50	4.55	4.59	4.59	4.61	4.61
4.75	4.76	4.78	4.82	4.82	4.83	4.86	4.93	4.94	4.94
4.94	4.96	4.97	5.00	5.01	5.02	5.05	5.06	5.08	5.09
5.10	5.12	5.13	5.15	5.15	5.15	5.16	5.16	5.16	5.18
5.19	5.23	5.24	5.29	5.32	5.33	5.35	5.37	5.37	5.39
5.41	5.43	5.44	5.46	5.46	5.47	5.50	5.51	5.53	5.55
5.55	5.56	5.61	5.62	5.64	5.65	5.65	5.66	5.67	5.67
5.68	5.69	5.70	5.75	5.75	5.75	5.76	5.76	5.79	5.80
5.81	5.81	5.81	5.81	5.85	5.85	5.90	5.90	6.00	6.03
6.03	6.04	6.04	6.05	6.06	6.07	6.09	6.13	6.21	6.34
6.43	6.61	6.62	6.65	6.81					

(a) Make a histogram of pH with 14 classes, using class boundaries 4.2, 4.4, ..., 7.0. How many modes does your histogram show? More than one mode suggests that the data contain groups that have different distributions.

(b) Make a second histogram, also with 14 classes, using class boundaries 4.14, 4.34, ..., 6.94. The classes are those

from (a) moved 0.06 to the left. How many modes does the new histogram show?

(c) Use your software's histogram function to make a histogram without specifying the number of classes or their boundaries. How does the software's default histogram compare with those in (a) and (b)?

1.39 Identify the histograms. A survey of a large college class asked the following questions:
1. Are you female or male? (In the data, male = 0, female = 1.)
2. Are you right-handed or left-handed? (In the data, right = 0, left = 1.)
3. What is your height in inches?
4. How many minutes do you study on a typical weeknight?

Figure 1.15 shows histograms of the student responses, in scrambled order and without scale markings. Which histogram goes with each variable? Explain your reasoning.

1.40 Sketch a skewed distribution. Sketch a histogram for a distribution that is skewed to the left. Suppose that you and your friends emptied your pockets of coins and recorded the year marked on each coin. The distribution of dates would be skewed to the left. Explain why.

1.41 Time spent studying. Do women study more than men? We asked the students in a large first-year college class how many minutes they studied on a typical weeknight. Here are the responses of random samples of 30 women and 30 men from the class: STUDYTIME

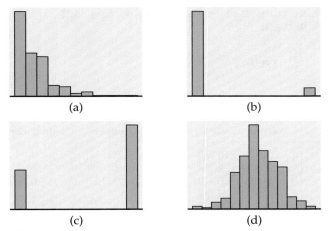

FIGURE 1.15 Match each histogram with its variable, for Exercise 1.39.

	Women					Men			
180	120	180	360	240	90	120	30	90	200
120	180	120	240	170	90	45	30	120	75
150	120	180	180	150	150	120	60	240	300
200	150	180	150	180	240	60	120	60	30
120	60	120	180	180	30	230	120	95	150
90	240	180	115	120	0	200	120	120	180

(a) Examine the data. Why are you not surprised that most responses are multiples of 10 minutes? We eliminated one student who claimed to study 30,000 minutes per night. Are there any other responses you consider suspicious?

(b) Make a back-to-back stemplot of these data. Report the approximate midpoints of both groups. Does it appear that women study more than men (or at least claim that they do)?

1.42 The density of the earth. In 1798 the English scientist Henry Cavendish measured the density of the earth by careful work with a torsion balance. The variable recorded was the density of the earth as a multiple of the density of water. Here are Cavendish's 29 measurements.[22]

5.50	5.61	4.88	5.07	5.26	5.55	5.36	5.29	5.58	5.65
5.57	5.53	5.62	5.29	5.44	5.34	5.79	5.10	5.27	5.39
5.42	5.47	5.63	5.34	5.46	5.30	5.75	5.68	5.85	

Present these measurements graphically by either a stemplot or a histogram and explain the reason for your choice. Then briefly discuss the main features of the distribution. In particular, what is your estimate of the density of the earth based on these measurements?

EARTHDENSITY

1.43 Grades and self-concept. Table 1.4 presents data on 78 seventh-grade students in a rural Midwestern school.[23] The researcher was interested in the relationship between the students' "self-concept" and their academic performance. The data we give here include each student's grade point average (GPA), score on a standard IQ test, and gender, taken from school records. Gender is coded as F for female and M for male. The students are identified only by an observation number (OBS). The missing OBS numbers show that some students dropped out of the study. The final variable is each student's score on the Piers-Harris Children's Self-Concept Scale, a psychological test administered by the researcher. SEVENTHGRADE

(a) How many variables does this data set contain? Which are categorical variables and which are quantitative variables?

(b) Make a stemplot of the distribution of GPA, after rounding to the nearest tenth of a point.

(c) Describe the shape, center, and spread of the GPA distribution. Identify any suspected outliers from the overall pattern.

(d) Make a back-to-back stemplot of the rounded GPAs for female and male students. Write a brief comparison of the two distributions.

1.44 Describe the IQ scores. Make a graph of the distribution of IQ scores for the seventh-grade students in Table 1.4. Describe the shape, center, and spread of the distribution, as well as any outliers. IQ scores are usually said to be centered at 100. Is the midpoint for these students close to 100, clearly above, or clearly below? SEVENTHGRADE

1.45 Describe the self-concept scores. Based on a suitable graph, briefly describe the distribution of self-concept scores for the students in Table 1.4. Be sure to identify any suspected outliers. SEVENTHGRADE

1.46 The Boston Marathon. Women were allowed to enter the Boston Marathon in 1972. Here are the times (in minutes, rounded to the nearest minute) for the winning women from 1972 to 2009:

Year	Time	Year	Time	Year	Time	Year	Time
1972	190	1982	150	1992	144	2002	141
1973	186	1983	143	1993	145	2003	145
1974	167	1984	149	1994	142	2004	144
1975	162	1985	154	1995	145	2005	145
1976	167	1986	145	1996	147	2006	143
1977	168	1987	146	1997	146	2007	149
1978	165	1988	145	1998	143	2008	145
1979	155	1989	144	1999	143	2009	152
1980	154	1990	145	2000	146		
1981	147	1991	144	2001	144		

Make a graph that shows change over time. What overall pattern do you see? Have times stopped improving in recent years? If so, when did improvement end?

MARATHON

TABLE 1.4

Educational data for 78 seventh-grade students

OBS	GPA	IQ	Gender	Self-concept	OBS	GPA	IQ	Gender	Self-concept
001	7.940	111	M	67	043	10.760	123	M	64
002	8.292	107	M	43	044	9.763	124	M	58
003	4.643	100	M	52	045	9.410	126	M	70
004	7.470	107	M	66	046	9.167	116	M	72
005	8.882	114	F	58	047	9.348	127	M	70
006	7.585	115	M	51	048	8.167	119	M	47
007	7.650	111	M	71	050	3.647	97	M	52
008	2.412	97	M	51	051	3.408	86	F	46
009	6.000	100	F	49	052	3.936	102	M	66
010	8.833	112	M	51	053	7.167	110	M	67
011	7.470	104	F	35	054	7.647	120	M	63
012	5.528	89	F	54	055	0.530	103	M	53
013	7.167	104	M	54	056	6.173	115	M	67
014	7.571	102	F	64	057	7.295	93	M	61
015	4.700	91	F	56	058	7.295	72	F	54
016	8.167	114	F	69	059	8.938	111	F	60
017	7.822	114	F	55	060	7.882	103	F	60
018	7.598	103	F	65	061	8.353	123	M	63
019	4.000	106	M	40	062	5.062	79	M	30
020	6.231	105	F	66	063	8.175	119	M	54
021	7.643	113	M	55	064	8.235	110	M	66
022	1.760	109	M	20	065	7.588	110	M	44
024	6.419	108	F	56	068	7.647	107	M	49
026	9.648	113	M	68	069	5.237	74	F	44
027	10.700	130	F	69	071	7.825	105	M	67
028	10.580	128	M	70	072	7.333	112	F	64
029	9.429	128	M	80	074	9.167	105	M	73
030	8.000	118	M	53	076	7.996	110	M	59
031	9.585	113	M	65	077	8.714	107	F	37
032	9.571	120	F	67	078	7.833	103	F	63
033	8.998	132	F	62	079	4.885	77	M	36
034	8.333	111	F	39	080	7.998	98	F	64
035	8.175	124	M	71	083	3.820	90	M	42
036	8.000	127	M	59	084	5.936	96	F	28
037	9.333	128	F	60	085	9.000	112	F	60
038	9.500	136	M	64	086	9.500	112	F	70
039	9.167	106	M	71	087	6.057	114	M	51
040	10.140	118	F	72	088	6.057	93	F	21
041	9.999	119	F	54	089	6.938	106	M	56

1.2 Describing Distributions with Numbers

We can begin with graphs, but numerical summaries make the comparisons more specific. A brief description of a distribution should include its *shape* and numbers describing its *center* and *spread*. We describe the shape of a distribution based on inspection of a histogram or a stemplot. Now we will learn specific ways to use numbers to measure the center and spread of a distribution. We can calculate these numerical measures for any quantitative variable. But to interpret measures of center and spread, and to choose among the several measures we will learn, you must think about the shape of the distribution and the meaning of the data. The numbers, like graphs, are aids to understanding, not "the answer" in themselves.

TIMETOSTART24

EXAMPLE

1.21   The distribution of business start times. An entrepreneur faces many bureaucratic and legal hurdles when starting a new business. The World Bank collects information about starting businesses throughout the world. They have determined the time, in days, to complete all of the procedures required to start a business.[24] Data for 195 countries are included in the data set (TimeToStart195) which is described in the data appendix. For this section we will examine data for a sample of 24 of these countries. Here are the data:

23	4	29	44	47	24	40	23	23	44	33	27
60	46	61	11	23	62	31	44	77	14	65	42

The stemplot in Figure 1.16 shows us the *shape, center,* and *spread* of the business start times. The stems are tens of days and the leaves are days. As is often the case when there are few observations, the shape of the distribution is irregular. There are peaks in the 20s and the 40s. the values range from 4 days to 77 days with a center somewhere in the middle of these two extremes. There do not appear to be any outliers.

```
0 | 4
1 | 1 4
2 | 3 3 3 3 4 7 9
3 | 1 3
4 | 0 2 4 4 4 6 7
5 |
6 | 0 1 2 5
7 | 7
```

FIGURE 1.16 Stemplot for sample of 24 business start times, for Example 1.21.

Measuring center: the mean

Numerical description of a distribution begins with a measure of its center or average. The two common measures of center are the *mean* and the *median*. The mean is the "average value" and the median is the "middle value." These are two different ideas for "center," and the two measures behave differently. We need precise recipes for the mean and the median.

THE MEAN $\overline{x}$

To find the **mean $\overline{x}$** of a set of observations, add their values and divide by the number of observations. If the n observations are $x_1, x_2, \ldots, x_n$, their mean is

$$\overline{x} = \frac{x_1 + x_2 + \cdots + x_n}{n}$$

or, in more compact notation,

$$\overline{x} = \frac{1}{n}\sum x_i$$

The $\sum$ (capital Greek sigma) in the formula for the mean is short for "add them all up." The bar over the x indicates the mean of all the x-values. Pronounce the mean $\overline{x}$ as "x-bar." This notation is so common that writers who are discussing data use $\overline{x}$, $\overline{y}$, etc. without additional explanation. The subscripts on the observations x_i are just a way of keeping the n observations separate. They do not necessarily indicate order or any other special facts about the data.

TIMETOSTART24

EXAMPLE

1.22 Mean time to start a business. The mean time to start a business is

$$\overline{x} = \frac{x_1 + x_2 + \cdots + x_n}{n}$$
$$= \frac{23 + 4 + \cdots + 42}{24}$$
$$= \frac{897}{24} = 37.375$$

The mean time to start a business for the 24 countries in our data set is 37.4 days. Note that we have rounded the answer. Our goal is using the mean to describe the center of a distribution; it is not to demonstrate that we can compute with great accuracy. The additional digits do not provide any additional useful information. In fact, they distract our attention from the important digits that are meaningful. Do you think it would be better to report the mean as 37 days?

TIMETOSTART25

USE YOUR KNOWLEDGE

1.47 Include the outlier. The complete business start time data set with 195 countries has a few with very large start times. In constructing the data set for Case 1.2, a random sample of 25 countries was selected. This sample included the South American country of Suriname where the start time is 694 days. This country was deleted for Example 1.21. Reconstruct the original random sample by including Suriname. Show

STATCOURSE

that the mean has increased to 64 days. (This is a rounded number. You should report the mean with two digits after the decimal.)

1.48 Find the mean. Here are the scores on the first exam in an introductory statistics course for 10 students:

80   73   92   85   75   98   93   55   80   90

Find the mean first-exam score for these students.

resistant measure

Exercise 1.48 illustrates an important weakness of the mean as a measure of center: *the mean is sensitive to the influence of a few extreme observations.* These may be outliers, but a skewed distribution that has no outliers will also pull the mean toward its long tail. Because the mean cannot resist the influence of extreme observations, we say that it is not a **resistant measure** of center. A measure that is resistant does more than limit the influence of outliers. Its value does not respond strongly to changes in a few observations, no matter how large those changes may be. The mean fails this requirement because we can make the mean as large as we wish by making a large enough increase in just one observation.

Measuring center: the median

We used the midpoint of a distribution as an informal measure of center in the previous section. The *median* is the formal version of the midpoint, with a specific rule for calculation.

THE MEDIAN *M*

The **median *M*** is the midpoint of a distribution. Half the observations are smaller than the median and the other half are larger than the median. Here is a rule for finding the median:

1. Arrange all observations in order of size, from smallest to largest.
2. If the number of observations n is odd, the median M is the center observation in the ordered list. Find the location of the median by counting $(n + 1)/2$ observations up from the bottom of the list.
3. If the number of observations n is even, the median M is the mean of the two center observations in the ordered list. The location of the median is again $(n + 1)/2$ from the bottom of the list.

Note that the formula $(n + 1)/2$ does *not* give the median, just the location of the median in the ordered list. Medians require little arithmetic, so they are easy to find by hand for small sets of data. Arranging even a moderate number of observations in order is tedious, however, so that finding the median by hand for larger sets of data is unpleasant. Even simple calculators have an $\bar{x}$ button, but you will need computer software or a graphing calculator to automate finding the median.

TIMETOSTART24

EXAMPLE

1.23 Median time to start a business. To find the median time to start a business for our 24 countries, we first arrange the data in order from smallest to largest.

4	11	14	23	23	23	23	24	27	29	31	33
40	42	44	44	44	46	47	60	61	62	65	77

The count of observations $n = 24$ is even. The median, then, is the average of the two center observations in the ordered list. To find the location of the center observations, we first compute

$$\text{location of } M = \frac{n+1}{2} = \frac{25}{2} = 12.5$$

Therefore, the center observations are the 12th and 13th observations in the ordered list. The median is

$$M = \frac{33 + 40}{2} = 36.5$$

Note that you can use the stemplot directly to compute the median. In the stemplot the cases are already ordered and you simply need to count from the top or the bottom to the desired location.

USE YOUR KNOWLEDGE

TIMETOSTART25

1.49 Include the outlier. Include Suriname where the start time is 694 days in the data set and show that the median is 40 days. Note that with this case included, the sample size is now 25 and the median is the 13th observation in the ordered list. Write out the ordered list and circle the outlier. Describe the effect of the outlier on the median for this set of data.

CALLCENTER80

1.50 Calls to a customer service center. The service times for 80 calls to a customer service center are given in Table 1.2 (page 15). Use these data to compute the median service time.

STATCOURSE

1.51 Find the median. Here are the scores on the first-exam in an introductory statistics course for 10 students:

80	73	92	85	75	98	93	55	80	90

Find the median first exam score for these students.

Mean versus median

Exercises 1.47 and 1.49 illustrate an important difference between the mean and the median. Suriname pulls the mean time to start a business up from 37 days to 64 days. The increase in the median is a lot less, from 36 days to 40 days.

The median is more *resistant* than the mean. If the largest starting time in the data set was 1200 days, the median for all 25 workers would still be 40 days. The largest observation just counts as one observation above the center, no matter how far above the center it lies. The mean uses the actual value of each observation and so will chase a single large observation upward.

The best way to compare the response of the mean and median to extreme observations is to use an interactive applet that allows you to place points on a line and then drag them with your computer's mouse. Exercises 1.83 to 1.85 use the *Mean and Median* applet on the Web site for this book, **www.whfreeman .com/ips7e**, to compare mean and median.

The median and mean are the most common measures of the center of a distribution. The mean and median of a symmetric distribution are close together. If the distribution is exactly symmetric, the mean and median are exactly the same. In a skewed distribution, the mean is farther out in the long tail than is the median. The endowment for a college or university is money set aside and invested. The income from the endowment is usually used to support various programs. The distribution of the sizes of the endowments of colleges and universities is strongly skewed to the right. Most institutions have modest endowments, but a few are very wealthy. The median endowment of colleges and universities in a recent year was $70 million—but the mean endowment was over $320 million. The few wealthy institutions pulled the mean up but did not affect the median. *Don't confuse the "average" value of a variable (the mean) with its "typical" value, which we might describe by the median.*

We can now give a better answer to the question of how to deal with outliers in data. First, look at the data to identify outliers and investigate their causes. You can then correct outliers if they are wrongly recorded, delete them for good reason, or otherwise give them individual attention. The outlier in Example 1.19 (page 19) can all be dropped from the data once we discover that it is an error. If you have no clear reason to drop outliers, you may want to use resistant methods, so that outliers have little influence over your conclusions. The choice is often a matter for judgment.

Measuring spread: the quartiles

A measure of center alone can be misleading. Two nations with the same median family income are very different if one has extremes of wealth and poverty and the other has little variation among families. A drug with the correct mean concentration of active ingredient is dangerous if some batches are much too high and others much too low. We are interested in the *spread* or *variability* of incomes and drug potencies as well as their centers. **The simplest useful numerical description of a distribution consists of both a measure of center and a measure of spread.**

We can describe the spread or variability of a distribution by giving several percentiles. The median divides the data in two; half of the observations are above the median and half are below the median. We could call the median the 50th percentile. The upper **quartile** is the median of the upper half of the data. Similarly, the lower quartile is the median of the lower half of the data. With the median, the quartiles divide the data into four equal parts; 25% of the data are in each part.

quartile

percentile   We can do a similar calculation for any percent. The **pth percentile** of a distribution is the value that has p percent of the observations fall at or below it. To calculate a percentile, arrange the observations in increasing order and count up the required percent from the bottom of the list. Our definition of percentiles is a bit inexact because there is not always a value with exactly p percent of the data at or below it. We will be content to take the nearest observation for most percentiles, but the quartiles are important enough to require an exact rule.

THE QUARTILES Q_1 AND Q_3

To calculate the quartiles:

1. Arrange the observations in increasing order and locate the median M in the ordered list of observations.

2. The **first quartile** Q_1 is the median of the observations whose position in the ordered list is to the left of the location of the overall median.

3. The **third quartile** Q_3 is the median of the observations whose position in the ordered list is to the right of the location of the overall median.

Here is an example.

TIMETOSTART24

EXAMPLE

1.24   Finding the quartiles.   Here is the ordered list of the times to start a business in our sample of 24 countries:

4	11	14	23	23	23	23	24	27	29	31	33
40	42	44	44	44	46	47	60	61	62	65	77

The count of observations $n = 24$ is even, so the median is at position $(24 + 1)/2 = 12.5$, that is, between the 12th and the 13th observation in the ordered list. There are 12 cases above this position and 12 below it. The first quartile is the median of the first 12 observations, and the third quartile is the median of the last 12 observations. Check that $Q_1 = 23$ and $Q_3 = 46.5$.

Notice that the quartiles are resistant. For example, Q_3 would have the same value if the highest start time was 770 days rather than 77 days.

There are slight differences in the methods used by software to compute percentiles. However, the results will generally quite similar, except in cases where the sample sizes are very small.

Be careful when several observations take the same numerical value. Write down all of the observations and apply the rules just as if they all had distinct values.

DATA FILE

STATCOURSE

USE YOUR KNOWLEDGE

1.52 Find the quartiles. Here are the scores on the first-exam in an introductory statistics course for 10 students:

80   73   92   85   75   98   93   55   80   90

Find the quartiles for these first exam scores.

EXAMPLE

1.25   Results from software. Statistical software often provides several numerical measures in response to a single command. Figure 1.17 displays such output from Minitab, JMP, and SPSS software for the data on the time to start a business. Examine the outputs carefully. Notice that they give different numbers of significant digits for some of these numerical summaries. Which output do you prefer?

There are several rules for calculating quartiles, which often give slightly different values. The differences are always small. For describing data, just report the values that your software gives.

The five-number summary and boxplots

In Section 1.1, we used the smallest and largest observations to indicate the spread of a distribution. These single observations tell us little about the

---

Minitab

Descriptive Statistics: TimeToStart

Variable	Total Count	Mean	StDev	Minimum	Q1	Median	Q3	Maximum
TimeToStart	24	37.38	18.57	4.00	23.00	36.50	46.75	77.00

(a)

JMP

Summary of Sheet1

	N Rows	N(TimeToStart)	Mean(TimeToStart)	Std Dev(TimeToStart)	Min(TimeToStart)	Max(TimeToStart)	Median(TimeToStart)	Quantiles25(TimeToStart)	Quantiles75(TimeToStart)
1	24	24	37.375	18.5749071	4	77	36.5	23	46.75

- Summary of Sheet1
- Source
- Columns (9/0)
  - N Rows
  - N(TimeToStart)
- Rows

(b)

SPSS

Descriptive Statistics

	N	Minimum	Maximum	Mean	Std. Deviation
TimeToStart	24	4	77	37.38	18.575

(c)

FIGURE 1.17 Descriptive statistics from Minitab (a), JMP (b), and SPSS (c) for the time to start a business, for Example 1.25.

distribution as a whole, but they give information about the tails of the distribution that is missing if we know only Q_1, M, and Q_3. To get a quick summary of both center and spread, combine all five numbers.

THE FIVE-NUMBER SUMMARY

The **five-number summary** of a set of observations consists of the smallest observation, the first quartile, the median, the third quartile, and the largest observation, written in order from smallest to largest. In symbols, the five-number summary is

$$\text{Minimum} \quad Q_1 \quad M \quad Q_3 \quad \text{Maximum}$$

EXAMPLE

1.26   Service center call lengths. Table 1.2 (page 15) gives the service center call lengths for the sample of 80 calls that we discussed in Example 1.14. The five-number summary for these data is 1.0, 54.5, 103.5, 200, and 2631. The distribution is highly skewed. The mean is 197 seconds, a value that is very close to the third quartile.

USE YOUR KNOWLEDGE

1.53   Verify the calculations. Refer to the five-number summary and the mean for service call lengths given in Example 1.26. Verify these results. Do not use software for this exercise and be sure to show all of your work.

1.54   Find the five-number summary. Here are the scores on the first exam in an introductory statistics course for 10 students:

$$80 \quad 73 \quad 92 \quad 85 \quad 75 \quad 98 \quad 93 \quad 55 \quad 80 \quad 90$$

Find the five-number summary for these first-exam scores.

The five-number summary leads to another visual representation of a distribution, the *boxplot*.

BOXPLOT

A **boxplot** is a graph of the five-number summary.

- A central box spans the quartiles Q_1 and Q_3.
- A line in the box marks the median M.
- Lines extend from the box out to the smallest and largest observations.

When you look at a boxplot, first locate the median, which marks the center of the distribution. Then look at the spread. The quartiles show the spread of the middle half of the data, and the extremes (the smallest and largest observations) show the spread of the entire data set.

CALLCENTER80

EXAMPLE

1.27 Service center call lengths. Table 1.2 (page 15) gives the call lengths for our sample of 80 service calls from our collection of 31,492 calls. In Exercise 1.54 you verified that the five-number summary for these data is 1.0, 54.5, 103.5, 200, and 2631. The boxplot is displayed in Figure 1.18. The skewness of the distribution is the major feature that we see in this plot. Note that the mean is marked with a "+" and appears very close to the third quartile at the upper edge of the box.

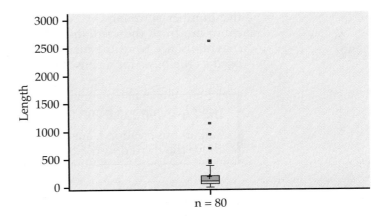

FIGURE 1.18 Modified boxplot for sample of 80 service center calls lengths, for Example 1.27. Suspected outliers identified by the 1.5 × *IQR* rule are plotted as individual points.

USE YOUR KNOWLEDGE

STATCOURSE

1.55 Make a boxplot. Here are the scores on the first exam in an introductory statistics course for 10 students:

80   73   92   85   75   98   93   55   80   90

Make a boxplot for these first-exam scores.

The 1.5 × *IQR* rule for suspected outliers

Look again at the boxplot for the 80 service center call lengths given in Figure 1.18 and the display of the data in Table 1.2 (page 15). There is a clear outlier, a call lasting 2631 seconds, more than twice the length of any other call. How shall we describe the spread of this distribution? The smallest and largest observations are extremes that do not describe the spread of the majority of the data. The distance between the quartiles (the range of the center half of the data) is a more resistant measure of spread. This distance is called the *interquartile range*.

> **THE INTERQUARTILE RANGE *IQR***
>
> The **interquartile range *IQR*** is the distance between the first and third quartiles,
>
> $$IQR = Q_3 - Q_1$$

For our data on service call lengths, $IQR = 200 - 54.5 = 145.5$. The quartiles and the IQR are not affected by changes in either tail of the distribution. They are therefore resistant, because changes in a few data points have no further effect once these points move outside the quartiles. However, *no single numerical measure of spread, such as IQR, is very useful for describing skewed distributions.* The two sides of a skewed distribution have different spreads, so one number can't summarize them. We can often detect skewness from the five-number summary by comparing how far the first quartile and the minimum are from the median (left tail) with how far the third quartile and the maximum are from the median (right tail). The interquartile range is mainly used as the basis for a rule of thumb for identifying suspected outliers.

> **THE 1.5 × *IQR* RULE FOR OUTLIERS**
>
> Call an observation a suspected outlier if it falls more than $1.5 \times IQR$ above the third quartile or below the first quartile.

EXAMPLE

1.28 Outliers for call length data. For the call length data in Table 1.2,

$$1.5 \times IQR = 1.5 \times 145.5 = 218.25$$

Any values below $54.5 - 218.25 = -163.75$ or above $200 + 218.25 = 418.25$ are flagged as possible outliers. There are no low outliers, but the 8 longest calls are flagged as possible high outliers. Their lengths are

438 465 479 700 700 951 1148 2631

Statistical software often uses the $1.5 \times IQR$ rule. For example, the stemplot in Figure 1.18 plots the 8 call lengths identified by the $1.5 \times IQR$ rule as separate points. A plot such as this where suspected outliers are identified individually is called a **modified boxplot.** In Figure 1.18, the lines extend out from the central box only to the smallest and largest observations that are not flagged by the $1.5 \times IQR$ rule.

The distribution of call lengths is very strongly skewed. We may well decide that only the longest call is truly an outlier in the sense of deviating from the overall pattern of the distribution. The other 7 calls are just part of the long right tail. The $1.5 \times IQR$ rule does not remove the need to look at the distribution and use judgment. It is useful mainly to call our attention to unusual observations.

modified boxplot

1.56 Find the *IQR*. Here are the scores on the first exam in an introductory statistics course for 10 students:

80   73   92   85   75   98   93   55   80   90

Find the interquartile range and use the $1.5 \times IQR$ rule to check for outliers. How low would the lowest score need to be for it to be an outlier according to this rule?

The modified boxplot in Figure 1.18 tells us much more about the distribution of call lengths than the five-number summary or other numerical measures. The routine methods of statistics compute numerical measures and draw conclusions based on their values. These methods are very useful, and we will study them carefully in later chapters. But they cannot be applied blindly, by feeding data to a computer program, because *statistical measures and methods based on them are generally meaningful only for distributions of sufficiently regular shape.* This principle will become clearer as we progress, but it is good to be aware at the beginning that quickly resorting to fancy calculations is the mark of a statistical amateur. Look, think, and choose your calculations selectively.

Measuring spread: the standard deviation

The five-number summary is not the most common numerical description of a distribution. That distinction belongs to the combination of the mean to measure center and the *standard deviation* to measure spread. The standard deviation measures spread by looking at how far the observations are from their mean.

THE STANDARD DEVIATION *s*

The **variance** s^2 of a set of observations is the average of the squares of the deviations of the observations from their mean. In symbols, the variance of n observations $x_1, x_2, \ldots, x_n$ is

$$s^2 = \frac{(x_1 - \bar{x})^2 + (x_2 - \bar{x})^2 + \cdots + (x_n - \bar{x})^2}{n - 1}$$

or, in more compact notation,

$$s^2 = \frac{1}{n - 1} \sum (x_i - \bar{x})^2$$

The **standard deviation** s is the square root of the variance s^2:

$$s = \sqrt{\frac{1}{n - 1} \sum (x_i - \bar{x})^2}$$

The idea behind the variance and the standard deviation as measures of spread is as follows: The deviations $x_i - \bar{x}$ display the spread of the values x_i

about their mean $\bar{x}$. Some of these deviations will be positive and some negative because some of the observations fall on each side of the mean. In fact, *the sum of the deviations of the observations from their mean will always be zero.* Squaring the deviations makes them all positive, so that observations far from the mean in either direction have large positive squared deviations. The variance is the average squared deviation. Therefore, s^2 and s will be large if the observations are widely spread about their mean, and small if the observations are all close to the mean.

EXAMPLE

METABOLIC

1.29 Metabolic rate. A person's metabolic rate is the rate at which the body consumes energy. Metabolic rate is important in studies of weight gain, dieting, and exercise. Here are the metabolic rates of 7 men who took part in a study of dieting. (The units are calories per 24 hours. These are the same calories used to describe the energy content of foods.)

$$1792 \quad 1666 \quad 1362 \quad 1614 \quad 1460 \quad 1867 \quad 1439$$

Enter these data into your calculator or software and verify that

$$\bar{x} = 1600 \text{ calories} \quad s = 189.24 \text{ calories}$$

Figure 1.19 plots these data as dots on the calorie scale, with their mean marked by an asterisk (∗). The arrows mark two of the deviations from the mean. If you were calculating s by hand, you would find the first deviation as

$$x_1 - \bar{x} = 1792 - 1600 = 192$$

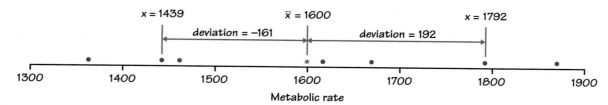

FIGURE 1.19 Metabolic rates for seven men, with the mean (*) and the deviations of two observations from the mean, for Example 1.29.

Exercise 1.78 asks you to calculate the seven deviations from Example 1.29, square them, and find s^2 and s directly from the deviations. Working one or two short examples by hand helps you understand how the standard deviation is obtained. In practice you will use either software or a calculator that will find s from keyed-in data. The software outputs in Figure 1.17 give both the variance and standard deviation for the data on the time to start a business.

USE YOUR KNOWLEDGE

STATCOURSE

1.57 Find the variance and the standard deviation. Here are the scores on the first exam in an introductory statistics course for 10 students:

$$80 \quad 73 \quad 92 \quad 85 \quad 75 \quad 98 \quad 93 \quad 55 \quad 80 \quad 90$$

Find the variance and the standard deviation for these first-exam scores.

The idea of the variance is straightforward: it is the average of the squares of the deviations of the observations from their mean. The details we have just presented, however, raise some questions.

Why do we square the deviations?

- First, the sum of the squared deviations of any set of observations from their mean is the smallest that the sum of squared deviations from any number can possibly be. This is not true of the unsquared distances. So squared deviations point to the mean as center in a way that distances do not.

- Second, the standard deviation turns out to be the natural measure of spread for a particularly important class of symmetric unimodal distributions, the *Normal distributions.* We will meet the Normal distributions in the next section. We commented earlier that the usefulness of many statistical procedures is tied to distributions of particular shapes. This is distinctly true of the standard deviation.

Why do we emphasize the standard deviation rather than the variance?

- One reason is that s, not s^2, is the natural measure of spread for Normal distributions, which are introduced in the next section.

- There is also a more general reason to prefer s to s^2. Because the variance involves squaring the deviations, it does not have the same unit of measurement as the original observations. The variance of the metabolic rates, for example, is measured in squared calories. Taking the square root remedies this. The standard deviation s measures spread about the mean in the original scale.

Why do we average by dividing by $n-1$ rather than n in calculating the variance?

- Because the sum of the deviations is always zero, the last deviation can be found once we know the other $n-1$. So we are not averaging n unrelated numbers. Only $n-1$ of the squared deviations can vary freely, and we average by dividing the total by $n-1$.

degrees of freedom
- The number $n-1$ is called the **degrees of freedom** of the variance or standard deviation. Many calculators offer a choice between dividing by n and dividing by $n-1$, so be sure to use $n-1$.

Properties of the standard deviation

Here are the basic properties of the standard deviation s as a measure of spread.

PROPERTIES OF THE STANDARD DEVIATION

- s measures spread about the mean and should be used only when the mean is chosen as the measure of center.

- $s = 0$ only when there is *no spread.* This happens only when all observations have the same value. Otherwise, $s > 0$. As the observations become more spread out about their mean, s gets larger.

- s, like the mean $\bar{x}$, is not resistant. A few outliers can make s very large.

1.58 A standard deviation of zero. Construct a data set with 5 cases that has a variable with $s = 0$.

The use of squared deviations renders s even more sensitive than $\bar{x}$ to a few extreme observations. For example, when we added Suriname to our sample of 24 countries for the analysis of the time to start a business (Example 1.22 and Exercise 1.48), we increase the standard deviation from 18.6 to 132.6! Distributions with outliers and strongly skewed distributions have large standard deviations. The number s does not give much helpful information about such distributions.

TIMETOSTART24

TIMETOSTART25

1.59 Effect of an outlier on the *IQR*. Find the *IQR* for the time to start a business with and without Suriname. What do you conclude about the sensitivity of this measure of spread to the inclusion of an outlier?

Choosing measures of center and spread

How do we choose between the five-number summary and $\bar{x}$ and s to describe the center and spread of a distribution? Because the two sides of a strongly skewed distribution have different spreads, no single number such as s describes the spread well. The five-number summary, with its two quartiles and two extremes, does a better job.

CHOOSING A SUMMARY

The five-number summary is usually better than the mean and standard deviation for describing a skewed distribution or a distribution with strong outliers. Use $\bar{x}$ and s only for reasonably symmetric distributions that are free of outliers.

Remember that a graph gives the best overall picture of a distribution. Numerical measures of center and spread report specific facts about a distribution, but they do not describe its entire shape. Numerical summaries do not disclose the presence of multiple modes or gaps, for example. **Always plot your data.**

Changing the unit of measurement

The same variable can be recorded in different units of measurement. Americans commonly record distances in miles and temperatures in degrees Fahrenheit, while the rest of the world measures distances in kilometers and temperatures in degrees Celsius. Fortunately, it is easy to convert numerical descriptions of a distribution from one unit of measurement to another. This is true because a change in the measurement unit is a *linear transformation* of the measurements.

LINEAR TRANSFORMATIONS

A **linear transformation** changes the original variable x into the new variable x_{new} given by an equation of the form

$$x_{new} = a + bx$$

Adding the constant a shifts all values of x upward or downward by the same amount. In particular, such a shift changes the origin (zero point) of the variable. Multiplying by the positive constant b changes the size of the unit of measurement.

EXAMPLE

1.30   Change the units

(a) If a distance x is measured in kilometers, the same distance in miles is

$$x_{new} = 0.62x$$

For example, a 10-kilometer race covers 6.2 miles. This transformation changes the units without changing the origin—a distance of 0 kilometers is the same as a distance of 0 miles.

(b) A temperature x measured in degrees Fahrenheit must be reexpressed in degrees Celsius to be easily understood by the rest of the world. The transformation is

$$x_{new} = \frac{5}{9}(x - 32) = -\frac{160}{9} + \frac{5}{9}x$$

Thus, the high of 95°F on a hot American summer day translates into 35°C. In this case

$$a = -\frac{160}{9} \quad \text{and} \quad b = \frac{5}{9}$$

This linear transformation changes both the unit size and the origin of the measurements. The origin in the Celsius scale (0°C, the temperature at which water freezes) is 32° in the Fahrenheit scale.

Linear transformations do not change the shape of a distribution. If measurements on a variable x have a right-skewed distribution, any new variable x_{new} obtained by a linear transformation $x_{new} = a + bx$ (for $b > 0$) will also have a right-skewed distribution. If the distribution of x is symmetric and unimodal, the distribution of x_{new} remains symmetric and unimodal.

Although a linear transformation preserves the basic shape of a distribution, the center and spread will change. Because linear changes of measurement scale are common, we must be aware of their effect on numerical descriptive measures of center and spread. Fortunately, the changes follow a simple pattern.

EXAMPLE

1.31 Use scores to find the points. In an introductory statistics course, homework counts for 300 points out of a total of 1000 possible points for all course requirements. During the semester there were 12 homework assignments and each was given a grade on a scale of 0 to 100. The maximum total score for the 12 homework assignments is therefore 1200. To convert the homework scores to final grade points, we need to convert the scale of 0 to 1200 to a scale of 0 to 300. We do this by multiplying the homework scores by 300/1200. In other words, we divide the homework scores by 4. Here are the homework scores and the corresponding final grade points for 5 students:

Student	1	2	3	4	5
Score	1056	1080	900	1164	1020
Points	264	270	225	291	255

These two sets of numbers measure the same performance on homework for the course. Since we obtained the points by dividing the scores by 4, the mean of the points will be the mean of the scores divided by 4. Similarly, the standard deviation of points will be the standard deviation of the scores divided by 4.

USE YOUR KNOWLEDGE

1.60 Calculate the points for a student. Use the setting of Example 1.31 to find the points for a student whose score is 950.

Here is a summary of the rules for linear transformations:

EFFECT OF A LINEAR TRANSFORMATION

To see the effect of a linear transformation on measures of center and spread, apply these rules:

- Multiplying each observation by a positive number b multiplies both measures of center (mean and median) and measures of spread (interquartile range and standard deviation) by b.

- Adding the same number a (either positive or negative) to each observation adds a to measures of center and to quartiles and other percentiles but does not change measures of spread.

In Example 1.31, when we converted from score to points, we described the transformation as dividing by 4. The multiplication part of the summary of the effect of a linear transformation applies to this case, because division by 4 is the same as multiplication by 0.25. Similarly, the second part of the summary applies to subtraction as well as addition, because subtraction is simply the addition of a negative number.

The measures of spread IQR and s do not change when we add the same number a to all of the observations because adding a constant changes the location of the distribution but leaves the spread unaltered. You can find the effect of a linear transformation $x_{new} = a + bx$ by combining these rules. For example, if x has mean $\bar{x}$, the transformed variable x_{new} has mean $a + b\bar{x}$.

SECTION 1.2 Summary

A numerical summary of a distribution should report its **center** and its **spread** or **variability.**

The **mean** $\bar{x}$ and the **median** M describe the center of a distribution in different ways. The mean is the arithmetic average of the observations, and the median is their midpoint.

When you use the median to describe the center of the distribution, describe its spread by giving the **quartiles.** The **first quartile Q_1** has one-fourth of the observations below it, and the **third quartile Q_3** has three-fourths of the observations below it.

The **interquartile range** is the difference between the quartiles. It is the spread of the center half of the data. The **1.5 $\times$ IQR rule** flags observations more than $1.5 \times IQR$ beyond the quartiles as possible outliers.

The **five-number summary** consisting of the median, the quartiles, and the smallest and largest individual observations provides a quick overall description of a distribution. The median describes the center, and the quartiles and extremes show the spread.

Boxplots based on the five-number summary are useful for comparing several distributions. The box spans the quartiles and shows the spread of the central half of the distribution. The median is marked within the box. Lines extend from the box to the extremes and show the full spread of the data. In a **modified boxplot,** points identified by the $1.5 \times IQR$ rule are plotted individually.

The **variance s^2** and especially its square root, the **standard deviation s,** are common measures of spread about the mean as center. The standard deviation s is zero when there is no spread and gets larger as the spread increases.

A **resistant measure** of any aspect of a distribution is relatively unaffected by changes in the numerical value of a small proportion of the total number of observations, no matter how large these changes are. The median and quartiles are resistant, but the mean and the standard deviation are not.

The mean and standard deviation are good descriptions for symmetric distributions without outliers. They are most useful for the Normal distributions introduced in the next section. The five-number summary is a better exploratory summary for skewed distributions.

Linear transformations have the form $x_{new} = a + bx$. A linear transformation changes the origin if $a \neq 0$ and changes the size of the unit of measurement if $b > 0$. Linear transformations do not change the overall shape of a distribution. A linear transformation multiplies a measure of spread by b and changes a percentile or measure of center m into $a + bm$.

Numerical measures of particular aspects of a distribution, such as center and spread, do not report the entire shape of most distributions. In some cases, particularly distributions with multiple peaks and gaps, these measures may not be very informative.

SECTION 1.2 Exercises

For Exercises 1.47 and 1.48,
see pages 30–31; for Exercises 1.49 to 1.51, see page 32;
for Exercise 1.52, see page 35; for Exercises 1.53 to 1.54, see
page 36; for Exercise 1.55, see page 37; for Exercise 1.56, see
page 39; for Exercise 1.57, see page 40; for Exercise 1.58 and
Exercise 1.59, see page 42; and for Exercise 1.60, see page 44.

1.61 The value of brands. A brand is a symbol or images that are associated with a company. An effective brand identifies the company and its products. Using a variety of measures, dollar values for brands can be calculated.[25] The most valuable brand is Coca-Cola with a value of $66,667,000. Coke is followed by IBM at $59.031 million, Microsoft at $59.007 million, GE at $53.086 million, and Toyota at $34.050 million. For this exercise you will use the brand values, reported in millions of dollars, for the top 100 brands. ⬤ BRANDS

(a) Graphically display the distribution of the values of these brands.

(b) Use numerical measures to summarize the distribution.

(c) Write a short paragraph discussing the dollar values of the top 100 brands. Include the results of your analysis.

1.62 Alcohol content of beer. Brewing beer involves a variety of steps that can affect the alcohol content. A Web site gives the percent alcohol for 86 domestic brands of beer.[26] ⬤ BEER

(a) Use graphical and numerical summaries of your choice to describe these data. Give reasons for your choice.

(b) The data set contains an outlier. Explain why this particular beer is unusual and how its outlier status is related to how it is marketed.

1.63 An outlier for alcohol content of beer. Refer to the previous exercise. ⬤ BEER

(a) Calculate the mean with and without the outlier. Do the same for the median. Explain how these statistics change when the outlier is excluded.

(b) Calculate the standard deviation with and without the outlier. Do the same for the quartiles. Explain how these statistics change when the outlier is excluded.

(c) Write a short paragraph summarizing what you have learned in this exercise.

1.64 Calories in beer. Refer to the previous two exercises. The data set also gives the calories per 12 ounces of beverage. ⬤ BEER

(a) Analyze the data and summarize the distribution of calories for these 86 brands of beer.

(b) In the previous exercise you identified one brand of beer as an outlier. To what extent is this brand an outlier in the distribution of calories? Explain your answer.

(c) The distribution of calories suggests that there may be two groups of beers which might be marketed differently. Examine the data file carefully and explain the characteristics of the two groups.

1.65 Create a data set. Create a data set for which the median would change by a large amount if the smallest observation is deleted.

1.66 Salaries of the New York Mets. The mean salary of the players on the 2008 New York Mets baseball team was $4,910,900 while the median salary was $2,375,000. What explains the difference between these two measures of center?

1.67 Potatoes. A quality product is one that is consistent and has very little variability in its characteristics. Controlling variability can be more difficult with agricultural products than with those that are manufactured. The following table gives the weights, in ounces, of the 25 potatoes sold in a 10-pound bag. ⬤ POTATOES

7.8	7.9	8.2	7.3	6.7	7.9	7.9	7.9	7.6	7.8	7.0	4.7	7.6
6.3	4.7	4.7	4.7	6.3	6.0	5.3	4.3	7.9	5.2	6.0	3.7	

(a) Summarize the data graphically and numerically. Give reasons for methods you chose to use in your summaries.

(b) Do you think that your numerical summaries do an effective job of describing these data? Why or why not?

(c) There appear to be two distinct clusters of weights for these potatoes. Divide the sample into two subsamples based on the clustering. Give the mean and standard deviation for each subsample. Do you think that this way of summarizing these data is better than a numerical summary that uses all of the data as a single sample? Give a reason for your answer.

1.68 Longleaf pine trees. The Wade Tract in Thomas County, Georgia, is an old-growth forest of longleaf pine trees (*Pinus palustris*) that has survived in a relatively undisturbed state since before the settlement of the area by Europeans. A study collected data about 584 of these trees.[27] One of the variables measured was the diameter at breast height (DBH). This is the diameter of the tree at

4.5 feet and the units are centimeters (cm). Only trees with DBH greater than 1.5 cm were sampled. Here are the diameters of a random sample of 40 of these trees: **LONGLEAF**

10.5	13.3	26.0	18.3	52.2	9.2	26.1	17.6	40.5	31.8
47.2	11.4	2.7	69.3	44.4	16.9	35.7	5.4	44.2	2.2
4.3	7.8	38.1	2.2	11.4	51.5	4.9	39.7	32.6	51.8
43.6	2.3	44.6	31.5	40.3	22.3	43.3	37.5	29.1	27.9

(a) Find the five-number summary for these data.

(b) Make a boxplot.

(c) Make a histogram.

(d) Write a short summary of the major features of this distribution. Do you prefer the boxplot or the histogram for these data?

1.69 Blood proteins in children from Papua New Guinea. C-reactive protein (CRP) is a substance that can be measured in the blood. Values increase substantially within 6 hours of an infection and reach a peak within 24 to 48 hours after. In adults, chronically high values have been linked to an increased risk of cardiovascular disease. In a study of apparently healthy children aged 6 to 60 months in Papua New Guinea, CRP was measured in 90 children.[28] The units are milligrams per liter (mg/l). Here are the data from a random sample of 40 of these children: **CRP**

0.00	3.90	5.64	8.22	0.00	5.62	3.92	6.81	30.61	0.00
73.20	0.00	46.70	0.00	0.00	26.41	22.82	0.00	0.00	3.49
0.00	0.00	4.81	9.57	5.36	0.00	5.66	0.00	59.76	12.38
15.74	0.00	0.00	0.00	0.00	9.37	20.78	7.10	7.89	5.53

(a) Find the five-number summary for these data.

(b) Make a boxplot.

(c) Make a histogram.

(d) Write a short summary of the major features of this distribution. Do you prefer the boxplot or the histogram for these data?

1.70 ⚠ Transform the blood proteins values. Refer to the previous exercise. With strongly skewed distributions such as this, we frequently reduce the skewness by taking a log transformation. We have a bit of a problem here, however, because some of the data are recorded as 0.00 and the logarithm of zero is not defined. For this variable, the value 0.00 is recorded whenever the amount of CRP in the blood is below the level that the measuring instrument is capable of detecting. The usual procedure in this circumstance is to add a small number to each observation before taking the logs. Transform

these data by adding 1 to each observation and then taking the logarithm. Use the questions in the previous exercise as a guide to your analysis and prepare a summary contrasting this analysis with the one that you performed in the previous exercise. **CRP**

1.71 ⚠ Vitamin A deficiency in children from Papua New Guinea. In the Papua New Guinea study that provided the data for the previous two exercises, the researchers also measured serum retinol. A low value of this variable can be an indicator of vitamin A deficiency. Here are the data on the same sample of 40 children from this study. The units are micromoles per liter (μmol/l).

1.15	1.36	0.38	0.34	0.35	0.37	1.17	0.97	0.97	0.67
0.31	0.99	0.52	0.70	0.88	0.36	0.24	1.00	1.13	0.31
1.44	0.35	0.34	1.90	1.19	0.94	0.34	0.35	0.33	0.69
0.69	1.04	0.83	1.11	1.02	0.56	0.82	1.20	0.87	0.41

Analyze these data. Use the questions in the previous two exercises as a guide. **VITAMINA**

1.72 Luck and puzzle solving. Children in a psychology study were asked to solve some puzzles and were then given feedback on their performance. They then were asked to rate how luck played a role in determining their scores.[29] This variable was recorded on a 1 to 10 scale with 1 corresponding to very lucky and 10 corresponding to very unlucky. Here are the scores for 60 children:

1	10	1	10	1	1	10	5	1	1	8	1	10	2	1
9	5	2	1	8	10	5	9	10	10	9	6	10	1	5
1	9	2	1	7	10	9	5	10	10	10	1	8	1	6
10	1	6	10	10	8	10	3	10	8	1	8	10	4	2

Use numerical and graphical methods to describe these data. Write a short report summarizing your work. **LUCK**

1.73 Median versus mean for net worth. A report on the assets of American households says that the median net worth of U.S. families is $120,300. The mean net worth of these families is $556,300.[30] What explains the difference between these two measures of center?

1.74 Carbon dioxide emissions. Table 1.3 (page 26) gives carbon dioxide (CO_2) emissions per person for countries with population at least 20 million. The distribution is strongly skewed to the right. The United States and several other countries appear to be high outliers.

(a) Give the five-number summary. Explain why this summary suggests that the distribution is right-skewed.

(b) Which countries are outliers according to the $1.5 \times IQR$ rule? Make a stemplot or histogram of the data.

Do you agree with the rule's suggestions about which countries are and are not outliers?

1.75 Mean versus median. A small accounting firm pays each of its six clerks $35,000, two junior accountants $80,000 each, and the firm's owner $320,000. What is the mean salary paid at this firm? How many of the employees earn less than the mean? What is the median salary?

1.76 Be careful about how you treat the zeros. In computing the median income of any group, some federal agencies omit all members of the group who had no income. Give an example to show that the reported median income of a group can go down even though the group becomes economically better off. Is this also true of the mean income?

1.77 How does the median change? The firm in Exercise 1.75 gives no raises to the clerks and junior accountants, while the owner's take increases to $455,000. How does this change affect the mean? How does it affect the median?

1.78 Metabolic rates. Calculate the mean and standard deviation of the metabolic rates in Example 1.29 (page 40), showing each step in detail. First find the mean $\bar{x}$ by summing the 7 observations and dividing by 7. Then find each of the deviations $x_i - \bar{x}$ and their squares. Check that the deviations have sum 0. Calculate the variance as an average of the squared deviations (remember to divide by $n - 1$). Finally, obtain s as the square root of the variance. METABOLIC

1.79 Hurricanes and losses. A discussion of extreme weather says, "In most states, hurricanes occur infrequently. Yet, when a hurricane hits, the losses can be catastrophic. Average annual losses are not a meaningful measure of damage from rare but potentially catastrophic events."[31] Why is this true?

1.80 Distributions for time spent studying. Exercise 1.41 (page 26) presented data on the nightly study time claimed by first-year college men and women. The most common methods for formal comparison of two groups use $\bar{x}$ and s to summarize the data. We wonder if this is appropriate here. Look at your back-to-back stemplot from Exercise 1.41, or make one now if you have not done so.

(a) What kinds of distributions are best summarized by $\bar{x}$ and s? It isn't easy to decide whether small data sets with irregular distributions fit the criteria. We will learn a better tool for making this decision in the next section.

(b) Each set of study times appears to contain a high outlier. Are these points flagged as suspicious by the

$1.5 \times IQR$ rule? How much does removing the outlier change $\bar{x}$ and s for each group? The presence of outliers makes us reluctant to use the mean and standard deviation for these data unless we remove the outliers on the grounds that these students were exaggerating.

1.81 The density of the earth. Many standard statistical methods that you will study in Part II of this book are intended for use with distributions that are symmetric and have no outliers. These methods start with the mean and standard deviation, $\bar{x}$ and s. Two examples of scientific data for which standard methods should work well are the pH measurements in Exercise 1.38 (page 26) and Cavendish's measurements of the density of the earth in Exercise 1.42 (page 27). EARTHDENSITY

(a) Summarize each of these data sets by giving $\bar{x}$ and s.

(b) Find the median for each data set. Is the median quite close to the mean, as we expect it to be in these examples?

1.82 IQ scores. Many standard statistical methods that you will study in Part II of this book are intended for use with distributions that are symmetric and have no outliers. These methods start with the mean and standard deviation, $\bar{x}$ and s. For example, standard methods would typically be used for the IQ and GPA data in Table 1.4 (page 28). IQGPA

(a) Find $\bar{x}$ and s for the IQ data. In large populations, IQ scores are standardized to have mean 100 and standard deviation 15. In what way does the distribution of IQ among these students differ from the overall population?

(b) Find the median IQ score. It is, as we expect, close to the mean.

(c) Find the mean and median for the GPA data. The two measures of center differ a bit. What feature of the data (see your stemplot in Exercise 1.43 or make a new stemplot) explains the difference?

1.83 Mean and median for two observations. The *Mean and Median* applet allows you to place observations on a line and see their mean and median visually. Place two observations on the line by clicking below it. Why does only one arrow appear?

1.84 Mean and median for three observations. In the *Mean and Median* applet, place three observations on the line by clicking below it, two close together near the center of the line and one somewhat to the right of these two.

(a) Pull the single rightmost observation out to the right. (Place the cursor on the point, hold down a mouse button, and drag the point.) How does the mean behave? How does the median behave? Explain briefly why each measure acts as it does.

(b) Now drag the rightmost point to the left as far as you can. What happens to the mean? What happens to the median as you drag this point past the other two (watch carefully)?

1.85 **Mean and median for five observations.** Place five observations on the line in the *Mean and Median* applet by clicking below it.

(a) Add one additional observation *without changing the median*. Where is your new point?

(b) Use the applet to convince yourself that when you add yet another observation (there are now seven in all), the median does not change no matter where you put the seventh point. Explain why this must be true.

1.86 Hummingbirds and flowers. Different varieties of the tropical flower *Heliconia* are fertilized by different species of hummingbirds. Over time, the lengths of the flowers and the form of the hummingbirds' beaks have evolved to match each other. Here are data on the lengths in millimeters of three varieties of these flowers on the island of Dominica:[32]

			H. bihai				
47.12	46.75	46.81	47.12	46.67	47.43	46.44	46.64
48.07	48.34	48.15	50.26	50.12	46.34	46.94	48.36

			H. caribaea red				
41.90	42.01	41.93	43.09	41.47	41.69	39.78	40.57
39.63	42.18	40.66	37.87	39.16	37.40	38.20	38.07
38.10	37.97	38.79	38.23	38.87	37.78	38.01	

			H. caribaea yellow				
36.78	37.02	36.52	36.11	36.03	35.45	38.13	37.1
35.17	36.82	36.66	35.68	36.03	34.57	34.63	

Make boxplots to compare the three distributions. Report the five-number summaries along with your graph. What are the most important differences among the three varieties of flower? HELICONIA

1.87 Compare the three varieties of flowers. The biologists who collected the flower length data in the previous exercise compared the three *Heliconia* varieties using statistical methods based on $\bar{x}$ and s. HELICONIA

(a) Find $\bar{x}$ and s for each variety.

(b) Make a stemplot of each set of flower lengths. Do the distributions appear suitable for use of $\bar{x}$ and s as summaries?

1.88 Imputation. Various problems with data collection can cause some observations to be missing. Suppose a data set has 20 cases. Here are the values of the variable x for 10 of these cases: IMPUTATION

17   6   12   14   20   23   9   12   16   21

The values for the other 10 cases are missing. One way to deal with missing data is called **imputation.** The basic idea is that missing values are replaced, or imputed, with values that are based on an analysis of the data that are not missing. For a data set with a single variable, the usual choice of a value for imputation is the mean of the values that are not missing. The mean for this data set is 15.

(a) Verify that the mean is 15 and find the standard deviation for the 10 cases for which x is not missing.

(b) Create a new data set with 20 cases by setting the values for the 10 missing cases to 15. Compute the mean and standard deviation for this data set.

(c) Summarize what you have learned about the possible effects of this type of imputation on the mean and the standard deviation.

1.89 **Shakespeare's plays.** Look at the histogram of lengths of words in Shakespeare's plays, Figure 1.13 (page 25). The heights of the bars tell us what percent of words have each length. What is the median length of words used by Shakespeare? Similarly, what are the quartiles? Give the five-number summary for Shakespeare's word lengths. SHAKESPEARE

1.90 **Create a data set.** Create a set of 5 positive numbers (repeats allowed) that have median 10 and mean 7. What thought process did you use to create your numbers?

1.91 Create another data set. Give an example of a small set of data for which the mean is larger than the third quartile.

1.92 **Deviations from the mean sum to zero.** Use the definition of the mean $\bar{x}$ to show that the sum of the deviations $x_i - \bar{x}$ of the observations from their mean is always zero. This is one reason why the variance and standard deviation use squared deviations.

1.93 **A standard deviation contest.** This is a standard deviation contest. You must choose four numbers from the whole numbers 0 to 20, with repeats allowed.

(a) Choose four numbers that have the smallest possible standard deviation.

(b) Choose four numbers that have the largest possible standard deviation.

(c) Is more than one choice possible in either (a) or (b)? Explain.

1.94 Does your software give incorrect answers? This exercise requires a calculator with a standard deviation button or statistical software on a computer. The observations

20,001   20,002   20,003

have mean $\bar{x} = 20{,}002$ and standard deviation $s = 1$. Adding a 0 in the center of each number, the next set becomes

200,001   200,002   200,003

The standard deviation remains $s = 1$ as more 0s are added. Use your calculator or computer to calculate the standard deviation of these numbers, adding extra 0s until you get an incorrect answer. How soon did you go wrong? This demonstrates that calculators and computers cannot handle an arbitrary number of digits correctly.

1.95 Compare three varieties of flowers. Exercise 1.86 reports data on the lengths in millimeters of flowers of three varieties of *Heliconia*. In Exercise 1.87 you found the mean and standard deviation for each variety. Starting from the $\bar{x}$- and s-values in millimeters, find the means and standard deviations in inches. (A millimeter is 1/1000 of a meter. A meter is 39.37 inches.)

1.96 ⚠ **The density of the earth.** Henry Cavendish (see Exercise 1.42, page 27) used $\bar{x}$ to summarize his 29 measurements of the density of the earth.

(a) Find $\bar{x}$ and s for his data.

(b) Cavendish recorded the density of the earth as a multiple of the density of water. The density of water is almost exactly 1 gram per cubic centimeter, so his measurements have these units. In American units, the density of water is 62.43 pounds per cubic foot. This is the weight of a cube of water measuring 1 foot (that is, 30.48 cm) on each side. Express Cavendish's first result for the earth (5.50 g/cm^3) in pounds per cubic foot. Then find $\bar{x}$ and s in pounds per cubic foot.

1.97 Weight gain. A study of diet and weight gain deliberately overfed 16 volunteers for eight weeks. The mean increase in fat was $\bar{x} = 2.42$ kilograms and the standard deviation was $s = 1.18$ kilograms. What are $\bar{x}$ and s in pounds? (A kilogram is 2.2 pounds.)

1.98 ⚠ **Changing units from inches to centimeters.** Changing the unit of length from inches to centimeters multiplies each length by 2.54 because there are 2.54 centimeters in an inch. This change of units multiplies our usual measures of spread by 2.54. This is true of *IQR* and the standard deviation. What happens to the variance when we change units in this way?

1.99 A different type of mean. The **trimmed mean** is a measure of center that is more resistant than the mean but uses more of the available information than the median. To compute the 10% trimmed mean, discard the highest 10% and the lowest 10% of the observations and compute the mean of the remaining 80%. Trimming eliminates the effect of a small number of outliers. Compute the 10% trimmed mean of the service time data in Table 1.2 (page 15). Then compute the 20% trimmed mean. Compare the values of these measures with the median and the ordinary untrimmed mean.

1.100 ⚠ **Changing units from centimeters to inches.** Refer to Exercise 1.68. Change the measurements from centimeters to inches by multiplying each value by 0.39. Answer the questions from the previous exercise and explain the effect of the transformation on these data.

1.3 Density Curves and Normal Distributions

We now have a kit of graphical and numerical tools for describing distributions. What is more, we have a clear strategy for exploring data on a single quantitative variable:

1. Always plot your data: make a graph, usually a stemplot or a histogram.

2. Look for the overall pattern and for striking deviations such as outliers.

3. Calculate an appropriate numerical summary to briefly describe center and spread.

Technology has expanded the set of graphs that we can choose for Step 1. It is possible, though painful, to make histograms by hand. Using software, clever algorithms can describe a distribution in a way that is not feasible by hand, by fitting a smooth curve to the data in addition to or instead of a histogram. The curves used are called **density curves.** Before we examine density curves in detail, here is an example of what software can do.

EXAMPLE

1.32 Density curves for pH and numbers of undergraduate students. Figure 1.20 illustrates the use of density curves along with histograms to describe distributions. Figure 1.20(a) shows the distribution of the acidity (pH) of rainwater, from Exercise 1.38 (page 26). That exercise illustrates how the choice of classes can change the shape of a histogram. The density curve and the software's default histogram agree that the distribution has a single peak and is approximately symmetric.

Figure 1.20(b) shows a strongly skewed distribution, numbers of undergraduate students in states from Example 1.17 (page 18). The histogram and density curve agree on the overall shape and on the long right tail.

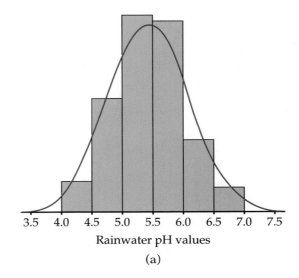

(a)

(b)

FIGURE 1.20 (a) The distribution of pH values measuring the acidity of 105 samples of rainwater, for Example 1.32. The roughly symmetric distribution is pictured with both a histogram and a density curve. (b) The distribution of the number of undergraduate students in states, for Example 1.32. This right-skewed distribution is pictured with both a histogram and a density curve.

In general, software that draws density curves describes the data in a way that is less arbitrary than choosing classes for a histogram. A smooth density curve is, however, an idealization that pictures the overall pattern of the data but ignores minor irregularities as well as any outliers. We will concentrate, not on general density curves, but on a special class, the bell-shaped Normal curves.

Density curves

One way to think of a density curve is as a smooth approximation to the irregular bars of a histogram. Figure 1.21 shows a histogram of the scores of all 947 seventh-grade students in Gary, Indiana, on the vocabulary part of the Iowa Test of Basic Skills. Scores of many students on this national test have a very regular distribution. The histogram is symmetric, and both tails fall off quite smoothly from a single center peak. There are no large gaps or obvious outliers. The curve drawn through the tops of the histogram bars in Figure 1.21 is a good description of the overall pattern of the data.

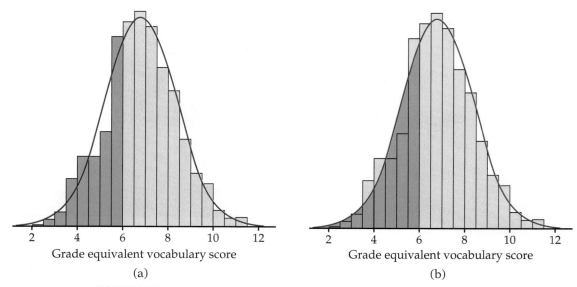

Grade equivalent vocabulary score

(a)

Grade equivalent vocabulary score

(b)

FIGURE 1.21   (a) The distribution of Iowa Test vocabulary scores for Gary, Indiana, seventh-graders. The shaded bars in the histogram represent scores less than or equal to 6.0. (b) The shaded area under the Normal density curve also represents scores less than or equal to 6.0. This area is 0.293, close to the true 0.303 for the actual data.

EXAMPLE

1.33   Vocabulary scores. In a histogram, the *areas* of the bars represent either counts or proportions of the observations. In Figure 1.21(a) we have shaded the bars that represent students with vocabulary scores 6.0 or lower. There are 287 such students, who make up the proportion $287/947 = 0.303$ of all Gary seventh-graders. The shaded bars in Figure 1.21(a) make up proportion 0.303 of the total area under all the bars. If we adjust the scale so that the total area of the bars is 1, the area of the shaded bars will be 0.303.

In Figure 1.21(b), we have shaded the *area under the curve* to the left of 6.0. Adjust the scale so that the total area under the curve is exactly 1.

Areas under the curve then represent proportions of the observations. That is, *area = proportion.* The curve is then a density curve. The shaded area under the density curve in Figure 1.21(b) represents the proportion of students with score 6.0 or lower. This area is 0.293, only 0.010 away from the histogram result. You can see that areas under the density curve give quite good approximations of areas given by the histogram.

DENSITY CURVE

A **density curve** is a curve that

- is always on or above the horizontal axis and

- has area exactly 1 underneath it.

A density curve describes the overall pattern of a distribution. The area under the curve and above any range of values is the proportion of all observations that fall in that range.

The density curve in Figure 1.21 is a *Normal curve.* Density curves, like distributions, come in many shapes. Figure 1.22 shows two density curves, a symmetric Normal density curve and a right-skewed curve. A density curve of an appropriate shape is often an adequate description of the overall pattern of a distribution. Outliers, which are deviations from the overall pattern, are not described by the curve.

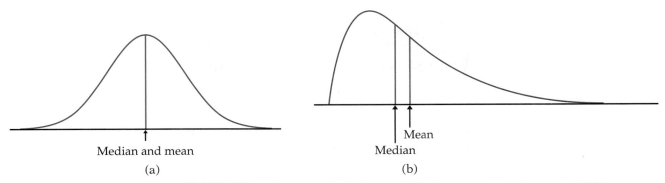

(a)   (b)

FIGURE 1.22   (a) A symmetric density curve with its mean and median marked. (b) A right-skewed density curve with its mean and median marked.

Measuring center and spread for density curves

Our measures of center and spread apply to density curves as well as to actual sets of observations, but only some of these measures are easily seen from the curve. A **mode** of a distribution described by a density curve is a peak point of the curve, the location where the curve is highest. Because areas under a density curve represent proportions of the observations, the **median** is the point with half the total area on each side. You can roughly locate the **quartiles** by dividing the area under the curve into quarters as accurately as possible by eye. The *IQR* is then the distance between the first and third quartiles. There

FIGURE 1.23 The mean of a density curve is the point at which it would balance.

are mathematical ways of calculating areas under curves. These allow us to locate the median and quartiles exactly on any density curve.

What about the mean and standard deviation? The mean of a set of observations is their arithmetic average. If we think of the observations as weights strung out along a thin rod, the mean is the point at which the rod would balance. This fact is also true of density curves. The mean is the point at which the curve would balance if it were made out of solid material. Figure 1.23 illustrates this interpretation of the mean. We have marked the mean and median on the density curves in Figure 1.22. A symmetric curve, such as the Normal curve in Figure 1.23(a), balances at its center of symmetry. Half the area under a symmetric curve lies on either side of its center, so this is also the median. For a right-skewed curve, such as that shown in Figure 1.23(b), the small area in the long right tail tips the curve more than the same area near the center. The mean (the balance point) therefore lies to the right of the median. It is hard to locate the balance point by eye on a skewed curve. There are mathematical ways of calculating the mean for any density curve, so we are able to mark the mean as well as the median in Figure 1.23(b). The standard deviation can also be calculated mathematically, but it can't be located by eye on most density curves.

MEDIAN AND MEAN OF A DENSITY CURVE

The **median** of a density curve is the equal-areas point, the point that divides the area under the curve in half.

The **mean** of a density curve is the balance point, at which the curve would balance if made of solid material.

The median and mean are the same for a symmetric density curve. They both lie at the center of the curve. The mean of a skewed curve is pulled away from the median in the direction of the long tail.

A density curve is an idealized description of a distribution of data. For example, the symmetric density curve in Figure 1.21 is exactly symmetric, but the histogram of vocabulary scores is only approximately symmetric. We therefore need to distinguish between the mean and standard deviation of the density curve and the numbers $\bar{x}$ and s computed from the actual observations. The

mean μ
standard deviation σ

usual notation for the mean of an idealized distribution is μ (the Greek letter mu). We write the standard deviation of a density curve as σ (the Greek letter sigma).

Normal distributions

One particularly important class of density curves has already appeared in Figures 1.21 and 1.22(a). These density curves are symmetric, unimodal, and bell-

Normal curves   shaped. They are called **Normal curves,** and they describe *Normal distributions*.

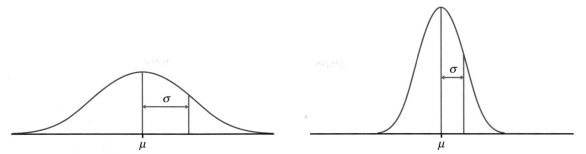

FIGURE 1.24   Two Normal curves, showing the mean μ and the standard deviation σ.

All Normal distributions have the same overall shape. The exact density curve for a particular Normal distribution is specified by giving its mean μ and its standard deviation σ. The mean is located at the center of the symmetric curve and is the same as the median. Changing μ without changing σ moves the Normal curve along the horizontal axis without changing its spread. The standard deviation σ controls the spread of a Normal curve. Figure 1.24 shows two Normal curves with different values of σ. The curve with the larger standard deviation is more spread out.

The standard deviation σ is the natural measure of spread for Normal distributions. Not only do μ and σ completely determine the shape of a Normal curve, but we can locate σ by eye on the curve. Here's how. As we move out in either direction from the center μ, the curve changes from falling ever more steeply

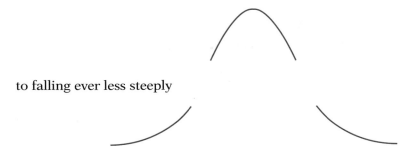

to falling ever less steeply

The points at which this change of curvature takes place are located at distance σ on either side of the mean μ. You can feel the change as you run your finger along a Normal curve, and so find the standard deviation. Remember that μ and σ alone do not specify the shape of most distributions, and that the shape of density curves in general does not reveal σ. These are special properties of Normal distributions.

There are other symmetric bell-shaped density curves that are not Normal. The Normal density curves are specified by a particular equation. The height of the density curve at any point x is given by

$$\frac{1}{\sigma\sqrt{2\pi}}e^{-\frac{1}{2}\left(\frac{x-\mu}{\sigma}\right)^2}$$

We will not make direct use of this fact, although it is the basis of mathematical work with Normal distributions. Notice that the equation of the curve is completely determined by the mean μ and the standard deviation σ.

Why are the Normal distributions important in statistics? Here are three reasons. First, Normal distributions are good descriptions for some distributions of *real data*. Distributions that are often close to Normal include scores on tests taken by many people (such as the Iowa Test of Figure 1.21), repeated careful measurements of the same quantity, and characteristics of biological populations (such as lengths of baby pythons and yields of corn). Second, Normal distributions are good approximations to the results of many kinds of *chance outcomes*, such as tossing a coin many times. Third, and most important, we will see that many *statistical inference* procedures based on Normal distributions work well for other roughly symmetric distributions. HOWEVER ... *even though many sets of data follow a Normal distribution, many do not.* Most income distributions, for example, are skewed to the right and so are not Normal. Non-Normal data, like nonnormal people, not only are common but are also sometimes more interesting than their Normal counterparts.

The 68–95–99.7 rule

Although there are many Normal curves, they all have common properties. Here is one of the most important.

THE 68–95–99.7 RULE

In the Normal distribution with mean μ and standard deviation σ:

- Approximately **68%** of the observations fall within σ of the mean μ.

- Approximately **95%** of the observations fall within 2σ of μ.

- Approximately **99.7%** of the observations fall within 3σ of μ.

Figure 1.25 illustrates the 68–95–99.7 rule. By remembering these three numbers, you can think about Normal distributions without constantly making detailed calculations.

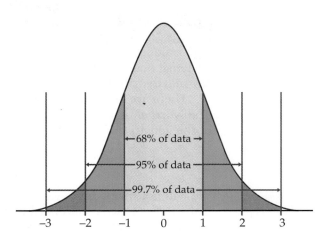

FIGURE 1.25 The 68–95–99.7 rule for Normal distributions.

EXAMPLE

1.34   Heights of young women. The distribution of heights of young women aged 18 to 24 is approximately Normal with mean $\mu = 64.5$ inches and standard deviation $\sigma = 2.5$ inches. Figure 1.26 shows what the 68–95–99.7 rule says about this distribution.

Two standard deviations is 5 inches for this distribution. The 95 part of the 68–95–99.7 rule says that the middle 95% of young women are between $64.5 - 5$ and $64.5 + 5$ inches tall, that is, between 59.5 inches and 69.5 inches. This fact is exactly true for an exactly Normal distribution. It is approximately true for the heights of young women because the distribution of heights is approximately Normal.

The other 5% of young women have heights outside the range from 59.5 to 69.5 inches. Because the Normal distributions are symmetric, half of these women are on the tall side. So the tallest 2.5% of young women are taller than 69.5 inches.

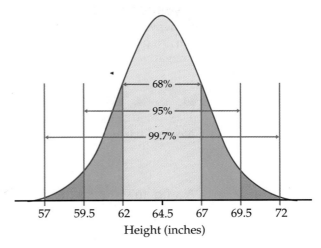

FIGURE 1.26 The 68–95–99.7 rule applied to the heights of young women, for Example 1.34.

Because we will mention Normal distributions often, a short notation is helpful. We abbreviate the Normal distribution with mean μ and standard deviation σ as **$N(\mu, \sigma)$**. For example, the distribution of young women's heights is $N(64.5, 2.5)$.

$N(\mu, \sigma)$

USE YOUR KNOWLEDGE

1.101   Test scores. Many states have programs for assessing the skills of students in various grades. The Indiana Statewide Testing for Educational Progress (ISTEP) is one such program.[33] In a recent year 76,531 tenth-grade Indiana students took the English/language arts exam. The mean score was 572 and the standard deviation was 51. Assuming that these scores are approximately Normally distributed, $N(572, 51)$, use the 68–95–99.7 rule to give a range of scores that includes 95% of these students.

1.102   Use the 68–95–99.7 rule. Refer to the previous exercise. Use the 68–95–99.7 rule to give a range of scores that includes 99.7% of these students.

Standardizing observations

As the 68–95–99.7 rule suggests, all Normal distributions share many properties. In fact, all Normal distributions are the same if we measure in units of size σ about the mean μ as center. Changing to these units is called *standardizing*. To standardize a value, subtract the mean of the distribution and then divide by the standard deviation.

STANDARDIZING AND z-SCORES

If x is an observation from a distribution that has mean μ and standard deviation σ, the **standardized value** of x is

$$z = \frac{x - \mu}{\sigma}$$

A standardized value is often called a **z-score.**

A z-score tells us how many standard deviations the original observation falls away from the mean, and in which direction. Observations larger than the mean are positive when standardized, and observations smaller than the mean are negative.

EXAMPLE

1.35 Find some z-scores. The heights of young women are approximately Normal with $\mu = 64.5$ inches and $\sigma = 2.5$ inches. The z-score for height is

$$z = \frac{\text{height} - 64.5}{2.5}$$

A woman's standardized height is the number of standard deviations by which her height differs from the mean height of all young women. A woman 68 inches tall, for example, has z-score

$$z = \frac{68 - 64.5}{2.5} = 1.4$$

or 1.4 standard deviations above the mean. Similarly, a woman 5 feet (60 inches) tall has z-score

$$z = \frac{60 - 64.5}{2.5} = -1.8$$

or 1.8 standard deviations less than the mean height.

USE YOUR KNOWLEDGE

1.103 Find the z-score. Consider the ISTEP scores (see Exercise 1.101), which we can assume are approximately Normal, $N(572, 51)$. Give the z-score for a student who received a score of 620.

1.104 Find another z-score. Consider the ISTEP scores which we can assume are approximately Normal, $N(572, 51)$. Give the z-score for a student who received a score of 510. Explain why your answer is negative even though all of the test scores are positive.

We need a way to write variables, such as "height" in Example 1.34, that follow a theoretical distribution such as a Normal distribution. We use capital letters near the end of the alphabet for such variables. If X is the height of a young woman, we can then shorten "the height of a young woman is less than 68 inches" to "$X < 68$." We will use lowercase x to stand for any specific value of the variable X.

We often standardize observations from symmetric distributions to express them in a common scale. We might, for example, compare the heights of two children of different ages by calculating their z-scores. The standardized heights tell us where each child stands in the distribution for his or her age group.

Standardizing is a linear transformation that transforms the data into the standard scale of z-scores. We know that a linear transformation does not change the shape of a distribution, and that the mean and standard deviation change in a simple manner. In particular, *the standardized values for any distribution always have mean 0 and standard deviation 1.*

If the variable we standardize has a Normal distribution, standardizing does more than give a common scale. It makes all Normal distributions into a single distribution, and this distribution is still Normal. Standardizing a variable that has any Normal distribution produces a new variable that has the *standard Normal distribution.*

THE STANDARD NORMAL DISTRIBUTION

The **standard Normal distribution** is the Normal distribution $N(0, 1)$ with mean 0 and standard deviation 1.

If a variable X has any Normal distribution $N(\mu, \sigma)$ with mean μ and standard deviation σ, then the standardized variable

$$Z = \frac{X - \mu}{\sigma}$$

has the standard Normal distribution.

Normal distribution calculations

Areas under a Normal curve represent proportions of observations from that Normal distribution. There is no formula for areas under a Normal curve. Calculations use either software that calculates areas or a table of areas. The table and most software calculate one kind of area: **cumulative proportions.** A cumulative proportion is the proportion of observations in a distribution that lie at or below a given value. When the distribution is given by a density curve, the cumulative proportion is the area under the curve to the left of a given value. Figure 1.27 shows the idea more clearly than words do.

cumulative proportion

The key to calculating Normal proportions is to match the area you want with areas that represent cumulative proportions. Then get areas for cumulative proportions either from software or (with an extra step) from a table. The following examples show the method in pictures.

EXAMPLE

1.36 The NCAA standard for SAT scores. The National Collegiate Athletic Association (NCAA) requires Division I athletes to get a combined score

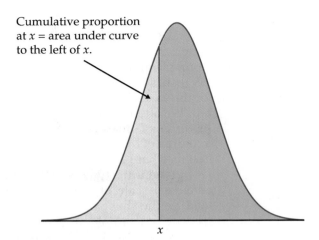

Cumulative proportion at x = area under curve to the left of x.

x

FIGURE 1.27 The *cumulative proportion* for a value x is the proportion of all observations from the distribution that are less than or equal to x. This is the area to the left of x under the Normal curve.

of at least 820 on the SAT Mathematics and Verbal tests to compete in their first college year. (Higher scores are required for students with poor high school grades.) The scores of the 1.4 million students in the class of 2003 who took the SATs were approximately Normal with mean 1026 and standard deviation 209. What proportion of all students had SAT scores of at least 820?

Here is the calculation in pictures: the proportion of scores above 820 is the area under the curve to the right of 820. That's the total area under the curve (which is always 1) minus the cumulative proportion up to 820.

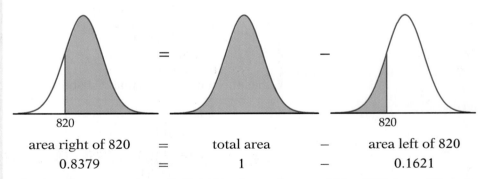

| area right of 820 | = | total area | − | area left of 820 |
| 0.8379 | = | 1 | − | 0.1621 |

That is, the proportion of all SAT takers who would be NCAA qualifiers is 0.8379, or about 84%.

There is *no* area under a smooth curve and exactly over the point 820. Consequently, the area to the right of 820 (the proportion of scores > 820) is the same as the area at or to the right of this point (the proportion of scores ≥ 820). The actual data may contain a student who scored exactly 820 on the SAT. That the proportion of scores exactly equal to 820 is 0 for a Normal distribution is a consequence of the idealized smoothing of Normal distributions for data.

EXAMPLE

1.37 NCAA partial qualifiers. The NCAA considers a student a "partial qualifier" eligible to practice and receive an athletic scholarship, but not to compete, if the combined SAT score is at least 720. What proportion of all students who take the SAT would be partial qualifiers? That is, what proportion have scores between 720 and 820? Here are the pictures:

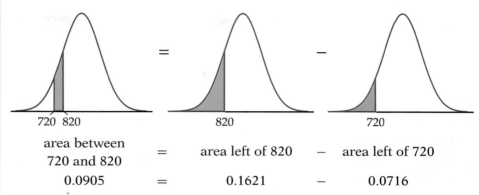

area between 720 and 820	=	area left of 820	−	area left of 720
0.0905	=	0.1621	−	0.0716

About 9% of all students who take the SAT have scores between 720 and 820.

How do we find the numerical values of the areas in Examples 1.36 and 1.37? If you use software, just plug in mean 1026 and standard deviation 209. Then ask for the cumulative proportions for 820 and for 720. (Your software will probably refer to these as "cumulative probabilities." We will learn in Chapter 4 why the language of probability fits.) If you make a sketch of the area you want, you will never go wrong.

You can use the *Normal Curve* applet on the text CD and Web site to find Normal proportions. The applet is more flexible than most software—it will find any Normal proportion, not just cumulative proportions. The applet is an excellent way to understand Normal curves. But, because of the limitations of Web browsers, the applet is not as accurate as statistical software.

If you are not using software, you can find cumulative proportions for Normal curves from a table. That requires an extra step, as we now explain.

Using the standard Normal table

The extra step in finding cumulative proportions from a table is that we must first standardize to express the problem in the standard scale of z-scores. This allows us to get by with just one table, a table of *standard Normal cumulative proportions*. Table A in the back of the book gives cumulative proportions for the standard Normal distribution. Table A also appears on the inside front cover. The pictures at the top of the table remind us that the entries are cumulative proportions, areas under the curve to the left of a value z.

> ### EXAMPLE
>
> **1.38   Find the proportion from z.** What proportion of observations on a standard Normal variable Z take values less than 1.47?
>
> *Solution:* To find the area to the left of 1.47, locate 1.4 in the left-hand column of Table A and then locate the remaining digit 7 as .07 in the top row. The entry opposite 1.4 and under .07 is 0.9292. This is the cumulative proportion we seek. Figure 1.28 illustrates this area.

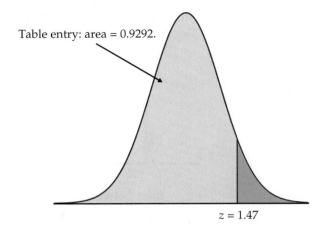

Table entry: area = 0.9292.

$z = 1.47$

FIGURE 1.28 The area under a standard Normal curve to the left of the point $z = 1.47$ is 0.9292, for Example 1.38.

Now that you see how Table A works, let's redo the NCAA Examples 1.36 and 1.37 using the table.

EXAMPLE

1.39 Find the proportion from x. What proportion of all students who take the SAT have scores of at least 820? The picture that leads to the answer is exactly the same as in Example 1.36. The extra step is that we first standardize in order to read cumulative proportions from Table A. If X is SAT score, we want the proportion of students for which $X \geq 820$.

1. *Standardize.* Subtract the mean, then divide by the standard deviation, to transform the problem about X into a problem about a standard Normal Z:

$$X \geq 820$$

$$\frac{X - 1026}{209} \geq \frac{820 - 1026}{209}$$

$$Z \geq -0.99$$

2. *Use the table.* Look at the pictures in Example 1.36. From Table A, we see that the proportion of observations less than -0.99 is 0.1611. The area to the right of -0.99 is therefore $1 - 0.1611 = 0.8389$. This is about 84%.

The area from the table in Example 1.39 (0.8389) is slightly less accurate than the area from software in Example 1.36 (0.8379) because we must round z to two places when we use Table A. The difference is rarely important in practice.

EXAMPLE

1.40 Proportion of partial qualifiers. What proportion of all students who take the SAT would be partial qualifiers in the eyes of the NCAA? That is, what proportion of students have SAT scores between 720 and 820? First, sketch the areas, exactly as in Example 1.37. We again use X as shorthand for an SAT score.

1. *Standardize.*

$$720 \leq \quad X \quad < 820$$

$$\frac{720 - 1026}{209} \leq \frac{X - 1026}{209} < \frac{820 - 1026}{209}$$

$$-1.46 \leq \quad Z \quad < -0.99$$

2. *Use the table.*

area between -1.46 and -0.99 = (area left of -0.99) $-$ (area left of -1.46)

$$= 0.1611 - 0.0721 = 0.0890$$

As in Example 1.37, about 9% of students would be partial qualifiers.

Sometimes we encounter a value of z more extreme than those appearing in Table A. For example, the area to the left of $z = -4$ is not given directly in the table. The z-values in Table A leave only area 0.0002 in each tail unaccounted for. For practical purposes, we can act as if there is zero area outside the range of Table A.

USE YOUR KNOWLEDGE

1.105 Find the proportion. Consider the ISTEP scores, which are approximately Normal, $N(572, 51)$. Find the proportion of students who have scores less than 620. Find the proportion of students who have scores greater than or equal to 620. Sketch the relationship between these two calculations using pictures of Normal curves similar to the ones given in Example 1.36.

1.106 Find another proportion. Consider the ISTEP scores, which are approximately Normal, $N(572, 51)$. Find the proportion of students who have scores between 620 and 660. Use pictures of Normal curves similar to the ones given in Example 1.37 to illustrate your calculations.

Inverse Normal calculations

Examples 1.34 to 1.40 illustrate the use of Normal distributions to find the proportion of observations in a given event, such as "SAT score between 720 and 820." We may instead want to find the observed value corresponding to a given proportion.

Statistical software will do this directly. Without software, use Table A backward, finding the desired proportion in the body of the table and then reading the corresponding z from the left column and top row.

EXAMPLE

1.41 How high for the top 10%? Scores on the SAT Verbal test in recent years follow approximately the $N(505, 110)$ distribution. How high must a student score in order to place in the top 10% of all students taking the SAT?

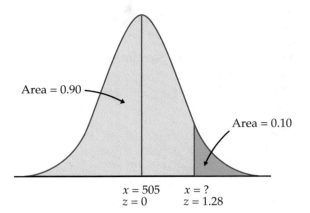

Area = 0.90

Area = 0.10

$x = 505$ $x = ?$
$z = 0$ $z = 1.28$

FIGURE 1.29   Locating the point on a Normal curve with area 0.10 to its right, for Example 1.41.

Again, the key to the problem is to draw a picture. Figure 1.29 shows that we want the score x with area above it 0.10. That's the same as area below x equal to 0.90.

Statistical software has a function that will give you the x for any cumulative proportion you specify. The function often has a name such as "inverse cumulative probability." Plug in mean 505, standard deviation 110, and cumulative proportion 0.9. The software tells you that $x = 645.97$. We see that a student must score at least 646 to place in the highest 10%.

Without software, first find the standard score z with cumulative proportion 0.9, then "unstandardize" to find x. Here is the two-step process:

1. *Use the table.* Look in the body of Table A for the entry closest to 0.9. It is 0.8997. This is the entry corresponding to $z = 1.28$. So $z = 1.28$ is the standardized value with area 0.9 to its left.

2. *Unstandardize* to transform the solution from z back to the original x scale. We know that the standardized value of the unknown x is $z = 1.28$. So x itself satisfies

$$\frac{x - 505}{110} = 1.28$$

Solving this equation for x gives

$$x = 505 + (1.28)(110) = 645.8$$

This equation should make sense: it finds the x that lies 1.28 standard deviations above the mean on this particular Normal curve. That is the "unstandardized" meaning of $z = 1.28$. The general rule for unstandardizing a z-score is

$$x = \mu + z\sigma$$

USE YOUR KNOWLEDGE

1.107   What score is needed to be in the top 25%? Consider the ISTEP scores, which are approximately Normal, $N(572, 51)$. How high a score is needed to be in the top 25% of students who take this exam?

1.108   Find the score that 80% of students will exceed. Consider the ISTEP scores, which are approximately Normal, $N(572, 51)$. Eighty percent of the students will score above x on this exam. Find x.

Normal quantile plots

The Normal distributions provide good descriptions of some distributions of real data, such as the Gary vocabulary scores. The distributions of some other common variables are usually skewed and therefore distinctly non-Normal. Examples include economic variables such as personal income and gross sales of business firms, the survival times of cancer patients after treatment, and the service lifetime of mechanical or electronic components. While experience can suggest whether or not a Normal distribution is plausible in a particular case, it is risky to assume that a distribution is Normal without actually inspecting the data.

A histogram or stemplot can reveal distinctly non-Normal features of a distribution, such as outliers, pronounced skewness, or gaps and clusters. If the stemplot or histogram appears roughly symmetric and unimodal, however, we need a more sensitive way to judge the adequacy of a Normal model. The most useful tool for assessing Normality is another graph, the **Normal quantile plot.**

Normal quantile plot

Here is the basic idea of a Normal quantile plot. The graphs produced by software use more sophisticated versions of this idea. It is not practical to make Normal quantile plots by hand.

1. Arrange the observed data values from smallest to largest. Record what percentile of the data each value occupies. For example, the smallest observation in a set of 20 is at the 5% point, the second smallest is at the 10% point, and so on.

2. Do Normal distribution calculations to find the values of z corresponding to these same percentiles. For example, $z = -1.645$ is the 5% point of the standard Normal distribution, and $z = -1.282$ is the 10% point. We call these values of Z **Normal scores.**

Normal scores

3. Plot each data point x against the corresponding Normal score. If the data distribution is close to any Normal distribution, the plotted points will lie close to a straight line.

Any Normal distribution produces a straight line on the plot because standardizing turns any Normal distribution into a standard Normal distribution. Standardizing is a linear transformation that can change the slope and intercept of the line in our plot but cannot turn a line into a curved pattern.

> **USE OF NORMAL QUANTILE PLOTS**
>
> If the points on a Normal quantile plot lie close to a straight line, the plot indicates that the data are Normal. Systematic deviations from a straight line indicate a non-Normal distribution. Outliers appear as points that are far away from the overall pattern of the plot.

Figures 1.30 to 1.32 are Normal quantile plots for data we have met earlier. The data x are plotted vertically against the corresponding standard Normal z-score plotted horizontally. The z-score scale generally extends from -3 to 3 because almost all of a standard Normal curve lies between these values. These figures show how Normal quantile plots behave.

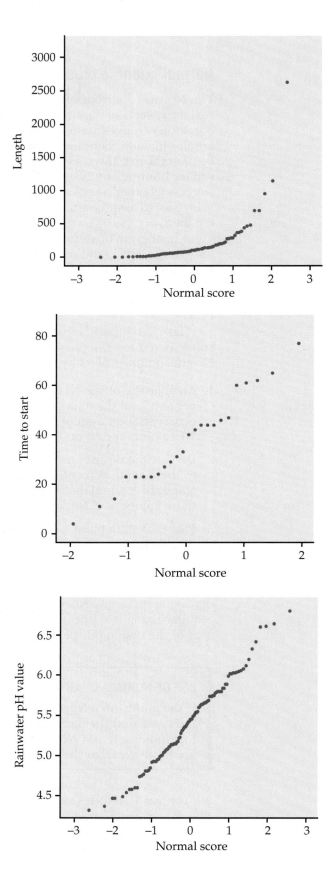

FIGURE 1.30 Normal quantile plot of the service center call lengths, for Example 1.42. This distribution is highly skewed.

FIGURE 1.31 Normal quantile plot of 24 times to start a business with the outlier, Suriname, excluded, for Example 1.42. This distribution is approximately normal.

FIGURE 1.32 Normal quantile plot of the acidity (pH) values of 105 samples of rainwater, for Example 1.44.

CALLCENTER80

EXAMPLE

1.42 Service center call lengths are not Normal. Figure 1.30 is a Normal quantile plot of the 80 call lengths in Table 1.2 (page 15). Because the plot is clearly curved we conclude that these data are not Normally distributed. The shape of the curve is what we typically see with a distribution that is strongly skewed to the right.

TIMETOSTART24

EXAMPLE

1.43 Times to start a business are approximately Normal. Figure 1.31 is a Normal quantile plot of the data on times to start a business from Example 1.21. We have excluded the Suriname, the outlier that you examined in Exercise 1.49. The plot is not particularly smooth but the overall pattern is approximately linear. Note that the sample size here is 24, much smaller than the 80 cases in the previous example. Because of the smaller sample size the z-score scale extends from -2 to 2 rather than from -3 to 3. With smaller sample sizes such as this, we have less information about the true shape of the distribution. Our examination of the normal quantile plot leads us to conclude that there is no clear deviation from Normality evident in the data.

ACIDRAIN

EXAMPLE

1.44 Acidity of rainwater is approximately Normal. Figure 1.32 is a Normal quantile plot of the 105 acidity (pH) measurements of rainwater from Exercise 1.38 (page 26). Histograms don't settle the question of approximate Normality of these data, because their shape depends on the choice of classes. The Normal quantile plot makes it clear that a Normal distribution is a good description—there are only minor wiggles in a generally straight-line pattern.

As Figure 1.32 illustrates, real data almost always show some departure from the theoretical Normal model. *When you examine a Normal quantile plot, look for shapes that show clear departures from Normality. Don't overreact to minor wiggles in the plot.* When we discuss statistical methods that are based on the Normal model, we will pay attention to the sensitivity of each method to departures from Normality. Many common methods work well as long as the data are approximately Normal and outliers are not present.

BEYOND THE BASICS

Density Estimation

density estimator

A density curve gives a compact summary of the overall shape of a distribution. Many distributions do not have the Normal shape. There are other families of density curves that are used as mathematical models for various distribution shapes. Modern software offers more flexible options. A **density estimator** does not start with any specific shape, such as the Normal shape. It looks at the data and draws a density curve that describes the overall shape of the data.

Density estimators join stemplots and histograms as useful graphical tools for exploratory data analysis.

Density estimates can capture other unusual features of a distribution. Here is an example.

STUBHUB

StubHub!
Where Fans Buy & Sell Tickets™

bimodal distribution

EXAMPLE

1.45 StubHub! StubHub! is a Web site where fans can buy and sell tickets to sporting events. Ticket holders wanting to sell their tickets provide the location of their seats and the selling price. People wanting to buy tickets can choose from among the tickets offered for a given event.[34]

On Saturday October 18, 2008, the eleventh-ranked Missouri football team played number one Texas in Austin. On Thursday October 16, 2008, StubHub! listed 64 pairs of tickets for the game. One pair was offered at $883 per ticket. It was noted that these seats were in a suite and that food and bar were included. We discarded this outlier and examined the distribution of the price per ticket for the remaining 63 pairs of tickets. The histogram with a density estimate is given in Figure 1.33. The distribution has two peaks, one around $160 and another around $360. This is the identifying characteristic of a **bimodal distribution.** Since the stadium has upper and lower level seats, we suspect that differences in prices between these two types of seats is responsible for the two peaks. (Texas won 56 to 31.)

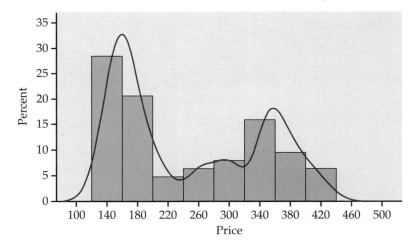

FIGURE 1.33 Histogram of StubHub! price per seat for tickets to the Missouri–Texas football game on October 18, 2008, with a density estimate, for Example 1.45. One outlier with a price per seat of $883 was deleted.

The previous exercise reminds of a continuing theme for data analysis. We looked at a histogram and a density estimate and saw something interesting. This led us to speculation. Additional data on the type and location of the seats may explain more about the prices than we see in Figure 1.33.

SECTION 1.3 Summary

The overall pattern of a distribution can often be described compactly by a **density curve.** A density curve has total area 1 underneath it. Areas under a density curve give proportions of observations for the distribution.

The **mean** μ (balance point), the **median** (equal-areas point), and the **quartiles** can be approximately located by eye on a density curve. The **standard**

deviation σ cannot be located by eye on most density curves. The mean and median are equal for symmetric density curves, but the mean of a skewed curve is located farther toward the long tail than is the median.

The **Normal distributions** are described by bell-shaped, symmetric, unimodal density curves. The mean μ and standard deviation σ completely specify the Normal distribution $N(\mu, \sigma)$. The mean is the center of symmetry, and σ is the distance from μ to the change-of-curvature points on either side.

To **standardize** any observation x, subtract the mean of the distribution and then divide by the standard deviation. The resulting **z-score** $z = (x - \mu)/\sigma$ says how many standard deviations x lies from the distribution mean. All Normal distributions are the same when measurements are transformed to the standardized scale. In particular, all Normal distributions satisfy the **68–95–99.7 rule.**

If X has the $N(\mu, \sigma)$ distribution, then the standardized variable $Z = (X - \mu)/\sigma$ has the **standard Normal distribution** $N(0, 1)$. Proportions for any Normal distribution can be calculated by software or from the **standard Normal table** (Table A), which gives the **cumulative proportions** of $Z < z$ for many values of z.

The adequacy of a Normal model for describing a distribution of data is best assessed by a **Normal quantile plot,** which is available in most statistical software packages. A pattern on such a plot that deviates substantially from a straight line indicates that the data are not Normal.

SECTION 1.3 Exercises

For Exercises 1.101 and 1.02, see page 57; for Exercises 1.103 and 1.104, see page 58; for Exercises 1.105 and 1.106, see page 63; and for Exercises 1.107 and 1.108, see page 64.

1.109 Sketch some normal curves.

(a) Sketch a normal curve that has mean 10 and standard deviation 3.

(b) On the same x axis, sketch a normal curve that has mean 20 and standard deviation 3.

(c) How does the normal curve change when the mean is varied but the standard deviation stays the same?

1.110 The effect of changing the standard deviation.

(a) Sketch a normal curve that has mean 10 and standard deviation 3.

(b) On the same x axis, sketch a normal curve that has mean 10 and standard deviation 1.

(c) How does the normal curve change when the standard deviation is varied but the mean stays the same?

1.111 Know your density.   Sketch density curves that might describe distributions with the following shapes:

(a) Symmetric, but with two peaks (that is, two strong clusters of observations).

(b) Single peak and skewed to the left.

1.112 Do women talk more?   Conventional wisdom suggests that women are more talkative than men. One study designed to examine this stereotype collected data on the speech of 42 women and 37 men in the United States.[35] TALK

(a) The mean number of words spoken per day by the women was 14,297 with a standard deviation of 6441. Use the 68–95–99.7 rule to describe this distribution.

(b) Do you think that applying the rule in this situation is reasonable? Explain your answer.

(c) The men averaged 14,060 words per day with a standard deviation of 9065. Answer the questions in parts (a) and (b) for the men.

(d) Do you think that the data support the conventional wisdom? Explain your answer. Note that in Section 7.2 we will learn formal statistical methods to answer this type of question.

1.113 Data from Mexico.   Refer to the previous exercise. A similar study in Mexico was conducted with

31 women and 20 men. The women averaged 14,704 words per day with a standard deviation of 6215. For men the mean was 15,022 and the standard deviation was 7864. TALKMEXICO

(a) Answer the questions from the previous exercise for the Mexican study.

(b) The means for both men and women are higher for the Mexican study than for the U.S. study. What conclusions can you draw from this observation?

1.114 Total scores. Here are the total scores of 10 students in an introductory statistics course. STATCOURSE

68   54   92   75   73   98   64   55   80   70

Previous experience with this course suggests that these scores should come from a distribution that is approximately Normal with mean 70 and standard deviation 10.

(a) Using these values for μ and σ, standardize the first exam scores of these 10 students.

(b) If the grading policy is to give grades of A to the top 15% of scores based on the Normal distribution with mean 70 and standard deviation 10, what is the cut-off for an A in terms of a standardized score?

(c) Which students earned a grade of A on the final exam for this course?

1.115 Assign more grades. Refer to the previous exercise. The grading policy says the cut-offs for the other grades correspond to the following: bottom 5% receive F, next 10% receive D, next 40% receive C, and next 30% receive B. These cut-offs are based on the $N(70, 10)$ distribution.

(a) Give the cut-offs for the grades in this course in terms of standardized scores.

(b) Give the cut-offs in terms of actual total scores.

(c) Do you think that this method of assigning grades is a good one? Give reasons for your answer.

1.116 A uniform distribution. If you ask a computer to generate "random numbers" between 0 and 1, you will get observations from a **uniform distribution**. Figure 1.34 graphs the density curve for a uniform distribution. Use areas under this density curve to answer the following questions.

(a) Why is the total area under this curve equal to 1?

(b) What proportion of the observations lie below 0.35?

(c) What proportion of the observations lie between 0.35 and 0.65?

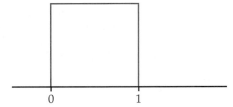

FIGURE 1.34 The density curve of a uniform distribution, for Exercise 1.116.

1.117 Use a different range for the uniform distribution. Many random number generators allow users to specify the range of the random numbers to be produced. Suppose that you specify that the outcomes are to be distributed uniformly between 0 and 4. Then the density curve of the outcomes has constant height between 0 and 4, and height 0 elsewhere.

(a) What is the height of the density curve between 0 and 4? Draw a graph of the density curve.

(b) Use your graph from (a) and the fact that areas under the curve are proportions of outcomes to find the proportion of outcomes that are less than 1.

(c) Find the proportion of outcomes that lie between 0.5 and 2.5.

1.118 Find the mean, the median, and the quartiles. What are the mean and the median of the uniform distribution in Figure 1.34? What are the quartiles?

1.119 Three density curves. Figure 1.35 displays three density curves, each with three points marked on it. At which of these points on each curve do the mean and the median fall?

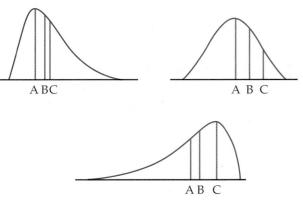

FIGURE 1.35 Three density curves, for Exercise 1.119.

1.120 Length of pregnancies. The length of human pregnancies from conception to birth varies according to a distribution that is approximately Normal with mean 266 days and standard deviation 16 days. Draw a density curve for this distribution on which the mean and standard deviation are correctly located.

1.121 (A) **Use the Normal Curve applet.** The 68–95–99.7 rule for Normal distributions is a useful approximation. You can use the *Normal Curve* applet on the text CD and Web site to see how accurate the rule is. Drag one flag across the other so that the applet shows the area under the curve between the two flags.

(a) Place the flags one standard deviation on either side of the mean. What is the area between these two values? What does the 68–95–99.7 rule say this area is?

(b) Repeat for locations two and three standard deviations on either side of the mean. Again compare the 68–95–99.7 rule with the area given by the applet.

1.122 Pregnancies and the 68–95–99.7 rule. The length of human pregnancies from conception to birth varies according to a distribution that is approximately Normal with mean 266 days and standard deviation 16 days. Use the 68–95–99.7 rule to answer the following questions.

(a) Between what values do the lengths of the middle 95% of all pregnancies fall?

(b) How short are the shortest 2.5% of all pregnancies? How long do the longest 2.5% last?

1.123 Horse pregnancies are longer. Bigger animals tend to carry their young longer before birth. The length of horse pregnancies from conception to birth varies according to a roughly Normal distribution with mean 336 days and standard deviation 3 days. Use the 68–95–99.7 rule to answer the following questions.

(a) Almost all (99.7%) horse pregnancies fall in what range of lengths?

(b) What percent of horse pregnancies are longer than 339 days?

1.124 (A) **Use the Normal Curve applet.** Use the *Normal Curve* applet for the standard Normal distribution to say how many standard deviations above and below the mean the quartiles of any Normal distribution lie.

1.125 Acidity of rainwater. The Normal quantile plot in Figure 1.32 (page 66) shows that the acidity (pH) measurements for rainwater samples in Exercise 1.36 are approximately Normal. How well do these scores satisfy the 68–95–99.7 rule? To find out, calculate the mean $\bar{x}$ and standard deviation s of the observations. Then calculate the percent of the 105 measurements that fall between

$\bar{x} - s$ and $\bar{x} + s$ and compare your result with 68%. Do the same for the intervals covering two and three standard deviations on either side of the mean. (The 68–95–99.7 rule is exact for any theoretical Normal distribution. It will hold only approximately for actual data.) (A) ACIDRAIN

1.126 Find some proportions. Using either Table A or your calculator or software, find the proportion of observations from a standard Normal distribution that satisfies each of the following statements. In each case, sketch a standard Normal curve and shade the area under the curve that is the answer to the question.

(a) $Z > 1.65$

(b) $Z < 1.65$

(c) $Z > -0.76$

(d) $-0.76 < Z < 1.65$

1.127 Find more proportions. Using either Table A or your calculator or software, find the proportion of observations from a standard Normal distribution for each of the following events. In each case, sketch a standard Normal curve and shade the area representing the proportion.

(a) $Z \leq -1.8$

(b) $Z \geq -1.8$

(c) $Z > 1.6$

(d) $-1.8 < Z < 1.6$

1.128 Find some values of z. Find the value z of a standard Normal variable Z that satisfies each of the following conditions. (If you use Table A, report the value of z that comes closest to satisfying the condition.) In each case, sketch a standard Normal curve with your value of z marked on the axis.

(a) 22% of the observations fall below z.

(b) 40% of the observations fall above z.

1.129 Find more values of z. The variable Z has a standard Normal distribution.

(a) Find the number z that has cumulative proportion 0.65.

(b) Find the number z such that the event $Z > z$ has proportion 0.45.

1.130 Find some values of z. The Wechsler Adult Intelligence Scale (WAIS) is the most common IQ test. The scale of scores is set separately for each age group and is approximately Normal with mean 100 and standard deviation 15. People with WAIS scores below 70 are considered mentally retarded when, for example, applying for Social Security disability benefits. What percent of adults are retarded by this criterion?

1.131 High IQ scores. The Wechsler Adult Intelligence Scale (WAIS) is the most common IQ test. The scale of scores is set separately for each age group and is approximately Normal with mean 100 and standard deviation 15. The organization MENSA, which calls itself "the high IQ society," requires a WAIS score of 130 or higher for membership. What percent of adults would qualify for membership?

There are two major tests of readiness for college, the ACT and the SAT. ACT scores are reported on a scale from 1 to 36. The distribution of ACT scores are approximately Normal with mean $\mu = 21.5$ and standard deviation $\sigma = 5.4$. SAT scores are reported on a scale from 600 to 2400. The SAT scores are approximately Normal with mean $\mu = 1509$ and standard deviation $\sigma = 321$. Exercises 1.132 to 1.141 are based on this information.

1.132 Compare an SAT score with an ACT score. Tonya scores 1820 on the SAT. Jermaine scores 29 on the ACT. Assuming that both tests measure the same thing, who has the higher score? Report the z-scores for both students.

1.133 Make another comparison. Jacob scores 16 on the ACT. Emily scores 1020 on the SAT. Assuming that both tests measure the same thing, who has the higher score? Report the z-scores for both students.

1.134 Find the ACT equivalent. Jose scores 2080 on the SAT. Assuming that both tests measure the same thing, what score on the ACT is equivalent to Jose's SAT score?

1.135 Find the SAT equivalent. Maria scores 30 on the ACT. Assuming that both tests measure the same thing, what score on the SAT is equivalent to Maria's ACT score?

1.136 Find an SAT percentile. Reports on a student's ACT or SAT usually give the percentile as well as the actual score. The percentile is just the cumulative proportion stated as a percent: the percent of all scores that were lower than this one. Maria scores 2090 on the SAT. What is her percentile?

1.137 Find an ACT percentile. Reports on a student's ACT or SAT usually give the percentile as well as the actual score. The percentile is just the cumulative proportion stated as a percent: the percent of all scores that were lower than this one. Jacob scores 19 on the ACT. What is his percentile?

1.138 How high is the top 10%? What SAT scores make up the top 10% of all scores?

1.139 How low is the bottom 20%? What SAT scores make up the bottom 20% of all scores?

1.140 Find the ACT quartiles. The quartiles of any distribution are the values with cumulative proportions 0.25 and 0.75. What are the quartiles of the distribution of ACT scores?

1.141 Find the SAT quintiles. The quintiles of any distribution are the values with cumulative proportions 0.20, 0.40, 0.60, and 0.80. What are the quintiles of the distribution of SAT scores?

1.142 Do you have enough "good cholesterol?" High-density lipoprotein (HDL) is sometimes called the "good cholesterol" because low values are associated with a higher risk of heart disease. According to the American Heart Association, people over the age of 20 years should have at least 40 mg/dL of HDL cholesterol.[36] U.S. women aged 20 and over have a mean HDL of 55 mg/dL with a standard deviation of 15.5 mg/dL. Assume that the distribution is Normal.

(a) What percent of women have low values of HDL (40 mg/dL or less)?

(b) HDL levels of 60 mg/dL are believed to protect people from heart disease. What percent of women have protective levels of HDL?

(c) Women with more than 40 mg/dL but less than 60 mg/dL of HDL are in the intermediate range, neither very good or very bad. What proportion are in this category?

1.143 Men and HDL cholesterol. HDL cholesterol levels for men have a mean of 46 mg/dL with a standard deviation of 13.6. Answer the questions given in the previous exercise for the population of men.

1.144 Diagnosing osteoporosis. Osteoporosis is a condition in which the bones become brittle due to loss of minerals. To diagnose osteoporosis, an elaborate apparatus measures bone mineral density (BMD). BMD is usually reported in standardized form. The standardization is based on a population of healthy young adults. The World Health Organization (WHO) criterion for osteoporosis is a BMD 2.5 standard deviations below the mean for young adults. BMD measurements in a population of people similar in age and sex roughly follow a Normal distribution.

(a) What percent of healthy young adults have osteoporosis by the WHO criterion?

(b) Women aged 70 to 79 are of course not young adults. The mean BMD in this age is about -2 on the standard scale for young adults. Suppose that the standard deviation is the same as for young adults. What percent of this older population has osteoporosis?

1.145 Length of pregnancies.   The length of human pregnancies from conception to birth varies according to a distribution that is approximately Normal with mean 266 days and standard deviation 16 days.

(a) What percent of pregnancies last less than 240 days (that's about 8 months)?

(b) What percent of pregnancies last between 240 and 270 days (roughly between 8 months and 9 months)?

(c) How long do the longest 20% of pregnancies last?

1.146 ⚠ Quartiles for Normal distributions.   The quartiles of any distribution are the values with cumulative proportions 0.25 and 0.75.

(a) What are the quartiles of the standard Normal distribution?

(b) Using your numerical values from (a), write an equation that gives the quartiles of the $N(\mu, \sigma)$ distribution in terms of μ and σ.

(c) The length of human pregnancies from conception to birth varies according to a distribution that is approximately Normal with mean 266 days and standard deviation 16 days. Apply your result from (b): what are the quartiles of the distribution of lengths of human pregnancies?

1.147 ⚠ IQR for Normal distributions.   Continue your work from the previous exercise. The interquartile range IQR is the distance between the first and third quartiles of a distribution.

(a) What is the value of the IQR for the standard Normal distribution?

(b) There is a constant c such that $IQR = c\sigma$ for any Normal distribution $N(\mu, \sigma)$. What is the value of c?

1.148 ⚠ Outliers for Normal distributions.   Continue your work from the previous two exercises. The percent of the observations that are suspected outliers according to the $1.5 \times IQR$ rule is the same for any Normal distribution. What is this percent?

1.149 Deciles of Normal distributions.   The **deciles** of any distribution are the 10th, 20th, ..., 90th percentiles. The first and last deciles are the 10th and 90th percentiles, respectively.

(a) What are the first and last deciles of the standard Normal distribution?

(b) The weights of 9-ounce potato chip bags are approximately Normal with mean 9.12 ounces and standard deviation 0.15 ounce. What are the first and last deciles of this distribution?

1.150 Carbon dioxide emissions.   Figure 1.36 is a Normal quantile plot of the emissions of carbon dioxide (CO_2) per person in 48 countries, from Table 1.3 (page 26). In what way is this distribution non-Normal? Comparing the plot with Table 1.3, which countries would you call outliers? 🌐 CO_2

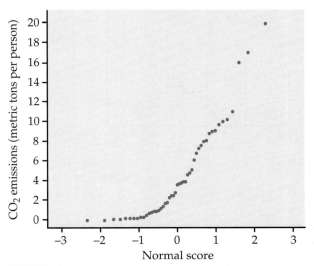

FIGURE 1.36 Normal quantile plot of CO_2 emissions in 48 countries, for Exercise 1.150.

1.151 ⚠ Three Normal quantile plots.   Figure 1.37 shows three Normal quantile plots for three distributions. Describe the three distributions.

The remaining exercises for this section require the use of software that will make Normal quantile plots.

1.152 Density of the earth.   We expect repeated careful measurements of the same quantity to be approximately Normal. Make a Normal quantile plot for Cavendish's measurements in Exercise 1.42 (page 27). Are the data approximately Normal? If not, describe any clear deviations from Normality. 🌐 EARTHDENSITY

1.153 Three varieties of flowers.   The study of tropical flowers and their hummingbird pollinators (Exercise 1.86, page 49) measured lengths for three varieties of *Heliconia* flowers. We expect that such biological measurements will have roughly Normal distributions. 🌐 HELICONIA

(a) Make Normal quantile plots for each of the three flower varieties. Which distribution is closest to Normal?

(b) The other two distributions show the same kind of mild deviation from Normality. In what way are these distributions non-Normal?

1.154 Use software to generate some data. Use software to generate 200 observations from the standard Normal distribution. Make a histogram of these observations. How does the shape of the histogram compare with a Normal density curve? Make a Normal quantile plot of the data. Does the plot suggest any

important deviations from Normality? (Repeating this exercise several times is a good way to become familiar with how histograms and Normal quantile plots look when data actually are close to Normal.)

1.155 Use software to generate more data. Use software to generate 200 observations from the uniform distribution described in Exercise 1.116. Make a histogram of these observations. How does the histogram compare with the density curve in Figure 1.34? Make a Normal quantile plot of your data. According to this plot, how does the uniform distribution deviate from Normality?

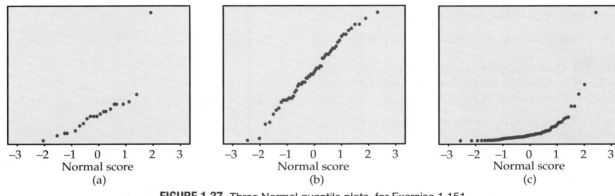

FIGURE 1.37 Three Normal quantile plots, for Exercise 1.151.

CHAPTER 1 Exercises

1.156 ⚔ **Fuel efficiency of hatchbacks and large sedans.** Let's compare the fuel efficiencies (MPG) of 2009 model hatchbacks and large sedans.[37] Here are the data:

Hatchbacks
30 29 28 27 27 27 27 27 26 25 25 25 24 24 24
24 24 23 23 22 22 21 21 21 21 21 21 21 20 20
20 20 20 20 20 20 19 19 19 19 18 16 16

Large sedans
19 19 18 18 18 18 17 17 17 17 17 17 17 17 17
17 16 16 16 16 16 16 16 16 15 15 13 13

Give a graphical and numerical descriptions of the fuel efficiencies for these two types of vehicles. What are the main features of the distributions? Compare the two distributions and summarize your results in a short paragraph. 📄 MPGHATCHLARGE

1.157 Binge drinking. The Behavioral Risk Factor Surveillance System (BRFSS) conducts a large survey of

health conditions and risk behaviors in the United States.[38] The BRFSS data set contains data on 29 demographic factors and risk factors for each state. Use the percentage of binge drinkers for this exercise. 📄 BRFSS

(a) Prepare a graphical display of the distribution and use your display to describe the major features of the distribution.

(b) Calculate numerical summaries. Give reasons for your choices.

(c) Write a short paragraph summarizing what the data tell us about binge drinking in the United States.

1.158 Eat your fruits and vegetables. Nutrition experts recommend that we eat five servings of fruits and vegetables each day. The BRFSS data set described in the previous exercise include a variable that gives the percent of people who regularly eat five or more servings of fruits and vegetables. Answer the questions given in the previous exercise for this variable. 📄 BRFSS

1.159 ⚠ **Vehicle colors.** Vehicle colors differ among types of vehicle in different regions. Here are data on the most popular colors in 2006 for several different regions of the world.[39]

Color	North America percent	South America percent	Europe percent	China percent	South Korea percent	Japan percent
Silver	19	26	28	24	21	27
White	16	11	4	16	18	24
Gray	13	14	16	3	19	12
Black	13	20	24	19	20	16
Blue	11	8	13	17	9	10
Red	11	10	6	9	6	3
Brown	7	7	4	1	6	2
Other	10	4	5	11	1	6

Use the methods you learned in this chapter to compare the vehicle color preferences for the regions of the world presented in this table. Write a report summarizing your findings with an emphasis on similarities and differences across regions. Include recommendations related to marketing and advertising of vehicles in these regions. ⬤ VCOLORSCOUNTRY

1.160 ⚠ **Balance of international payments for Canada.** Visit the Web page **www40.statcan.ca/l01/ cst01/econ01a.htm** that provides data on Canada's balance of international payments. Select some data from this Web page and use the methods that you learned in this chapter to create graphical and numerical summaries. Write a report summarizing your findings that includes supporting evidence from your analyses.

1.161 ⚠ **Canadian government revenue and expenditures by province and territory.** Visit the Web pages **www40.statcan.ca/l01/cst01/govt08a.htm**, **www40.statcan.ca/l01/cst01/govt08b.htm**, and **www40.statcan.ca/l01/cst01/govt08c.htm** You need to look at the three pages to obtain data for all provinces and territories. Select some data from these Web page and use the methods that you learned in this chapter to create graphical and numerical summaries. Write a report summarizing your findings that includes supporting evidence from your analyses.

1.162 Internet use. The World Bank collects data on many variables related to development for countries throughout the world.[40] One of these is Internet use, expressed as the number of users per 100 people. The data file for this exercise gives this variable for 182 countries. Use graphical and numerical methods to describe this distribution. Write a short report summarizing what the data tell about worldwide Internet use. ⬤ INTERNETUSE

1.163 Internet use in Europe. Refer to the previous exercise. Now examine the data only for the countries in Europe. Answer the questions given in the previous exercise and compare your results with those that you found for all 182 countries. ⬤ INTERNETEUROPE

1.164 Park space and population. Below are data on park and open space in several U.S. cities with high population density.[41] In this table, population is reported in thousands of people, and park and open space is called open space, with units of acres. ⬤ PARKSPACE

City	Population	Open space
Baltimore	651	5,091
Boston	589	4,865
Chicago	2,896	11,645
Long Beach	462	2,887
Los Angeles	3,695	29,801
Miami	362	1,329
Minneapolis	383	5,694
New York	8,008	49,854
Oakland	399	3,712
Philadelphia	1,518	10,685
San Francisco	777	5,916
Washington, D.C.	572	7,504

(a) Make a bar graph for population. Describe what you see in the graph.

(b) Do the same for open space.

(c) For each city, divide the open space by population. This gives rates: acres of open space per thousand residents.

(d) Make a bar graph of the rates.

(e) Redo the bar graph that you made in part (d) by ordering the cities by their open space to population rate.

(f) Which of the two bar graphs in (d) and (e) do you prefer? Give reasons for your answer.

1.165 Compare two Normal curves. In Exercise 1.101, we worked with the distribution of ISTEP scores on the English/language arts portion of the exam for tenth-graders. We used the fact that the distribution of scores for the 76,531 students who took the exam was approximately $N(572, 51)$. These students were classified in a variety of ways, and summary statistics were reported for these different subgroups. When classified by gender, the scores for the women are approximately $N(579, 49)$, and the scores for the men are approximately $N(565, 55)$. Figure 1.38 gives the Normal density curves for these two distributions. Here is a possible description of these data: women score about 14 points higher than men on the ISTEP English/language arts exam. Critically evaluate this statement and then write your own summary based on the distributions displayed in Figure 1.38.

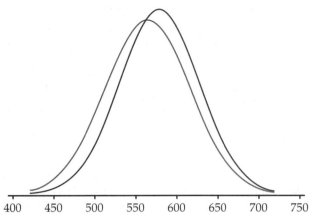

FIGURE 1.38 Normal density curves for ISTEP scores of women and men, for Exercise 1.165.

1.166 Leisure time for college students. You want to measure the amount of "leisure time" that college students enjoy. Write a brief discussion of two issues:

(a) How will you define "leisure time"?

(b) Once you have defined leisure time, how will you measure Sally's leisure time this week?

1.167 Biological clocks. Many plants and animals have "biological clocks" that coordinate activities with the time of day. When researchers looked at the length of the biological cycle in the plant *Arabidopsis* by measuring leaf movements, they found that the length of the cycle is not always 24 hours. Further study discovered that cycle length changes systematically with north-south location. Table 1.5 contains cycle lengths for 149 locations around

the world.[42] Describe the distribution of cycle lengths with a histogram and numerical summaries. In particular, how much variation is there among locations? **BIOCLOCK**

1.168 Product preference. Product preference depends in part on the age, income, and gender of the consumer. A market researcher selects a large sample of potential car buyers. For each consumer, she records gender, age, household income, and automobile preference. Which of these variables are categorical and which are quantitative?

1.169 Internet service. Late in 2008, there were over 96 million residential subscribers to Internet service in the United States. The numbers of subscribers claimed by the top 10 providers of service were as follows:[43]

Service provider	Subscribers (millions)	Service provider	Subscribers (millions)
AT&T	14.8	EarthLink	3.0
Comcast	14.7	Charter	2.9
Road Runner	8.6	Qwest	2.8
Verizon	8.5	Cablevision	2.4
America Online	7.5	United Online	1.5

Display these data in a graph. How many subscribers do the many smaller providers have? Add an "Other" entry in your graph. Business people looking at this graph see an industry that offers opportunities for larger companies to take over. **INTERNETPROVIDE**

1.170 Weights are not Normal. The heights of people of the same sex and similar ages follow Normal distributions reasonably closely. Weights, on the other hand, are not Normally distributed. The weights of

---

TABLE 1.5

Biological clock cycle lengths for a plant species in different locations

23.89	23.72	23.74	24.35	25.05	24.56	23.69	22.33	23.79	22.12
25.39	23.08	25.64	23.98	25.84	25.46	24.37	24.13	24.40	24.74
24.44	24.82	23.56	24.96	24.21	23.85	24.57	23.44	23.64	24.23
24.01	24.58	25.57	23.73	24.11	23.21	25.08	24.03	24.62	23.51
23.21	23.41	23.69	22.97	24.65	24.65	24.29	23.89	25.08	23.89
24.95	23.09	23.21	24.66	23.88	25.33	24.38	24.68	25.34	25.22
23.45	23.39	25.43	23.16	23.95	23.25	24.72	24.89	24.88	24.71
23.58	25.98	24.28	24.25	23.16	24.19	27.22	23.77	26.21	24.33
24.34	24.89	24.32	24.14	24.00	23.48	25.81	24.99	24.18	22.73
24.18	23.95	24.48	23.89	24.24	24.96	24.58	24.29	24.31	23.64
23.87	23.68	24.87	23.00	23.48	24.26	23.34	25.11	24.69	24.97
24.64	24.49	23.61	24.07	26.60	24.91	24.76	25.09	26.56	25.13
24.81	25.63	25.63	24.69	24.41	23.79	22.88	22.00	23.33	25.12
24.00	24.31	23.03	24.51	28.55	22.96	23.61	24.72	24.04	25.18
24.30	24.22	24.39	24.73	24.68	24.14	24.57	24.42	25.62	

women aged 20 to 29 have mean 141.7 pounds and median 133.2 pounds. The first and third quartiles are 118.3 pounds and 157.3 pounds. What can you say about the shape of the weight distribution? Explain your reasoning.

1.171 What graph would you use?   What type of graph or graphs would you plan to make in a study of each of the following issues?

(a) What makes of cars do students drive? How old are their cars?

(b) How many hours per week do students study? How does the number of study hours change during a semester?

(c) Which radio stations are most popular with students?

(d) When many students measure the concentration of the same solution for a chemistry course laboratory assignment, do their measurements follow a Normal distribution?

1.172 Spam filters.   A university department installed a spam filter on its computer system. During a 21-day period, 6693 messages were tagged as spam. How much spam you get depends on what your online habits are. Here are the counts for some students and faculty in this department (with log-in IDs changed, of course):

ID	Count	ID	Count	ID	Count	ID	Count
AA	1818	BB	1358	CC	442	DD	416
EE	399	FF	389	GG	304	HH	251
II	251	JJ	178	KK	158	LL	103

All other department members received fewer than 100 spam messages. How many did the others receive in total? Make a graph and comment on what you learn from these data.   SPAMFILTERS

1.173 Two distributions.   If two distributions have exactly the same mean and standard deviation, must their histograms have the same shape? If they have the same five-number summary, must their histograms have the same shape? Explain.

1.174 Babe Ruth and Mark McGwire.   Babe Ruth hit 60 home runs in 1927, a record that stood until Mark McGwire hit 70 in 1998. A proper comparison of Ruth and McGwire should include their historical context. Here are the number of home runs by the major league leader for each year in baseball history, 1876 to 2003, in order from left to right. Make a time plot. (Be sure to add the scale of years.)   RUTHMCGWIRE

5	3	4	9	6	7	7	10	27	11	11	17	14	20	14	16	13
19	18	17	13	11	15	25	12	16	16	13	10	9	12	10	12	9
10	21	14	19	19	24	12	12	**11**	**29**	**54**	**59**	42	**41**	**46**	39	**47**
60	**54**	**46**	56	**46**	58	48	49	36	49	46	58	35	43	37	36	34
33	28	44	51	40	54	47	42	37	47	49	51	52	44	47	46	41
61	49	45	49	52	49	44	44	49	45	48	40	44	36	38	38	52
46	48	48	31	39	40	43	40	40	**49**	42	47	51	44	43	46	43
50	**52**	**58**	**70**	**65**	50	73	57	47								

(a) Describe the effect of World War II (1942 to 1945 seasons).

(b) Ruth led in the 11 years in boldface between 1918 and 1931. McGwire led in the 5 boldface years between 1987 and 1999. Briefly compare the achievements of Ruth and McGwire in the context of their times.

1.175 Barry Bonds.   The single-season home run record was broken by Barry Bonds of the San Francisco Giants in 2001 when he hit 73 home runs. Here are Bonds's home run totals from 1986 (his first year) to 2003:

16 25 24 19 33 25 34 46 37 33 42 40 37 34 49 73 46 45

Make a stemplot of these data. Bonds's record year is a high outlier. How do his career mean and median number of home runs change when we drop the record 73? What general fact about the mean and median does your result illustrate?   BONDS

1.176 Norms for reading scores.   Raw scores on behavioral tests are often transformed for easier comparison. A test of reading ability has mean 70 and standard deviation 10 when given to third-graders. Sixth-graders have mean score 80 and standard deviation 11 on the same test. To provide separate "norms" for each grade, we want scores in each grade to have mean 100 and standard deviation 20.

(a) What linear transformation will change third-grade scores x into new scores $x_{new} = a + bx$ that have the desired mean and standard deviation? (Use $b > 0$ to preserve the order of the scores.)

(b) Do the same for the sixth-grade scores.

(c) David is a third-grade student who scores 72 on the test. Find David's transformed score. Nancy is a sixth-grade student who scores 78. What is her transformed score? Who scores higher within his or her grade?

(d) Suppose that the distribution of scores in each grade is Normal. Then both sets of transformed scores have the $N(100, 20)$ distribution. What percent of third-graders have scores less than 75? What percent of sixth-graders have scores less than 75?

1.177 Use software to generate some data. Most statistical software packages have routines for generating values of variables having specified distributions. Use your statistical software to generate 30 observations from the $N(25, 8)$ distribution. Compute the mean and standard deviation $\bar{x}$ and s of the 30 values you obtain. How close are $\bar{x}$ and s to the μ and σ of the distribution from which the observations were drawn? Repeat 19 more times the process of generating 30 observations from the $N(25, 8)$ distribution and recording $\bar{x}$ and s. Make a stemplot of the 20 values of $\bar{x}$ and another stemplot of the 20 values of s. Make Normal quantile plots of both sets of data. Briefly describe each of these distributions. Are they symmetric or skewed? Are they roughly Normal? Where are their centers? (The distributions of measures like $\bar{x}$ and s when repeated sets of observations are made from the same theoretical distribution will be very important in later chapters.)

Looking at Data— Relationships

Introduction

In Chapter 1 we learned to use graphical and numerical methods to describe the distribution of a single variable. Many of the interesting examples of the use of statistics involve relationships between pairs of variables. Learning ways to describe relationships with graphical and numerical methods is the focus of this chapter.

> **EXAMPLE**
>
> **2.1 Stress and lack of sleep.** Stress is a common problem for college students. Exploring factors that are associated with stress may lead to strategies that will help students to relieve some of the stress that they experience. Recent studies have suggested that a lack of sleep is associated with stress.[1] The two variables involved in the relationship here are lack of sleep and stress. The cases are the students who are the subjects for a particular study.

When we study relationships between two variables, it is not sufficient to collect data on both variables. A key idea for this chapter is that both variables must be measured on the same cases.

USE YOUR KNOWLEDGE

2.1 Relationship between first test and final exam. You want to study the relationship between the score on the first exam and the score on the final for the 30 students enrolled in an elementary statistics class. Who are the cases for your study?

We use the term *associated* to describe the relationship between two variables, such as the stress and lack of sleep in Example 2.1. Here is another example where two variables are associated.

EXAMPLE

2.2 Size and price of a coffee beverage. You visit a local Starbucks to buy a Mocha Frappuccino®. The barista explains that this blended coffee beverage comes in three sizes and asks if you want a Tall, a Grande, or a Venti. The prices are $3.50, $4.00, and $4.50, respectively. There is a clear association between the size of the Mocha Frappuccino and its price.

ASSOCIATION BETWEEN VARIABLES

Two variables measured on the same cases are **associated** if knowing the value of one of the variables tells you something about the values of the other variable that you would not know without this information.

In the Mocha Frappuccino example, knowing the size tells you the exact price, so the association here is very strong. Many statistical associations, however, are simply overall tendencies that allow exceptions. Some people get adequate sleep and are highly stressed. Others get little sleep and do not experience much stress. The association here is much weaker than the one in the Mocha Frappuccino example.

Examining relationships

When you examine the relationship between two or more variables, first ask the preliminary questions that are familiar from Chapter 1:

- What *cases* do the data describe?
- What *variables* are present? How are they measured?
- Which variables are *quantitative* and which are *categorical*?

EXAMPLE

2.3 Stress and lack of sleep. A study of stress and lack of sleep collected data on 1125 students from an urban Midwestern university. Two of the variables measured were the Pittsburgh Sleep Quality Index (PSQI) and the Subjective Units of Distress Scale (SUDS). In this study the cases are the 1125 students studied.[2] The PSQI is based on responses to a large number of questions which are summarized in a single variable that has a value between 0 and 21 for each subject. Therefore, we will treat the PSQI as a quantitative variable. SUDS is a similar scale with values between 0 and 100 for each subject. We will treat SUDS as a quantitative variable also.

In many situations, we measure a collection of categorical variables and then combine them in a scale that can be viewed as a quantitative variable. The PSQI is an example. We can also turn the tables in the other direction. Here is an example:

EXAMPLE

2.4 Hemoglobin and anemia. Hemoglobin is a measure of iron in the blood. The units are grams of hemoglobin per deciliter of blood (g/dL). Normal values depend on age and gender. Adult women are typically have values between 12 g/dL and 16 g/dL.

Anemia is a major problem in developing countries and many studies have been designed to address the problem. In these studies computing the mean hemoglobin is not particularly useful. For studies like this, using a definition of severe anemia as hemoglobin of less than 8 g/dL is more appropriate. Thus, for example, researchers can compare the proportions of subjects who are severely anemic for two treatments rather than the difference in the mean hemoglobin levels. In this situation, the categorical variable, severely anemic or not, is much more useful than the quantitative variable, hemoglobin.

When analyzing data to draw conclusions it is important to carefully consider the best way to summarize the data. Just because a variable is measured as a quantitative variable, it does not necessarily follow that the best summary is based on the mean (or the median). As the previous example illustrates, converting a quantitative variable to a categorical variable is a very useful option to keep in mind.

USE YOUR KNOWLEDGE

2.2 Create a categorical variable from a quantitative variable. Consider the study described in Example 2.3. Some analyses compared three groups of students. The students were classified as having optimal sleep quality (a PSQI of 5 or less), borderline sleep quality (a PSQI of 6 and 7), and poor (a PSQI of 8 or more). When the three groups of students are compared, is the PSQI being used as a quantitative variable or as a categorical variable? Explain your answer and describe some advantages to using the optimal, borderline, and poor categories in explaining the results of a study such as this.

2.3 Replace names by ounces. In the Mocha Frappuccino example, the variable size is categorical, with Tall, Grande, and Venti as the possible values. Suppose you converted these values to the number of ounces: Tall is 12 ounces, Grande is 16 ounces, and Venti is 24 ounces. For studying the relationship between ounces and price, describe the cases and the variables, and state whether each is quantitative or categorical.

When you examine the relationship between two variables, a new question becomes important:

- Is your purpose simply to explore the nature of the relationship, or do you hope to show that one of the variables can explain variation in the other? That is, are some of the variables *response variables* and others *explanatory variables?*

RESPONSE VARIABLE, EXPLANATORY VARIABLE

A **response variable** measures an outcome of a study. An **explanatory variable** explains or causes changes in the response variables.

EXAMPLE

2.5 Stress and lack of sleep. Refer to the study of stress and lack of sleep in Example 2.3. Here, the explanatory variable is the Pittsburgh Sleep Quality Index and the response variable is the Subjective Units of Distress Scale.

USE YOUR KNOWLEDGE

2.4 Sleep and stress or stress and sleep? Consider the scenario described in the previous example. Make an argument for treating the Subjective Units of Distress Scale as the explanatory variable and the Pittsburgh Sleep Quality Index as the response variable.

In some studies it is easy to identify explanatory and response variables. The following example illustrates one situation where this is true: when we actually set values of one variable to see how it affects another variable.

EXAMPLE

2.6 How much calcium do you need? Adolescence is a time when bones are growing very actively. If young people do not have enough calcium, their bones will not grow properly. How much calcium is enough? Research designed to answer this question has been performed for many years at events called "Camp Calcium."[3] At these camps subjects eat a controlled diet that are identical except for the amount of calcium. The amount of calcium retained by the body is the major response variable of interest. Since the amount of calcium consumed is controlled by the researchers, this variable is the explanatory variable.

When you don't set the values of either variable but just observe both variables, there may or may not be explanatory and response variables. Whether there are depends on how you plan to use the data.

EXAMPLE

2.7 Student loans. A college student aid officer looks at the findings of the National Student Loan Survey. She notes data on the amount of debt of recent graduates, their current income, and how stressful they feel about college debt. She isn't interested in predictions but is simply trying to understand the situation of recent college graduates.

A sociologist looks at the same data with an eye to using amount of debt and income, along with other variables, to explain the stress caused by college debt. Now amount of debt and income are explanatory variables, and stress level is the response variable.

In many studies, the goal is to show that changes in one or more explanatory variables actually *cause* changes in a response variable. But many explanatory-response relationships do not involve direct causation. The SAT scores of high school students help predict the students' future college grades, but high SAT scores certainly don't cause high college grades.

Some of the statistical techniques in this chapter require us to distinguish explanatory from response variables; others make no use of this distinction.

independent variable
dependent variable

You will often see explanatory variables called **independent variables** and response variables called **dependent variables.** The idea behind this language is that response variables depend on explanatory variables. Because the words "independent" and "dependent" have other meanings in statistics that are unrelated to the explanatory-response distinction, we prefer to avoid those words.

Most statistical studies examine data on more than one variable. Fortunately, statistical analysis of several-variable data builds on the tools used for examining individual variables. The principles that guide our work also remain the same:

- Start with a graphical display of the data.

- Look for overall patterns and deviations from those patterns.

- Based on what you see, use numerical summaries to describe specific aspects of the data.

2.1 Scatterplots

EXAMPLE

2.8 Spam botnets. A botnet is a remotely and silently controlled collection of networked computers. Botnets are illicitly created through the use of viruses, Trojans, and other malware to assimilate computers, or bots, into the botnet, generally without the knowledge of the computer owner. Some botnets can grow to many thousands of bots located all over the world. A botnet that is used to send unwanted commercial emails, called spam, is called a spam botnet.[4] About 120 billion spam messages are sent per day and the cost of dealing with spam messages is estimated to be $140 billion per year.[5] Here is some information about 10 large botnets.

Botnet	Bots (thousands)	Spams per day (billions)	Botnet	Bots (thousands)	Spams per day (billions)
Srizbi	315	60	Grum	50	2
Bobax	185	9	Ozdok	35	10
Rustock	150	30	Nucrypt	20	5
Cutwail	125	16	Wopla	20	0.6
Storm	85	3	Spamthru	12	0.35

The variables are the number of bots operated by the botnet and the number of spam messages per day produced by these bots. The first botnet listed is a botnet called Srizbi which was discovered on June 30, 2007.[6] Srizbi has 315,000 bots that generate 60 billion spams per day.

USE YOUR KNOWLEDGE

SPAM

2.5   Make a data set.

(a) Create a spreadsheet that contains the spam botnet data.

(b) How many cases are in your data set?

(c) Describe the labels, variables, and values that you used.

(d) Which columns give quantitative variables?

SPAM

2.6   Use your data set. Using the data set that you created in the previous exercise, find graphical and numerical summaries for bots and spam messages per day.

The most common way to display the relation between two quantitative variables is a *scatterplot*.

SCATTERPLOT

A **scatterplot** shows the relationship between two quantitative variables measured on the same individuals. The values of one variable appear on the horizontal axis, and the values of the other variable appear on the vertical axis. Each individual in the data appears as the point in the plot fixed by the values of both variables for that individual.

EXAMPLE

SPAM

2.9   Bots and spam messages. We think that a netbot that has a large number of bots would be capable of generating a large amount of spam messages, relative to a netbot that has a smaller number of bots. Therefore, we think of the number of bots as an explanatory variable and the number of spam messages as a response variable. We begin our study of this relationship with a graphical display of the two variables.

Figure 2.1 gives a scatterplot that displays the relationship between the response variable, spam messages per day, and the explanatory variable, number of bots. Notice that 6 of the 10 botnets are clustered in the lower-left part of the plot with relatively low values for both bots and spam messages per day. On the other hand, the botnet Srizbi stands out with the highest values for both variables.

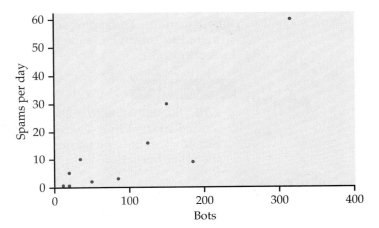

FIGURE 2.1 Scatterplot of spams per day (in billions) versus bots (in thousands), for Example 2.9.

Always plot the explanatory variable, if there is one, on the horizontal axis (the x axis) of a scatterplot. We usually call the explanatory variable x and the response variable y. If there is no explanatory-response distinction, either variable can go on the horizontal axis. The time plots in Section 1.1 (page 7) are special scatterplots where the explanatory variable x is a measure of time.

USE YOUR KNOWLEDGE

SPAM

2.7 Make a scatterplot.

    (a) Make a scatterplot similar to Figure 2.1 for the spam botnet data.

    (b) Mark the location of the botnet Bobax on your plot.

SPAM

2.8 Change the units.

    (a) Create a spreadsheet with the spam botnet data using the actual values. In other words, for Srizbi use 315,000 for the number of bots and 60,000,000,000 for the number of spam messages per day.

    (b) Make a scatterplot for the data coded in this way.

    (c) Describe how this scatterplot differs from Figure 2.1.

Interpreting scatterplots

To look more closely at a scatterplot such as Figure 2.1, apply the strategies of exploratory analysis learned in Chapter 1.

> **EXAMINING A SCATTERPLOT**
>
> In any graph of data, look for the **overall pattern** and for striking **deviations** from that pattern.
>
> You can describe the overall pattern of a scatterplot by the **form, direction,** and **strength** of the relationship.
>
> An important kind of deviation is an **outlier,** an individual value that falls outside the overall pattern of the relationship.

linear relationship

Figure 2.1 shows a clear *form:* the data lie in a roughly straight-line, or **linear,** pattern. To help us see this relationship, we can use software to put a straight line through the data. We will see more details about how this is done in Section 2.3.

    SPAM

cluster

> ### EXAMPLE
>
> **2.10 Scatterplot with a straight line.** Figure 2.2 plots the botnet data with a fitted straight line. There is a **cluster** of points in the lower left. Although Srizbi appears to be an outlier, it roughly lies in the same linear pattern as the other botnets.

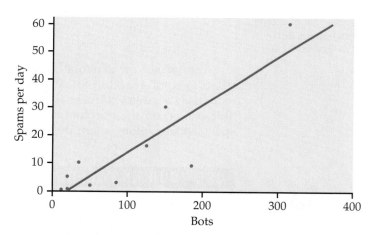

FIGURE 2.2 Scatterplot of spams per day (in billions) versus bots (in thousands) with a fitted straight line, for Example 2.10.

The relationship in Figure 2.2 also has a clear *direction:* botnets with more bots generate more spam messages than botnets that have fewer bots. This is a *positive association* between the two variables.

> **POSITIVE ASSOCIATION, NEGATIVE ASSOCIATION**
>
> Two variables are **positively associated** when above-average values of one tend to accompany above-average values of the other and below-average values also tend to occur together.
>
> Two variables are **negatively associated** when above-average values of one tend to accompany below-average values of the other, and vice versa.

The *strength* of a relationship in a scatterplot is determined by how closely the points follow a clear form. The overall relationship in Figure 2.1 is fairly moderate. Botnets with similar numbers of bots have a fair amount of scatter in the number of spam messages per day that they produce. Here is an example of a stronger linear relationship.

DEBT

> ### EXAMPLE
>
> **2.11 Debt for 24 countries.** The amount of debt owed by a country is a measure of its economic health. The Organization for Economic Co-Operation and Development (OECD) collects data on the central government debt for many countries. One of their tables gives the debt for 30 countries for the years 1998 to 2007.[7] Since there are a few countries with a very large amount of debt, let's concentrate on the 24 countries with debt less than US$ trillion in 2006. The 6 countries excluded are the United Kingdom, France, Germany, Italy, the United States, and Japan. The data are given in Figure 2.3.

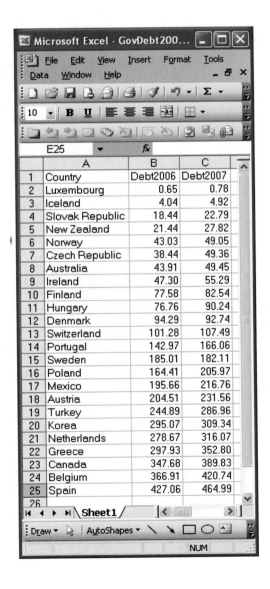

FIGURE 2.3 Central government debt in 2006 and 2007 for 24 countries with debt in 2006 less than US$1 trillion, in US$ billions, for Example 2.11.

Figure 2.4 is a scatterplot of the central government debt in 2007 versus the central government debt in 2006. The scatterplot shows a strong positive relationship between the debt in these two years.

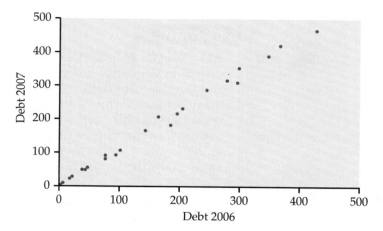

FIGURE 2.4   Scatterplot of debt in 2007 (US$ billion) versus debt in 2006 (US$ billions) for 24 countries with less than US$1 trillion debt in 2006, for Example 2.11.

USE YOUR KNOWLEDGE

2.9   Make a scatterplot. In our Mocha Frappuccino example, the 12-ounce drink costs $3.50, the 16-ounce drink costs $4.00, and the 24-ounce drink costs $4.50. Explain which variable should be used as the explanatory variable and make a scatterplot. Describe the scatterplot and the association between these two variables.

It is tempting to conclude that the strong linear relationship that we have found between the central government debt in 2006 and 2007 for the 24 countries is evidence that the amount of debt for each country is approximately the same in the two years. The first exercise below asks you to explore this temptation.

DEBT

DEBT29

USE YOUR KNOWLEDGE

2.10   Are the debts in 2006 and 2007 approximately the same? Use the methods you learned in Chapter 1 to examine whether or not the central government debts in 2006 and 2007 are approximately the same. (*Hint:* think about creating a new variable that would help you to answer this question.)

2.11   What about the countries with very large debts? In Example 2.11 we excluded six countries. The original data file did not list a value for the debt in 2007 for Japan. Here are the debts, in US$ billions, for the other five countries:

Country	2006 debt	2007 debt
United Kingdom	1168	1231
France	1240	1454
Germany	1252	1409
Italy	1892	2167
United States	4848	5055

Add the data for these five countries to your data set and make a scatterplot that includes the data for all 29 countries. Summarize the relationship. Does the additional data change the relationship? Explain your answer.

Of course, not all relationships are linear. Here is an example where the relationship is described by a nonlinear curve.

BESTCOUNTRIES

> ### EXAMPLE
>
> **2.12 Forbes.com Best Countries for Business.** Forbes.com analyzes business climates in 120 countries and determines an ordered list of these countries called *The Best Countries for Business*.[8] Let's look at two of the variables that they use to determine the ranks in their list, gross domestic product per capita and unemployment rate. We exclude the data from a few countries with very extreme values or missing values on one or more of the variables used in the rankings. Figure 2.5 is a scatterplot of gross domestic product per capita versus unemployment rate for 99 countries.
>
> The scatterplot suggests that there is a negative relationship between these two variables. The relationship is approximately linear for small values of unemployment, particularly for values less that about 15%. However, after that point the curve decreases more slowly; so overall, we have a nonlinear relationship.

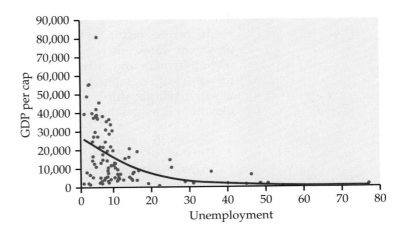

FIGURE 2.5 Scatterplot of gross domestic product per capita versus unemployment for 99 countries, for Example 2.12. There is a negative nonlinear relationship between these two variables.

Look at Figure 2.5 carefully. Notice that the two variables that we are examining are both skewed toward large values. This shows up in the plot with much of the data in the lower left part of the scatterplot. Skewed data are quite common in business applications of statistics, particularly when the measured variable is some kind of count or amount. In these situations, we often apply

transformation a **transformation** to the data. This means that we replace the original values by the transformed values and then use the transformed values for our analysis.

Transforming data is common in statistical practice. There are systematic principles that describe how transformations behave and guide the search for transformations that will, for example, make a distribution more Normal or a curved relationship more linear. You can read about these principles in the extra material entitled *Transforming Relationships* available on the text CD and Web site.

The log transformation

log transformation

The most important transformation that we will use is the **log transformation.** This transformation can only be used for variables that have strictly positive values. Occasionally, we use it when there are zeros, but in this case we first replace the zero values by some small value, often one-half of the smallest positive value in the data set.

You have probably encountered logarithms in one of your mathematics courses as a way to do certain kinds of arithmetic. Logarithms are a lot more fun when used in statistical analyses. We will use natural logarithms. Statistical software and statistical calculators generally provide easy ways to perform this transformation.

BESTCOUNTRIES

EXAMPLE

2.13 Gross domestic product per capita and unemployment with logarithms.
Figure 2.6 is a scatterplot of the log of the gross domestic product per capita versus the log of the unemployment rate. Notice how the data now fill up much of the central part of the scatterplot in contrast to the clustering that we noticed in Figure 2.5. Here we see that the relationship is essentially flat for values of log unemployment that are less than about 1.6. This point corresponds to an unemployment rate of about 5%. So for low unemployment rates, there appears to be little or no relationship between unemployment and gross domestic product per capita. On the other hand, for values of unemployment greater than 5%, we see an approximate linear negative relationship with gross domestic product per capita. High unemployment is associated with low gross domestic product per capita.

FIGURE 2.6 Scatterplot of log gross domestic product versus log unemployment for 99 countries with a smooth curve, for Example 2.13. There appears to be essentially no relationship for low unemployment values, up to about 5% (1.6 on the log scale), and a negative relationship for unemployment values greater that 5%.

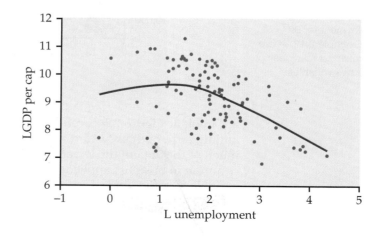

Use of transformations and the interpretation of scatterplots is an art that requires judgment and knowledge about the variables that we are studying. Always ask yourself if the relationship that you see makes sense. If it does not, then additional analyses are needed to understand the data.

Adding categorical variables to scatterplots*

In Examples 2.12 and 2.13, we looked at two of the variables used by Forbes .com to construct their Best Countries for Business list. There are several more variables that they use, but they do not give the details about exactly how their list is constructed. Let's take a look at how the two variables we examined relate to whether or not a country ranks high or low on the list.

CATEGORICAL VARIABLES IN SCATTERPLOTS

To add a categorical variable to a scatterplot, use a different plot color or symbol for each category.

 **BESTCOUNTRIES**

> **EXAMPLE**
>
> **2.14   Gross domestic product per capita and the rankings.** We start by creating a categorical variable that indicates whether or not a country ranks in the top half of the Best Countries for Business list. In our scatterplot, we will use the symbol "H" for countries that rank in the top half of the list and "L" for countries that rank in the bottom half of the list. Examine the scatterplot in Figure 2.7 carefully. Notice that the countries in the top part of the

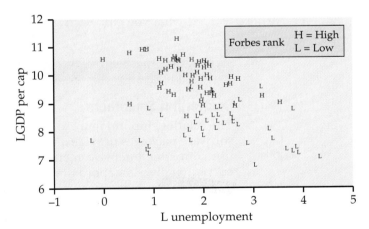

FIGURE 2.7 Scatterplot of log gross domestic product versus log unemployment for 99 countries with different plotting symbols. Countries with Forbes.com in the top half of their Best Countries for Business rankings are plotted with the symbol "H" and those in the lower half are plotted with the symbol "L," for Example 2.14. Countries with high gross domestic product tend to be rated high, while the unemployment rate does not appear to be a major factor in the distinguishing high versus low rankings.

---

*This section is optional.

plot, those with relatively high gross domestic product per capita, tend to rank high in the Best Countries for Business list. On the other hand, countries in the bottom part, those with relatively low gross domestic product per capita, tend to rank low. What about unemployment? No clear pattern is evident. Although the countries with the six highest unemployment rates are all ranked low, they also have very low values for gross domestic product per capita.

In this example, we used a quantitative variable, rank in the Best Countries for Business list, to create a categorical variable that indicated whether a country ranked high or low in list. Of course, if the variable that you want for a plotting symbol is categorical, no conversion is needed. Careful judgment is needed in applying using this graphical method. Don't be discouraged if your first attempt is not very successful. In performing a good data analysis, you will often produce several plots before you find the one that you believe to be most effective in describing the data.[9]

USE YOUR KNOWLEDGE

 **BESTCOUNTRIES**

2.12 Change the plotting symbol. In Example 2.14 we used the plotting symbols "H" and "L" to distinguish countries that ranked high and low on the Best Countries for Business list. Let's see if we can learn anything more about these variables by refining our categorical variable further. Define a new categorical variable that has three distinct values corresponding to a rank in the top third (ranks 1 to 40), the middle third (ranks 41 to 80), and the bottom third (ranks 81 to 120). Choose appropriate plotting symbols and make a scatterplot similar to Figure 2.7 using this categorical variable. Describe your scatterplot and compare what you can learn from it with what we learned in Example 2.14.

BEYOND THE BASICS

Scatterplot Smoothers

The relationship in Figure 2.4 (page 88) appears to be linear. Some statistical software packages provide a tool to help us make this kind of judgment. These

algorithms use computer intensive methods called **algorithms** that calculate a smooth curve that gives an approximate fit to the points in a scatterplot. This is called

smoothing **smoothing** a scatterplot. Usually, these methods use a smoothing parameter that determines how smooth the fit will be. You can vary it until you have a fit that you judge to suitable for your data. Here is an example.

 DEBT

EXAMPLE

2.15 Debt for 24 countries with a smooth fit. Figure 2.8 gives the scatterplot that we examined in Figure 2.4 with a smooth fit. Notice that the smooth curve fits almost all of the points. The curve is too wavy and does not provide a good summary of the relationship.

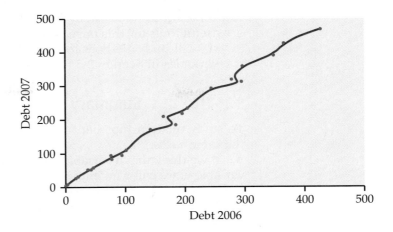

FIGURE 2.8 Scatterplot of debt in 2007 (US$ billion) versus debt in 2006 (US$ billion) for 24 countries with less than US$100 billion debt in 2006, with a smooth curve fit to the data, for Example 2.15. This smooth curve fits the data too well and does not provide a good summary of the relationship.

Our first attempt at smoothing the data was not very successful. This scenario happens frequently when we use data analysis methods to learn something from our data. Don't be discouraged when your first attempt at summarizing data produces unsatisfactory results. Take what you learn and refine your analysis until you are satisfied that you have found a good summary. It is your last attempt, not your first, that is most important.

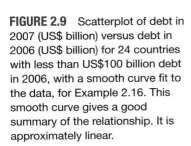 **DEBT**

EXAMPLE

2.16   A better smooth fit for the debt data. By varying the smoothing parameter, we can make the curve more or less smooth. Figure 2.9 gives the same data in the previous figure with a better smooth fit. The smooth curve is very close to a straight line. In this way we have confirmed our original impression that the relationship between these two variables is approximately linear.

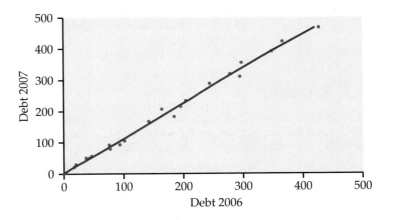

FIGURE 2.9 Scatterplot of debt in 2007 (US$ billion) versus debt in 2006 (US$ billion) for 24 countries with less than US$100 billion debt in 2006, with a smooth curve fit to the data, for Example 2.16. This smooth curve gives a good summary of the relationship. It is approximately linear.

Categorical explanatory variables

Scatterplots display the association between two quantitative variables. To display a relationship between a categorical variable and a quantitative variable, make a side-by-side comparison of the distributions of the response for each category. Back-to-back stemplots (page 10) and side-by-side boxplots (page 36) are useful tools for this purpose.

We will study methods for describing the association between two categorical variables in Section 2.5 (page 136).

SECTION 2.1 Summary

To study relationships between variables, we must measure the variables on the same cases.

If we think that a variable x may explain or even cause changes in another variable y, we call x an **explanatory variable** and y a **response variable.**

A **scatterplot** displays the relationship between two quantitative variables. Mark values of one variable on the horizontal axis (x axis) and values of the other variable on the vertical axis (y axis). Plot each individual's data as a point on the graph.

Always plot the explanatory variable, if there is one, on the x axis of a scatterplot. Plot the response variable on the y axis.

Plot points with different colors or symbols to see the effect of a categorical variable in a scatterplot.

In examining a scatterplot, look for an overall pattern showing the **form, direction,** and **strength** of the relationship, and then for **outliers** or other deviations from this pattern.

Form: Linear relationships, where the points show a straight-line pattern, are an important form of relationship between two variables. Curved relationships and **clusters** are other forms to watch for.

Direction: If the relationship has a clear direction, we speak of either **positive association** (high values of the two variables tend to occur together) or **negative association** (high values of one variable tend to occur with low values of the other variable).

Strength: The **strength** of a relationship is determined by how close the points in the scatterplot lie to a simple form such as a line.

To display the relationship between a categorical explanatory variable and a quantitative response variable, make a graph that compares the distributions of the response for each category of the explanatory variable.

SECTION 2.1 Exercises

For Exercise 2.1, see page 80; for 2.2 and 2.4, see pages 81–82 for 2.5 and 2.6, see page 84; for 2.7 and 2.8, see page 85; for 2.9 to 2.11, see pages 88–89, and for 2.12, see page 92.

2.13 What's wrong? Explain what is wrong with each of the following:

(a) A boxplot can be used to examine the relationship between two variables.

(b) In a scatterplot we put the response variable on the y axis and the explanatory variable on the x axis.

(c) If two variables are positively associated, then high values of one variable are associated with low values of the other variable.

2.14 Make some sketches. For each of the following situations, make a scatterplot that illustrates the given relationship between two variables.

(a) A strong negative linear relationship.

(b) No apparent relationship.

(c) A weak positive relationship.

(d) A more complicated relationship. Give the sketch and explain the relationship.

2.15 Who does not have health insurance? The lack of adequate health insurance coverage is a major problem for many Americans. The Current Population Survey collected data on the characteristics of the uninsured.[10] The numbers of uninsured and the total number of people classified by age for 2006 area as follows. The units are thousands of people. HEALTHINSURANCE

Age group	Number uninsured	Total number
Under 18 years	8,661	74,101
18 to 24 years	8,323	28,405
25 to 34 years	10,713	39,868
35 to 44 years	8,018	42,762
45 to 64 years	10,738	75,653
65 years and older	541	36,035

(a) Plot the number of uninsured versus age group.

(b) Find the total number of uninsured persons and use this total to compute the percent of the uninsured who are in each age group.

(c) Plot the percents versus age group.

(d) Explain how the plot you produced in part (c) differs from the plot that you made in part (a).

(e) Summarize what you can conclude from these plots.

2.16 Which age groups have the larger percent uninsured? Refer to the previous exercise. Let's take a look at the data from a different point of view. HEALTHINSURANCE

(a) For each age group calculate the percent that are uninsured using the number of uninsured persons and the total number of persons in each group.

(b) Make a plot of the percent uninsured versus age group.

(c) Summarize the information in your plot and write a short summary of what you conclude from your analysis.

2.17 Compare the two percents. In the previous two exercises, you computed percents in two different ways and generated plots versus age group. Describe the difference between the two ways with an emphasis on what kinds of conclusions can be drawn from each. HEALTHINSURANCE

2.18 What's in the beer? Beer100.com advertises itself as "Your Place for All Things Beer." One of their things is a list of 86 domestic beer brands with the percent alcohol, calories per 12 ounces, and carbohydrates per 12 ounces (in grams).[11] BEER

(a) Figure 2.10 gives a scatterplot of carbohydrates versus percent alcohol. Give a short summary of what can be learned from the plot.

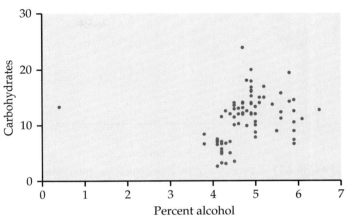

FIGURE 2.10 Scatterplot of carbohydrates versus percent alcohol for 86 brands of beer, for Exercise 2.18.

(b) One of the points is an outlier. Use the data file to find the outlier brand of beer. How is this brand of beer marketed as compared with the other brands?

(c) Remove the outlier from the data set and generate a scatterplot of the remaining data.

(d) Describe the relationship between carbohydrates and percent alcohol based on what you see in your scatterplot.

2.19 More beer. Refer to the previous exercise. BEER

(a) Make a scatterplot of calories versus percent alcohol using the data set without the outlier.

(b) Describe the relationship between these two variables.

2.20 Will you live longer if you use the Internet? The World Bank collects data on many variables related to world development for countries throughout the world. Two of these are Internet use, in number of users per 100 people, and life expectancy, in years.[12] Figure 2.11 is a scatterplot of life expectancy versus Internet use. INTERNETANDLIFE

(a) Describe the relationship between these two variables.

(b) A friend looks at this plot and concludes that using the Internet will increase the length of your life. Write a short paragraph explaining why the association seen in the scatterplot does not provide a reason to draw this conclusion.

2.21 ⚠ **Let's look at Europe.** Refer to the previous exercise. Figure 2.12 gives a scatterplot for the same data

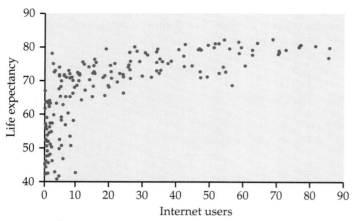

FIGURE 2.11 Scatterplot of life expectancy (in years) versus Internet users (per 100 people) for 181 countries, for Exercise 2.20.

for the 48 European countries in the data set. Compare this figure with Figure 2.11 which plots that data for all 181 countries in the data set. Write a paragraph summarizing the relationship between life expectancy and Internet use for European countries with an emphasis on how the European countries compare with the entire set of 181 countries. Be sure to take into account the fact the software used here automatically chooses the range of values for each axis so that the space in the plot is used efficiently. In this case, the range of values for Internet use is the same for both scatterplots but the range of values for life expectancy is quite different. ● INTERNETANDLIFEE

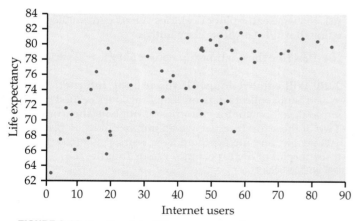

FIGURE 2.12 Scatterplot of life expectancy (in years) versus Internet users (per 100 people) for 48 European countries, for Exercise 2.21.

2.22 ⚔ **How would you make a better plot?** In the previous two exercises, we looked at the relationship between life expectancy and Internet use. First, we made a scatterplot for all 181 countries in the data set. Then we made one for the subset of 48 European countries. Explain how you would construct a single plot to make a comparison between the European countries and the other countries in the data set. (*Optional:* Make the plot if you have software that can do what you need.) ● INTERNETANDLIFE

2.23 Average temperatures. Here are the average temperatures in degrees for Lafayette, Indiana, during the months of February through May: ● WLAFTEMPS

Month	February	March	April	May
Temperature (degrees F)	30	41	51	62

(a) Explain why month should be the explanatory variable for examining this relationship.

(b) Make a scatterplot and describe the relationship.

2.24 Relationship between first test and final exam. How strong is the relationship between the score on the first exam and the score on the final exam in an elementary statistics course? Here are data for eight students from such a course: ● STATCOURSE8

First-test score	153	144	162	149	127	118	158	153
Final-exam score	145	140	145	170	145	175	170	160

(a) Which variable should play the role of the explanatory variable in describing this relationship?

(b) Make a scatterplot and describe the relationship.

(c) Give some possible reasons why this relationship is so weak.

2.25 Relationship between second test and final exam. Refer to the previous exercise. Here are the data for the second test and the final exam for the same students: ● STATCOURSE8

Second-test score	158	162	144	162	136	158	175	153
Final-exam score	145	140	145	170	145	175	170	160

(a) Explain why you should use the second-test score as the explanatory variable.

(b) Make a scatterplot and describe the relationship.

(c) Why do you think the relationship between the second-test score and the final-exam score is stronger than the relationship between the first-test score and the final-exam score?

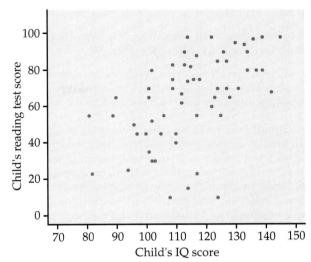

2.26 Add an outlier to the plot. Refer to the previous exercise. Add a ninth student whose scores on the second test and final exam would lead you to classify the additional data point as an outlier. Highlight the outlier on your scatterplot and describe the performance of the student on the second exam and final exam and why that leads to the conclusion that the result is an outlier. Give a possible reason for the performance of this student.

2.27 Explanatory and response variables. In each of the following situations, is it more reasonable to simply explore the relationship between the two variables or to view one of the variables as an explanatory variable and the other as a response variable? In the latter case, which is the explanatory variable and which is the response variable?

(a) The weight of a child and the age of the child from birth to 10 years.

(b) High school English grades and high school math grades.

(c) The rental price of apartments and the number of bedrooms in the apartment.

(d) The amount of sugar added to a cup of coffee and how sweet the coffee tastes.

(e) The student evaluation scores for an instructor and the student evaluation scores for the course.

2.28 Parents' income and student loans. How well does the income of a college student's parents predict how much the student will borrow to pay for college? We have data on parents' income and college debt for a sample of 1200 recent college graduates. What are the explanatory and response variables? Are these variables categorical or quantitative? Do you expect a positive or negative association between these variables? Why?

2.29 Reading ability and IQ. A study of reading ability in school children chose 60 fifth-grade children at random from a school. The researchers had the children's scores on an IQ test and on a test of reading ability.[13] Figure 2.13 plots reading test score (response) against IQ score (explanatory).

(a) Explain why we should expect a positive association between IQ and reading score for children in the same grade. Does the scatterplot show a positive association?

(b) A group of four points appear to be outliers. In what way do these children's IQ and reading scores deviate from the overall pattern?

(c) Ignoring the outliers, is the association between IQ and reading scores roughly linear? Is it very strong? Explain your answers.

FIGURE 2.13 IQ and reading test scores for 60 fifth-grade children, for Exercise 2.29.

2.30 Can children estimate their reading ability? The main purpose of the study cited in Exercise 2.29 was to ask whether school children can estimate their own reading ability. The researchers had the children's scores on a test of reading ability. They asked each child to estimate his or her reading level, on a scale from 1 (low) to 5 (high). Figure 2.14 is a scatterplot of the children's estimates (response) against their reading scores (explanatory).

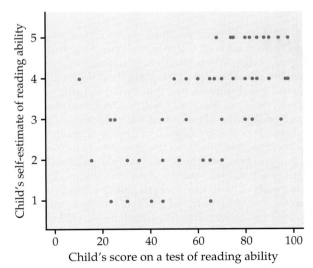

FIGURE 2.14 Reading test scores for 60 fifth-grade children and the children's estimates of their own reading levels, for Exercise 2.30.

(a) What explains the "stair-step" pattern in the plot?

(b) Is there an overall positive association between reading score and self-estimate?

(c) There is one clear outlier. What is this child's self-estimated reading level? Does this appear to over- or underestimate the level as measured by the test?

2.31 Small falcons in Sweden. Often the percent of an animal species in the wild that survive to breed again is lower following a successful breeding season. This is part of nature's self-regulation, tending to keep population size stable. A study of merlins (small falcons) in northern Sweden observed the number of breeding pairs in an isolated area and the percent of males (banded for identification) who returned the next breeding season. Here are data for nine years:[14] 🔵 FALCONS

Pairs	28	29	29	29	30	32	33	38	38
Percent	82	83	70	61	69	58	43	50	47

(a) Why is the response variable the *percent* of males that return rather than the *number* of males that return?

(b) Make a scatterplot. To emphasize the pattern, also plot the mean response for years with 29 and 38 breeding pairs and draw lines connecting the mean responses for the six values of the explanatory variable.

(c) Describe the pattern. Do the data support the theory that a smaller percent of birds survive following a successful breeding season?

2.32 Biological clocks. Many plants and animals have "biological clocks" that coordinate activities with the time of day. When researchers looked at the length of the biological cycles in the plant *Arabidopsis* by measuring leaf movements, they found that the length of the cycle is not always 24 hours. The researchers suspected that the plants adapt their clocks to their north-south position. Plants don't know geography, but they do respond to light, so the researchers looked at the relationship between the plants' cycle lengths and the length of the day on June 21 at their locations. The data file includes data on cycle length and day length, both in hours, for 146 plants.[15] Plot cycle length as the response variable against day length as the explanatory variable. Does there appear to be a positive association? Is it a strong association? Explain your answers. 🔵 BIOCLOCKS

2.33 Social rejection and pain. We often describe our emotional reaction to social rejection as "pain." A clever study asked whether social rejection causes activity in areas of the brain that are known to be activated by physical pain. If it does, we really do experience social and physical pain in similar ways. Subjects were first included and then deliberately excluded from a social activity while increases in blood flow in their brains were measured. After each activity, the subjects filled out questionnaires that assessed how excluded they felt.

Here are data for 13 subjects.[16] The explanatory variable is "social distress" measured by each subject's questionnaire score after exclusion relative to the score after inclusion (values greater than 1 show the degree of distress caused by exclusion). The response variable is activity in the anterior cingulate cortex, a region of the brain that is activated by physical pain. 🔵 SOCIALREJECTION

Subject	Social distress	Brain activity	Subject	Social distress	Brain activity
1	1.26	−0.055	8	2.18	0.025
2	1.85	−0.040	9	2.58	0.027
3	1.10	−0.026	10	2.75	0.033
4	2.50	−0.017	11	2.75	0.064
5	2.17	−0.017	12	3.33	0.077
6	2.67	0.017	13	3.65	0.124
7	2.01	0.021			

Plot brain activity against social distress. Describe the direction, form, and strength of the relationship, as well as any outliers. Do the data suggest that brain activity in the "pain" region is directly related to the distress from social exclusion?

2.34 Business revenue and team value in the NBA. Management theory says that the value of a business should depend on its operating income, the income produced by the business after taxes. (Operating income excludes income from sales of assets and investments, which don't reflect the actual business.) Total revenue, which ignores costs, should be less important. Debt includes borrowing for the construction of a new arena. Table 2.1 shows the value, operating income, debt, and revenue of the teams in the National Basketball Association (NBA).[17] Professional sports teams are generally privately owned, often by very wealthy individuals who may treat their team as a source of prestige rather than as a business. 🔵 NBA

(a) Plot team value against revenue. Describe the relationship.

(b) Plot team value against debt. Describe the relationship.

TABLE 2.1

NBA teams as businesses

Team	Value ($millions)	Revenue ($millions)	Debt ($millions)	Income ($millions)
New York Knicks	613	0	208	29.6
Los Angeles Lakers	584	22	191	47.9
Chicago Bulls	504	11	165	55.4
Detroit Pistons	480	0	160	40.4
Cleveland Cavaliers	477	42	159	13.1
Houston Rockets	469	15	156	31.2
Dallas Mavericks	466	26	153	−13.6
Phoenix Suns	452	39	148	28.9
Boston Celtics	447	40	149	20.1
San Antonio Spurs	415	13	138	19.0
Toronto Raptors	400	40	138	27.7
Miami Heat	393	43	131	−1.1
Philadelphia 76ers	360	18	116	0.3
Utah Jazz	358	3	119	8.8
Washington Wizards	353	47	118	14.9
Sacramento Kings	350	24	117	7.0
Orlando Magic	349	29	100	6.2
Golden State Warriors	335	22	112	14.2
Denver Nuggets	329	14	112	−26.3
Portland Trail Blazers	307	34	114	−0.9
Atlanta Hawks	306	23	102	6.7
Indiana Pacers	303	16	101	−6.5
Minnesota Timberwolves	301	17	100	−5.7
Oklahoma City Thunder	300	47	82	−9.4
Los Angeles Clippers	297	0	99	10.7
New Jersey Nets	295	71	98	−0.9
Memphis Grizzlies	294	51	95	−3.2
New Orleans Hornets	285	35	95	3.2
Charlotte Bobcats	284	53	95	−4.9
Milwaukee Bucks	278	20	94	5.4

(c) Plot team value against income. Describe the relationship. In your description be sure to pay attention to the teams that have negative income, that is, to the teams that lost money.

(d) Write a short summary comparing the relationships that you described in parts (a), (b), and (c) of this exercise.

2.35 🔺 **Body mass and metabolic rate.** Metabolic rate, the rate at which the body consumes energy, is important in studies of weight gain, dieting, and exercise. The following table gives data on the lean body mass and resting metabolic rate for 12 women and 7 men who are subjects in a study of dieting. Lean body mass, given in kilograms, is a person's weight leaving out all fat. Metabolic rate is measured in calories burned per 24 hours, the same calories used to describe the energy

content of foods. The researchers believe that lean body mass is an important influence on metabolic rate. 🔵 BODYMASS

Subject	Sex	Mass	Rate	Subject	Sex	Mass	Rate
1	M	62.0	1792	11	F	40.3	1189
2	M	62.9	1666	12	F	33.1	913
3	F	36.1	995	13	M	51.9	1460
4	F	54.6	1425	14	F	42.4	1124
5	F	48.5	1396	15	F	34.5	1052
6	F	42.0	1418	16	F	51.1	1347
7	M	47.4	1362	17	F	41.2	1204
8	F	50.6	1502	18	M	51.9	1867
9	F	42.0	1256	19	M	46.9	1439
10	M	48.7	1614				

(a) Make a scatterplot of the data, using different symbols or colors for men and women.

(b) Is the association between these variables positive or negative? What is the form of the relationship? How strong is the relationship? Does the pattern of the relationship differ for women and men? How do the male subjects as a group differ from the female subjects as a group?

2.36 **How do icicles grow?** How fast do icicles grow? Japanese researchers measured the growth of icicles in a cold chamber under various conditions of temperature, wind, and water flow.[18] Table 2.2 contains data produced under two sets of conditions. In both cases, there was no wind and the temperature was set at $-11°C$.

TABLE 2.2

Growth of icicles over time

Run 8903				Run 8905			
Time (min)	Length (cm)	Time (min)	Length (cm)	Time (min)	Length (cm)	Time (min)	Length (cm)
10	0.6	130	18.1	10	0.3	130	10.4
20	1.8	140	19.9	20	0.6	140	11.0
30	2.9	150	21.0	30	1.0	150	11.9
40	4.0	160	23.4	40	1.3	160	12.7
50	5.0	170	24.7	50	3.2	170	13.9
60	6.1	180	27.8	60	4.0	180	14.6
70	7.9			70	5.3	190	15.8
80	10.1			80	6.0	200	16.2
90	10.9			90	6.9	210	17.9
100	12.7			100	7.8	220	18.8
110	14.4			110	8.3	230	19.9
120	16.6			120	9.6	240	21.1

TABLE 2.3

World record times for the 10,000-meter run

Men				Women	
Record year	Time (seconds)	Record year	Time (seconds)	Record year	Time (seconds)
1912	1880.8	1963	1695.6	1967	2286.4
1921	1840.2	1965	1659.3	1970	2130.5
1924	1835.4	1972	1658.4	1975	2100.4
1924	1823.2	1973	1650.8	1975	2041.4
1924	1806.2	1977	1650.5	1977	1995.1
1937	1805.6	1978	1642.4	1979	1972.5
1938	1802.0	1984	1633.8	1981	1950.8
1939	1792.6	1989	1628.2	1981	1937.2
1944	1775.4	1993	1627.9	1982	1895.3
1949	1768.2	1993	1618.4	1983	1895.0
1949	1767.2	1994	1612.2	1983	1887.6
1949	1761.2	1995	1603.5	1984	1873.8
1950	1742.6	1996	1598.1	1985	1859.4
1953	1741.6	1997	1591.3	1986	1813.7
1954	1734.2	1997	1587.8	1993	1771.8
1956	1722.8	1998	1582.7		
1956	1710.4	2004	1580.3		
1960	1698.8	2005	1577.3		
1962	1698.2				

Water flowed over the icicle at a higher rate (29.6 milligrams per second) in run 8905 and at a slower rate (11.9 mg/s) in Run 8903. ICICLES

(a) Make a scatterplot of the length of the icicle in centimeters versus time in minutes, using separate symbols for the two runs.

(b) Write a careful explanation of what your plot shows about the growth of icicles.

2.37 **Records for men and women in the 10K.** Table 2.3 shows the progress of world record times (in seconds) for the 10,000-meter run for both men and women.[19] TENK

(a) Make a scatterplot of world record time against year, using separate symbols for men and women. Describe the pattern for each sex. Then compare the progress of men and women.

(b) Women began running this long distance later than men, so we might expect their improvement to be more rapid. Moreover, it is often said that men have little advantage over women in distance running as opposed to sprints, where muscular strength plays a greater role. Do the data appear to support these claims?

2.2 Correlation

A scatterplot displays the form, direction, and strength of the relationship between two quantitative variables. Linear (straight-line) relations are particularly important because a straight line is a simple pattern that is quite common. We say a linear relationship is strong if the points lie close to a straight line, and weak if they are widely scattered about a line. Our eyes are not good judges of how strong a relationship is. The two scatterplots in Figure 2.15 depict exactly the same data, but the plot on the right is drawn smaller in a large field. The plot on the right seems to show a stronger relationship. Our eyes can be fooled by changing the plotting scales or the amount of white space around the cloud of points in a scatterplot.[20] We need to follow our strategy for data analysis by using a numerical measure to supplement the graph. *Correlation* is the measure we use.

The correlation *r*

We have data on variables x and y for n individuals. Think, for example, of measuring height and weight for n people. Then x_1 and y_1 are your height and

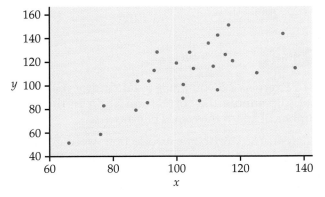
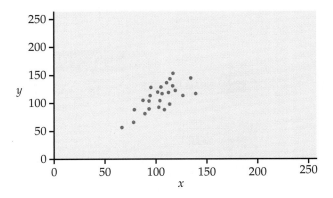

FIGURE 2.15 Two scatterplots of the same data. The linear pattern in the plot on the right appears stronger because of the surrounding space.

your weight, x_2 and y_2 are my height and my weight, and so on. For the ith individual, height x_i goes with weight y_i. Here is the definition of correlation.

CORRELATION

The **correlation** measures the **direction** and **strength** of the linear relationship between two quantitative variables. Correlation is usually written as r.

Suppose that we have data on variables x and y for n individuals. The means and standard deviations of the two variables are $\bar{x}$ and s_x for the x-values, and $\bar{y}$ and s_y for the y-values. The correlation r between x and y is

$$r = \frac{1}{n-1} \sum \left(\frac{x_i - \bar{x}}{s_x} \right) \left(\frac{y_i - \bar{y}}{s_y} \right)$$

As always, the summation sign $\sum$ means "add these terms for all the individuals." The formula for the correlation r is a bit complex. It helps us see what correlation is but is not convenient for actually calculating r. In practice you should use software or a calculator that finds r from keyed-in values of two variables x and y. Exercise 2.47 asks you to calculate a correlation step-by-step from the definition to solidify its meaning.

The formula for r begins by standardizing the observations. Suppose, for example, that x is height in centimeters and y is weight in kilograms and that we have height and weight measurements for n people. Then $\bar{x}$ and s_x are the mean and standard deviation of the n heights, both in centimeters. The value

$$\frac{x_i - \bar{x}}{s_x}$$

is the standardized height of the ith person, familiar from Chapter 1. The standardized height says how many standard deviations above or below the mean a person's height lies. Standardized values have no units—in this example, they are no longer measured in centimeters. Standardize the weights also. The correlation r is an average of the products of the standardized height and the standardized weight for the n people.

USE YOUR KNOWLEDGE

SPAM Case 2.1

2.38 Spam botnets. In Exercise 2.5 you made a data set for the botnet data. Use that data set to compute the correlation between the number of bots and the number of spam messages per day.

SPAM

2.39 Change the units. In the previous exercise bots were given in thousands and spam messages per day were recorded in billions. In Exercise 2.8 you created a data set using the actual values. For example, Srizbi has 315,000 bots and generates 60,000,000,000 spam messages per day.

(a) Find the correlation between bots and spam messages using this data set.

(b) Compare this correlation with the one that you computed in the previous exercise.

(c) What can you say in general about the effect of changing units in this way on the size of the correlation?

Properties of correlation

The formula for correlation helps us see that r is positive when there is a positive association between the variables. Height and weight, for example, have a positive association. People who are above average in height tend to also be above average in weight. Both the standardized height and the standardized weight for such a person are positive. People who are below average in height tend also to have below-average weight. Then both standardized height and standardized weight are negative. In both cases, the products in the formula for r are mostly positive and so r is positive. In the same way, we can see that r is negative when the association between x and y is negative. More detailed study of the formula gives more detailed properties of r. Here is what you need to know in order to interpret correlation:

- Correlation makes no use of the distinction between explanatory and response variables. It makes no difference which variable you call x and which you call y in calculating the correlation.

- *Correlation requires that both variables be quantitative, so that it makes sense to do the arithmetic indicated by the formula for r.* For example, we cannot calculate a correlation between the incomes of a group of people and what city they live in, because city is a categorical variable.

- Because r uses the standardized values of the observations, r does not change when we change the units of measurement of x, y, or both. Measuring height in inches rather than centimeters and weight in pounds rather than kilograms does not change the correlation between height and weight. The correlation r itself has no unit of measurement; it is just a number.

- Positive r indicates positive association between the variables, and negative r indicates negative association.

- The correlation r is always a number between -1 and 1. Values of r near 0 indicate a very weak linear relationship. The strength of the relationship increases as r moves away from 0 toward either -1 or 1. Values of r close to -1 or 1 indicate that the points lie close to a straight line. The extreme values $r = -1$ and $r = 1$ occur only when the points in a scatterplot lie exactly along a straight line.

- Correlation measures the strength of only the linear relationship between two variables. *Correlation does not describe curved relationships between variables, no matter how strong they are.*

• *Like the mean and standard deviation, the correlation is not resistant: r is strongly affected by a few outlying observations.* Use *r* with caution when outliers appear in the scatterplot.

The scatterplots in Figure 2.16 illustrate how values of *r* closer to 1 or −1 correspond to stronger linear relationships. To make the essential meaning of *r* clear, the standard deviations of both variables in these plots are equal and the horizontal and vertical scales are the same. In general, it is not so easy to guess the value of *r* from the appearance of a scatterplot. Remember that changing the plotting scales in a scatterplot may mislead our eyes, but it does not change the standardized values of the variables and therefore cannot change the correlation. To explore how extreme observations can influence *r*, use the *Correlation and Regression* applet available on the text CD and Web site.

Finally, remember that **correlation is not a complete description of two-variable data,** even when the relationship between the variables is linear. You should give the means and standard deviations of both *x* and *y* along with the correlation. (Because the formula for correlation uses the means and standard deviations, these measures are the proper choices to accompany a correlation.) Conclusions based on correlations alone may require rethinking in the light of a more complete description of the data.

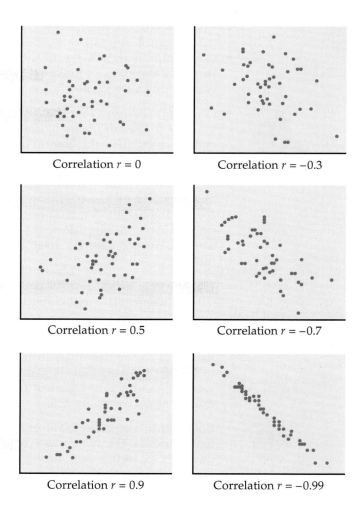

FIGURE 2.16 How the correlation *r* measures the direction and strength of a linear association.

EXAMPLE

2.17 Scoring of figure skating in the Olympics. Until a scandal at the 2002 Olympics brought change, figure skating was scored by judges on a scale from 0.0 to 6.0. The scores were often controversial. We have the scores awarded by two judges, Pierre and Elena, to many skaters. How well do they agree? We calculate that the correlation between their scores is $r = 0.9$. But the mean of Pierre's scores is 0.8 point lower than Elena's mean.

These facts in the example above do not contradict each other. They are simply different kinds of information. The mean scores show that Pierre awards lower scores than Elena. But because Pierre gives *every* skater a score about 0.8 point lower than Elena, the correlation remains high. Adding the same number to all values of either x or y does not change the correlation. If both judges score the same skaters, the competition is scored consistently because Pierre and Elena agree on which performances are better than others. The high r shows their agreement. But if Pierre scores some skaters and Elena others, we must add 0.8 points to Pierre's scores to arrive at a fair comparison.

SECTION 2.2 Summary

The **correlation** r measures the direction and strength of the linear (straight line) association between two quantitative variables x and y. Although you can calculate a correlation for any scatterplot, r measures only linear relationships.

Correlation indicates the direction of a linear relationship by its sign: $r > 0$ for a positive association and $r < 0$ for a negative association.

Correlation always satisfies $-1 \leq r \leq 1$ and indicates the strength of a relationship by how close it is to -1 or 1. Perfect correlation, $r = \pm 1$, occurs only when the points lie exactly on a straight line.

Correlation ignores the distinction between explanatory and response variables. The value of r is not affected by changes in the unit of measurement of either variable. Correlation is not resistant, so outliers can greatly change the value of r.

SECTION 2.2 Exercises

2.40 Thinking about correlation. Figure 2.4 (page 88) is a scatterplot of 2007 debt versus 2006 debt for 24 countries. Is the correlation r for these data near -1, clearly negative but not near -1, near 0, clearly positive but not near 1, or near 1? Explain your answer. DEBT

2.41 Brand names and generic products.

(a) If a store always prices its generic "store brand" products at 90% of the brand name products' prices, what would be the correlation between the prices of the brand name products and the store brand products? (*Hint:* Draw a scatterplot for several prices.)

(b) If the store always prices its generic products $1 less than the corresponding brand name products, then what would be the correlation between the prices of the brand name products and the store brand products?

2.42 Strong association but no correlation. Here is a data set that illustrates an important point about correlation: CORRELATION

X	20	30	40	50	60
Y	10	30	50	30	10

(a) Make a scatterplot of Y versus X.

(b) Describe the relationship between Y and X. Is it weak or strong? Is it linear?

(c) Find that the correlation between Y and X.

(d) What important point about correlation does this exercise illustrate?

2.43 Alcohol and carbohydrates in beer. Figure 2.10 (page 95) gives a scatterplot of carbohydrates versus percent alcohol in 86 brands of beer. Compute the correlation for these data. BEER

2.44 Alcohol and carbohydrates in beer revisited. Refer to the previous exercise. The data that you used to compute the correlation includes an outlier. BEER

(a) Remove the outlier and recompute the correlation.

(b) Write a short paragraph about the possible effects of outliers on a correlation using this example to illustrate your ideas.

2.45 Will you live longer if you use the Internet? Figure 2.11 (page 96) is a scatterplot of the number of life expectancy versus Internet users per 100 people for 181 countries. In Exercise 2.20 you described this relationship. Make a plot of the data similar to Figure 2.11 and report the correlation. INTERNETANDLIFEE

2.46 Let's look at Europe. Refer to the previous exercise. Figure 2.12 (page 96) gives a scatterplot for the same data for the 48 European countries in the data set. INTERNETANDLIFEE

(a) Make a plot of the data similar to Figure 2.12.

(b) Report the correlation.

(c) Summarize the differences and similarities between the relationship for all 181 countries and the results that you found in this exercise for the European countries only.

2.47 Second test and final exam. In Exercise 2.25 you looked at the relationship between the score on the second test and the score on the final exam in an elementary statistics course. Here are the data: STATCOURSE8

Second-test score	158	162	144	162	136	158	175	153
Final-exam score	145	140	145	170	145	175	170	160

(a) Find the correlation between these two variables.

(b) Do you think that the correlation between the first test and the final exam should be higher than, approximately equal to, or lower than the correlation between the second test and the final exam? Give a reason for your answer.

2.48 First test and final exam. Refer to the previous exercise. Here are the data for the first test and the final exam. STATCOURSE8

First-exam score	153	144	162	149	127	118	158	153
Final-exam score	145	140	145	170	145	175	170	160

(a) Find the correlation between these two variables.

(b) In Exercise 2.24 we noted that the relationship between these two variables is weak. Does your calculation of the correlation support this statement? Explain your answer.

(c) Examine part (b) of the previous exercise. Does your calculation agree with your prediction?

2.49 The effect of a different point. Examine the data in the Exercise 2.47 and add a ninth student who has low scores on the second test and the final exam and fits the overall pattern of the other scores in the data set. Calculate the correlation and compare it with the correlation that you calculated in Exercise 2.47. Write a short summary of your findings.

2.50 The effect of an outlier. Refer to the Exercise 2.47. Add a ninth student whose scores on the second test and final exam would lead you to classify the additional data point as an outlier. Recalculate the correlation with this additional case and summarize the effect it as on the value of the correlation.

2.51 NBA teams. Table 2.1 (page 99) gives the values of the 30 teams in the National Basketball Association, along with their total revenues, debt, and operating incomes. You made scatterplots of value against the three explanatory variables in Exercise 2.34. Find the correlations of team value with revenue, with debt, and with operating income. Do you think that the values of r provide a good first comparison of what the plots show about predicting value? NBA

2.52 Correlations measure strong and weak linear associations. Your scatterplots for Exercises 2.32 (page 98) and 2.36 (Table 2.2, page 100) illustrate a quite weak linear association and a very strong linear association. Find the correlations that go with these plots. It isn't surprising that a laboratory experiment on physical behavior (the icicles) gives a much stronger correlation than field data on living things (the biological clock). How strong a correlation must be to interest scientists depends on the field of study. BIOCLOCKS and ICICLES

2.53 Heights of people who date each other. A student wonders if tall women tend to date taller men than do short women. She measures herself, her dormitory roommate, and the women in the adjoining rooms; then she measures the next man each woman dates. Here are the data (heights in inches): DATEHEIGHTS

Women (x)	66	64	66	65	70	65
Men (y)	72	68	70	68	71	65

(a) Make a scatterplot of these data. Based on the scatterplot, do you expect the correlation to be positive or negative? Near ±1 or not?

(b) Find the correlation r between the heights of the men and women.

(c) How would r change if all the men were 6 inches shorter than the heights given in the table? Does the correlation tell us whether women tend to date men taller than themselves?

(d) If heights were measured in centimeters rather than inches, how would the correlation change? (There are 2.54 centimeters in an inch.)

(e) If every woman dated a man exactly 3 inches taller than herself, what would be the correlation between male and female heights?

2.54 An interesting set of data. Make a scatterplot of the following data:

x	1	2	3	4	10	10
y	1	3	3	5	1	11

Use your calculator to show that the correlation is about 0.5. What feature of the data is responsible for reducing the correlation to this value despite a strong straight-line association between x and y in most of the observations?
🌐 INTERESTING

2.55 🕐 ⚠️ **Use the applet.** You are going to use the *Correlation and Regression* applet to make different scatterplots with 10 points that have correlation close to 0.8. *Many patterns can have the same correlation. Always plot your data before you trust a correlation.*

(a) Stop after adding the first 2 points. What is the value of the correlation? Why does it have this value no matter where the 2 points are located?

(b) Make a lower-left to upper-right pattern of 10 points with correlation about $r = 0.8$. (You can drag points up or down to adjust r after you have 10 points.) Make a rough sketch of your scatterplot.

(c) Make another scatterplot, this time with 9 points in a vertical stack at the left of the plot. Add one point far to the right and move it until the correlation is close to 0.8. Make a rough sketch of your scatterplot.

(d) Make yet another scatterplot, this time with 10 points in a curved pattern that starts at the lower left, rises to the right, then falls again at the far right. Adjust the points up or down until you have a quite smooth curve with correlation close to 0.8. Make a rough sketch of this scatterplot also.

2.56 🕐 **Use the applet.** Go to the *Correlation and Regression* applet. Click on the scatterplot to create a group of 10 points in the lower-right corner of the scatterplot with a strong straight-line negative pattern (correlation about −0.9).

(a) Add one point at the upper left that is in line with the first 10. How does the correlation change?

(b) ⚠️ Drag this last point down until it is opposite the group of 10 points. How small can you make the correlation? Can you make the correlation positive? *A single outlier can greatly strengthen or weaken a correlation. Always plot your data to check for outlying points.*

2.57 What is the correlation? Suppose that women always married men 2 years older than themselves. Draw a scatterplot of the ages of 5 married couples, with the wife's age as the explanatory variable. What is the correlation r for your data? Why?

2.58 🔺 **High correlation does not mean that the values are the same.** Investment reports often include correlations. Following a table of correlations among mutual funds, a report adds, "Two funds can have perfect correlation, yet different levels of risk. For example, Fund A and Fund B may be perfectly correlated, yet Fund A moves 20% whenever Fund B moves 10%." Write a brief explanation, for someone who knows no statistics, of how this can happen. Include a sketch to illustrate your explanation.

2.59 Student ratings of teachers. A college newspaper interviews a psychologist about student ratings of the teaching of faculty members. The psychologist says, "The evidence indicates that the correlation between the research productivity and teaching rating of faculty members is close to zero." The paper reports this as "Professor McDaniel said that good researchers tend to be poor teachers, and vice versa." Explain why the paper's report is wrong. Write a statement in plain language (don't use the word "correlation") to explain the psychologist's meaning.

2.60 What's wrong? Each of the following statements contains a blunder. Explain in each case what is wrong.

(a) "There is a high correlation between the age of American workers and their occupation."

(b) "We found a high correlation ($r = 1.19$) between students' ratings of faculty teaching and ratings made by other faculty members."

(c) "The correlation between the gender of a group of students and the color of their cell phone was $r = 0.23$."

2.61 **IQ and GPA.** Table 1.4 (page 28) reports data on 78 seventh-grade students. We expect a positive association between IQ and GPA. Moreover, some people think that self-concept is related to school performance. Examine in detail the relationships between GPA and the two explanatory variables IQ and self-concept. Are the relationships roughly linear? How strong are they? Are there unusual points? What is the effect of removing these points? 🔵 SEVENTHGRADE

2.3 Least-Squares Regression

Correlation measures the direction and strength of the linear (straight-line) relationship between two quantitative variables. If a scatterplot shows a linear relationship, we would like to summarize this overall pattern by drawing a line on the scatterplot. A *regression line* summarizes the relationship between two variables, but only in a specific setting: when one of the variables helps explain or predict the other. That is, regression describes a relationship between an explanatory variable and a response variable.

REGRESSION LINE

A **regression line** is a straight line that describes how a response variable *y* changes as an explanatory variable *x* changes. We often use a regression line to **predict** the value of *y* for a given value of *x*. Regression, unlike correlation, requires that we have an explanatory variable and a response variable.

🔵 DATA FILE
FIDGET

EXAMPLE

2.18 Fidgeting and fat gain. Does fidgeting keep you slim? Some people don't gain weight even when they overeat. Perhaps fidgeting and other "nonexercise activity" (NEA) explains why—the body might spontaneously increase nonexercise activity when fed more. Researchers deliberately overfed 16 healthy young adults for 8 weeks. They measured fat gain (in kilograms) and, as an explanatory variable, increase in energy use (in calories) from activity other than deliberate exercise—fidgeting, daily living, and the like. Here are the data:[21]

NEA increase (cal)	−94	−57	−29	135	143	151	245	355
Fat gain (kg)	4.2	3.0	3.7	2.7	3.2	3.6	2.4	1.3

NEA increase (cal)	392	473	486	535	571	580	620	690
Fat gain (kg)	3.8	1.7	1.6	2.2	1.0	0.4	2.3	1.1

Figure 2.17 is a scatterplot of these data. The plot shows a moderately strong negative linear association with no outliers. The correlation is $r = -0.7786$. People with larger increases in nonexercise activity do indeed gain less fat. A line drawn through the points will describe the overall pattern well.

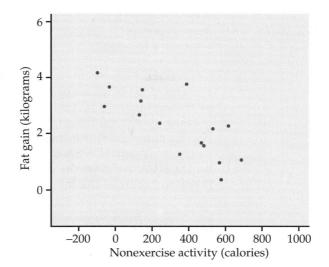

FIGURE 2.17 Fat gain after 8 weeks of overeating plotted against the increase in nonexercise activity over the same period, for Example 2.18.

Fitting a line to data

fitting a line

When a scatterplot displays a linear pattern, we can describe the overall pattern by drawing a straight line through the points. Of course, no straight line passes exactly through all of the points. **Fitting a line** to data means drawing a line that comes as close as possible to the points. The equation of a line fitted to the data gives a compact description of the dependence of the response variable y on the explanatory variable x.

> **STRAIGHT LINES**
>
> Suppose that y is a response variable (plotted on the vertical axis) and x is an explanatory variable (plotted on the horizontal axis). A straight line relating y to x has an equation of the form
>
> $$y = b_0 + b_1 x$$
>
> In this equation, b_1 **is the slope**, the amount by which y changes when x increases by one unit. The number b_0 **is the intercept**, the value of y when $x = 0$.

In practice, we will use software to obtain values of b_0 and b_1 for a given set of data.

EXAMPLE

2.19 Regression line for fat gain. Any straight line describing the nonexercise activity data has the form

$$\text{fat gain} = b_0 + (b_1 \times \text{NEA increase})$$

In Figure 2.17 we have drawn the regression line with the equation

$$\text{fat gain} = 3.505 - (0.00344 \times \text{NEA increase})$$

The figure shows that this line fits the data well. The slope $b_1 = -0.00344$ tells us that fat gained goes down by 0.00344 kilogram for each added calorie of NEA increase.

The slope b_1 of a line $y = b_0 + b_1 x$ is the *rate of change* in the response y as the explanatory variable x changes. The slope of a regression line is an important numerical description of the relationship between the two variables. For Example 2.19, the intercept, $b_0 = 3.505$ kilograms. This value is the estimated fat gain if NEA does not change. When we substitute the value zero for the NEA increase, the regression equation gives 3.505 (the intercept) as the predicted value of the fat gain.

USE YOUR KNOWLEDGE

2.62 Plot the data with the line. Make a sketch of the data in Example 2.18 and plot the line

$$\text{fat gain} = 4.505 - (0.00344 \times \text{NEA increase})$$

on your sketch. Explain why this line does not give a good fit to the data.

Prediction

prediction

We can use a regression line to **predict** the response y for a specific value of the explanatory variable x.

EXAMPLE

2.20 Prediction for fat gain. Based on the linear pattern, we want to predict the fat gain for an individual whose NEA increases by 400 calories when she overeats. To use the fitted line to predict fat gain, go "up and over" on the graph in Figure 2.18. From 400 calories on the x axis, go up to the fitted line and over to the y axis. The graph shows that the predicted gain in fat is a bit more than 2 kilograms.

If we have the equation of the line, it is faster and more accurate to substitute $x = 400$ in the equation. The predicted fat gain is

$$\text{fat gain} = 3.505 - (0.00344 \times 400) = 2.13 \text{ kilograms}$$

The accuracy of predictions from a regression line depends on how much scatter about the line the data show. In Figure 2.18, fat gains for similar increases in NEA show a spread of 1 or 2 kilograms. The regression line summarizes the pattern but gives only roughly accurate predictions.

USE YOUR KNOWLEDGE

2.63 Predict the fat gain. Use the regression equation in Example 2.19 to predict the fat gain for a person whose NEA increases by 600 calories.

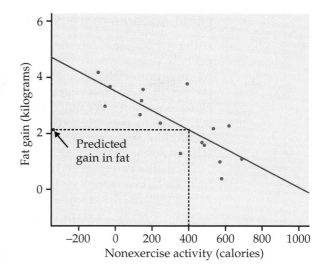

FIGURE 2.18   A regression line fitted to the nonexercise activity data and used to predict fat gain for an NEA increase of 400 calories, for Example 2.20.

EXAMPLE

2.21   Is this prediction reasonable?   Can we predict the fat gain for someone whose nonexercise activity increases by 1500 calories when she overeats? We can certainly substitute 1500 calories into the equation of the line. The prediction is

$$\text{fat gain} = 3.505 - (0.00344 \times 1500) = -1.66 \text{ kilograms}$$

That is, we predict that this individual loses fat when she overeats. This prediction is not trustworthy. Look again at Figure 2.18. An NEA increase of 1500 calories is far outside the range of our data. We can't say whether increases this large ever occur, or whether the relationship remains linear at such extreme values. Predicting fat gain when NEA increases by 1500 calories *extrapolates* the relationship beyond what the data show.

EXTRAPOLATION

Extrapolation is the use of a regression line for prediction far outside the range of values of the explanatory variable x used to obtain the line. Such predictions are often not accurate and should be avoided.

USE YOUR KNOWLEDGE

2.64   Would you use the regression equation to predict?   Consider the following values for NEA increase: −400, 200, 500, 1000. For each, decide whether you would use the regression equation in Example 2.19 to predict fat gain or whether you would be concerned that the prediction would not be trustworthy because of extrapolation. Give reasons for your answers.

Least-squares regression

Different people might draw different lines by eye on a scatterplot. This is especially true when the points are widely scattered. We need a way to draw a regression line that doesn't depend on our guess as to where the line should go. No line will pass exactly through all the points, but we want one that is as close as possible. We will use the line to predict y from x, so we want a line that is as close as possible to the points in the *vertical* direction. That's because the prediction errors we make are errors in y, which is the vertical direction in the scatterplot.

The line in Figure 2.18 predicts 2.13 kilograms of fat gain for an increase in nonexercise activity of 400 calories. If the actual fat gain turns out to be 2.3 kilograms, the error is

$$\text{error} = \text{observed gain} - \text{predicted gain}$$
$$= 2.3 - 2.13 = 0.17 \text{ kilograms}$$

Errors are positive if the observed response lies above the line, and negative if the response lies below the line. We want a regression line that makes these prediction errors as small as possible. Figure 2.19 illustrates the idea. For clarity, the plot shows only three of the points from Figure 2.18, along with the line, on an expanded scale. The line passes below two of the points and above one of them. The vertical distances of the data points from the line appear as vertical line segments. A "good" regression line makes these distances as small as possible. There are many ways to make "as small as possible" precise. The most common is the *least-squares* idea. The line in Figures 2.18 and 2.19 is in fact the least-squares regression line.

LEAST-SQUARES REGRESSION LINE

The **least-squares regression line of y on x** is the line that makes the sum of the squares of the vertical distances of the data points from the line as small as possible.

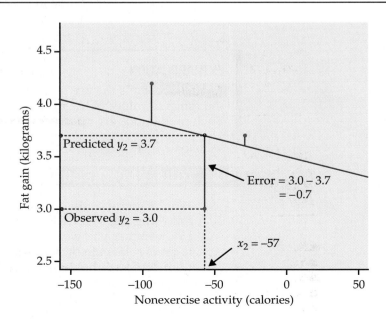

FIGURE 2.19 The least-squares idea: make the errors in predicting y as small as possible by minimizing the sum of their squares.

Here is the least-squares idea expressed as a mathematical problem. We represent n observations on two variables x and y as

$$(x_1, y_1), \ (x_2, y_2), \dots, \ (x_n, \ y_n)$$

If we draw a line $y = b_0 + b_1 x$ through the scatterplot of these observations, the line predicts the value of y corresponding to x_i as $\hat{y}_i = b_0 + b_1 x_i$. We write $\hat{y}$ (read "y-hat") in the equation of a regression line to emphasize that the line gives a *predicted* response $\hat{y}$ for any x. The predicted response will usually not be exactly the same as the actually *observed* response y. The method of least squares chooses the line that makes the sum of the squares of these errors as small as possible. To find this line, we must find the values of the intercept b_0 and the slope b_1 that minimize

$$\sum (\text{error})^2 = \sum (y_i - b_0 - b_1 x_i)^2$$

for the given observations x_i and y_i. For the NEA data, for example, we must find the b_0 and b_1 that minimize

$$(4.2 - b_0 + 94 b_1)^2 + (3.0 - b_0 + 57 b_1)^2 + \cdots + (1.1 - b_0 - 690 b_1)^2$$

These values are the intercept and slope of the least-squares line.

You will use software or a calculator with a regression function to find the equation of the least-squares regression line from data on x and y. We will therefore give the equation of the least-squares line in a form that helps our understanding but is not efficient for calculation.

EQUATION OF THE LEAST-SQUARES REGRESSION LINE

We have data on an explanatory variable x and a response variable y for n individuals. The means and standard deviations of the sample data are $\bar{x}$ and s_x for x and $\bar{y}$ and s_y for y, and the correlation between x and y is r. The equation of the least-squares regression line of y on x is

$$\hat{y} = b_0 + b_1 x$$

with **slope**

$$b_1 = r \frac{s_y}{s_x}$$

and **intercept**

$$b_0 = \bar{y} - b_1 \bar{x}$$

EXAMPLE

2.22 Check the calculations. Verify from the data in Example 2.18 that the mean and standard deviation of the 16 increases in NEA are

$$\bar{x} = 324.8 \text{ calories} \quad \text{and} \quad s_x = 257.66 \text{ calories}$$

The mean and standard deviation of the 16 fat gains are

$$\bar{y} = 2.388 \text{ kg} \quad \text{and} \quad s_y = 1.1389 \text{ kg}$$

The correlation between fat gain and NEA increase is $r = -0.7786$. The least-squares regression line of fat gain y on NEA increase x therefore has slope

$$b_1 = r\frac{s_y}{s_x} = -0.7786\frac{1.1389}{257.66}$$
$$= -0.00344 \text{ kg per calorie}$$

and intercept

$$b_0 = \bar{y} - b_1\bar{x} = 2.388 - (-0.00344)(324.8)$$
$$= 3.505 \text{ kg}$$

The equation of the least-squares line is

$$\hat{y} = 3.505 - 0.00344x$$

When doing calculations like this by hand, you may need to carry extra decimal places in the preliminary calculations to get accurate values of the slope and intercept. Using software or a calculator with a regression function eliminates this worry.

Interpreting the regression line

The slope $b_1 = -0.00344$ kilograms per calorie in Example 2.22 is the change in fat gain as NEA increases. The units "kilograms of fat gained per calorie of NEA" come from the units of y (kilograms) and x (calories). Although the correlation does not change when we change the units of measurement, the equation of the least-squares line does change. The slope in grams per calorie would be 1000 times as large as the slope in kilograms per calorie, because there are 1000 grams in a kilogram. The small value of the slope, $b_1 = -0.00344$, does not mean that the effect of increased NEA on fat gain is small—it just reflects the choice of kilograms as the unit for fat gain. *The slope and intercept of the least-squares line depend on the units of measurement—you can't conclude anything from their size.*

EXAMPLE

2.23 Regression using software. Figure 2.20 displays the basic regression output for the nonexercise activity data from two statistical software packages. Other software produces very similar output. You can find the slope and intercept of the least-squares line, calculated to more decimal places than we need, in both outputs. The software also provides information that we do not yet need, including some that we trimmed from Figure 2.20.

Part of the art of using software is to ignore the extra information that is almost always present. Look for the results that you need. Once you understand a statistical method, you can read output from almost any software.

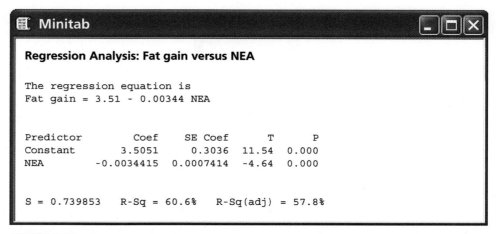

Linear Regression – Select Fields: NEA, Fat gain

Confidence Level 0.9500

Estimates	Estimate	Std. Error	t Statistic	Pr(>|t|)	CI Lower Bound	CI Upper Bound
(Intercept)	3.5051	0.3036	11.5446	1.5344e–8	2.8539	4.1563
NEA	−0.0034	0.0007	−4.6418	0.0004	−0.0050	−0.0019

N	16
r-Squared	0.6061
Adjusted r-Squared	0.5780
s	0.7399
Correlation Coefficient	−0.7786

```
Minitab

Regression Analysis: Fat gain versus NEA

The regression equation is
Fat gain = 3.51 - 0.00344 NEA


Predictor          Coef      SE Coef        T        P
Constant         3.5051       0.3036    11.54    0.000
NEA           -0.0034415    0.0007414    -4.64    0.000


S = 0.739853    R-Sq = 60.6%    R-Sq(adj) = 57.8%
```

FIGURE 2.20 Regression results for the nonexercise activity data from two statistical software packages. Other software produces similar output.

Facts about least-squares regression

Regression is one of the most common statistical settings, and least squares is the most common method for fitting a regression line to data. Here are some facts about least-squares regression lines.

Fact 1. There is a close connection between correlation and the slope of the least-squares line. The slope is

$$b_1 = r\frac{s_y}{s_x}$$

This equation says that along the regression line, **a change of one standard deviation in x corresponds to a change of r standard deviations in y.** When the variables are perfectly correlated ($r = 1$ or $r = -1$), the change in the predicted response $\hat{y}$ is the same (in standard deviation units) as the change in x. Otherwise, because $-1 \leq r \leq 1$, the change in $\hat{y}$ is less than the change in x. As the correlation grows less strong, the prediction $\hat{y}$ moves less in response to changes in x.

Fact 2. The least-squares regression line always passes through the point $(\overline{x}, \overline{y})$ on the graph of y against x. So the least-squares regression line of y on x is the line with slope rs_y/s_x that passes through the point $(\overline{x}, \overline{y})$. We can describe regression entirely in terms of the basic descriptive measures $\overline{x}$, s_x, $\overline{y}$, s_y, and r.

Fact 3. The distinction between explanatory and response variables is essential in regression. Least-squares regression looks at the distances of the data points from the line only in the y direction. If we reverse the roles of the two variables, we get a different least-squares regression line.

Correlation and regression

Least-squares regression looks at the distances of the data points from the line only in the y direction. So the two variables x and y play different roles in regression.

SPAM

> ### EXAMPLE
>
> **2.24 Bots and spam messages per day.** Figure 2.21 is a scatterplot of the data spam botnet data described in Example 2.8 (page 83). There is a positive linear relationship. The two lines on the plot are the two least-squares regression lines. The regression line using bots to predict spam messages per day is red. The regression line using spam messages per day to predict bots is blue. *Regression of spams per day on bots and regression of bots on spam messages per day give different lines.* In the regression setting, you must decide which variable is explanatory.

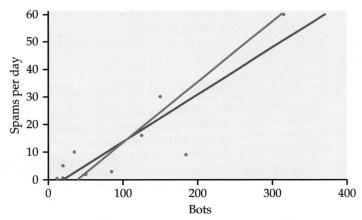

FIGURE 2.21 Scatterplot of spams per day versus the number of bots for 10 botnets, from Example 2.8. The two lines are the two least-squares regression lines: using bots to predict spams per day (red) and using spams per day to predict bots (blue), for Example 2.24.

Even though the correlation r ignores the distinction between explanatory and response variables, there is a close connection between correlation and regression. We saw that the slope of the least-squares line involves r. Another connection between correlation and regression is even more important. In fact, the numerical value of r as a measure of the strength of a linear relationship is best interpreted by thinking about regression. Here is the fact we need.

---

r^2 IN REGRESSION

The **square of the correlation, r^2,** is the fraction of the variation in the values of y that is explained by the least-squares regression of y on x.

The correlation between NEA increase and fat gain for the 16 subjects in Example 2.18 (page 108) is $r = -0.7786$. Because $r^2 = 0.606$, the straight-line relationship between NEA and fat gain explains about 61% of the vertical scatter in fat gains in Figure 2.18 (page 111). When you report a regression, give r^2 as a measure of how successfully the regression explains the response. Both software outputs in Figure 2.20 include r^2, either in decimal form or as a percent. When you see a correlation, square it to get a better feel for the strength of the association. Perfect correlation ($r = -1$ or $r = 1$) means the points lie exactly on a line. Then $r^2 = 1$ and all of the variation in one variable is accounted for by the linear relationship with the other variable. If $r = -0.7$ or $r = 0.7$, $r^2 = 0.49$ and about half the variation is accounted for by the linear relationship. In the r^2 scale, correlation ± 0.7 is about halfway between 0 and ± 1.

USE YOUR KNOWLEDGE

2.65 What fraction of the variation is explained? Consider the following correlations: -0.9, -0.5, -0.3, 0, 0.3, 0.5, and 0.9. For each give the fraction of the variation in y that is explained by the least-squares regression of y on x. Summarize what you have found from performing these calculations.

The use of r^2 to describe the success of regression in explaining the response y is very common. It rests on the fact that there are two sources of variation in the responses y in a regression setting. Figure 2.16 gives a rough visual picture of the two sources. The first reason for the variation in fat gains is that there is a relationship between fat gain y and increase in NEA x. As x increases from -94 calories to 690 calories among the 16 subjects, it pulls fat gain y with it along the regression line in the figure. The linear relationship explains this part of the variation in fat gains.

The fat gains do not lie exactly on the line, however, but are scattered above and below it. This is the second source of variation in y, and the regression line tells us nothing about how large it is. The vertical dashed lines in Figure 2.16 show a rough average for the spread in y when we fix a value of x. We use r^2 to measure variation along the line as a fraction of the total variation in the fat gains. In Figure 2.18, about 61% of the variation in fat gains among the 16 subjects is due to the straight-line tie between y and x. The remaining 39% is vertical scatter in the observed responses remaining after the line has fixed the predicted responses.

*Understanding r^2

Here is a more specific interpretation of r^2. The fat gains y in Figure 2.18 range from 0.4 kilograms to 4.2 kilograms. The variance of these responses, a measure of how variable they are, is

$$\text{variance of observed values } y = 1.297$$

Much of this variability is due to the fact that as x increases from -94 calories to 690 calories it pulls height y along with it. If the only variability in the observed responses were due to the straight-line dependence of fat gain on NEA, the observed gains would lie exactly on the regression line. That is, they would be the same as the predicted gains $\hat{y}$. We can compute the predicted gains by substituting the NEA values for each subject into the equation of the least-squares line. Their variance describes the variability in the predicted responses. The result is

$$\text{variance of predicted values } \hat{y} = 0.786$$

This is what the variance would be if the responses fell exactly on the line, that is, if the linear relationship explained 100% of the observed variation in y. Because the responses don't fall exactly on the line, the variance of the predicted values is smaller than the variance of the observed values. Here is the fact we need:

$$r^2 = \frac{\text{variance of predicted values } \hat{y}}{\text{variance of observed values } y}$$

$$= \frac{0.786}{1.297} = 0.606$$

This fact is always true. The squared correlation gives the variance the responses would have if there were no scatter about the least-squares line as a fraction of the variance of the actual responses. This is the exact meaning of "fraction of variation explained" as an interpretation of r^2.

These connections with correlation are special properties of least-squares regression. They are not true for other methods of fitting a line to data. One reason that least squares is the most common method for fitting a regression line to data is that it has many convenient special properties.

SECTION 2.3 Summary

A **regression line** is a straight line that describes how a response variable y changes as an explanatory variable x changes.

The most common method of fitting a line to a scatterplot is least squares. The **least-squares regression line** is the straight line $\hat{y} = b_0 + b_1 x$ that minimizes the sum of the squares of the vertical distances of the observed y-values from the line.

You can use a regression line to **predict** the value of y for any value of x by substituting this x into the equation of the line. **Extrapolation** beyond the range of x-values spanned by the data is risky.

---

*This explanation is optional reading.

The **slope** b_1 of a regression line $\hat{y} = b_0 + b_1 x$ is the rate at which the predicted response $\hat{y}$ changes along the line as the explanatory variable x changes. Specifically, b_1 is the change in $\hat{y}$ when x increases by 1. The numerical value of the slope depends on the units used to measure x and y.

The **intercept** b_0 of a regression line $\hat{y} = b_0 + b_1 x$ is the predicted response $\hat{y}$ when the explanatory variable $x = 0$. This prediction is not particularly useful unless x can actually take values near 0.

The least-squares regression line of y on x is the line with slope $b_1 = rs_y/s_x$ and intercept $b_0 = \bar{y} - b_1\bar{x}$. This line always passes through the point $(\bar{x}, \bar{y})$.

Correlation and regression are closely connected. The correlation r is the slope of the least-squares regression line when we measure both x and y in standardized units. The square of the correlation r^2 is the fraction of the variance of one variable that is explained by least-squares regression on the other variable.

SECTION 2.3 Exercises

For Exercise 2.64, see page 111; and for Exercise 2.65, see page 117. Read about residuals on pages 122–123 before attempting Exercises 2.67, 2.74, 2.75, 2.88, and 2.89.

2.66 Open space and population. The New York City Open Accessible Space Information System Cooperative (OASIS) is an organization of public and private sector representatives that has developed an information system designed to enhance the stewardship of open space.[22] Data from the OASIS Web site for 12 large U.S. cities follow. The variables are population in thousands and open total park or open space within city limits in acres. 🌐 OASIS

City	Population	Open space
Baltimore	651	5,091
Boston	589	4,865
Chicago	2,896	11,645
Long Beach	462	2,887
Los Angeles	3,695	29,801
Miami	362	1,329
Minneapolis	383	5,694
New York	8,008	49,854
Oakland	399	3,712
Philadelphia	1,518	10,685
San Francisco	777	5,916
Washington, D.C.	572	7,504

(a) Make a scatterplot of the data using population as the explanatory variable and open space as the response variable.

(b) Is is reasonable to fit a straight line to these data? Explain your answer.

(c) Find the least squares regression line. Report the equation of the line and draw the line on your scatterplot.

(d) What proportion of the variation in open space is explained by population?

2.67 Prepare the report card. Refer to the previous exercise. One way to compare cities with respect to the amount of open space that they have is to use the residuals from the regression analysis that you performed in the previous exercise. Cities with positive residuals are doing better than predicted while those with negative residuals are doing worse. Find the residual for each city and make a table with the city name and the residual, ordered from best to worst by the size of the residual. 🌐 OASIS

2.68 Is New York an outlier? Refer to Exercises 2.66 and 2.67. Write a short paragraph about the data point corresponding to New York City. Is this point an outlier? If it were deleted from the data set, would the least squares regression line change very much? Compare the analysis results with and without this observation. 🌐 OASIS

2.69 Open space per person. Refer to Exercises 2.66, 2.67 and 2.68. Open space in acres per person is an alternative way to report open space. Divide open space by population to compute the value of this variable for each city. Using this new variable as the response variable and population as the explanatory variable, answer the questions given in Exercise 2.66. How do your new results compare with those that you found in that exercise? 🌐 OASIS

2.70 A different report card. Refer to Exercise 2.67. Prepare a report card based on the analysis of open space per person that you performed in the previous exercise. Write a short paragraph comparing this report card with the one that you prepared in Exercise 2.67. Which do you prefer? Give reasons for your answer. 🌐 OASIS

2.71 Alcohol and carbohydrates in beer. Figure 2.10 (page 95) gives a scatterplot of carbohydrates versus percent alcohol in 86 brands of beer. In Exercise 2.43 you calculated the correlation between these two variables. Find the equation of the least-squares regression line for these data. 🔵 BEER

2.72 Alcohol and carbohydrates in beer revisited. Refer to the previous exercise. The data that you used to compute the least-squares regression line includes an outlier. 🔵 BEER

(a) Remove the outlier and recompute the least-squares regression line.

(b) Write a short paragraph about the possible effects of outliers on a least-squares regression line using this example to illustrate your ideas.

2.73 Always plot your data! Table 2.4 presents four sets of data prepared by the statistician Frank Anscombe to illustrate the dangers of calculating without first plotting the data.[23] 🔵 ANSCOMBE

(a) Without making scatterplots, find the correlation and the least-squares regression line for all four data sets. What do you notice? Use the regression line to predict y for $x = 10$.

(b) Make a scatterplot for each of the data sets and add the regression line to each plot.

(c) In which of the four cases would you be willing to use the regression line to describe the dependence of y on x? Explain your answer in each case.

2.74 Data generated by software. The following 20 observations on Y and X were generated by a computer program. 🔵 GENERATEDDATA

Y	X	Y	X
34.38	22.06	27.07	17.75
30.38	19.88	31.17	19.96
26.13	18.83	27.74	17.87
31.85	22.09	30.01	20.20
26.77	17.19	29.61	20.65
29.00	20.72	31.78	20.32
28.92	18.10	32.93	21.37
26.30	18.01	30.29	17.31
29.49	18.69	28.57	23.50
31.36	18.05	29.80	22.02

(a) Make a scatterplot and describe the relationship between Y and X.

(b) Find the equation of the least-squares regression line and add the line to your plot.

(c) Plot the residuals versus X.

(d) What percent of the variability in Y is explained by X?

(e) Summarize your analysis of these data in a short paragraph.

2.75 Add an outlier. Refer to the previous exercise. Add an additional observation with $Y = 50$ and $X = 30$ to the data set. Repeat the analysis that you performed in the previous exercise and summarize your results paying particular attention to the effect of this outlier. 🔵 GENDATA21A

2.76 Add a different outlier. Refer to the previous two exercises. Add an additional observation with $Y = 29$ and $X = 50$ to the original data set. 🔵 GENDATA21B

TABLE 2.4

Four data sets for exploring correlation and regression

Data Set A

x	10	8	13	9	11	14	6	4	12	7	5
y	8.04	6.95	7.58	8.81	8.33	9.96	7.24	4.26	10.84	4.82	5.68

Data Set B

x	10	8	13	9	11	14	6	4	12	7	5
y	9.14	8.14	8.74	8.77	9.26	8.10	6.13	3.10	9.13	7.26	4.74

Data Set C

x	10	8	13	9	11	14	6	4	12	7	5
y	7.46	6.77	12.74	7.11	7.81	8.84	6.08	5.39	8.15	6.42	5.73

Data Set D

x	8	8	8	8	8	8	8	8	8	8	19
y	6.58	5.76	7.71	8.84	8.47	7.04	5.25	5.56	7.91	6.89	12.50

(a) Repeat the analysis that you performed in the first exercise and summarize your results paying particular attention to the effect of this outlier.

(b) In this exercise and in the previous one, you added an outlier to the original data set and reanalyzed the data. Write a short summary of the changes in correlations that can result from different kinds of outliers.

2.77 The regression equation. The equation of a least-squares regression line is $y = 12 + 6x$.

(a) What is the value of y for $x = 5$?

(b) If x increases by one unit, what is the corresponding increase in y?

(c) What is the intercept for this equation?

2.78 Progress in math scores. Every few years, the National Assessment of Educational Progress asks a national sample of eighth-graders to perform the same math tasks. The goal is to get an honest picture of progress in math. Here are the last few national mean scores, on a scale of 0 to 500:[24] NAEPMATH

Year	1990	1992	1996	2000	2003	2005	2008
Score	263	268	272	273	278	279	281

(a) Make a time plot of the mean scores, by hand. This is just a scatterplot of score against year. There is a slow linear increasing trend.

(b) Find the regression line of mean score on time step-by-step. First calculate the mean and standard deviation of each variable and their correlation (use a calculator with these functions). Then find the equation of the least-squares line from these. Draw the line on your scatterplot. What percent of the year-to-year variation in scores is explained by the linear trend?

(c) Now use software or the regression function on your calculator to verify your regression line.

2.79 Social exclusion and pain. Exercise 2.33 (page 98) gives data from a study that shows that social exclusion causes "real pain." That is, activity in the area of the brain that responds to physical pain goes up as distress from social exclusion goes up. Your scatterplot in Exercise 2.33 shows a moderately strong linear relationship. SOCIALREJECTION

(a) What is the equation of the least-squares regression line for predicting brain activity from social distress score? Make a scatterplot with this line drawn on it.

(b) On your plot, show the "up and over" lines that predict brain activity for social distress score 2.0. Use the equation of the regression line to get the predicted brain activity level. Verify that it agrees with your plot.

(c) What percent of the variation in brain activity among these subjects is explained by the straight-line relationship with social distress score?

2.80 Growth of icicles. Table 2.2 (page 100) gives data on the growth of icicles at two rates of water flow. You examined these data in Exercise 2.36. Use least-squares regression to estimate the rate (centimeters per minute) at which icicles grow at these two flow rates. How does flow rate affect growth? ICICLES

2.81 Icicle growth. Find the mean and standard deviation of the times and icicle lengths for the data on Run 8903 in Table 2.2 (page 100). Find the correlation between the two variables. Use these five numbers to find the equation of the regression line for predicting length from time. Verify that your result agrees with that in the previous exercise. Use the same five numbers to find the equation of the regression line for predicting the time an icicle has been growing from its length. What units does the slope of each of these lines have?

2.82 Metabolic rate and lean body mass. Compute the mean and the standard deviation of the metabolic rates and lean body masses in Exercise 2.35 (page 99) and the correlation between these two variables. Use these values to find the slope of the regression line of metabolic rate on lean body mass. Also find the slope of the regression line of lean body mass on metabolic rate. What are the units for each of the two slopes? BODYMASS

2.83 IQ and self-concept. Table 1.4 (page 28) reports data on 78 seventh-grade students. We want to know how well each of IQ score and self-concept score predicts GPA using least-squares regression. We also want to know which of these explanatory variables predicts GPA better. Give numerical measures that answer these questions, and explain your answers. SEVENTHGRADE

2.84 Heights of husbands and wives. The mean height of American women in their early twenties is about 64.5 inches and the standard deviation is about 2.5 inches. The mean height of men the same age is about 68.5 inches, with standard deviation about 2.7 inches. If the correlation between the heights of husbands and wives is about $r = 0.5$, what is the equation of the regression line of the husband's height on the wife's height in young couples? Draw a graph of this regression line. Predict the height of the husband of a woman who is 67 inches tall.

2.85 A property of the least-squares regression line. Use the equation for the least-squares regression line to show that this line always passes through the point $(\bar{x}, \bar{y})$.

2.86 Icicle growth. The data for Run 8903 in Table 2.2 (page 100) describe how the length y in centimeters of an icicle increases over time x. Time is measured in minutes. **ICICLES**

(a) What are the numerical values and units of measurement for each of $\bar{x}$, s_x, $\bar{y}$, s_y, and the correlation r between x and y?

(b) There are 2.54 centimeters in an inch. If we measure length y in inches rather than in centimeters, what are the new values of $\bar{y}$, s_y, and the correlation r?

(c) If we measure length y in inches rather than in centimeters, what is the new value of the slope b_1 of the least-squares line for predicting length from time?

2.87 Class attendance and grades. A study of class attendance and grades among first-year students at a state university showed that in general students who attended a higher percent of their classes earned higher grades. Class attendance explained 16% of the variation in grade index among the students. What is the numerical value of the correlation between percent of classes attended and grade index?

2.88 Best countries for business. Figure 2.5 (page 89) gives a scatterplot of the gross domestic product per capita versus the unemployment rate for 99 countries. **BESTCOUNTRIES**

(a) Plot the data and add the least-squares regression line to the plot.

(b) Is it appropriate to use this least-squares regression line to describe the relationship shown in your plot? Explain your answer.

(c) Plot the residuals versus unemployment rate. Interpret the plot and explain how it helps you to understand this data set.

2.89 Best countries for business with logs. Refer to the previous exercise. Figure 2.6 (page 90) gives a scatterplot of the log gross domestic product per capita versus the log unemployment rate for 99 countries. **BESTCOUNTRIES**

(a) Plot the data and add the least-squares regression line to the plot.

(b) Is it appropriate to use this least-squares regression line to describe the relationship shown in your plot? Explain your answer.

(c) Plot the residuals versus log unemployment rate. Interpret the plot and explain how it helps you to understand this data set.

2.90 Delete data for countries with low unemployment rates. Refer to the previous exercise. Delete the countries with log unemployment rates lower than 1.6. This corresponds to an unemployment rate of about 5%. Answer the questions given for the previous exercise and explain the effect of deleting the countries with low unemployment rates. **BESTCOUNTRIES**

2.91 Revenue and value of NBA teams. Table 2.1 (page 99) gives the values of the 30 teams in the National Basketball Association, along with their operating incomes, debt, and revenues. Treat value as the response variable. Perform separate analyses using income, debt, and revenue as explanatory variables. Write a short report summarizing your findings. Include plots, correlations, and the least-squares regression line. **NBA**

2.4 Cautions about Correlation and Regression

Correlation and regression are among the most common statistical tools. They are used in more elaborate form to study relationships among many variables, a situation in which we cannot see the essentials by studying a single scatterplot. We need a firm grasp of the use and limitations of these tools, both now and as a foundation for more advanced statistics.

Residuals

A regression line describes the overall pattern of a linear relationship between an explanatory variable and a response variable. Deviations from the overall pattern are also important. In the regression setting, we see deviations by looking at the scatter of the data points about the regression line. The vertical distances from the points to the least-squares regression line are as small as possible in the sense that they have the smallest possible sum of squares. Because they represent "left-over" variation in the response after fitting the regression line, these distances are called *residuals*.

RESIDUALS

A **residual** is the difference between an observed value of the response variable and the value predicted by the regression line. That is,

$$\text{residual} = \text{observed } y - \text{predicted } y$$
$$= y - \hat{y}$$

EXAMPLE

2.25 Residuals for fat gain. Example 2.18 (page 108) describes measurements on 16 young people who volunteered to overeat for 8 weeks. Those whose nonexercise activity (NEA) spontaneously rose substantially gained less fat than others. Figure 2.22(a) is a scatterplot of these data. The pattern is linear. The least-squares line is

$$\text{fat gain} = 3.505 - (0.00344 \times \text{NEA increase})$$

One subject's NEA rose by 135 calories. That subject gained 2.7 kilograms of fat. The predicted gain for 135 calories is

$$\hat{y} = 3.505 - (0.00344 \times 135) = 3.04 \text{ kg}$$

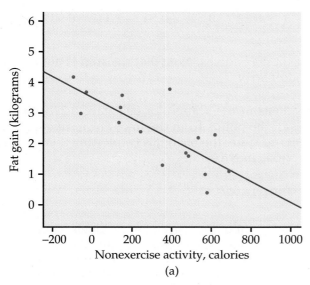

(a)

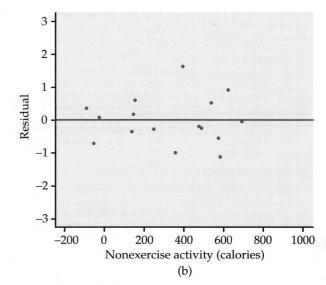

(b)

FIGURE 2.22 (a) Scatterplot of fat gain versus increase in nonexercise activity, with the least squares regression line, for Example 2.25. (b) Residual plot for the regression displayed in Figure 2.20 (a). The line at $y = 0$ marks the mean of the residuals.

The residual for this subject is therefore

$$\text{residual} = \text{observed } y - \text{predicted } y$$
$$= y - \hat{y}$$
$$= 2.7 - 3.04 = -0.34 \text{ kg}$$

Most regression software will calculate and store residuals for you.

USE YOUR KNOWLEDGE

2.92 Find the predicted value and the residual. Another individual in the NEA data set has NEA increase equal to 143 calories and fat gain equal to 3.2 kg. Find the predicted value of fat gain for this individual and then calculate the residual. Explain why this residual is positive.

Because the residuals show how far the data fall from our regression line, examining the residuals helps assess how well the line describes the data. Although residuals can be calculated from any model fitted to the data, the residuals from the least-squares line have a special property: **the mean of the least-squares residuals is always zero.**

USE YOUR KNOWLEDGE

FIDGET

2.93 Find the sum of the residuals. Here are the 16 residuals for the NEA data rounded to two decimal places:

0.37	−0.70	0.10	−0.34	0.19	0.61	−0.26	−0.98
1.64	−0.18	−0.23	0.54	−0.54	−1.11	0.93	−0.03

Find the sum of these residuals. Note that the sum is not exactly zero because of roundoff error.

You can see the residuals in the scatterplot of Figure 2.22(a) by looking at the vertical deviations of the points from the line. The *residual plot* in Figure 2.22(b) makes it easier to study the residuals by plotting them against the explanatory variable, increase in NEA.

RESIDUAL PLOTS

A **residual plot** is a scatterplot of the regression residuals against the explanatory variable. Residual plots help us assess the fit of a regression line.

Because the mean of the residuals is always zero, the horizontal line at zero in Figure 2.22(b) helps orient us. This line (residual = 0) corresponds to the fitted line in Figure 2.22(a). The residual plot magnifies the deviations from the line to make patterns easier to see. If the regression line catches the overall pattern of the data, there should be *no pattern* in the residuals. That is, the residual plot should show an unstructured horizontal band centered at zero. The residuals in Figure 2.22(b) do have this irregular scatter.

You can see the same thing in the scatterplot of Figure 2.22(a) and the residual plot of Figure 2.22(b). It's just a bit easier in the residual plot. Deviations from an irregular horizontal pattern point out ways in which the regression line fails to catch the overall pattern. Here is an example:

EXAMPLE

2.26 Patterns in the life expectancy and Internet residuals. Figure 2.11 (page 96) gives a scatterplot of life expectancy versus Internet use for 181 countries. The first thing that we see in the plot is the large number of countries with low Internet use. In addition, the relationship between life expectancy and Internet use appears to be curved. For low values of Internet use, there is a clear relationship, while for higher values, the curve appears to become fairly flat. Figure 2.23 gives the residuals for the regression when we predict life expectancy using Internet use. Look at the left part of the plot where the values of Internet use are low. Here we see that there are many more negative residuals than positive ones. Now look at the right part of the plot where the values of Internet use are high. Here we see that the residuals tend to be negative.

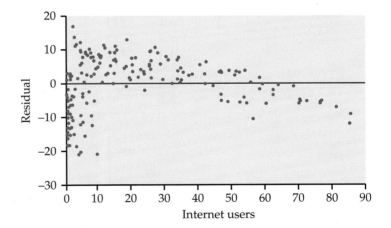

FIGURE 2.23 Residual plot for the regression of life expectancy on Internet users, for Example 2.26.

The residual pattern in Figure 2.23 is characteristic of a simple curved relationship. There are many ways in which a relationship can deviate from a linear pattern. We now have an important tool for examining these deviations. Use it frequently and carefully when you study relationships.

Outliers and influential observations

When you look at scatterplots and residual plots, look for striking individual points as well as for an overall pattern. Here is an example of data that contain some unusual cases.

EXAMPLE

2.27 Diabetes and blood sugar. People with diabetes must manage their blood sugar levels carefully. They measure their fasting plasma glucose (FPG) several times a day with a glucose meter. Another measurement, made at regular medical checkups, is called HbA1c. This is roughly the percent of red blood cells that have a glucose molecule attached. It measures average exposure to glucose over a period of several months. Table 2.5 gives data on both HbA1c and FPG for 18 diabetics five months after they had completed a diabetes education class.[25]

Because both FPG and HbA1c measure blood glucose, we expect a positive association. The scatterplot in Figure 2.24 shows a surprisingly weak relationship, with correlation $r = 0.4819$. The line on the plot is the least-squares regression line for predicting FPG from HbA1c. Its equation is

$$\hat{y} = 66.4 + 10.41x$$

It appears that one-time measurements of FPG can vary quite a bit among people with similar long-term levels, as measured by HbA1c.

TABLE 2.5

Two measures of glucose level in diabetics

Subject	HbA1c (%)	FPG (mg/ml)	Subject	HbA1c (%)	FPG (mg/ml)	Subject	HbA1c (%)	FPG (mg/ml)
1	6.1	141	7	7.5	96	13	10.6	103
2	6.3	158	8	7.7	78	14	10.7	172
3	6.4	112	9	7.9	148	15	10.7	359
4	6.8	153	10	8.7	172	16	11.2	145
5	7.0	134	11	9.4	200	17	13.7	147
6	7.1	95	12	10.4	271	18	19.3	255

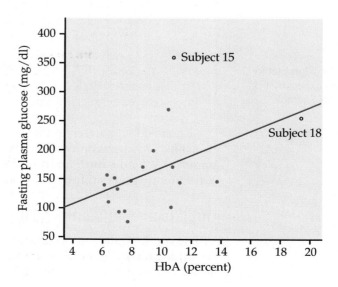

FIGURE 2.24 Scatterplot of fasting plasma glucose against HbA (which measures long-term blood glucose), with the least-squares line, for Example 2.27.

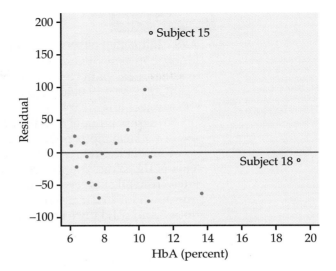

FIGURE 2.25 Residual plot for the regression of fasting plasma glucose on HbA. Subject 15 is an outlier in fasting plasma glucose. Subject 18 is an outlier in HbA that may be influential but does not have a large residual.

Two unusual cases are marked in Figure 2.24. Subjects 15 and 18 are unusual in different ways. Subject 15 has dangerously high FPG and lies far from the regression line in the y direction. Subject 18 is close to the line but far out in the x direction. The residual plot in Figure 2.25 confirms that Subject 15 has a large residual and that Subject 18 does not.

Points that are outliers in the x direction, like Subject 18, can have a strong influence on the position of the regression line. Least-squares lines make the sum of squares of the vertical distances to the points as small as possible. A point that is extreme in the x direction with no other points near it pulls the line toward itself.

OUTLIERS AND INFLUENTIAL OBSERVATIONS IN REGRESSION

An **outlier** is an observation that lies outside the overall pattern of the other observations. Points that are outliers in the y direction of a scatterplot have large regression residuals, but other outliers need not have large residuals.

An observation is **influential** for a statistical calculation if removing it would markedly change the result of the calculation. Points that are outliers in the x direction of a scatterplot are often influential for the least-squares regression line.

Influence is a matter of degree—how much does a calculation change when we remove an observation? It is difficult to assess influence on a regression line without actually doing the regression both with and without the suspicious observation. A point that is an outlier in x is often influential. But if the point happens to lie close to the regression line calculated from the other observations, then its presence will move the line only a little and the point will not be influential. The influence of a point that is an outlier in y depends on whether there are many other points with similar values of x that hold the line in place. Figures 2.24 and 2.25 identify two unusual observations. How influential are they?

EXAMPLE

2.28 Influential observations. Subjects 15 and 18 both influence the correlation between FPG and HbA1c, in opposite directions. Subject 15 weakens the linear pattern; if we drop this point, the correlation increases from $r = 0.4819$ to $r = 0.5684$. Subject 18 extends the linear pattern; if we omit this subject, the correlation drops from $r = 0.4819$ to $r = 0.3837$.

To assess influence on the least-squares line, we recalculate the line leaving out a suspicious point. Figure 2.26 shows three least-squares lines. The solid line is the regression line of FPG on HbA1c based on all 18 subjects. This is the same line that appears in Figure 2.24. The dotted line is calculated from all subjects except Subject 18. You see that point 18 does pull the line down toward itself. But the influence of Subject 18 is not very large—the dotted and solid lines are close together for HbA1c values between 6 and 14, the range of all except Subject 18.

The dashed line omits Subject 15, the outlier in y. Comparing the solid and dashed lines, we see that Subject 15 pulls the regression line up. The influence is again not large, but it exceeds the influence of Subject 18.

FIGURE 2.26 Three regression lines for predicting fasting plasma glucose from HbA, for Example 2.28. The solid line uses all 18 subjects. The dotted line leaves out Subject 18. The dashed line leaves out Subject 15. "Leaving one out" calculations are the surest way to assess influence.

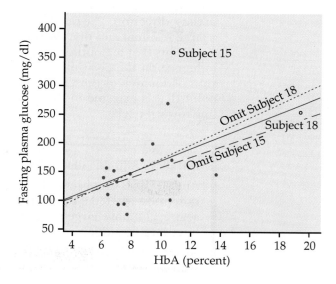

The best way to see how points that are extreme in x can influence the regression line is to use the *Correlation and Regression* applet on the text CD and Web site. As Exercise 2.101 demonstrates, moving one point can pull the line to almost any position on the graph.

We did not need the distinction between outliers and influential observations in Chapter 1. A single large salary that pulls up the mean salary $\bar{x}$ for a group of workers is an outlier because it lies far above the other salaries. It is also influential because the mean changes when it is removed. In the regression setting, however, not all outliers are influential. Because influential observations draw the regression line toward themselves, we may not be able to spot them by looking for large residuals.

Beware the lurking variable

Correlation and regression are powerful tools for measuring the association between two variables and for expressing the dependence of one variable on the other. These tools must be used with an awareness of their limitations. We have seen that

- Correlation measures *only linear association,* and fitting a straight line makes sense only when the overall pattern of the relationship is linear. Always plot your data before calculating.

- *Extrapolation* (using a fitted model far outside the range of the data that we used to fit it) often produces unreliable predictions.

- Correlation and least-squares regression are *not resistant.* Always plot your data and look for potentially influential points.

Another caution is even more important: the relationship between two variables can often be understood only by taking other variables into account. *Lurking variables* can make a correlation or regression misleading.

LURKING VARIABLE

A **lurking variable** is a variable that is not among the explanatory or response variables in a study and yet may influence the interpretation of relationships among those variables.

EXAMPLE

2.29   Discrimination in medical treatment? Studies show that men who complain of chest pain are more likely to get detailed tests and aggressive treatment such as bypass surgery than are women with similar complaints. Is this association between gender and treatment due to discrimination?

Perhaps not. Men and women develop heart problems at different ages—women are on the average between 10 and 15 years older than men. Aggressive treatments are more risky for older patients, so doctors may hesitate to recommend them. Lurking variables—the patient's age and condition—may explain the relationship between gender and doctors' decisions.

Here is an example of a different type of lurking variable.

EXAMPLE

2.30   Gas and electricity bills. A single-family household receives bills for gas and electricity each month. The 12 observations for a recent year are plotted with the least-squares regression line in Figure 2.27. We have arbitrarily chosen to put the electricity bill on the x axis and the gas bill on the y axis. There is a clear negative association. Does this mean that a high electricity bill causes the gas bill to be low and vice versa?

To understand the association in this example, we need to know a little more about the two variables. In this household, heating is done by gas and

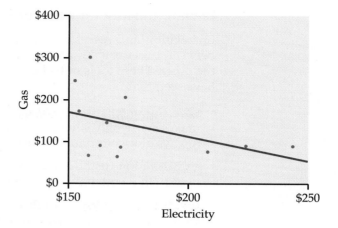

FIGURE 2.27 Scatterplot with least-squares regression line for predicting monthly charges for gas using monthly charges for electricity for a household, for Example 2.30.

cooling is done by electricity. Therefore, in the winter months the gas bill will be relatively high and the electricity bill will be relatively low. The pattern is reversed in the summer months. The association that we see in this example is due to a lurking variable: time of year.

Correlations that are due to lurking variables are sometimes called "nonsense correlations." The correlation is real. What is nonsense is the suggestion that the variables are directly related so that changing one of the variables *causes* changes in the other. The question of causation is important enough to merit separate treatment in Section 2.6. For now, just remember that an association between two variables x and y can reflect many types of relationship among x, y, and one or more lurking variables.

ASSOCIATION DOES NOT IMPLY CAUSATION

An association between an explanatory variable x and a response variable y, even if it is very strong, is not by itself good evidence that changes in x actually cause changes in y.

Lurking variables sometimes create a correlation between x and y, as in Examples 2.29 and 2.30. When you observe an association between two variables, always ask yourself if the relationship that you see might be due to a lurking variable. As in Example 2.30, time is often a likely candidate.

Beware correlations based on averaged data

Regression or correlation studies sometimes work with averages or other measures that combine information from many individuals. For example, if we plot the average height of young children against their age in months, we will see a very strong positive association with correlation near 1. But individual children of the same age vary a great deal in height. A plot of height against age for individual children will show much more scatter and lower correlation than the plot of average height against age.

A correlation based on averages over many individuals is usually higher than the correlation between the same variables based on data for individuals. This fact reminds us again of the importance of noting exactly what variables a statistical study involves.

The restricted-range problem

A regression line is often used to predict the response y to a given value x of the explanatory variable. Successful prediction does not require a cause-and-effect relationship. If both x and y respond to the same underlying unmeasured variables, x may help us predict y even though x has no direct influence on y. For example, the scores of SAT exams taken in high school help predict college grades. There is no cause-and-effect tie between SAT scores and college grades. Rather, both reflect a student's ability and knowledge.

How well do SAT scores, perhaps with the help of high school grades, predict college GPA? We can use the correlation r and its square to get a rough answer. There is, however, a subtle difficulty.

EXAMPLE

2.31   SAT scores and GPA. Combining several studies for students graduating from college since 1980, the College Board reports these correlations between explanatory variables and the overall GPA of college students:

SAT Math and Verbal	High school grades	SAT plus grades
$r = 0.36$	$r = 0.42$	$r = 0.52$

Because $0.52^2 = 0.27$, we see that SAT scores plus students' high school records explain about 27% of the variation in GPA among college students.

The subtle problem? Colleges differ greatly in the range of students they attract. Almost all students at highly selective colleges have high SAT scores and did well in high school. At other colleges, many students are in the middle range of SAT scores and high school performance. Both sets of students receive the full spread of grades. We suspect that if a highly selective college admitted weaker students they would get lower grades, and that the typical student at a highly selective college would get very high grades at a college with a broader range of students. This is the *restricted-range problem:* the data do not contain information on the full range of both explanatory and response variables. *When data suffer from restricted range, r and r^2 are lower than they would be if the full range could be observed.*

Thus, $r = 0.52$ understates the actual ability of SAT scores and high school grades to predict college GPA. One investigator found 21 colleges that enrolled the full range of high school graduates. Sure enough, for these colleges, $r = 0.65$.[26]

Note that the correlations in Example 2.31 involve more than one explanatory variable. We often use several explanatory variables together to predict a response. This is called *multiple regression.* Each r in the example is a *multiple correlation coefficient*, whose square is the proportion of variation in the response explained by the multiple regression. Chapter 11 introduces multiple regression.

Data Mining

Chapters 1 and 2 of this book are devoted to the important aspect of statistics called *exploratory data analysis* (EDA). We use graphs and numerical summaries to examine data, searching for patterns and paying attention to striking deviations from the patterns we find. In discussing regression, we advanced to using the pattern we find (in this case, a linear pattern) for prediction.

Suppose now that we have a truly enormous database, such as all purchases recorded by the cash register scanners of a national retail chain during the past week. Surely this treasure chest of data contains patterns that might guide business decisions. If we could see clearly the types of activewear preferred in large California cities and compare the preferences of small Midwest cities—right now, not at the end of the season—we might improve profits in both parts of the country by matching stock with demand. This sounds much like EDA, and indeed it is. Exploring really large databases in the hope of finding useful patterns is called **data mining.** Here are some distinctive features of data mining:

- When you have terabytes of data, even straightforward calculations and graphics become very time-consuming. So efficient algorithms are very important.

- The structure of the database and the process of storing the data (the fashionable term is *data warehousing*), perhaps by unifying data scattered across many departments of a large corporation, require careful consideration.

- Data mining requires automated tools that work based on only vague queries by the user. The process is too complex to do step-by-step as we have done in EDA.

All these features point to the need for sophisticated computer science as a basis for data mining. Indeed, data mining is often viewed as a part of computer science. Yet many statistical ideas and tools—mostly tools for dealing with multidimensional data, not the sort of thing that appears in a first statistics course—are very helpful. Like many other modern developments, data mining crosses the boundaries of traditional fields of study.

Do remember that the perils we associate with blind use of correlation and regression are yet more perilous in data mining, where the fog of an immense data base can prevent clear vision. Extrapolation, ignoring lurking variables, and confusing association with causation are traps for the unwary data miner.

SECTION 2.4 Summary

You can examine the fit of a regression line by plotting the **residuals,** which are the differences between the observed and predicted values of y. Be on the lookout for points with unusually large residuals and also for nonlinear patterns and uneven variation about the line.

Also look for **influential observations,** individual points that substantially change the regression line. Influential observations are often outliers in the x direction, but they need not have large residuals.

Correlation and regression must be **interpreted with caution.** Plot the data to be sure that the relationship is roughly linear and to detect outliers and influential observations.

Lurking variables may explain the relationship between the explanatory and response variables. Correlation and regression can be misleading if you ignore important lurking variables.

We cannot conclude that there is a cause-and-effect relationship between two variables just because they are strongly associated. **High correlation does not imply causation.**

A correlation based on averages is usually higher than if we used data for individuals. A correlation based on data with a **restricted range** is often lower than would be the case if we could observe the full range of the variables.

SECTION 2.4 Exercises

For Exercises 2.92 and 2.93, see page 124.

2.94   What's wrong? Each of the following statements contains an error. Describe each error and explain why the statement is wrong.

(a) If the residuals are all positive, this implies that there is a positive relationship between the response variable and the explanatory variable.

(b) A negative relationship can never be due to causation.

(c) A lurking variable is always a response variable.

2.95   What's wrong? Each of the following statements contains an error. Describe each error and explain why the statement is wrong.

(a) High correlation implies causation.

(b) An outlier will always have a small residual.

(c) If we have data at values of X equal to 1, 2, 3, 4, and 5, and we try to predict the value of Y using a least-squares regression line, we are doing an extrapolation.

2.96   Use of the Internet and a long life. Exercise 2.20 (page 95) asks the question, "Will you live longer if you use the Internet?" Figure 2.11 (page 95) is a scatterplot of life expectancy in years versus Internet use for 181 countries. The scatterplot shows a positive association between these two variables. Do you think that this plot indicates that Internet use causes people to live longer? Give another possible explanation for why these two variables are positively associated.   🌐 INTERNETANDLIFE

2.97   How's your self-esteem? People who do well tend to feel good about themselves. Perhaps helping people feel good about themselves will help them do better in their jobs and in life. For a time, raising self-esteem became a goal in many schools and companies. Can you think of explanations for the association between high self-esteem and good performance other than "Self-esteem causes better work"?

2.98   Are big hospitals bad for you? A study shows that there is a positive correlation between the size of a hospital (measured by its number of beds x) and the median number of days y that patients remain in the hospital. Does this mean that you can shorten a hospital stay by choosing a small hospital? Why?

2.99   Does herbal tea help nursing-home residents? A group of college students believes that herbal tea has remarkable powers. To test this belief, they make weekly visits to a local nursing home, where they visit with the residents and serve them herbal tea. The nursing-home staff reports that after several months many of the residents are healthier and more cheerful. We should commend the students for their good deeds but doubt that herbal tea helped the residents. Identify the explanatory and response variables in this informal study. Then explain what lurking variables account for the observed association.

2.100   Price and ounces. In Example 2.2 (page 80) and Exercise 2.3 (page 82) we examined the relationship between the price and the size of a Mocha Frappuccino©. The 12-ounce Tall drink costs \$3.50, the 16-ounce Grande is \$4.00, and the 24-ounce Venti is \$4.50.

(a) Plot the data and describe the relationship. (Explain why you should plot size in ounces on the x axis.)

(b) Find the least-squares regression line for predicting the price using size. Add the line to your plot.

(c) Draw a vertical line from the least-squares line to each data point. This gives a graphical picture of the residuals.

(d) Find the residuals and verify that they sum to zero.

(e) Plot the residuals versus size. Interpret this plot.

2.101 Average monthly temperatures. Here are the average monthly temperatures for Chicago, Illinois:
🌑 CHICAGOTEMPS

Month	1	2	3	4	5	6
Temperature (°F)	21.0	25.4	37.2	48.6	58.9	68.6

Month	7	8	9	10	11	12
Temperature (°F)	73.2	71.7	64.4	52.8	40.0	26.6

In this table, months are coded as integers, with January corresponding to 1 and December corresponding to 12.

(a) Plot the data with month on the x axis and temperature on the y axis. Describe the relationship.

(b) Find the least-squares regression line and add it to the plot. Does the line give a good fit to the data? Explain your answer.

(c) Calculate the residuals and plot them versus month. Describe the pattern and explain what the residual plot tells you about the relationship between temperature and month in Chicago.

(d) Do you think you would find a similar pattern if you plotted the same kind of data for another city?

(e) Would your answer to part (d) change if the other city was Melbourne, Australia? Explain why or why not.

2.102 Growth of infants in Egypt. A study of nutrition in developing countries collected data from the Egyptian village of Nahya. Here are the mean weights (in kilograms) for 170 infants in Nahya who were weighed each month during their first year of life:[27] 🌑 INFANTGROWTH

Age (months)	1	2	3	4	5	6	7	8	9	10	11	12
Weight (kg)	4.3	5.1	5.7	6.3	6.8	7.1	7.2	7.2	7.2	7.2	7.5	7.8

(a) Plot weight against time.

(b) A hasty user of statistics enters the data into software and computes the least-squares line without plotting the data. The result is

```
The regression equation is
weight = 4.88 + 0.267 age
```

Plot this line on your graph. Is it an acceptable summary of the overall pattern of growth? Remember that you can calculate the least-squares line for *any* set of two-variable data. It's up to you to decide if it makes sense to fit a line.

(c) Fortunately, the software also prints out the residuals from the least-squares line. In order of age along the rows, they are

−0.85	−0.31	0.02	0.35	0.58	0.62
0.45	0.18	−0.08	−0.35	−0.32	−0.28

Verify that the residuals have sum zero (except for roundoff error). Plot the residuals against age and add a horizontal line at zero. Describe carefully the pattern that you see.

2.103 Effect of using means. Your plot for Exercise 2.102 shows that the increase of the mean weight of children in Nahya is very linear during the first 5 months of life. The correlation of age and weight is $r = 0.9964$ for the first 5 months. Weight in these data is the mean for 170 children. Explain why the correlation between age and weight for the 170 individual children would surely be much smaller.

2.104 A test for job applicants. Your company gives a test of cognitive ability to job applicants before deciding whom to hire. Your boss has asked you to use company records to see if this test really helps predict the performance ratings of employees. Explain carefully to your boss why the restricted-range problem may make it difficult to see a strong relationship between test scores and performance ratings.

2.105 🔺 ⚠️ **A lurking variable.** The effect of a lurking variable can be surprising when individuals are divided into groups. In recent years, the mean SAT score of all high school seniors has increased. But the mean SAT score has decreased for students at each level of high school grades (A, B, C, and so on). Explain how grade inflation in high school (the lurking variable) can account for this pattern. *A relationship that holds for each group within a population need not hold for the population as a whole. In fact, the relationship can even change direction.*

2.106 🔺 **Another example.** Here is another example of the group effect cautioned about in the previous exercise. Explain how as a nation's population grows older mean income can go down for workers in each age group, yet still go up for all workers.

2.107 Basal metabolic rate. Careful statistical studies often include examination of potential lurking variables. This was true of the study of the effect of nonexercise activity (NEA) on fat gain (Example 2.18, page 108), our lead example in Section 2.3. Overeating may lead our bodies to spontaneously increase NEA (fidgeting and the like). Our bodies might also spontaneously increase their basal metabolic rate (BMR), which measures energy use while resting. If both energy uses increased, regressing fat gain on NEA alone would be misleading. Here are data on BMR and fat gain for the same 16 subjects whose NEA we examined earlier: 🖕 FIDGET

BMR increase (cal)	117	352	244	−42	−3	134	136	−32
Fat gain (kg)	4.2	3.0	3.7	2.7	3.2	3.6	2.4	1.3

BMR increase (cal)	−99	9	−15	−70	165	172	100	35
Fat gain (kg)	3.8	1.7	1.6	2.2	1.0	0.4	2.3	1.1

The correlation between NEA and fat gain is $r = -0.7786$. The slope of the regression line for predicting fat gain from NEA is $b_1 = -0.00344$ kilogram per calorie. What are the correlation and slope for BMR and fat gain? Explain why these values show that BMR has much less effect on fat gain than does NEA.

2.108 Gas chromatography. Gas chromatography is a technique used to detect very small amounts of a substance, for example, a contaminant in drinking water. Laboratories use regression to calibrate such techniques. The following data show the results of five measurements for each of four amounts of the substance being investigated.[28] The explanatory variable x is the amount of substance in the specimen, measured in nanograms (ng), units of 10^{-9} gram. The response variable y is the reading from the gas chromatograph. 🖕 GASCHROMO

Amount (ng)	Response				
0.25	6.55	7.98	6.54	6.37	7.96
1.00	29.7	30.0	30.1	29.5	29.1
5.00	211	204	212	213	205
20.00	929	905	922	928	919

(a) Make a scatterplot of these data. The relationship appears to be approximately linear, but the wide variation in the response values makes it hard to see detail in this graph.

(b) Compute the least-squares regression line of y on x, and plot this line on your graph.

(c) Now compute the residuals and make a plot of the residuals against x. It is much easier to see deviations from linearity in the residual plot. Describe carefully the pattern displayed by the residuals.

2.109 Golf scores. Here are the golf scores of 6 members of a women's golf team in first two rounds of NCAA Tournament play:[29] 🖕 GOLFNCAA

Player	Gulyanamitta	Hernandez	Hoffmeister
Round 1	80	74	76
Round 2	76	72	76

Player	LeBlanc	Mess	Sinha
Round 1	76	85	88
Round 2	77	83	83

(a) Plot the data with the Round 1 scores on the x axis and the Round 2 scores on the y axis.

(b) Describe the relationship.

(c) Calculate the least-squares regression line and add it to your plot.

(d) Circle the observation for Maria Hernandez. She was the NCAA champion in this tournament.

2.110 Climate change. Drilling down beneath a lake in Alaska yields chemical evidence of past changes in climate. Biological silicon, left by the skeletons of single-celled creatures called diatoms, measures the abundance of life in the lake. A rather complex variable based on the ratio of certain isotopes relative to ocean water gives an indirect measure of moisture, mostly from snow. As we drill down, we look farther into the past. Here are data from 2300 to 12,000 years ago:[30] 🖕 SILICONISOTOPE

Isotope (%)	Silicon (mg/g)	Isotope (%)	Silicon (mg/g)	Isotope (%)	Silicon (mg/g)
−19.90	97	−20.71	154	−21.63	224
−19.84	106	−20.80	265	−21.63	237
−19.46	118	−20.86	267	−21.19	188
−20.20	141	−21.28	296	−19.37	337

(a) Make a scatterplot of silicon (response) against isotope (explanatory). Ignoring the outlier, describe the direction, form, and strength of the relationship. The researchers say that this and relationships among other variables they measured are evidence for cyclic changes in climate that are linked to changes in the sun's activity.

(b) The researchers single out one point: "The open circle in the plot is an outlier that was excluded in the correlation analysis." Circle this outlier on your graph. What is the correlation with and without this point? The point strongly influences the correlation.

(c) Is the outlier also strongly influential for the regression line? Calculate and draw on your graph two regression lines, and discuss what you see.

2.111 🖕 ⚠️ **Use the applet.** It isn't easy to guess the position of the least-squares line by eye. Use the *Correlation and Regression* applet to compare a line you draw with the least-squares line. Click on the scatterplot to create a group of 15 to 20 points from lower left to upper right with a clear positive straight-line pattern (correlation around 0.7). Click the "Draw line" button and use the mouse to draw a line through the middle of the cloud of points from lower left to upper right. Note the "thermometer" that appears above the plot. The red portion is the sum of the squared vertical distances from the points in the plot to the least-squares line. The green portion is the "extra" sum of squares for your line—it shows by how much your line misses the smallest possible sum of squares.

(a) You drew a line by eye through the middle of the pattern. Yet the right-hand part of the bar is probably almost entirely green. What does that tell you?

(b) Now click the "Show least-squares line" box. Is the slope of the least-squares line smaller (the new line is less steep) or larger (line is steeper) than that of your line? If you repeat this exercise several times, you will consistently get the same result. *The least-squares line minimizes the vertical distances of the points from the line. It is not the line through the "middle" of the cloud of points.* This is one reason why it is hard to draw a good regression line by eye.

2.112 Use the applet. Go to the *Correlation and Regression* applet. Click on the scatterplot to create a group of 10 points in the lower-right corner of the scatterplot with a strong straight-line pattern (correlation about −0.9). In Exercise 2.56 (page 107) you started here to see that correlation r is not resistant. Now click the "Show least-squares line" box to display the regression line.

(a) Add one point at the upper left that is far from the other 10 points but exactly on the regression line. Why does this outlier have no effect on the line even though it changes the correlation?

(b) Now drag this last point down until it is opposite the group of 10 points. You see that one end of the least-squares line chases this single point, while the other end remains near the middle of the original group of 10. What makes the last point so influential?

2.113 Education and income. There is a strong positive correlation between years of education and income for economists employed by business firms. (In particular, economists with doctorates earn more than economists with only a bachelor's degree.) There is also a strong positive correlation between years of education and income for economists employed by colleges and universities. But when all economists are considered, there is a *negative* correlation between education and income. The explanation for this is that business pays high salaries and employs mostly economists with bachelor's degrees, while colleges pay lower salaries and employ mostly economists with doctorates. Sketch a scatterplot with two groups of cases (business and academic) that illustrates how a strong positive correlation within each group and a negative overall correlation can occur together.

2.114 Dangers of not looking at a plot. Table 2.4 (page 120) presents four sets of data prepared by the statistician Frank Anscombe to illustrate the dangers of calculating without first plotting the data.[31] ANSCOMBE

(a) Use x to predict y for each of the four data sets. Find the predicted values and residuals for each of the four regression equations.

(b) Plot the residuals versus x for each of the four data sets.

(c) Write a summary of what the residuals tell you for each data set and explain how the residuals help you to understand these data.

2.5 Data Analysis for Two-Way Tables

LOOK BACK
quantitative and categorical
variables p. 2

When we study relationships between two variables, one of the first questions we ask is whether each variable is quantitative or categorical. For two quantitative variables, we use a scatterplot to examine the relationship, and we fit a line to the data if the relationship is approximately linear. If one of the variables is quantitative and the other is categorical, we can use the methods in Chapter 1 to describe the distribution of the quantitative variable for each value of the categorical variable. This leaves us with the situation where both variables are categorical. In this section we discuss methods for studying these relationships.

Some variables—such as gender, race, and occupation—are inherently categorical. Other categorical variables are created by grouping values of a quantitative variable into classes. Published data are often reported in grouped form to save space. To describe categorical data, we use the *counts* (frequencies) or *percents* (relative frequencies) of individuals that fall into various categories.

The two-way table

two-way table

A key idea in studying relationships between two variables is that both variables must be measured on the same individuals or cases. When both variables are categorical, the raw data are summarized in a **two-way table** that gives counts of observations for each combination of values of the two categorical variables. Here is an example.

> ### EXAMPLE
>
> **2.32 Binge drinking by college students.** Alcohol abuse has been described by college presidents as the number one problem on campus, and it is an important cause of death in young adults. How common is it? A survey of 17,096 students in U.S. four-year colleges collected information on drinking behavior and alcohol-related problems.[32] The researchers defined "frequent binge drinking" as having five or more drinks in a row three or more times in the past two weeks. Here is the two-way table classifying students by gender and whether or not they are frequent binge drinkers:

Two-way table for frequent binge drinking and gender

	Gender	
Frequent binge drinker	**Men**	**Women**
Yes	1630	1684
No	5550	8232

BINGEGENDER

We see that there are 1630 male students who are frequent binge drinkers and 5550 male students who are not.

BINGEGENDER

> ### USE YOUR KNOWLEDGE
>
> **2.115 Read the table.** How many female students are binge drinkers? How many are not?

row and column variables

cell

For the binge-drinking example, we could view gender as an explanatory variable and frequent binge drinking as a response variable. This is why we put gender in the columns (like the x axis in a regression) and frequent binge drinking in the rows (like the y axis in a regression). We call binge drinking the **row variable** because each horizontal row in the table describes the drinking behavior. Gender is the **column variable** because each vertical column describes one gender group. Each combination of values for these two variables is called a **cell.** For example, the cell corresponding to women who are not frequent binge drinkers contains the number 8232. This table is called a 2×2 table because there are 2 rows and 2 columns.

To describe relationships between two categorical variables, we compute different types of percents. Our job is easier if we expand the basic two-way table by adding various totals. We illustrate the idea with our binge-drinking example.

BINGEGENDER

EXAMPLE

2.33 Add the margins to the table. We expand the table in Example 2.32 by adding the totals for each row, for each column, and the total number of all of the observations. Here is the result:

Two-way table for frequent binge drinking and gender

Frequent binge drinker	Gender		Total
	Men	Women	
Yes	1,630	1,684	3,314
No	5,550	8,232	13,782
Total	7,180	9,916	17,096

In this study there are 7180 male students. The total number of binge drinkers is 3314 and the total number of individuals in the study is 17,096.

USE YOUR KNOWLEDGE

BINGEGENDER

2.116 Read the margins of the table. How many women are subjects in the binge-drinking study? What is the total number of students who are not binge drinkers?

In this example, be sure that you understand how the table is obtained from the raw data. Think about a data file with one line per subject. There would be 17,096 lines or records in this data set. In the two-way table, each individual is counted once and only once. As a result, the sum of the counts in the table is the total number of individuals in the data set. *Most errors in the use of categorical-data methods come from a misunderstanding of how these tables are constructed.*

Joint distribution

We are now ready to compute some proportions that help us understand the data in a two-way table. Suppose that we are interested in the men who are binge drinkers. The proportion of these is simply 1630 divided by 17,096, or 0.095. We would estimate that 9.5% of college students are male frequent binge drinkers. For each cell, we can compute a proportion by dividing the cell entry by the total sample size. The collection of these proportions is the **joint distribution** of the two categorical variables.

joint distribution

BINGEGENDER

EXAMPLE

2.34 The joint distribution. For the binge-drinking example, the joint distribution of binge drinking and gender is

Joint distribution of frequent binge drinking and gender

Frequent binge drinker	Gender	
	Men	Women
Yes	0.095	0.099
No	0.325	0.482

Because this is a distribution, the sum of the proportions should be 1. For this example the sum is 1.001. The difference is due to roundoff error.

USE YOUR KNOWLEDGE

BINGEGENDER

2.117 Explain the computation. Explain how the entry for the women who are not binge drinkers in Example 2.34 is computed from the table in Example 2.33.

From the joint distribution we see that the proportions of men and women who are frequent binge drinkers are similar in the population of college students. For the men we have 9.5%; the women are slightly higher at 9.9%. Note, however, that the proportion of women who are not frequent binge drinkers is also higher than the proportion of men. One reason for this is that there are more women in the sample than men. To understand this set of data we will need to do some additional calculations. Let's look at the distribution of gender.

Marginal distributions

marginal distribution

When we examine the distribution of a single variable in a two-way table, we are looking at a **marginal distribution.** There are two marginal distributions, one for each categorical variable in the two-way table. They are very easy to compute.

BINGEGENDER

EXAMPLE

2.35 The marginal distribution of gender. Look at the table in Example 2.33. The total numbers of men and women are given in the bottom row, labeled "Total." Our sample has 7180 men and 9916 women. To find the marginal distribution of gender we simply divide these numbers by the total sample size, 17,096. The marginal distribution of gender is

Marginal distribution of gender

	Men	Women
Proportion	0.420	0.580

Note that the proportions sum to 1; there is no roundoff error.

Often we prefer to use percents rather than proportions. Here is the marginal distribution of gender described with percents:

BINGEGENDER

Marginal distribution of gender		
	Men	**Women**
Percent	42.0%	58.0%

Which form do you prefer?

The other marginal distribution for this example is the distribution of binge drinking.

BINGEGENDER

EXAMPLE

2.36   The marginal distribution in percents. Here is the marginal distribution of the frequent-binge-drinking variable (in percents):

Marginal distribution of frequent binge drinking		
	Yes	**No**
Percent	19.4%	80.6%

USE YOUR KNOWLEDGE

2.118   Explain the marginal distribution. Explain how the marginal distribution of frequent binge drinking given in Example 2.36 is computed from the entries in the table given in Example 2.33.

Each marginal distribution from a two-way table is a distribution for a single categorical variable. We can use a bar graph or a pie chart to display such a distribution. For our two-way table, we will be content with numerical summaries: for example, 58% of these college students are women, and 19.4% of the students are frequent binge drinkers. When we have more rows or columns, the graphical displays are particularly useful.

LOOK BACK
bar graphs and pie charts p. 7

Describing relations in two-way tables

The table in Example 2.34 contains much more information than the two marginal distributions of gender alone and frequent binge drinking alone. We need to do a little more work to examine the relationship. *Relationships among categorical variables are described by calculating appropriate percents from the counts given.* What percents do you think we should use to describe the relationship between gender and frequent binge drinking?

EXAMPLE

2.37   Women who are frequent binge drinkers. What percent of the women in our sample are frequent binge drinkers? This is the count of the women who are frequent binge drinkers as a percent of the number of women in the sample:

$$\frac{1684}{9916} = 0.170 = 17.0\%$$

BINGEGENDER

USE YOUR KNOWLEDGE

2.119 Find the percent. Show that the percent of men who are frequent binge drinkers is 22.7%.

Recall that when we looked at the joint distribution of gender and binge drinking, we found that among all college students in the sample, 9.5% were male frequent binge drinkers and 9.9% were female frequent binge drinkers. The percents are fairly similar because the counts for these two groups, 1630 and 1684, are close. The calculations that we just performed, however, give us a different view. When we look separately at women and men, we see that the proportions of frequent binge drinkers are somewhat different, 17.0% for women versus 22.7% for men.

Conditional distributions

conditional distribution

In Example 2.37 we looked at the women alone and examined the distribution of the other categorical variable, frequent binge drinking. Another way to say this is that we conditioned on the value of gender being female. Similarly, we can condition on the value of gender being male. When we condition on the value of one variable and calculate the distribution of the other variable, we obtain a **conditional distribution.** Note that in Example 2.37 we calculated only the percent for frequent binge drinking. The complete conditional distribution gives the proportions or percents for all possible values of the conditioning variable.

BINGEGENDER

EXAMPLE

2.38 Conditional distribution of binge drinking for women. For women, the conditional distribution of the binge-drinking variable in terms of percents is

Conditional distribution of binge drinking for women		
	Yes	**No**
Percent	17.0%	83.0%

Note that we have included the percents for both of the possible values, Yes and No, of the binge-drinking variable. These percents sum to 100%.

BINGEGENDER

USE YOUR KNOWLEDGE

2.120 A conditional distribution. Perform the calculations to show that the conditional distribution of binge drinking for men is

	Yes	**No**
Conditional distribution of binge drinking for men		
Percent	22.7%	77.3%

Comparing the conditional distributions (Example 2.38 and Exercise 2.120) reveals the nature of the association between gender and frequent binge drinking. In this set of data the men are more likely to be frequent binge drinkers than the women.

Bar graphs can help us to see relationships between two categorical variables. No single graph (such as a scatterplot) portrays the form of the relationship between categorical variables, and no single numerical measure (such as the correlation) summarizes the strength of an association. Bar graphs are flexible enough to be helpful, but you must think about what comparisons you want to display. For numerical measures, we must rely on well-chosen percents or on more advanced statistical methods.[33]

A two-way table contains a great deal of information in compact form. Making that information clear almost always requires finding percents. You must decide which percents you need. Of course, we prefer to use software to compute the joint, marginal, and conditional distributions.

BINGEGENDER

EXAMPLE

2.39 Software output. Figure 2.28 gives computer output for the data in Example 2.32 using SPSS, Minitab, and SAS. There are minor variations among software packages, but these are typical of what is usually produced.

```
  SPSS                                                          _ □ X

                BINGE * GENDER Crosstabulation
                                      GENDER          Total
                                   Male    Female
    BINGE  Yes           Count     1630      1684     3314
                  % within BINGE   49.2%     50.8%   100.0%
                  % within GENDER  22.7%     17.0%    19.4%
                      % of Total    9.5%      9.9%    19.4%
           No             Count    5550      8232    13782
                  % within BINGE   40.3%     59.7%   100.0%
                  % within GENDER  77.3%     83.0%    80.6%
                      % of Total   32.5%     48.2%    80.6%
    Total              Count       7180      9916    17096
                  % within BINGE   42.0%     58.0%   100.0%
                  % within GENDER 100.0%    100.0%   100.0%
                      % of Total   42.0%     58.0%   100.0%
```

FIGURE 2.28 Computer output for the binge-drinking study in Example 2.39.

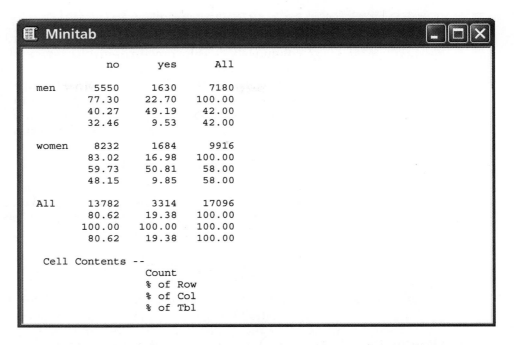

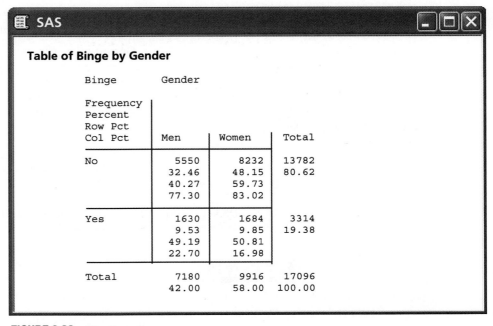

FIGURE 2.28 (*Continued*)

Each cell in the 2 × 2 table has four entries. These are the count (the number of observations in the cell), the conditional distributions for rows and columns, and the joint distribution. Note that all of these are expressed as percents rather than proportions. Marginal totals and distributions are given in the rightmost column and the bottom row.

Most software packages order the row and column labels numerically or alphabetically. In general, it is better to use words rather than numbers for the column labels. This sometimes involves some additional work, but it avoids the kind of confusion that can result when you forget the real values associated with each numerical value. You should verify that the entries in Figure 2.28 correspond to the calculations that we performed in Examples 2.34 to 2.38. In addition, verify the calculations for the conditional distributions of gender for each value of the frequent-binge-drinking variable.

Simpson's paradox

As is the case with quantitative variables, the effects of lurking variables can strongly influence relationships between two categorical variables. Here is an example that demonstrates the surprises that can await the unsuspecting consumer of data.

CUSTSERVREP

EXAMPLE

2.40 Which customer service representative is better? A customer service center has a goal of resolving customer questions in 10 minutes or less. Here are the records for two representatives:

Goal met	Representative	
	Alexis	Peyton
Yes	172	118
No	28	82
Total	200	200

Alexis has met the goal 172 times out of 200, a success rate of 86%. For Peyton, the success rate is 118 out of 200, or 59%. Alexis clearly has the better success rate.

Let's look at the data in a little more detail. The data summarized come from two different weeks in the year.

CUSTSERVICEREP

EXAMPLE

2.41 Lets look at the data more carefully. Here are the counts broken down by week:

Goal met	Week 1		Week 2	
	Alexis	Peyton	Alexis	Peyton
Yes	162	19	10	99
No	18	1	10	81
Total	180	20	20	180

For week 1, Alexis met the goal 90% of the time (162/180) while Peyton met the goal 95% of the time (19/20). Peyton had the better performance in week 1. What about week 2? Here Alexis met the goal 50% of the time (10/20) while the success rate for Peyton was 55% (99/180). Peyton again had the better performance. How does this analysis compare with the analysis that combined the counts for the two weeks? That analysis clearly showed that Alexis had the better performance, 59% versus 86%.

These results can be explained by a lurking variable related to week. The first week was during a period where the product had been in use for several months. Most of the calls to the customer service center concerned problems that had been encountered before. The representatives were trained to answer these questions and usually had no trouble in meeting the goal of resolving the problems quickly. On the other hand, the second week occurred shortly after the release of a new version of the product. Most of the calls during this week concerned new problems that the representatives had not yet encountered. Many more of these questions took longer than the 10-minute goal to resolve.

Look at the total in the bottom row of the detailed table. During the first week, when calls were easy to resolve, Alexis handled 180 calls and Peyton handled 20. The situation was exactly the opposite during the second week when the calls were difficult to resolve. There were 20 calls for Alexis and 180 for Peyton.

The original two-way table, which did not take account of week, was misleading. This example illustrates *Simpson's paradox*.

SIMPSON'S PARADOX

An association or comparison that holds for all of several groups can reverse direction when the data are combined to form a single group. This reversal is called **Simpson's paradox.**

The lurking variables in our Simpson's paradox example are categorical. That is, they break the observations into groups, work week. *Simpson's paradox is an extreme form of the fact that observed associations can be misleading when there are lurking variables.*

three-way table

The data in Example 2.41 are given in a **three-way table** that reports counts for each combination of three categorical variables: week, representative, and whether or not the goal was met. In our example, we looked at the three-way table by constructing two two-way tables for representative by goal, one for each week. The original table can be obtained by adding the corresponding

aggregation

counts for these two tables. This process is called **aggregating** the data. When we aggregated data in our example we ignored the variable week, which then became a lurking variable. *Conclusions that seem obvious when we look only at aggregated data can become quite different when the data are examined in more detail.*

SECTION 2.5 Summary

A **two-way table** of counts organizes data about two categorical variables. Values of the **row variable** label the rows that run across the table, and values of the **column variable** label the columns that run down the table. Two-way tables are often used to summarize large amounts of data by grouping outcomes into categories.

The **joint distribution** of the row and column variables is found by dividing the count in each cell by the total number of observations.

The **row totals** and **column totals** in a two-way table give the **marginal distributions** of the two variables separately. It is clearer to present these distributions as percents of the table total. Marginal distributions do not give any information about the relationship between the variables.

To find the **conditional distribution** of the row variable for one specific value of the column variable, look only at that one column in the table. Find each entry in the column as a percent of the column total.

There is a conditional distribution of the row variable for each column in the table. Comparing these conditional distributions is one way to describe the association between the row and the column variables. It is particularly useful when the column variable is the explanatory variable. When the row variable is explanatory, find the conditional distribution of the column variable for each row and compare these distributions.

Bar graphs are a flexible means of presenting categorical data. There is no single best way to describe an association between two categorical variables.

We present data on three categorical variables in a **three-way table,** printed as separate two-way tables for each level of the third variable. A comparison between two variables that holds for each level of a third variable can be changed or even reversed when the data are **aggregated** by summing over all levels of the third variable. **Simpson's paradox** refers to the reversal of a comparison by aggregation. It is an example of the potential effect of lurking variables on an observed association.

SECTION 2.5 Exercises

For Exercise 2.115, see page 137; for 2.116, see page 138; for 2.117, see page 139; for 2.118, see page 140; for 2.119, see page 141; and for 2.120, see page 142.

2.121 Exercise and adequate sleep. A survey of 656 boys and girls who were 13 to 18 years old asked about adequate sleep and other health-related behaviors. The recommended amount of sleep is six to eight hours per night.[34] In the survey 54% of the respondents reported that they got less than this amount of sleep on school nights. An exercise scale was developed and was used to classify the students as above or below the median in this domain. Here is the 2 × 2 table of counts with students classified as getting or not getting adequate sleep and by the exercise variable: SLEEP

	Exercise	
Enough sleep	**High**	**Low**
Yes	151	115
No	148	242

(a) Find the distribution of adequate sleep for the high exercisers.

(b) Do the same for the low exercisers.

(c) Summarize the relationship between adequate sleep and exercise using the results of parts (a) and (b).

2.122 Adequate sleep and exercise. Refer to the previous exercise. 🔊 SLEEP

(a) Find the distribution of exercise for those who get adequate sleep.

(b) Do the same for those who do not get adequate sleep.

(c) Write a short summary of the relationship between adequate sleep and exercise using the results of parts (a) and (b).

(d) Compare this summary with the summary that you obtained in part (c) of the previous exercise. Which do you prefer? Give a reason for your answer.

2.123 Which hospital is safer? Insurance companies and consumers are interested in the performance of hospitals. The government releases data about patient outcomes in hospitals that can be useful in making informed health care decisions. Here is a two-way table of data on the survival of patients after surgery in two hospitals. All patients undergoing surgery in a recent time period are included. "Survived" means that the patient lived at least 6 weeks following surgery. 🔊 HOSPITALS

	Hospital A	Hospital B
Died	63	16
Survived	2037	784
Total	2100	800

What percent of Hospital A patients died? What percent of Hospital B patients died? These are the numbers one might see reported in the media.

2.124 Patients in "poor" or "good" condition. Refer to the previous exercise. Not all surgery cases are equally serious, however. Patients are classified as being in either "poor" or "good" condition before surgery. Here are the data broken down by patient condition. The entries in the original two-way table are just the sums of the "poor" and "good" entries in this pair of tables. 🔊 HOSPITALS

Good Condition			Poor Condition		
	Hospital A	Hospital B		Hospital A	Hospital B
Died	6	8	Died	57	8
Survived	594	592	Survived	1443	192
Total	600	600	Total	1500	200

(a) Find the death rate for Hospital A patients who were classified as "poor" before surgery. Do the same for

Hospital B. In which hospital do "poor" patients fare better?

(b) Repeat (a) for patients classified as "good" before surgery.

(c) What is your recommendation to someone facing surgery and choosing between these two hospitals?

(d) How can Hospital A do better in both groups, yet do worse overall? Look at the data and carefully explain how this can happen.

2.125 Full-time and part-time college students. The Census Bureau provides estimates of numbers of people in the United States classified in various ways.[35] Let's look at college students. The following table gives us data to examine the relation between age and full-time or part-time status. The numbers in the table are expressed as thousands of U.S. college students. 🔊 USCOLLEGESTUDENTS

U.S. college students by age and status

	Status	
Age	Full-time	Part-time
15–19	3388	389
20–24	5238	1164
25–34	1703	1699
35 and over	762	2045

(a) What is the U.S. Census Bureau estimate of the number of full-time college students aged 15 to 19?

(b) Give the joint distribution of age and status for this table.

(c) What is the marginal distribution of age? Display the results graphically.

(d) What is the marginal distribution of status? Display the results graphically.

2.126 Condition on age. Refer to the previous exercise. Find the conditional distribution of status for each of the four age categories. Display the distributions graphically and summarize their differences and similarities. 🔊 USCOLLEGESTUDENTS

2.127 Condition on status. Refer to the previous two exercises. Compute the conditional distribution of age for each of the two status categories. Display the distributions graphically and write a short paragraph describing the distributions and how they differ. 🔊 USCOLLEGESTUDENTS

2.128 Complete the table. Here are the row and column totals for a two-way table with two rows and two columns:

a	b	200
c	d	200
200	200	400

Find *two different* sets of counts *a*, *b*, *c*, and *d* for the body of the table that give these same totals. This shows that the relationship between two variables cannot be obtained from the two individual distributions of the variables.

2.129 Construct a table with no association. Construct a 3 × 3 table of counts where there is no apparent association between the row and column variables.

2.130 Survey response rates. A market research firm conducted a survey of companies in its state. They mailed a questionnaire to 300 small companies, 300 medium-sized companies, and 300 large companies. The rate of nonresponse is important in deciding how reliable survey results are. Here are the data on response to this survey: RESPONSERATES

Size of company	Response		Total
	Yes	**No**	
Small	175	125	300
Medium	145	155	300
Large	120	180	300

(a) What was the overall percent of nonresponse?

(b) Describe how nonresponse is related to the size of the business. (Use percents to make your statements precise.)

(c) Draw a bar graph to compare the nonresponse percents for the three size categories.

(d) Using the total number of responses as a base, compute the percent of responses that come from each of small, medium, and large businesses.

(e) The sampling plan was designed to obtain equal numbers of responses from small, medium, and large companies. In preparing an analysis of the survey results, do you think it would be reasonable to proceed as if the responses represented companies of each size equally?

2.131 Career plans of young women and men. A study of the career plans of young women and men sent questionnaires to all 722 members of the senior class in the College of Business Administration at the University of Illinois. One question asked which major within the business program the student had chosen.[36] Here are the data from the students who responded: CAREERPLANS

Major	Gender	
	Female	**Male**
Accounting	68	56
Administration	91	40
Economics	5	6
Finance	61	59

(a) Describe the differences between the distributions of majors for women and men with percents, with a graph, and in words.

(b) What percent of the students did not respond to the questionnaire? The nonresponse weakens conclusions drawn from these data.

2.132 Treatment for cocaine addiction. Cocaine addiction can be difficult to overcome. Since addicts derive pleasure from the drug, one proposed aid is to provide an antidepressant drug. A 3-year study with 72 chronic cocaine users compared an antidepressant drug called desipramine with lithium and a placebo. (Lithium is a standard drug to treat cocaine addiction. A placebo is a tablet with no effects that tastes and looks like the antidepressant drug. It is used so that the effect of being in the study but not taking the antidepressant drug can be seen.) One-third of the subjects, chosen at random, received each treatment.[37] Here are the results: COCAINE

Treatment	Cocaine relapse?	
	Yes	**No**
Desipramine	10	14
Lithium	18	6
Placebo	20	4

Compare the effectiveness of the three treatments in preventing relapse. Use percents and draw a bar graph. Write a brief summary of your conclusions.

2.6 The Question of Causation*

In many studies of the relationship between two variables, the goal is to establish that changes in the explanatory variable *cause* changes in the response variable. Even when a strong association is present, the conclusion that this association is due to a causal link between the variables is often hard to find. What ties between two variables (and others lurking in the background) can

---

*This section is optional.

explain an observed association? What constitutes good evidence for causation? We begin our consideration of these questions with a set of examples. In each case, there is a clear association between an explanatory variable x and a response variable y. Moreover, the association is positive whenever the direction makes sense.

Explaining Association

EXAMPLE

2.42 Observed associations. Here are some examples of observed association between x and y:

1. x = mother's body mass index
   y = daughter's body mass index

2. x = amount of the artificial sweetener saccharin in a rat's diet
   y = count of tumors in the rat's bladder

3. x = a student's SAT score as a high school senior
   y = a student's first-year college grade point average

4. x = monthly flow of money into stock mutual funds
   y = monthly rate of return for the stock market

5. x = whether a person regularly attends religious services
   y = how long the person lives

6. x = the number of years of education a worker has
   y = the worker's income

Explaining association: causation

Figure 2.29 shows in outline form how a variety of underlying links between variables can explain association. The dashed double-arrow line represents an observed association between the variables x and y. Some associations are explained by a direct cause-and-effect link between these variables. The first diagram in Figure 2.28 shows "x causes y" by a solid arrow running from x to y.

Items 1 and 2 in Example 2.42 are examples of direct causation. *Even when direct causation is present, very often it is not a complete explanation of an association between two variables.* The best evidence for causation comes from experiments that actually change x while holding all other factors fixed. If y changes, we have good reason to think that x caused the change in y.

FIGURE 2.29 Possible explanations for an observed association. The dashed double-arrow lines show an association. The solid arrows show a cause-and-effect link. The variable x is explanatory, y is a response variable, and z is a lurking variable.

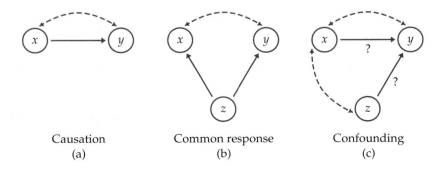

Explaining association: common response

common response

"Beware the lurking variable" is good advice when thinking about an association between two variables. The second diagram in Figure 2.28 illustrates **common response.** The observed association between the variables x and y is explained by a lurking variable z. Both x and y change in response to changes in z. This common response creates an association even though there may be no direct causal link between x and y.

The third and fourth items in Example 2.42 illustrate how common response can create an association.

Explaining association: confounding

For item 1 in Example 2.42 we expect that inheritance explains part of the association between the body mass indexes (BMIs) of daughters and their mothers. Can we use r or r^2 to say how much inheritance contributes to the daughters' BMIs? No. It may well be that mothers who are overweight also set an example of little exercise, poor eating habits, and lots of television. Their daughters pick up these habits to some extent, so the influence of heredity is mixed up with influences from the girls' environment. We call this mixing of influences *confounding.*

> **CONFOUNDING**
>
> Two variables are **confounded** when their effects on a response variable cannot be distinguished from each other. The confounded variables may be either explanatory variables or lurking variables.

When many uncontrolled variables are related to a response variable, you should always ask whether or not confounding of several variables prevents you from drawing conclusions about causation. The third diagram in Figure 2.29 illustrates confounding. Both the explanatory variable x and the lurking variable z may influence the response variable y. Because x is confounded with z, we cannot distinguish the influence of x from the influence of z. We cannot say how strong the direct effect of x on y is. In fact, it can be hard to say if x influences y at all.

The last two associations in Example 2.42 (Items 5 and 6) are explained in part by confounding.

Many observed associations are at least partly explained by lurking variables. Both common response and confounding involve the influence of a lurking variable (or variables) z on the response variable y. The distinction between these two types of relationship is less important than the common element, the influence of lurking variables. The most important lesson of these examples is one we have already emphasized: **even a very strong association between two variables is not by itself good evidence that there is a cause-and-effect link between the variables.**

Establishing causation

How can a direct causal link between x and y be established? The best method—indeed, the only fully compelling method—of establishing causation is to conduct a carefully designed experiment in which the effects of possible lurking variables are controlled. Chapter 3 explains how to design convincing experiments.

Many of the sharpest disputes in which statistics plays a role involve questions of causation that cannot be settled by experiment. Does gun control reduce violent crime? Does living near power lines cause cancer? Has "outsourcing" work to overseas locations reduced overall employment in the United States? All of these questions have become public issues. All concern associations among variables. And all have this in common: they try to pinpoint cause and effect in a setting involving complex relations among many interacting variables. Common response and confounding, along with the number of potential lurking variables, make observed associations misleading. Experiments are not possible for ethical or practical reasons. We can't assign some people to live near power lines or compare the same nation with and without strong gun controls.

EXAMPLE

2.43 Power lines and leukemia. Electric currents generate magnetic fields. So living with electricity exposes people to magnetic fields. Living near power lines increases exposure to these fields. Really strong fields can disturb living cells in laboratory studies. Some people claim that the weaker fields we experience if we live near power lines cause leukemia in children.

It isn't ethical to do experiments that expose children to magnetic fields. It's hard to compare cancer rates among children who happen to live in more and less exposed locations because leukemia is rare and locations vary in many ways other than magnetic fields. We must rely on studies that compare children who have leukemia with children who don't.

A careful study of the effect of magnetic fields on children took five years and cost $5 million. The researchers compared 638 children who had leukemia and 620 who did not. They went into the homes and actually measured the magnetic fields in the children's bedrooms, in other rooms, and at the front door. They recorded facts about nearby power lines for the family home and also for the mother's residence when she was pregnant. Result: no evidence of more than a chance connection between magnetic fields and childhood leukemia.[38]

"No evidence" that magnetic fields are connected with childhood leukemia doesn't prove that there is no risk. It says only that a careful study could not find any risk that stands out from the play of chance that distributes leukemia cases across the landscape. Critics continue to argue that the study failed to measure some lurking variables, or that the children studied don't fairly represent all children. Nonetheless, a carefully designed study comparing children with and without leukemia is a great advance over haphazard and sometimes emotional counting of cancer cases.

EXAMPLE

2.44 Smoking and lung cancer. Despite the difficulties, it is sometimes possible to build a strong case for causation in the absence of experiments. The evidence that smoking causes lung cancer is about as strong as nonexperimental evidence can be.

Doctors had long observed that most lung cancer patients were smokers. Comparison of smokers and similar nonsmokers showed a very strong association between smoking and death from lung cancer. Could the association be due to common response? Might there be, for example, a genetic factor that predisposes people both to nicotine addiction and to lung cancer? Smoking and lung cancer would then be positively associated even if smoking had no direct effect on the lungs. Or perhaps confounding is to blame. It might be that smokers live unhealthy lives in other ways (diet, alcohol, lack of exercise) and that some other habit confounded with smoking is a cause of lung cancer. How were these objections overcome?

Let's answer this question in general terms: What are the criteria for establishing causation when we cannot do an experiment?

- *The association is strong.* The association between smoking and lung cancer is very strong.

- *The association is consistent.* Many studies of different kinds of people in many countries link smoking to lung cancer. That reduces the chance that a lurking variable specific to one group or one study explains the association.

- *Higher doses are associated with stronger responses.* People who smoke more cigarettes per day or who smoke over a longer period get lung cancer more often. People who stop smoking reduce their risk.

- *The alleged cause precedes the effect in time.* Lung cancer develops after years of smoking.

- *The alleged cause is plausible.* Experiments show that tars from cigarette smoke cause cancer when applied to the backs of mice.

Medical authorities do not hesitate to say that smoking causes lung cancer. The U.S. surgeon general states that cigarette smoking is "the largest avoidable cause of death and disability in the United States."[39] The evidence for causation is strong—but it is not as strong as the evidence provided by well-designed experiments.

SECTION 2.6 Summary

Some observed associations between two variables are due to a **cause-and-effect** relationship between these variables, but others are explained by **lurking variables.**

The effect of lurking variables can operate through **common response** if changes in both the explanatory and response variables are caused by changes in lurking variables. **Confounding** of two variables (either explanatory or lurking variables) means that we cannot distinguish their effects on the response variable.

That an association is due to causation is best established by an **experiment** that changes the explanatory variable while controlling other influences on the response.

In the absence of experimental evidence, be cautious in accepting claims of causation. Good evidence of causation requires a strong association that appears consistently in many studies, a clear explanation for the alleged causal link, and careful examination of possible lurking variables.

SECTION 2.6 Exercises

2.133 Iron and anemia. A lack of adequate iron in the diet is associated with anemia, a condition where the body does not have enough red blood cells. However, anemia is also associated with malaria and infections with worms called helminths. Discuss these observed associations using the framework of Figure 2.29.

2.134 Stress and lack of sleep in college students. Studies of college students have shown that stress and lack of sleep are associated. Do you think that lack of sleep causes stress or that stress causes a lack of sleep? Write a short paragraph summarizing your opinions.

2.135 Online courses. Many colleges offer online versions of some courses that are also taught in the classroom. It often happens that the students who enroll in the online version do better than the classroom students on the course exams. This does not show that online instruction is more effective than classroom teaching, because the kind of people who sign up for online courses are often quite different from the classroom students. Suggest some student characteristics that you think could be confounded with online versus classroom. Use a diagram like Figure 2.29(c) to illustrate your ideas.

2.136 Marriage and income. Data show that men who are married, and also divorced or widowed men, earn quite a bit more than men who have never been married. This does not mean that a man can raise his income by getting married. Suggest several lurking variables that you think are confounded with marital status and that help explain the association between marital status and income. Use a diagram like Figure 2.29(c) to illustrate your ideas.

2.137 Exercise and self-confidence. A college fitness center offers an exercise program for staff members who choose to participate. The program assesses each participant's fitness, using a treadmill test, and also administers a personality questionnaire. There is a moderately strong positive correlation between fitness score and score for self-confidence. Is this good evidence that improving fitness increases self-confidence? Explain why or why not.

2.138 Music and academic performance. The Kalamazoo (Michigan) Symphony once advertised a "Mozart for Minors" program with this statement: "Question: Which students scored 51 points higher in verbal skills and 39 points higher in math? Answer: Students who had experience in music."[40] In fact, good academic performance and early exposure to classical music are in part common responses to lurking variables.

What background information about students could explain the association? Use a diagram like Figure 2.29(b) to show the situation.

2.139 Coaching for the SAT. A study finds that high school students who take the SAT, enroll in an SAT coaching course, and then take the SAT a second time raise their SAT Mathematics scores from a mean of 521 to a mean of 561.[41] What factors other than "taking the course causes higher scores" might explain this improvement?

2.140 Computer chip manufacturing and miscarriages. A study showed that women who work in the production of computer chips have abnormally high numbers of miscarriages. The union claimed that exposure to chemicals used in production caused the miscarriages. Another possible explanation is that these workers spend most of their work time standing up. Illustrate these relationships in a diagram like those in Figure 2.29.

2.141 Hospital size and length of stay. A study shows that there is a positive correlation between the size of a hospital (measured by its number of beds x) and the median number of days y that patients remain in the hospital. Does this mean that you can shorten a hospital stay by choosing a small hospital? Use a diagram like one of those in Figure 2.29 to explain the association.

2.142 Watching TV and low grades. Children who watch many hours of television get lower grades in school on the average than those who watch less TV. Explain clearly why this fact does not show that watching TV *causes* poor grades. In particular, suggest some other variables that may be confounded with heavy TV viewing and may contribute to poor grades.

2.143 Artificial sweeteners. People who use artificial sweeteners in place of sugar tend to be heavier than people who use sugar. Does this mean that artificial sweeteners cause weight gain? Give a more plausible explanation for this association.

2.144 Exercise and mortality. A sign in a fitness center says, "Mortality is halved for men over 65 who walk at least 2 miles a day."

(a) Mortality is eventually 100% for everyone. What do you think "mortality is halved" means?

(b) Assuming that the claim is true, explain why this fact does not show that exercise *causes* lower mortality.

2.145 Effect of a math skills refresher initiative.
Students enrolling in an elementary statistics course take a pretest that assesses their math skills. Those who receive low scores are given the opportunity to take three one-hour refresher sessions designed to review the basic math skills needed for the statistics course. Those who took the refresher sessions performed worse than those who did not on the final exam in the statistics course. Can you conclude that the refresher course has a negative impact on performance in the statistics course? Explain your answer.

CHAPTER 2 Exercises

2.146 Popularity of a first name. The Social Security Administration maintains lists of the top 1000 names for boys and girls born each year since 1879.[42] The name "Atticus" made the list in five recent years. Here are the ranks for those years: 🔵 ATTICUS

Year	2004	2005	2006	2007	2008	2009
Rank	935	792	767	682	689	609

(a) Plot rank versus year.

(b) Find the equation of the least-squares regression line and add it to your plot.

(c) Do these data suggest that this "Atticus" is becoming more popular, less popular, or staying the same in popularity over this period of time? Give reasons for your answer.

2.147 You select the name. Refer to the previous exercise. Choose a first name and find the rank of this name for the past several years from the Social Security Web site, ssa.gov/OACT/babynames Answer the questions from the previous exercise for this name.

2.148 Salaries and raises. For this exercise we consider a hypothetical employee who starts working in year 1 with a salary of $50,000. Each year her salary increases by approximately 5%. By year 20, she is earning $226,000. The following table gives her salary for each year (in thousands of dollars): 🔵 RAISES

Year	Salary	Year	Salary	Year	Salary	Year	Salary
1	50	6	63	11	81	16	104
2	53	7	67	12	85	17	109
3	56	8	70	13	90	18	114
4	58	9	74	14	93	19	120
5	61	10	78	15	99	20	126

(a) Figure 2.30 is a scatterplot of salary versus year with the least-squares regression line. Describe the relationship between salary and year for this person.

(b) The value of r^2 for these data is 0.9832. What percent of the variation in salary is explained by year? Would you say that this is an indication of a strong linear relationship? Explain your answer.

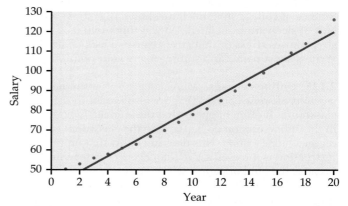

FIGURE 2.30 Plot of salary versus year for an individual who receives approximately a 5% raise each year for 20 years with the least-squares regression line, for Exercise 2.148.

2.149 Look at the residuals. Refer to the previous exercise. Figure 2.31 is a plot of the residuals versus year. 🔵 RAISES

(a) Interpret the residual plot.

(b) Explain how this plot highlights the deviations from the least-squares regression line that you can see in Figure 2.30.

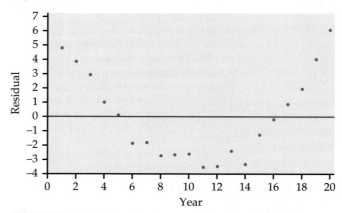

FIGURE 2.31 Plot of residuals versus year for an individual who receives approximately a 5% raise each year for 20 years, for Exercise 2.149.

2.150 Try logs. Refer to the previous two exercises. Figure 2.32 is a scatterplot with the least-squares regression line for log salary versus year. For this model, $r^2 = 0.9995$. ⬤ RAISES

(a) Compare this plot with Figure 2.30. Write a short summary of the similarities and the differences.

(b) Figure 2.33 is a plot of the residuals for the model using year to predict log salary. Compare this plot with Figure 2.31 and summarize your findings.

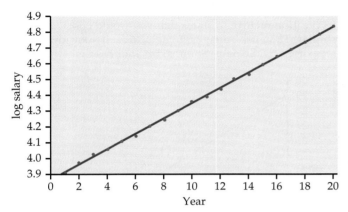

FIGURE 2.32 Plot of log salary versus year for an individual who receives approximately a five percent raise each year for twenty years with the least-squares regression line, for Exercise 2.150.

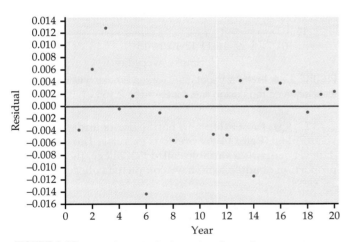

FIGURE 2.33 Plot of residuals, based on log salary, versus year for an individual who receives approximately a five percent raise each year for twenty years, for Exercise 2.150.

2.151 Do some predictions. The individual whose salary we have been studying wants to do some financial planning. Specifically, she would like to predict her salary 5 years into the future, that is for year 25. She is willing to assume that her employment situation will be stable for the next 5 years and that it will be similar to the last 20 years. ⬤ RAISES

(a) Predict her salary for year 25 using the least-squares regression equation constructed to predict salary from year.

(b) Predict her salary for year 25 using the least-squares regression equation constructed to predict log salary from year. Note that you will need to take the predicted log salary and convert this value back to the predicted salary. Many calculators have a function that will perform this operation.

(c) Which prediction do you prefer? Explain your answer.

(d) Someone looking at the numerical summaries and not the plots for these analyses, says that because both models have very high values of r^2, they should perform equally well in doing this prediction. Write a response to this comment.

(e) Write a short summary about the value of graphical summaries and the problems of extrapolation using what you have learned in studying this salary data.

2.152 Faculty salaries. Here are the salaries for a sample of professors in a mathematics department at a large Midwestern university for the academic years 2007–2008 and 2008–2009.
⬤ FACULTYSALARIES

2007–2008 salary ($)	2008–2009 salary ($)	2007–2008 salary ($)	2008–2009 salary ($)
141,800	142,900	133,650	136,350
109,800	113,600	129,160	132,485
106,000	110,500	71,972	76,072
95,700	99,800	72,000	76,000
109,000	111,180	79,500	82,700
108,790	111,240	138,850	141,830
100,500	105,100	119,506	122,906
146,000	150,080	112,100	115,200

(a) Construct a scatterplot with the 2008–2009 salaries on the vertical axis and the 2007–2008 salaries on the horizontal axis.

(b) Comment on the form, direction, and strength of the relationship in your scatterplot.

(c) What proportion of the variation in 2008–2009 salaries is explained by 2007–2008 salaries?

2.153 Find the line and examine the residuals. Refer to the previous exercise. ⬤ FACULTYSALARIES

(a) Find the least squares regression line for prediction 2008–2009 salaries from 2007–2008 salaries.

(b) Analyze the residuals paying attention to any outliers or influential observations. Write a summary of your findings.

2.154 Bigger raises for those earning less. Refer to the previous two exercises. The 2007–2008 salaries do an excellent job of predicting the 2008–2009 salaries. Is there anything more that we can learn from these data? In this department there is a tradition of giving higher than average percent raises to those whose salaries are lower. Let's see if we can find evidence to support this idea in the data. 🖟 FACULTYSALARIES

(a) Compute the percentage raise for each faculty member. Take the difference between the 2008–2009 salary and the 2007–2008 salary, dividing by the 2007–2008 salary, and then multiply by 100. Make a scatterplot with raise as the response variable and the 2007–2008 salary as the explanatory variable. Describe the relationship that you see in your plot.

(b) Find the least-squares regression line and add it to your plot.

(c) Analyze the residuals. Are there any outliers or influential cases? Make a graphical display and include this in a short summary of what you conclude.

(d) Is there evidence in the data to support the idea that greater percentage raises are given to those with lower salaries? Write a summary of your answer and include numerical and graphical summaries to support your conclusion.

2.155 Graduation rates. One of the factors used to evaluate undergraduate programs is the proportion of incoming students who graduate. This quantity, called the graduation rate, can be predicted by other variables such as the SAT or ACT scores and the high school record of the incoming students. One of the components that *U.S. News & World Report* uses when evaluating colleges is the difference between the actual graduation rate and the rate predicted by a regression equation.[43] In this chapter, we call this quantity the residual. Explain why the residual is a better measure to evaluate college graduation rates than the raw graduation rate.

2.156 🔺 **Eating fruits and vegetables and smoking.** The Centers for Disease Prevention and Control (CDC) Behavior Risk Factor Surveillance System (BRFSS) collects data related to health conditions and risk behaviors.[44] Aggregated data by state are in the BRFSS data set described in the Data Appendix. Figure 2.34 is a plot of two of the BRFSS variables. Fruits and Vegetables is the percent of adults in the state who report eating at least five servings of fruits and vegetables per day; Smoking is the percent who smoke every day. 🖟 BRFSS

(a) Describe the relationship between Fruits and Vegetables and Smoking. Explain why you might expect this type of association.

(b) Find the correlation between the two variables.

(c) For Utah, 22.8% eat at least five servings of fruits and vegetables per day and 6.8% smoke every day. Find Utah on the plot and describe its position relative to the other states.

(d) For California, the percents are 28.9% for Fruits and vegetables and 8.9% for Smoking. Find California on the plot and describe its position relative to the other states.

(e) Pick your favorite state and write a short summary of its position relative to states that you would consider to be similar. Then use Table 2.6 to determine if your idea is supported by the data. Summarize your results.

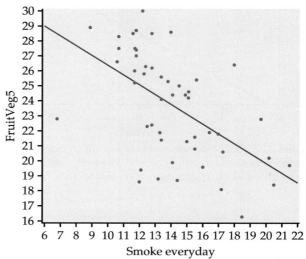

FIGURE 2.34 Fruits & Vegetables versus Smoking with least-squares regression line, for Exercise 2.156.

2.157 🔺 **Education and eating fruits and vegetables.** Refer to the previous exercise. The RBFSS data set contains a variable called EdCollege, the proportion of adults who have completed college. 🖟 BRFSS

(a) Plot the data with Fruits and Vegetables on the *x* axis and EdCollege on the *y* axis. Describe the overall pattern of the data.

(b) Add the least squares regression line to your plot. Does the line give a summary of the overall pattern? Explain your answer.

(c) Pick out a few states and use their position in the graph to write a short summary of how they compare with other states.

(d) Can you conclude that earning a college degree will cause you to eat five servings of fruits and vegetables per day? Explain your answer.

TABLE 2.6

Fruit and vegetable consumption and smoking

State	Fruits & Vegetables (percent)	Smoking (percent)	State	Fruits & Vegetables (percent)	Smoking (percent)
Alabama	20.6	17.3	Montana	25.3	13.8
Alaska	24.2	15.1	Nebraska	24.1	13.4
Arizona	28.3	10.7	Nevada	21.9	16.4
Arkansas	21.8	17.0	New Hampshire	28.5	12.8
California	28.9	8.9	New Jersey	27.5	10.7
Colorado	25.8	12.3	New Mexico	22.4	12.8
Connecticut	28.5	11.6	New York	27.4	11.8
Delaware	21.4	13.4	North Carolina	21.6	15.5
Florida	26.2	12.8	North Dakota	21.9	13.3
Georgia	25.0	14.5	Ohio	20.8	15.5
Hawaii	28.7	11.8	Oklahoma	16.3	18.5
Idaho	22.3	12.5	Oregon	27.0	11.8
Illinois	24.6	15.1	Pennsylvania	25.4	15.6
Indiana	22.8	19.7	Rhode Island	25.6	13.4
Iowa	19.9	14.1	South Carolina	18.7	14.4
Kansas	18.8	13.2	South Dakota	18.6	12.0
Kentucky	18.4	20.5	Tennessee	26.4	18.0
Louisiana	19.6	16.0	Texas	25.2	11.7
Maine	28.6	14.0	Utah	22.8	6.8
Maryland	26.6	10.6	Vermont	30.0	12.2
Massachusetts	27.5	11.7	Virginia	26.3	12.4
Michigan	21.3	15.0	Washington	26.0	11.7
Minnesota	19.4	12.1	West Virginia	19.7	21.5
Mississippi	18.1	17.2	Wisconsin	24.4	14.1
Missouri	20.2	20.2	Wyoming	24.4	14.9

2.158 Predicting text pages. The editor of a statistics text would like to plan for the next edition. A key variable is the number of pages that will be in the final version. Text files are prepared by the authors using a word processor called LaTeX, and separate files contain figures and tables. For the previous edition of the text, the number of pages in the LaTeX files can easily be determined, as well as the number of pages in the final version of the text. Here are the data: 🌀 TEXTPAGES

Chapter	1	2	3	4	5	6	7	8	9	10	11	12	13
LaTeX pages	77	73	59	80	45	66	81	45	47	43	31	46	26
Text pages	99	89	61	82	47	68	87	45	53	50	36	52	19

(a) Plot the data and describe the overall pattern.

(b) Find the equation of the least-squares regression line and add the line to your plot.

(c) Find the predicted number of pages for the next edition if the number of LaTeX pages is 62.

(d) Write a short report for the editor explaining to her how you constructed the regression equation and how she could use it to estimate the number of pages in the next edition of the text.

2.159 🛡️ **Points scored in women's basketball games.** Use the Internet to find the scores for the past season's women's basketball team at a college of your choice. Is there a relationship between the points scored by your chosen team and the points scored by the opponent? Summarize the data and write a report on your findings.

2.160 🛡️ **Look at the data for men.** Refer to the previous exercise. Analyze the data for the men's team from the same college and compare your results with those for the women.

2.161 Endangered animals and habitat. Endangered animal species often live in isolated patches of habitat. If the population size in a patch varies a lot (due to weather, for example), the species is more likely to disappear from that patch in a bad year. Here is a general question: Is there less variation in population size when a patch of habitat has more diverse vegetation? If so, maintaining habitat diversity can help protect endangered species.

A researcher measured the variation over time in the population of a cricket species in 45 habitat patches. He also measured the diversity of each patch.[45] He reported his results by giving the least-squares equation

$$\text{population variation} = 84.4 - 0.13 \times \text{diversity}$$

along with the fact that $r^2 = 0.34$. Do these results support the idea that more diversity goes with less variation in population size? Is the relationship very strong or only moderately strong?

2.162 ⊕ **Monkey calls.** The usual way to study the brain's response to sounds is to have subjects listen to "pure tones." The response to recognizable sounds may differ. To compare responses, researchers anesthetized macaque monkeys. They fed pure tones and also monkey calls directly to their brains by inserting electrodes. Response to the stimulus was measured by the firing rate (electrical spikes per second) of neurons in various areas of the brain. Table 2.7 contains the responses for 37 neurons.[46] ⊛ MONKEYCALLS

(a) One notable finding is that responses to monkey calls are generally stronger than responses to pure tones. Give a numerical measure that supports this finding.

(b) Make a scatterplot of monkey call response against pure tone response (explanatory variable). Find the least-squares line and add it to your plot. Mark on your plot the point with the largest residual (positive or negative) and also a point that is an outlier in the x direction.

(c) How influential are each of these points for the correlation r?

(d) How influential are each of these points for the regression line?

2.163 ⊕ **Plywood strength.** How strong is a building material such as plywood? To be specific, support a 24-inch by 2-inch strip of plywood at both ends and apply force in the middle until the strip breaks. The modulus of rupture (MOR) is the force needed to break the strip. We would like to be able to predict MOR without actually breaking the wood. The modulus of elasticity (MOE) is found by bending the wood without breaking it. Both MOE and MOR are measured in pounds per square inch. Here are data for 32 specimens of the same type of plywood:[47] ⊛ MOEMOR

MOE	MOR	MOE	MOR
2,005,400	11,591	2,181,910	12,702
1,166,360	8,542	1,559,700	11,209
1,842,180	12,750	2,372,660	12,799
2,088,370	14,512	1,580,930	12,062
1,615,070	9,244	1,879,900	11,357
1,938,440	11,904	1,594,750	8,889
2,047,700	11,208	1,558,770	11,565
2,037,520	12,004	2,212,310	15,317
1,774,850	10,541	1,747,010	11,794
1,457,020	10,314	1,791,150	11,413
1,959,590	11,983	2,535,170	13,920
1,720,930	10,232	1,355,720	9,286
1,355,960	8,395	1,646,010	8,814
1,411,210	10,654	1,472,310	6,326
1,842,630	10,223	1,488,440	9,214
1,984,690	13,499	2,349,090	13,645

Can we use MOE to predict MOR accurately? Use the data to write a discussion of this question.

TABLE 2.7

Neuron response to tones and monkey calls

Tone	Call	Tone	Call	Tone	Call	Tone	Call
474	500	145	42	71	134	35	103
256	138	141	241	68	65	31	70
241	485	129	194	59	182	28	192
226	338	113	123	59	97	26	203
185	194	112	182	57	318	26	135
174	159	102	141	56	201	21	129
176	341	100	118	47	279	20	193
168	85	74	62	46	62	20	54
161	303	72	112	41	84	19	66
150	208						

2.164 Distribution of the residuals. Some statistical methods require that the residuals from a regression line have a Normal distribution. The residuals for the nonexercise activity example, are given in Exercise 2.93 (page 124). Is their distribution close to Normal? Make a Normal quantile plot to find out.

2.165 Asian culture and thinness. Asian culture does not emphasize thinness, but young Asians are often influenced by Western culture. In a study of concerns about weight among young Korean women, researchers administered the Drive for Thinness scale (a questionnaire) to 264 female college students in Seoul, South Korea. This scale measures excessive concern with weight and dieting and fear of weight gain. In Exercise 1.37 (page 25), you examined the distribution of Drive for Thinness scores among these college students. The study looked at several explanatory variables. One was "Body Dissatisfaction," also measured by a questionnaire. ⬤ THINNESS

(a) Make a scatterplot of Drive for Thinness (response) against Body Dissatisfaction. The appearance of the plot is a result of the fact that both variables take only whole-number values. Such variables are common in the social and behavioral sciences.

(b) Add the least-squares line to your plot. The line shows a linear relationship. How strong is this relationship? Body Dissatisfaction was more strongly correlated with Drive for Thinness than any of the other explanatory variables examined. Rather weak relationships are common in social and behavioral sciences, because individuals vary a great deal. Using several explanatory variables together improves prediction of the response. This is *multiple regression*, discussed in Chapter 11.

2.166 Solar heating panels and gas consumption. To study the energy savings due to adding solar heating panels to a house, researchers measured the natural-gas consumption of the house for more than a year and then installed solar panels and observed the natural-gas consumption for almost two years. The explanatory variable x is degree-days per day during the several weeks covered by each observation, and the response variable y is gas consumption (in hundreds of cubic feet) per day during the same period. Figure 2.35 plots y against x, with separate symbols for observations taken before and after the installation of the solar panels.[48] The least-squares regression lines were computed separately for the before and after data and are drawn on the plot. The regression lines are

$$\text{Before:} \quad \hat{y} = 1.089 + 0.189x$$
$$\text{After:} \quad \hat{y} = 0.853 + 0.157x$$

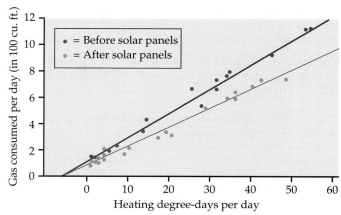

FIGURE 2.35 The regression of residential natural-gas consumption on heating degree-days before and after installation of solar heating panels, for Exercise 2.166.

(a) Does the scatterplot suggest that a straight line is an appropriate description of the relationship between degree-days and natural-gas consumption? Do any individual observations appear to have large residuals or to be highly influential?

(b) About how much additional natural gas was consumed per day for each additional degree-day before the panels were added? After the panels were added?

(c) The daily average temperature during January in this location is about 30°, which corresponds to 35 degree- days per day. Use the regression lines to predict daily gas usage for a day with 35 degree-days before and after installation of the panels.

(d) The Energy Information Agency says that natural gas cost consumers about $1.20 per 100 cubic feet in the fall of 2003. At this rate, how much money do the solar panels save in the 31 days of January?

2.167 ⬤ Marine bacteria and X-rays. Expose marine bacteria to X-rays for time periods from 1 to 15 minutes. Here are the number of surviving bacteria (in hundreds) on a culture plate after each exposure time:[49] ⬤ MARINEBACTERIA

Time	Count	Time	Count	Time	Count
1	355	6	106	11	36
2	211	7	104	12	32
3	197	8	60	13	21
4	166	9	56	14	19
5	142	10	38	15	15

Previous experience suggests that the relationship between time and the *logarithm* of the count of surviving bacteria is linear.

(a) Find the regression line of count on time and make plots of the data and the residuals. In what way is this relationship nonlinear?

(b) Repeat your work for the regression of the logarithm of count on time. Make a numerical comparison of the linearity of the two relationships.

2.168 ⚔ **Midterm-exam scores and final-exam scores.** We expect that students who do well on the midterm exam in a course will usually also do well on the final exam. Gary Smith of Pomona College looked at the exam scores of all 346 students who took his statistics class over a 10-year period.[50] The least-squares line for predicting final-exam score from midterm-exam score was $\hat{y} = 46.6 + 0.41x$.

Octavio scores 10 points above the class mean on the midterm. How many points above the class mean do you predict that he will score on the final? (*Hint:* What is the predicted final-exam score for the class mean midterm score $\bar{x}$?) This is an example of **regression to the mean,** the phenomenon that gave "regression" its name: students who do well on the midterm will on the average do less well on the final, but still above the class mean.)

2.169 Firefighters and fire damage. Someone says, "There is a strong positive correlation between the number of firefighters at a fire and the amount of damage the fire does. So sending lots of firefighters just causes more damage." Explain why this reasoning is wrong.

2.170 University degrees in Asia. Asia has become a major competitor of the United States and Western Europe in education as well as economics. Here are counts of first university degrees in science and engineering in the three regions:[51] 🌐 UNIVDEGREES

	Region		
	United	Western	
Field	States	Europe	Asia
Engineering	61,941	158,931	280,772
Natural science	111,158	140,126	242,879
Social science	182,166	116,353	236,018

Direct comparison of counts of degrees would require us to take into account Asia's much larger population. We can, however, compare the distribution of degrees by field of study in the three regions. Do this using calculations and graphs, and write a brief summary of your findings.

2.171 ⚔ **Motivation to participate in volunteer service.** A study examined patterns and characteristics of volunteer service for young people from high school through early adulthood.[52] Here are some data that can be used to compare males and females on participation in

unpaid volunteer service or community service and motivation for participation:

	Participants			
	Motivation			
	Strictly	Court-		
Gender	voluntary	ordered	Other	Nonparticipants
Men	31.9%	2.1%	6.3%	59.7%
Women	43.7%	1.1%	6.5%	48.7%

Note that the percents in each row sum to 100%. Graphically compare the volunteer service profiles for men and women. Describe any differences that are striking.

2.172 ⚔ **Look at volunteers only.** Refer to the previous exercise. Recompute the table for volunteers only. To do this, take the entries for each motivation and divide by the percent of volunteers. Do this separately for each gender. Verify that the percents sum to 100% for each gender. Give a graphical summary to compare the motivation of men and women who are volunteers. Compare this with your summary in the previous exercise, and write a short paragraph describing similarities and differences in these two views of the data.

2.173 An example of Simpson's paradox. Mountain View University has professional schools in business and law. Here is a three-way table of applicants to these professional schools, categorized by gender, school, and admission decision:[53]

	Business			Law		
	Admit			Admit	
Gender	Yes	No	Gender	Yes	No
Male	400	200	Male	90	110
Female	200	100	Female	200	200

(a) Make a two-way table of gender by admission decision for the combined professional schools by summing entries in the three-way table.

(b) From your two-way table, compute separately the percents of male and female applicants admitted. Male applicants are admitted to Mountain View's professional schools at a higher rate than female applicants.

(c) Now compute separately the percents of male and female applicants admitted by the business school and by the law school.

(d) Explain carefully, as if speaking to a skeptical reporter, how it can happen that Mountain View appears to favor males when this is not true within each of the professional schools.

2.174 Construct an example with four schools. Refer to the previous exercise. Make up a similar table for a hypothetical university having four different schools that illustrates the same point. Carefully summarize your table with the appropriate percents.

2.175 Class size and class level. A university classifies its classes as either "small" (fewer than 40 students) or "large." A dean sees that 62% of Department A's classes are small, while Department B has only 40% small classes. She wonders if she should cut Department A's budget and insist on larger classes. Department A responds to the dean by pointing out that classes for third- and fourth-year students tend to be smaller than classes for first- and second-year students. The following three-way table gives the counts of classes by department, size, and student audience. Write a short report for the dean that summarizes these data. Start by computing the percents of small classes in the two departments and include other numerical and graphical comparisons as needed. Here are the numbers of classes to be analyzed: CLASSSIZE

	Department A			Department B		
Year	Large	Small	Total	Large	Small	Total
First	2	0	2	18	2	20
Second	9	1	10	40	10	50
Third	5	15	20	4	16	20
Fourth	4	16	20	2	14	16

2.176 Identity theft. A study of identity theft looked at how well consumers protect themselves from this increasingly prevalent crime. The behaviors of 61 college students were compared with the behaviors of 59 nonstudents.[54] One of the questions was, "When asked to create a password, I have used either my mother's maiden name, or my pet's name, or my birth date, or the last four digits of my social security number, or a series of consecutive numbers." For the students, 22 agreed with this statement while 30 of the nonstudents agreed.

(a) Display the data in a two-way table and analyze the data. Write a short summary of your results.

(b) The students in this study were junior and senior college students from two sections of a course in Internet marketing at a large northeastern university. The nonstudents were a group of individuals who were recruited to attend commercial focus groups on the West Coast conducted by a lifestyle marketing organization. Discuss how the method of selecting the subjects in this study relates to the conclusions that can be drawn from it.

2.177 Athletes and gambling. A survey of student athletes that asked questions about gambling behavior classified students according to the National Collegiate Athletic Association (NCAA) division.[55] For male student athletes, the percents who reported wagering on collegiate sports are given here along with the numbers of respondents in each division:

Division	I	II	III
Percent	17.2%	21.0%	24.4%
Number	5619	2957	4089

(a) Analyze the data. Give details and a short summary of your conclusion.

(b) The percents in the preceding table are given in the NCAA report, but the numbers of male student athletes in each division who responded to the survey question are estimated based on other information in the report. To what extent do you think this has an effect on the results?

(c) Some student athletes may be reluctant to provide this kind of information, even in a survey where there is no possibility that they can be identified. Discuss how this fact may affect your conclusions.

2.178 Health conditions and risk behaviors. The data set BRFSS described in the Data Appendix gives several variables related to health conditions and risk behaviors as well as demographic information for the 50 states and the District of Columbia. Pick at least three pairs of variables to analyze. Write a short report on your findings. BRFSS

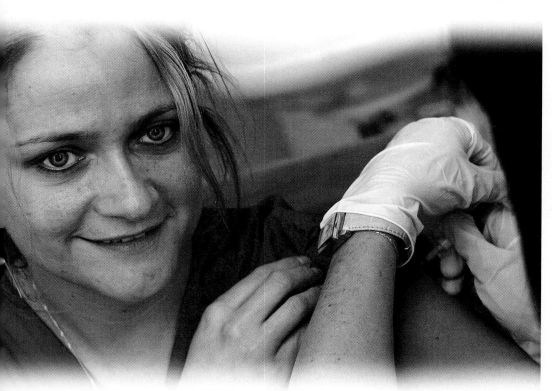

CHAPTER 3

3.1 Design of Experiments

3.2 Sampling Design

3.3 Toward Statistical Inference

3.4 Ethics

Producing Data

Introduction

In Chapters 1 and 2 we learned some basic tools of *data analysis*. We used graphs and numbers to describe data. When we do **exploratory data analysis,** we rely heavily on plotting the data. We look for patterns that suggest interesting conclusions or questions for further study. However, *exploratory analysis alone can rarely provide convincing evidence for its conclusions, because striking patterns we find in data can arise from many sources.*

exploratory data analysis

Anecdotal data

It is tempting to simply draw conclusions from our own experience, making no use of more broadly representative data. A magazine article about Pilates says that men need this form of exercise even more than women. The article describes the benefits that two men received from taking Pilates classes. A newspaper ad states that a particular brand of windows are "considered to be the best" and says that "now is the best time to replace your windows and doors." These types of stories, or *anecdotes*, sometimes provide quantitative data. However, this type of data does not give us a sound basis for drawing conclusions.

ANECDOTAL EVIDENCE

Anecdotal evidence is based on haphazardly selected individual cases, which often come to our attention because they are striking in some way. These cases need not be representative of any larger group of cases.

USE YOUR KNOWLEDGE

3.1 Wizards of Waverly Place. Your friends are big fans of "Waverly Place," a Disney Channel show about a family with three children who are training to be wizards. To what extent do you think you can generalize your preferences for this show to all students at your college?

3.2 Describe an anecdote. Find an example from some recent experience where anecdotal evidence is used to draw a conclusion that is not justified. Describe the example and explain why it cannot be used in this way.

3.3 Preference for Powerade. Ashley is a hard-core runner. She and all her friends prefer Powerade Ion$^{4®}$ to Gatorade$^®$. Explain why Ashley's experience is not good evidence that most young people prefer Powerade to Gatorade.

3.4 Reliability of a product. A friend has driven a Toyota Camry for more than 200,000 miles and with only the usual service maintenance expenses. Explain why not all Camry owners can expect this kind of performance.

Available data

available data   Occasionally, data are collected for a particular purpose but can also serve as the basis for drawing sound conclusions about other research questions. We use the term **available data** for this type of data.

AVAILABLE DATA

Available data are data that were produced in the past for some other purpose but that may help answer a present question.

The library and the Internet can be good sources of available data. Because producing new data is expensive, we all use available data whenever possible. However, the clearest answers to present questions often require that data be produced to answer those specific questions. Here are two examples:

> ### EXAMPLE
>
> **3.1 International manufacturing productivity.** If you visit the U.S. Bureau of Labor Statistics Web site, `bls.gov`, you can find many interesting sets of data and statistical summaries. One recent study compared the economies of 17 countries. The study showed that from 2007 to 2008, the Republic of Korea and the United States each had the highest productivity increases (tied at 1.2%) while Singapore had the largest decline (-6.6%).
>
> **3.2 Math skills of children.** At the Web site of the National Center for Education Statistics, `nces.ed.gov`, you will find full details about the math skills of schoolchildren in the latest National Assessment of Educational Progress (Figure 3.1). Mathematics scores have slowly but steadily increased since 1990. All racial/ethnic groups, both men and women, and students in most states are getting better in math.

Many nations have a single national statistical office, such as Statistics Canada (`statcan.gc.ca`) or Mexico's INEGI (`inegi.org.mx`). More than 70 different U.S. agencies collect data. You can reach most of them through the government's FedStats site (`fedstats.gov`).

> ### USE YOUR KNOWLEDGE
>
> **3.5 A popular product.** A Web site claims that the Ikea Billy is a bookcase that is very popular among college students. What kind of information would you like to know about the background of the statement before you would consider it to be reliable?

A survey of college athletes is designed to estimate the percent who gamble. Do restaurant patrons give higher tips when their server repeats their order carefully? The validity of our conclusions from the analysis of data collected to address these issues rests on a foundation of carefully collected data. In this chapter, we will develop the skills needed to produce trustworthy data and to judge the quality of data produced by others. The techniques for producing data we will study require no formulas, but they are among the most important ideas in statistics. Statistical designs for producing data rely on either *sampling* or *experiments*.

Sample surveys and experiments

How have the attitudes of Americans, on issues ranging from abortion to work, changed over time? **Sample surveys** are the usual tool for answering questions like these.

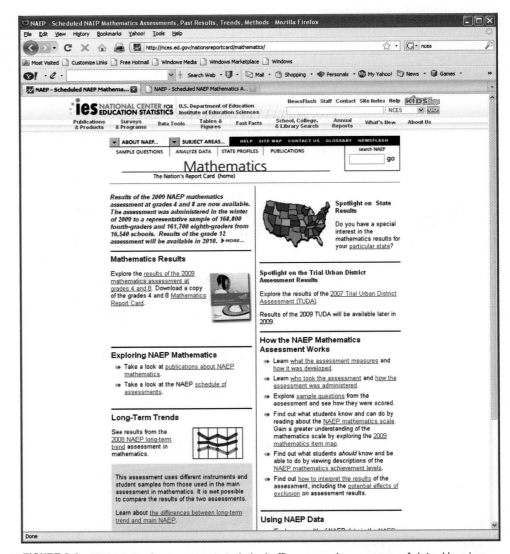

FIGURE 3.1   Web sites of government statistical offices are prime sources of data. Here is a page about mathematics scores of children in grades 4 and 8 from the National Center of Education Statistics Web pages.

EXAMPLE

3.3   The General Social Survey.   One of the most important sample surveys is the General Social Survey (GSS) conducted by the NORC, a national organization for research and computing affiliated with the University of Chicago.[1] The GSS interviews about 3000 adult residents of the United States every second year.

sample
population

The GSS selects a **sample** of adults to represent the larger **population** of all English-speaking adults living in the United States. The idea of *sampling* is to study a part in order to gain information about the whole. Data are often produced by sampling a population of people or things. Opinion polls, for example, report the views of the entire country based on interviews with a sample of about 1000 people. Government reports on employment and unemployment are produced from a monthly sample of about 60,000 households. The quality of manufactured items is monitored by inspecting small samples each hour or each shift.

USE YOUR KNOWLEDGE

> **3.6 Find a sample survey.** Use the Internet or some printed material to find an example of a sample survey that interests you. Describe the population, how the sample was collected, and some of the conclusions.

census   In all of our examples, the expense of examining every item in the population makes sampling a practical necessity. Timeliness is another reason for preferring a sample to a **census,** which is an attempt to contact every individual in the entire population. We want information on current unemployment and public opinion next week, not next year. Moreover, a carefully conducted sample is often more accurate than a census. Accountants, for example, sample a firm's inventory to verify the accuracy of the records. Attempting to count every last item in the warehouse would be not only expensive, but also inaccurate. Bored people do not count carefully.

If conclusions based on a sample are to be valid for the entire population, a sound design for selecting the sample is required. Sampling designs are the topic of Section 3.2.

A sample survey collects information about a population by selecting and measuring a sample from the population. The goal is a picture of the population, disturbed as little as possible by the act of gathering information. Sample surveys are one kind of *observational study.*

OBSERVATION VERSUS EXPERIMENT

In an **observational study** we observe individuals and measure variables of interest but do not attempt to influence the responses.

In an **experiment** we deliberately impose some treatment on individuals and we observe their responses.

USE YOUR KNOWLEDGE

> **3.7 H1N1 vaccines.** A report issued by the Centers for Disease Control and Prevention stated that among 120 adults who received an injection of a monovalent H1N1 influenza A vaccine, 116, or 97%, had an effective response by three weeks after the vaccination. They also reported that the rates of adverse events such as headaches were not significantly different from a control group.[2] Is this an observational study or an experiment? Why? What are the explanatory and response variables?

3.8 **Violent acts on prime-time TV.** A typical hour of prime-time television shows three to five violent acts. Linking family interviews and police records shows a clear association between time spent watching TV as a child and later aggressive behavior.[3]

(a) Explain why this is an observational study rather than an experiment. What are the explanatory and response variables?

(b) Suggest several lurking variables describing a child's home life that may be confounded with how much TV he or she watches. Explain why confounding makes it difficult to conclude that more TV *causes* more aggressive behavior.

intervention

An observational study, even one based on a statistical sample, is a poor way to determine what will happen if we change something. The best way to see the effects of a change is to do an **intervention**—where we actually impose the change. When our goal is to understand cause and effect, experiments are the only source of fully convincing data.

> ### EXAMPLE
>
> **3.4 Child care and and behavior.** A study of child care enrolled 1364 infants in 1991 and planned to follow them through their sixth year in school. Twelve years later, the researchers published an article finding that "the more time children spent in child care from birth to age four-and-a-half, the more adults tended to rate them, both at age four-and-a-half and at kindergarten, as less likely to get along with others, as more assertive, as disobedient, and as aggressive."[4]

What can we conclude from this study? If parents choose to use child care, are they more likely to see these undesirable behaviors in their children?

> ### EXAMPLE
>
> **3.5 Is there a cause and effect relationship?** Example 3.4 describes an observational study. Parents made all child care decisions and the study did not attempt to influence them. A summary of the study stated, "The study authors noted that their study was not designed to prove a cause and effect relationship. That is, the study cannot prove whether spending more time in child care causes children to have more problem behaviors."[5] Perhaps employed parents who use child care are under stress and the children react to their parents' stress. Perhaps single parents are more likely to use child care. Perhaps parents are more likely to place in child care children who already have behavior problems.

We can imagine an experiment that would remove these difficulties. From a large group of young children, choose some to be placed in child care and others to remain at home. This is an experiment because the treatment (child care or not) is imposed on the children. Of course, this particular experiment is neither practical nor ethical.

confounded

In Examples 3.4 and 3.5, we say that the effect of child care on behavior is **confounded** with (mixed up with) other characteristics of families who use child care. Observational studies that examine the effect of a single variable on an outcome can be misleading when the effects of the explanatory variable are confounded with those of other variables. Because experiments allow us to isolate the effects of specific variables, we generally prefer them. Here is an example.

EXAMPLE

3.6 Which Web design sells more? A company that sells products on the Internet wants to decide which of two possible Web designs to use. During a two-week period they will use both designs and collect data on sales. They randomly select one of the designs to be used on the first day and then alternate the two designs on each of the following days. At the end of this period they compare the sales for the two designs.

Experiments usually require some sort of randomization, as in this example. We begin the discussion of statistical designs for data collection in Section 3.1 with the principles underlying the design of experiments.

USE YOUR KNOWLEDGE

3.9 Software for teaching creative writing. An educational software company wants to compare the effectiveness of its computer animation for teaching creative writing with that of a textbook presentation. The company tests the creative writing of each of a group of second-year college students and then randomly divides them into two groups. One group uses the animation, and the other studies the text. The company retests all the students and compares the increase in creative writing in the two groups. Is this an experiment? Why or why not? What are the explanatory and response variables?

3.10 Apples or apple juice. Food rheologists study different forms of foods and how the form of a food affects how full we feel when we eat it. One study prepared samples of apple juice and apples with the same amount of calories. Half of the subjects were fed apples on one day followed by apple juice on a later day; the other half received the apple juice followed by the apples. After eating, the subjects were asked about how full they felt. Is this an experiment? Why or why not? What are the explanatory and response variables?

statistical inference

Statistical techniques for producing data are the foundation for **statistical inference,** which answers specific questions with a known degree of confidence. In Section 3.3, we discuss some basic ideas related to inference.

Should an experiment or sample survey that could possibly provide interesting and important information always be performed? How can we safeguard the privacy of subjects in a sample survey? What constitutes the mistreatment of people or animals who are studied in an experiment? These are questions of ethics

ethics. In Section 3.4, we address ethical issues related to the design of studies and the analysis of data.

3.1 Design of Experiments

A study is an experiment when we actually do something to people, animals, or objects in order to observe the response. Here is the basic vocabulary of experiments.

EXPERIMENTAL UNITS, SUBJECTS, TREATMENT

The individuals on which the experiment is done are the **experimental units.** When the units are human beings, they are called **subjects.** A specific experimental condition applied to the units is called a **treatment.**

Because the purpose of an experiment is to reveal the response of one variable to changes in other variables, the distinction between explanatory and response variables is important. The explanatory variables in an experiment are often called **factors.** Many experiments study the joint effects of several factors. In such an experiment, each treatment is formed by combining a specific value (often called a **level**) of each of the factors.

factors

level of a factor

EXAMPLE

3.7 Are smaller class sizes better? Do smaller classes in elementary school really benefit students in areas such as scores on standard tests, staying in school, and going on to college? We might do an observational study that compares students who happened to be in smaller and larger classes in their early school years. Small classes are expensive, so they are more common in schools that serve richer communities. Students in small classes tend to also have other advantages: their schools have more resources, their parents are better educated, and so on. Confounding makes it impossible to isolate the effects of small classes.

The Tennessee STAR program was an experiment on the effects of class size. It has been called "one of the most important educational investigations ever carried out." The *subjects* were 6385 students who were beginning kindergarten. Each student was assigned to one of three *treatments:* regular class (22 to 25 students) with one teacher, regular class with a teacher and a full-time teacher's aide, and small class (13 to 17 students). These treatments are levels of a single *factor,* the type of class. The students stayed in the same type of class for four years, then all returned to regular classes. In later years, students from the small classes had higher scores on standard tests, were less likely to fail a grade, had better high school grades, and so on. The benefits of small classes were greatest for minority students.[6]

Example 3.7 illustrates the big advantage of experiments over observational studies. **In principle, experiments can give good evidence for causation.**

In an experiment, we study the specific factors we are interested in, while controlling the effects of lurking variables. All the students in the Tennessee STAR program followed the usual curriculum at their schools. Because students were assigned to different class types within their schools, school resources and family backgrounds were not confounded with class type. The only systematic difference was the type of class. When students from the small classes did better than those in the other two types, we can be confident that class size made the difference.

EXAMPLE

3.8 Repeated exposure to advertising. What are the effects of repeated exposure to an advertising message? The answer may depend both on the length of the ad and on how often it is repeated. An experiment investigated this question using undergraduate students as *subjects.* All subjects viewed a 40-minute television program that included ads for a digital camera. Some subjects saw a 30-second commercial; others, a 90-second version. The same commercial was shown either 1, 3, or 5 times during the program.

This experiment has two *factors:* length of the commercial, with 2 levels, and repetitions, with 3 levels. The 6 combinations of one level of each factor form 6 *treatments.* Figure 3.2 shows the layout of the treatments. After viewing, all the subjects answered questions about their recall of the ad, their attitude toward the camera, and their intention to purchase it. These are the *response variables.*[7]

		Factor B Repetitions		
		1 time	3 times	5 times
Factor A Length	30 seconds	1	2	3
	90 seconds	4	5	6

FIGURE 3.2 The treatments in the study of advertising, for Example 3.8. Combining the levels of the two factors forms six treatments.

Example 3.8 shows how experiments allow us to study the combined effects of several factors. The interaction of several factors can produce effects that could not be predicted from looking at the effects of each factor alone. Perhaps longer commercials increase interest in a product, and more commercials also increase interest, but if we both make a commercial longer and show it more often, viewers get annoyed and their interest in the product drops. The two-factor experiment in Example 3.8 will help us find out.

USE YOUR KNOWLEDGE

3.11 Food for a trip beyond the moon. Storing food for long periods of time is a major challenge for those planning for human space travel beyond the moon. One problem is that exposure to radiation decreases the length of time that food can be stored. One experiment examined the effects of nine different levels of radiation on a particular type of fat, or lipid.[8] The amount of oxidation of the lipid is the measure of the extent of the damage due to the radiation. Three samples are exposed to each radiation level. Give the experimental units, the treatments, and the response variable. Describe the factor and its levels. There are many different types of lipids. To what extent do you think the results of this experiment can be generalized to other lipids?

3.12 Learning how to draw. A course in computer graphics technology requires students to learn multiview drawing concepts. This topic is traditionally taught using supplementary material printed on paper. The instructor of the course believes that a Web-based interactive drawing program will be more effective in increasing the drawing skills of the students.[9] The 50 students who are enrolled in the course will be randomly assigned to either the paper-based instruction or the Web-based instruction. A standardized drawing test will be given before and after the instruction. Explain why this study is an experiment and give the experimental units, the treatments, and the response variable. Describe the factor and its levels. To what extent do you think the results of this experiment can be generalized to other settings?

Comparative experiments

Laboratory experiments in science and engineering often have a simple design with only a single treatment, which is applied to all experimental units. The design of such an experiment can be outlined as

$$\text{Treatment} \longrightarrow \text{Observe response}$$

For example, we may subject a beam to a load (treatment) and measure its deflection (observation). We rely on the controlled environment of the laboratory to protect us from lurking variables. When experiments are conducted in the field or with living subjects, such simple designs often yield invalid data. That is, we cannot tell whether the response was due to the treatment or to lurking variables.

EXAMPLE

3.9 T-shirt logos and sales. A student organization sells t-shirts to raise money that supports its charitable activities. This year they decide to change the logo of the t-shirt in the hope that a more attractive logo will increase sales. They also organized a sales team that made personal contacts with other students who might buy a t-shirt. Many people say that they like the new logo and sales this year are higher than last year. The design of the experiment was

$$\text{New logo} \longrightarrow \text{Observe sales}$$

The t-shirt logo experiment was poorly designed to evaluate the effect of the new logo. Perhaps sales would have increased even with the old logo. Or, the increase in sales could have been due to the personal contacts that were made by the sales team. In medical settings this phenomenon is called a **placebo effect.** A placebo is a dummy treatment. People respond favorably to personal attention or to any treatment that they hope will help them. On the other hand, the new logo may have been very effective in increasing sales. From this experiment we don't know if the increase was due to the new logo, to the personal contacts with potential customers, or to a general environment where more students were inclined to buy t-shirts.

placebo effect

The t-shirt logo experiment gave inconclusive results because the effect of the logo was confounded with other factors that could have had an effect on sales. The best way to avoid confounding is to do a comparative experiment. T-shirts with both logos could be sold and the sales of each could then be compared.

In medical settings, it is standard practice to randomize patients to a **control group,** or a **treatment group.** All patients are treated the same in every way except that the treatment group receives the product that is being evaluated.

control group
treatment group

Uncontrolled experiments in medicine and the behavioral sciences can be dominated by such influences as the details of the experimental arrangement, the selection of subjects, and the placebo effect. The result is often *bias.*

> ### BIAS
>
> The design of a study is **biased** if it systematically favors certain outcomes.

An uncontrolled study of a new medical therapy, for example, is biased in favor of finding the treatment effective because of the placebo effect. It should not surprise you to learn that uncontrolled studies in medicine give new therapies a much higher success rate than proper comparative experiments. Well-designed experiments usually compare several treatments.

USE YOUR KNOWLEDGE

3.13 Does using statistical software improve exam scores? An instructor in an elementary statistics course wants to know if using a new statistical software package will improve students' final-exam scores. He asks for volunteers and about half the class agrees to work with the new software. He compares the final-exam scores of the students who used the new software with the scores of those who did not. Discuss possible sources of bias in this study.

Randomization

experimental design

The **design of an experiment** first describes the response variable or variables, the factors (explanatory variables), and the layout of the treatments, with comparison as the leading principle. Figure 3.2 illustrates this aspect of the design of a study of response to advertising. The second aspect of design is the rule used to assign the experimental units to the treatments. Comparison of the effects of several treatments is valid only when all treatments are applied to similar groups of experimental units. If one corn variety is planted on more fertile ground, or if one cancer drug is given to more seriously ill patients, comparisons among treatments are meaningless. Systematic differences among the groups of experimental units in a comparative experiment cause bias. How can we assign experimental units to treatments in a way that is fair to all of the treatments?

Experimenters often attempt to match groups by elaborate balancing acts. Medical researchers, for example, try to match the patients in a "new drug" experimental group and a "standard drug" control group by age, sex, physical condition, smoker or not, and so on. Matching is helpful but not adequate—there are too many lurking variables that might affect the outcome. The experimenter is unable to measure some of these variables and will not think of others until after the experiment. Some important variables, such as how advanced a cancer patient's disease is, are so subjective that an experimenter might bias the study by, for example, assigning more advanced cancer cases to a promising new treatment in the unconscious hope that it will help them.

The statistician's remedy is to rely on chance to make an assignment that does not depend on any characteristic of the experimental units and that does not rely on the judgment of the experimenter in any way. The use of chance can be combined with matching, but the simplest design creates groups by chance alone. Here is an example.

EXAMPLE

3.10 Which new cell phone should be marketed? Two teams have prepared prototypes for a new cell phone. Before deciding which one will be marketed, they will be evaluated by college students. Forty students will receive a new phone. They will use it for two weeks and then answer some questions about how well they like the phone. The 40 students will be randomized with 20 receiving each phone.

This experiment has a single factor (commercial) with two levels. The researchers must divide the 40 student subjects into two groups of 20. To do this in a completely unbiased fashion, put the names of the 40 students in a hat, mix them up, and draw 20. These students will receive phone 1 and the remaining 20 will receive phone 2. Figure 3.3 outlines the design of this experiment.

randomization   The use of chance to divide experimental units into groups is called **randomization.** The design in Figure 3.3 combines comparison and randomization to arrive at the simplest randomized comparative design. This "flowchart" outline presents all the essentials: randomization, the sizes of the

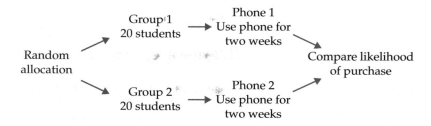

FIGURE 3.3 Outline of a randomized comparative experiment, for Example 3.10.

groups and which treatment they receive, and the response variable. There are, as we will see later, statistical reasons for generally using treatment groups about equal in size.

USE YOUR KNOWLEDGE

3.14 Diagram the drawing experiment. Refer to Exercise 3.12 (page 172). Draw a diagram similar to Figure 3.3 that describes the computer graphics drawing experiment.

3.15 Diagram the food for moon experiment. Refer to Exercise 3.11 (page 171). Draw a diagram similar to Figure 3.3 that describes the food for space travel experiment.

Randomized comparative experiments

The logic behind the randomized comparative design in Figure 3.3 is as follows:

- Randomization produces two groups of subjects that we expect to be similar in all respects before the treatments are applied.

- Comparative design helps ensure that influences other than the characteristics of the cell phone operate equally on both groups.

- Therefore, differences in the satisfaction with the phone must be due either to the characteristics of the cell phone or to the chance assignment of subjects to the two groups.

That "either-or" deserves more comment. We cannot say that *any* difference in the satisfaction for the two cell phones is caused by the characteristics of the cell phone. There would be some difference even if both groups used the same phone. Some students would be more likely to be highly favorable with any new cell phone. Chance can assign more of these students to one of the phones, so that there is a chance difference between the groups. We would not trust an experiment with just one subject in each group, for example. The results would depend too much on which phone got lucky and received the subject who was more likely to be highly satisfied. If we assign many students to each group, however, the effects of chance will average out. There will be little difference in the satisfaction in the two groups unless the phone characteristics causes a difference. "Use enough subjects to reduce chance variation" is the third big idea of statistical design of experiments.

PRINCIPLES OF EXPERIMENTAL DESIGN

The basic principles of statistical design of experiments are

1. **Compare** two or more treatments. This will control the effects of lurking variables on the response.

2. **Randomize**—use impersonal chance to assign experimental units to treatments.

3. **Repeat** each treatment on many units to reduce chance variation in the results.

We hope to see a difference in the responses so large that it is unlikely to happen just because of chance variation. We can use the laws of probability, which give a mathematical description of chance behavior, to learn if the treatment effects are larger than we would expect to see if only chance were operating. If they are, we call them *statistically significant*.

STATISTICAL SIGNIFICANCE

An observed effect so large that it would rarely occur by chance is called **statistically significant.**

You will often see the phrase "statistically significant" in reports of investigations in many fields of study. It tells you that the investigators found good evidence for the effect they were seeking.

How to randomize

The idea of randomization is to assign subjects to treatments by drawing names from a hat. In practice, experimenters use software to carry out randomization. Most statistical software will choose 20 out of a list of 40 at random, for example. The list might contain the names of 40 human subjects. The 20 chosen form one group, and the 20 that remain form the second group. The *Simple Random Sample* applet on the text Web site makes it particularly easy to choose treatment groups at random.

You can randomize without software by using a *table of random digits*. Thinking about random digits helps you to understand randomization even if you will use software in practice. Table B at the back of the book and on the back endpaper is a table of random digits.

RANDOM DIGITS

A **table of random digits** is a list of the digits 0, 1, 2, 3, 4, 5, 6, 7, 8, 9 that has the following properties:

1. The digit in any position in the list has the same chance of being any one of 0, 1, 2, 3, 4, 5, 6, 7, 8, 9.

2. The digits in different positions are independent in the sense that the value of one has no influence on the value of any other.

You can think of Table B as the result of asking an assistant (or a computer) to mix the digits 0 to 9 in a hat, draw one, then replace the digit drawn, mix again, draw a second digit, and so on. The assistant's mixing and drawing saves us the work of mixing and drawing when we need to randomize. Table B begins with the digits 19223950340575628713. To make the table easier to read, the digits appear in groups of five and in numbered rows. The groups and rows have no meaning—the table is just a long list of digits having the properties 1 and 2 described above.

Our goal is to use random digits for experimental randomization. We need the following facts about random digits, which are consequences of the basic properties 1 and 2:

- Any *pair* of random digits has the same chance of being any of the 100 possible pairs: 00, 01, 02, ..., 98, 99.
- Any *triple* of random digits has the same chance of being any of the 1000 possible triples: 000, 001, 002, ..., 998, 999.
- ...and so on for groups of four or more random digits.

EXAMPLE

3.11 Randomize the students. In the cell phone experiment of Example 3.10, we must divide 40 students at random into two groups of 20 students each.

Step 1: *Label.* Give each student a numerical label, using as few digits as possible. Two digits are needed to label 40 students, so we use labels

$$01, 02, 03, \ldots, 39, 40$$

It is also correct to use labels 00 to 39 or some other choice of 40 two-digit labels.

Step 2: *Table.* Start anywhere in Table B and read two-digit groups. Suppose we begin at line 130, which is

$$69051 \ 64817 \ 87174 \ 09517 \ 84534 \ 06489 \ 87201 \ 97245$$

The first 10 two-digit groups in this line are

$$69 \ 05 \ 16 \ 48 \ 17 \ 87 \ 17 \ 40 \ 95 \ 17$$

Each of these two-digit groups is a label. The labels 00 and 41 to 99 are not used in this example, so we ignore them. The first 20 labels between 01 and 40 that we encounter in the table choose students for the first phone. Of the first 10 labels in line 130, we ignore four because they are too high (over 40). The others are 05, 16, 17, 17, 40, and 17. The students labeled 05, 16, 17, and 40 will evaluate the first phone. Ignore the second and third 17s because that student is already in the group. Run your finger across line 130 (and continue to the following lines) until you have chosen 20 students. They are the students labeled

$$05, 16, 17, 40, 20, 19, 32, 04, 25, 29, 37, 39, 31, 18, 07, 13, 33, 02, 36, 23$$

You should check at least the first few of these. These students will receive the first phone. The remaining 20 will evaluate the second phone.

As Example 3.11 illustrates, randomization requires two steps: assign labels to the experimental units and then use Table B to select labels at random. Be sure that all labels are the same length so that all have the same chance to be chosen. Use the shortest possible labels—one digit for 10 or fewer individuals, two digits for 11 to 100 individuals, and so on. Don't try to scramble the labels as you assign them. Table B will do the required randomizing, so assign labels in any convenient manner, such as in alphabetical order for human subjects. You can read digits from Table B in any order—along a row, down a column, and so on—because the table has no order. As an easy standard practice, we recommend reading along rows.

It is easy to use statistical software or Excel to randomize. Here are the steps:

Step 1: *Label.* The first step of assigning labels to the experimental units is similar to the procedure we described previously. One difference, however, is that we are not restricted to using numerical labels. Any system where each experimental unit has a unique label identifier will work.

Step 2: *Use the computer.* Once we have the labels, we then create a data file with the labels and generate a random number for each label. In Excel, this can be done with the RAND() function. Finally, we sort the entire data set based on the random numbers. Groups are formed by selecting units in order from the sorted list.

This process is essentially the same as writing the labels on a deck of cards, shuffling the cards, and dealing them out one at a time.

EXAMPLE

3.12 Using software for the randomization. Let's do a randomization similar to the one we did in Example 3.11, but this time using Excel. Here we will use 10 experimental units. We will assign 5 to the treatment group and 5 to the control group. We first create a data set with the numbers 1 to 10 in the first column. See Figure 3.4(a). Then we use RAND() to generate 10 random numbers in the second column. See Figure 3.4(b). Finally, we sort the data set based on the numbers in the second column. See Figure 3.4(c). The first 5 labels (8, 5, 9, 4, and 6) are assigned to the experimental group. The remaining 5 labels (3, 10, 7, 2, and 1) correspond to the control group.

FIGURE 3.4 Randomization of 10 experimental units using a spreadsheet, for Example 3.12. (a) Labels. (b) Random numbers. (c) Sorted list of labels.

(a) A	(b) A	B	(c) A	B
1	1	0.925672	8	0.077044
2	2	0.893959	5	0.118440
3	3	0.548247	9	0.348467
4	4	0.349591	4	0.349591
5	5	0.118440	6	0.390180
6	6	0.390180	3	0.548247
7	7	0.760262	10	0.601167
8	8	0.077044	7	0.760262
9	9	0.348467	2	0.893959
10	10	0.601167	1	0.925672

completely randomized design

When all experimental units are allocated at random among all treatments, as in Example 3.11, the experimental design is **completely randomized.** Completely randomized designs can compare any number of treatments. The treatments can be formed by levels of a single factor or by more than one factor.

EXAMPLE

3.13 Randomization for the TV commercial experiment. Figure 3.2 (page 171) displays six treatments formed by the two factors in an experiment on response to a TV commercial. Suppose that we have 150 students who are willing to serve as subjects. We must assign 25 students at random to each group. Figure 3.5 outlines the completely randomized design.

To carry out the random assignment, label the 150 students 001 to 150. (Three digits are needed to label 150 subjects.) Enter Table B and read three-digit groups until you have selected 25 students to receive Treatment 1 (a 30-second ad shown once). If you start at line 140, the first few labels for Treatment 1 subjects are 129, 048, and 003.

Continue in Table B to select 25 more students to receive Treatment 2 (a 30-second ad shown 3 times). Then select another 25 for Treatment 3 and so on until you have assigned 125 of the 150 students to Treatments 1 through 5. The 25 students who remain get Treatment 6. The randomization is straightforward, but very tedious to do by hand. We recommend the *Simple Random Sample* applet. Exercise 3.37 shows how to use the applet to do the randomization for this example.

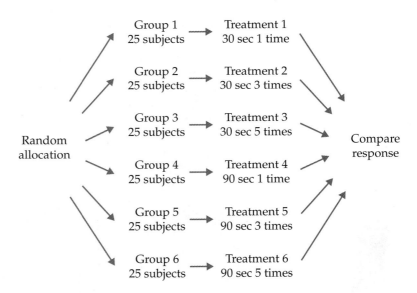

FIGURE 3.5 Outline of a completely randomized design comparing six treatments, for Example 3.13.

USE YOUR KNOWLEDGE

3.16 Do the randomization. Use computer software to carry out the randomization in Example 3.13.

Cautions about experimentation

The logic of a randomized comparative experiment depends on our ability to treat all the experimental units identically in every way except for the

double-blind

actual treatments being compared. Good experiments therefore require careful attention to details. The ideal situation is where a study is **double-blind**—neither the subjects themselves nor the experimenters know which treatment any subject had received. The double-blind method avoids unconscious bias by, for example, a doctor who doesn't think that "just a placebo" can benefit a patient.

Many—perhaps most—experiments have some weaknesses in detail. The environment of an experiment can influence the outcomes in unexpected ways. Although experiments are the gold standard for evidence of cause and effect, really convincing evidence usually requires that a number of studies in different places with different details produce similar results. Here are some brief examples of what can go wrong.

EXAMPLE

3.14 Placebo for a marijuana experiment. A study of the effects of marijuana recruited young men who used marijuana. Some were randomly assigned to smoke marijuana cigarettes, while others were given placebo cigarettes. This failed: the control group recognized that their cigarettes were phony and complained loudly. It may be quite common for blindness to fail because the subjects can tell which treatment they are receiving.[10]

lack of realism

The most serious potential weakness of experiments is **lack of realism.** The subjects or treatments or setting of an experiment may not realistically duplicate the conditions we really want to study. Here is an example.

EXAMPLE

3.15 Layoffs and feeling bad. How do layoffs at a workplace affect the workers who remain on the job? Psychologists asked student subjects to proofread text for extra course credit, then "let go" some of the workers (who were actually accomplices of the experimenters). Some subjects were told that those let go had performed poorly (Treatment 1). Others were told that not all could be kept and that it was just luck that they were kept and others let go (Treatment 2). We can't be sure that the reactions of the students are the same as those of workers who survive a layoff in which other workers lose their jobs. Many behavioral science experiments use student subjects in a campus setting. Do the conclusions apply to the real world?

Lack of realism can limit our ability to apply the conclusions of an experiment to the settings of greatest interest. Most experimenters want to generalize their conclusions to some setting wider than that of the actual experiment. *Statistical analysis of an experiment cannot tell us how far the results will generalize to other settings.* Nonetheless, the randomized comparative experiment, because of its ability to give convincing evidence for causation, is one of the most important ideas in statistics.

Matched pairs designs

matched pairs design

Completely randomized designs are the simplest statistical designs for experiments. They illustrate clearly the principles of control, randomization, and repetition. However, completely randomized designs are often inferior to more elaborate statistical designs. In particular, matching the subjects in various ways can produce more precise results than simple randomization.

The simplest use of matching is a **matched pairs design,** which compares just two treatments. The subjects are matched in pairs. For example, an experiment to compare two advertisements for the same product might use pairs of subjects with the same age, sex, and income. The idea is that matched subjects are more similar than unmatched subjects, so that comparing responses within a number of pairs is more efficient than comparing the responses of groups of randomly assigned subjects. Randomization remains important: which one of a matched pair sees the first ad is decided at random. One common variation of the matched pairs design imposes both treatments on the same subjects, so that each subject serves as his or her own control. Here is an example.

> ### EXAMPLE
>
> **3.16 Matched pairs for the cell phone prototype experiment.** Example 3.10 describes an experiment to compare two prototypes of a new cell phone. The experiment compared two treatments, phone 1 and phone 2. The response variable is the satisfaction of the college student participant with the new cell phone. In Example 3.10, 40 student subjects were assigned at random, 20 students to each phone. This is a completely randomized design, outlined in Figure 3.3. Subjects differ in how satisfied they are with cell phones in general. The completely randomized design relies on chance to create two similar groups of subjects.
>
> If we wanted to do a matched pairs version of this experiment, we would have each college student use each phone for two weeks. An effective design would randomize the *order* in which the phones are evaluated for each student. This will eliminate bias due to the possibility that the first phone evaluated will be systematically evaluated higher or lower than the second phone evalated.

The completely randomized design uses chance to decide which subjects will evaluate each cell phone prototype. The matched pairs design uses chance to decide which 20 subjects will evaluate phone 1 first. The other 20 will evaluate the phone 2 first.

Block designs

The matched pairs design of Example 3.16 uses the principles of comparison of treatments, randomization, and repetition on several experimental units. However, the randomization is not complete (all subjects randomly assigned to treatment groups) but restricted to assigning the order of the treatments for each subject. *Block designs* extend the use of "similar subjects" from pairs to larger groups.

BLOCK DESIGN

A **block** is a group of experimental units or subjects that are known before the experiment to be similar in some way that is expected to affect the response to the treatments. In a **block design,** the random assignment of units to treatments is carried out separately within each block.

Block designs can have blocks of any size. A block design combines the idea of creating equivalent treatment groups by matching with the principle of forming treatment groups at random. Blocks are another form of *control.* They control the effects of some outside variables by bringing those variables into the experiment to form the blocks. Here are some typical examples of block designs.

EXAMPLE

3.17 Blocking in a cancer experiment. The progress of a type of cancer differs in women and men. A clinical experiment to compare three therapies for this cancer therefore treats sex as a blocking variable. Two separate randomizations are done, one assigning the female subjects to the treatments and the other assigning the male subjects. Figure 3.6 outlines the design of this experiment. Note that there is no randomization involved in making up the blocks. They are groups of subjects who differ in some way (sex in this case) that is apparent before the experiment begins.

3.18 Blocking in an agriculture experiment. The soil type and fertility of farmland differ by location. Because of this, a test of the effect of tillage type (two types) and pesticide application (three application schedules) on soybean yields uses small fields as blocks. Each block is divided into six plots, and the six treatments are randomly assigned to plots separately within each block.

3.19 Blocking in an education experiment. The Tennessee STAR class size experiment (Example 3.7) used a block design. It was important to compare different class types in the same school because the children in a school come from the same neighborhood, follow the same curriculum, and have

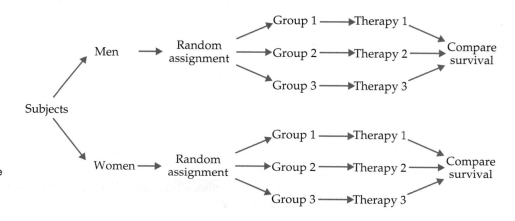

FIGURE 3.6 Outline of a block design, for Example 3.17. The blocks consist of male and female subjects. The treatments are the three therapies for cancer.

the same school environment outside class. In all, 79 schools across Tennessee participated in the program. That is, there were 79 blocks. New kindergarten students were randomly placed in the three types of class separately within each school.

Blocks allow us to draw separate conclusions about each block, for example, about men and women in the cancer study in Example 3.17. Blocking also allows more precise overall conclusions because the systematic differences between men and women can be removed when we study the overall effects of the three therapies. The idea of blocking is an important additional principle of statistical design of experiments. A wise experimenter will form blocks based on the most important unavoidable sources of variability among the experimental units. Randomization will then average out the effects of the remaining variation and allow an unbiased comparison of the treatments.

SECTION 3.1 **Summary**

In an experiment, one or more **treatments** are imposed on the **experimental units** or **subjects.** Each treatment is a combination of **levels** of the explanatory variables, which we call **factors.**

The **design** of an experiment refers to the choice of treatments and the manner in which the experimental units or subjects are assigned to the treatments.

The basic principles of statistical design of experiments are **control, randomization,** and **repetition.**

The simplest form of control is **comparison.** Experiments should compare two or more treatments in order to prevent **confounding** the effect of a treatment with other influences, such as lurking variables.

Randomization uses chance to assign subjects to the treatments. Randomization creates treatment groups that are similar (except for chance variation) before the treatments are applied. Randomization and comparison together prevent **bias,** or systematic favoritism, in experiments.

You can carry out randomization by giving numerical labels to the experimental units and using a **table of random digits** to choose treatment groups.

Repetition of the treatments on many units reduces the role of chance variation and makes the experiment more sensitive to differences among the treatments.

Good experiments require attention to detail as well as good statistical design. Many behavioral and medical experiments are **double-blind. Lack of realism** in an experiment can prevent us from generalizing its results.

In addition to comparison, a second form of control is to restrict randomization by forming **blocks** of experimental units that are similar in some way that is important to the response. Randomization is then carried out separately within each block.

Matched pairs are a common form of blocking for comparing just two treatments. In some matched pairs designs, each subject receives both treatments in a random order. In others, the subjects are matched in pairs as closely as possible, and one subject in each pair receives each treatment.

SECTION 3.1 Exercises

For Exercises 3.1 to 3.4, see page 164; for Exercise 3.5, see page 165; for Exercise 3.6, see page 167; for Exercises 3.7 and 3.8, see pages 167–168; for Exercises 3.9 and 3.10, see page 169; for Exercises 3.11 and 3.12, see pages 171–172; for Exercise 3.13, see page 173; for Exercises 3.14 and 3.15, see page 175; and for Exercise 3.16, see page 179.

3.17 What is needed? Explain what is deficient in each of the following proposed experiments and explain how you would improve the experiment.

(a) Two product promotion offers are to be compared. The first, which offers two items for $2, will be used in a store on Friday. The second, which offers three items for $3, will be used in the same store on Saturday.

(b) A study compares two marketing campaigns to encourage individuals to eat more fruits and vegetables. The first campaign is launched in Florida at the same time that the second campaign is launched in Minnesota.

(c) You want to evaluate the effectiveness of a new investment strategy. You try the strategy for one year and evaluate the performance of the strategy.

3.18 What is wrong? Explain what is wrong with each of the following randomization procedures and describe how you would do the randomization correctly.

(a) Twenty students are to be used to evaluate a new treatment. Ten men are assigned to receive the treatment and 10 women are assigned to be the controls.

(b) Ten subjects are to be assigned to two treatments, 5 to each. For each subject, a coin is tossed. If the coin comes up heads, the subject is assigned to the first treatment; if the coin comes up tails, the subject is assigned to the second treatment.

(c) An experiment will assign 40 rats to four different treatment conditions. The rats arrive from the supplier in batches of 10 and the treatment lasts two weeks. The first batch of 10 rats is randomly assigned to one of the four treatments, and data for these rats are collected. After a one-week break, another batch of 10 rats arrives and is assigned to one of the three remaining treatments. The process continues until the last batch of rats is given the treatment that has not been assigned to the three previous batches.

3.19 Evaluate an online version of a course. An online version of a traditional course is to be evaluated by randomly assigning students to either the online version or the traditional course. The change in a standardized test score is the response variable. Explain why this experiment cannot be done in a double-blind fashion.

3.20 Can you change attitudes toward binge drinking? A experiment designed to change attitudes about binge drinking is to be performed using college students as subjects. Discuss some variables that you might use if you were to use a block design for this experiment.

3.21 Evaluate a new employee orientation program. Your company runs a two-day orientation program Monday and Tuesday each week for new employees. A new program is to be compared with the current one. Set up an experiment to compare the new program with the old. Be sure to provide details regarding randomization and what outcome variables you will measure.

3.22 Medical magnets. Some claim that magnets can be used to reduce pain. Design a double-blind experiment to test this claim. Write a proposal requesting funding for your study giving all of the important details including the number of subjects, issues concerning randomization, and how you will make the study double-blind.

3.23 Calcium and vitamin D. Vitamin D is needed for the body to use calcium. An experiment is designed the effects of calcium and vitamin D supplements on the bones of first-year college students. The outcome measure is the total body bone mineral content (TBBMC), a measure of bone health. Three doses of calcium will be used: 0, 200, and 400 mg/day. The doses of vitamin D will be 0, 50, and 100 international units (IU) per day. The calcium and vitamin D will be given in a single tablet. All tablets including those with no calcium and no vitamin D will look identical. Subjects for the study will be 90 men and 90 women.

(a) What are the factors and the treatments for this experiment?

(b) Draw a picture explaining how you would randomize the 180 college students to the treatments.

(c) Use a spreadsheet to carry out the randomization.

3.24 🔺 The interaction of calcium and vitamin D. Refer to the previous exercise. It is expected that the effect of vitamin D on TBBMC will be greatest when there is an adequate intake of calcium. In other words, the effect of vitamin D is expected to be greater with the higher doses of calcium. It is also expected that the men in the study will generally have adequate calcium and vitamin D in their diets, so the effect of the supplements will be very small at best. Draw pictures that illustrate these expected interactions.

3.25 **Compare two versions of a product.** A coffee house wants to compare two new varieties of coffee.

(a) Describe an experiment where different customers evaluate each variety. Be sure to provide details including how many customers you will use, issues related to randomization, and what evaluation data you will collect.

(b) Do the same for an experiment where each customer evaluates both varieties of coffee.

(c) Which experiment do you prefer? Give reasons for your answer.

3.26 The *Sports Illustrated* jinx. Some people believe that teams or individual athletes who appear on the cover of *Sports Illustrated* magazine will experience bad luck soon after they appear. Can you evaluate this belief with an experiment? Explain your answer.

For each of the experimental situations described in Exercises 3.27 to 3.29, identify the experimental units or subjects, the factors, the treatments, and the response variables.

3.27 How well do pine trees grow in shade? Ability to grow in shade may help pines in the dry forests of Arizona resist drought. How well do these pines grow in shade? Investigators planted pine seedlings in a greenhouse in either full light or light reduced to 5% of normal by shade cloth. At the end of the study, they dried the young trees and weighed them.

3.28 Will the students exercise more and eat better? Most American adolescents don't eat well and don't exercise enough. Can middle schools increase physical activity among their students? Can they persuade students to eat better? Investigators designed a "physical activity intervention" to increase activity in physical education classes and during leisure periods throughout the school day. They also designed a "nutrition intervention" that improved school lunches and offered ideas for healthy home-packed lunches. Each participating school was randomly assigned to one of the interventions, both interventions, or no intervention. The investigators observed physical activity and lunchtime consumption of fat.

3.29 Refusals in telephone surveys. How can we reduce the rate of refusals in telephone surveys? Most people who answer at all listen to the interviewer's introductory remarks and then decide whether to continue. One study made telephone calls to randomly selected households to ask opinions about the next election. In some calls, the interviewer gave her name, in others she identified the university she was representing, and in still others she identified both herself and the university. For each type of call, the interviewer either did or did not offer to send a copy of the final survey results to the person interviewed. Do these differences in the introduction affect whether the interview is completed?

3.30 Does aspirin prevent strokes and heart attacks? The Bayer Aspirin Web site claims that "Nearly five decades of research now link aspirin to the prevention of stroke and heart attacks." The most important evidence for this claim comes from the Physicians' Health Study, a large medical experiment involving 22,000 male physicians. One group of about 11,000 physicians took an aspirin every second day, while the rest took a placebo. After several years the study found that subjects in the aspirin group had significantly fewer heart attacks than subjects in the placebo group.

(a) Identify the experimental subjects, the factor and its levels, and the response variable in the Physicians' Health Study.

(b) Use a diagram to outline a completely randomized design for the Physicians' Health Study.

(c) What does it mean to say that the aspirin group had "significantly fewer heart attacks"?

3.31 Chronic tension headaches. Doctors identify "chronic tension-type headaches" as headaches that occur almost daily for at least six months. Can antidepressant medications or stress management training reduce the number and severity of these headaches? Are both together more effective than either alone? Investigators compared four treatments: antidepressant alone, placebo alone, antidepressant plus stress management, and placebo plus stress management. Outline the design of the experiment. The headache sufferers named in the table have agreed to participate in the study. Use software or Table B at line 151 to randomly assign the subjects to the treatments.

Anderson	Archberger	Bezawada	Cetin	Cheng	Chronopoulou
Codrington	Daggy	Daye	Engelbrecht	Guha	Hatfield
Hua	Kim	Kumar	Leaf	Li	Lipka
Lu	Martin	Mehta	Mi	Nolan	Olbricht
Park	Paul	Rau	Saygin	Shu	Tang
Towers	Tyner	Vassilev	Wang	Watkins	Xu

3.32 Smoking marijuana and willingness to work. How does smoking marijuana affect willingness to work? Canadian researchers persuaded people who used marijuana to live for 98 days in a "planned environment." The subjects earned money by weaving belts. They used their earnings to pay for meals and other consumption and could keep any money left over. One group smoked two potent marijuana cigarettes every evening. The other group smoked two weak marijuana cigarettes. All subjects could buy more cigarettes but were given strong or weak cigarettes, depending on their group. Did the weak and strong groups differ in work output and earnings?[11]

(a) Outline the design of this experiment.

(b) Here are the names of the 20 subjects. Use software or Table B at line 101 to carry out the randomization your design requires.

Becker	Brifcani	Chen	Crabill	Cunningham
Dicklin	Fein	Gorman	Knapp	Lucas
McCarty	Merkulyeva	Mitchell	Ponder	Roe
Saeed	Seele	Truong	Wayman	Woodley

3.33 CHALLENGE **Measuring water quality in streams and lakes.** Water quality of streams and lakes is an issue of concern to the public. Although trained professionals typically are used to take reliable measurements, many volunteer groups are gathering and distributing information based on data that they collect.[12]

You are part of a team to train volunteers to collect accurate water quality data. Design an experiment to evaluate the effectiveness of the training. Write a summary of your proposed design to present to your team. Be sure to include all the details that they will need to evaluate your proposal.

3.34 Guilt among workers who survive a layoff. Workers who survive a layoff of other employees at their location may suffer from "survivor guilt." A study of survivor guilt and its effects used as subjects 90 students who were offered an opportunity to earn extra course credit by doing proofreading. Each subject worked in the same cubicle as another student, who was an accomplice of the experimenters. At a break midway through the work, one of three things happened:

Treatment 1: The accomplice was told to leave; it was explained that this was because she performed poorly.

Treatment 2: It was explained that unforeseen circumstances meant there was only enough work for one person. By "chance," the accomplice was chosen to be laid off.

Treatment 3: Both students continued to work after the break.

The subjects' work performance after the break was compared with performance before the break.[13]

(a) Outline the design of this completely randomized experiment.

(b) If you are using software, randomly assign the 90 students to the treatments. If not, use Table B at line 153 to choose the first four subjects for Treatment 1.

3.35 Diagram the exercise and eating experiment. Twenty-four public middle schools agree to participate in the experiment described in Exercise 3.28. Use a diagram to outline a completely randomized design for this experiment. Then do the randomization required to assign schools to treatments. If you use Table B, start at line 160.

3.36 Price cuts on athletic shoes. Stores advertise price reductions to attract customers. What type of price cut is most attractive? Market researchers prepared ads for athletic shoes announcing different levels of discounts (20%, 40%, 60%, or 80%). The student subjects who read the ads were also given "inside information" about the fraction of shoes on sale (25%, 50%, 75%, or 100%). Each subject then rated the attractiveness of the sale on a scale of 1 to 7.[14]

(a) There are two factors. Make a sketch like Figure 3.2 (page 171) that displays the treatments formed by all combinations of levels of the factors.

(b) Outline a completely randomized design using 96 student subjects. Use software or Table B at line 111 to choose the subjects for the first treatment.

3.37 APPLET **Use the simple random sample applet.** The *Simple Random Sample* applet allows you to randomly assign experimental units to more than two groups without difficulty. Example 3.13 (page 179) describes a randomized comparative experiment in which 150 students are randomly assigned to six groups of 25.

(a) Use the applet to randomly choose 25 out of 150 students to form the first group. Which students are in this group?

(b) The population hopper now contains the 125 students who were not chosen, in scrambled order. Click "Sample" again to choose 25 of these remaining students to make up the second group. Which students were chosen?

(c) Click "Sample" three more times to choose the third, fourth, and fifth groups. Don't take the time to write down these groups. Check that there are only 25 students remaining in the population hopper. These subjects get Treatment 6. Which students are they?

3.38 APPLET **Use the simple random sample applet.** You can use the *Simple Random Sample* applet to choose a treatment group at random once you have labeled the subjects. Example 3.11 (page 177) uses Table B to choose 20 students from a group of 40 for the treatment group in a study of the effect of cell phones on driving. Use the applet to choose the 20 students for the experimental group. Which students did you choose? The remaining 20 students make up the control group.

3.39 🔵 **Health benefits of bee pollen.** "Bee pollen is effective for combating fatigue, depression, cancer, and colon disorders." So says a Web site that offers the pollen for sale. We wonder if bee pollen really does prevent colon disorders. Here are two ways to study this question. Explain why the first design will produce more trustworthy data.

(a) Find 400 women who do not have colon disorders. Randomly assign 200 to take bee pollen capsules and the other 200 to take placebo capsules that are identical in appearance. Follow both groups for 5 years.

(b) Find 200 women who take bee pollen regularly. Match each with a woman of the same age, race, and occupation who does not take bee pollen. Follow both groups for 5 years.

3.40 🔵 **Effectiveness of price discounts.** Experiments with more than one factor allow insight into interactions between the factors. A study of the attractiveness of advertised price discounts had two factors: percent of all goods on sale (25%, 50%, 75%, or 100%) and whether the discount was stated precisely as 60% off or as a range, 50% to 70% off. Subjects rated the attractiveness of the sale on a scale of 1 to 7. Figure 3.7 shows the mean ratings for the eight treatments formed from the two factors.[15] Based on these results, write a careful description of how percent on sale and precise discount versus range of discounts influence the attractiveness of a sale.

3.41 Five-digit zip codes and delivery time of mail. Does adding the five-digit postal zip code to an address

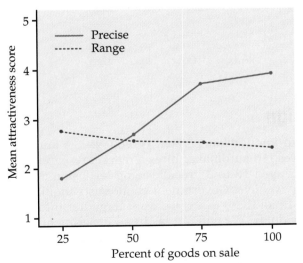

FIGURE 3.7 Mean responses to eight treatments in an experiment with two factors, showing interaction between the factors, for Exercise 3.40.

really speed up delivery of letters? Does adding the four more digits that make up "zip + 4" speed delivery yet more? What about mailing a letter on Monday, Thursday, or Saturday? Describe the design of an experiment on the speed of first-class mail delivery. For simplicity, suppose that all letters go from you to a friend, so that the sending and receiving locations are fixed.

3.42 🔵 **Use the simple random sample applet.** The *Simple Random Sample* applet can demonstrate how randomization works to create similar groups for comparative experiments. Suppose that (unknown to the experimenters) the 20 even-numbered students among the 40 subjects for the cell phone study in Example 3.11 (page 177) have fast reactions, and that the odd-numbered students have slow reactions. We would like the experimental and control groups to contain similar numbers of the fast reactors. Use the applet to choose 10 samples of size 20 from the 40 students. (Be sure to click "Reset" after each sample.) Record the counts of even-numbered students in each of your 10 samples. You see that there is considerable chance variation but no systematic bias in favor of one or the other group in assigning the fast-reacting students. Larger samples from larger populations will on the average do a better job of making the two groups equivalent.

3.43 Does oxygen help football players? We often see players on the sidelines of a football game inhaling oxygen. Their coaches think this will speed their recovery. We might measure recovery from intense exercise as follows: Have a football player run 100 yards three times in quick succession. Then allow three minutes to rest before running 100 yards again. Time the final run. Because players vary greatly in speed, you plan a matched pairs experiment using 20 football players as subjects. Describe the design of such an experiment to investigate the effect of inhaling oxygen during the rest period. Why should each player's two trials be on different days? Use Table B at line 140 to decide which players will get oxygen on their first trial.

3.44 Carbon dioxide in the atmosphere. The concentration of carbon dioxide (CO_2) in the atmosphere is increasing rapidly due to our use of fossil fuels. Because plants use CO_2 to fuel photosynthesis, more CO_2 may cause trees and other plants to grow faster. An elaborate apparatus allows researchers to pipe extra CO_2 to a 30-meter circle of forest. We want to compare the growth in base area of trees in treated and untreated areas to see if extra CO_2 does in fact increase growth. We can afford to treat six circular areas.[16]

(a) Describe the design of a completely randomized experiment using six well-separated 30-meter circular areas in a pine forest. Sketch the forest area with the six circles and carry out the randomization your design calls for.

(b) Regions within the forest may differ in soil fertility. Describe a matched pairs design using three pairs of circles that will reduce the extra variation due to different fertility. Sketch the forest area with the new arrangement of circles and carry out the randomization your design calls for.

3.45 🔺 **Calcium and the bones of young girls.** Calcium is important to the bone development of young girls. To study how the bodies of young girls process calcium, investigators used the setting of a summer camp. Calcium was given in punch at either a high or a low level. The camp diet was otherwise the same for all girls. Suppose that there are 50 campers.

(a) Outline a completely randomized design for this experiment.

(b) Describe a matched pairs design in which each girl receives both levels of calcium (with a "washout period" where they eat their normal diet, between). What is the advantage of the matched pairs design over the completely randomized design?

(c) The same randomization can be used in different ways for both designs. Label the subjects 01 to 50. You must choose 25 of the 50. Use Table B at line 110 to choose just the first 5 of the 25. How are the 25 subjects chosen treated in the completely randomized design? How are they treated in the matched pairs design?

3.46 🔺 **Random digits.** Table B is a table of random digits. Which of the following statements are true of a table of random digits, and which are false? Explain your answers.

(a) There are exactly four 0s in each row of 40 digits.

(b) Each pair of digits has chance 1/100 of being 00.

(c) The digits 0000 can never appear as a group, because this pattern is not random.

3.47 Vitamin C for ultramarathon runners. An ultramarathon, as you might guess, is a footrace longer than the 26.2 miles of a marathon. Runners commonly develop respiratory infections after an ultramarathon. Will taking 600 milligrams of vitamin C daily reduce these infections? Researchers randomly assigned ultramarathon runners to receive either vitamin C or a placebo. Separately, they also randomly assigned these treatments to a group of nonrunners the same age as the runners. All subjects were watched for 14 days after the big race to see if infections developed.[17]

(a) What is the name for this experimental design?

(b) Use a diagram to outline the design.

(c) The report of the study said

Sixty-eight percent of the runners in the placebo group reported the development of symptoms of upper respiratory tract infection after the race; this was significantly more than that reported by the vitamin C-supplemented group (33%).

Explain to someone who knows no statistics why "significantly more" means there is good reason to think that vitamin C works.

3.2 Sampling Design

A political scientist wants to know what percent of college-age adults consider themselves conservatives. An automaker hires a market research firm to learn what percent of adults aged 18 to 35 recall seeing television advertisements for a new sport utility vehicle. Government economists inquire about average household income. In all these cases, we want to gather information about a large group of individuals. We will not, as in an experiment, impose a treatment in order to observe the response. Also, time, cost, and inconvenience forbid contacting every individual. In such cases, we gather information about only part of the group—a *sample*—in order to draw conclusions about the whole. **Sample surveys** are an important kind of observational study.

sample survey

POPULATION AND SAMPLE

The entire group of individuals that we want information about is called the **population.**

A **sample** is a part of the population that we actually examine in order to gather information.

Notice that "population" is defined in terms of our desire for knowledge. If we wish to draw conclusions about all U.S. college students, that group is our population even if only local students are available for questioning. The sample is the part from which we draw conclusions about the whole. The **design** of a sample survey refers to the method used to choose the sample from the population.

sample design

EXAMPLE

3.20 The Reading Recovery program. The Reading Recovery (RR) program has specially trained teachers work one-on-one with at-risk first-grade students to help them learn to read. A study was designed to examine the relationship between the RR teachers' beliefs about their ability to motivate students and the progress of the students whom they teach.[18] The Reading Recovery International Data Evaluation Center Web site (`idecweb.us`) says that there are 13,823 RR teachers. The researchers send a questionnaire to a random sample of 200 of these. The population consists of all 13,823 RR teachers, and the sample is the 200 that were randomly selected.

Unfortunately, our idealized framework of population and sample does not exactly correspond to the situations that we face in many cases. In Example 3.20, the list of teachers was prepared at a particular time in the past. It is very likely that some of the teachers on the list are no longer working as RR teachers today. New teachers have been trained in RR methods and are not on the list. In spite of these difficulties, we still view the list as the population. We may have out-of-date addresses for some who are still working as RR teachers, and some teachers may choose not to respond to our survey questions.

In reporting the results of a sample survey it is important to include all details regarding the procedures used. Follow-up mailings or phone calls to those who do not initially respond can help increase the response rate. The proportion of the original sample who actually provide usable data is called the **response rate** and should be reported for all surveys. If only 150 of the teachers who were sent questionnaires provided usable data, the response rate would be 150/200, or 75%.

response rate

USE YOUR KNOWLEDGE

3.48 Job satisfaction in Mongolian universities. An educational research team wanted to examine the relationship between faculty participation in decision making and job satisfaction in Mongolian public universities. They are planning to randomly select 300 faculty members from a list of 2500 faculty members in these universities. The Job Descriptive Index (JDI) will be used to measure job satisfaction, and the Conway Adaptation of the Alutto-Belasco Decisional Participation Scale will be used to measure decision participation. Describe the population and the sample for this study. Can you determine the response rate?

3.49 Taxes and forestland usage. A study was designed to assess the impact of taxes on forestland usage in part of the Upper Wabash River Watershed in Indiana.[19] A survey was sent to 772 forest owners from this region and 348 were returned. Consider the population, the sample, and the response rate for this study. Describe these based on the information given and indicate any additional information that you would need to give a complete answer.

Poor sample designs can produce misleading conclusions. Here is an example.

EXAMPLE

3.21 Sampling pieces of steel. A mill produces large coils of thin steel for use in manufacturing home appliances. The quality engineer wants to submit a sample of 5-cm squares to detailed laboratory examination. She asks a technician to cut a sample of 10 such squares. Wanting to provide "good" pieces of steel, the technician carefully avoids the visible defects in the coil material when cutting the sample. The laboratory results are wonderful, but the customers complain about the material they are receiving.

Online opinion polls are particularly vulnerable to bias because the sample who respond are not representative of the population at large. People who take the trouble to respond to an open invitation are not representative of the entire adult population.

In Example 3.21, the sample was selected in a manner that guaranteed that it would not be representative of the entire population. This sampling scheme displays *bias*, or systematic error, in favoring some parts of the population over others. Online polls use *voluntary response samples,* a particularly common form of biased sample.

VOLUNTARY RESPONSE SAMPLE

A **voluntary response sample** consists of people who choose themselves by responding to a general appeal. Voluntary response samples are biased because people with strong opinions, especially negative opinions, are most likely to respond.

The remedy for bias in choosing a sample is to allow impersonal chance to do the choosing, so that there is neither favoritism by the sampler nor voluntary response. Random selection of a sample eliminates bias by giving all individuals an equal chance to be chosen, just as randomization eliminates bias in assigning experimental subjects.

Simple random samples

The simplest sampling design amounts to placing names in a hat (the population) and drawing out a handful (the sample). This is *simple random sampling*.

SIMPLE RANDOM SAMPLE

A **simple random sample (SRS)** of size n consists of n individuals from the population chosen in such a way that every set of n individuals has an equal chance to be the sample actually selected.

Each treatment group in a completely randomized experimental design is an SRS drawn from the available experimental units. We select an SRS by labeling all the individuals in the population and using software or a table of random digits to select a sample of the desired size, just as in experimental randomization. Notice that an SRS not only gives each individual an equal chance to be chosen (thus avoiding bias in the choice) but also gives every possible sample an equal chance to be chosen. There are other random sampling designs that give each individual, but not each sample, an equal chance. One such design, systematic random sampling, is described in Exercise 3.66.

SPRINGBREAK

EXAMPLE

3.22 Spring break destinations. A campus newspaper plans a major article on spring break destinations. The authors intend to call a few randomly chosen resorts at each destination to ask about their attitudes toward groups of students as guests. Here are the resorts listed in one city. The first step is to label the members of this population as shown.

01	Aloha Kai	08	Captiva	15	Palm Tree	22	Sea Shell
02	Anchor Down	09	Casa del Mar	16	Radisson	23	Silver Beach
03	Banana Bay	10	Coconuts	17	Ramada	24	Sunset Beach
04	Banyan Tree	11	Diplomat	18	Sandpiper	25	Tradewinds
05	Beach Castle	12	Holiday Inn	19	Sea Castle	26	Tropical Breeze
06	Best Western	13	Lime Tree	20	Sea Club	27	Tropical Shores
07	Cabana	14	Outrigger	21	Sea Grape	28	Veranda

Now enter Table B, and read two-digit groups until you have chosen three resorts. If you enter at line 185, Banana Bay (03), Palm Tree (15), and Cabana (07) will be called.

Most statistical software will select an SRS for you, eliminating the need for Table B. The *Simple Random Sample* applet on the text CD and Web site is a convenient way to automate this task.

Excel can do the job in a way similar to what we used when we randomized experimental units to treatments in designed experiments. There are four steps:

1. Create a data set with all of the elements of the population in the first column.

2. Assign a random number to each element of the population; put these in the second column.

3. Sort the data set by the random number column.

4. The simple random sample is obtained by taking elements in the sorted list until the desired sample size is reached.

We illustrate the procedure with a simplified version of Example 3.23.

EXAMPLE

3.23 Select a random sample. Suppose that the population from Example 3.22 is only the first two rows of the display given there:

Aloha Kai	Captiva	Palm Tree	Sea Shell
Anchor Down	Casa del Mar	Radisson	Silver Beach

Note that we do not need the numerical labels to identify the individuals in the population. Suppose that we want to select a simple random sample of three resorts from this population. Figure 3.8(a) gives the spreadsheet with the population names. The random numbers generated by the RAND() function are given in the second column in Figure 3.8(b). The sorted data set is given in Figure 3.8(c). We have added a third column to the speadsheet to indicate which resorts were selected for our random sample. They are Captiva, Radisson, and Silver Beach.

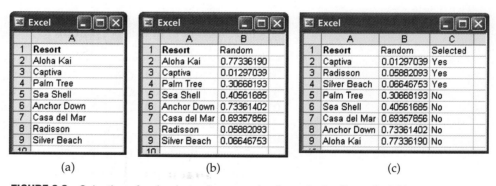

FIGURE 3.8 Selection of a simple random sample of resorts, for Example 3.23.

USE YOUR KNOWLEDGE

3.50 Ringtones for cell phones. You decide to change the ringtones for your cell phone by choosing 2 from a list of the 10 most popular ringtones.[20] Here is the list:

Forever	Empire State of Mind	Party in the USA	Need You Now
Fireflies	Down	Whatcha Say	Halloween
Big Green Tractor	Thriller		

Select your two ringtones using a simple random sample.

3.51 Listen to three songs. The walk to your statistics class takes about 10 minutes, about the amount of time needed to listen to three songs on your iPod. You decide to take a simple random sample of songs from a Billboard List of Rock Songs.[21] Here is the list:

Wheels	Check My Brain	I Will Not Bow	Break
Jars	Uprising	If You're Wondering	The Fixer
Savior	You're Going Down		

Select the three songs for your iPod using a simple random sample.

Stratified samples

The general framework for designs that use chance to choose a sample is a *probability sample.*

> **PROBABILITY SAMPLE**
>
> A **probability sample** is a sample chosen by chance. We must know what samples are possible and what chance, or probability, each possible sample has.

Some probability sampling designs (such as an SRS) give each member of the population an *equal* chance to be selected. This may not be true in more elaborate sampling designs. In every case, however, the use of chance to select the sample is the essential principle of statistical sampling.

Designs for sampling from large populations spread out over a wide area are usually more complex than an SRS. For example, it is common to sample important groups within the population separately, then combine these samples. This is the idea of a *stratified sample.*

> **STRATIFIED RANDOM SAMPLE**
>
> To select a **stratified random sample,** first divide the population into groups of similar individuals, called **strata.** Then choose a separate SRS in each stratum and combine these SRSs to form the full sample.

Choose the strata based on facts known before the sample is taken. For example, a population of election districts might be divided into urban, suburban, and rural strata. A stratified design can produce more exact information than an SRS of the same size by taking advantage of the fact that individuals in the same stratum are similar to one another. Think of the extreme case in which all individuals in each stratum are identical: just one individual from each stratum is then enough to completely describe the population. Strata for sampling are similar to blocks in experiments. We have two names because the idea of grouping similar units before randomizing arose separately in sampling and in experiments.

> **EXAMPLE**
>
> **3.24 A stratified sample of dental claims.** A dentist is suspected of defrauding insurance companies by describing some dental procedures incorrectly on claim forms and overcharging for them. An investigation begins by examining a sample of his bills for the past three years. Because there are five suspicious types of procedures, the investigators take a stratified sample. That is, they randomly select bills for each of the five types of procedures separately.

Multistage samples

Another common means of restricting random selection is to choose the sample in stages. This is common practice for national samples of households or people. For example, data on employment and unemployment are gathered by the government's Current Population Survey, which conducts interviews in about 60,000 households each month. The cost of sending interviewers to the widely scattered households in an SRS would be too high. Moreover, the government wants data broken down by states and large cities. The Current Population Survey therefore uses a **multistage sampling design.** The final sample consists of clusters of nearby households that an interviewer can easily visit. Most opinion polls and other national samples are also multistage, though interviewing in most national samples today is done by telephone rather than in person, eliminating the economic need for clustering. The Current Population Survey sampling design is roughly as follows:[22]

multistage sample

Stage 1. Divide the United States into 2007 geographical areas called Primary Sampling Units, or PSUs. PSUs do not cross state lines. Select a sample of 754 PSUs. This sample includes the 428 PSUs with the largest population and a stratified sample of 326 of the others.

Stage 2. Divide each PSU selected into smaller areas called "blocks." Stratify the blocks using ethnic and other information and take a stratified sample of the blocks in each PSU.

Stage 3. Sort the housing units in each block into clusters of four nearby units. Interview the households in a probability sample of these clusters.

Analysis of data from sampling designs more complex than an SRS takes us beyond basic statistics. But the SRS is the building block of more elaborate designs, and analysis of other designs differs more in complexity of detail than in fundamental concepts.

Cautions about sample surveys

Random selection eliminates bias in the choice of a sample from a list of the population. Sample surveys of large human populations, however, require much more than a good sampling design.[23] To begin, we need an accurate and complete list of the population. Because such a list is rarely available, most samples suffer from some degree of *undercoverage*. A sample survey of households, for example, will miss not only homeless people, but also prison inmates and students in dormitories. An opinion poll conducted by telephone will miss the large number of American households without residential phones. The results of national sample surveys therefore have some bias if the people not covered—who most often are poor people—differ from the rest of the population.

A more serious source of bias in most sample surveys is *nonresponse,* which occurs when a selected individual cannot be contacted or refuses to cooperate. Nonresponse to sample surveys often reaches 50% or more, even with careful planning and several callbacks. Because nonresponse is higher in urban areas, most sample surveys substitute other people in the same area to avoid favoring rural areas in the final sample. If the people contacted differ from those who are rarely at home or who refuse to answer questions, some bias remains.

UNDERCOVERAGE AND NONRESPONSE

Undercoverage occurs when some groups in the population are left out of the process of choosing the sample.

Nonresponse occurs when an individual chosen for the sample can't be contacted or does not cooperate.

EXAMPLE

3.25 Nonresponse in the Current Population Survey. How bad is nonresponse? The Current Population Survey (CPS) has the lowest nonresponse rate of any poll we know: only about 4% of the households in the CPS sample refuse to take part and another 3% or 4% can't be contacted. People are more likely to respond to a government survey such as the CPS, and the CPS contacts its sample in person before doing later interviews by phone.

The General Social Survey (Figure 3.9) is the nation's most important social science research survey. The GSS also contacts its sample in person, and it is run by a university. Despite these advantages, its most recent survey had a 30% rate of nonresponse.

What about polls done by the media and by market research and opinion-polling firms? We don't know their rates of nonresponse, because they won't say. That itself is a bad sign. The Pew Research Center for People and the Press designed a careful telephone survey and published the results: out of 2879 households called, 1658 were never at home, refused, or would not finish the interview. That's a nonresponse rate of 58%.[24]

Most sample surveys, and almost all opinion polls, are now carried out by telephone. This and other details of the interview method can affect the results. When presented with several options for a reply, such as completely agree, mostly agree, mostly disagree, and completely disagree, people tend to be a little more likely to respond to the first one or two options presented.

response bias   The behavior of the respondent or of the interviewer can cause **response bias** in sample results. Respondents may lie, especially if asked about illegal or unpopular behavior. The race or sex of the interviewer can influence responses to questions about race relations or attitudes toward feminism. Answers to questions that ask respondents to recall past events are often inaccurate because of faulty memory. For example, many people "telescope" events in the

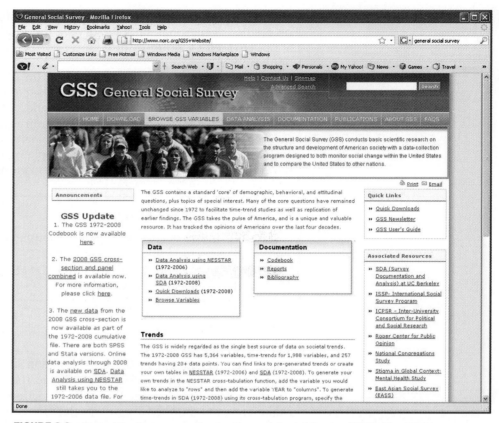

FIGURE 3.9 Part of the home page for the General Social Survey (GSS). The GSS has assessed attitudes on a wide variety of topics since 1972. Its continuity over time makes the GSS a valuable source for studies of changing attitudes.

past, bringing them forward in memory to more recent time periods. "Have you visited a dentist in the last 6 months?" will often elicit a "Yes" from someone who last visited a dentist 8 months ago.[25]

EXAMPLE

3.26 Overreporting of voter behavior. "One of the most frequently observed survey measurement errors is the overreporting of voting behavior."[26] People know they should vote, so those who didn't vote tend to save face by saying that they did. Here are the data from a typical sample of 663 people after an election:

		What they said:	
		I voted	I didn't
What they did:	Voted	358	13
	Didn't vote	120	172

You can see that 478 people (72%) said that they voted, but only 371 people (56%) actually did vote.

The **wording of questions** is the most important influence on the answers given to a sample survey. Confusing or leading questions can introduce strong bias, and even minor changes in wording can change a survey's outcome. Here are some examples.

> ### EXAMPLE
>
> **3.27 The form of the question is important.** In response to the question "Are you heterosexual, homosexual, or bisexual?" in a social science research survey, one woman answered, "It's just me and my husband, so bisexual." The issue is serious, even if the example seems silly: reporting about sexual behavior is difficult because people understand and misunderstand sexual terms in many ways.
>
> How do Americans feel about government help for the poor? Only 13% think we are spending too much on "assistance to the poor," but 44% think we are spending too much on "welfare." How do the Scots feel about the movement to become independent from England? Well, 51% would vote for "independence for Scotland," but only 34% support "an independent Scotland separate from the United Kingdom." It seems that "assistance to the poor" and "independence" are nice, hopeful words. "Welfare" and "separate" are negative words.[27]

The statistical design of sample surveys is a science, but this science is only part of the art of sampling. Because of nonresponse, response bias, and the difficulty of posing clear and neutral questions, you should hesitate to fully trust reports about complicated issues based on surveys of large human populations. *Insist on knowing the exact questions asked, the rate of nonresponse, and the date and method of the survey before you trust a poll result.*

SECTION 3.2 Summary

A sample survey selects a **sample** from the **population** of all individuals about which we desire information. We base conclusions about the population on data about the sample.

The **design** of a sample refers to the method used to select the sample from the population. **Probability sampling designs** use impersonal chance to select a sample.

The basic probability sample is a **simple random sample (SRS).** An SRS gives every possible sample of a given size the same chance to be chosen.

Choose an SRS by labeling the members of the population and using a **table of random digits** to select the sample. Software can automate this process.

To choose a **stratified random sample,** divide the population into **strata,** groups of individuals that are similar in some way that is important to the response. Then choose a separate SRS from each stratum and combine them to form the full sample.

Multistage samples select successively smaller groups within the population in stages, resulting in a sample consisting of clusters of individuals. Each stage may employ an SRS, a stratified sample, or another type of sample.

Failure to use probability sampling often results in **bias,** or systematic errors in the way the sample represents the population. **Voluntary response** samples, in which the respondents choose themselves, are particularly prone to large bias.

In human populations, even probability samples can suffer from bias due to **undercoverage** or **nonresponse,** from **response bias** due to the behavior of the interviewer or the respondent, or from misleading results due to **poorly worded questions.**

SECTION 3.2   Exercises

For Exercises 3.48 and 3.49, see pages 189–190; and for Exercises 3.50 and 3.51, see pages 192–193.

3.52 What kind of sample? In each of the following situations, describe the sample as an SRS, a stratified random sample, a multistage sample, or a voluntary response sample. Explain your answers.

(a) There are seven sections of an introductory statistics course. A random sample of three sections is chosen and then random samples of 8 students from each of these sections are chosen.

(b) A student organization has 55 members. A table of random numbers is used to select a sample of 5.

(c) An online poll asks people who visit this site to choose their favorite television show.

(d) Separate random samples of male and female first-year college students in an introductory psychology are selected to receive a one-week alternate instructional method.

3.53 What's wrong? Explain what is wrong in each of the following scenarios.

(a) The population consists of all individuals selected in a simple random sample.

(b) In a poll of an SRS of residents in a local community, respondents are asked to indicate the level of their concern about the dangers of dihydrogen monoxide, a substance that is a major component of acid rain and in its gaseous state can cause severe burns. (*Hint*: Ask a friend who is majoring in chemistry about this substance or search the Internet for information about it.)

(c) Students in a class are asked to raise their hands if they have cheated on an exam one or more times within the past year.

3.54 What's wrong? Explain what is wrong with each of the following random selection procedures and explain how you would do the randomization correctly.

(a) To determine the reading level of an introductory statistics text, you evaluate all the written material in the third chapter.

(b) You want to sample student opinions about a proposed change in procedures for changing majors. You hand out questionnaires to 100 students as they arrive for class at 7:30 A.M.

(c) A population of subjects is put in alphabetical order and a simple random sample of size 10 is taken by selecting the first 10 subjects in the list.

3.55 Importance of students as customers. A committee on community relations in a college town plans to survey local businesses about the importance of students as customers. From telephone book listings, the committee chooses 150 businesses at random. Of these, 73 return the questionnaire mailed by the committee. What is the population for this sample survey? What is the sample? What is the rate (percent) of nonresponse?

3.56 ⚔ **Consumer spending.** A Gallup Poll used telephone interviews to collect data on consumer spending on different days of the week.[28] Here are the averages (in dollars) for each day of the week:

Monday	59	Friday	63
Tuesday	56	Saturday	73
Wednesday	55	Sunday	76
Thursday	59		

(a) Display the data graphically and write a short paragraph describing these averages.

(b) The data were collected between January 2 and October 21, 2009. Discuss how this choice may have affected the results.

3.57 ⚔ **Which channel do you watch for news?** A Pew Research Center survey asked people about what channel they regularly watch for news and their party identification.[29] For one analysis they focused on those who regularly watch the Fox News Channel, CNN, MSNBC, and nightly network news. Here are the political profiles (in percents) for each of these news sources:

Party	Fox	CNN	MSNBC	Network
Republican	39	18	18	22
Democrat	33	51	45	45
Independent	22	23	27	26
Other/Don't know	6	8	10	7

Display the data graphically and write a report summarizing the results.

3.58 Identify the populations. For each of the following sampling situations, identify the population as exactly as possible. That is, say what kind of individuals the population consists of and say exactly which individuals fall in the population. If the information given is not complete, complete the description of the population in a reasonable way. RESIDENTS

(a) A college has changed its core curriculum and wants to obtain detailed feedback information from the students during each of the first 12 weeks of the coming semester. Each week, a random sample of 5 students will be selected to be interviewed.

(b) The American Community Survey (ACS) will replace the census "long form" starting with the 2010 census. The main part of the ACS contacts 250,000 addresses by mail each month, with follow-up by phone and in person if there is no response. Each household answers questions about their housing, economic, and social status.

(c) An opinion poll contacts 1161 adults and asks them, "Which political party do you think has better ideas for leading the country in the twenty-first century?"

3.59 Interview residents of apartment complexes. You are planning a report on apartment living in a college town. You decide to select 5 apartment complexes at random for in-depth interviews with residents. Select a simple random sample of 5 of the following apartment complexes. If you use Table B, start at line 137.

Ashley Oaks	Country View	Mayfair Village
Bay Pointe	Country Villa	Nobb Hill
Beau Jardin	Crestview	Pemberly Courts
Bluffs	Del-Lynn	Peppermill
Brandon Place	Fairington	Pheasant Run
Briarwood	Fairway Knolls	Richfield
Brownstone	Fowler	Sagamore Ridge
Burberry	Franklin Park	Salem Courthouse
Cambridge	Georgetown	Village Manor
Chauncey Village	Greenacres	Waterford Court
Country Squire	Lahr House	Williamsburg

3.60 Using GIS to identify mint field conditions. A Geographic Information System (GIS) is to be used to distinguish different conditions in mint fields. Ground observations will be used to classify regions of each field as either healthy mint, diseased mint, or weed-infested mint. The GIS divides mint-growing areas into regions called pixels. An experimental area contains 200 pixels. For a random sample of 25 pixels, ground measurements will be made to determine the status of the mint, and these observations will be compared with information obtained by the GIS. Select the random sample. If you use Table B, start at line 112 and choose only the first 5 pixels in the sample.

3.61 **Use the simple random sample applet.** After you have labeled the individuals in a population, the *Simple Random Sample* applet automates the task of choosing an SRS. Use the applet to choose the sample in the previous exercise.

3.62 **Use the simple random sample applet.** There are approximately 371 active telephone area codes covering Canada, the United States, and some Caribbean areas. (More are created regularly.) You want to choose an SRS of 25 of these area codes for a study of available telephone numbers. Label the codes 001 to 371 and use the *Simple Random Sample* applet to choose your sample. (If you use Table B, start at line 120 and choose only the first 5 codes in the sample.)

3.63 Census tracts. The Census Bureau divides the entire country into "census tracts" that contain about 4000 people. Each tract is in turn divided into small "blocks," which in urban areas are bounded by local streets. An SRS of blocks from a census tract is often the next-to-last stage in a multistage sample. Figure 3.10 shows part of census tract 8051.12, in Cook County, Illinois, west of

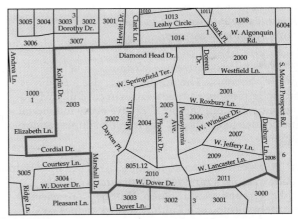

FIGURE 3.10 Census blocks in Cook County, Illinois, for Exercises 3.63 and 3.65. The outlined area is a block group.

Chicago. The 44 blocks in this tract are divided into three "block groups." Group 1 contains 6 blocks numbered 1000 to 1005; Group 2 (outlined in Figure 3.10) contains 12 blocks numbered 2000 to 2011; Group 3 contains 26 blocks numbered 3000 to 3025. Use Table B, beginning at line 135, to choose an SRS of 5 of the 44 blocks in this census tract. Explain carefully how you labeled the blocks.

3.64 Repeated use of Table B. In using Table B repeatedly to choose samples or do randomization for experiments, you should not always begin at the same place, such as line 101. Why not?

3.65 A stratified sample. Exercise 3.63 asks you to choose an SRS of blocks from the census tract pictured in Figure 3.10. You might instead choose a stratified sample of one block from the 6 blocks in Group 1, two from the 12 blocks in Group 2, and three from the 26 blocks in Group 3. Choose such a sample, explaining carefully how you labeled blocks and used Table B.

3.66 Systematic random samples. Systematic random samples are often used to choose a sample of apartments in a large building or dwelling units in a block at the last stage of a multistage sample. An example will illustrate the idea of a systematic sample. Suppose that we must choose 4 addresses out of 100. Because $100/4 = 25$, we can think of the list as four lists of 25 addresses. Choose 1 of the first 25 at random, using Table B. The sample contains this address and the addresses 25, 50, and 75 places down the list from it. If 13 is chosen, for example, then the systematic random sample consists of the addresses numbered 13, 38, 63, and 88.

(a) A study of dating among college students wanted a sample of 200 of the 9000 single male students on campus. The sample consisted of every 45th name from a list of the 9000 students. Explain why the survey chooses every 45th name.

(b) Use Table B at line 125 to choose the starting point for this systematic sample.

3.67 Systematic random samples versus simple random samples. The previous exercise introduces systematic random samples. Explain carefully why a systematic random sample *does* give every individual the same chance to be chosen but is *not* a simple random sample.

3.68 Random digit telephone dialing. An opinion poll in California uses random digit dialing to choose telephone numbers at random. Numbers are selected separately within each California area code. The size of the sample in each area code is proportional to the population living there. AREACODES

(a) What is the name for this kind of sampling design?

(b) California area codes, in rough order from north to south, are

530 707 916 209 415 925 510 650 408 831 805 559 760
661 818 213 626 323 562 709 310 949 909 858 619

Another California survey does not call numbers in all area codes but starts with an SRS of 10 area codes. Choose such an SRS. If you use Table B, start at line 122.

3.69 Stratified samples of forest areas. Stratified samples are widely used to study large areas of forest. Based on satellite images, a forest area in the Amazon basin is divided into 14 types. Foresters studied the four most commercially valuable types: alluvial climax forests of quality levels 1, 2, and 3, and mature secondary forest. They divided the area of each type into large parcels, chose parcels of each type at random, and counted tree species in a 20- by 25-meter rectangle randomly placed within each parcel selected. Here is some detail:

Forest type	Total parcels	Sample size
Climax 1	36	4
Climax 2	72	7
Climax 3	31	3
Secondary	42	4

Choose the stratified sample of 18 parcels. Be sure to explain how you assigned labels to parcels. If you use Table B, start at line 140.

3.70 Select club members to go to a convention. A club has 30 student members and 10 faculty members. The students are

Abel	Fisher	Huber	Moran	Reinmann
Carson	Golomb	Jimenez	Moskowitz	Santos
Chen	Griswold	Jones	Neyman	Shaw
David	Hein	Kiefer	O'Brien	Thompson
Deming	Hernandez	Klotz	Pearl	Utts
Elashoff	Holland	Liu	Potter	Vlasic

and the faculty members are

Andrews	Fernandez	Kim	Moore	Rabinowitz
Besicovitch	Gupta	Lightman	Phillips	Yang

The club can send 5 students and 3 faculty members to a convention and decides to choose those who will go by random selection. Select a stratified random sample of 5 students and 3 faculty members.

3.71 ⚠️ **Stratified samples for alcohol attitudes.** At a party there are 30 students over age 21 and 20 students under age 21. You choose at random 3 of those over 21 and separately choose at random 2 of those under 21 to interview about attitudes toward alcohol. You have given every student at the party the same chance to be interviewed: what is that chance? Why is your sample not an SRS?

3.72 Stratified samples for accounting audits. Accountants use stratified samples during audits to verify a company's records of such things as accounts receivable. The stratification is based on the dollar amount of the item and often includes 100% sampling of the largest items. One company reports 5000 accounts receivable. Of these, 100 are in amounts over $50,000; 500 are in amounts between $1000 and $50,000; and the remaining 4400 are in amounts under $1000. Using these groups as strata, you decide to verify all of the largest accounts and to sample 5% of the midsize accounts and 1% of the small accounts. How would you label the two strata from which you will sample? Use Table B, starting at line 115, to select the first 5 accounts from each of these strata.

3.73 The sampling frame. The list of individuals from which a sample is actually selected is called the **sampling frame.** Ideally, the frame should list every individual in the population, but in practice this is often difficult. A frame that leaves out part of the population is a common source of undercoverage.

(a) Suppose that a sample of households in a community is selected at random from the telephone directory. What households are omitted from this frame? What types of people do you think are likely to live in these households? These people will probably be underrepresented in the sample.

(b) It is usual in telephone surveys to use random digit dialing equipment that selects the last four digits of a telephone number at random after being given the area code and the exchange. The exchange is the first three digits of the telephone number. Which of the households that you mentioned in your answer to (a) will be included in the sampling frame by random digit dialing?

3.74 Survey questions. Comment on each of the following as a potential sample survey question. Is the question clear? Is it slanted toward a desired response?

(a) "Some cell phone users have developed brain cancer. Should all cell phones come with a warning label explaining the danger of using cell phones?"

(b) "Do you agree that a national system of health insurance should be favored because it would provide health insurance for everyone and would reduce administrative costs?"

(c) "In view of escalating environmental degradation and incipient resource depletion, would you favor economic incentives for recycling of resource-intensive consumer goods?"

3.75 Use of a budget surplus. In 2000, when the federal budget showed a large surplus, the Pew Research Center asked two questions of random samples of adults. Both questions stated that Social Security would be "fixed." Here are the uses suggested for the remaining surplus:

> *Should the money be used for a tax cut, or should it be used to fund new government programs?*
> *Should the money be used for a tax cut, or should it be spent on programs for education, the environment, health care, crime fighting and military defense?*

One of these questions drew 60% favoring a tax cut; the other, only 22%. Which wording pulls respondents toward a tax cut? Why?

3.76 ⚠️ **How many children are in your family?** A teacher asks her class, "How many children are there in your family, including yourself?" The mean response is about 3 children. According to the U.S. Census Bureau, in 2008, families that have children average 1.86 children. Why is a sample like this biased toward higher outcomes?

3.77 ⚠️ **Bad survey questions.** Write your own examples of bad sample survey questions.

(a) Write a biased question designed to get one answer rather than another.

(b) Write a question that is confusing, so that it is hard to answer.

3.78 ⚠️ **Economic attitudes of Spaniards.** Spain's Centro de Investigaciones Sociológicos carried out a sample survey on the economic attitudes of Spaniards.[30] Of the 2496 adults interviewed, 72% agreed that "Employees with higher performance must get higher pay." On the other hand, 71% agreed that "Everything a society produces should be distributed among its members as equally as possible and there should be no major differences." Use these conflicting results as an example in a short explanation of why opinion polls often fail to reveal public attitudes clearly.

3.3 Toward Statistical Inference

A market research firm interviews a random sample of 2500 adults. Result: 66% find shopping for clothes frustrating and time consuming. That's the truth about the 2500 people in the sample. What is the truth about the almost 220 million American adults who make up the population? Because the sample was chosen at random, it's reasonable to think that these 2500 people represent the entire population fairly well. So the market researchers turn the *fact* that 66% of the *sample* find shopping frustrating into an *estimate* that about 66% of *all adults* feel this way. That's a basic move in statistics: use a fact about a sample to estimate the truth about the whole population. We call this **statistical inference** because we infer conclusions about the wider population from data on selected individuals. To think about inference, we must keep straight whether a number describes a sample or a population. Here is the vocabulary we use.

statistical inference

PARAMETERS AND STATISTICS

A **parameter** is a number that describes the **population.** A parameter is a fixed number, but in practice we do not know its value.

A **statistic** is a number that describes a **sample.** The value of a statistic is known when we have taken a sample, but it can change from sample to sample. We often use a statistic to estimate an unknown parameter.

EXAMPLE

3.28   Building loyalty in your customers. Henley Centre HeadlightVision and Yankelovich market uses *Dollars & Consumer Sense* to provide clients with research about maintaining and improving their business. One phase of the research focuses on customer loyalty. They use a Web interface to collect data from between 1000 and 2500 potential customers using 30- to 40-minute surveys.[31] Let's assume that 1650 out of 2500 potential customers in a sample show strong interest in a product. The proportion of the sample who are interested is

$$\hat{p} = \frac{1650}{2500} = 0.66 = 66\%$$

The number $\hat{p} = 0.66$ is a *statistic*. The corresponding *parameter* is the proportion (call it p) of all potential customers who would have expressed interest in this product if they had been asked. We don't know the value of the parameter p, so we use the statistic $\hat{p}$ to estimate it.

USE YOUR KNOWLEDGE

3.79   Sexual harassment of college students. A recent survey of 2036 undergraduate college students aged 18 to 24 reports that 62% of college students say they have encountered some type of sexual harassment while at college.[32] Describe the sample and the population for this setting.

3.80 Web polls. If you connect to the Web site `wnd.com/polls`, you will be given the opportunity to give your opinion about a different question of public interest each day. Can you apply the ideas about populations and samples that we have just discussed to this poll? Explain why or why not.

Sampling variability

sampling variability

If Yankelovich took a second random sample of 2500 adults, the new sample would have different people in it. It is almost certain that there would not be exactly 1650 positive responses. That is, the value of the statistic $\hat{p}$ will vary from sample to sample. This basic fact is called **sampling variability:** the value of a statistic varies in repeated random sampling. Could it happen that one random sample finds that 66% of potential customers are interested in this product and a second random sample finds that only 42% would express interest? Random samples eliminate *bias* from the act of choosing a sample, but they can still be wrong because of the *variability* that results when we choose at random. If the variation when we take repeat samples from the same population is too great, we can't trust the results of any one sample.

We are saved by the second great advantage of random samples. The first advantage is that choosing at random eliminates favoritism. That is, random sampling attacks bias. The second advantage is that if we take lots of random samples of the same size from the same population, the variation from sample to sample will follow a predictable pattern. **All of statistical inference is based on one idea: to see how trustworthy a procedure is, ask what would happen if we repeated it many times.**

To understand why sampling variability is not fatal, we ask, "What would happen if we took many samples?" Here's how to answer that question:

- Take a large number of samples from the same population.
- Calculate the sample proportion $\hat{p}$ for each sample.
- Make a histogram of the values of $\hat{p}$.
- Examine the distribution displayed in the histogram for shape, center, and spread, as well as outliers or other deviations.

simulation

In practice it is too expensive to take many samples from a large population such as all adult U.S. residents. But we can imitate many samples by using random digits. Using random digits from a table or computer software to imitate chance behavior is called **simulation.**

EXAMPLE

3.29 Simulate a random sample. We will simulate drawing simple random samples (SRSs) of size 100 from the population of potential customers. Suppose that in fact 60% of the population have interest in the product. Then the true value of the parameter we want to estimate is $p = 0.6$. (Of course, we would not sample in practice if we already knew that $p = 0.6$. We are sampling here to understand how sampling behaves.)

We can imitate the population by a table of random digits, with each entry standing for a person. Six of the 10 digits (say 0 to 5) stand for people who have interest in the product. The remaining four digits, 6 to 9, stand for those who do not. Because all digits in a random number table are equally likely, this assignment produces a population proportion of frustrated shoppers equal to $p = 0.6$. We then imitate an SRS of 100 people from the population by taking 100 consecutive digits from Table B. The statistic $\hat{p}$ is the proportion of 0s to 5s in the sample.

Here are the first 100 entries in Table B with digits 0 to 5 highlighted:

19223	95034	05756	28713	96409	12531	42544	82853
73676	47150	99400	01927	27754	42648	82425	36290
45467	71709	77558	00095				

There are 64 digits between 0 and 5, so $\hat{p} = 64/100 = 0.64$. A second SRS based on the second 100 entries in Table B gives a different result, $\hat{p} = 0.55$. The two sample results are different, and neither is equal to the true population value $p = 0.6$. That's sampling variability.

Sampling distributions

Simulation is a powerful tool for studying chance. Now that we see how simulation works, it is faster to abandon Table B and to use a computer programmed to generate random numbers.

EXAMPLE

3.30   Take many random samples. Figure 3.11 illustrates the process of choosing many samples and finding the sample proportion $\hat{p}$ for each one. Follow the flow of the figure from the population at the left, to choosing an SRS and finding the $\hat{p}$ for this sample, to collecting together the $\hat{p}$'s from many samples. The histogram at the right of the figure shows the distribution of the

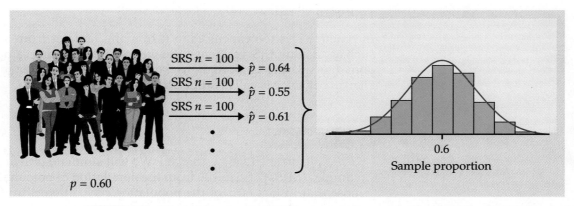

FIGURE 3.11   The results of many SRSs have a regular pattern. Here we draw 1000 SRSs of size 100 from the same population. The population proportion is $p = 0.60$. The histogram shows the distribution of 1000 sample proportions.

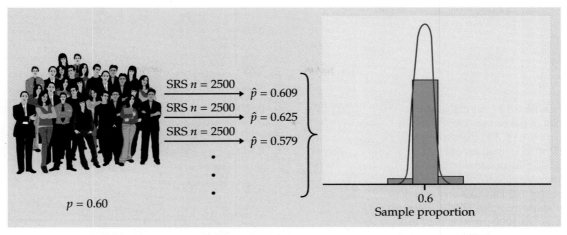

FIGURE 3.12 The distribution of sample proportions for 1000 SRSs of size 2500 drawn from the same population as in Figure 3.11. The two histograms have the same scale. The statistic from the larger sample is less variable.

values of $\hat{p}$ from 1000 separate SRSs of size 100 drawn from a population with $p = 0.6$.

Of course, Yankelovich samples 2500 people, not just 100. Figure 3.12 is parallel to Figure 3.11. It shows the process of choosing 1000 SRSs, each of size 2500, from a population in which the true proportion is $p = 0.6$. The 1000 values of $\hat{p}$ from these samples form the histogram at the right of the figure. Figures 3.11 and 3.12 are drawn on the same scale. Comparing them shows what happens when we increase the size of our samples from 100 to 2500. These histograms display the *sampling distribution* of the statistic $\hat{p}$ for two sample sizes.

SAMPLING DISTRIBUTION

The **sampling distribution** of a statistic is the distribution of values taken by the statistic in all possible samples of the same size from the same population.

Strictly speaking, the sampling distribution is the ideal pattern that would emerge if we looked at all possible samples of size 100 from our population. A distribution obtained from a fixed number of trials, like the 1000 trials in Figure 3.11, is only an approximation to the sampling distribution. We will see that probability theory, the mathematics of chance behavior, can sometimes describe sampling distributions exactly. The interpretation of a sampling distribution is the same, however, whether we obtain it by simulation or by the mathematics of probability.

We can use the tools of data analysis to describe any distribution. Let's apply those tools to Figures 3.11 and 3.12.

• **Shape:** The histograms look Normal. Figure 3.13 is a Normal quantile plot of the values of $\hat{p}$ for our samples of size 100. It confirms that the

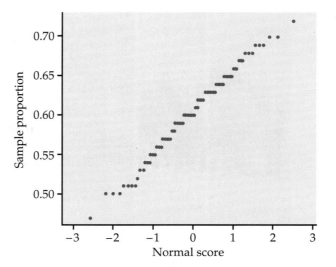

FIGURE 3.13 Normal quantile plot of the sample proportions in Figure 3.11. The distribution is close to Normal except for some granularity due to the fact that sample proportions from a sample size of 100 can take only values that are multiples of 0.01. Because a plot of 1000 points is hard to read, this plot presents only every 10th value.

distribution in Figure 3.11 is close to Normal. The 1000 values for samples of size 2500 in Figure 3.12 are even closer to Normal. The Normal curves drawn through the histograms describe the overall shape quite well.

- **Center:** In both cases, the values of the sample proportion $\hat{p}$ vary from sample to sample, but the values are centered at 0.6. Recall that $p = 0.6$ is the true population parameter. Some samples have a $\hat{p}$ less than 0.6 and some greater, but there is no tendency to be always low or always high. That is, $\hat{p}$ has no **bias** as an estimator of p. This is true for both large and small samples. (Want the details? The mean of the 1000 values of $\hat{p}$ is 0.598 for samples of size 100 and 0.6002 for samples of size 2500. The median value of $\hat{p}$ is exactly 0.6 for samples of both sizes.)

- **Spread:** The values of $\hat{p}$ from samples of size 2500 are much less spread out than the values from samples of size 100. In fact, the standard deviations are 0.051 for Figure 3.11 and 0.0097, or about 0.01, for Figure 3.12.

Although these results describe just two sets of simulations, they reflect facts that are true whenever we use random sampling.

USE YOUR KNOWLEDGE

3.81 Effect of sample size on the sampling distribution. You are planning a study and are considering taking an SRS of either 200 or 400 observations. Explain how the sampling distribution would differ for these two scenarios.

Bias and variability

Our simulations show that a sample of size 2500 will almost always give an estimate $\hat{p}$ that is close to the truth about the population. Figure 3.12 illustrates this fact for just one value of the population proportion, but it is true for any population. Samples of size 100, on the other hand, might give an estimate of 50% or 70% when the truth is 60%.

Thinking about Figures 3.11 and 3.12 helps us restate the idea of bias when we use a statistic like $\hat{p}$ to estimate a parameter like p. It also reminds us that variability matters as much as bias.

BIAS AND VARIABILITY

Bias concerns the center of the sampling distribution. A statistic used to estimate a parameter is **unbiased** if the mean of its sampling distribution is equal to the true value of the parameter being estimated.

The **variability of a statistic** is described by the spread of its sampling distribution. This spread is determined by the sampling design and the sample size n. Statistics from larger probability samples have smaller spreads.

We can think of the true value of the population parameter as the bull's-eye on a target, and of the sample statistic as an arrow fired at the bull's-eye. Bias and variability describe what happens when an archer fires many arrows at the target. *Bias* means that the aim is off, and the arrows land consistently off the bull's-eye in the same direction. The sample values do not center about the population value. Large *variability* means that repeated shots are widely scattered on the target. Repeated samples do not give similar results but differ widely among themselves. Figure 3.14 shows this target illustration of the two types of error.

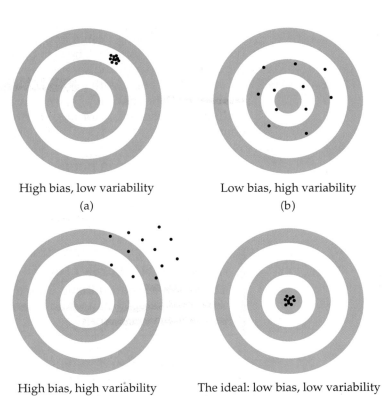

High bias, low variability
(a)

Low bias, high variability
(b)

High bias, high variability
(c)

The ideal: low bias, low variability
(d)

FIGURE 3.14 Bias and variability in shooting arrows at a target. Bias means the archer systematically misses in the same direction. Variability means that the arrows are scattered.

Notice that small variability (repeated shots are close together) can accompany large bias (the arrows are consistently away from the bull's-eye in one direction). And small bias (the arrows center on the bull's-eye) can accompany large variability (repeated shots are widely scattered). A good sampling scheme, like a good archer, must have both small bias and small variability. Here's how we do this.

MANAGING BIAS AND VARIABILITY

To reduce bias, use random sampling. When we start with a list of the entire population, simple random sampling produces unbiased estimates—the values of a statistic computed from an SRS neither consistently overestimate nor consistently underestimate the value of the population parameter.

To reduce the variability of a statistic from an SRS, use a **larger sample**. You can make the variability as small as you want by taking a large enough sample.

In practice, Yankelovich takes only one sample. We don't know how close to the truth an estimate from this one sample is because we don't know what the truth about the population is. But *large random samples almost always give an estimate that is close to the truth.* Looking at the pattern of many samples shows that we can trust the result of one sample. The Current Population Survey's sample of 60,000 households estimates the national unemployment rate very accurately. Of course, only probability samples carry this guarantee. Using a probability sampling design and taking care to deal with practical difficulties reduce bias in a sample. The size of the sample then determines how close to the population truth the sample result is likely to fall. Results from a sample survey usually come with a **margin of error** that sets bounds on the size of the likely error. The margin of error directly reflects the variability of the sample statistic, so it is smaller for larger samples. We will describe the details in later chapters.

margin of error

Sampling from large populations

Yankelovich's sample of 2500 adults is only about 1 out of every 90,000 adults in the United States. Does it matter whether we sample 1-in-100 individuals in the population or 1-in-90,000?

POPULATION SIZE DOESN'T MATTER

The variability of a statistic from a random sample does not depend on the size of the population, as long as the population is at least 100 times larger than the sample.

Why does the size of the population have little influence on the behavior of statistics from random samples? To see why this is plausible, imagine sampling harvested corn by thrusting a scoop into a lot of corn kernels. The scoop doesn't know whether it is surrounded by a bag of corn or by an entire truckload. As long as the corn is well mixed (so that the scoop selects a random sample), the variability of the result depends only on the size of the scoop.

The fact that the variability of sample results is controlled by the size of the sample has important consequences for sampling design. An SRS of size 2500 from the 220 million adult residents of the United States gives results as precise as an SRS of size 2500 from the 665,000 adult inhabitants of San Francisco. This is good news for designers of national samples but bad news for those who want accurate information about the citizens of San Francisco. If both use an SRS, both must use the same size sample to obtain equally trustworthy results.

Why randomize?

Why randomize? The act of randomizing guarantees that the results of analyzing our data are subject to the laws of probability. The behavior of statistics is described by a sampling distribution. The form of the distribution is known, and in many cases is approximately Normal. Often the center of the distribution lies at the true parameter value, so that the notion that randomization eliminates bias is made more precise. The spread of the distribution describes the variability of the statistic and can be made as small as we wish by choosing a large enough sample. In a randomized experiment, we can reduce variability by choosing larger groups of subjects for each treatment.

These facts are at the heart of formal statistical inference. Later chapters will have much to say in more technical language about sampling distributions and the way statistical conclusions are based on them. What any user of statistics must understand is that all the technical talk has its basis in a simple question: *What would happen if the sample or the experiment were repeated many times?* The reasoning applies not only to an SRS but also to the complex sampling designs actually used by opinion polls and other national sample surveys. The same conclusions hold as well for randomized experimental designs. The details vary with the design but the basic facts are true whenever randomization is used to produce data.

Remember that proper statistical design is not the only aspect of a good sample or experiment. *The sampling distribution shows only how a statistic varies due to the operation of chance in randomization. It reveals nothing about possible bias due to undercoverage or nonresponse in a sample, or to lack of realism in an experiment.* The actual error in estimating a parameter by a statistic can be much larger than the sampling distribution suggests. What is worse, there is no way to say how large the added error is. The real world is less orderly than statistics textbooks imply.

BEYOND THE BASICS

Capture-recapture sampling

Sockeye salmon return to reproduce in the river where they were hatched four years earlier. How many salmon survived natural perils and heavy fishing to make it back this year? How many mountain sheep are there in Colorado? Are migratory songbird populations in North America decreasing or holding their own? These questions concern the size of animal populations. Biologists address them with a special kind of repeated sampling, called *capture-recapture sampling.*

EXAMPLE

3.31 Estimate the number of least flycatchers. You are interested in the number of least flycatchers migrating along a major route in the north-central United States. You set up "mist nets" that capture the birds but do not harm them. The birds caught in the net are fitted with a small aluminum leg band and released. Last year you banded and released 200 least flycatchers. This year you repeat the process. Your net catches 120 least flycatchers, 12 of which have tags from last year's catch.

The proportion of your second sample that have bands should estimate the proportion in the entire population that are banded. So if N is the unknown number of least flycatchers, we should have approximately

proportion banded in sample = proportion banded in population

$$\frac{12}{120} = \frac{200}{N}$$

Solve for N to estimate that the total number of flycatchers migrating while your net was up this year is approximately

$$N = 200 \times \frac{120}{12} = 2000$$

The capture-recapture idea extends the use of a sample proportion to estimate a population proportion. The idea works well if both samples are SRSs from the population and the population remains unchanged between samples. In practice, complications arise because, for example, some of the birds tagged last year died before this year's migration. Variations on capture-recapture samples are widely used in wildlife studies and are now finding other applications. One way to estimate the census undercount in a district is to consider the census as "capturing and marking" the households that respond. Census workers then visit the district, take an SRS of households, and see how many of those counted by the census show up in the sample. Capture-recapture estimates the total count of households in the district. As with estimating wildlife populations, there are many practical pitfalls. Our final word is as before: the real world is less orderly than statistics textbooks imply.

SECTION 3.3 Summary

A number that describes a population is a **parameter.** A number that can be computed from the data is a **statistic.** The purpose of sampling or experimentation is usually **inference:** use sample statistics to make statements about unknown population parameters.

A statistic from a probability sample or randomized experiment has a **sampling distribution** that describes how the statistic varies in repeated data production. The sampling distribution answers the question "What would happen if we repeated the sample or experiment many times?" Formal statistical inference is based on the sampling distributions of statistics.

A statistic as an estimator of a parameter may suffer from **bias** or from high **variability.** Bias means that the center of the sampling distribution is not equal to the true value of the parameter. The variability of the statistic is described by the spread of its sampling distribution. Variability is usually reported by giving a **margin of error** for conclusions based on sample results.

Properly chosen statistics from randomized data production designs have no bias resulting from the way the sample is selected or the way the experimental units are assigned to treatments. We can reduce the variability of the statistic by increasing the size of the sample or the size of the experimental groups.

SECTION 3.3 **Exercises**

For Exercises 3.79 and 3.80, see pages 202–203; and for Exercise 3.81, see page 206.

3.82 What's wrong? State what is wrong in each of the following scenarios.

(a) A parameter describes a sample.

(b) Bias and variability are two names for the same thing.

(c) Large samples are always better than small samples.

(d) A sampling distribution is something generated by a computer.

3.83 Describe the population and the sample. For each of the following situations, describe the population and the sample.

(a) A survey of 17,096 students in U.S. four-year colleges reported that 19.4% were binge drinkers.

(b) In a study of work stress, 100 restaurant workers were asked about the impact of work stress on their personal lives.

(c) A tract of forest has 584 longleaf pine trees. The diameters of 40 of these trees were measured.

3.84 Bias and variability. Figure 3.15 shows histograms of four sampling distributions of statistics intended to estimate the same parameter. Label each distribution relative to the others as high or low bias and as high or low variability.

3.85 Sampling invoices. We will illustrate the idea of a sampling distribution in the case of a very small sample from a very small population. The population contains 10 past due invoices. Here are the number of days each invoice is past due:

Invoice	0	1	2	3	4	5	6	7	8	9
Days past due	8	12	10	5	7	3	15	9	7	6

The parameter of interest is the mean of this population, which is 8.2. The sample is an SRS of $n = 4$ invoices

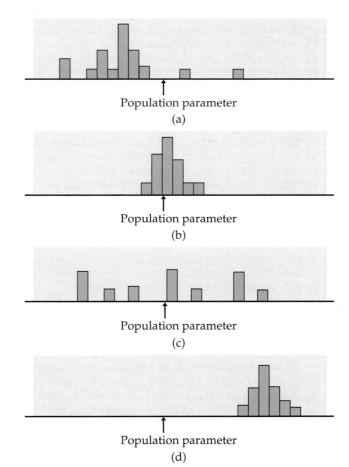

FIGURE 3.15 Determine which of these sampling distributions displays high or low bias and high or low variability, for Exercise 3.84.

drawn from the population. Because the invoices are labeled 0 to 9, a single random digit from Table B chooses one invoice for the sample.

(a) Use Table B to draw an SRS of size 4 from this population. Write the past due values for the days in your sample and calculate their mean $\bar{x}$. This statistic is an estimate of the population parameter.

(b) Repeat this process 10 times. Make a histogram of the 10 values of $\bar{x}$. You are constructing the sampling distribution of $\bar{x}$. Is the center of your histogram close to 8.2? (Ten repetitions give only a crude approximation to the sampling distribution. If possible, pool your work with that of other students—using different parts of Table B— to obtain at least 100 repetitions. A histogram of these values of $\bar{x}$ is a better approximation to the sampling distribution.)

3.86 Gallup Canada polls. Gallup Canada bases its polls of Canadian public opinion on telephone samples of about 1000 adults, the same sample size as Gallup uses in the United States. Canada's population is about one-ninth as large as that of the United States, so the percent of adults that Gallup interviews in Canada is nine times as large as in the United States. Does this mean that the margin of error for a Gallup Canada poll is smaller? Explain your answer.

3.87 People want to move. The headline of a report on a Gallup Poll says "700 million worldwide desire to migrate permanently." They estimate that about 16% of the world's adults would like to move to another country.[33]

(a) This poll took samples of adults aged 15 and over in 135 countries. What is this type of sampling procedure called?

(b) These countries represent 93% of the world's adult population. Do you think that this means that the survey is seriously biased? Give reasons for your answer.

(c) The report says that the margin of error for individual countries ranges from ±3% to ±6%. Explain what this means in simple terms.

3.88 The health care system in Ontario. The Ministry of Health in the Canadian province of Ontario wants to know whether the national health care system is achieving its goals in the province. The ministry conducted the Ontario Health Survey, which interviewed a probability sample of 61,239 adults who live in Ontario.[34]

(a) What is the population for this sample survey? What is the sample?

(b) The survey found that 76% of males and 86% of females in the sample had visited a general practitioner at least once in the past year. Do you think these estimates are close to the truth about the entire population? Why?

3.89 Use the probability applet. The *Probability* applet simulates tossing a coin, with the advantage that you can choose the true long-term proportion, or probability, of a head. Suppose we have a population in which proportion $p = 0.6$ (the parameter) approve of legal

gambling. Tossing a coin with probability $p = 0.6$ of a head simulates this situation: each head is a person who approves of legal gambling, and each tail is a person who does not. Set the "Probability of heads" in the applet to 0.6 and the number of tosses to 25. This simulates an SRS of size 25 from this population. By alternating between "Toss" and "Reset" you can take many samples quickly.

(a) Take 50 samples, recording the number of heads in each sample. Make a histogram of the 50 sample proportions (count of heads divided by 25). You are constructing the sampling distribution of this statistic.

(b) Another population contains only 20% who approve of legal gambling. Take 50 samples of size 25 from this population, record the number in each sample who approve, and make a histogram of the 50 sample proportions. How do the centers of your two histograms reflect the differing truths about the two populations?

3.90 Use the statistical software for simulations. Statistical software can speed simulations. We are interested in the sampling distribution of the proportion $\hat{p}$ of people who find shopping frustrating in an SRS from a population in which proportion p find shopping frustrating. Here, p is a parameter and $\hat{p}$ is a statistic used to estimate p. We will see in Chapter 5 that "binomial" is the key word to look for in the software menus.

(a) Set $n = 50$ and $p = 0.6$ and generate 100 binomial observations. These are the counts for 100 SRSs of size 50 when 60% of the population finds shopping frustrating. Save these counts and divide them by 50 to get values of $\hat{p}$ from 100 SRSs. Make a stemplot of the 100 values of $\hat{p}$.

(b) Repeat this process with $p = 0.3$, representing a population in which only 30% of people find shopping frustrating. Compare your two stemplots. How does changing the parameter p affect the center and spread of the sampling distribution?

(c) Now generate 100 binomial observations with $n = 200$ and $p = 0.6$. This simulates 100 SRSs, each of size 200. Obtain the 100 sample proportions $\hat{p}$ and make a stemplot. Compare this with your stemplot from (a). How does changing the sample size n affect the center and spread of the sampling distribution?

3.91 Use Table B for a simulation. We can construct a sampling distribution by hand in the case of a very small sample from a very small population. The population contains 10 students. Here are their scores on an exam:

Student	0	1	2	3	4	5	6	7	8	9
Score	82	62	80	58	72	73	65	66	74	62

The parameter of interest is the mean score, which is 69.4. The sample is an SRS of $n = 4$ students drawn from this population. The students are labeled 0 to 9 so that a single random digit from Table B chooses one student for the sample.

(a) Use Table B to draw an SRS of size 4 from this population. Write the four scores in your sample and calculate the mean $\bar{x}$ of the sample scores. This statistic is an estimate of the population parameter.

(b) Repeat this process 9 more times. Make a histogram of the 10 values of $\bar{x}$. You are constructing the sampling distribution of $\bar{x}$. Is the center of your histogram close to 69.4? (Ten repetitions give only a crude approximation to the sampling distribution. If possible, pool your work with that of other students—using different parts of Table B— to obtain several hundred repetitions and make a histogram of the values of $\bar{x}$. This histogram is a better approximation to the sampling distribution.)

3.92 ⏣ **Use the simple random sample applet.** The *Simple Random Sample* applet can illustrate the idea of a sampling distribution. Form a population labeled 1 to 100. We will choose an SRS of 10 of these numbers. That is, in this exercise, the numbers themselves are the population, not just labels for 100 individuals. The mean of the whole numbers 1 to 100 is 50.5. This is the parameter, the mean of the population.

(a) Use the applet to choose an SRS of size 10. Which 10 numbers were chosen? What is their mean? This is a statistic, the sample mean $\bar{x}$.

(b) Although the population and its mean 50.5 remain fixed, the sample mean changes as we take more samples. Take another SRS of size 10. (Use the "Reset" button to return to the original population before taking the second sample.) What are the 10 numbers in your sample? What is their mean? This is another value of $\bar{x}$.

(c) Take 8 more SRSs from this same population and record their means. You now have 10 values of the sample mean $\bar{x}$ from 10 SRSs of the same size from the same population. Make a histogram of the 10 values and mark the population mean 50.5 on the horizontal axis. Are your 10 sample values roughly centered at the population value? (If you kept going forever, your $\bar{x}$-values would form the sampling distribution of the sample mean; the population mean would indeed be the center of this distribution.)

3.93 Analyze simple random samples. The CSDATA data set contains the college grade point averages (GPAs) of all 224 students in a university entering class who planned to major in computer science. This is our population. Statistical software can take repeated samples to illustrate sampling variability. ⬤ CSDATA

(a) Using software, describe this population with a histogram and with numerical summaries. In particular, what is the mean GPA in the population? This is a parameter.

(b) Choose an SRS of 20 members from this population. Make a histogram of the GPAs in the sample and find their mean. The sample mean is a statistic. Briefly compare the distributions of GPA in the sample and in the population.

(c) Repeat the process of choosing an SRS of size 20 four more times (five in all). Record the five histograms of your sample GPAs. Does it seem reasonable to you from this small trial that an SRS will usually produce a sample that is generally representative of the population?

3.94 Simulate the sampling distribution of the mean. Continue the previous exercise, using software to illustrate the idea of a sampling distribution.

(a) Choose 20 more SRSs of size 20 in addition to the 5 you have already chosen. Don't make histograms of these latest samples—just record the mean GPA for each sample. Make a histogram of the 25 sample means. This histogram is a rough approximation to the sampling distribution of the mean.

(b) One sign of bias would be that the distribution of the sample means was systematically on one side of the true population mean. Mark the population mean GPA on your histogram of the 25 sample means. Is there a clear bias?

(c) Find the mean and standard deviation of your 25 sample means. We expect that the mean will be close to the true mean of the population. Is it? We also expect that the standard deviation of the sampling distribution will be smaller than the standard deviation of the population. Is it?

3.95 Toss a coin. Coin tossing can illustrate the idea of a sampling distribution. The population is all outcomes (heads or tails) we would get if we tossed a coin forever. The parameter p is the proportion of heads in this population. We suspect that p is close to 0.5. That is, we think the coin will show about one-half heads in the long run. The sample is the outcomes of 20 tosses, and the statistic $\hat{p}$ is the proportion of heads in these 20 tosses (count of heads divided by 20).

(a) Toss a coin 20 times and record the value of $\hat{p}$.

(b) Repeat this sampling process 9 more times. Make a stemplot of the 10 values of $\hat{p}$. You are constructing the sampling distribution of $\hat{p}$. Is the center of this distribution close to 0.5? (Ten repetitions give only a crude approximation to the sampling distribution. If possible, pool your work with that of other students to obtain several hundred repetitions and make a histogram of the values of $\hat{p}$.)

3.4 Ethics

The production and use of data, like all human endeavors, raise ethical questions. We won't discuss the telemarketer who begins a telephone sales pitch with "I'm conducting a survey." Such deception is clearly unethical. It enrages legitimate survey organizations, which find the public less willing to talk with them. Neither will we discuss those few researchers who, in the pursuit of professional advancement, publish fake data. There is no ethical question here—faking data to advance your career is just wrong. It will end your career when uncovered. But just how honest must researchers be about real, unfaked data? Here is an example that suggests the answer is "More honest than they often are."

> ### EXAMPLE
>
> **3.32 Provide all of the critical information.** Papers reporting scientific research are supposed to be short, with no extra baggage. Brevity can allow the researchers to avoid complete honesty about their data. Did they choose their subjects in a biased way? Did they report data on only some of their subjects? Did they try several statistical analyses and report only the ones that looked best? The statistician John Bailar screened more than 4000 medical papers in more than a decade as consultant to the *New England Journal of Medicine*. He says, "When it came to the statistical review, it was often clear that critical information was lacking, and the gaps nearly always had the practical effect of making the authors' conclusions look stronger than they should have."[35] The situation is no doubt worse in fields that screen published work less carefully.

The most complex issues of data ethics arise when we collect data from people. The ethical difficulties are more severe for experiments that impose some treatment on people than for sample surveys that simply gather information. Trials of new medical treatments, for example, can do harm as well as good to their subjects. Here are some basic standards of data ethics that must be obeyed by any study that gathers data from human subjects, whether sample survey or experiment.

BASIC DATA ETHICS

The organization that carries out the study must have an **institutional review board** that reviews all planned studies in advance in order to protect the subjects from possible harm.

All individuals who are subjects in a study must give their **informed consent** before data are collected.

All individual data must be kept **confidential.** Only statistical summaries for groups of subjects may be made public.

The law requires that studies funded by the federal government obey these principles. But neither the law nor the consensus of experts is completely clear about the details of their application.

Institutional review boards

The purpose of an institutional review board is not to decide whether a proposed study will produce valuable information or whether it is statistically sound. The board's purpose is, in the words of one university's board, "to protect the rights and welfare of human subjects (including patients) recruited to participate in research activities." The board reviews the plan of the study and can require changes. It reviews the consent form to be sure that subjects are informed about the nature of the study and about any potential risks. Once research begins, the board monitors its progress at least once a year.

The most pressing issue concerning institutional review boards is whether their workload has become so large that their effectiveness in protecting subjects drops. When the government temporarily stopped human-subject research at Duke University Medical Center in 1999 due to inadequate protection of subjects, more than 2000 studies were going on. That's a lot of review work. There are shorter review procedures for projects that involve only minimal risks to subjects, such as most sample surveys. When a board is overloaded, there is a temptation to put more proposals in the minimal-risk category to speed the work.

USE YOUR KNOWLEDGE

The exercises in this section on ethics are designed to help you think about the issues that we are discussing and to formulate some opinions. In general there are no wrong or right answers, but you need to give reasons for your answers.

3.96 Do these proposals involve minimal risk? You are a member of your college's institutional review board. You must decide whether several research proposals qualify for lighter review because they involve only minimal risk to subjects. Federal regulations say that "minimal risk" means the risks are no greater than "those ordinarily encountered in daily life or during the performance of routine physical or psychological examinations or tests." That's vague. Which of these do you think qualifies as "minimal risk"?

(a) Draw a drop of blood by pricking a finger in order to measure blood sugar.

(b) Draw blood from the arm for a full set of blood tests.

(c) Insert a tube that remains in the arm, so that blood can be drawn regularly.

3.97 Who should be on an institutional review board? Government regulations require that institutional review boards consist of at least five people, including at least one scientist, one nonscientist, and one person from outside the institution. Most boards are larger, but many contain just one outsider.

(a) Why should review boards contain people who are not scientists?

(b) Do you think that one outside member is enough? How would you choose that member? (For example, would you prefer a medical doctor? A member of the clergy? An activist for patients' rights?)

Informed consent

Both words in the phrase "informed consent" are important, and both can be controversial. Subjects must be *informed* in advance about the nature of a study and any risk of harm it may bring. In the case of a sample survey, physical harm is not possible. The subjects should be told what kinds of questions the survey will ask and about how much of their time it will take. Experimenters must tell subjects the nature and purpose of the study and outline possible risks. Subjects must then *consent* in writing.

EXAMPLE

3.33   Who can give informed consent? Are there some subjects who can't give informed consent? It was once common, for example, to test new vaccines on prison inmates who gave their consent in return for good-behavior credit. Now we worry that prisoners are not really free to refuse, and the law forbids medical experiments in prisons.

Young children can't give fully informed consent, so the usual procedure is to ask their parents. A study of new ways to teach reading is about to start at a local elementary school, so the study team sends consent forms home to parents. Many parents don't return the forms. Can their children take part in the study because the parents did not say "No," or should we allow only children whose parents returned the form and said "Yes"?

What about research into new medical treatments for people with mental disorders? What about studies of new ways to help emergency room patients who may be unconscious or have suffered a stroke? In most cases, there is not time even to get the consent of the family. Does the principle of informed consent bar realistic trials of new treatments for unconscious patients?

These are questions without clear answers. Reasonable people differ strongly on all of them. There is nothing simple about informed consent.[36]

The difficulties of informed consent do not vanish even for capable subjects. Some researchers, especially in medical trials, regard consent as a barrier to getting patients to participate in research. They may not explain all possible risks; they may not point out that there are other therapies that might be better than those being studied; they may be too optimistic in talking with patients even when the consent form has all the right details. On the other hand, mentioning every possible risk leads to very long consent forms that really are barriers. "They are like rental car contracts," one lawyer said. Some subjects don't read forms that run five or six printed pages. Others are frightened by the large number of possible (but unlikely) disasters that might happen and so refuse to participate. Of course, unlikely disasters sometimes happen. When they do, lawsuits follow and the consent forms become yet longer and more detailed.

Confidentiality

Ethical problems do not disappear once a study has been cleared by the review board, has obtained consent from its subjects, and has actually collected data about the subjects. It is important to protect the subjects' privacy by keeping all data about individuals confidential. The report of an opinion poll may say what percent of the 1500 respondents felt that legal immigration should be reduced. It may not report what *you* said about this or any other issue.

anonymity

Confidentiality is not the same as **anonymity.** Anonymity means that subjects are anonymous—their names are not known even to the director of the study. Anonymity is rare in statistical studies. Even where anonymity is possible (mainly in surveys conducted by mail), it prevents any follow-up to improve nonresponse or inform subjects of results.

Any breach of confidentiality is a serious violation of data ethics. The best practice is to separate the identity of the subjects from the rest of the data at once. Sample surveys, for example, use the identification only to check on who did or did not respond. In an era of advanced technology, however, it is no longer enough to be sure that each individual set of data protects people's privacy. The government, for example, maintains a vast amount of information about citizens in many separate data bases—census responses, tax returns, Social Security information, data from surveys such as the Current Population Survey, and so on. Many of these data bases can be searched by computers for statistical studies. A clever computer search of several data bases might be able, by combining information, to identify you and learn a great deal about you even if your name and other identification have been removed from the data available for search. A colleague from Germany once remarked that "female full professor of statistics with a PhD from the United States" was enough to identify her among all the citizens of Germany. Privacy and confidentiality of data are hot issues among statisticians in the computer age.

EXAMPLE

3.34 Data collected by the government. Citizens are required to give information to the government. Think of tax returns and Social Security contributions. The government needs these data for administrative purposes—to see if we paid the right amount of tax and how large a Social Security benefit we are owed when we retire. Some people feel that individuals should be able to forbid any other use of their data, even with all identification removed. This would prevent using government records to study, say, the ages, incomes, and household sizes of Social Security recipients. Such a study could well be vital to debates on reforming Social Security.

USE YOUR KNOWLEDGE

3.98 How can we obtain informed consent? A researcher suspects that traditional religious beliefs tend to be associated with an authoritarian personality. She prepares a questionnaire that measures authoritarian tendencies and also asks many religious questions. Write a description of the purpose of this research to be read by subjects in order to obtain their informed consent. You must balance the conflicting goals of not deceiving the subjects as to what the questionnaire will tell about them and of not biasing the sample by scaring off religious people.

3.99 Should we allow this personal information to be collected? In which of the following circumstances would you allow collecting personal information without the subjects' consent?

(a) A government agency takes a random sample of income tax returns to obtain information on the average income of people in different occupations. Only the incomes and occupations are recorded from the returns, not the names.

(b) A social psychologist attends public meetings of a religious group to study the behavior patterns of members.

(c) A social psychologist pretends to be converted to membership in a religious group and attends private meetings to study the behavior patterns of members.

Clinical trials

Clinical trials are experiments that study the effectiveness of medical treatments on actual patients. Medical treatments can harm as well as heal, so clinical trials spotlight the ethical problems of experiments with human subjects. Here are the starting points for a discussion:

- Randomized comparative experiments are the only way to see the true effects of new treatments. Without them, risky treatments that are no better than placebos will become common.

- Clinical trials produce great benefits, but most of these benefits go to future patients. The trials also pose risks, and these risks are borne by the subjects of the trial. So we must balance future benefits against present risks.

- Both medical ethics and international human rights standards say that "the interests of the subject must always prevail over the interests of science and society."

The quoted words are from the 1964 Helsinki Declaration of the World Medical Association, the most respected international standard. The most outrageous examples of unethical experiments are those that ignore the interests of the subjects.

EXAMPLE

3.35 The Tuskegee study. In the 1930s, syphilis was common among black men in the rural South, a group that had almost no access to medical care. The Public Health Service Tuskegee study recruited 399 poor black sharecroppers with syphilis and 201 others without the disease in order to observe how syphilis progressed when no treatment was given. Beginning in 1943, penicillin became available to treat syphilis. The study subjects were not treated. In fact, the Public Health Service prevented any treatment until word leaked out and forced an end to the study in the 1970s.

The Tuskegee study is an extreme example of investigators following their own interests and ignoring the well-being of their subjects. A 1996 review said, "It has come to symbolize racism in medicine, ethical misconduct in human research, paternalism by physicians, and government abuse of vulnerable people." In 1997, President Clinton formally apologized to the surviving participants in a White House ceremony.[37]

Because "the interests of the subject must always prevail," medical treatments can be tested in clinical trials only when there is reason to hope that they will help the patients who are subjects in the trials. Future benefits aren't enough to justify experiments with human subjects. Of course, if there is already strong evidence that a treatment works and is safe, it is unethical *not* to give it. Here are the words of Dr. Charles Hennekens of the Harvard Medical School, who directed the large clinical trial that showed that aspirin reduces the risk of heart attacks:[38]

> There's a delicate balance between when to do or not do a randomized trial. On the one hand, there must be sufficient belief in the agent's potential to justify exposing half the subjects to it. On the other hand, there must be sufficient doubt about its efficacy to justify withholding it from the other half of subjects who might be assigned to placebos.

Why is it ethical to give a control group of patients a placebo? Well, we know that placebos often work. What is more, placebos have no harmful side effects. So in the state of balanced doubt described by Dr. Hennekens, the placebo group may be getting a better treatment than the drug group. If we *knew* which treatment was better, we would give it to everyone. When we don't know, it is ethical to try both and compare them.

The idea of using a control or placebo is a fundamental principle to be considered in designing experiments. In many situations deciding what to use as an appropriate control requires some careful thought. *The choice of control can have a substantial impact on the how the results of an experiment are interpreted.* Here is an example:

EXAMPLE

3.36 Attentiveness improves by nearly 20%. The manufacturer of a breakfast cereal designed for children claims that eating this cereal has been clinically shown to improve attentiveness by nearly 20%. The study used two groups of children who were tested before and after breakfast. One group received the cereal for breakfast while breakfast for the control group was water. The results of the tests taken three hours after breakfast were used in the claim.

The Federal Trade Commission investigated the marketing of this product. They charged that the claim was false and violated federal law. The charges were settled and the company agreed to not use misleading claims in their advertising.[39]

It is not sufficient to obtain appropriate controls. They data must be collected from all groups in the same way. Here is an example of this type of flawed design:

EXAMPLE

3.37 Accurate identification of ovarian cancer. Two scientists published a paper claiming to have developed a very exciting new method to detect ovarian cancer using blood samples. When other scientists were unable to reproduce the results in different labs, the original work was examined more carefully. In the original study there were samples for women with ovarian cancer and healthy controls. The serum samples were all analyzed using a mass spectrometer. The control samples were analyzed on one day and the cancer samples were analyzed on the next day. This design was flawed in that it could not control for changes over time in the measuring instrument.[40]

USE YOUR KNOWLEDGE

3.100 Is this study ethical? Researchers on aging proposed to investigate the effect of supplemental health services on the quality of life of older people. Eligible patients on the rolls of a large medical clinic were to be randomly assigned to treatment and control groups. The treatment group would be offered hearing aids, dentures, transportation, and other services not available without charge to the control group. The review board felt that providing these services to some but not other persons in the same institution raised ethical questions. Do you agree?

3.101 Should the treatments be given to everyone? Effective drugs for treating AIDS are very expensive, so most African nations cannot afford to give them to large numbers of people. Yet AIDS is more common in parts of Africa than anywhere else. Several clinical trials are looking at ways to prevent pregnant mothers infected with HIV from passing the infection to their unborn children, a major source of HIV infections in Africa. Some people say these trials are unethical because they do not give effective AIDS drugs to their subjects, as would be required in rich nations. Others reply that the trials are looking for treatments that can work in the real world in Africa and that they promise benefits at least to the children of their subjects. What do you think?

Behavioral and social science experiments

When we move from medicine to the behavioral and social sciences, the direct risks to experimental subjects are less acute, but so are the possible benefits to the subjects. Consider, for example, the experiments conducted by psychologists in their study of human behavior.

EXAMPLE

3.38 Personal space. Psychologists observe that people have a "personal space" and get annoyed if others come too close to them. We don't like strangers to sit at our table in a coffee shop if other tables are available, and we see people move apart in elevators if there is room to do so. Americans tend to require more personal space than people in most other cultures. Can violations of personal space have physical, as well as emotional, effects?

Investigators set up shop in a men's public rest room. They blocked off urinals to force men walking in to use either a urinal next to an experimenter (treatment group) or a urinal separated from the experimenter (control group). Another experimenter, using a periscope from a toilet stall, measured how long the subject took to start urinating and how long he kept at it.[41]

This personal space experiment illustrates the difficulties facing those who plan and review behavioral studies.

- There is no risk of harm to the subjects, although they would certainly object to being watched through a periscope. What should we protect subjects from when physical harm is unlikely? Possible emotional harm? Undignified situations? Invasion of privacy?

- What about informed consent? The subjects in Example 3.38 did not even know they were participating in an experiment. Many behavioral experiments rely on hiding the true purpose of the study. The subjects would change their behavior if told in advance what the investigators were looking for. Subjects are asked to consent on the basis of vague information. They receive full information only after the experiment.

The "Ethical Principles" of the American Psychological Association require consent unless a study merely observes behavior in a public place. They allow deception only when it is necessary to the study, does not hide information that might influence a subject's willingness to participate, and is explained to subjects as soon as possible. The personal space study (from the 1970s) does not meet current ethical standards.

We see that the basic requirement for informed consent is understood differently in medicine and psychology. Here is an example of another setting with yet another interpretation of what is ethical. The subjects get no information and give no consent. They don't even know that an experiment may be sending them to jail for the night.

EXAMPLE

3.39 Domestic violence. How should police respond to domestic-violence calls? In the past, the usual practice was to remove the offender and order him to stay out of the household overnight. Police were reluctant to make arrests because the victims rarely pressed charges. Women's groups argued that arresting offenders would help prevent future violence even if no charges were filed. Is there evidence that arrest will reduce future offenses? That's a question that experiments have tried to answer.

A typical domestic-violence experiment compares two treatments: arrest the suspect and hold him overnight, or warn the suspect and release him. When police officers reach the scene of a domestic-violence call, they calm the participants and investigate. Weapons or death threats require an arrest. If the facts permit an arrest but do not require it, an officer radios headquarters for instructions. The person on duty opens the next envelope in a file prepared in advance by a statistician. The envelopes contain the treatments in random order. The police either arrest the suspect or warn and release him, depending on the contents of the envelope. The researchers then watch police records and visit the victim to see if the domestic violence reoccurs.

The first such experiment appeared to show that arresting domestic-violence suspects does reduce their future violent behavior. As a result of this evidence, arrest has become the common police response to domestic violence.

The domestic-violence experiments shed light on an important issue of public policy. Because there is no informed consent, the ethical rules that govern clinical trials and most social science studies would forbid these experiments. They were cleared by review boards because, in the words of one domestic-violence researcher, "These people became subjects by committing acts that allow the police to arrest them. You don't need consent to arrest someone."

SECTION 3.4 Summary

Approval of an **institutional review board** is required for studies that involve human or animals as subjects.

Human subjects must give **informed consent** if they are to participate in experiments.

Data on human subjects must be kept **confidential.**

SECTION 3.4 Exercises

For Exercises 3.96 and 3.97, see page 215; for Exercises 3.98 and 3.99, see pages 217–218; and for Exercises 3.100 and 3.101, see page 220.

3.102 What is wrong? Explain what is wrong in each of the following scenarios.

(a) Clinical trials are always ethical as long as they randomly assign patients to the treatments.

(b) The job of an institutional review board is complete when they decide to allow a study to be conducted.

(c) A treatment that has no risk of physical harm to subjects is always ethical.

3.103 How should the samples have been analyzed? Refer to the ovarian cancer diagnostic test study in Example 3.37 (page 220). Describe how you would process the samples through the mass spectrometer.

3.104 The Vytorin controversy. Vytorin is a combination pill designed to lower cholesterol. It consists of a relatively inexpensive and widely used drug, Zocor, and a newer drug called Zetia. Early study results suggested that Vytorin was no more effective than Zetia. Critics claimed that the makers of the drugs tried to change the response variable for the study and two congressional panels investigated why there was a two-year delay in the release of the results. Use the Web to search for more information about this controversy and write a report of what you have found. Include an evaluation in the framework of ethical use of experiments and data. A good place to start your search would be to look for the phrase "Vytorin's Shortcomings."

3.105 Facebook and academic performance. *First Monday* is a peer-reviewed journal on the Internet. They recently published two articles concerning Facebook and academic performance. Visit their Web site, firstmonday.org, look at the first three articles in Volume 14, Number 5–4, May 2009. Identify the key controversial issues that involve the use of statistics addressed in these articles and write a report summarizing the facts as you see them. Be sure to include your opinions regarding ethical issues related to this work.

3.106 Select a random sample. A data set contains information about 14,959 people aged 25 to 64 whose highest level of education is a bachelor's degree.

(a) To select an SRS of these people, how would you assign labels?

(b) Use Table B at line 165 to choose the first 3 members of the SRS.

3.107 Examples of designs. Give a detailed example of each of the following designs.

(a) A matched pairs design.

(b) A stratified random sample.

(c) A completely randomized design.

3.108 What type of study? What is the best way to answer each of the following questions: an experiment, a sample survey, or an observational study that is not a sample survey? Explain your choices.

(a) Are people generally satisfied with how things are going in the country right now?

(b) Do college students learn basic accounting better in a classroom or using an online course?

(c) How long do your teachers wait on the average after they ask their class a question?

3.109 Choose the type of study. Give an example of a question about college students, their behavior, or their opinions that would best be answered by

(a) a sample survey.

(b) an observational study that is not a sample survey.

(c) an experiment.

3.110 Informed consent to take blood samples. Researchers from Yale, working with medical teams in Tanzania, wanted to know how common infection with the AIDS virus is among pregnant women in that country. To do this, they planned to test blood samples drawn from pregnant women.

Yale's institutional review board insisted that the researchers get the informed consent of each woman and tell her the results of the test. This is the usual procedure in developed nations. The Tanzanian government did not want to tell the women why blood was drawn or tell them the test results. The government feared panic if many people turned out to have an incurable disease for which the country's medical system could not provide care. The study was canceled. Do you think that Yale was right to apply its usual standards for protecting subjects?

3.111 The General Social Survey. One of the most important nongovernment surveys in the United States is the National Opinion Research Center's General Social Survey. The GSS regularly monitors public opinion on a wide variety of political and social issues. Interviews are conducted in person in the subject's home. Are a subject's responses to GSS questions anonymous, confidential, or both? Explain your answer.

3.112 Anonymity and confidentiality in health screening. Texas A&M, like many universities, offers free screening for HIV, the virus that causes AIDS. The announcement says, "Persons who sign up for the HIV Screening will be assigned a number so that they do not have to give their name." They can learn the results of the test by telephone, still without giving their name. Does this practice offer *anonymity* or just *confidentiality?*

3.113 Anonymity and confidentiality in mail surveys. Some common practices may appear to offer anonymity while actually delivering only confidentiality. Market researchers often use mail surveys that do not ask the respondent's identity but contain hidden codes on the questionnaire that identify the respondent. A false claim of anonymity is clearly unethical. If only confidentiality is promised, is it also unethical to say nothing about the identifying code, perhaps causing respondents to believe their replies are anonymous?

3.114 Use of stored blood. Long ago, doctors drew a blood specimen from you as part of treating minor anemia. Unknown to you, the sample was stored. Now researchers plan to use stored samples from you and many other people to look for genetic factors that may influence anemia. It is no longer possible to ask your consent. Modern technology can read your entire genetic makeup from the blood sample.

(a) Do you think it violates the principle of informed consent to use your blood sample if your name is on it but you were not told that it might be saved and studied later?

(b) Suppose that your identity is not attached. The blood sample is known only to come from (say) "a 20-year-old white female being treated for anemia." Is it now OK to use the sample for research?

(c) Perhaps we should use biological materials such as blood samples only from patients who have agreed to allow the material to be stored for later use in research. It isn't possible to say in advance what kind of research, so this falls short of the usual standard for informed consent. Is it nonetheless acceptable, given complete confidentiality and the fact that using the sample can't physically harm the patient?

3.115 Testing vaccines. One of the most important goals of AIDS research is to find a vaccine that will protect against HIV. Because AIDS is so common in parts of Africa, that is the easiest place to test a vaccine. It is likely, however, that a vaccine would be so expensive that it could not (at least at first) be widely used in Africa. Is it ethical to test in Africa if the benefits go mainly to rich countries? The treatment group of subjects would get the vaccine, and the placebo group would later be given the vaccine if it proved effective. So the actual subjects would benefit—it is the future benefits that would go elsewhere. What do you think?

3.116 Political polls. The presidential election campaign is in full swing, and the candidates have hired polling organizations to take regular polls to find out what the voters think about the issues. What information should the pollsters be required to give out?

(a) What does the standard of informed consent require the pollsters to tell potential respondents?

(b) The standards accepted by polling organizations also require giving respondents the name and address of the organization that carries out the poll. Why do you think this is required?

(c) The polling organization usually has a professional name such as "Samples Incorporated," so respondents don't know that the poll is being paid for by a political party or candidate. Would revealing the sponsor to respondents bias the poll? Should the sponsor always be announced whenever poll results are made public?

3.117 Should poll results be made public? Some people think that the law should require that all political poll results be made public. Otherwise, the possessors of poll results can use the information to their own advantage. They can act on the information, release only selected parts of it, or time the release for best effect. A candidate's organization replies that they are paying for the poll in order to gain information for their own use, not to amuse the public. Do you favor requiring complete disclosure of political poll results? What about other private surveys, such as market research surveys of consumer tastes?

CHAPTER 3 Exercises

3.118 Experiments and surveys. Write a short report describing the differences and similarities between experiments and surveys. Include a discussion of the advantages and disadvantages of each.

3.119 Online behavioral advertising. The Federal Trade Commission (FTC) Staff Report, "Self-Regulatory Principles for Online Behavioral Advertising," defines behavioral advertising as *the tracking of a consumer's online activities over time—including the searches the consumer has conducted, the Web pages visited and the content viewed—in order to deliver advertising targeted to the individual consumer's interests.* The report suggests four governing concepts for their proposals. These are (1) transparency and control: when companies collect information from consumers for advertising, they should tell the consumer about how the data will be collected and customers should be given a choice about whether to allow the data to be collected; (2) security and data retention: data should be kept secure and should be retained only as long as it is needed; (3) privacy: before data is used in a way that is different from promises made when it is collected, it should obtain consent from the consumer; and (4) sensitive data: affirmative express consent should be obtained before using any sensitive data.[42] Write a report discussing your opinions concerning online behavioral advertising and the four governing concepts. Pay particular attention to issues related to the ethical collection and use of statistical data.

3.120 Confidentiality at NORC. The National Opinion Research Center conducts a large number of surveys and has established procedures for protecting the confidentiality of their survey participants. For their Survey of Consumer Finances, they provide a pledge to participants regarding confidentiality. This pledge is available at norc.org. Review the pledge and summarize its key parts. Do you think that the pledge adequately addresses issues related to the ethical collection and use of data? Explain your answer.

3.121 Make it an experiment! In the following observational studies, describe changes that could be made to the data collection process that would result in an experiment rather than an observational study. Also, offer suggestions about unseen biases or lurking variables that may be present in the studies as they are described here.

(a) A friend of yours likes to play Texas Hold Em. Every time that he tells you about his playing, he says that he won.

(b) In an introductory statistics class you notice that the students who sit in the first two rows of seats had a higher score on the first exam in the class than the other students in the class.

3.122 Name the designs. What is the name for each of these study designs?

(a) A study to compare two methods of preserving wood started with boards of southern white pine. Each board was ripped from end to end to form two edge-matched specimens. One was assigned to Method A, the other to Method B.

(b) A survey on youth and smoking contacted by telephone 300 smokers and 300 nonsmokers, all 14 to 22 years of age.

(c) Does air pollution induce DNA mutations in mice? Starting with 40 male and 40 female mice, 20 of each sex were housed in a polluted industrial area downwind from a steel mill. The other 20 of each sex were housed at an unpolluted rural location 30 kilometers away.

3.123 Price promotions and consumer's expectations. A researcher studying the effect of price promotions on consumers' expectations makes up two different histories of the store price of a hypothetical brand of laundry detergent for the past year. Students in a marketing course view one or the other price history on a computer. Some students see a steady price, while others see regular promotions that temporarily cut the price. Then the students are asked what price they would expect to pay for the detergent. Is this study an experiment? Why? What are the explanatory and response variables?

3.124 Calcium and healthy bones. Adults need to eat foods or supplements that contain enough calcium to maintain healthy bones. Calcium intake is generally measured in milligrams per day (mg/d) and one measure of healthy bones is total body bone mineral density measured in grams per centimeter squared (TBBMD, g/cm^2). Suppose that you want to study the relationship between calcium intake and TBBMD.

(a) Design an observational study to study the relationship.

(b) Design an experiment to study the relationship.

(c) Compare the relative merits of your two designs. Which do you prefer? Give reasons for your answer.

3.125 Choose the type of study. Give an example of a question about pets and their owners, their behavior, or their opinions that would best be answered by

(a) a sample survey.

(b) an observational study that is not a sample survey.

(c) an experiment.

3.126 Compare the fries. Do consumers prefer the fries from Burger King or from McDonald's? Design a blind test in which neither source of the fries is identified.

Describe briefly the design of a matched pairs experiment to investigate this question. How will you use randomization?

3.127 Bicycle gears. How does the time it takes a bicycle rider to travel 100 meters depend on which gear is used and how steep the course is? It may be, for example, that higher gears are faster on the level but lower gears are faster on steep inclines. Discuss the design of a two-factor experiment to investigate this issue, using one bicycle with three gears and one rider. How will you use randomization?

3.128 ⚠ **Design an experiment.** The previous two exercises illustrate the use of statistically designed experiments to answer questions that arise in everyday life. Select a question of interest to you that an experiment might answer and carefully discuss the design of an appropriate experiment.

3.129 ⚠ **Design a survey.** You want to investigate the attitudes of students at your school about the faculty's commitment to teaching. The student government will pay the costs of contacting about 500 students.

(a) Specify the exact population for your study; for example, will you include part-time students?

(b) Describe your sample design. Will you use a stratified sample?

(c) Briefly discuss the practical difficulties that you anticipate; for example, how will you contact the students in your sample?

3.130 Compare two doses of a drug. A drug manufacturer is studying how a new drug behaves in patients. Investigators compare two doses: 5 milligrams (mg) and 10 mg. The drug can be administered by injection, by a skin patch, or by intravenous drip. Concentration in the blood after 30 minutes (the response variable) may depend both on the dose and on the method of administration.

(a) Make a sketch that describes the treatments formed by combining dosage and method. Then use a diagram to outline a completely randomized design for this two-factor experiment.

(b) "How many subjects?" is a tough issue. We will explain the basic ideas in Chapter 6. What can you say now about the advantage of using larger groups of subjects?

3.131 Would the results be different for men and women? The drug that is the subject of the experiment in Exercise 3.130 may behave differently in men and women.

How would you modify your experimental design to take this into account?

3.132 ⚔ **Informed consent.** The requirement that human subjects give their informed consent to participate in an experiment can greatly reduce the number of available subjects. For example, a study of new teaching methods asks the consent of parents for their children to be randomly assigned to be taught by either a new method or the standard method. Many parents do not return the forms, so their children must continue to be taught by the standard method. Why is it not correct to consider these children as part of the control group along with children who are randomly assigned to the standard method?

3.133 ⚔ **Two ways to ask sensitive questions.** Sample survey questions are usually read from a computer screen. In a Computer Aided Personal Interview (CAPI), the interviewer reads the questions and enters the responses. In a Computer Aided Self Interview (CASI), the interviewer stands aside and the respondent reads the questions and enters responses. One method almost always shows a higher percent of subjects admitting use of illegal drugs. Which method? Explain why.

3.134 Your institutional review board. Your college or university has an institutional review board that screens all studies that use human subjects. Get a copy of the document that describes this board (you can probably find it online).

(a) According to this document, what are the duties of the board?

(b) How are members of the board chosen? How many members are not scientists? How many members are not employees of the college? Do these members have some special expertise, or are they simply members of the "general public"?

3.135 Use of data produced by the government. Data produced by the government are often available free or at low cost to private users. For example, satellite weather data produced by the U.S. National Weather Service are available free to TV stations for their weather reports and to anyone on the Web. *Opinion 1:* Government data should be available to everyone at minimal cost. European governments, on the other hand, charge TV stations for weather data. *Opinion 2:* The satellites are expensive, and the TV stations are making a profit from their weather services, so they should share the cost. Which opinion do you support, and why?

3.136 Should we ask for the consent of the parents? The Centers for Disease Control and Prevention, in a survey of teenagers, asked the subjects if they were sexually active. Those who said "Yes" were then asked,

> *How old were you when you had sexual intercourse for the first time?*

Should consent of parents be required to ask minors about sex, drugs, and other such issues, or is consent of the minors themselves enough? Give reasons for your opinion.

3.137 A theft experiment. Students sign up to be subjects in a psychology experiment. When they arrive, they are told that interviews are running late and are taken to a waiting room. The experimenters then stage a theft of a valuable object left in the waiting room. Some subjects are alone with the thief, and others are in pairs—these are the treatments being compared. Will the subject report the theft? The students had agreed to take part in an unspecified study, and the true nature of the experiment is explained to them afterward. Do you think this study is ethically OK?

3.138 A cheating experiment. A psychologist conducts the following experiment: she measures the attitude of subjects toward cheating and then has them play a game rigged so that winning without cheating is impossible. The computer that organizes the game also records—unknown to the subjects—whether or not they cheat. Then attitude toward cheating is retested. Subjects who cheat tend to change their attitudes to find cheating more acceptable. Those who resist the temptation to cheat tend to condemn cheating more strongly on the second test of attitude. These results confirm the psychologist's theory. This experiment tempts subjects to cheat. The subjects are led to believe that they can cheat secretly when in fact they are observed. Is this experiment ethically objectionable? Explain your position.

Probability: The Study of Randomness

Introduction

The reasoning of statistical inference rests on asking, "How often would this method give a correct answer if I used it very many times?" When we produce data by random sampling or randomized comparative experiments, the laws of probability answer the question "What would happen if we did this many times?" Games of chance like Texas hold 'em are exciting because the outcomes are determined by the rules of probability.

4.1 Randomness

LOOK BACK
sampling distributions p. 205

Toss a coin, or choose an SRS. The result can't be predicted in advance, because the result will vary when you toss the coin or choose the sample repeatedly. But there is nonetheless a regular pattern in the results, a pattern that emerges clearly only after many repetitions. This remarkable fact is the basis for the idea of probability.

EXAMPLE

4.1 Toss a coin 5000 times. When you toss a coin, there are only two possible outcomes, heads or tails. Figure 4.1 shows the results of tossing a coin 5000 times twice. For each number of tosses from 1 to 5000, we have plotted the proportion of those tosses that gave a head. Trial A (red line) begins tail, head, tail, tail. You can see that the proportion of heads for Trial A starts at 0 on the first toss, rises to 0.5 when the second toss gives a head, then falls to 0.33 and 0.25 as we get two more tails. Trial B (blue dotted line), on the other hand, starts with five straight heads, so the proportion of heads is 1 until the sixth toss.

The proportion of tosses that produce heads is quite variable at first. Trial A starts low and Trial B starts high. As we make more and more tosses, however, the proportions of heads for both trials get close to 0.5 and stay there. If we made yet a third trial at tossing the coin a great many times, the proportion of heads would again settle down to 0.5 in the long run. We say that 0.5 is the *probability* of a head. The probability 0.5 appears as a horizontal line on the graph.

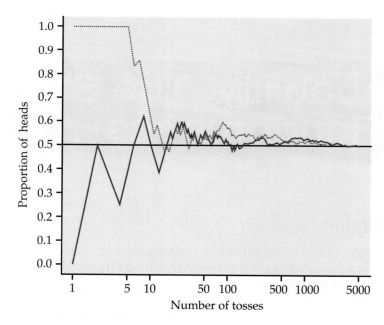

FIGURE 4.1 The proportion of tosses of a coin that give a head varies as we make more tosses. Eventually, however, the proportion approaches 0.5, the probability of a head. This figure shows the results of two trials of 5000 tosses each.

The *Probability* applet on the text Web site animates Figure 4.1. It allows you to choose the probability of a head and simulate any number of tosses of a coin with that probability. Try it. You will see that the proportion of heads gradually settles down close to the chosen probability. Equally important, you will also see that the proportion in a small or moderate number of tosses can be far from the probability. *Probability describes only what happens in the long run. Most people expect chance outcomes to show more short-term regularity than is actually true.*

EXAMPLE

4.2 Significance testing and Type I errors. In Chapter 6 we will learn about significance testing and Type I errors. When we perform a significance test, we have the possibility of making a Type I error under certain circumstances. The significance testing procedure is set up so that the probability of making this kind of error is small, usually five percent. If we perform a large number of significance tests under this set of circumstances, the proportion of times that we will make a Type I error is 0.05.

fair coin   In the coin toss setting, the probability of a head is a characteristic of the coin being tosses. A coin is called **fair** if the probability of a head is 0.5, that is, it is equally likely to come up heads or tails. In the same way, when the Type I error of a statistical significance procedure is set at 0.05, this probability is a characteristic of the procedure. If we toss a fair coin once, we do not know if it will come up heads or tails. Similarly, if we perform a significance test once, we do not know if we will make a Type I error or not. However, if the procedure is designed to have a Type I error probability of 0.05, then we are much less likely than not to make a Type I error.

The language of probability

"Random" in statistics is not a synonym for "haphazard," but a description of a kind of order that emerges in the long run. We often encounter the unpredictable side of randomness in our everyday experience, but we rarely see enough repetitions of the same random phenomenon to observe the long-term regularity that probability describes. You can see that regularity emerging in Figure 4.1. In the very long run, the proportion of tosses that give a head is 0.5. This is the intuitive idea of probability. Probability 0.5 means "occurs half the time in a very large number of trials."

RANDOMNESS AND PROBABILITY

We call a phenomenon **random** if individual outcomes are uncertain but there is nonetheless a regular distribution of outcomes in a large number of repetitions.

The **probability** of any outcome of a random phenomenon is the proportion of times the outcome would occur in a very long series of repetitions.

Not all coins are fair. In fact most real coins have bumps and imperfections that make the probability of heads a little different from 0.5. The probability might be 0.499999 or 0.500002. For our study of probability in this chapter, we will assume that we know the actual values of probabilities. Thus, we assume things like fair coins, even though we know that real coins are not exactly fair. We do this to learn what kinds of outcomes we are likely to see when we make such assumptions. When we study statistical inference in later chapters, we look at the situation from the opposite point of view: given that we have observed certain outcomes, what can we say about the probabilities that generated these outcomes?

USE YOUR KNOWLEDGE

4.1 Use Table B. We can use the random digits in Table B in the back of the text to simulate tossing a fair coin. Start at line 119 and read the numbers from left to right. If the number is 0, 1, 2, 3, or 4, you will say that the coin toss resulted in a head; if the number is a 5, 6, 7, 8, or 9, the outcome is tails. Use the first 20 random digits on line 119 to simulate 20 tosses of a fair coin. What is the actual proportion of heads in your simulated sample? Explain why you did not get exactly 10 heads.

Probability describes what happens in very many trials, and we must actually observe many trials to pin down a probability. In the case of tossing a coin, some diligent people have in fact made thousands of tosses.

EXAMPLE

4.3 Many tosses of a coin. The French naturalist Count Buffon (1707–1788) tossed a coin 4040 times. Result: 2048 heads, or proportion 2048/4040 = 0.5069 for heads.

Around 1900, the English statistician Karl Pearson heroically tossed a coin 24,000 times. Result: 12,012 heads, a proportion of 0.5005.

While imprisoned by the Germans during World War II, the South African statistician John Kerrich tossed a coin 10,000 times. Result: 5067 heads, proportion of heads 0.5067.

Thinking about randomness

That some things are random is an observed fact about the world. The outcome of a coin toss, the time between emissions of particles by a radioactive source, and the sexes of the next litter of lab rats are all random. So is the outcome of a random sample or a randomized experiment. Probability theory is the branch of mathematics that describes random behavior. Of course, we can never observe a probability exactly. We could always continue tossing the coin, for example. Mathematical probability is an idealization based on imagining what would happen in an indefinitely long series of trials.

The best way to understand randomness is to observe random behavior—not only the long-run regularity but the unpredictable results of short runs. You can do this with physical devices such as coins and dice, but software simulations of random behavior allow faster exploration. As you explore randomness, remember:

independence

- You must have a long series of **independent** trials. That is, the outcome of one trial must not influence the outcome of any other. Imagine a crooked gambling house where the operator of a roulette wheel can stop it where she chooses—she can prevent the proportion of "red" from settling down to a fixed number. These trials are not independent.

- The idea of probability is empirical. Simulations start with given probabilities and imitate random behavior, but we can estimate a real-world probability only by actually observing many trials.

- Nonetheless, simulations are very useful because we need long runs of trials. In situations such as coin tossing, the proportion of an outcome often requires several hundred trials to settle down to the probability of that outcome. The kinds of physical random devices suggested in the exercises are too slow for this. Short runs give only rough estimates of a probability.

The uses of probability

Probability theory originated in the study of games of chance. Tossing dice, dealing shuffled cards, and spinning a roulette wheel are examples of deliberate randomization. In that respect, they are similar to random sampling. Although games of chance are ancient, they were not studied by mathematicians until the sixteenth and seventeenth centuries. It is only a mild simplification to say that probability as a branch of mathematics arose when seventeenth-century French gamblers asked the mathematicians Blaise Pascal and Pierre de Fermat for help. Gambling is still with us, in casinos and state lotteries. We will make use of games of chance as simple examples that illustrate the principles of probability.

Careful measurements in astronomy and surveying led to further advances in probability in the eighteenth and nineteenth centuries because the results of repeated measurements are random and can be described by distributions much like those arising from random sampling. Similar distributions appear in data on human life span (mortality tables) and in data on lengths or weights in a population of skulls, leaves, or cockroaches.[1] Now, we employ the mathematics of probability to describe the flow of traffic through a highway system, the Internet, or a computer processor; the genetic makeup of individuals or populations; the energy states of subatomic particles; the spread of epidemics or rumors; and the rate of return on risky investments. Although we are interested in probability because of its usefulness in statistics, the mathematics of chance is important in many fields of study.

SECTION 4.1 Summary

A **random phenomenon** has outcomes that we cannot predict but that nonetheless have a regular distribution in very many repetitions.

The **probability** of an event is the proportion of times the event occurs in many repeated trials of a random phenomenon.

SECTION 4.1   Exercises

For Exercise 4.1, see page 230.

4.2   Graduation rates for bachelor's students. In the United States about 56% of students enrolled in college for a bachelor's degree graduate within six years. If you select a U.S. college student at random, the probability is 0.56 that you have selected someone who will graduate within six years. Do you think that this probability would apply to all colleges? Write a short paragraph explaining your answer.

4.3   Simulate free throws. The professional basketball player Diana Taurasi made about 90% of her free throws in a recent season when she was named Most Valuable Player in the league. Use Table B or the *Probability* applet to simulate 100 free throws shot independently by a player who has probability 0.9 of making each shot.

(a) What percent of the 100 shots did she hit?

(b) Examine the sequence of hits and misses. How long was the longest run of shots made? Of shots missed? (Sequences of random outcomes often show runs longer than our intuition thinks likely.)

4.4   Is music playing on the radio? Turn on your favorite music radio station 8 times at least 10 minutes apart. Each time record whether or not music is playing. Calculate the number of times music is playing divided by 8. This number is an estimate of the probability that music is playing when you turn on this station. It is also an estimate of the proportion of time that music is playing on this station.

4.5   Wait 5 seconds between each observation. Refer to the previous exercise. Explain why you would not want to wait only 5 seconds between each time you turn the radio station on.

4.6   Winning at craps. The game of craps starts with a "come-out" roll where the shooter rolls a pair of dice. If the total is 7 or 11, the shooter wins immediately (there are ways that the shooter can win on later rolls if other numbers are rolled on the come-out roll). Roll a pair of dice 25 times and estimate the probability that the shooter wins immediately on the come-out roll. For a pair of perfectly made dice, the probability is 0.2222.

4.7   Side effects of eye drops. You go to the doctor and she prescribes a medicine for an eye infection that you have. Suppose that the probability of a serious side effect from the medicine is 0.00001. Explain in simple terms what this number means.

4.8   Use the *Probability* applet. The idea of probability is that the *proportion* of heads in many tosses of a balanced coin eventually gets close to 0.5. But does the actual *count* of heads get close to one-half the number of tosses? Let's find out. Set the "Probability of heads" in the *Probability* applet to 0.5 and the number of tosses to 50. You can extend the number of tosses by clicking "Toss" again to get 50 more. Don't click "Reset" during this exercise.

(a) After 50 tosses, what is the proportion of heads? What is the count of heads? What is the difference between the count of heads and 25 (one-half the number of tosses)?

(b) Keep going to 150 tosses. Again record the proportion and count of heads and the difference between the count and 75 (half the number of tosses).

(c) Keep going. Stop at 300 tosses and again at 600 tosses to record the same facts. Although it may take a long time, the laws of probability say that the proportion of heads will always get close to 0.5 and also that the difference between the count of heads and half the number of tosses will always grow without limit.

4.9   A question about dice. Here is a question that a French gambler asked the mathematicians Fermat and Pascal at the very beginning of probability theory: what is the probability of getting at least one six in rolling four dice? The *Law of Large Numbers* applet allows you to roll several dice and watch the outcomes. (Ignore the title of the applet for now.) Because simulation—just like real random phenomena—often takes very many trials to estimate a probability accurately, let's simplify the question: is this probability clearly greater than 0.5, clearly less than 0.5, or quite close to 0.5? Use the applet to roll four dice until you can confidently answer this question. You will have to set "Rolls" to 1 so that you have time to look at the four up-faces. Keep clicking "Roll dice" to roll again and again. How many times did you roll four dice? What percent of your rolls produced at least one six?

4.2 Probability Models

The idea of probability as a proportion of outcomes in very many repeated trials guides our intuition but is hard to express in mathematical form. A description of a random phenomenon in the language of mathematics is called a **probability model.** To see how to proceed, think first about a very simple random phenomenon, tossing a coin once. When we toss a coin, we cannot know the outcome in advance. What do we know? We are willing to say that the outcome will be either heads or tails. Because the coin appears to be balanced, we believe that each of these outcomes has probability 1/2. This description of coin tossing has two parts:

probability model

- A list of possible outcomes

- A probability for each outcome

This two-part description is the starting point for a probability model. We will begin by describing the outcomes of a random phenomenon and then learn how to assign probabilities to the outcomes.

Sample spaces

A probability model first tells us what outcomes are possible.

SAMPLE SPACE

The **sample space** S of a random phenomenon is the set of all possible outcomes.

The name "sample space" is natural in random sampling, where each possible outcome is a sample and the sample space contains all possible samples. To specify S, we must state what constitutes an individual outcome and then state which outcomes can occur. We often have some freedom in defining the sample space, so the choice of S is a matter of convenience as well as correctness. The idea of a sample space, and the freedom we may have in specifying it, are best illustrated by examples.

EXAMPLE

4.4   Sample space for tossing a coin. Toss a coin. There are only two possible outcomes, and the sample space is

$$S = \{\text{heads, tails}\}$$

or, more briefly, $S = \{\text{H, T}\}$.

EXAMPLE

4.5   Sample space for random digits. Let your pencil point fall blindly into Table B of random digits. Record the value of the digit it lands on. The possible outcomes are

$$S = \{0, 1, 2, 3, 4, 5, 6, 7, 8, 9\}$$

EXAMPLE

4.6 Sample space for tossing a coin four times. Toss a coin four times and record the results. That's a bit vague. To be exact, record the results of each of the four tosses in order. A typical outcome is then HTTH. Counting shows that there are 16 possible outcomes. The sample space S is the set of all 16 strings of four H's and T's.

Suppose that our only interest is the number of heads in four tosses. Now we can be exact in a simpler fashion. The random phenomenon is to toss a coin four times and count the number of heads. The sample space contains only five outcomes:

$$S = \{0, 1, 2, 3, 4\}$$

This example illustrates the importance of carefully specifying what constitutes an individual outcome.

Although these examples seem remote from the practice of statistics, the connection is surprisingly close. Suppose that in conducting an opinion poll you select four people at random from a large population and ask each if he or she favors reducing federal spending on low-interest student loans. The answers are "Yes" or "No." The possible outcomes—the sample space—are exactly as in Example 4.4 if we replace heads by "Yes" and tails by "No." Similarly, the possible outcomes of an SRS of 1500 people are the same in principle as the possible outcomes of tossing a coin 1500 times. One of the great advantages of mathematics is that the essential features of quite different phenomena can be described by the same mathematical model.

USE YOUR KNOWLEDGE

4.10   What color is your hair? A student is asked what color hair he or she has. Set up an appropriate sample space for this setting. Note that there is not a single correct answer to this exercise, so give reasons for your choice.

The sample spaces described above previously correspond to categorical variables where we can list all the possible values. Other sample spaces correspond to quantitative variables. Here is an example.

EXAMPLE

4.7   Using software. Most statistical software has a function that will generate a random number between 0 and 1. The sample space is

$$S = \{\text{all numbers between 0 and 1}\}$$

This S is a mathematical idealization. Any specific random number generator produces numbers with some limited number of decimal places so that, strictly speaking, not all numbers between 0 and 1 are possible outcomes. For example, Minitab generates random numbers like 0.736891, with six decimal places. The entire interval from 0 to 1 is easier to think about. It also has the advantage of being a suitable sample space for different software systems that produce random numbers with different numbers of digits.

4.11 How many hours do you study? You record the number of hours per week that a randomly selected student spends studying. What is the sample space?

A sample space S lists the possible outcomes of a random phenomenon. To complete a mathematical description of the random phenomenon, we must also give the probabilities with which these outcomes occur. The true long-term proportion of any outcome—say, "exactly 2 heads in four tosses of a coin"—can be found only empirically, and then only approximately. How then can we describe probability mathematically? Rather than immediately attempting to give "correct" probabilities, let's confront the easier task of laying down rules that any assignment of probabilities must satisfy. We need to assign probabilities not only to single outcomes but also to sets of outcomes.

EVENT

An **event** is an outcome or a set of outcomes of a random phenomenon. That is, an event is a subset of the sample space.

EXAMPLE

4.8 Exactly 2 heads in four tosses. Take the sample space S for four tosses of a coin to be the 16 possible outcomes in the form HTHH. Then "exactly 2 heads" is an event. Call this event A. The event A expressed as a set of outcomes is

$$A = \{\text{HHTT, HTHT, HTTH, THHT, THTH, TTHH}\}$$

In a probability model, events have probabilities. What properties must any assignment of probabilities to events have? Here are some basic facts about any probability model. These facts follow from the idea of probability as "the long-run proportion of repetitions on which an event occurs."

1. **Any probability is a number between 0 and 1.** Any proportion is a number between 0 and 1, so any probability is also a number between 0 and 1. An event with probability 0 never occurs, and an event with probability 1 occurs on every trial. An event with probability 0.5 occurs in half the trials in the long run.

2. **All possible outcomes together must have probability 1.** Because every trial will produce an outcome, the sum of the probabilities for all possible outcomes must be exactly 1.

3. **If two events have no outcomes in common, the probability that one or the other occurs is the sum of their individual probabilities.** If one event occurs in 40% of all trials, a different event occurs in 25% of all

trials, and the two can never occur together, then one or the other occurs on 65% of all trials because 40% + 25% = 65%.

4. **The probability that an event does not occur is 1 minus the probability that the event does occur.** If an event occurs in (say) 70% of all trials, it fails to occur in the other 30%. The probability that an event occurs and the probability that it does not occur always add to 100%, or 1.

Probability rules

Formal probability uses mathematical notation to state Facts 1 to 4 more concisely. We use capital letters near the beginning of the alphabet to denote events. If A is any event, we write its probability as $P(A)$. Here are our probability facts in formal language. As you apply these rules, remember that they are just another form of intuitively true facts about long-run proportions.

PROBABILITY RULES

Rule 1. The probability $P(A)$ of any event A satisfies $0 \leq P(A) \leq 1$.

Rule 2. If S is the sample space in a probability model, then $P(S) = 1$.

Rule 3. Two events A and B are **disjoint** if they have no outcomes in common and so can never occur together. If A and B are disjoint,

$$P(A \text{ or } B) = P(A) + P(B)$$

This is the **addition rule for disjoint events.**

Rule 4. The **complement** of any event A is the event that A does not occur, written as A^c. The **complement rule** states that

$$P(A^c) = 1 - P(A)$$

You may find it helpful to draw a picture to remind yourself of the meaning of complements and disjoint events. A picture like Figure 4.2 that shows the sample space S as a rectangular area and events as areas within S is called a **Venn diagram.** The events A and B in Figure 4.2 are disjoint because they do not overlap. As Figure 4.3 shows, the complement A^c contains exactly the outcomes that are not in A.

Venn diagram

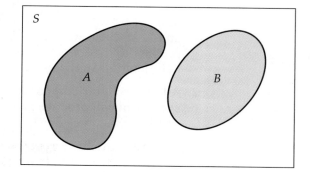

FIGURE 4.2 Venn diagram showing disjoint events A and B. Disjoint events have no common outcomes.

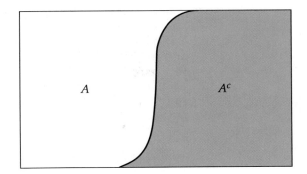

FIGURE 4.3 Venn diagram showing the complement A^c of an event A. The complement consists of all outcomes that are not in A.

EXAMPLE

4.9 Favorite colors. What is your favorite color? Our preferences can be related to our personality, our moods, or particular objects. Here is a probability model for color preferences.[2]

Color	Blue	Green	Purple	Red	Black
Probability	0.42	0.14	0.14	0.08	0.07

Color	Orange	Yellow	Brown	Grey	White
Probability	0.05	0.03	0.03	0.02	0.02

Each probability is between 0 and 1. The probabilities add to 1 because these outcomes together make up the sample space S. Our probability model corresponds to selecting a person at random and asking what is their favorite color.

Let's use the probability Rules 3 and 4 to find some probabilities for favorite colors.

EXAMPLE

4.10 Is it blue or green? What is the probability that a person's favorite color is blue or green? If the favorite is blue, it cannot be green, so these two events are disjoint. Using Rule 3, we find

$$P(\text{Blue or Green}) = P(\text{Blue}) + P(\text{Green})$$
$$= 0.42 + 0.14 = 0.56$$

There is a 56% chance that a randomly selected person will have blue or green as their favorite color. Suppose we want to find the probability that the favorite is not purple.

EXAMPLE

4.11 Use the complement rule. To solve this problem, we could use Rule 3 and add the probabilities for Blue, Green, Red, Black, Orange, Yellow, Brown, Grey, and White. However, it is easier to use the probability that we have for Purple and Rule 4. The event that the favorite is not Purple is the complement of the event that the favorite is Purple. Using our notation for events, we have

$$P(\text{not Purple}) = 1 - P(\text{Purple})$$
$$= 1 - 0.14 = 0.86$$

We see that 86% of people have a favorite color that is not purple.

USE YOUR KNOWLEDGE

4.12 Favorites of Black or White. Find the probability that the favorite color is Black or White.

4.13 Blue, Green, Black, Brown, Grey, or White. Find the probability that the favorite color is Blue, Green, Black, Brown, Grey, or White using Rule 4. Explain why this calculation is easier than finding the answer using Rule 3.

Assigning probabilities: finite number of outcomes

The individual outcomes of a random phenomenon are always disjoint. So the addition rule provides a way to assign probabilities to events with more than one outcome: start with probabilities for individual outcomes and add to get probabilities for events. This idea works well when there are only a finite (fixed and limited) number of outcomes.

> **PROBABILITIES IN A FINITE SAMPLE SPACE**
>
> Assign a probability to each individual outcome. These probabilities must be numbers between 0 and 1 and must have sum 1.
>
> The probability of any event is the sum of the probabilities of the outcomes making up the event.

EXAMPLE

Benford's law

4.12 Benford's law. Faked numbers in tax returns, payment records, invoices, expense account claims, and many other settings often display patterns that aren't present in legitimate records. Some patterns, like too many round numbers, are obvious and easily avoided by a clever crook. Others are more subtle. It is a striking fact that the first digits of numbers in legitimate records often follow a distribution known as **Benford's law.** Here it is (note that a first digit can't be 0):[3]

First digit	1	2	3	4	5	6	7	8	9
Probability	0.301	0.176	0.125	0.097	0.079	0.067	0.058	0.051	0.046

Benford's law usually applies to the first digits of the sizes of similar quantities, such as invoices, expense account claims, and county populations. Investigators can detect fraud by comparing the first digits in records such as invoices paid by a business with these probabilities.

EXAMPLE

4.13 Find some probabilities for Benford's law. Consider the events

$$A = \{\text{first digit is 1}\}$$
$$B = \{\text{first digit is 6 or greater}\}$$

From the table of probabilities,

$$P(A) = P(1) = 0.301$$
$$P(B) = P(6) + P(7) + P(8) + P(9)$$
$$= 0.067 + 0.058 + 0.051 + 0.046 = 0.222$$

Note that $P(B)$ is not the same as the probability that a first digit is strictly greater than 6. The probability $P(6)$ that a first digit is 6 is included in "6 or greater" but not in "greater than 6."

USE YOUR KNOWLEDGE

4.14 Benford's law. Using the probabilities for Benford's law, find the probability that a first digit is anything other than 1.

4.15 Use the addition rule. Use the addition rule with the probabilities for the events A and B from Example 4.13 to find the probability that a first digit is either 1 or 6 or greater.

Be careful to apply the addition rule only to disjoint events.

EXAMPLE

4.14 Apply the addition rule to Benford's law. Check that the probability of the event C that a first digit is odd is

$$P(C) = P(1) + P(3) + P(5) + P(7) + P(9) = 0.609$$

The probability

$$P(B \text{ or } C) = P(1) + P(3) + P(5) + P(6) + P(7) + P(8) + P(9) = 0.727$$

is *not* the sum of $P(B)$ and $P(C)$, because events B and C are not disjoint. Outcomes 7 and 9 are common to both events.

Assigning probabilities: equally likely outcomes

Assigning correct probabilities to individual outcomes often requires long observation of the random phenomenon. In some circumstances, however, we are willing to assume that individual outcomes are equally likely because of some balance in the phenomenon. Ordinary coins have a physical balance that should make heads and tails equally likely, for example, and the table of random digits comes from a deliberate randomization.

EXAMPLE

4.15 First digits that are equally likely. You might think that first digits are distributed "at random" among the digits 1 to 9 in business records. The 9 possible outcomes would then be equally likely. The sample space for a single digit is

$$S = \{1, 2, 3, 4, 5, 6, 7, 8, 9\}$$

Because the total probability must be 1, the probability of each of the 9 outcomes must be 1/9. That is, the assignment of probabilities to outcomes is

First digit	1	2	3	4	5	6	7	8	9
Probability	1/9	1/9	1/9	1/9	1/9	1/9	1/9	1/9	1/9

The probability of the event B that a randomly chosen first digit is 6 or greater is

$$P(B) = P(6) + P(7) + P(8) + P(9)$$
$$= \frac{1}{9} + \frac{1}{9} + \frac{1}{9} + \frac{1}{9} = \frac{4}{9} = 0.444$$

Compare this with the Benford's law probability in Example 4.13. A crook who fakes data by using "random" digits will end up with too many first digits 6 or greater and too few 1s and 2s.

In Example 4.15 all outcomes have the same probability. Because there are 9 equally likely outcomes, each must have probability 1/9. Because exactly 4 of the 9 equally likely outcomes are 6 or greater, the probability of this event is 4/9. In the special situation where all outcomes are equally likely, we have a simple rule for assigning probabilities to events.

EQUALLY LIKELY OUTCOMES

If a random phenomenon has k possible outcomes, all equally likely, then each individual outcome has probability $1/k$. The probability of any event A is

$$P(A) = \frac{\text{count of outcomes in } A}{\text{count of outcomes in } S}$$
$$= \frac{\text{count of outcomes in } A}{k}$$

Most random phenomena do not have equally likely outcomes, so the general rule for finite sample spaces is more important than the special rule for equally likely outcomes.

USE YOUR KNOWLEDGE

4.16 Possible outcomes for rolling a die. A die has six sides with 1 to 6 "spots" on the sides. Give the probability distribution for the six possible outcomes that can result when a perfect die is rolled.

Independence and the multiplication rule

Rule 3, the addition rule for disjoint events, describes the probability that *one or the other* of two events A and B will occur in the special situation when A and B cannot occur together because they are disjoint. Our final rule describes the probability that *both* events A and B occur, again only in a special situation. More general rules appear in Section 4.5, but in our study of statistics we will need only the rules that apply to special situations.

Suppose that you toss a fair coin twice. You are counting heads, so two events of interest are

$$A = \{\text{first toss is a head}\}$$
$$B = \{\text{second toss is a head}\}$$

The events A and B are not disjoint. They occur together whenever both tosses give heads. We want to compute the probability of the event $\{A$ and $B\}$ that *both* tosses are heads. The Venn diagram in Figure 4.4 illustrates the event $\{A$ and $B\}$ as the overlapping area that is common to both A and B.

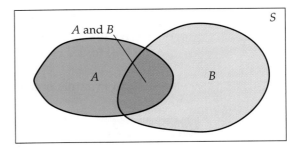

FIGURE 4.4 Venn diagram showing the event $\{A$ and $B\}$. This event consists of outcomes common to A and B.

The coin tossing of Buffon, Pearson, and Kerrich described in Example 4.3 makes us willing to assign probability 1/2 to a head when we toss a coin. So

$$P(A) = 0.5$$
$$P(B) = 0.5$$

What is $P(A$ and $B)$? Our common sense says that it is 1/4. The first coin will give a head half the time and then the second will give a head on half of those trials, so both coins will give heads on $1/2 \times 1/2 = 1/4$ of all trials in the long run. This reasoning assumes that the second coin still has probability 1/2 of a head after the first has given a head. This is true—we can verify it by tossing two coins many times and observing the proportion of heads on the second toss after the first toss has produced a head. We say that the events "head on the first toss" and "head on the second toss" are *independent*. Here is our final probability rule.

THE MULTIPLICATION RULE FOR INDEPENDENT EVENTS

Rule 5. Two events A and B are **independent** if knowing that one occurs does not change the probability that the other occurs. If A and B are independent,

$$P(A \text{ and } B) = P(A)P(B)$$

This is the **multiplication rule for independent events**.

Our definition of independence is rather informal. We will make this informal idea precise in Section 4.5. In practice, though, we rarely need a precise definition of independence, because independence is usually *assumed* as part of a probability model when we want to describe random phenomena that seem to be physically unrelated to each other. Here is an example of independence.

EXAMPLE

4.16   Coins do not have memory. Because a coin has no memory and most coin tossers cannot influence the fall of the coin, it is safe to assume that successive coin tosses are independent. For a fair coin, this means that after we see the outcome of the first toss, we still assign probability 1/2 to heads on the second toss.

USE YOUR KNOWLEDGE

4.17   Two tails in two tosses. What is the probability of obtaining two tails on two tosses of a fair coin?

Here is an example of a situation where there are dependent events.

EXAMPLE

4.17   Dependent events in cards. The colors of successive cards dealt from the same deck are not independent. A standard 52-card deck contains 26 red and 26 black cards. For the first card dealt from a shuffled deck, the probability of a red card is $26/52 = 0.50$ because the 52 possible cards are equally likely. Once we see that the first card is red, we know that there are only 25 reds among the remaining 51 cards. The probability that the second card is red is therefore only $25/51 = 0.49$. Knowing the outcome of the first deal changes the probabilities for the second.

USE YOUR KNOWLEDGE

4.18   The probability of a second ace. A deck of 52 cards contains 4 aces, so the probability that a card drawn from this deck is an ace is 4/52. If we know that the first card drawn is an ace, what is the probability that the second card drawn is also an ace? Using the idea of independence, explain why this probability is not 4/52.

Here is another example of a situation where events are dependent.

EXAMPLE

4.18 Taking a test twice. If you take an IQ test or other mental test twice in succession, the two test scores are not independent. The learning that occurs on the first attempt influences your second attempt. If you learn a lot, then your second test score might be a lot higher than your first test score.

When independence is part of a probability model, the multiplication rule applies. Here is an example.

EXAMPLE

4.19 Mendel's peas. Gregor Mendel used garden peas in some of the experiments that revealed that inheritance operates randomly. The seed color of Mendel's peas can be either green or yellow. Two parent plants are "crossed" (one pollinates the other) to produce seeds. Each parent plant carries two genes for seed color, and each of these genes has probability 1/2 of being passed to a seed. The two genes that the seed receives, one from each parent, determine its color. The parents contribute their genes independently of each other.

Suppose that both parents carry the G and the Y genes. The seed will be green if both parents contribute a G gene; otherwise it will be yellow. If M is the event that the male contributes a G gene and F is the event that the female contributes a G gene, then the probability of a green seed is

$$P(M \text{ and } F) = P(M)P(F)$$
$$= (0.5)(0.5) = 0.25$$

In the long run, 1/4 of all seeds produced by crossing these plants will be green.

The multiplication rule applies only to independent events; you cannot use it if events are not independent. Here is a distressing example of misuse of the multiplication rule.

EXAMPLE

4.20 Sudden infant death syndrome. Sudden infant death syndrome (SIDS) causes babies to die suddenly (often in their cribs) with no explanation. Deaths from SIDS have been greatly reduced by placing babies on their backs, but as yet no cause is known.

When more than one SIDS death occurs in a family, the parents are sometimes accused. One "expert witness" popular with prosecutors in England told juries that there is only a 1 in 73 million chance that two children in the same family could have died naturally. Here's his calculation: the rate of SIDS in a nonsmoking middle-class family is 1 in 8500. So the probability of two deaths is

$$\frac{1}{8500} \times \frac{1}{8500} = \frac{1}{72,250,000}$$

Several women were convicted of murder on this basis, without any direct evidence that they harmed their children.

As the Royal Statistical Society said, this reasoning is nonsense. It assumes that SIDS deaths in the same family are independent events. The cause of SIDS is unknown: "There may well be unknown genetic or environmental factors that predispose families to SIDS, so that a second case within the family becomes much more likely."[4] The British government decided to review the cases of 258 parents convicted of murdering their babies.

The multiplication rule $P(A \text{ and } B) = P(A)P(B)$ holds if A and B are independent but not otherwise. The addition rule $P(A \text{ or } B) = P(A) + P(B)$ holds if A and B are disjoint but not otherwise. Resist the temptation to use these simple formulas when the circumstances that justify them are not present. *You must also be certain not to confuse disjointness and independence. Disjoint events cannot be independent.* If A and B are disjoint, then the fact that A occurs tells us that B cannot occur—look again at Figure 4.2. Unlike disjointness or complements, independence cannot be pictured by a Venn diagram, because it involves the probabilities of the events rather than just the outcomes that make up the events.

Applying the probability rules

If two events A and B are independent, then their complements A^c and B^c are also independent and A^c is independent of B. Suppose, for example, that 75% of all registered voters in a suburban district are Republicans. If an opinion poll interviews two voters chosen independently, the probability that the first is a Republican and the second is not a Republican is $(0.75)(0.25) = 0.1875$. The multiplication rule also extends to collections of more than two events, provided that all are independent. Independence of events A, B, and C means that no information about any one or any two can change the probability of the remaining events. The formal definition is a bit messy. Fortunately, independence is usually assumed in setting up a probability model. We can then use the multiplication rule freely, as in this example.

By combining the rules we have learned, we can compute probabilities for rather complex events. Here is an example.

EXAMPLE

4.21 HIV testing. Many people who come to clinics to be tested for HIV, the virus that causes AIDS, don't come back to learn the test results. Clinics now use "rapid HIV tests" that give a result in a few minutes. The false positive rate for a diagnostic test is the probability that a person with no disease will have a positive test result. For the rapid HIV tests, the Food and Drug Administration (FDA) has established 2% as the maximum false positive rate for a rapid HIV test.[5] If a clinic uses a test that meets the FDA standard and tests 50 people who are free of HIV antibodies, what is the probability that at least one false-positive will occur?

It is reasonable to assume as part of the probability model that the test results for different individuals are independent. The probability that the test is positive for a single person is 0.02, so the probability of a negative result is $1 - 0.02 = 0.98$ by the complement rule. The probability of at least one false-positive among the 50 people tested is therefore

$$P(\text{at least one positive}) = 1 - P(\text{no positives})$$
$$= 1 - P(100 \text{ negatives})$$
$$= 1 - 0.98^{50}$$
$$= 1 - 0.3642 = 0.6358$$

There is approximately a 64% chance that at least one of the 50 people will test positive for HIV, even though no one has the virus.

Concern about excessive numbers of false positives led the New York City Department of Health and Mental Hygiene to suspend the use of one particular rapid HIV test.[6]

SECTION 4.2 Summary

A **probability model** for a random phenomenon consists of a sample space S and an assignment of probabilities P.

The **sample space** S is the set of all possible outcomes of the random phenomenon. Sets of outcomes are called **events.** P assigns a number $P(A)$ to an event A as its probability.

The **complement** A^c of an event A consists of exactly the outcomes that are not in A. Events A and B are **disjoint** if they have no outcomes in common. Events A and B are **independent** if knowing that one event occurs does not change the probability we would assign to the other event.

Any assignment of probability must obey the rules that state the basic properties of probability:

> **Rule 1.** $0 \leq P(A) \leq 1$ for any event A.
> **Rule 2.** $P(S) = 1$.
> **Rule 3. Addition rule:** If events A and B are **disjoint,** then $P(A \text{ or } B) = P(A) + P(B)$.
> **Rule 4. Complement rule:** For any event A, $P(A^c) = 1 - P(A)$.
> **Rule 5. Multiplication rule:** If events A and B are **independent,** then $P(A \text{ and } B) = P(A)P(B)$.

SECTION 4.2 Exercises

For Exercise 4.10, see page 234; for Exercise 4.11, see page 235; for Exercises 4.12 and 4.13, see page 238; for Exercises 4.14 and 4.15, see page 239; for Exercise 4.16, see page 241; for Exercise 4.17, see page 242; and for Exercise 4.18, see page 242.

4.19 What's wrong? In each of the following scenarios, there is something wrong. Describe what is wrong and give a reason for your answer.

(a) If two events are disjoint, we can multiply their probabilities to determine the probability that they will both occur.

(b) If the probability of A is 0.6 and the probability of B is 0.5, the probability of both A and B happening is 1.1.

(c) If the probability of A is 0.35, then the probability of the complement of A is -0.35.

4.20 What's wrong? In each of the following scenarios, there is something wrong. Describe what is wrong and give a reason for your answer.

(a) If the sample space consists of two outcomes, then each outcome has probability 0.5.

(b) If we select a digit at random, then the probability of selecting a 2 is 0.2.

(c) If the probability of A is 0.2, the probability of B is 0.3, and the probability of A and B is 0.5, then A and B are independent.

4.21 Evaluating Web page designs. You are a Web page designer and you set up a page with five different links. A user of the page can click on one of the links or he or she can leave that page. Describe the sample space for the outcome of a visitor to your Web page.

4.22 Record the length of time spent on the page. Refer to the previous exercise. You also decide to measure the length of time a visitor spends on your page. Give the sample space for this measure.

4.23 Ringtones. What are the popular ringtones? The web site `funtonia.com` updates its list of top ringtones frequently. Here are probabilities for the top 10 ringtones listed by the site recently:[7]

Ringtone	Probability	Ringtone	Probability
Empire State of Mind	0.180	Bad Romance	0.081
Baby By Me	0.136	I Can Transform Ya	0.075
Forever	0.114	Down	0.070
Party in the USA	0.107	I Gotta Feeling	0.068
Fireflies	0.103	Money To Blow	0.066

(a) What is the probability that a randomly selected ringtone from this list is either Empire State of Mind or I Gotta Feeling?

(b) What is the probability that a randomly selected ringtone from this list is not Empire State of Mind and not I Gotta Feeling? Be sure to show how you computed this answer.

4.24 More ringtones. Refer to the previous exercise.

(a) If two ringtones are selected independently, what is the probability that both are Party in the USA?

(b) Describe in words the complement of the event described in part (a) of this exercise. Find the probability of this event.

4.25 Distribution of blood types. All human blood can be "ABO-typed" as one of O, A, B, or AB, but the distribution of the types varies a bit among groups of people. Here is the distribution of blood types for a randomly chosen person in the United States:[8]

Blood type	A	B	AB	O
U.S. probability	0.42	0.11	?	0.44

(a) What is the probability of type AB blood in the United States?

(b) Maria has type B blood. She can safely receive blood transfusions from people with blood types O and B. What is the probability that a randomly chosen person from the United States can donate blood to Maria?

4.26 Blood types in Ireland. The distribution of blood types in Ireland differs from the U.S. distribution given in the previous exercise:

Blood type	A	B	AB	O
Ireland probability	0.35	0.10	0.03	0.52

Choose a person from the United States and a person from Ireland at random, independently of each other. What is the probability that both have type O blood? What is the probability that both have the same blood type?

4.27 Are the probabilities legitimate? In each of the following situations, state whether or not the given assignment of probabilities to individual outcomes is legitimate, that is, satisfies the rules of probability. If not, give specific reasons for your answer.

(a) Choose a college student at random and record gender and enrollment status: P(female full-time) = 0.44, P(female part-time) = 0.56, P(male full-time) = 0.46, P(male part-time) = 0.54.

(b) Deal a card from a shuffled deck: P(clubs) = 16/52, P(diamonds) = 12/52, P(hearts) = 12/52, P(spades) = 12/52.

(c) Roll a die and record the count of spots on the up-face: $P(1) = 1/3$, $P(2) = 0$, $P(3) = 1/6$, $P(4) = 1/3$, $P(5) = 1/6$, $P(6) = 0$.

4.28 French and English in Canada. Canada has two official languages, English and French. Choose a Canadian at random and ask, "What is your mother tongue?" Here is the distribution of responses, combining many separate languages from the broad Asian/Pacific region:[9]

Language	English	French	Asian/Pacific	Other
Probability	0.59	?	0.07	0.11

(a) What probability should replace "?" in the distribution?

(b) What is the probability that a Canadian's mother tongue is not English? Explain how you computed your answer.

4.29 Education levels of young adults. Choose a young adult (age 25 to 34 years) at random. The probability is 0.12 that the person chosen did not complete high school, 0.31 that the person has a high school diploma but no further education, and 0.29 that the person has at least a bachelor's degree.

(a) What must be the probability that a randomly chosen young adult has some education beyond high school but does not have a bachelor's degree?

(b) What is the probability that a randomly chosen young adult has at least a high school education?

4.30 🔺 **Loaded dice.** There are many ways to produce crooked dice. To *load* a die so that 6 comes up too often and 1 (which is opposite 6) comes up too seldom, add a bit of lead to the filling of the spot on the 1 face. Because the spot is solid plastic, this works even with transparent dice. If a die is loaded so that 6 comes up with probability 0.21 and the probabilities of the 2, 3, 4, and 5 faces are not affected, what is the assignment of probabilities to the six faces?

4.31 Rh blood types. Human blood is typed as O, A, B, or AB and also as Rh-positive or Rh-negative. ABO type and Rh-factor type are independent because they are governed by different genes. In the American population, 84% of people are Rh-positive. Use the information about ABO type in Exercise 4.25 to give the probability distribution of blood type (ABO and Rh) for a randomly chosen person.

4.32 Roulette. A roulette wheel has 38 slots, numbered 0, 00, and 1 to 36. The slots 0 and 00 are colored green, 18 of the others are red, and 18 are black. The dealer spins the wheel and at the same time rolls a small ball along the wheel in the opposite direction. The wheel is carefully balanced so that the ball is equally likely to land in any slot when the wheel slows. Gamblers can bet on various combinations of numbers and colors.

(a) What is the probability that the ball will land in any one slot?

(b) If you bet on "red," you win if the ball lands in a red slot. What is the probability of winning?

(c) The slot numbers are laid out on a board on which gamblers place their bets. One column of numbers on the board contains all multiples of 3, that is, 3, 6, 9, ..., 36. You place a "column bet" that wins if any of these numbers comes up. What is your probability of winning?

4.33 Winning the lottery. A state lottery's Pick 3 game asks players to choose a three-digit number, 000 to 999. The state chooses the winning three-digit number at random, so that each number has probability 1/1000. You

win if the winning number contains the digits in your number, in any order.

(a) Your number is 491. What is your probability of winning?

(b) Your number is 222. What is your probability of winning?

4.34 PINs. The personal identification numbers (PINs) for automatic teller machines usually consist of four digits. You notice that most of your PINs have at least one 0, and you wonder if the issuers use lots of 0s to make the numbers easy to remember. Suppose that PINs are assigned at random, so that all four-digit numbers are equally likely.

(a) How many possible PINs are there?

(b) What is the probability that a PIN assigned at random has at least one 0?

4.35 Universal blood donors. People with type O-negative blood are universal donors. That is, any patient can receive a transfusion of O-negative blood. Only 7% of the American population have O-negative blood. If 10 people appear at random to give blood, what is the probability that at least 1 of them is a universal donor?

4.36 Disappearing Internet sites. Internet sites often vanish or move, so that references to them can't be followed. In fact, 13% of Internet sites referenced in papers in major scientific journals are lost within two years after publication.[10] If a paper contains seven Internet references, what is the probability that all seven are still good two years later? What specific assumptions did you make in order to calculate this probability?

4.37 Is this calculation correct? Government data show that 6% of the American population are at least 75 years of age and that about 51% are women. Explain why it is wrong to conclude that because $(0.06)(0.51) = 0.0306$ about 3% of the population are women aged 75 or over.

4.38 Colored dice. Here's more evidence that our intuition about chance behavior is not very accurate. A six-sided die has four green and two red faces, all equally probable. Psychologists asked students to say which of these color sequences is most likely to come up at the beginning of a long set of rolls of this die:

RGRRR

RGRRRG

GRRRR

More than 60% chose the second sequence.[11] What is the correct probability of each sequence?

4.39 Random walks and stock prices. The "random walk" theory of securities prices holds that price movements in disjoint time periods are independent of each other. Suppose that we record only whether the price is up or down each year, and that the probability that our portfolio rises in price in any one year is 0.65. (This probability is approximately correct for a portfolio containing equal dollar amounts of all common stocks listed on the New York Stock Exchange.)

(a) What is the probability that our portfolio goes up for three consecutive years?

(b) If you know that the portfolio has risen in price two years in a row, what probability do you assign to the event that it will go down next year?

(c) What is the probability that the portfolio's value moves in the same direction in both of the next two years?

4.40 ⚠ **Axioms of probability.** Show that any assignment of probabilities to events that obeys Rules 2 and 3 on page 236 automatically obeys the complement rule (Rule 4). This implies that a mathematical treatment of probability can start from just Rules 1, 2, and 3. These rules are sometimes called *axioms* of probability.

4.41 ⚠ **Independence of complements.** Show that if events A and B obey the multiplication rule, $P(A \text{ and } B) = P(A)P(B)$, then A and the complement B^c of B also obey the multiplication rule, $P(A \text{ and } B^c) = P(A)P(B^c)$. That is, if events A and B are independent, then A and B^c are also independent. (*Hint:* Start by drawing a Venn diagram and noticing that the events "A and B" and "A and B^c" are disjoint.)

Mendelian inheritance. *Some traits of plants and animals depend on inheritance of a single gene. This is called Mendelian inheritance, after Gregor Mendel (1822--1884). Exercises 4.42 to 4.45 are based on the following information about Mendelian inheritance of blood type.*

Each of us has an ABO blood type, which describes whether two characteristics called A and B are present. Every human being has two blood type alleles (gene forms), one inherited from our mother and one from our father.

Each of these alleles can be A, B, or O. Which two we inherit determines our blood type. Here is a table that shows what our blood type is for each combination of two alleles:

Alleles inherited	Blood type
A and A	A
A and B	AB
A and O	A
B and B	B
B and O	B
O and O	O

We inherit each of a parent's two alleles with probability 0.5. We inherit independently from our mother and father.

4.42 Blood types of children. Hannah and Jacob both have alleles A and B.

(a) What blood types can their children have?

(b) What is the probability that their next child has each of these blood types?

4.43 Parents with alleles B and O. Nancy and David both have alleles B and O.

(a) What blood types can their children have?

(b) What is the probability that their next child has each of these blood types?

4.44 Two children. Jennifer has alleles A and O. José has alleles A and B. They have two children. What is the probability that both children have blood type A? What is the probability that both children have the same blood type?

4.45 Three children. Jasmine has alleles A and O. Joshua has alleles B and O.

(a) What is the probability that a child of these parents has blood type O?

(b) If Jasmine and Joshua have three children, what is the probability that all three have blood type O? What is the probability that the first child has blood type O and the next two do not?

4.3 Random Variables

Sample spaces need not consist of numbers. When we toss a coin four times, we can record the outcome as a string of heads and tails, such as HTTH. In statistics, however, we are most often interested in numerical outcomes such as the count of heads in the four tosses. It is convenient to use a shorthand notation: Let X be the number of heads. If our outcome is HTTH, then $X = 2$. If the next outcome is TTTH, the value of X changes to $X = 1$. The possible values of X are 0, 1, 2, 3, and 4. Tossing a coin four times will give X one of these possible values. Tossing four more times will give X another and probably different value. We call X a *random variable* because its values vary when the coin tossing is repeated.

RANDOM VARIABLE

A **random variable** is a variable whose value is a numerical outcome of a random phenomenon.

We usually denote random variables by capital letters near the end of the alphabet, such as X or Y. Of course, the random variables of greatest interest to us are outcomes such as the mean $\bar{x}$ of a random sample, for which we will keep the familiar notation.[12] As we progress from general rules of probability toward statistical inference, we will concentrate on random variables. When a random variable X describes a random phenomenon, the sample space S just lists the possible values of the random variable. We usually do not mention S separately. There remains the second part of any probability model, the assignment of probabilities to events. There are two main ways of assigning probabilities to the values of a random variable. The two types of probability models that result will dominate our application of probability to statistical inference.

Discrete random variables

We have learned several rules of probability, but only one method of assigning probabilities: state the probabilities of the individual outcomes and assign probabilities to events by summing over the outcomes. The outcome probabilities must be between 0 and 1 and have sum 1. When the outcomes are numerical, they are values of a random variable. We will now attach a name to random variables having probability assigned in this way.[13]

DISCRETE RANDOM VARIABLE

A **discrete random variable** X has a finite number of possible values. The **probability distribution** of X lists the values and their probabilities:

Value of X	x_1	x_2	x_3	$\cdots$	x_k
Probability	p_1	p_2	p_3	$\cdots$	p_k

The probabilities p_i must satisfy two requirements:
1. Every probability p_i is a number between 0 and 1.
2. $p_1 + p_2 + \cdots + p_k = 1$.

Find the probability of any event by adding the probabilities p_i of the particular values x_i that make up the event.

EXAMPLE

4.22 Grade distributions. North Carolina State University posts the grade distributions for its courses online.[14] Students in one section of English 210 in the spring 2006 semester received 31% A's, 40% B's, 20% C's, 4% D's, and 5% F's. Choose an English 210 student at random. To "choose at random" means to give every student the same chance to be chosen. The student's grade on a four-point scale (with A = 4) is a random variable X.

The value of X changes when we repeatedly choose students at random, but it is always one of 0, 1, 2, 3, or 4. Here is the distribution of X:

Value of X	0	1	2	3	4
Probability	0.05	0.04	0.20	0.40	0.31

The probability that the student got a B or better is the sum of the probabilities of an A and a B. In the language of random variables,

$$P(X \geq 3) = P(X = 3) + P(X = 4)$$
$$= 0.40 + 0.31 = 0.71$$

USE YOUR KNOWLEDGE

4.46 Will the course satisfy the requirement? Refer to Example 4.22. Suppose that a grade of D or F in English 210 will not count as satisfying a requirement for a major in linguistics. What is the probability that a randomly selected student will not satisfy this requirement?

probability histogram   We can use histograms to show probability distributions as well as distributions of data. Figure 4.5 displays **probability histograms** that compare

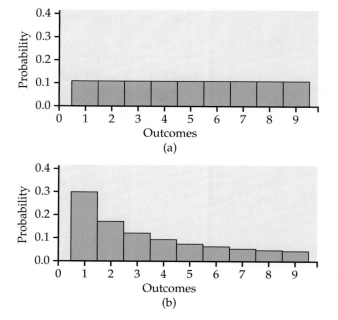

FIGURE 4.5 Probability histograms for (a) random digits 1 to 9 and (b) Benford's law. The height of each bar shows the probability assigned to a single outcome.

the probability model for random digits for business records (Example 4.15) with the model given by Benford's law (Example 4.12). The height of each bar shows the probability of the outcome at its base. Because the heights are probabilities, they add to 1. As usual, all the bars in a histogram have the same width. So the areas also display the assignment of probability to outcomes. Think of these histograms as idealized pictures of the results of very many trials. The histograms make it easy to quickly compare the two distributions.

EXAMPLE

4.23 Number of heads in four tosses of a coin. What is the probability distribution of the discrete random variable X that counts the number of heads in four tosses of a coin? We can derive this distribution if we make two reasonable assumptions:

- The coin is balanced, so it is fair and each toss is equally likely to give H or T.

- The coin has no memory, so tosses are independent.

The outcome of four tosses is a sequence of heads and tails such as HTTH. There are 16 possible outcomes in all. Figure 4.6 lists these outcomes along with the value of X for each outcome. The multiplication rule for independent events tells us that, for example,

$$P(\text{HTTH}) = \frac{1}{2} \times \frac{1}{2} \times \frac{1}{2} \times \frac{1}{2} = \frac{1}{16}$$

Each of the 16 possible outcomes similarly has probability 1/16. That is, these outcomes are equally likely.

The number of heads X has possible values 0, 1, 2, 3, and 4. These values are *not* equally likely. As Figure 4.6 shows, there is only one way that $X = 0$ can occur: namely, when the outcome is TTTT. So

$$P(X = 0) = \frac{1}{16} = 0.0625$$

The event $\{X = 2\}$ can occur in six different ways, so that

$$P(X = 2) = \frac{\text{count of ways } X = 2 \text{ can occur}}{16}$$

$$= \frac{6}{16} = 0.375$$

FIGURE 4.6 Possible outcomes in four tosses of a coin, for Example 4.23. The outcomes are arranged by the values of the random variable X, the number of heads.

		HTTH		
		HTHT		
	HTTT	THTH	HHHT	
	THTT	HHTT	HHTH	
	TTHT	THHT	HTHH	
TTTT	TTTH	TTHH	THHH	HHHH
$X = 0$	$X = 1$	$X = 2$	$X = 3$	$X = 4$

We can find the probability of each value of X from Figure 4.6 in the same way. Here is the result:

Value of X	0	1	2	3	4
Probability	0.0625	0.25	0.375	0.25	0.0625

Figure 4.7 is a probability histogram for the distribution in Example 4.23. The probability distribution is exactly symmetric. The probabilities (bar heights) are idealizations of the proportions after very many tosses of four coins. The actual distribution of proportions observed would be nearly symmetric but is unlikely to be exactly symmetric.

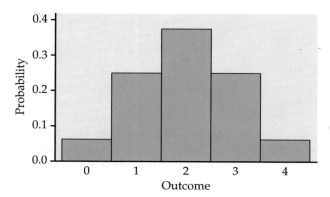

FIGURE 4.7 Probability histogram for the number of heads in four tosses of a coin.

EXAMPLE

4.24 Probability of at least two heads. Any event involving the number of heads observed can be expressed in terms of X, and its probability can be found from the distribution of X. For example, the probability of tossing at least two heads is

$$P(X \geq 2) = 0.375 + 0.25 + 0.0625 = 0.6875$$

The probability of at least one head is most simply found by use of the complement rule:

$$P(X \geq 1) = 1 - P(X = 0)$$
$$= 1 - 0.0625 = 0.9375$$

Recall that tossing a coin n times is similar to choosing an SRS of size n from a large population and asking a yes-or-no question. We will extend the results of Example 4.23 when we return to sampling distributions in the next chapter.

USE YOUR KNOWLEDGE

4.47 Two tosses of a fair coin. Find the probability distribution for the number of heads that appear in two tosses of a fair coin.

Continuous random variables

When we use the table of random digits to select a digit between 0 and 9, the result is a discrete random variable. The probability model assigns probability 1/10 to each of the 10 possible outcomes. Suppose that we want to choose a number at random between 0 and 1, allowing *any* number between 0 and 1 as the outcome. Software random number generators will do this. You can visualize such a random number by thinking of a spinner (Figure 4.8) that turns freely on its axis and slowly comes to a stop. The pointer can come to rest anywhere on a circle that is marked from 0 to 1. The sample space is now an entire interval of numbers:

$$S = \{\text{all numbers } x \text{ such that } 0 \leq x \leq 1\}$$

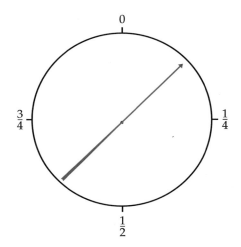

FIGURE 4.8 A spinner that generates a random number between 0 and 1.

How can we assign probabilities to events such as $\{0.3 \leq x \leq 0.7\}$? As in the case of selecting a random digit, we would like all possible outcomes to be equally likely. But we cannot assign probabilities to each individual value of x and then sum, because there are infinitely many possible values. Instead, we use a new way of assigning probabilities directly to events—as *areas under a density curve*. Any density curve has area exactly 1 underneath it, corresponding to total probability 1.

EXAMPLE

4.25 Uniform random numbers. The random number generator will spread its output uniformly across the entire interval from 0 to 1 as we allow it to generate a long sequence of numbers. The results of many trials are represented by the density curve of a **uniform distribution.** This density curve appears in red in Figure 4.9. It has height 1 over the interval from 0 to 1, and height 0 everywhere else. The area under the density curve is 1: the area of a square with base 1 and height 1. The probability of any event is the area under the density curve and above the event in question.

As Figure 4.9(a) illustrates, the probability that the random number generator produces a number X between 0.3 and 0.7 is

$$P(0.3 \leq X \leq 0.7) = 0.4$$

uniform distribution

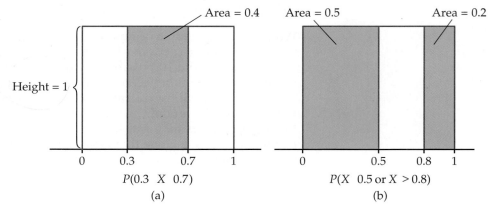

FIGURE 4.9 Assigning probabilities for generating a random number between 0 and 1, for Example 4.25. The probability of any interval of numbers is the area above the interval and under the density curve.

because the area under the density curve and above the interval from 0.3 to 0.7 is 0.4. The height of the density curve is 1, and the area of a rectangle is the product of height and length, so the probability of any interval of outcomes is just the length of the interval.

Similarly,

$$P(X \leq 0.5) = 0.5$$
$$P(X > 0.8) = 0.2$$
$$P(X \leq 0.5 \text{ or } X > 0.8) = 0.7$$

Notice that the last event consists of two nonoverlapping intervals, so the total area above the event is found by adding two areas, as illustrated by Figure 4.9(b). This assignment of probabilities obeys all of our rules for probability.

Probability as area under a density curve is a second important way of assigning probabilities to events. Figure 4.10 illustrates this idea in general form. We call X in Example 4.25 a *continuous random variable* because its values are not isolated numbers but an entire interval of numbers.

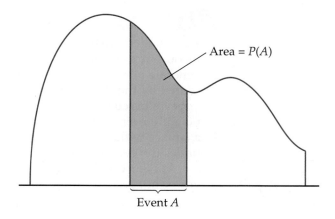

FIGURE 4.10 The probability distribution of a continuous random variable assigns probabilities as areas under a density curve. The total area under any density curve is 1.

USE YOUR KNOWLEDGE

4.48 Find the probability. For the uniform distribution described in Example 4.25, find the probability that X is between 0.1 and 0.4.

CONTINUOUS RANDOM VARIABLE

A **continuous random variable** X takes all values in an interval of numbers. The **probability distribution** of X is described by a density curve. The probability of any event is the area under the density curve and above the values of X that make up the event.

The probability model for a continuous random variable assigns probabilities to intervals of outcomes rather than to individual outcomes. In fact, **all continuous probability distributions assign probability 0 to every individual outcome.** Only intervals of values have positive probability. To see that this is true, consider a specific outcome such as $P(X = 0.8)$ in the context of Example 4.25. The probability of any interval is the same as its length. The point 0.8 has no length, so its probability is 0.

Although this fact may seem odd, it makes intuitive, as well as mathematical, sense. The random number generator produces a number between 0.79 and 0.81 with probability 0.02. An outcome between 0.799 and 0.801 has probability 0.002. A result between 0.799999 and 0.800001 has probability 0.000002. You see that as we approach 0.8 the probability gets closer to 0. To be consistent, the probability of outcome *exactly* equal to 0.8 must be 0. Because there is no probability exactly at $X = 0.8$, the two events $\{X > 0.8\}$ and $\{X \geq 0.8\}$ have the same probability. We can ignore the distinction between > and ≥ when finding probabilities for continuous (but not discrete) random variables.

Normal distributions as probability distributions

The density curves that are most familiar to us are the Normal curves. Because any density curve describes an assignment of probabilities, *Normal distributions are probability distributions*. Recall that $N(\mu, \sigma)$ is our shorthand for the Normal distribution having mean μ and standard deviation σ. In the language of random variables, if X has the $N(\mu, \sigma)$ distribution, then the standardized variable

$$Z = \frac{X - \mu}{\sigma}$$

is a standard Normal random variable having the distribution $N(0, 1)$.

EXAMPLE

4.26 Texting while driving. Texting while driving can be dangerous, but young people want to remain connected. Suppose that 26% of teen drivers text while driving. If we take a sample of 500 teen drivers, what percent would we expect to say that they text while driving?[15]

The proportion $p = 0.26$ is a *parameter* that describes the population of teen drivers. The proportion $\hat{p}$ of the sample who answer say that they text while driving is a *statistic* used to estimate p. The statistic $\hat{p}$ is a random variable because repeating the SRS would give a different sample of 500 teen drivers and a different value of $\hat{p}$. We will see in the next chapter that $\hat{p}$ has approximately the $N(0.26, 0.0196)$ distribution. The mean 0.26 of this distribution is the same as the population parameter because $\hat{p}$ is an unbiased estimate of p. The standard deviation is controlled mainly by the size of the sample.

LOOK BACK
normal distribution calculations
p. 59

What is the probability that the survey result differs from the truth about the population by no more than 3 percent points? We can use what we learned about Normal distribution calculations (page 59) to answer this question. Because $p = 0.26$, the survey misses by no more than 3 percent points if the sample proportion is within 0.23 and 0.29. Figure 4.11 shows this probability as an area under a Normal density curve. You can find it by software or by standardizing and using Table A. From Table A,

$$P(0.23 \leq \hat{p} \leq 0.29) = P\left(\frac{0.23 - 0.26}{0.0196} \leq \frac{\hat{p} - 0.26}{0.0196} \leq \frac{0.29 - 0.26}{0.0196}\right)$$

$$= P(-1.53 \leq Z \leq 1.53)$$

$$= 0.9370 - 0.0630 = 0.8740$$

About 87% of the time, the sample $\hat{p}$ will be within 3 percentage points of the parameter p.

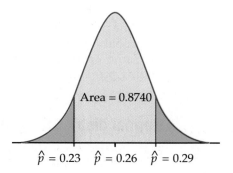

Area = 0.8740

$\hat{p} = 0.23$ $\hat{p} = 0.26$ $\hat{p} = 0.29$

FIGURE 4.11 Probability in Example 4.26 as area under a Normal density curve.

We began this chapter with a general discussion of the idea of probability and the properties of probability models. Two very useful specific types of probability models are distributions of discrete and continuous random variables. In our study of statistics we will employ only these two types of probability models.

SECTION 4.3 Summary

A **random variable** is a variable taking numerical values determined by the outcome of a random phenomenon. The **probability distribution** of a random variable X tells us what the possible values of X are and how probabilities are assigned to those values.

A random variable X and its distribution can be **discrete** or **continuous.**

A **discrete random variable** has finitely many possible values. The probability distribution assigns each of these values a probability between 0 and 1 such that the sum of all the probabilities is exactly 1. The probability of any event is the sum of the probabilities of all the values that make up the event.

A **continuous random variable** takes all values in some interval of numbers. A **density curve** describes the probability distribution of a continuous random variable. The probability of any event is the area under the curve and above the values that make up the event.

Normal distributions are one type of continuous probability distribution. You can picture a probability distribution by drawing a **probability histogram** in the discrete case or by graphing the density curve in the continuous case.

SECTION 4.3 Exercises

For Exercise 4.46, see page 250; for Exercise 4.47, see page 252; and for Exercise 4.48, see page 255.

4.49 What's wrong? In each of the following scenarios, there is something wrong. Describe what is wrong and give a reason for your answer.

(a) The probabilities for a discrete statistic always add to one.

(b) A continuous random variable can take any value between zero and one.

(c) Normal distributions are discrete random variables.

4.50 Use of Twitter. Suppose that the population proportion of Internet users who say that they use Twitter or another service to post updates about themselves or to see updates about others is 19%.[16] Think about selecting random samples from a population in which 19% are Twitter users.

(a) Describe the sample space for selecting a single person.

(b) If you select three people, describe the sample space.

(c) Using the results of (b), define the sample space for the random variable that expresses the number of Twitter users in the sample of size 3.

(d) What information is contained in the sample space for part (b) that is not contained in the sample space for part (c)? Do you think this information is important? Explain your answer.

4.51 Use of Twitter. Find the probabilities for parts (a), (b), and (c) of the previous exercise.

4.52 Households and families in government data. In government data, a household consists of all occupants of a dwelling unit, while a family consists of two or more persons who live together and are related by blood or

marriage. So all families form households, but some households are not families. Here are the distributions of household size and of family size in the United States:

Number of persons	1	2	3	4	5	6	7
Household probability	0.27	0.33	0.16	0.14	0.06	0.03	0.01
Family probability	0	0.44	0.22	0.20	0.09	0.03	0.02

Make probability histograms for these two discrete distributions, using the same scales. What are the most important differences between the sizes of households and families?

4.53 Discrete or continuous. In each of the following situations decide if the random variable is discrete or continuous and give a reason for your answer.

(a) Your Web page has five different links and a user can click on one of the links or can leave the page. You record the length of time that a user spends on the Web page before clicking one of the links or leaving the page.

(b) The number of hits on your Web page.

(c) The yearly income of a visitor to your Web page.

4.54 Texas hold 'em. The game of Texas hold 'em starts with each player receiving two cards. Here is the probability distribution for the number of aces in two-card hands:

Number of aces	0	1	2
Probability	0.8507	0.1448	0.0045

(a) Verify that this assignment of probabilities satisfies the requirement that the sum of the probabilities for a discrete distribution must be 1.

(b) Make a probability histogram for this distribution.

(c) What is the probability that a hand contains at least one ace? Show two different ways to calculate this probability.

4.55 Spell-checking software. Spell-checking software catches "nonword errors," which result in a string of letters that is not a word, as when "the" is typed as "teh." When undergraduates are asked to write a 250-word essay (without spell-checking), the number X of nonword errors has the following distribution:

Value of X	0	1	2	3	4
Probability	0.1	0.3	0.3	0.2	0.1

(a) Sketch the probability distribution for this random variable.

(b) Write the event "at least one nonword error" in terms of X. What is the probability of this event?

(c) Describe the event $X \leq 2$ in words. What is its probability? What is the probability that $X < 2$?

4.56 Length of human pregnancies. The length of human pregnancies from conception to birth varies according to a distribution that is approximately Normal with mean 266 days and standard deviation 16 days. Call the length of a randomly chosen pregnancy Y.

(a) Make a sketch of the density curve for this random variable.

(b) What is $P(Y \leq 280)$?

4.57 Tossing two dice. Some games of chance rely on tossing two dice. Each die has six faces, marked with $1, 2, \ldots, 6$ spots called pips. The dice used in casinos are carefully balanced so that each face is equally likely to come up. When two dice are tossed, each of the 36 possible pairs of faces is equally likely to come up. The outcome of interest to a gambler is the sum of the pips on the two up-faces. Call this random variable X.

(a) Write down all 36 possible pairs of faces.

(b) If all pairs have the same probability, what must be the probability of each pair?

(c) Write the value of X next to each pair of faces and use this information with the result of (b) to give the probability distribution of X. Draw a probability histogram to display the distribution.

(d) One bet available in craps wins if a 7 or an 11 comes up on the next roll of two dice. What is the probability of rolling a 7 or an 11 on the next roll?

(e) Several bets in craps lose if a 7 is rolled. If any outcome other than 7 occurs, these bets either win or continue to the next roll. What is the probability that anything other than a 7 is rolled?

4.58 ⚠ **Nonstandard dice.** Nonstandard dice can produce interesting distributions of outcomes. You have two balanced, six-sided dice. One is a standard die, with faces having 1, 2, 3, 4, 5, and 6 spots. The other die has three faces with 0 spots and three faces with 6 spots. Find the probability distribution for the total number of spots Y on the up-faces when you roll these two dice.

4.59 ⚠ **Dungeons & Dragons.** Role-playing games like Dungeons & Dragons use many different types of dice, usually having either 4, 6, 8, 10, 12, or 20 sides. Roll a balanced 8-sided die and a balanced 6-sided die and add the spots on the up-faces. Call the sum X. What is the probability distribution of the random variable X?

4.60 Foreign-born residents of California. The Census Bureau reports that 27% of California residents are foreign-born. Suppose that you choose three Californians at random, so that each has probability 0.27 of being foreign-born and the three are independent of each other. Let the random variable W be the number of foreign-born people you chose.

(a) What are the possible values of W?

(b) Look at your three people in order. There are eight possible arrangements of foreign (F) and domestic (D) birth. For example, FFD means the first two are foreign-born and the third is not. All eight arrangements are equally likely. What is the probability of each one?

(c) What is the value of W for each arrangement in (b)? What is the probability of each possible value of W? (This is the distribution of a Yes/No response for an SRS of size 3. In principle, the same idea works for an SRS of any size.)

4.61 Uniform random numbers. Let X be a random number between 0 and 1 produced by the idealized uniform random number generator described in Example 4.25 and Figure 4.9. Find the following probabilities:

(a) $P(X < 0.6)$

(b) $P(X \leq 0.6)$

(c) What important fact about continuous random variables does comparing your answers to parts (a) and (b) illustrate?

4.62 Find the probabilities. Let the random variable X be a random number with the uniform density curve in Figure 4.9. Find the following probabilities:

(a) $P(X \geq 0.30)$

(b) $P(X = 0.30)$

(c) $P(0.30 < X < 1.30)$

(d) $P(0.20 \leq X \leq 0.25 \text{ or } 0.7 \leq X \leq 0.9)$

(e) The probability that X is not in the interval 0.4 to 0.7.

4.63 Uniform numbers between 0 and 2. Many random number generators allow users to specify the range of the random numbers to be produced. Suppose that you specify that the range is to be all numbers between 0 and 2. Call the random number generated Y. Then the density curve of the random variable Y has constant height between 0 and 2, and height 0 elsewhere.

(a) What is the height of the density curve between 0 and 2? Draw a graph of the density curve.

(b) Use your graph from (a) and the fact that probability is area under the curve to find $P(Y \leq 1.6)$.

(c) Find $P(0.5 < Y < 1.7)$.

(d) Find $P(Y \geq 0.95)$.

4.64 The sum of two uniform random numbers. Generate *two* random numbers between 0 and 1 and take Y to be their sum. Then Y is a continuous random variable that can take any value between 0 and 2. The density curve of Y is the triangle shown in Figure 4.12.

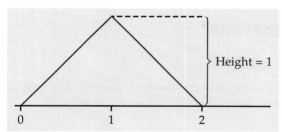

FIGURE 4.12 The density curve for the sum Y of two random numbers, for Exercise 4.64.

(a) Verify by geometry that the area under this curve is 1.

(b) What is the probability that Y is less than 1? (Sketch the density curve, shade the area that represents the probability, then find that area. Do this for (c) also.)

(c) What is the probability that Y is greater than 0.6?

4.65 How many close friends? How many close friends do you have? Suppose that the number of close friends adults claim to have varies from person to person with mean $\mu = 9$ and standard deviation $\sigma = 2.4$. An opinion poll asks this question of an SRS of 1100 adults. We will see in the next chapter that in this situation the sample mean response $\bar{x}$ has approximately the Normal distribution with mean 9 and standard deviation 0.0724. What is $P(8 \leq \bar{x} \leq 10)$, the probability that the statistic $\bar{x}$ estimates the parameter μ to within ± 1?

4.66 Normal approximation for a sample proportion. A sample survey contacted an SRS of 700 registered voters in Oregon shortly after an election and asked respondents whether they had voted. Voter records show that 56% of registered voters had actually voted. We will see in the next chapter that in this situation the proportion $\hat{p}$ of the sample who voted has approximately the Normal distribution with mean $\mu = 0.56$ and standard deviation $\sigma = 0.019$.

(a) If the respondents answer truthfully, what is $P(0.52 \leq \hat{p} \leq 0.60)$? This is the probability that the statistic $\hat{p}$ estimates the parameter 0.56 within plus or minus 0.04.

(b) In fact, 72% of the respondents said they had voted ($\hat{p} = 0.72$). If respondents answer truthfully, what is $P(\hat{p} \geq 0.72)$? This probability is so small that it is good evidence that some people who did not vote claimed that they did vote.

4.4 Means and Variances of Random Variables

The probability histograms and density curves that picture the probability distributions of random variables resemble our earlier pictures of distributions of data. In describing data, we moved from graphs to numerical measures such as means and standard deviations. Now we will make the same move to expand our descriptions of the distributions of random variables. We can speak of the mean winnings in a game of chance or the standard deviation of the randomly varying number of calls a travel agency receives in an hour. In this section we will learn more about how to compute these descriptive measures and about the laws they obey.

The mean of a random variable

The mean $\bar{x}$ of a set of observations is their ordinary average. The mean of a random variable X is also an average of the possible values of X, but with an essential change to take into account the fact that not all outcomes need be equally likely. An example will show what we must do.

EXAMPLE

4.27 The Tri-State Pick 3 lottery. Most states and Canadian provinces have government-sponsored lotteries. Here is a simple lottery wager, from the Tri-State Pick 3 game that New Hampshire shares with Maine and Vermont. You choose a three-digit number, 000 to 999. The state chooses a three-digit winning number at random and pays you $500 if your number is chosen. Because there are 1000 three-digit numbers, you have probability 1/1000 of winning. Taking X to be the amount your ticket pays you, the probability distribution of X is

Payoff X	$0	$500
Probability	0.999	0.001

What is your average payoff from many tickets? The ordinary average of the two possible outcomes $0 and $500 is $250, but that makes no sense as the average because $500 is much less likely than $0. In the long run you receive $500 once in every 1000 tickets and $0 on the remaining 999 of 1000 tickets. The long-run average payoff is

$$\$500\frac{1}{1000} + \$0\frac{999}{1000} = \$0.50$$

or 50 cents. That number is the mean of the random variable X. (Tickets cost $1, so in the long run the state keeps half the money you wager.)

If you play Tri-State Pick 3 several times, we would as usual call the mean of the actual amounts you win $\bar{x}$. The mean in Example 4.27 is a different quantity—it is the long-run average winnings you expect if you play a very large number of times.

USE YOUR KNOWLEDGE

4.67 Find the mean of the probability distribution. You toss a fair coin. If the outcome is heads, you win $1.00; if the outcome is tails, you win nothing. Let X be the amount that you win in a single toss of a coin. Find the probability distribution of this random variable and its mean.

Just as probabilities are an idealized description of long-run proportions, the mean of a probability distribution describes the long-run average outcome. We can't call this mean $\bar{x}$, so we need a different symbol. The common symbol for the **mean of a probability distribution** is μ, the Greek letter mu. We used μ in Chapter 1 for the mean of a Normal distribution, so this is not a new notation. We will often be interested in several random variables, each having

mean μ

a different probability distribution with a different mean. To remind ourselves that we are talking about the mean of X, we often write μ_X rather than simply μ. In Example 4.27, $\mu_X = \$0.50$. Notice that, as often happens, the mean is not a possible value of X. You will often find the mean of a random variable *expected value* X called the **expected value** of X. This term can be misleading, for we don't necessarily expect one observation on X to be close to its expected value.

The mean of any discrete random variable is found just as in Example 4.27. It is an average of the possible outcomes, but a weighted average in which each outcome is weighted by its probability. Because the probabilities add to 1, we have total weight 1 to distribute among the outcomes. An outcome that occurs half the time has probability one-half and gets one-half the weight in calculating the mean. Here is the general definition.

MEAN OF A DISCRETE RANDOM VARIABLE

Suppose that X is a discrete random variable whose distribution is

Value of X	x_1	x_2	x_3	$\cdots$	x_k
Probability	p_1	p_2	p_3	$\cdots$	p_k

To find the **mean** of X, multiply each possible value by its probability, then add all the products:

$$\mu_X = x_1 p_1 + x_2 p_2 + \cdots + x_k p_k$$
$$= \sum x_i p_i$$

EXAMPLE

4.28 The mean of equally likely first digits. If first digits in a set of data all have the same probability, the probability distribution of the first digit X is then

First digit X	1	2	3	4	5	6	7	8	9
Probability	1/9	1/9	1/9	1/9	1/9	1/9	1/9	1/9	1/9

The mean of this distribution is

$$\mu_X = 1 \times \frac{1}{9} + 2 \times \frac{1}{9} + 3 \times \frac{1}{9} + 4 \times \frac{1}{9} + 5 \times \frac{1}{9} + 6 \times \frac{1}{9} + 7 \times \frac{1}{9}$$
$$+ 8 \times \frac{1}{9} + 9 \times \frac{1}{9}$$
$$= 45 \times \frac{1}{9} = 5$$

Suppose the random digits in Example 4.28 had a different probability distribution. In Example 4.12 (page 238) we described Benford's law as a probability distribution that describes first digits of numbers in many real situations. Let's calculate the mean for Benford's law.

EXAMPLE

4.29 The mean of first digits following Benford's law. Here is the distribution of the first digit for data that follow Benford's law. We use the letter V for this random variable to distinguish it from the one that we studied in Example 4.28. The distribution of V is

First digit V	1	2	3	4	5	6	7	8	9
Probability	0.301	0.176	0.125	0.097	0.079	0.067	0.058	0.051	0.046

The mean of V is

$$\mu_V = (1)(0.301) + (2)(0.176) + (3)(0.125) + (4)(0.097) + (5)(0.079)$$
$$+ (6)(0.067) + (7)(0.058) + (8)(0.051) + (9)(0.046)$$
$$= 3.441$$

The mean reflects the greater probability of smaller first digits under Benford's law than when first digits 1 to 9 are equally likely.

Figure 4.13 locates the means of X and V on the two probability histograms. Because the discrete uniform distribution of Figure 4.13(a) is symmetric, the mean lies at the center of symmetry. We can't locate the mean of the right-skewed distribution of Figure 4.13(b) by eye—calculation is needed.

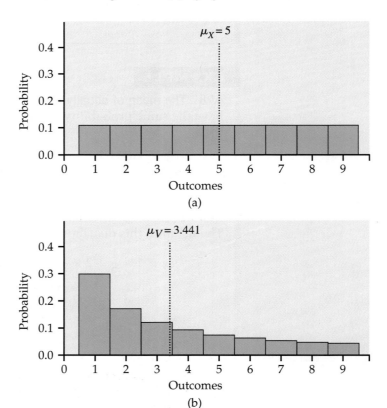

FIGURE 4.13 Locating the mean of a discrete random variable on the probability histogram for (a) digits between 1 and 9 chosen at random; (b) digits between 1 and 9 chosen from records that obey Benford's law.

What about continuous random variables? The probability distribution of a continuous random variable X is described by a density curve. Chapter 1 (page 54) showed how to find the mean of the distribution: it is the point at which the area under the density curve would balance if it were made out of solid material. The mean lies at the center of symmetric density curves such as the Normal curves. Exact calculation of the mean of a distribution with a skewed density curve requires advanced mathematics.[17] The idea that the mean is the balance point of the distribution applies to discrete random variables as well, but in the discrete case we have a formula that gives us this point.

Statistical estimation and the law of large numbers

We would like to estimate the mean height μ of the population of all American women between the ages of 18 and 24 years. This μ is the mean μ_X of the random variable X obtained by choosing a young woman at random and measuring her height. To estimate μ, we choose an SRS of young women and use the sample mean $\bar{x}$ to estimate the unknown population mean μ. In the language of Section 3.3 (page 202), μ is a *parameter* and $\bar{x}$ is a *statistic*. Statistics obtained from probability samples are random variables because their values vary in repeated sampling. The sampling distributions of statistics are just the probability distributions of these random variables.

LOOK BACK
sampling distributions p. 204

It seems reasonable to use $\bar{x}$ to estimate μ. An SRS should fairly represent the population, so the mean $\bar{x}$ of the sample should be somewhere near the mean μ of the population. Of course, we don't expect $\bar{x}$ to be exactly equal to μ, and we realize that if we choose another SRS, the luck of the draw will probably produce a different $\bar{x}$.

If $\bar{x}$ is rarely exactly right and varies from sample to sample, why is it nonetheless a reasonable estimate of the population mean μ? We gave one answer in Section 3.4: $\bar{x}$ is unbiased and we can control its variability by choosing the sample size. Here is another answer: if we keep on adding observations to our random sample, the statistic $\bar{x}$ is *guaranteed* to get as close as we wish to the parameter μ and then stay that close. We have the comfort of knowing that if we can afford to keep on measuring more women, eventually we will estimate the mean height of all young women very accurately. This remarkable fact is called the *law of large numbers*. It is remarkable because it holds for *any* population, not just for some special class such as Normal distributions.

LAW OF LARGE NUMBERS

Draw independent observations at random from any population with finite mean μ. Decide how accurately you would like to estimate μ. As the number of observations drawn increases, the mean $\bar{x}$ of the observed values eventually approaches the mean μ of the population as closely as you specified and then stays that close.

The behavior of $\bar{x}$ is similar to the idea of probability. In the long run, the *proportion* of outcomes taking any value gets close to the *probability* of that value, and the *average outcome* gets close to the distribution *mean*. Figure 4.1 (page 228) shows how proportions approach probability in one example. Here is an example of how sample means approach the distribution mean.

EXAMPLE

4.30 Heights of young women. The distribution of the heights of all young women is close to the Normal distribution with mean 64.5 inches and standard deviation 2.5 inches. Suppose that $\mu = 64.5$ were exactly true. Figure 4.14 shows the behavior of the mean height $\bar{x}$ of n women chosen at random from a population whose heights follow the $N(64.5, 2.5)$ distribution. The graph plots the values of $\bar{x}$ as we add women to our sample. The first woman drawn had height 64.21 inches, so the line starts there. The second had height 64.35 inches, so for $n = 2$ the mean is

$$\bar{x} = \frac{64.21 + 64.35}{2} = 64.28$$

This is the second point on the line in the graph.

At first, the graph shows that the mean of the sample changes as we take more observations. Eventually, however, the mean of the observations gets close to the population mean $\mu = 64.5$ and settles down at that value. The law of large numbers says that this *always* happens.

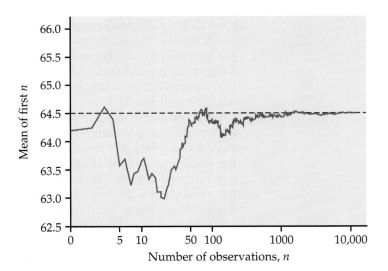

FIGURE 4.14 The law of large numbers in action. As we take more observations, the sample mean always approaches the mean of the population.

USE YOUR KNOWLEDGE

4.68 Use the *Law of Large Numbers* applet. The *Law of Large Numbers* applet animates a graph like Figure 4.14. Use it to better understand the law of large numbers by making a similar graph.

The mean μ of a random variable is the average value of the variable in two senses. By its definition, μ is the average of the possible values, weighted by their probability of occurring. The law of large numbers says that μ is also the long-run average of many independent observations on the variable. The law of large numbers can be proved mathematically starting from the basic laws of probability.

Thinking about the law of large numbers

The law of large numbers says broadly that the average results of many independent observations are stable and predictable. The gamblers in a casino may win or lose, but the casino will win in the long run because the law of large numbers says what the average outcome of many thousands of bets will be. An insurance company deciding how much to charge for life insurance and a fast-food restaurant deciding how many beef patties to prepare also rely on the fact that averaging over many individuals produces a stable result. It is worth the effort to think a bit more closely about so important a fact.

The "law of small numbers"   Both the rules of probability and the law of large numbers describe the regular behavior of chance phenomena *in the long run*. Psychologists have discovered that our intuitive understanding of randomness is quite different from the true laws of chance.[18] For example, most people believe in an incorrect "law of small numbers." That is, we expect even short sequences of random events to show the kind of average behavior that in fact appears only in the long run.

Some teachers of statistics begin a course by asking students to toss a coin 50 times and bring the sequence of heads and tails to the next class. The teacher then announces which students just wrote down a random-looking sequence rather than actually tossing a coin. The faked tosses don't have enough "runs" of consecutive heads or consecutive tails. Runs of the same outcome don't look random to us but are in fact common. For example, the probability of a run of three or more consecutive heads or tails in just 10 tosses is greater than 0.8.[19] The runs of consecutive heads or consecutive tails that appear in real coin tossing (and that are predicted by the mathematics of probability) seem surprising to us. Because we don't expect to see long runs, we may conclude that the coin tosses are not independent or that some influence is disturbing the random behavior of the coin.

> **EXAMPLE**
>
> **4.31   The "hot hand" in basketball.**   Belief in the law of small numbers influences behavior. If a basketball player makes several consecutive shots, both the fans and her teammates believe that she has a "hot hand" and is more likely to make the next shot. This is doubtful. Careful study suggests that runs of baskets made or missed are no more frequent in basketball than would be expected if each shot were independent of the player's previous shots. Baskets made or missed are just like heads and tails in tossing a coin. (Of course, some players make 30% of their shots in the long run and others make 50%, so a coin-toss model for basketball must allow coins with different probabilities of a head.) Our perception of hot or cold streaks simply shows that we don't perceive random behavior very well.[20]

Our intuition doesn't do a good job of distinguishing random behavior from systematic influences. This is also true when we look at data. We need statistical inference to supplement exploratory analysis of data because probability calculations can help verify that what we see in the data is more than a random pattern.

How large is a large number?   The law of large numbers says that the actual mean outcome of many trials gets close to the distribution mean μ as more trials are made. It doesn't say how many trials are needed to guarantee a mean outcome close to μ. That depends on the *variability* of the random outcomes. The more variable the outcomes, the more trials are needed to ensure that the mean outcome $\bar{x}$ is close to the distribution mean μ. Casinos understand this: the outcomes of games of chance are variable enough to hold the interest of gamblers. Only the casino plays often enough to rely on the law of large numbers. Gamblers get entertainment; the casino has a business.

BEYOND THE BASICS

More laws of large numbers

The law of large numbers is one of the central facts about probability. It helps us understand the mean μ of a random variable. It explains why gambling casinos and insurance companies make money. It assures us that statistical estimation will be accurate if we can afford enough observations. The basic law of large numbers applies to independent observations that all have the same distribution. Mathematicians have extended the law to many more general settings. Here are two of these.

Is there a winning system for gambling?   Serious gamblers often follow a system of betting in which the amount bet on each play depends on the outcome of previous plays. You might, for example, double your bet on each spin of the roulette wheel until you win—or, of course, until your fortune is exhausted. Such a system tries to take advantage of the fact that you have a memory even though the roulette wheel does not. Can you beat the odds with a system based on the outcomes of past plays? No. Mathematicians have established a stronger version of the law of large numbers that says that, if you do not have an infinite fortune to gamble with, your long-run average winnings μ remain the same as long as successive trials of the game (such as spins of the roulette wheel) are independent.

What if observations are not independent?   You are in charge of a process that manufactures video screens for computer monitors. Your equipment measures the tension on the metal mesh that lies behind each screen and is critical to its image quality. You want to estimate the mean tension μ for the process by the average $\bar{x}$ of the measurements. Alas, the tension measurements are not independent. If the tension on one screen is a bit too high, the tension on the next is more likely to also be high. Many real-world processes are like this—the process stays stable in the long run, but observations made close together are likely to be both above or both below the long-run mean. Again the mathematicians come to the rescue: as long as the dependence dies out fast enough as we take measurements farther and farther apart in time, the law of large numbers still holds.

Rules for means

You are studying flaws in the painted finish of refrigerators made by your firm. Dimples and paint sags are two kinds of surface flaw. Not all refrigerators have the same number of dimples: many have none, some have one, some two, and so on. You ask for the average number of imperfections on a refrigerator. The inspectors report finding an average of 0.7 dimples and 1.4 sags per refrigerator. How many total imperfections of both kinds (on the average) are there on a refrigerator? That's easy: if the average number of dimples is 0.7 and the average number of sags is 1.4, then counting both gives an average of $0.7 + 1.4 = 2.1$ flaws.

In more formal language, the number of dimples on a refrigerator is a random variable X that varies as we inspect one refrigerator after another. We know only that the mean number of dimples is $\mu_X = 0.7$. The number of paint sags is a second random variable Y having mean $\mu_Y = 1.4$. (As usual, the subscripts keep straight which variable we are talking about.) The total number of both dimples and sags is another random variable, the sum $X + Y$. Its mean μ_{X+Y} is the average number of dimples and sags together. It is just the sum of the individual means μ_X and μ_Y. That's an important rule for how means of random variables behave.

Here's another rule. The crickets living in a field have mean length 1.2 inches. What is the mean in centimeters? There are 2.54 centimeters in an inch, so the length of a cricket in centimeters is 2.54 times its length in inches. If we multiply every observation by 2.54, we also multiply their average by 2.54. The mean in centimeters must be 2.54×1.2, or about 3.05 centimeters. More formally, the length in inches of a cricket chosen at random from the field is a random variable X with mean μ_X. The length in centimeters is $2.54X$, and this new random variable has mean $2.54\mu_X$.

The point of these examples is that means behave like averages. Here are the rules we need.

RULES FOR MEANS

Rule 1. If X is a random variable and a and b are fixed numbers, then

$$\mu_{a+bX} = a + b\mu_X$$

Rule 2. If X and Y are random variables, then

$$\mu_{X+Y} = \mu_X + \mu_Y$$

EXAMPLE

4.32 Sales of cars, trucks, and SUVs. Linda is a sales associate at a large auto dealership. At her commission rate of 25% of gross profit on each vehicle she sells, Linda expects to earn \$350 for each car sold and \$400 for each truck or SUV sold. Linda motivates herself by using probability estimates of her sales. For a sunny Saturday in April, she estimates her car sales as follows:

Cars sold	0	1	2	3
Probability	0.3	0.4	0.2	0.1

Linda's estimate of her truck or SUV sales is

Vehicles sold	0	1	2
Probability	0.4	0.5	0.1

Take X to be the number of cars Linda sells and Y the number of trucks or SUVs. The means of these random variables are

$$\mu_X = (0)(0.3) + (1)(0.4) + (2)(0.2) + (3)(0.1)$$
$$= 1.1 \ \text{cars}$$
$$\mu_Y = (0)(0.4) + (1)(0.5) + (2)(0.1)$$
$$= 0.7 \ \text{trucks or SUVs}$$

Linda's earnings, at \$350 per car and \$400 per truck or SUV, are

$$Z = 350X + 400Y$$

Combining Rules 1 and 2, her mean earnings are

$$\mu_Z = 350\mu_X + 400\mu_Y$$
$$= (350)(1.1) + (400)(0.7) = \$665$$

This is Linda's best estimate of her earnings for the day. It's a bit unusual for individuals to use probability estimates, but they are a common tool for business planners.

personal probability

The probabilities in Example 4.32 are **personal probabilities** that describe Linda's informed opinion about her sales in the coming weekend. Although personal probabilities need not be based on observing many repetitions of a random phenomenon, they must obey the rules of probability if they are to make sense. Personal probability extends the usefulness of probability models to one-time events, but remember that they are subject to the follies of human opinion. Overoptimism is common: 40% of college students think that they will eventually reach the top 1% in income.

USE YOUR KNOWLEDGE

4.69 Find μ_Y. The random variable X has mean $\mu_X = 10$. If $Y = 15 + 8X$, what is μ_Y?

4.70 Find μ_W. The random variable U has mean $\mu_U = 20$ and the random variable V has mean $\mu_V = 20$. If $W = 0.5U + 0.5V$, find μ_W.

The variance of a random variable

The mean is a measure of the center of a distribution. A basic numerical description requires in addition a measure of the spread or variability of the distribution. The variance and the standard deviation are the measures of spread that accompany the choice of the mean to measure center. Just as for the mean, we need a distinct symbol to distinguish the variance of a random variable from the variance s^2 of a data set. We write the variance of a random variable X as σ_X^2. Once again the subscript reminds us which variable we have in mind. The definition of the variance σ_X^2 of a random variable is similar to the definition of the sample variance s^2 given in Chapter 1. That is, the variance is an average value of the squared deviation $(X - \mu_X)^2$ of the variable X from its mean μ_X. As for the mean, the average we use is a weighted average in which each outcome is weighted by its probability in order to take account of outcomes that are not equally likely. Calculating this weighted average is straightforward for discrete random variables but requires advanced mathematics in the continuous case. Here is the definition.

VARIANCE OF A DISCRETE RANDOM VARIABLE

Suppose that X is a discrete random variable whose distribution is

Value of X	x_1	x_2	x_3	$\cdots$	x_k
Probability	p_1	p_2	p_3	$\cdots$	p_k

and that μ_X is the mean of X. The **variance** of X is

$$\sigma_X^2 = (x_1 - \mu_X)^2 p_1 + (x_2 - \mu_X)^2 p_2 + \cdots + (x_k - \mu_X)^2 p_k$$
$$= \sum (x_i - \mu_X)^2 p_i$$

The **standard deviation** σ_X of X is the square root of the variance.

EXAMPLE

4.33 Find the mean and the variance. In Example 4.32 we saw that the number X of cars that Linda hopes to sell has distribution

Cars sold	0	1	2	3
Probability	0.3	0.4	0.2	0.1

We can find the mean and variance of X by arranging the calculation in the form of a table. Both μ_X and σ_X^2 are sums of columns in this table.

x_i	p_i	$x_i p_i$	$(x_i - \mu_X)^2 p_i$		
0	0.3	0.0	$(0 - 1.1)^2 (0.3)$	=	0.363
1	0.4	0.4	$(1 - 1.1)^2 (0.4)$	=	0.004
2	0.2	0.4	$(2 - 1.1)^2 (0.2)$	=	0.162
3	0.1	0.3	$(3 - 1.1)^2 (0.1)$	=	0.361
		$\mu_X = 1.1$		$\sigma_X^2 =$	0.890

We see that $\sigma_X^2 = 0.89$. The standard deviation of X is $\sigma_X = \sqrt{0.89} = 0.943$. The standard deviation is a measure of the variability of the number of cars Linda sells. As in the case of distributions for data, the standard deviation of a probability distribution is easiest to understand for Normal distributions.

USE YOUR KNOWLEDGE

4.71 Find the variance and the standard deviation. The random variable X has the following probability distribution:

Value of X	0	2
Probability	0.5	0.5

Find the variance σ_X^2 and the standard deviation σ_X for this random variable.

Rules for variances and standard deviations

What are the facts for variances that parallel Rules 1 and 2 for means? *The mean of a sum of random variables is always the sum of their means, but this addition rule is true for variances only in special situations.* To understand why, take X to be the percent of a family's after-tax income that is spent and Y the percent that is saved. When X increases, Y decreases by the same amount. Though X and Y may vary widely from year to year, their sum $X + Y$ is always 100% and does not vary at all. It is the association between the variables X and Y that prevents their variances from adding. If random variables are independent, this kind of association between their values is ruled out and their variances do add. Two
independence   random variables X and Y are **independent** if knowing that any event involving X alone did or did not occur tells us nothing about the occurrence of any event involving Y alone. Probability models often assume independence when the random variables describe outcomes that appear unrelated to each other. You should ask in each instance whether the assumption of independence seems reasonable.

When random variables are not independent, the variance of their sum
correlation   depends on the **correlation** between them as well as on their individual variances. In Chapter 2, we met the correlation r between two observed variables measured on the same individuals. We defined (page 102) the correlation r as an average of the products of the standardized x and y observations. The correlation between two random variables is defined in the same way, once again using a weighted average with probabilities as weights. We won't give the details—it is enough to know that the correlation between two random variables has the same basic properties as the correlation r calculated from data. We use ρ, the Greek letter rho, for the correlation between two random variables. The correlation ρ is a number between -1 and 1 that measures the direction and strength of the linear relationship between two variables. **The correlation between two independent random variables is zero.**

Returning to family finances, if X is the percent of a family's after-tax income that is spent and Y the percent that is saved, then $Y = 100 - X$. This is a perfect linear relationship with a negative slope, so the correlation between X and Y is $\rho = -1$. With the correlation at hand, we can state the rules for manipulating variances.

RULES FOR VARIANCES AND STANDARD DEVIATIONS

Rule 1. If X is a random variable and a and b are fixed numbers, then

$$\sigma^2_{a+bX} = b^2 \sigma^2_X$$

Rule 2. If X and Y are independent random variables, then

$$\sigma^2_{X+Y} = \sigma^2_X + \sigma^2_Y$$
$$\sigma^2_{X-Y} = \sigma^2_X + \sigma^2_Y$$

This is the **addition rule** for variances of independent random variables.

Rule 3. If X and Y have correlation ρ, then

$$\sigma^2_{X+Y} = \sigma^2_X + \sigma^2_Y + 2\rho\sigma_X\sigma_Y$$
$$\sigma^2_{X-Y} = \sigma^2_X + \sigma^2_Y - 2\rho\sigma_X\sigma_Y$$

This is the **general addition rule** for variances of random variables.

To find the standard deviation, take the square root of the variance.

Because a variance is the average of squared deviations from the mean, multiplying X by a constant b multiplies σ^2_X by the square of the constant. Adding a constant a to a random variable changes its mean but does not change its variability. The variance of $X + a$ is therefore the same as the variance of X. Because the square of -1 is 1, the addition rule says that the variance of a difference of independent random variables is the *sum* of the variances. For independent random variables, the difference $X - Y$ is more variable than either X or Y alone because variations in both X and Y contribute to variation in their difference.

As with data, we prefer the standard deviation to the variance as a measure of the variability of a random variable. *Rule 2 for variances implies that standard deviations of independent random variables do not add.* To combine standard deviations, use the rules for variances. For example, the standard deviations of $2X$ and $-2X$ are both equal to $2\sigma_X$ because this is the square root of the variance $4\sigma^2_X$.

EXAMPLE

4.34 Payoff in the Tri-State Pick 3 lottery. The payoff X of a $1 ticket in the Tri-State Pick 3 game is $500 with probability 1/1000 and 0 the rest of the time. Here is the combined calculation of mean and variance:

x_i	p_i	$x_i p_i$	$(x_i - \mu_X)^2 p_i$	
0	0.999	0	$(0 - 0.5)^2(0.999)$ =	0.24975
500	0.001	0.5	$(500 - 0.5)^2(0.001)$ =	249.50025
		$\mu_X = 0.5$	σ_X^2 =	249.75

The mean payoff is 50 cents. The standard deviation is $\sigma_X = \sqrt{249.75} = \$15.80$. It is usual for games of chance to have large standard deviations, because large variability makes gambling exciting.

If you buy a Pick 3 ticket, your winnings are $W = X - 1$ because the dollar you paid for the ticket must be subtracted from the payoff. Let's find the mean and variance for this random variable.

EXAMPLE

4.35 Winnings in the Tri-State Pick 3 lottery. By the rules for means, the mean amount you win is

$$\mu_W = \mu_X - 1 = -\$0.50$$

That is, you lose an average of 50 cents on a ticket. The rules for variances remind us that the variance and standard deviation of the winnings $W = X - 1$ are the same as those of X. Subtracting a fixed number changes the mean but not the variance.

Suppose now that you buy a $1 ticket on each of two different days. The payoffs X and Y on the two tickets are independent because separate drawings are held each day. Your total payoff is $X + Y$. Let's find the mean and standard deviation for this payoff.

EXAMPLE

4.36 Two tickets. The mean for the payoff for the two tickets is

$$\mu_{X+Y} = \mu_X + \mu_Y = \$0.50 + \$0.50 = \$1.00$$

Because X and Y are independent, the variance of $X + Y$ is

$$\sigma_{X+Y}^2 = \sigma_X^2 + \sigma_Y^2 = 249.75 + 249.75 = 499.5$$

The standard deviation of the total payoff is

$$\sigma_{X+Y} = \sqrt{499.5} = \$22.35$$

This is not the same as the sum of the individual standard deviations, which is $\$15.80 + \$15.80 = \$31.60$. Variances of independent random variables add; standard deviations do not.

When we add random variables that are correlated, we need to use the correlation for the calculation of the variance, but not for the calculation of the mean. Here is an example.

EXAMPLE

4.37 Utility bills. Consider a household where the monthly bill for natural gas averages $125 with a standard deviation of $75, while the monthly bill for electricity averages $174 with a standard deviation of $41. The correlation between the two bills is −0.55.

Let's compute the mean and standard deviation of the sum of the natural gas bill and the electricity bill. We let X stand for the natural gas bill and Y stand for the electricity bill. Then the total is $X+Y$. Using the rules for means, we have

$$\mu_{X+Y} = \mu_X + \mu_Y = 125 + 174 = 299$$

To find the standard deviation we first find the variance and then take the square root to determine the standard deviation. From the general addition rule for variances of random variables,

$$\sigma_{X+Y}^2 = \sigma_X^2 + \sigma_Y^2 + 2\rho\sigma_X\sigma_Y$$
$$= (75)^2 + (41)^2 + (2)(-0.55)(75)(41)$$
$$= 3923$$

Therefore the standard deviation is

$$\sigma_{X+Y} = \sqrt{3923} = 63$$

The total of the natural gas bill and the electricity bill has mean $299 and standard deviation $63.

The negative correlation in Example 4.37 is due to the fact that, in this household, natural gas is used for heating and electricity is used for air conditioning. So, when it is warm, the electric charges are high and the natural gas charges are low. When it is cool, the reverse is true. This causes the standard deviation of the sum to be less than it would be if the two bills were uncorrelated (see Exercise 4.79, on page 276).

There are situations where we need to combine several of our rules to find means and standard deviations. Here is an example.

EXAMPLE

4.38 Calcium intake. To get enough calcium for optimal bone health, tablets containing calcium are often recommend to supplement the calcium in the diet. One study designed to evaluate the effectiveness of a supplement followed a group of young people for seven years. Each subject was assigned to take a tablet containing 1000 mg of calcium per day (mg/d) or a placebo tablet which was identical except that it had no calcium.[21] A major problem with studies like this one is compliance: subjects do not always take the treatments assigned to them.

In this study, the compliance rate declined to about 47% toward the end of the seven-year period. The standard deviation was 22%. Calcium from the diet averaged 850 mg/d with a standard deviation of 330 mg/d. The correlation between compliance and diet intake is 0.68. Let's find the mean and standard deviation for the total calcium intake. We let S stand for the intake from the supplement and D stand for the intake from the diet.

We start with the intake from the supplement. Since the compliance is 47% and the amount in each tablet is 1000 mg, the mean for S is

$$\mu_S = 1000(0.47) = 470$$

Since the standard deviation of the compliance is 22%, the variance of S is

$$\sigma_S^2 = 1000^2(0.22)^2 = 48400$$

The standard deviation is

$$\sigma_S = \sqrt{48400} = 220$$

Be sure to verify which rules for means and variances are used in these calculations.

We can now find the mean and standard deviation for the total intake. The mean is

$$\mu_{S+D} = \mu_S + \mu_D = 470 + 850 = 1320$$

and the variance is

$$\sigma_{S+D}^2 = \sigma_S^2 + \sigma_D^2 + 2\rho\sigma_S\sigma_D = 220^2 + 330^2 + 2(0.68)(220)(330) = 256036$$

and the standard deviation is

$$\sigma_{S+D} = \sqrt{256036} = 506$$

The mean of the total calcium intake is 1320 mg/d and the standard deviation is 506 mg/d.

The correlation in this example illustrates an unfortunate fact about compliance and having an adequate diet. Some of the subjects in this study have diets that provide an adequate amount of calcium while others do not. The positive correlation between compliance and dietary intake tells us that those who have relatively high dietary intakes are more likely to take the assigned supplements. On the other hand, those subjects with relatively low dietary intakes, the ones who need the supplement the most, are less likely to take the assigned supplements.

SECTION 4.4 Summary

The probability distribution of a random variable X, like a distribution of data, has a **mean μ_X** and a **standard deviation σ_X.**

The **law of large numbers** says that the average of the values of X observed in many trials must approach μ.

The **mean μ** is the balance point of the probability histogram or density curve. If X is discrete with possible values x_i having probabilities p_i, the mean is the average of the values of X, each weighted by its probability:

$$\mu_X = x_1 p_1 + x_2 p_2 + \cdots + x_k p_k$$

The **variance** σ_X^2 is the average squared deviation of the values of the variable from their mean. For a discrete random variable,

$$\sigma_X^2 = (x_1 - \mu)^2 p_1 + (x_2 - \mu)^2 p_2 + \cdots + (x_k - \mu)^2 p_k$$

The **standard deviation** σ_X is the square root of the variance. The standard deviation measures the variability of the distribution about the mean. It is easiest to interpret for Normal distributions.

The mean and variance of a continuous random variable can be computed from the density curve, but to do so requires more advanced mathematics.

The means and variances of random variables obey the following rules. If a and b are fixed numbers, then

$$\mu_{a+bX} = a + b\mu_X$$
$$\sigma_{a+bX}^2 = b^2 \sigma_X^2$$

If X and Y are any two random variables having correlation ρ, then

$$\mu_{X+Y} = \mu_X + \mu_Y$$
$$\sigma_{X+Y}^2 = \sigma_X^2 + \sigma_Y^2 + 2\rho\sigma_X\sigma_Y$$
$$\sigma_{X-Y}^2 = \sigma_X^2 + \sigma_Y^2 - 2\rho\sigma_X\sigma_Y$$

If X and Y are **independent**, then $\rho = 0$. In this case,

$$\sigma_{X+Y}^2 = \sigma_X^2 + \sigma_Y^2$$
$$\sigma_{X-Y}^2 = \sigma_X^2 + \sigma_Y^2$$

To find the standard deviation, take the square root of the variance.

SECTION 4.4 Exercises

For Exercise 4.67, see page 260; for Exercise 4.68, see page 264; for Exercises 4.69 and 4.70, see page 268; and for Exercise 4.71, see page 270.

4.72 What's wrong? In each of the following scenarios, there is something wrong. Describe what is wrong and give a reason for your answer.

(a) If you toss a fair coin three times and get heads all three times then the probability of getting a tail on the next toss is much greater than a half.

(b) If you multiply a random variable by 10, then the mean is multiplied by 10 and the variance is multiplied by 10.

(c) When finding the mean of the sum of two random variables, you need to know the correlation between them.

4.73 Servings of fruits and vegetables. The following table gives the distribution of the number of servings of fruits and vegetables per day in a population.

Number of servings X	0	1	2	3	4	5
Probability	0.3	0.1	0.1	0.2	0.1	0.2

Find the mean and the standard deviation for this random variable.

4.74 Mean of the distribution for the number of aces. In Exercise 4.54 you examined the probability distribution for the number of aces when you are dealt two cards in the game of Texas hold 'em. Let X represent the number of aces in a randomly selected deal of two cards in this game. Here is the probability distribution for the random variable X:

Value of X	0	1	2
Probability	0.8507	0.1448	0.0045

Find μ_X, the mean of the probability distribution of X.

4.75 Mean of the grade distribution. Example 4.22 gives the distribution of grades (A = 4, B = 3, and so on) in English 210 at North Carolina State University as

Value of X	0	1	2	3	4
Probability	0.05	0.04	0.20	0.40	0.31

Find the average (that is, the mean) grade in this course.

4.76 Mean of the distributions of errors. Typographical and spelling errors can be either "nonword errors" or "word errors." A nonword error is not a real word, as when "the" is typed as "teh." A word error is a real word, but not the right word, as when "lose" is typed as "loose." When undergraduates are asked to write a 250-word essay (without spell-checking), the number of nonword errors has the following distribution:

Errors	0	1	2	3	4
Probability	0.1	0.3	0.3	0.2	0.1

The number of word errors has this distribution:

Errors	0	1	2	3
Probability	0.4	0.3	0.2	0.1

What are the mean numbers of nonword errors and word errors in an essay?

4.77 Standard deviation of the number of aces. Refer to Exercise 4.74. Find the standard deviation of the number of aces.

4.78 Standard deviation of the grades. Refer to Exercise 4.75. Find the standard deviation of the grade distribution.

4.79 Suppose that the correlation is zero. Refer to Example 4.37 (page 273).

(a) Recompute the standard deviation for the total of the natural gas bill and the electricity bill assuming that the correlation is zero.

(b) Is this standard deviation larger or smaller that the standard deviation computed in Example 4.37 (page 273)? Explain why.

4.80 Find the mean of the sum. Figure 4.12 (page 259) displays the density curve of the sum $Y = X_1 + X_2$ of two independent random numbers, each uniformly distributed between 0 and 1.

(a) The mean of a continuous random variable is the balance point of its density curve. Use this fact to find the mean of Y from Figure 4.12.

(b) Use the same fact to find the means of X_1 and X_2. (They have the density curve pictured in Figure 4.9, page 254.) Verify that the mean of Y is the sum of the mean of X_1 and the mean of X_2.

4.81 Calcium supplements and calcium in the diet. Refer to Example 4.38 (page 273). Suppose that people who have high intakes of calcium in their diets are more compliant than those who have low intakes. What effect would this have on the calculation of the standard deviation for the total calcium intake? Explain your answer.

4.82 The effect of correlation. Find the mean and standard deviation of the total number of errors (nonword errors plus word errors) in an essay if the error counts have the distributions given in Exercise 4.76 and

(a) the counts of nonword and word errors are independent.

(b) students who make many nonword errors also tend to make many word errors, so that the correlation between the two error counts is 0.5.

4.83 Means and variances of sums. The rules for means and variances allow you to find the mean and variance of a sum of random variables without first finding the distribution of the sum, which is usually much harder to do.

(a) A single toss of a balanced coin has either 0 or 1 head, each with probability 1/2. What are the mean and standard deviation of the number of heads?

(b) Toss a coin four times. Use the rules for means and variances to find the mean and standard deviation of the total number of heads.

(c) Example 4.23 (page 251) finds the distribution of the number of heads in four tosses. Find the mean and standard deviation from this distribution. Your results in parts (b) and (c) should agree.

4.84 ⚠ **Toss a four-sided die twice.** Role-playing games like Dungeons & Dragons use many different types of dice. Suppose that a four-sided die has faces marked 1, 2, 3, 4. The intelligence of a character is determined by rolling this die twice and adding 1 to the sum of the spots. The faces are equally likely and the two rolls are independent. What is the average (mean) intelligence for such characters? How spread out are their intelligences, as measured by the standard deviation of the distribution?

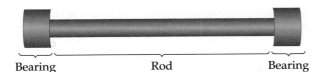

Bearing   Rod   Bearing

FIGURE 4.15 Sketch of a mechanical assembly, for Exercise 4.85.

4.85 A mechanical assembly. A mechanical assembly (Figure 4.15) consists of a rod with a bearing on each end. The three parts are manufactured independently, and all vary a bit from part to part. The length of the rod has mean 12 centimeters (cm) and standard deviation 0.004 millimeters (mm). The length of a bearing has mean 2 cm and standard deviation 0.001 mm. What are the mean and standard deviation of the total length of the assembly?

4.86 Sums of Normal random variables. Continue your work in the previous exercise. Dimensions of mechanical parts are often roughly Normal. According to the 68–95–99.7 rule, 95% of rods have lengths within $\pm d_1$ of 12 cm and 95% of bearings have lengths within $\pm d_2$ of 2 cm.

(a) What are the values of d_1 and d_2? These are often called the "natural tolerances" of the parts.

(b) Statistical theory says that any sum of independent Normal random variables has a Normal distribution. So the total length of the assembly is roughly Normal. What is the natural tolerance for the total length? It is *not* $d_1 + 2d_2$, because standard deviations don't add.

4.87 Will you assume independence? In which of the following games of chance would you be willing to assume independence of X and Y in making a probability model? Explain your answer in each case.

(a) In blackjack, you are dealt two cards and examine the total points X on the cards (face cards count 10 points). You can choose to be dealt another card and compete based on the total points Y on all three cards.

(b) In craps, the betting is based on successive rolls of two dice. X is the sum of the faces on the first roll, and Y the sum of the faces on the next roll.

4.88 Transform the distribution of heights from centimeters to inches. A report of the National Center for Health Statistics says that the heights of 20-year-old men have mean 176.8 centimeters (cm) and standard deviation 7.2 cm. There are 2.54 centimeters in an inch. What are the mean and standard deviation in inches?

4.89 What happens when the correlation is 1? We know that variances add if the random variables involved are uncorrelated $(\rho = 0)$, but not otherwise. The

opposite extreme is perfect positive correlation $(\rho = 1)$. Show by using the general addition rule for variances that in this case the standard deviations add. That is, $\sigma_{X+Y} = \sigma_X + \sigma_Y$ if $\rho_{XY} = 1$.

4.90 A random variable with given mean and standard deviation. Here is a simple way to create a random variable X that has mean μ and standard deviation σ: X takes only the two values $\mu - \sigma$ and $\mu + \sigma$, each with probability 0.5. Use the definition of the mean and variance for discrete random variables to show that X does have mean μ and standard deviation σ.

Insurance. The business of selling insurance is based on probability and the law of large numbers. Consumers buy insurance because we all face risks that are unlikely but carry high cost. Think of a fire destroying your home. So we form a group to share the risk: we all pay a small amount, and the insurance policy pays a large amount to those few of us whose homes burn down. The insurance company sells many policies, so it can rely on the law of large numbers. Exercises 4.91 to 4.94 explore aspects of insurance.

4.91 Fire insurance. An insurance company looks at the records for millions of homeowners and sees that the mean loss from fire in a year is $\mu = \$300$ per person. (Most of us have no loss, but a few lose their homes. The $300 is the average loss.) The company plans to sell fire insurance for $300 plus enough to cover its costs and profit. Explain clearly why it would be stupid to sell only 10 policies. Then explain why selling thousands of such policies is a safe business.

4.92 Mean and standard deviation for 10 and for 12 policies. In fact, the insurance company sees that in the entire population of homeowners, the mean loss from fire is $\mu = \$300$ and the standard deviation of the loss is $\sigma = \$400$. What are the mean and standard deviation of the average loss for 10 policies? (Losses on separate policies are independent.) What are the mean and standard deviation of the average loss for 12 policies?

4.93 Life insurance. According to the current Commissioners' Standard Ordinary mortality table, adopted by state insurance regulators in December 2002, a 25-year-old man has these probabilities of dying during the next five years:[22]

Age at death	25	26	27	28	29
Probability	0.00039	0.00044	0.00051	0.00057	0.00060

(a) What is the probability that the man does not die in the next five years?

(b) An online insurance site offers a term insurance policy that will pay $100,000 if a 25-year-old man dies within the next five years. The cost is $175 per year. So the insurance company will take in $875 from this policy if the man does not die within five years. If he does die, the company must pay $100,000. Its loss depends on how many premiums were paid, as follows:

Age at death	25	26	27	28	29
Loss	$99,825	$99,650	$99,475	$99,300	$99,125

What is the insurance company's mean cash intake from such polices?

4.94 Risk for one versus thousands of life insurance policies. It would be quite risky for you to insure the life of a 25-year-old friend under the terms of Exercise 4.93. There is a high probability that your friend would live and you would gain $875 in premiums. But if he were to die, you would lose almost $100,000. Explain carefully why selling insurance is not risky for an insurance company that insures many thousands of 25-year-old men.

4.5 General Probability Rules*

Our study of probability has concentrated on random variables and their distributions. Now we return to the laws that govern any assignment of probabilities. The purpose of learning more laws of probability is to be able to give probability models for more complex random phenomena. We have already met and used five rules.

RULES OF PROBABILITY

Rule 1. $0 \leq P(A) \leq 1$ for any event A

Rule 2. $P(S) = 1$

Rule 3. Addition rule: If A and B are **disjoint** events, then

$$P(A \text{ or } B) = P(A) + P(B)$$

Rule 4. Complement rule: For any event A,

$$P(A^c) = 1 - P(A)$$

Rule 5. Multiplication rule: If A and B are **independent** events, then

$$P(A \text{ and } B) = P(A)P(B)$$

General addition rules

Probability has the property that if A and B are disjoint events, then $P(A \text{ or } B) = P(A) + P(B)$. What if there are more than two events, or if the events are not disjoint? These circumstances are covered by more general addition rules for probability.

UNION

The **union** of any collection of events is the event that at least one of the collection occurs.

---

*This section extends the rules of probability discussed in Section 4.2. This material is not needed for understanding the statistical methods in later chapters. It can therefore be omitted if desired.

For two events A and B, the union is the event $\{A \text{ or } B\}$ that A or B or both occur. From the addition rule for two disjoint events we can obtain rules for more general unions. Suppose first that we have several events—say A, B, and C—that are disjoint in pairs. That is, no two can occur simultaneously. The Venn diagram in Figure 4.16 illustrates three disjoint events. The addition rule for two disjoint events extends to the following law:

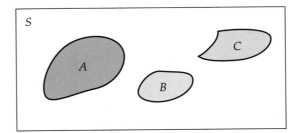

FIGURE 4.16 The addition rule for disjoint events: $P(A \text{ or } B \text{ or } C) = P(A) + P(B) + P(C)$ when events A, B, and C are disjoint.

ADDITION RULE FOR DISJOINT EVENTS

If events A, B, and C are disjoint in the sense that no two have any outcomes in common, then

$$P(\text{one or more of } A, B, C) = P(A) + P(B) + P(C)$$

This rule extends to any number of disjoint events.

EXAMPLE

4.39 Probabilities as areas. Generate a random number X between 0 and 1. What is the probability that the first digit after the decimal point will be odd? The random number X is a continuous random variable whose density curve has constant height 1 between 0 and 1 and is 0 elsewhere. The event that the first digit of X is odd is the union of five disjoint events. These events are

$$0.10 \le X < 0.20$$

$$0.30 \le X < 0.40$$

$$0.50 \le X < 0.60$$

$$0.70 \le X < 0.80$$

$$0.90 \le X < 1.00$$

Figure 4.17 illustrates the probabilities of these events as areas under the density curve. Each area is 0.1. The union of the five therefore has probability equal to the sum, or 0.5. As we should expect, a random number is equally likely to begin with an odd or an even digit.

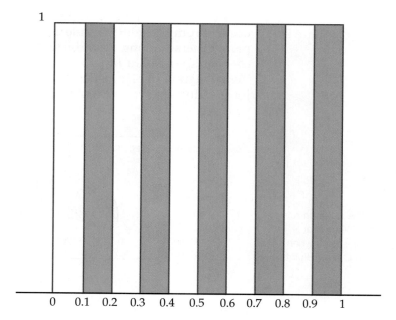

FIGURE 4.17 The probability that the first digit after the decimal point of a random number is odd is the sum of the probabilities of the 5 disjoint events shown. See Example 4.39.

USE YOUR KNOWLEDGE

4.95 Probability that you roll a 3 or a 5. If you roll a die, the probability of each of the six possible outcomes (1, 2, 3, 4, 5, 6) is 1/6. What is the probability that you roll a 3 or a 5?

If events A and B are not disjoint, they can occur simultaneously. The probability of their union is then *less* than the sum of their probabilities. As Figure 4.18 suggests, the outcomes common to both are counted twice when we add probabilities, so we must subtract this probability once. Here is the addition rule for the union of any two events, disjoint or not.

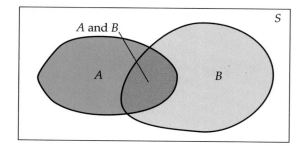

FIGURE 4.18 The union of two events that are not disjoint. The general addition rule says that $P(A \text{ or } B) = P(A) + P(B) - P(A \text{ and } B)$.

GENERAL ADDITION RULE FOR UNIONS OF TWO EVENTS

For any two events A and B,

$$P(A \text{ or } B) = P(A) + P(B) - P(A \text{ and } B)$$

If *A* and *B* are disjoint, the event {*A* and *B*} that both occur has no outcomes in it. This *empty event* is the complement of the sample space *S* and must have probability 0. So the general addition rule includes Rule 3, the addition rule for disjoint events.

EXAMPLE

4.40 Adequate sleep and exercise. Suppose that 40% of adults get enough sleep and 46% exercise regularly. What is the probability that an adult gets enough sleep or exercises regularly? To find this probability, we also need to know the percent who get enough sleep and exercise. Let's assume that 24% do both.

We will use the notation of the general addition rule for unions of two events. Let *A* be the event that an adult gets enough sleep and let *B* be the event that a person exercises regularly. We are given that $P(A) = 0.40$, $P(B) = 0.46$, and $P(A \text{ and } B) = 0.24$. Therefore,

$$P(A \text{ or } B) = P(A) + P(B) - P(A \text{ and } B)$$
$$= 0.40 + 0.46 - 0.24$$
$$= 0.62$$

The probability that an adult gets enough sleep or exercises regularly is 0.62, or 62%.

USE YOUR KNOWLEDGE

4.96 Probability that your roll is even or greater than 4. If you roll a die, the probability of each of the six possible outcomes (1, 2, 3, 4, 5, 6) is 1/6. What is the probability that your roll is even or greater than 4?

Venn diagrams are a great help in finding probabilities for unions because you can just think of adding and subtracting areas. Figure 4.19 shows some events and their probabilities for Example 4.40. What is the probability that an adult gets adequate sleep and exercises? The Venn diagram shows that this is the probability that an individual gets adequate sleep minus the probability that an adult gets adequate sleep and exercises regularly, $0.40 - 0.24 = 0.16$.

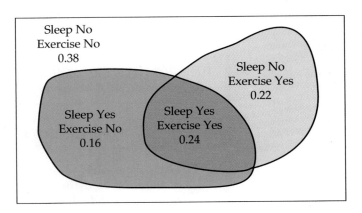

FIGURE 4.19 Venn diagram and probabilities for Example 4.40.

Similarly, the probability that an adult does not get adequate sleep and exercises regularly is $0.46 - 0.24 = 0.22$. The four probabilities that appear in the figure add to 1 because they refer to four disjoint events whose union is the entire sample space.

Conditional probability

The probability we assign to an event can change if we know that some other event has occurred. This idea is the key to many applications of probability.

EXAMPLE

4.41   Probability of being dealt an ace. Slim is a professional poker player. He stares at the dealer, who prepares to deal. What is the probability that the card dealt to Slim is an ace? There are 52 cards in the deck. Because the deck was carefully shuffled, the next card dealt is equally likely to be any of the cards that Slim has not seen. Four of the 52 cards are aces. So

$$P(\text{ace}) = \frac{4}{52} = \frac{1}{13}$$

This calculation assumes that Slim knows nothing about any cards already dealt. Suppose now that he is looking at 4 cards already in his hand, and that one of them is an ace. He knows nothing about the other 48 cards except that exactly 3 aces are among them. Slim's probability of being dealt an ace *given what he knows* is now

$$P(\text{ace} \mid 1 \text{ ace in 4 visible cards}) = \frac{3}{48} = \frac{1}{16}$$

Knowing that there is 1 ace among the 4 cards Slim can see changes the probability that the next card dealt is an ace.

conditional probability   The new notation $P(A \mid B)$ is a **conditional probability.** That is, it gives the probability of one event (the next card dealt is an ace) under the condition that we know another event (exactly 1 of the 4 visible cards is an ace). You can read the bar | as "given the information that."

MULTIPLICATION RULE

The probability that both of two events A and B happen together can be found by

$$P(A \text{ and } B) = P(A)P(B \mid A)$$

Here $P(B \mid A)$ is the conditional probability that B occurs, given the information that A occurs.

USE YOUR KNOWLEDGE

4.97   The probability of another ace. Refer to Example 4.41. Suppose two of the four cards in Slim's hand are aces. What is the probability that the next card dealt to him is an ace?

EXAMPLE

4.42 Downloading music from the Internet. The multiplication rule is just common sense made formal. For example, 29% of Internet users download music files, and 67% of downloaders say they don't care if the music is copyrighted.[23] So the percent of Internet users who download music (event A) *and* don't care about copyright (event B) is 67% of the 29% who download, or

$$(0.67)(0.29) = 0.1943 = 19.43\%$$

The multiplication rule expresses this as

$$P(A \text{ and } B) = P(A) \times P(B \mid A)$$
$$= (0.29)(0.67) = 0.1943$$

4.43 Probability of a favorable draw. Slim is still at the poker table. At the moment, he wants very much to draw two diamonds in a row. As he sits at the table looking at his hand and at the upturned cards on the table, Slim sees 11 cards. Of these, 4 are diamonds. The full deck contains 13 diamonds among its 52 cards, so 9 of the 41 unseen cards are diamonds. To find Slim's probability of drawing two diamonds, first calculate

$$P(\text{first card diamond}) = \frac{9}{41}$$

$$P(\text{second card diamond} \mid \text{first card diamond}) = \frac{8}{40}$$

Slim finds both probabilities by counting cards. The probability that the first card drawn is a diamond is 9/41 because 9 of the 41 unseen cards are diamonds. If the first card is a diamond, that leaves 8 diamonds among the 40 remaining cards. So the *conditional* probability of another diamond is 8/40. The multiplication rule now says that

$$P(\text{both cards diamonds}) = \frac{9}{41} \times \frac{8}{40} = 0.044$$

Slim will need luck to draw his diamonds.

USE YOUR KNOWLEDGE

4.98 The probability that the next two cards are diamonds. In the setting of Exercise 4.43, suppose Slim sees 25 cards and the only diamonds are the 3 in his hand. What is the probability that the next 2 cards dealt to Slim will be diamonds? This outcome would give him 5 cards from the same suit, a hand that is called a flush.

If $P(A)$ and $P(A \text{ and } B)$ are given, we can rearrange the multiplication rule to produce a *definition* of the conditional probability $P(B \mid A)$ in terms of unconditional probabilities.

DEFINITION OF CONDITIONAL PROBABILITY

When $P(A) > 0$, the **conditional probability** of B given A is

$$P(B \mid A) = \frac{P(A \text{ and } B)}{P(A)}$$

Be sure to keep in mind the distinct roles in $P(B \mid A)$ of the event B whose probability we are computing and the event A that represents the information we are given. The conditional probability $P(B \mid A)$ makes no sense if the event A can never occur, so we require that $P(A) > 0$ whenever we talk about $P(B \mid A)$.

EXAMPLE

4.44 College students. Here is the distribution of U.S. college students classified by age and full-time or part-time status:[24]

Age	Status	
	Full-time	**Part-time**
15 to 19	0.21	0.02
20 to 24	0.32	0.07
25 to 34	0.10	0.10
30 and over	0.05	0.13

Let's compute the probability that a full-time student is aged 15 to 19. We know that the probability that a student is full-time *and* aged 15 to 19 is 0.21 from the table of probabilities. But, what we want here is a conditional probability, given that a student is full-time. Rather than asking about age among all students, we restrict our attention to the subpopulation of students who are full-time. Let

$A =$ the student is between 15 and 19 years of age

$B =$ the student is a full-time student

Our formula is

$$P(A \mid B) = \frac{P(A \text{ and } B)}{P(B)}$$

We read $P(A \text{ and } B) = 0.21$ from the table as we mentioned previously. What about $P(B)$? This is the probability that a student is full-time. Notice that there are four groups of students in our table that fit this description. To find the probability needed, we add the entries:

$$P(B) = 0.21 + 0.32 + 0.10 + 0.05 = 0.68$$

We are now ready to complete the calculation of the conditional probability:

$$P(A \mid B) = \frac{P(A \text{ and } B)}{P(B)}$$

$$= \frac{0.21}{0.68}$$

$$= 0.31$$

The probability that a student is 15 to 19 years of age given that the student is full-time is 0.31.

Here is another way to give the information in the last sentence of this example: 31% of full-time college students are 15 to 19 years old. Which way do you prefer?

USE YOUR KNOWLEDGE

4.99   **What rule did we use?** In Example 4.44, we calculated $P(B)$. What rule did we use for this calculation? Explain why this rule applies in this setting.

4.100   **Find the conditional probability.** Refer to Example 4.44. What is the probability that a student is full-time given that the student is 15 to 19 years old? Explain in your own words the difference between this calculation and the one that we did in Example 4.44.

General multiplication rules

The definition of conditional probability reminds us that in principle all probabilities, including conditional probabilities, can be found from the assignment of probabilities to events that describe random phenomena. More often, however, conditional probabilities are part of the information given to us in a probability model, and the multiplication rule is used to compute $P(A \text{ and } B)$. This rule extends to more than two events.

The union of a collection of events is the event that *any* of them occur. Here is the corresponding term for the event that *all* of them occur.

INTERSECTION

The **intersection** of any collection of events is the event that *all* of the events occur.

To extend the multiplication rule to the probability that all of several events occur, the key is to condition each event on the occurrence of *all* of the preceding events. For example, the intersection of three events A, B, and C has probability

$$P(A \text{ and } B \text{ and } C) = P(A)P(B \mid A)P(C \mid A \text{ and } B)$$

EXAMPLE

4.45 High school athletes and professional careers. Only 5% of male high school basketball, baseball, and football players go on to play at the college level. Of these, only 1.7% enter major league professional sports. About 40% of the athletes who compete in college and then reach the pros have a career of more than three years.[25] Define these events:

$$A = \{\text{competes in college}\}$$
$$B = \{\text{competes professionally}\}$$
$$C = \{\text{pro career longer than 3 years}\}$$

What is the probability that a high school athlete competes in college and then goes on to have a pro career of more than three years? We know that

$$P(A) = 0.05$$
$$P(B \mid A) = 0.017$$
$$P(C \mid A \text{ and } B) = 0.4$$

The probability we want is therefore

$$P(A \text{ and } B \text{ and } C) = P(A)P(B \mid A)P(C \mid A \text{ and } B)$$
$$= 0.05 \times 0.017 \times 0.4 = 0.00034$$

Only about 3 of every 10,000 high school athletes can expect to compete in college and have a professional career of more than three years. High school students would be wise to concentrate on studies rather than on unrealistic hopes of fortune from pro sports.

Tree diagrams

Probability problems often require us to combine several of the basic rules into a more elaborate calculation. Here is an example that illustrates how to solve problems that have several stages.

EXAMPLE

4.46 Online chat rooms. Online chat rooms are dominated by the young. Teens are the biggest users. If we look only at adult Internet users (aged 18 and over), 47% of the 18 to 29 age group chat, as do 21% of the 30 to 49 age group and just 7% of those 50 and over. To learn what percent of all Internet users participate in chat, we also need the age breakdown of users. Here it is: 29% of adult Internet users are 18 to 29 years old (event A_1), another 47% are 30 to 49 (event A_2), and the remaining 24% are 50 and over (event A_3).[26]

tree diagram   What is the probability that a randomly chosen user of the Internet participates in chat rooms (event C)? To find out, use the **tree diagram** in Figure 4.20 to organize your thinking. Each segment in the tree is one stage of the problem. Each complete branch shows a path through the two stages. The probability written on each segment is the conditional probability of an Internet user following that segment, given that he or she has reached the node from which it branches.

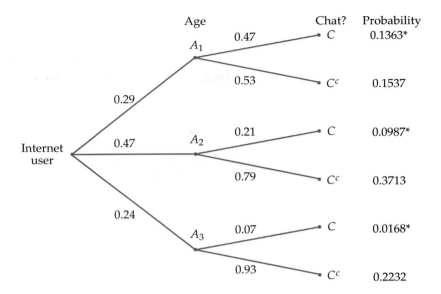

FIGURE 4.20 Tree diagram for Example 4.46. The probability $P(C)$ is the sum of the probabilities of the three branches marked with asterisks (*).

Starting at the left, an Internet user falls into one of the three age groups. The probabilities of these groups

$$P(A_1) = 0.29 \qquad P(A_2) = 0.47 \qquad P(A_3) = 0.24$$

mark the leftmost branches in the tree. Conditional on being 18 to 29 years old, the probability of participating in chat is $P(C \mid A_1) = 0.47$. So the conditional probability of *not* participating is

$$P(C^c \mid A_1) = 1 - 0.47 = 0.53$$

These conditional probabilities mark the paths branching out from the A_1 node in Figure 4.20. The other two age group nodes similarly lead to two branches marked with the conditional probabilities of chatting or not. The probabilities on the branches from any node add to 1 because they cover all possibilities, given that this node was reached.

There are three disjoint paths to C, one for each age group. By the addition rule, $P(C)$ is the sum of their probabilities. The probability of reaching C through the 18 to 29 age group is

$$P(C \text{ and } A_1) = P(A_1)P(C \mid A_1)$$
$$= 0.29 \times 0.47 = 0.1363$$

Follow the paths to C through the other two age groups. The probabilities of these paths are

$$P(C \text{ and } A_2) = P(A_2)P(C \mid A_2) = (0.47)(0.21) = 0.0987$$

$$P(C \text{ and } A_3) = P(A_3)P(C \mid A_3) = (0.24)(0.07) = 0.0168$$

The final result is

$$P(C) = 0.1363 + 0.0987 + 0.0168 = 0.2518$$

About 25% of all adult Internet users take part in chat rooms.

It takes longer to explain a tree diagram than it does to use it. Once you have understood a problem well enough to draw the tree, the rest is easy. Tree diagrams combine the addition and multiplication rules. The multiplication rule says that the probability of reaching the end of any complete branch is the product of the probabilities written on its segments. The probability of any outcome, such as the event C that an adult Internet user takes part in chat rooms, is then found by adding the probabilities of all branches that are part of that event.

USE YOUR KNOWLEDGE

4.101 Draw a tree diagram. Refer to Slim's chances of a flush in Exercise 4.98. Draw a tree diagram to describe the outcomes for the two cards that he will be dealt. At the first stage, his draw can be a diamond or a nondiamond. At the second stage, he has the same possible outcomes but the probabilities are different.

Bayes's rule

There is another kind of probability question that we might ask in the context of thinking about online chat. What percent of adult chat room participants are aged 18 to 29?

EXAMPLE

4.47 Conditional versus unconditional probabilities. In the notation of Example 4.46 this is the conditional probability $P(A_1 \mid C)$. Start from the definition of conditional probability and then apply the results of Example 4.46:

$$P(A_1 \mid C) = \frac{P(A_1 \text{ and } C)}{P(C)}$$

$$= \frac{0.1363}{0.2518} = 0.5413$$

Over half of adult chat room participants are between 18 and 29 years old. Compare this conditional probability with the original information (unconditional) that 29% of adult Internet users are between 18 and 29 years old. Knowing that a person chats increases the probability that he or she is young.

We know the probabilities $P(A_1)$, $P(A_2)$, and $P(A_3)$ that give the age distribution of adult Internet users. We also know the conditional probabilities $P(C \mid A_1)$, $P(C \mid A_2)$, and $P(C \mid A_3)$ that a person from each age group chats. Example 4.46 shows how to use this information to calculate $P(C)$. The method can be summarized in a single expression that adds the probabilities of the three paths to C in the tree diagram:

$$P(C) = P(A_1)P(C \mid A_1) + P(A_2)P(C \mid A_2) + P(A_3)P(C \mid A_3)$$

In Example 4.47 we calculated the "reverse" conditional probability $P(A_1 \mid C)$. The denominator 0.2518 in that example came from the previous expression. Put in this general notation, we have another probability law.

BAYES'S RULE

Suppose that A_1, A_2, ..., A_k are disjoint events whose probabilities are not 0 and add to exactly 1. That is, any outcome is in exactly one of these events. Then if C is any other event whose probability is not 0 or 1,

$$P(A_i \mid C) = \frac{P(C \mid A_i)P(A_i)}{P(C \mid A_1)P(A_1) + P(C \mid A_2)P(A_2) + \cdots + P(A_k)P(C \mid A_k)}$$

The numerator in Bayes's rule is always one of the terms in the sum that makes up the denominator. The rule is named after Thomas Bayes, who wrestled with arguing from outcomes like C back to the A_i in a book published in 1763. It is far better to think your way through problems like Examples 4.46 and 4.47 than to memorize these formal expressions.

Independence again

The conditional probability $P(B \mid A)$ is generally not equal to the unconditional probability $P(B)$. That is because the occurrence of event A generally gives us some additional information about whether or not event B occurs. If knowing that A occurs gives no additional information about B, then A and B are independent events. The formal definition of independence is expressed in terms of conditional probability.

INDEPENDENT EVENTS

Two events A and B that both have positive probability are **independent** if

$$P(B \mid A) = P(B)$$

This definition makes precise the informal description of independence given in Section 4.2. We now see that the multiplication rule for independent events, $P(A \text{ and } B) = P(A)P(B)$, is a special case of the general multiplication rule, $P(A \text{ and } B) = P(A)P(B \mid A)$, just as the addition rule for disjoint events is a special case of the general addition rule.

SECTION 4.5 Summary

The **complement** A^c of an event A contains all outcomes that are not in A. The **union** $\{A \text{ or } B\}$ of events A and B contains all outcomes in A, in B, or in both A and B. The **intersection** $\{A \text{ and } B\}$ contains all outcomes that are in both A and B, but not outcomes in A alone or B alone.

The **conditional probability** $P(B \mid A)$ of an event B, given an event A, is defined by

$$P(B \mid A) = \frac{P(A \text{ and } B)}{P(A)}$$

when $P(A) > 0$. In practice, conditional probabilities are most often found from directly available information.

The essential general rules of elementary probability are

Legitimate values: $0 \le P(A) \le 1$ for any event A

Total probability 1: $P(S) = 1$

Complement rule: $P(A^c) = 1 - P(A)$

Addition rule: $P(A \text{ or } B) = P(A) + P(B) - P(A \text{ and } B)$

Multiplication rule: $P(A \text{ and } B) = P(A)P(B \mid A)$

If A and B are **disjoint,** then $P(A \text{ and } B) = 0$. The general addition rule for unions then becomes the special addition rule, $P(A \text{ or } B) = P(A) + P(B)$.

A and B are **independent** when $P(B \mid A) = P(B)$. The multiplication rule for intersections then becomes $P(A \text{ and } B) = P(A)P(B)$.

In problems with several stages, draw a **tree diagram** to organize use of the multiplication and addition rules.

SECTION 4.5   Exercises

For Exercise 4.95, see page 280; for Exercise 4.96, see page 281; for Exercise 4.97, see page 282; for Exercise 4.98, see page 283; for Exercises 4.99 and 4.100, see page 285; and for Exercise 4.101, see page 288.

4.102 What's wrong? In each of the following scenarios, there is something wrong. Describe what is wrong and give a reason for your answer.

(a) $P(A \text{ or } B$ is always equal to the sum of $P(A)$ and $P(B)$.

(b) The probability of an event minus the probability of its complement is always equal to one.

(c) Two events are disjoint if $P(B \mid A) = P(B)$.

4.103 Exercise and sleep. Suppose 40% of adults get enough sleep, 46% get enough exercise, and 24% do both. Find the probabilities of the following events:

(a) enough sleep and not enough exercise

(b) not enough sleep and enough exercise

(c) not enough sleep and not enough exercise

(d) For each of parts (a), (b), and (c), state the rule that you used to find your answer.

4.104 Exercise and sleep. Refer to the previous exercise. Draw a Venn diagram showing the probabilities for exercise and sleep.

4.105 Lying to a teacher. Suppose that 48% of high school students would admit to lying at least once to a

teacher during the past year and 25% of students are male and would admit to lying at least once to a teacher during the past year.[27] Assume that 50% of the students are male. What is the probability that a randomly selected student is either male or is a liar? Be sure to show your work and indicate all of the rules that you use to find your answer.

4.106 Lying to a teacher. Refer to the previous exercise. Suppose that you select a student from the subpopulation of liars. What is the probability that the student is female? Be sure to show your work and indicate all of the rules that you use to find your answer.

4.107 Binge drinking and gender. In a college population, students are classified by gender and whether or not they are frequent binge drinkers. Here are the probabilities:

	Men	Women
Binge drinker	0.11	0.12
Not binge drinker	0.32	0.45

(a) Verify that the sum of the probabilities is 1.

(b) What is the probability that a randomly selected student is not a binge drinker?

(c) What is the probability that a randomly selected male student is not a binge drinker?

(d) Explain why your answers to (b) and (c) are different. Use language that would be understood by someone who has not studied the material in this chapter.

4.108 Find some probabilities. Refer to the previous exercise.

(a) Find the probability that a randomly selected student is a male binge drinker, and find the probability that a randomly selected student is a female binge drinker.

(b) Find the probability that a student is a binge drinker, given that the student is male and find the probability that a student is a binge drinker, given that the student is female.

(c) Your answer for part (a) gives a higher probability for females, while your answer for part (b) gives a higher probability for males. Interpret your answers in terms of the question of whether there are gender differences in binge-drinking behavior. Decide which comparison you prefer and explain the reasons for your preference.

4.109 Attendance at two-year and four-year colleges. In a large national population of college students, 61% attend four-year institutions and the rest attend two-year institutions. Males make up 44% of the students in the four-year institutions and 41% of the students in the two-year institutions.

(a) Find the four probabilities for each combination of gender and type of institution in the following table. Be sure that your probabilities sum to 1.

	Men	Women
Four-year institution		
Two-year institution		

(b) Consider randomly selecting a female student from this population. What is the probability that she attends a four-year institution?

4.110 Draw a tree diagram. Refer to the previous exercise. Draw a tree diagram to illustrate the probabilities in a situation where you first identify the type of institution attended and then identify the gender of the student.

4.111 Draw a different tree diagram for the same setting. Refer to the previous two exercises. Draw a tree diagram to illustrate the probabilities in a situation where you first identify the gender of the student and then identify the type of institution attended. Explain why the probabilities in this tree diagram are different from those that you used in the previous exercise.

4.112 Education and income. Call a household prosperous if its income exceeds $100,000. Call the household educated if the householder completed college. Select an American household at random, and let A be the event that the selected household is prosperous and B the event that it is educated. According to the Current Population Survey, $P(A) = 0.138$, $P(B) = 0.261$, and the probability that a household is both prosperous and educated is $P(A \text{ and } B) = 0.082$. What is the probability $P(A \text{ or } B)$ that the household selected is either prosperous or educated?

4.113 Find a conditional probability. In the setting of the previous exercise, what is the conditional probability that a household is prosperous, given that it is educated? Explain why your result shows that events A and B are not independent.

4.114 Draw a Venn diagram. Draw a Venn diagram that shows the relation between the events A and B in Exercise 4.106. Indicate each of the following events on your diagram and use the information in Exercise 4.112 to calculate the probability of each event. Finally, describe in words what each event is.

(a) $\{A \text{ and } B\}$

(b) $\{A^c \text{ and } B\}$

(c) $\{A \text{ and } B^c\}$

(d) $\{A^c \text{ and } B^c\}$

4.115 Sales of cars and light trucks. Motor vehicles sold to individuals are classified as either cars or light trucks (including SUVs) and as either domestic or imported. In a recent year, 69% of vehicles sold were light trucks, 78% were domestic, and 55% were domestic light trucks. Let A be the event that a vehicle is a car and B the event that it is imported. Write each of the following events in set notation and give its probability.

(a) The vehicle is a light truck.

(b) The vehicle is an imported car.

4.116 Income tax returns. In a recent year, the Internal Revenue Service received 142,978,806 individual tax returns. Of these, 17,993,498 reported an adjusted gross income of at least $100,000, and 392,220 reported at least $1 million.[28] If you know that a randomly chosen return shows an income of $100,000 or more, what is the conditional probability that the income is at least $1 million?

4.117 Conditional probabilities and independence. Using the information in Exercise 4.115, answer these questions.

(a) Given that a vehicle is imported, what is the conditional probability that it is a light truck?

(b) Are the events "vehicle is a light truck" and "vehicle is imported" independent? Justify your answer.

4.118 Job offers. Julie is graduating from college. She has studied biology, chemistry, and computing and hopes to work as a forensic scientist applying her science background to crime investigation. Late one night she thinks about some jobs she has applied for. Let A, B, and C be the events that Julie is offered a job by

A = the Connecticut Office of the Chief Medical Examiner

B = the New Jersey Division of Criminal Justice

C = the federal Disaster Mortuary Operations Response Team

Julie writes down her personal probabilities for being offered these jobs:

$P(A) = 0.7$ $P(B) = 0.5$ $P(C) = 0.3$
$P(A \text{ and } B) = 0.3$ $P(A \text{ and } C) = 0.1$ $P(B \text{ and } C) = 0.1$
$P(A \text{ and } B \text{ and } C) = 0$

Make a Venn diagram of the events A, B, and C. As in Figure 4.19 (page 281), mark the probabilities of every intersection involving these events and their complements. Use this diagram for Exercises 4.119 to 4.121.

4.119 Find the probability of at least one offer. What is the probability that Julie is offered at least one of the three jobs?

4.120 Find the probability of another event. What is the probability that Julie is offered both the Connecticut and New Jersey jobs, but not the federal job?

4.121 Find a conditional probability. If Julie is offered the federal job, what is the conditional probability that she is also offered the New Jersey job? If Julie is offered the New Jersey job, what is the conditional probability that she is also offered the federal job?

4.122 Academic degrees and gender. Here are the projected numbers (in thousands) of earned degrees in the United States in the 2010–2011 academic year, classified by level and by the sex of the degree recipient:[29]

	Bachelor's	Master's	Professional	Doctorate
Female	933	402	51	26
Male	661	260	44	26

(a) Convert this table to a table giving the probabilities for selecting a degree earned and classifying the recipient by gender and the degree by the levels given above.

(b) If you choose a degree recipient at random, what is the probability that the person you choose is a woman?

(c) What is the conditional probability that you choose a woman, given that the person chosen received a professional degree?

(d) Are the events "choose a woman" and "choose a professional degree recipient" independent? How do you know?

4.123 Find some probabilities. The previous exercise gives the projected number (in thousands) of earned degrees in the United States in the 2010–2011 academic year. Use these data to answer the following questions.

(a) What is the probability that a randomly chosen degree recipient is a man?

(b) What is the conditional probability that the person chosen received a bachelor's degree, given that he is a man?

(c) Use the multiplication rule to find the probability of choosing a male bachelor's degree recipient. Check your result by finding this probability directly from the table of counts.

Working. In the language of government statistics, you are "in the labor force" if you are available for work and either working or actively seeking work. The unemployment rate is the proportion of the labor force (not of the entire population) who are unemployed. Here are data from the Current Population Survey for the civilian population aged 25 years and over. The table entries are counts in thousands of people.[30] Exercises 4.124 to 4.127 concern these data.

Highest education	Total population	In labor force	Employed
Did not finish high school	28,021	12,623	11,552
High school but no college	59,844	38,210	36,249
Some college, but no bachelor's degree	46,777	33,928	32,429
College graduate	51,568	40,414	39,250

4.124 Find the unemployment rates. Find the unemployment rate for people with each level of education. How does the unemployment rate change with education? Explain carefully why your results show that level of education and being employed are not independent.

4.125 Conditional probabilities and independence.

(a) What is the probability that a randomly chosen person 25 years of age or older is in the labor force?

(b) If you know that the person chosen is a college graduate, what is the conditional probability that he or she is in the labor force?

(c) Are the events "in the labor force" and "college graduate" independent? How do you know?

4.126 Find some conditional probabilities. You know that a person is employed. What is the conditional probability that he or she is a college graduate? You know that a second person is a college graduate. What is the conditional probability that he or she is employed?

4.127 ⚠ **A lurking variable.** Beware the lurking variable. The low labor force participation rate of people who did not finish high school is explained by the confounding of education level with a variable that lurks behind the "aged 25 years and over" restriction for these data. Explain this confounding.

4.128 Spelling errors. As explained in Exercise 4.76 (page 276), spelling errors in a text can be either nonword errors or word errors. Nonword errors make up 25% of all errors. A human proofreader will catch 90% of nonword errors and 70% of word errors. What percent of all errors will the proofreader catch? (Draw a tree diagram to organize the information given.)

Genetic counseling. Conditional probabilities and Bayes's rule are a basis for counseling people who may have genetic defects that can be passed to their children. Exercises 4.129 to 4.131 concern genetic counseling settings.

4.129 Albinism. People with albinism have little pigment in their skin, hair, and eyes. The gene that governs albinism has two forms (called alleles), which we denote by a and A. Each person has a pair of these genes, one inherited from each parent. A child inherits one of each parent's two alleles, independently with probability 0.5. Albinism is a recessive trait, so a person is albino only if the inherited pair is aa.

(a) Beth's parents are not albino but she has an albino brother. This implies that both of Beth's parents have type Aa. Why?

(b) Which of the types aa, Aa, AA could a child of Beth's parents have? What is the probability of each type?

(c) Beth is not albino. What are the conditional probabilities for Beth's possible genetic types, given this fact? (Use the definition of conditional probability.)

4.130 Find some conditional probabilities. Beth knows the probabilities for her genetic types from part (c) of the previous exercise. She marries Bob, who is albino. Bob's genetic type must be aa.

(a) What is the conditional probability that a child of Beth and Bob is non-albino if Beth has type Aa? What is the conditional probability of a non-albino child if Beth has type AA?

(b) Beth and Bob's first child is non-albino. What is the conditional probability that Beth is a carrier, type Aa?

4.131 Muscular dystrophy. Muscular dystrophy is an incurable muscle-wasting disease. The most common and serious type, called DMD, is caused by a sex-linked recessive mutation. Specifically: women can be carriers but do not get the disease; a son of a carrier has probability 0.5 of having DMD; a daughter has probability 0.5 of being a carrier. As many as one-third of DMD cases, however, are due to spontaneous mutations in sons of mothers who are not carriers. Toni has one son, who has DMD.

In the absence of other information, the probability is 1/3 that the son is the victim of a spontaneous mutation and 2/3 that Toni is a carrier. There is a screening test called the CK test that is positive with probability 0.7 if a woman is a carrier and with probability 0.1 if she is not. Toni's CK test is positive. What is the probability that she is a carrier?

CHAPTER 4 Exercises

4.132 Repeat the experiment many times. Here is a probability distribution for a random variable X:

Value of X	−1	2
Probability	0.3	0.7

A single experiment generates a random value from this distribution. If the experiment is repeated many times, what will be the approximate proportion of times that the value is −1? Give a reason for your answer.

4.133 Repeat the experiment many times and take the mean. Here is a probability distribution for a random variable X:

Value of X	−1	2
Probability	0.3	0.7

A single experiment generates a random value from this distribution. If the experiment is repeated many times, what will be the approximate value of the mean of these random variables? Give a reason for your answer.

4.134 Work with a transformation. Here is a probability distribution for a random variable X:

Value of X	1	2
Probability	0.4	0.6

(a) Find the mean and the standard deviation of this distribution.

(b) Let $Y = 3X - 2$. Use the rules for means and variances to find the mean and the standard deviation of the distribution of Y.

(c) For part (b) give the rules that you used to find your answer.

4.135 ⚑ **A different transformation.** Refer to the previous exercise. Now let $Y = 3X^2 - 2$.

(a) Find the distribution of Y.

(b) Find the mean and standard deviation for the distribution of Y.

(c) Explain why the rules that you used for part (b) of the previous exercise do not work for this transformation.

4.136 Toss a pair of dice two times. Consider tossing a pair of fair dice two times. For each of the following pairs of events, tell whether they are disjoint, independent, or neither.

(a) $A = 6$ on the first roll, $B = 5$ or less on the first roll.

(b) $A = 6$ on the first roll, $B = 7$ or less on the second roll.

(c) $A = 6$ or less on the second roll, $B = 5$ or less on the first roll.

(d) $A = 6$ or less on the second roll, $B = 5$ or less on the second roll.

4.137 Find the probabilities. Refer to the previous exercise. Find the probabilities for each event.

4.138 Some probability distributions. Here is a probability distribution for a random variable X:

Value of X	2	3	4
Probability	0.3	0.4	0.3

(a) Find the mean and standard deviation for this distribution.

(b) Construct a different probability distribution with the same possible values, the same mean, and a larger standard deviation. Show your work and report the standard deviation of your new distribution.

(c) Construct a different probability distribution with the same possible values, the same mean, and a smaller standard deviation. Show your work and report the standard deviation of your new distribution.

4.139 A fair bet at craps. Almost all bets made at gambling casinos favor the house. In other words, the difference between the amount bet and the mean of the distribution of the payoff is a positive number. An exception is "taking the odds" at the game of craps, a bet that a player can make under certain circumstances. The bet becomes available when a shooter throws a 4, 5, 6, 8, 9, or 10 on the initial roll. This number is called the "point"; when a point is rolled, we say that a point has been established. If a 4 is the point, an odds bet can be made that wins if a 4 is rolled before a 7 is rolled. The probability of winning this bet is 1/3 and the payoff for a $10 bet is $20 (you keep the $10 you bet and you receive an additional $20). The same probability of winning and payoff apply for an odds bet on a 10. For an initial roll of 5 or 9, the odds bet has a winning probability of 2/5 and the payoff for a $10 bet is $15. Similarly, when the initial roll is 6 or 8, the odds bet has a winning probability of 5/11 and the payoff for a $10 bet is $12. Find the mean of the payoff distribution for each of these bets. Then confirm that the bets are fair by showing that the difference between amount bet and the mean of the distribution of the payoff is zero.

4.140 An ancient Korean drinking game. An ancient Korean drinking game involves a 14-sided die. The players roll the die in turn and must submit to whatever humiliation is written on the up-face: something like "Keep still when tickled on face." Six of the 14 faces are squares. Let's call them A, B, C, D, E, and F for short. The other eight faces are triangles, which we will call 1, 2, 3, 4, 5, 6, 7, and 8. Each of the squares is equally likely. Each of the triangles is also equally likely, but the triangle probability differs from the square probability. The probability of getting a square is 0.72. Give the probability model for the 14 possible outcomes.

4.141 Wine tasters. Two wine tasters rate each wine they taste on a scale of 1 to 5. From data on their ratings of a large number of wines, we obtain the following probabilities for both tasters' ratings of a randomly chosen wine:

Taster 1	Taster 2				
	1	2	3	4	5
1	0.03	0.02	0.01	0.00	0.00
2	0.02	0.07	0.06	0.02	0.01
3	0.01	0.05	0.25	0.05	0.01
4	0.00	0.02	0.05	0.20	0.02
5	0.00	0.01	0.01	0.02	0.06

(a) Why is this a legitimate assignment of probabilities to outcomes?

(b) What is the probability that the tasters agree when rating a wine?

(c) What is the probability that Taster 1 rates a wine higher than 3? What is the probability that Taster 2 rates a wine higher than 3?

4.142 SAT scores. The College Board finds that the distribution of students' SAT scores depends on the level of education their parents have. Children of parents who did not finish high school have SAT Math scores X with mean 445 and standard deviation 106. Scores Y of children of parents with graduate degrees have mean 566 and standard deviation 109. Perhaps we should standardize to a common scale for equity. Find positive numbers a, b, c, and d such that $a + bX$ and $c + dY$ both have mean 500 and standard deviation 100.

4.143 Lottery tickets. Joe buys a ticket in the Tri-State Pick 3 lottery every day, always betting on 956. He will win something if the winning number contains 9, 5, and 6 in any order. Each day, Joe has probability 0.006 of winning, and he wins (or not) independently of other days because a new drawing is held each day. What is the probability that Joe's first winning ticket comes on the 20th day?

4.144 Slot machines. Slot machines are now video games, with winning determined by electronic random number generators. In the old days, slot machines were like this: you pull the lever to spin three wheels; each wheel has 20 symbols, all equally likely to show when the wheel stops spinning; the three wheels are independent of each other. Suppose that the middle wheel has 8 bells among its 20 symbols, and the left and right wheels have 1 bell each.

(a) You win the jackpot if all three wheels show bells. What is the probability of winning the jackpot?

(b) What is the probability that the wheels stop with exactly 2 bells showing?

The following exercises require familiarity with the material presented in the optional Section 4.5.

4.145 Mathematics degrees and gender. Of the 21,189 degrees in mathematics given by U.S. colleges and universities in a recent year, 71% were bachelor's degrees, 23% were master's degrees, and the rest were doctorates. Moreover, women earned 44% of the bachelor's degrees, 41% of the master's degrees, and 30% of the doctorates.[31] You choose a mathematics degree at random and find that

it was awarded to a woman. What is the probability that it is a bachelor's degree?

4.146 Higher education at two-year and four-year institutions. The following table gives the counts of U.S. institutions of higher education classified as public or private and as two-year or four-year:[32]

	Public	Private
Two-year	639	1894
Four-year	1061	622

Convert the counts to probabilities and summarize the relationship between these two variables using conditional probabilities.

4.147 Odds bets at craps. Refer to the odds bets at craps in Exercise 4.139. Suppose that whenever the shooter has an initial roll of 4, 5, 6, 8, 9, or 10, you take the odds. Here are the probabilities for these initial rolls:

Point	4	5	6	8	9	10
Probability	3/36	4/36	5/36	5/36	4/36	3/36

Draw a tree diagram with the first stage showing the point rolled and the second stage showing whether the point is again rolled before a 7 is rolled. Include a first-stage branch showing the outcome that a point is not established. In this case, the amount bet is zero and the distribution of the winnings is the special random variable that has $P(X = 0) = 1$. For the combined betting system where the player always makes a $10 odds bet when it is available, show that the game is fair.

4.148 Weights and heights of children adjusted for age. The idea of conditional probabilities has many interesting applications, including the idea of a conditional distribution. For example, the National Center for Health Statistics produces distributions for weight and height for children while conditioning on other variables. Visit the Web site cdc.gov/growthcharts/ and describe the different ways that weight and height distributions are conditioned on other variables.

4.149 Wine tasting. In the setting of Exercise 4.141, Taster 1's rating for a wine is 3. What is the conditional probability that Taster 2's rating is higher than 3?

4.150 An interesting case of independence. Independence of events is not always obvious. Toss two balanced coins independently. The four possible combinations of heads and tails in order each have probability 0.25. The events

$$A = \text{head on the first toss}$$

$$B = \text{both tosses have the same outcome}$$

may seem intuitively related. Show that $P(B \mid A) = P(B)$, so that A and B are in fact independent.

4.151 Find some conditional probabilities. Choose a point at random in the square with sides $0 \le x \le 1$ and $0 \le y \le 1$. This means that the probability that the point falls in any region within the square is the area of that region. Let X be the x coordinate and Y the y coordinate of the point chosen. Find the conditional probability $P(Y < 1/2 \mid Y > X)$. (*Hint:* Sketch the square and the events $Y < 1/2$ and $Y > X$.)

4.152 Sample surveys for sensitive issues. It is difficult to conduct sample surveys on sensitive issues because many people will not answer questions if the answers might embarrass them. **Randomized response** is an effective way to guarantee anonymity while collecting information on topics such as student cheating or sexual behavior. Here is the idea. To ask a sample of students whether they have plagiarized a term paper while in college, have each student toss a coin in private. If the coin lands heads *and* they have not plagiarized, they are to answer "No." Otherwise, they are to give "Yes" as their answer. Only the student knows whether the answer reflects the truth or just the coin toss, but the researchers can use a proper random sample with follow-up for nonresponse and other good sampling practices.

Suppose that in fact the probability is 0.3 that a randomly chosen student has plagiarized a paper. Draw a tree diagram in which the first stage is tossing the coin and the second is the truth about plagiarism. The outcome at the end of each branch is the answer given to the randomized-response question. What is the probability of a "No" answer in the randomized-response poll? If the probability of plagiarism were 0.2, what would be the probability of a "No" response on the poll? Now suppose that you get 39% "No" answers in a randomized-response poll of a large sample of students at your college. What do you estimate to be the percent of the population who have plagiarized a paper?

CHAPTER

5.1 The Sampling
    Distribution of
    a Sample Mean

5.2 Sampling Distributions
    for Counts and
    Proportions

Sampling Distributions

Introduction

Statistical inference draws conclusions about a population or process on the basis of data. The data are summarized by *statistics* such as means and proportions. When the data are produced by random sampling or randomized experimentation, a statistic is a random variable that obeys the laws of probability theory. *Sampling distributions* of statistics provide the link between probability and data. A sampling distribution shows how a statistic would vary in repeated data productions. That is, a sampling distribution is a probability distribution that answers the question "What would happen if we did this experiment or sampling many times?" Suppose you plan to survey 1000 college students at your university about their sleeping habits. The sampling distribution of the average hours of sleep per night describes what this average would be if many

LOOK BACK
sampling distribution p. 204

simple random samples of 1000 subjects were drawn from the population of college students at your university. In other words, it gives you an idea of what you are likely to see from your survey.

> **THE DISTRIBUTION OF A STATISTIC**
>
> A statistic from a random sample or randomized experiment is a random variable. The probability distribution of the statistic is its **sampling distribution.**

Probability distributions also play a second role in statistical inference. Any quantity that can be measured for each member of a population is described by the distribution of its values for all members of the population. This is the context in which we first met distributions as density curves that provide models for the overall pattern of data. Imagine choosing one individual at random from the population. The results of repeated choices have a probability distribution that is the distribution of the population.

◄ **LOOK BACK**
density curves p. 52

> **EXAMPLE**
>
> **5.1 Total sleep time of college students.** A recent survey describes the distribution of total sleep time among college students as approximately Normal with a mean of 7.02 hours and standard deviation of 1.15 hours.[1] Select a college student at random and obtain his or her sleep time. The result is a random variable X. Prior to the random sampling, we don't know the sleep time of the chosen college student, but we do know that in repeated sampling X will have the same $N(7.02, 1.15)$ distribution that describes the pattern of sleep time in the entire population. We call $N(7.02, 1.15)$ the *population distribution*.

> **POPULATION DISTRIBUTION**
>
> The **population distribution** of a variable is the distribution of its values for all members of the population. The population distribution is also the probability distribution of the variable when we choose one individual at random from the population.

◄ **LOOK BACK**
SRS p. 191

The population of all college students actually exists, so that we can in principle draw an SRS from it. Sometimes our population of interest does not actually exist. For example, suppose we are interested in studying final-exam scores in a statistics course and we have the scores of the 34 students who took the course last semester. For the purposes of statistical inference, we might want to consider these 34 students as part of a hypothetical population of similar students who would take this course. In this sense, these students represent not only themselves, but also a larger population of similar students. The key idea is to think of the observations that you have as coming from a population with a probability distribution.

USE YOUR KNOWLEDGE

5.1 Number of apps on an iPhone. AppsFire is a service that shares the names of the apps on your iPhone with everyone else using the service. This, in a sense, creates an iPhone app recommendation system. Recently, the service drew a sample of 1200 AppsFire users and reported an average of 65 apps per device.[2] State the population of this survey, the statistic, and some likely values from the population distribution.

To progress from discussing probability as a topic in itself to probability as a foundation for inference, we start by studying the sampling distributions of some common statistics. In each case, the sampling distribution depends on both the population distribution and the way we collect the data from the population.

5.1 The Sampling Distribution of a Sample Mean

A variety of statistics are used to describe quantitative data. The sample mean, percentiles, and standard deviation are all examples of statistics based on quantitative data. Statistical theory describes the sampling distributions of these statistics. The general framework to construct a sampling distribution is the same for all statistics. In this section we will concentrate on the sample mean. Because sample means are just averages of observations, they are among the most frequently used statistics.

EXAMPLE

5.2 Sample means are approximately Normal. Figure 5.1 illustrates two striking facts about the sampling distribution of a sample mean. Figure 5.1(a)

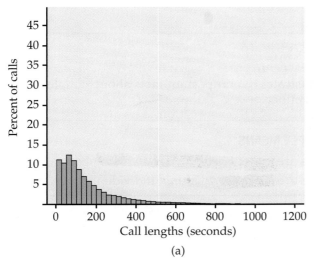

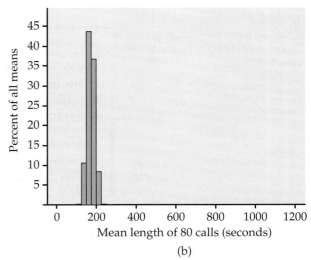

(a) (b)

FIGURE 5.1 (a) The distribution of lengths of all customer service calls received by a bank in a month. (b) The distribution of the sample means $\bar{x}$ for 500 random samples of size 80 from this population. The scales and histogram classes are exactly the same in both panels.

displays the distribution of customer service call lengths for a bank service center for a month. There are more than 30,000 calls in this population.[3] (We omitted a few extreme outliers, calls that lasted more than 20 minutes.) The distribution is extremely skewed to the right. The population mean is $\mu = 173.95$ seconds.

Table 1.2 (page 15) contains the lengths of a sample of 80 calls from this population. The mean of these 80 calls is $\bar{x} = 196.6$ seconds. If we were to take another sample of size 80, we would likely get a different value of $\bar{x}$. This is because this new sample would contain a different set of calls. To find the sampling distribution of $\bar{x}$, we take many SRSs of size 80 and calculate $\bar{x}$ for each sample. Figure 5.1(b) is the distribution of the values of $\bar{x}$ for 500 random samples. The scales and choice of classes are exactly the same as in Figure 5.1(a), so that we can make a direct comparison.

The sample means are much less spread out than the individual call lengths. What is more, the distribution in Figure 5.1(b) is roughly symmetric rather than skewed. The Normal quantile plot in Figure 5.2 confirms that the distribution is close to Normal.

CALL_LENGTH

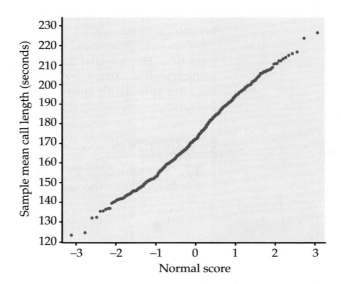

FIGURE 5.2 Normal quantile plot of the 500 sample means in Figure 5.1(b). The distribution is close to Normal.

This example illustrates two important facts about sample means that we will discuss in this section.

FACTS ABOUT SAMPLE MEANS

1. Sample means are less variable than individual observations.
2. Sample means are more Normal than individual observations.

These two facts contribute to the popularity of sample means in statistical inference.

The mean and standard deviation of $\bar{x}$

The sample mean $\bar{x}$ from a sample or an experiment is an estimate of the mean μ of the underlying population. The sampling distribution of $\bar{x}$ is determined by the design used to produce the data, the sample size n, and the population distribution.

Select an SRS of size n from a population, and measure a variable X on each individual in the sample. The n measurements are values of n random variables $X_1, X_2, \ldots, X_n$. A single X_i is a measurement on one individual selected at random from the population and therefore has the distribution of the population. If the population is large relative to the sample, we can consider $X_1, X_2, \ldots, X_n$ to be independent random variables each having the same distribution. This is our probability model for measurements on each individual in an SRS.

The **sample mean of an SRS of size n is**

$$\bar{x} = \frac{1}{n}(X_1 + X_2 + \cdots + X_n)$$

If the population has mean μ, then μ is the mean of the distribution of each observation X_i. To get the mean of $\bar{x}$, we use the **addition rule** for means of random variables. Specifically,

$$\mu_{\bar{x}} = \frac{1}{n}(\mu_{X_1} + \mu_{X_2} + \cdots + \mu_{X_n})$$

$$= \frac{1}{n}(\mu + \mu + \cdots + \mu) = \mu$$

LOOK BACK
addition rule for means p. 267

LOOK BACK
unbiased estimator p. 207

That is, *the mean of $\bar{x}$ is the same as the mean of the population.* The sample mean $\bar{x}$ is therefore an **unbiased estimator** of the unknown population mean μ.

The observations are independent, so the **addition rule for variances** also applies:

LOOK BACK
addition rule for variances
p. 271

$$\sigma_{\bar{x}}^2 = \left(\frac{1}{n}\right)^2 \left(\sigma_{X_1}^2 + \sigma_{X_2}^2 + \cdots + \sigma_{X_n}^2\right)$$

$$= \left(\frac{1}{n}\right)^2 \left(\sigma^2 + \sigma^2 + \cdots + \sigma^2\right)$$

$$= \frac{\sigma^2}{n}$$

With n in the denominator, the variability of $\bar{x}$ about its mean decreases as the sample size grows. Thus, a sample mean from a large sample will usually be very close to the true population mean μ. Because the standard deviation of $\bar{x}$ is $\sigma/\sqrt{n}$, the standard deviation of the statistic decreases in proportion to the square root of the sample size. This means that a sample size must be multiplied by 4 in order to divide the statistic's standard deviation in half. The sample size must be multiplied by 100 in order to reduce the standard deviation by a factor of 10. Here is a summary of these facts.

MEAN AND STANDARD DEVIATION OF A SAMPLE MEAN

Let $\bar{x}$ be the mean of an SRS of size n from a population having mean μ and standard deviation σ. The mean and standard deviation of $\bar{x}$ are

$$\mu_{\bar{x}} = \mu$$

$$\sigma_{\bar{x}} = \frac{\sigma}{\sqrt{n}}$$

How accurately does a sample mean $\bar{x}$ estimate a population mean μ? Because the values of $\bar{x}$ vary from sample to sample, we must give an answer in terms of the sampling distribution. We know that $\bar{x}$ is an unbiased estimator of μ, so its values in repeated samples are not systematically too high or too low. Most samples will give an $\bar{x}$-value close to μ if the sampling distribution is concentrated close to its mean μ. So the accuracy of estimation depends on the spread of the sampling distribution.

EXAMPLE

5.3 Standard deviations for sample means of service call lengths. The standard deviation of the population of service call lengths in Figure 5.1(a) is $\sigma = 184.81$ seconds. The length of a single call will often be far from the population mean. If we choose an SRS of 20 calls, the standard deviation of their mean length is

$$\sigma_{\bar{x}} = \frac{184.81}{\sqrt{20}} = 41.32 \text{ seconds}$$

Averaging over more calls reduces the variability and makes it more likely that $\bar{x}$ is close to μ. Our sample size of 80 calls is 4 times 20, so the standard deviation will be half as large:

$$\sigma_{\bar{x}} = \frac{184.81}{\sqrt{80}} = 20.66 \text{ seconds}$$

USE YOUR KNOWLEDGE

5.2 Find the mean and the standard deviation of the sampling distribution. You take an SRS of size 36 from a population with mean 240 and standard deviation 18. Find the mean and standard deviation of the sampling distribution of your sample mean.

5.3 The effect of increasing the sample size. In the setting of the previous exercise, repeat the calculations for a sample size of 144. Explain the effect of the increase on the sample mean and standard deviation.

The central limit theorem

We have described the center and spread of the probability distribution of a sample mean $\bar{x}$, but not its shape. The shape of the distribution of $\bar{x}$ depends on the shape of the population distribution. Here is one important case: if the population distribution is Normal, then so is the distribution of the sample mean.

SAMPLING DISTRIBUTION OF A SAMPLE MEAN

If a population has the $N(\mu, \sigma)$ distribution, then the sample mean $\bar{x}$ of n independent observations has the $N(\mu, \sigma/\sqrt{n})$ distribution.

This is a somewhat special result. Many population distributions are not Normal. The service call lengths in Figure 5.1(a), for example, are strongly skewed. Yet Figures 5.1(b) and 5.2 show that means of samples of size 80 are close to Normal. One of the most famous facts of probability theory says that, for large sample sizes, the distribution of $\bar{x}$ is close to a Normal distribution. This is true no matter what shape the population distribution has, as long as the population has a finite standard deviation σ. This is the **central limit theorem.** It is much more useful than the fact that the distribution of $\bar{x}$ is exactly Normal if the population is exactly Normal.

central limit theorem

CENTRAL LIMIT THEOREM

Draw an SRS of size n from any population with mean μ and finite standard deviation σ. When n is large, the sampling distribution of the sample mean $\bar{x}$ is approximately Normal:

$$\bar{x} \text{ is approximately } N\left(\mu, \frac{\sigma}{\sqrt{n}}\right)$$

EXAMPLE

5.4 How close will the sample mean be to the population mean? With the Normal distribution to work with, we can better describe how accurately a random sample of 80 calls estimates the mean length of all the calls in the population. The population standard deviation for the more than 30,000 calls in the population of Figure 5.1(a) is $\sigma = 184.81$ seconds. From Example 5.3 we know $\sigma_{\bar{x}} = 20.66$ seconds. By the 95 part of the 68–95–99.7 rule, about 95% of all samples will have mean $\bar{x}$ within two standard deviations of μ, that is, within ± 41.32 seconds of μ.

◄ LOOK BACK
68–95–99.7 rule p. 56

USE YOUR KNOWLEDGE

5.4 Use the 68–95–99.7 rule. You take an SRS of size 144 from a population with mean 240 and standard deviation 18. According to the central limit theorem, what is the approximate sampling distribution of the sample mean? Use the 95 part of the 68–95–99.7 rule to describe the variability of this sample mean.

For the sample size of $n = 80$ in Example 5.4, the sample mean is not very accurate. The population of service call lengths is very spread out, so the sampling distribution of $\bar{x}$ has a large standard deviation.

EXAMPLE

5.5 How can we reduce the standard deviation? In the setting of Example 5.4, if we want to reduce the standard deviation of $\bar{x}$ by a factor of 4, we must take a sample 16 times as large, $n = 16 \times 80$, or 1280. Then

$$\sigma_{\bar{x}} = \frac{184.81}{\sqrt{1280}} = 5.166 \text{ seconds}$$

For samples of size 1280, about 95% of the sample means will be within twice 5.166, or 10.33 seconds, of the population mean μ.

5.5 The effect of increasing the sample size. In the setting of Exercise 5.3, suppose we increase the sample size to 1296. Use the 95 part of the 68–95–99.7 rule to describe the variability of this sample mean. Compare your results with those you found in Exercise 5.3.

Example 5.5 reminds us that if the population is very spread out, the $\sqrt{n}$ in the standard deviation of $\bar{x}$ implies very large samples are needed to estimate the population mean accurately. The main point of the example, however, is that the central limit theorem allows us to use Normal probability calculations to answer questions about sample means even when the population distribution is not Normal. How large a sample size n is needed for $\bar{x}$ to be close to Normal depends on the population distribution. More observations are required if the shape of the population distribution is far from Normal. Even for the very skewed call length population, however, samples of size 80 are large enough. Here is a more detailed example.

EXAMPLE

5.6 The central limit theorem in action. Figure 5.3 shows the central limit theorem in action for another very non-Normal population. Figure 5.3(a) displays the density curve of a single observation from the population. The distribution is strongly right-skewed, and the most probable outcomes are near 0. The mean μ of this distribution is 1, and its standard deviation σ is also 1. This particular continuous distribution is called an **exponential distribution.** Exponential distributions are used as models for how long an

exponential distribution

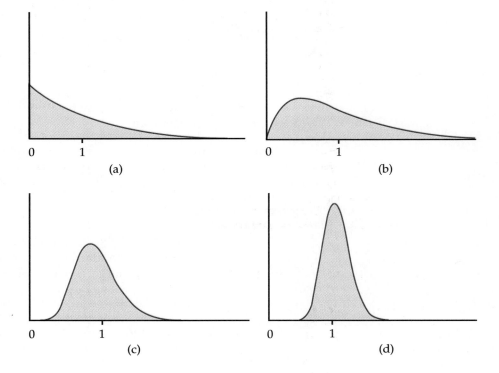

FIGURE 5.3 The central limit theorem in action: the distribution of sample means from a strongly non-Normal population becomes more Normal as the sample size increases. (a) The distribution of 1 observation. (b) The distribution of $\bar{x}$ for 2 observations. (c) The distribution of $\bar{x}$ for 10 observations. (d) The distribution of $\bar{x}$ for 25 observations.

electronic component will last and for the time between text messages arriving on your cell phone.

Figures 5.3(b), (c), and (d) are the density curves of the sample means of 2, 10, and 25 observations from this population. As n increases, the shape becomes more Normal. The mean remains at $\mu = 1$, but the standard deviation decreases, taking the value $1/\sqrt{n}$. The density curve for 10 observations is still somewhat skewed to the right but already resembles a Normal curve having $\mu = 1$ and $\sigma = 1/\sqrt{10} = 0.32$. The density curve for $n = 25$ is yet more Normal. The contrast between the shapes of the population distribution and of the distribution of the mean of 10 or 25 observations is striking.

The *Central Limit Theorem* applet animates Figure 5.3. You can slide the sample size n from 1 to 100 and watch both the exact density curve of $\bar{x}$ and the Normal approximation. As you increase n, the two curves move closer together.

EXAMPLE

5.7 Time between text message arrivals. Suppose the time X between text messages arriving on your cell phone is governed by the exponential distribution with mean $\mu = 25$ minutes and standard deviation $\sigma = 25$ minutes. You record the times between your next 50 messages. What is the probability that their average exceeds 21 minutes?

The central limit theorem says that the sample mean time $\bar{x}$ (in minutes) between text messages has approximately the Normal distribution with mean equal to the population mean $\mu = 25$ minutes and standard deviation

$$\frac{\sigma}{\sqrt{50}} = \frac{25}{\sqrt{50}} = 3.54 \text{ minutes}$$

The distribution of $\bar{x}$ is therefore approximately $N(25, 3.54)$. Figure 5.4 shows this Normal curve (solid) and also the actual density curve of $\bar{x}$ (dashed).

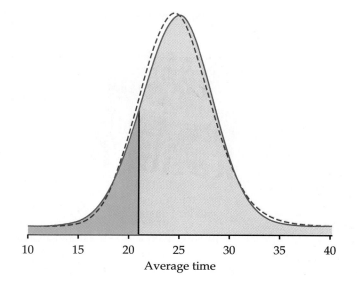

FIGURE 5.4 The exact distribution (dashed) and the Normal approximation from the central limit theorem (solid) for the average time between text messages arriving on your cell phone, for Example 5.7.

The probability we want is $P(\bar{x} > 21.0)$. A Normal distribution calculation gives this probability as 0.8708. This is the area to the right of 21 under the solid Normal curve in Figure 5.4. The exactly correct probability is the area under the dashed density curve in the figure. It is 0.8750. The central limit theorem Normal approximation is off by only about 0.0042.

USE YOUR KNOWLEDGE

5.6 Find a probability. Refer to the example above. Find the probability that the mean time between 50 text messages is less than 28 minutes.

EXAMPLE

5.8   Convert the results to the total time. In Example 5.7 what can we say about the total time between 50 text messages? According to the central limit theorem

$$P(\bar{x} > 21.0) = 0.8708$$

We know that sample mean is the total time divided by 50, so the event $\{\bar{x} > 21.0\}$ is the same as the event $\{50\bar{x} > 50(21.0)\}$. We can say that the probability is 0.8708 that the total time is $50(21.0) = 1{,}050$ minutes (17.5 hours) or greater.

Figure 5.5 summarizes the facts about the sampling distribution of $\bar{x}$ in a way that emphasizes the big idea of a sampling distribution.

- **Keep taking random samples of size n from a population with mean μ.**
- **Find the sample mean $\bar{x}$ for each sample.**
- **Collect all the $\bar{x}$'s and display their distribution.**

That's the sampling distribution of $\bar{x}$. Sampling distributions are the key to understanding statistical inference. Keep this figure in mind as you go forward.

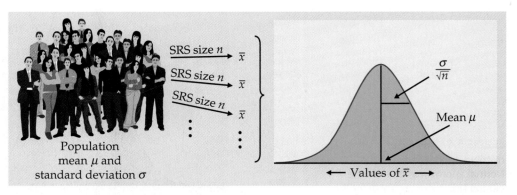

FIGURE 5.5   The sampling distribution of a sample mean $\bar{x}$ has mean μ and standard deviation $\sigma/\sqrt{n}$. The distribution is Normal if the population distribution is Normal; it is approximately Normal for large samples in any case.

A few more facts

The central limit theorem is the big fact of this section. Here are three useful smaller facts related to our topic.

The fact that the sample mean of an SRS from a Normal population has a Normal distribution is a special case of a more general fact: **any linear combination of independent Normal random variables is also Normally distributed.** That is, if X and Y are independent Normal random variables and a and b are any fixed numbers, $aX + bY$ is also Normally distributed, and so it is for any number of Normal variables. In particular, the sum or difference of independent Normal random variables has a Normal distribution. The mean and standard deviation of $aX + bY$ are found as usual from the addition rules for means and variances. These facts are often used in statistical calculations. Here is an example.

LOOK BACK
rules for means, p. 267
rules for variances, p. 271

EXAMPLE

5.9 Getting to and from campus. You live off campus and take the shuttle, provided by your apartment complex, to and from campus. Your time on the shuttle in minutes varies day to day. The time going to campus X has the $N(20, 4)$ distribution, and the time returning from campus Y varies according to the $N(18, 8)$ distribution. If they vary independently, what is the probability that you will be on the shuttle for less time going to campus?

The difference in times $X - Y$ is Normally distributed, with mean and variance

$$\mu_{X-Y} = \mu_X - \mu_Y = 20 - 18 = 2$$
$$\sigma^2_{X-Y} = \sigma^2_X + \sigma^2_Y = 4^2 + 8^2 = 80$$

Because $\sqrt{80} = 8.94$, $X - Y$ has the $N(2, 8.94)$ distribution. Figure 5.6 illustrates the probability computation:

$$P(X < Y) = P(X - Y < 0)$$
$$= P\left(\frac{(X - Y) - 2}{8.94} < \frac{0 - 2}{8.94}\right)$$
$$= P(Z < -0.22) = 0.4129$$

Although on average it takes longer to go to campus than return, the trip to campus will be shorter in roughly two of every five days.

The second useful fact is that **more general versions of the central limit theorem say that the distribution of a sum or average of many small random quantities is close to Normal.** This is true even if the quantities are not independent (as long as they are not too highly correlated) and even if they have different distributions (as long as no single random quantity is so large that it dominates the others). The central limit theorem suggests why the Normal distributions are common models for observed data. Any variable that is a sum of many small influences will have approximately a Normal distribution.

Finally, **the central limit theorem also applies to discrete random variables.** An average of discrete random variables will never result in a continuous sampling distribution but the Normal distribution often serves as a good approximation. In the next section, we will discuss the sampling distribution and

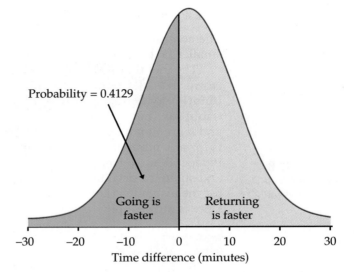

FIGURE 5.6 The Normal probability calculation for Example 5.9. The difference in times going to campus and returning from campus $(X - Y)$ is Normal with mean 2 minutes and standard deviation 8.94 minutes.

Probability = 0.4129

Going is faster

Returning is faster

Time difference (minutes)

Normal approximation for counts and proportions. This Normal approximation is just an example of the central limit theorem applied to these discrete random variables.

SECTION 5.1 Summary

The **sample mean** $\bar{x}$ of an SRS of size n drawn from a large population with mean μ and standard deviation σ has a sampling distribution with mean and standard deviation

$$\mu_{\bar{x}} = \mu$$
$$\sigma_{\bar{x}} = \frac{\sigma}{\sqrt{n}}$$

The sample mean $\bar{x}$ is therefore an unbiased estimator of the population mean μ and is less variable than a single observation.

Linear combinations of independent Normal random variables have Normal distributions. In particular, if the population has a Normal distribution, so does $\bar{x}$.

The **central limit theorem** states that for large n the sampling distribution of $\bar{x}$ is approximately $N(\mu, \sigma/\sqrt{n})$ for any population with mean μ and finite standard deviation σ.

SECTION 5.1 Exercises

For Exercise 5.1, see page 299; for Exercises 5.2 and 5.3, see pag 302; for Exercise 5.4, see page 303; for Exercise 5.5, see page 304; and for Exercise 5.6, see page 306.

5.7   What is wrong? Explain what is wrong in each of the following scenarios.

(a) If the variance of a population is 10, then the variance of the mean for an SRS of 30 observations from this population will be $10/\sqrt{30}$.

(b) When taking SRS's from a population, larger sample sizes will result in larger standard deviations of the sample mean.

(c) The mean of a sampling distribution of $\bar{x}$ changes when the sample size changes.

5.8   What is wrong? Explain what is wrong in each of the following statements.

(a) For large n, the distribution of observed values will be approximately Normal.

(b) The 68–95–99.7 rule says that $\bar{x}$ should be within $\mu \pm 2\sigma$ about 95% of the time.

(c) The central limit theorem states that for large n, μ is approximately Normal.

5.9 Generating a sampling distribution. Let's illustrate the idea of a sampling distribution in the case of a very small sample from a very small population. The population is the 10 scholarship players currently on your men's basketball team. For convenience, the 10 players have been labeled with the integers 0 to 9. For each player, the total amount of time spent (in minutes) on Facebook during the last month is recorded in the table below.

Player	0	1	2	3	4	5	6	7	8	9
Total Time (min)	370	290	358	366	323	319	358	309	327	368

The parameter of interest is the average amount of time on Facebook. The sample is an SRS of size $n = 3$ drawn from this population of players. Because the players are labeled 0 to 9, a single random digit from Table B chooses one player for the sample.

(a) Find the mean of the 10 players in the population. This is the population mean μ.

(b) Use Table B to draw an SRS of size 3 from this population (Note: you may sample the same player's time more than once). Write down the three times in your sample and calculate the sample mean $\bar{x}$. This statistic is an estimate of μ.

(c) Repeat this process 10 times using different parts of Table B. Make a histogram of the 10 values of $\bar{x}$. You are constructing the sampling distribution of $\bar{x}$.

(d) Is the center of your histogram close to μ? Would it get closer to μ the more times you repeated this sampling process? Explain.

5.10 Total sleep time of college students. In Example 5.1, the total sleep time per night among college students was approximately Normally distributed with mean $\mu = 7.02$ hours and standard deviation $\sigma = 1.15$ hours. Suppose you plan to take an SRS of size $n = 200$ and compute the average total sleep time.

(a) What is the standard deviation for the average time?

(b) Use the 95 part of the 68–95–99.7 rule to describe the variability of this sample mean.

(c) What is the probability that your average will be below 6.9 hours?

5.11 Determining sample size. Recall the previous exercise. Suppose you want to use a sample size such that about 95% of the averages fall within ±5 minutes of the true mean $\mu = 7.02$.

(a) Based on your answer to part (b) in Exercise 5.8, should the sample size be larger or smaller than 200? Explain.

(b) What standard deviation of the average do you need such that about 95% of all samples will have a mean within 5 minutes of μ?

(c) Using the standard deviation calculated in part (b), determine the number of students you need to sample.

5.12 Songs on an iPod. An iPod has about 10,000 songs. The distribution of the play time for these songs is highly skewed. Assume that the standard deviation for the population is 280 seconds.

(a) What is the standard deviation of the average time when you take an SRS of 10 songs from this population?

(b) How many songs would you need to sample if you wanted the standard deviation of $\bar{x}$ to be 15 seconds?

5.13 Bottling an energy drink. A bottling company uses a filling machine to fill cans with an energy drink. The cans are supposed to contain 250 milliliters (ml). The machine, however, has some variability, so the standard deviation of the size is $\sigma = 3$ ml. A sample of 6 cans is inspected each hour for process control purposes, and records are kept of the sample mean volume. If the process mean is exactly equal to the target value, what will be the mean and standard deviation of the numbers recorded?

5.14 Play times for songs on an iPod. Averages of several measurements are less variable than individual measurements. Suppose the true mean duration of the play time for the songs in the iPod of Exercise 5.12 is 350 seconds.

(a) Assuming the play times to be Normally distributed, sketch on the same graph the two Normal curves, one for sampling a single song and one for the mean of 10 songs.

(b) What is the probability that the sample mean differs from the population mean by more than 19 seconds when only 1 song is sampled?

(c) How does the probability that you calculated in part (b) change for the mean of an SRS of 10 songs?

5.15 Can volumes. Averages are less variable than individual observations. Suppose that the can volumes in Exercise 5.13 vary according to a Normal distribution. In that case, the mean $\bar{x}$ of an SRS of cans also has a Normal distribution.

(a) Make a sketch of the Normal curve for a single can. Add the Normal curve for the mean of an SRS of 6 cans on the same sketch.

(b) What is the probability that the volume of a single randomly chosen can differs from the target value by 1 ml or more?

(c) What is the probability that the mean volume of an SRS of 6 cans differs from the target value by 1 ml or more?

5.16 Number of friends on Facebook. Facebook provides a variety of statistics on their Web site that detail the growth and popularity of the site.[4] One such statistic is that the average user has 130 friends. This distribution only takes integer values, so it is certainly not Normal. We'll also assume it is skewed to the right with a standard deviation $\sigma = 85$. Consider an SRS of 30 Facebook users.

(a) What are the mean and standard deviation of the total number of friends in this sample?

(b) What are the mean and standard deviation of the mean number of friends per user?

(c) Use the central limit theorem to find the probability that the average number of friends in 30 Facebook users is greater than 140.

5.17 ⚠ **Cholesterol levels of teenagers.** A study of the health of teenagers plans to measure the blood cholesterol level of an SRS of 13- to 16-year olds. The researchers will report the mean $\bar{x}$ from their sample as an estimate of the mean cholesterol level μ in this population.

(a) Explain to someone who knows no statistics what it means to say that $\bar{x}$ is an "unbiased" estimator of μ.

(b) The sample result $\bar{x}$ is an unbiased estimator of the population truth μ no matter what size SRS the study chooses. Explain to someone who knows no statistics why a large sample gives more trustworthy results than a small sample.

5.18 ACT scores of high school seniors. The scores of high school seniors on the ACT college entrance examination in a recent year had mean $\mu = 19.2$ and standard deviation $\sigma = 5.1$. The distribution of scores is only roughly Normal.

(a) What is the approximate probability that a single student randomly chosen from all those taking the test scores 23 or higher?

(b) Now take an SRS of 25 students who took the test. What are the mean and standard deviation of the sample mean score $\bar{x}$ of these 25 students?

(c) What is the approximate probability that the mean score $\bar{x}$ of these students is 23 or higher?

(d) Which of your two Normal probability calculations in parts (a) and (c) is more accurate? Why?

5.19 Gypsy moths threaten oak and aspen trees. The gypsy moth is a serious threat to oak and aspen trees. A state agriculture department places traps throughout the state to detect the moths. When traps are checked periodically, the mean number of moths trapped is only 0.5, but some traps have several moths. The distribution of moth counts is discrete and strongly skewed, with standard deviation 0.7.

(a) What are the mean and standard deviation of the average number of moths $\bar{x}$ in 50 traps?

(b) Use the central limit theorem to find the probability that the average number of moths in 50 traps is greater than 0.6.

5.20 Grades in an English course. North Carolina State University posts the grade distributions for its courses online.[5] Students in one section of English 210 in the Fall 2008 semester received 33% A's, 24% B's, 18% C's, 16% D's, and 9% F's.

(a) Using the common scale A = 4, B = 3, C = 2, D = 1, F = 0, take X to be the grade of a randomly chosen English 210 student. Use the definitions of the mean (page 261) and standard deviation (page 269) for discrete random variables to find the mean μ and the standard deviation σ of grades in this course.

(b) English 210 is a large course. We can take the grades of an SRS of 50 students to be independent of each other. If $\bar{x}$ is the average of these 50 grades, what are the mean and standard deviation of $\bar{x}$?

(c) What is the probability $P(X \geq 3)$ that a randomly chosen English 210 student gets a B or better? What is the approximate probability $P(\bar{x} \geq 3)$ that the grade point average for 50 randomly chosen English 210 students is a B or better?

5.21 Diabetes during pregnancy. Sheila's doctor is concerned that she may suffer from gestational diabetes (high blood glucose levels during pregnancy). There is variation in both the actual glucose level and blood test that measures the level. A patient is classified as having gestational diabetes if the glucose level is above 140 milligrams per deciliter (mg/dl) one hour after a sugary drink is ingested. Sheila's measured glucose level one hour after ingesting the sugary drink varies according to the Normal distribution with $\mu = 125$ mg/dl and $\sigma = 10$ mg/dl.

(a) If a single glucose measurement is made, what is the probability that Sheila is diagnosed as having gestational diabetes?

(b) If measurements are made instead on three separate days and the mean result is compared with the criterion 140 mg/dl, what is the probability that Sheila is diagnosed as having gestational diabetes?

5.22 A lottery payoff. A $1 bet in a state lottery's Pick 3 game pays $500 if the three-digit number you choose exactly matches the winning number, which is drawn at random. Here is the distribution of the payoff X:

Payoff X	$0	$500
Probability	0.999	0.001

Each day's drawing is independent of other drawings.

(a) What are the mean and standard deviation of X?

(b) Joe buys a Pick 3 ticket twice a week. What does the law of large numbers say about the average payoff Joe receives from his bets?

(c) What does the central limit theorem say about the distribution of Joe's average payoff after 104 bets in a year?

(d) Joe comes out ahead for the year if his average payoff is greater than $1 (the amount he spent each day on a ticket). What is the probability that Joe ends the year ahead?

5.23 Defining a high glucose reading. In Exercise 5.21, Sheila's measured glucose level one hour after ingesting the sugary drink varies according to the Normal distribution with $\mu = 125$ mg/dl and $\sigma = 10$ mg/dl. What is the level L such that there is probability only 0.05 that the mean glucose level of three test results falls above L for Sheila's glucose level distribution?

5.24 Flaws in carpets. The number of flaws per square yard in a type of carpet material varies with mean 1.3 flaws per square yard and standard deviation 1.5 flaws per square yard. This population distribution cannot be Normal, because a count takes only whole-number values. An inspector studies 200 square yards of the material, records the number of flaws found in each square yard, and calculates $\bar{x}$, the mean number of flaws per square yard inspected. Use the central limit theorem to find the approximate probability that the mean number of flaws exceeds 2 per square yard.

5.25 Weights of airline passengers. In response to the increasing weight of airline passengers, the Federal Aviation Administration told airlines to assume that passengers average 190 pounds in the summer, including clothing and carry-on baggage. But passengers vary: the FAA gave a mean but not a standard deviation. A reasonable standard deviation is 35 pounds. Weights are not Normally distributed, especially when the population includes both men and women, but they are not very non-Normal. A commuter plane carries 25 passengers. What is the approximate probability that the total weight of the passengers exceeds 5200 pounds? (*Hint:* To apply the central limit theorem, restate the problem in terms of the mean weight.)

5.26 Risks and insurance. The idea of insurance is that we all face risks that are unlikely but carry high cost. Think of a fire destroying your home. So we form a group to share the risk: we all pay a small amount, and the insurance policy pays a large amount to those few of us whose homes burn down. An insurance company looks at the records for millions of homeowners and sees that the mean loss from fire in a year is $\mu = \$250$ per house and that the standard deviation of the loss is $\sigma = \$1000$. (The distribution of losses is extremely right-skewed: most people have $0 loss, but a few have large losses.) The company plans to sell fire insurance for $250 plus enough to cover its costs and profit.

(a) Explain clearly why it would be unwise to sell only 12 policies. Then explain why selling many thousands of such policies is a safe business.

(b) If the company sells 10,000 policies, what is the approximate probability that the average loss in a year will be greater than $275?

5.27 Treatment of cotton fabrics. "Durable press" cotton fabrics are treated to improve their recovery from wrinkles after washing. Unfortunately, the treatment also reduces the strength of the fabric. The breaking strength of untreated fabric is Normally distributed with mean 57 pounds and standard deviation 2.2 pounds. The same type of fabric after treatment has Normally distributed breaking strength with mean 30 pounds and standard deviation 1.6 pounds.[6] A clothing manufacturer tests 6 specimens of each fabric. All 12 strength measurements are independent.

(a) What is the probability that the mean breaking strength of the 6 untreated specimens exceeds 50 pounds?

(b) What is the probability that the mean breaking strength of the 6 untreated specimens is at least 25 pounds greater than the mean strength of the 6 treated specimens?

5.28 Advertisements and brand image. Many companies place advertisements to improve the image of their brand rather than to promote specific products. In a randomized comparative experiment, business students read ads that cited either the *Wall Street Journal* or the *National Enquirer* for important facts about a fictitious company. The students then rated the trustworthiness of the source on a 7-point scale. Suppose that in the population of all students scores for the *Journal* have mean 4.8 and standard deviation 1.5, while scores for the *Enquirer* have mean 2.4 and standard deviation 1.6.[7]

(a) There are 28 students in each group. Although individual scores are discrete, the mean score for a group of 28 will be close to Normal. Why?

(b) What are the means and standard deviations of the sample mean scores $\bar{y}$ for the *Journal* group and $\bar{x}$ for the *Enquirer* group?

(c) We can take all 56 scores to be independent because students are not told each other's scores. What is the distribution of the difference $\bar{y} - \bar{x}$ between the mean scores in the two groups?

(d) Find $P(\bar{y} - \bar{x} \geq 1)$.

5.29 Ⓐ **Treatment and control groups.** The two previous exercises illustrate a common setting for statistical inference. This exercise gives the general form of the sampling distribution needed in this setting. We have a sample of n observations from a treatment group and an independent sample of m observations from a control group. Suppose that the response to the treatment has the $N(\mu_X, \sigma_X)$ distribution and that the response of control subjects has the $N(\mu_Y, \sigma_Y)$ distribution. Inference about the difference $\mu_Y - \mu_X$ between the population means is based on the difference $\bar{y} - \bar{x}$ between the sample means in the two groups.

(a) Under the assumptions given, what is the distribution of $\bar{y}$? Of $\bar{x}$?

(b) What is the distribution of $\bar{y} - \bar{x}$?

5.30 Ⓐ **Investments in two funds.** Linda invests her money in a portfolio that consists of 70% Fidelity 500 Index Fund and 30% Fidelity Diversified International Fund. Suppose that in the long run the annual real return X on the 500 Index Fund has mean 9% and standard deviation 19%, the annual real return Y on the Diversified International Fund has mean 11% and standard deviation 17%, and the correlation between X and Y is 0.6.

(a) The return on Linda's portfolio is $R = 0.7X + 0.3Y$. What are the mean and standard deviation of R?

(b) The distribution of returns is typically roughly symmetric but with more extreme high and low observations than a Normal distribution. The average return over a number of years, however, is close to Normal. If Linda holds her portfolio for 20 years, what is the approximate probability that her average return is less than 5%?

(c) The calculation you just made is not overly helpful, because Linda isn't really concerned about the mean return $\bar{R}$. To see why, suppose that her portfolio returns 12% this year and 6% next year. The mean return for the two years is 9%. If Linda starts with $1000, how much does she have at the end of the first year? At the end of the second year? How does this amount compare with what she would have if both years had the mean return, 9%? Over 20 years, there may be a large difference between the ordinary mean $\bar{R}$ and the *geometric mean*, which reflects the fact that returns in successive years multiply rather than add.

5.31 Concrete blocks and mortar. You are building a wall from precast concrete blocks. Standard 8-inch blocks are $7\frac{5}{8}$ inches high to allow for a $\frac{3}{8}$-inch layer of mortar under each row of blocks. In practice, the height of a block-plus-mortar row varies according to a Normal distribution with mean 8 inches and standard deviation 0.1 inch. Heights of successive rows are independent. Your wall has four rows of blocks. What is the distribution of the height of the wall? What is the probability that the height differs from the design height of 32 inches by more than half an inch?

5.2 Sampling Distributions for Counts and Proportions

LOOK BACK
categorical variable p. 2

In the previous section, we discussed the probability distribution of the sample mean, which meant a focus on population values that were quantitative. We will now shift our focus to population values that are categorical. Counts and proportions are discrete statistics that describe categorical data. We focus our discussion on the simplest case of a random variable with only two possible categories. Here is an example.

EXAMPLE

5.10 Parents put too much pressure on their children. A sample survey asks 2000 college students whether they think that parents put too much pressure on their children. We would like to view the responses of these students as representative of a larger population of students who hold similar beliefs. That is, we will view the responses of the sampled students as an **SRS** from a population.

When there are only two possible outcomes for a random variable, we can summarize the results by giving the count for one of the possible outcomes. We let n represent the sample size and we use X to represent the random variable that gives the count for the outcome of interest.

EXAMPLE

5.11 The random variable of interest. In our sample survey of college students, $n = 2000$. We will ask each student in our sample whether he or she feels parents put too much pressure on their children. The variable X is the number of students who think that parents do put too much pressure on their children. Suppose we observe $X = 840$.

In our example, we chose the random variable X to be the number of students who think that parents put too much pressure on their children. We could have chosen X to be the number of students who do not think that parents put too much pressure on their children. The choice is yours. Often we make the choice based on how we would like to describe the results in a written summary. Which choice do you prefer in this example?

When a random variable has only two possible outcomes, we can also use **sample proportion** the **sample proportion**, $\hat{p} = X/n$, as a summary.

EXAMPLE

5.12 The sample proportion. The sample proportion of students surveyed who think that parents put too much pressure on their children is

$$\hat{p} = \frac{840}{2000} = 0.42$$

Notice that this summary takes into account the sample size n. We need to know n in order to properly interpret the meaning of the random variable X. For example, the conclusion we would draw about student opinions in our survey would be quite different if we had observed $X = 840$ from a sample half as small, $n = 1000$.

USE YOUR KNOWLEDGE

5.32 Seniors who have taken a statistics course. In a random sample of 250 senior students from your college, 45% reported that they had taken a statistics course. Give n, X, and $\hat{p}$ for this setting.

5.33 Use of the Internet to find a place to live. A poll of 1500 college students asked whether or not they have used the Internet to find a place to live sometime within the past year. There were 825 students who answered "Yes"; the other 675 answered "No."

(a) What is n?

(b) Choose one of the two possible outcomes to define the random variable, X. Give a reason for your choice.

(c) What is the value of X?

(d) Find the sample proportion, $\hat{p}$.

Just like the sample mean, sample counts and sample proportions are commonly used statistics and understanding their sampling distributions is important for statistical inference. These statistics, however, are discrete random variables and thus introduce us to a new family of probability distributions.

The binomial distributions for sample counts

The distribution of a count X depends on how the data are produced. Here is a simple but common situation.

THE BINOMIAL SETTING

1. There is a fixed number of observations n.
2. The n observations are all independent.
3. Each observation falls into one of just two categories, which for convenience we call "success" and "failure."
4. The probability of a success, call it p, is the same for each observation.

Think of tossing a coin n times as an example of the binomial setting. Each toss gives either heads or tails and the outcomes of successive tosses are independent. If we call heads a success, then p is the probability of a head and remains the same as long as we toss the same coin. The number of heads we count is a random variable X. The distribution of X, and more generally the distribution of the count of successes in any binomial setting, is completely determined by the number of observations n and the success probability p.

BINOMIAL DISTRIBUTIONS

The distribution of the count X of successes in the binomial setting is called the **binomial distribution** with parameters n and p. The parameter n is the number of observations, and p is the probability of a success on any one observation. The possible values of X are the whole numbers from 0 to n. As an abbreviation, we say that X is $B(n, p)$.

The binomial distributions are an important class of discrete probability distributions. Later in this section we will learn how to assign probabilities to outcomes and how to find the mean and standard deviation of binomial distributions. That said, *the most important skill for using binomial distributions is the ability to recognize situations to which they do and don't apply.* This can be done by checking all the facets of the Binomial setting.

EXAMPLE

5.13 Two binomial examples. (a) Genetics says that children receive genes from their parents independently. Each child of a particular pair of parents has probability 0.25 of having type O blood. If these parents have 3 children, the number who have type O blood is the count X of successes in 3 independent trials with probability 0.25 of a success on each trial. So X has the $B(3, 0.25)$ distribution.

(b) Engineers define reliability as the probability that an item will perform its function under specific conditions for a specific period of time. Replacement heart valves made of animal tissue, for example, have probability 0.77 of performing well for 15 years.[8] The probability of failure is therefore 0.23. It is reasonable to assume that valves in different patients fail (or not) independently of each other. The number of patients in a group of 500 who will need another valve replacement within 15 years has the $B(500, 0.23)$ distribution.

USE YOUR KNOWLEDGE

5.34 Genetics and blood types. Genetics says that children receive genes from their parents independently. Suppose each child of a particular pair of parents has probability 0.25 of having type O blood. If these parents have 4 children, what is the distribution of the number who have type O blood? Explain your answer.

5.35 Toss a coin. Toss a fair coin 15 times. Give the distribution of X, and the number of heads that you observe.

Binomial distributions in statistical sampling

The binomial distributions are important in statistics when we wish to make inferences about the proportion p of "successes" in a population. Here is a typical example.

EXAMPLE

5.14 Audits of financial records. The financial records of businesses may be audited by state tax authorities to test compliance with tax laws. It is too time-consuming to examine all sales and purchases made by a company during the period covered by the audit. Suppose the auditor examines an SRS of 150 sales records out of 10,000 available. One issue is whether each sale was correctly classified as subject to state sales tax or not. Suppose that 800 of the 10,000 sales are incorrectly classified. Is the count X of misclassified records in the sample a binomial random variable?

Choosing an SRS from a population is not quite a binomial setting. Removing one record in Example 5.14 changes the proportion of bad records in the remaining population, so the state of the second record chosen is not independent of the first. Because the population is large, however, removing a few items has a very small effect on the composition of the remaining population. Successive inspection results are very nearly independent. The population proportion of misclassified records is

$$p = \frac{800}{10,000} = 0.08$$

If the first record chosen is bad, the proportion of bad records remaining is $799/9999 = 0.079908$. If the first record is good, the proportion of bad records left is $800/9999 = 0.080008$. These proportions are so close to 0.08 that for practical purposes we can act as if removing one record has no effect on the proportion of misclassified records remaining. We act as if the count X of misclassified sales records in the audit sample has the binomial distribution $B(150, 0.08)$.

Populations like the one described in Example 5.14 often contain a relatively small number of items with very large values. For this example, these values would be very large sale amounts and likely represent an important group of items to the auditor. An SRS taken from such a population will likely include very few items of this type. Therefore, it is common to use a stratified sample in settings like this. Strata are defined based on dollar value of the sale, and within each stratum, an SRS is taken. The results are then combined to obtain an estimate for the entire population.

◄ **LOOK BACK**
stratified sample p. 193

SAMPLING DISTRIBUTION OF A COUNT

A population contains proportion p of successes. If the population is much larger than the sample, the count X of successes in an SRS of size n has approximately the binomial distribution $B(n, p)$.

The accuracy of this approximation improves as the size of the population increases relative to the size of the sample. As a rule of thumb, we will use the binomial sampling distribution for counts when the population is at least 20 times as large as the sample.

Finding binomial probabilities

We will later give a formula for the probability that a binomial random variable takes any of its values. In practice, you will rarely have to use this formula for calculations. Some calculators and most statistical software packages calculate binomial probabilities.

EXAMPLE

5.15 The probability of exactly 10 misclassified sales records. In the audit setting of Example 5.14, what is the probability that the audit finds exactly 10 misclassified sales records? What is the probability that the audit finds no more than 10 misclassified records? Figure 5.7 shows the output from one

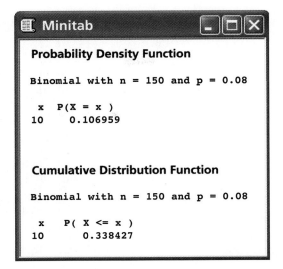

FIGURE 5.7 Binomial probabilities: output from the Minitab statistical software, for Example 5.15.

statistical software system. You see that if the count X has the $B(150, 0.08)$ distribution,

$$P(X = 10) = 0.106959$$
$$P(X \le 10) = 0.338427$$

It was easy to request these calculations in the software's menus. For the TI 83/84 calculator, the functions binompdf and binomcdf would be used. Typically, the output supplies more decimal places than we need and uses labels that may not be helpful (for example, "Probability Density Function" when the distribution is discrete, not continuous). But, as usual with software, we can ignore distractions and find the results we need.

If you do not have suitable computing facilities, you can still shorten the work of calculating binomial probabilities for some values of n and p by looking up probabilities in Table C in the back of this book. The entries in the table are the probabilities $P(X = k)$ of individual outcomes for a binomial random variable X.

EXAMPLE

5.16 The probability histogram. Suppose that the audit in Example 5.14 chose just 15 sales records. What is the probability that no more than 1 of the 15 is misclassified? The count X of misclassified records in the sample has approximately the $B(15, 0.08)$ distribution. Figure 5.8 is a probability histogram for this distribution. The distribution is strongly skewed. Although X can take any whole-number value from 0 to 15, the probabilities of values larger than 5 are so small that they do not appear in the histogram.

We want to calculate

$$P(X \le 1) = P(X = 0) + P(X = 1)$$

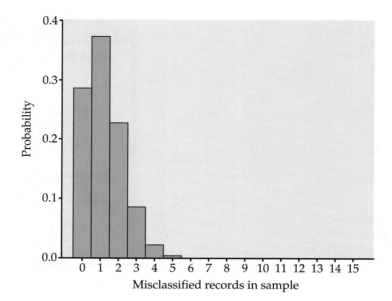

FIGURE 5.8 Probability histogram for the binomial distribution with $n = 15$ and $p = 0.08$, for Example 5.16.

		p
n	k	**.08**
15	0	.2863
	1	.3734
	2	.2273
	3	.0857
	4	.0223
	5	.0043
	6	.0006
	7	.0001
	8	
	9	

when X has the $B(15, 0.08)$ distribution. To use Table C for this calculation, look opposite $n = 15$ and under $p = 0.08$. This part of the table appears at the left. The entry opposite each k is $P(X = k)$. Blank entries are 0 to four decimal places, so we have omitted most of them here. You see that

$$P(X \leq 1) = P(X = 0) + P(X = 1)$$
$$= 0.2863 + 0.3734 = 0.6597$$

About two-thirds of all samples will contain no more than 1 bad record. In fact, almost 29% of the samples will contain no bad records. The sample of size 15 cannot be trusted to provide adequate evidence about misclassified sales records. A larger number of observations is needed.

The values of p that appear in Table C are all 0.5 or smaller. When the probability of a success is greater than 0.5, restate the problem in terms of the number of failures. The probability of a failure is less than 0.5 when the probability of a success exceeds 0.5. When using the table, always stop to ask whether you must count successes or failures.

EXAMPLE

5.17 Poor-quality sleepers. In the survey of 1125 college students described in Example 5.1, 65% of the respondents were classified as poor-quality sleepers. You randomly sample 12 students in your dormitory and classify 10 of them as poor-quality sleepers. Given the results in Example 5.1, is this an unusually high number of students?

To answer this question, assume that the student classifications are independent with the probability of a poor-quality outcome of 0.65. This independence assumption may not be reasonable if the students study and socialize together. We'll assume this is not an issue here. Because the probability of being classified as a poor-quality sleeper is greater than 0.5, we count those students classified as a quality sleeper to use Table C. The probability of being a quality sleeper is $1 - 0.65$, or 0.35. The number X of quality sleepers out of 12 students has the $B(12, 0.35)$ distribution.

We want the probability of classifying at most 2 students. This is

$$P(X \leq 2) = P(X = 0) + P(X = 1) + P(X = 2)$$
$$= 0.0057 + 0.0368 + 0.1088 = 0.1513$$

We would expect to classify 10 or more students as poor-quality sleepers about 15% of the time, or roughly three of every twenty surveys. While this seems like a high number of poor-quality sleepers, this outcome is well within the range of the usual chance variation due to random sampling.

USE YOUR KNOWLEDGE

5.36 Free throw shooting. Katie is a basketball player who makes 80% of her free throws. In a recent game, she had 10 free throws and missed 6 of them. How unusual is this outcome? Using software, calculator, or Table C, compute $P(X \leq 4)$, where X is the number of free throws made in 10 shots.

5.37 Find the probabilities.

(a) Suppose X has the $B(5, 0.4)$ distribution. Use software, calculator, or Table C to find $P(X = 0)$ and $P(X \geq 3)$.

(b) Suppose X has the $B(5, 0.6)$ distribution. Use software, calculator, or Table C to find $P(X = 5)$ and $P(X \leq 2)$.

(c) Explain the relationship between your answers to parts (a) and (b) of this exercise.

Binomial mean and standard deviation

If a count X has the $B(n, p)$ distribution, what are the mean μ_X and the standard deviation σ_X? We can guess the mean. If we expect 65% of the students to be classified as poor-quality sleepers, the mean number in 12 students should be 65% of 12, or 7.8. That's μ_X when X has the $B(12, 0.65)$ distribution. Intuition suggests more generally that the mean of the $B(n, p)$ distribution should be np. Can we show that this is correct and also obtain a short formula for the standard deviation? Because binomial distributions are discrete probability distributions, we could find the mean and variance by using the definitions in Section 4.4. Here is an easier way.

◀ **LOOK BACK**
means and variances of random
variables p. 259

A binomial random variable X is the count of successes in n independent observations that each have the same probability p of success. Let the random variable S_i indicate whether the ith observation is a success or failure by taking the values $S_i = 1$ if a success occurs and $S_i = 0$ if the outcome is a failure. The S_i are independent because the observations are, and each S_i has the same simple distribution:

Outcome	1	0
Probability	p	$1 - p$

← **LOOK BACK**
mean and variance of a
discrete random variable
pp. 261 and 269

From the definition of the mean of a discrete random variable, we know that the mean of each S_i is

$$\mu_S = (1)(p) + (0)(1-p) = p$$

Similarly, the definition of the variance shows that $\sigma_S^2 = p(1-p)$. Because each S_i is 1 for a success and 0 for a failure, to find the total number of successes X we add the S_i's:

$$X = S_1 + S_2 + \cdots + S_n$$

Apply the addition rules for means and variances to this sum. To find the mean of X we add the means of the S_i's:

$$\mu_X = \mu_{S_1} + \mu_{S_2} + \cdots + \mu_{S_n}$$
$$= n\mu_S = np$$

Similarly, the variance is n times the variance of a single S, so that $\sigma_X^2 = np(1-p)$. The standard deviation σ_X is the square root of the variance. Here is the result.

BINOMIAL MEAN AND STANDARD DEVIATION

If a count X has the binomial distribution $B(n, p)$, then

$$\mu_X = np$$
$$\sigma_X = \sqrt{np(1-p)}$$

EXAMPLE

5.18   The Helsinki Heart Study. The Helsinki Heart Study asked whether the anticholesterol drug gemfibrozil reduces heart attacks. In planning such an experiment, the researchers must be confident that the sample sizes are large enough to enable them to observe enough heart attacks. The Helsinki study planned to give gemfibrozil to about 2000 men aged 40 to 55 and a placebo to another 2000. The probability of a heart attack during the five-year period of the study for men this age is about 0.04. What are the mean and standard deviation of the number of heart attacks that will be observed in one group if the treatment does not change this probability?

There are 2000 independent observations, each having probability $p = 0.04$ of a heart attack. The count X of heart attacks has the $B(2000, 0.04)$ distribution, so that

$$\mu_X = np = (2000)(0.04) = 80$$
$$\sigma_X = \sqrt{np(1-p)} = \sqrt{(2000)(0.04)(0.96)} = 8.76$$

The expected number of heart attacks is large enough to permit conclusions about the effectiveness of the drug. In fact, there were 84 heart attacks among the 2035 men actually assigned to the placebo, quite close to the mean. The gemfibrozil group of 2046 men suffered only 56 heart attacks.

This is evidence that the drug reduces the chance of a heart attack. In a later chapter we will learn how to determine if this is strong enough evidence to conclude the drug is effective.

Sample proportions

What proportion of a company's sales records have an incorrect sales tax classification? What percent of adults favor stronger laws restricting firearms? In statistical sampling we often want to estimate the **proportion** p of "successes" in a population. Our estimator is the sample proportion of successes:

$$\hat{p} = \frac{\text{count of successes in sample}}{\text{size of sample}}$$

$$= \frac{X}{n}$$

Be sure to distinguish between the proportion $\hat{p}$ and the count X. The count takes whole-number values between 0 and n, but a proportion is always a number between 0 and 1. In the binomial setting, the count X has a binomial distribution. The proportion $\hat{p}$ does *not* have a binomial distribution. We can, however, do probability calculations about $\hat{p}$ by restating them in terms of the count X and using binomial methods. In Example 5.8 (page 306) we took a similar approach for the sum, restating the problem in terms of the sample mean and then using the Normal distribution to calculate the probability.

EXAMPLE

5.19 Buying clothes online. A recent survey by the Consumer Reports National Research Center revealed that 85% of all respondents were very or completely satisfied with their online clothes-shopping experience.[9] It was also reported, however, that people over the age of 40 were generally more satisfied than younger respondents. You decide to take a nationwide random sample of 2500 college students and ask if they agree or disagree that "I am very or completely satisfied with my online clothes-shopping experience." Suppose that 60% of all college students would agree if asked this question. What is the probability that the sample proportion who agree is at least 58%? The count X who agree has the binomial distribution $B(2500, 0.6)$. The sample proportion $\hat{p} = X/2500$ does *not* have a binomial distribution, because it is not a count. We can translate any question about a sample proportion $\hat{p}$ into a question about the count X. Because 58% of 2500 is 1450,

$$P(\hat{p} \geq 0.58) = P(X \geq 1450)$$
$$= P(X = 1450) + P(X = 1451) + \cdots + P(X = 2500)$$

This is a rather elaborate calculation. We must add more than 1000 binomial probabilities. Software tells us that $P(\hat{p} \geq 0.58) = 0.9802$. Because some software packages cannot handle an n as large as 2500, we need another way to do this calculation.

LOOK BACK
rules for means, p. 267
rules for variances, p. 271

As a first step, find the mean and standard deviation of a sample proportion. We know the mean and standard deviation of a sample count, so apply the rules from Section 4.4 for the mean and variance of a constant times a random variable. Here is the result.

MEAN AND STANDARD DEVIATION OF A SAMPLE PROPORTION

Let $\hat{p}$ be the sample proportion of successes in an SRS of size n drawn from a large population having population proportion p of successes. The mean and standard deviation of $\hat{p}$ are

$$\mu_{\hat{p}} = p$$

$$\sigma_{\hat{p}} = \sqrt{\frac{p(1-p)}{n}}$$

The formula for $\sigma_{\hat{p}}$ is exactly correct in the binomial setting. It is approximately correct for an SRS from a large population. We will use it when the population is at least 20 times as large as the sample.

EXAMPLE

5.20 The mean and the standard deviation. The mean and standard deviation of the proportion of the survey respondents in Example 5.19 who are satisfied with their online clothes-shopping experience are

$$\mu_{\hat{p}} = p = 0.6$$

$$\sigma_{\hat{p}} = \sqrt{\frac{p(1-p)}{n}} = \sqrt{\frac{(0.6)(0.4)}{2500}} = 0.0098$$

USE YOUR KNOWLEDGE

5.38 Find the mean and the standard deviation. If we toss a fair coin 100 times the number of heads is a random variable that is binomial.

(a) Find the mean and the standard deviation of the sample proportion.

(b) Is your answer to part (a) the same as the mean and the standard deviation of the sample count? Explain your answer.

LOOK BACK
unbiased statistic p. 207

The fact that the mean of $\hat{p}$ is p states in statistical language that the sample proportion $\hat{p}$ in an SRS is an *unbiased estimator* of the population proportion p. When a sample is drawn from a new population having a different value of the population proportion p, the sampling distribution of the unbiased estimator $\hat{p}$ changes so that its mean moves to the new value of p. We observed this fact empirically in Section 3.4 and have now verified it from the laws of probability.

The variability of $\hat{p}$ about its mean, as described by the variance or standard deviation, gets smaller as the sample size increases. So a sample proportion from a large sample will usually lie quite close to the population proportion p. We observed this in the simulation experiment on page 203 in Section 3.3. Now we have discovered exactly how the variability decreases: the standard deviation is $\sqrt{p(1-p)/n}$. Similar to what we observed in the previous section, the $\sqrt{n}$ in the denominator means that the sample size must be multiplied by 4 if we wish to divide the standard deviation in half.

Normal approximation for counts and proportions

Using simulation, we discovered in Section 3.4 that the sampling distribution of a sample proportion $\hat{p}$ is close to Normal. Now we know that the distribution of $\hat{p}$ is that of a binomial count divided by the sample size n. This seems at first to be a contradiction. To clear up the matter, look at Figure 5.9. This is a probability histogram of the exact distribution of the proportion of frustrated shoppers $\hat{p}$, based on the binomial distribution $B(2500, 0.6)$. There are hundreds of narrow bars, one for each of the 2501 possible values of $\hat{p}$. Most have probabilities too small to show in a graph. *The probability histogram looks very Normal!* In fact, both the count X and the sample proportion $\hat{p}$ are approximately Normal in large samples.

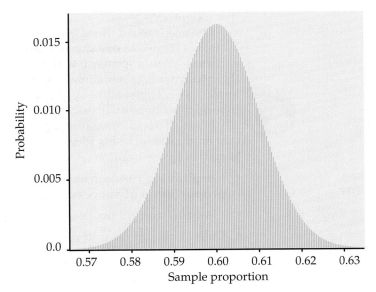

FIGURE 5.9 Probability histogram of the sample proportion $\hat{p}$ based on a binomial count with $n = 2500$ and $p = 0.6$. The distribution is very close to Normal.

LOOK BACK
central limit theorem p. 303

We also know this to be true as a result of the central limit theorem discussed in the previous section. Recall that we can consider the count X as a sum

$$X = S_1 + S_2 + \cdots + S_n$$

of independent random variables S_i that take the value 1 if a success occurs on the ith trial and the value 0 otherwise. The proportion of successes $\hat{p} = X/n$ can then be thought of as the sample mean of the S_i and, like all sample means, is approximately Normal when n is large. Given that $\hat{p}$ is approximately Normal, the count will also be approximately Normal since it is just a constant n times $\hat{p}$, an approximately Normal random variable.

NORMAL APPROXIMATION FOR COUNTS AND PROPORTIONS

Draw an SRS of size n from a large population having population proportion p of successes. Let X be the count of successes in the sample and $\hat{p} = X/n$ be the sample proportion of successes. When n is large, the sampling distributions of these statistics are approximately Normal:

$$X \text{ is approximately } N\left(np, \sqrt{np(1-p)}\right)$$

$$\hat{p} \text{ is approximately } N\left(p, \sqrt{\frac{p(1-p)}{n}}\right)$$

As a rule of thumb, we will use this approximation for values of n and p that satisfy $np \geq 10$ and $n(1-p) \geq 10$.

These Normal approximations are easy to remember because they say that $\hat{p}$ and X are Normal, with their usual means and standard deviations. Whether or not you use the Normal approximations should depend on how accurate your calculations need to be. For most statistical purposes great accuracy is not required. Our "rule of thumb" for use of the Normal approximations reflects this judgment.

The accuracy of the Normal approximations improves as the sample size n increases. They are most accurate for any fixed n when p is close to 1/2, and least accurate when p is near 0 or 1. You can compare binomial distributions with their Normal approximations by using the *Normal Approximation to Binomial* applet. This applet allows you to change n or p while watching the effect on the binomial probability histogram and the Normal curve that approximates it.

Figure 5.10 summarizes the distribution of a sample proportion in a form that emphasizes the big idea of a sampling distribution. Just as with Figure 5.5, keep this figure in mind as you move toward statistical inference.

- **Keep taking random samples of size n from a population that contains proportion p of successes.**

- **Find the sample proportion $\hat{p}$ for each sample.**

- **Collect all the $\hat{p}$'s and display their distribution.**

That's the sampling distribution of $\hat{p}$.

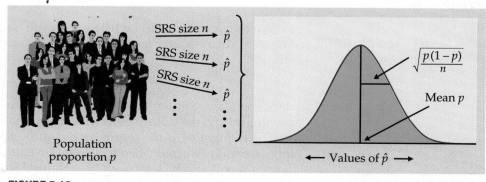

FIGURE 5.10 The sampling distribution of a sample proportion $\hat{p}$ is approximately Normal with mean p and standard deviation $\sqrt{p(1-p)/n}$.

EXAMPLE

5.21   Compare the Normal approximation with the exact calculation. Let's compare the Normal approximation for the calculation of Example 5.19 with the exact calculation from software. We want to calculate $P(\hat{p} \geq 0.58)$ when the sample size is $n = 2500$ and the population proportion is $p = 0.6$. Example 5.20 shows that

$$\mu_{\hat{p}} = p = 0.6$$

$$\sigma_{\hat{p}} = \sqrt{\frac{p(1-p)}{n}} = 0.0098$$

Act as if $\hat{p}$ were Normal with mean 0.6 and standard deviation 0.0098. The approximate probability, as illustrated in Figure 5.11, is

$$P(\hat{p} \geq 0.58) = P\left(\frac{\hat{p} - 0.6}{0.0098} \geq \frac{0.58 - 0.6}{0.0098}\right)$$

$$\doteq P(Z \geq -2.04) = 0.9793$$

That is, about 98% of all samples have a sample proportion that is at least 0.58. Because the sample was large, this Normal approximation is quite accurate. It misses the software value 0.9802 by only 0.0009.

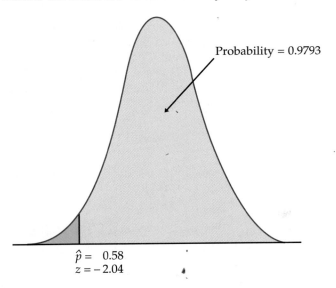

Probability = 0.9793

$\hat{p} = 0.58$
$z = -2.04$

FIGURE 5.11 The Normal probability calculation for Example 5.21.

EXAMPLE

5.22   Using the Normal approximation. The audit described in Example 5.14 examined an SRS of 150 sales records for compliance with sales tax laws. In fact, 8% of all the company's sales records have an incorrect sales tax classification. The count X of bad records in the sample has approximately the $B(150, 0.08)$ distribution.

According to the Normal approximation to the binomial distributions, the count X is approximately Normal with mean and standard deviation

$$\mu_X = np = (150)(0.08) = 12$$

$$\sigma_X = \sqrt{np(1-p)} = \sqrt{(150)(0.08)(0.92)} = 3.3226$$

The Normal approximation for the probability of no more than 10 misclassified records is the area to the left of $X = 10$ under the Normal curve. Using Table A,

$$P(X \leq 10) = P\left(\frac{X - 12}{3.3226} \leq \frac{10 - 12}{3.3226}\right)$$
$$\doteq P(Z \leq -0.60) = 0.2743$$

Software tells us that the actual binomial probability that no more than 10 of the records in the sample are misclassified is $P(X \leq 10) = 0.3384$. The Normal approximation is only roughly accurate. Because $np = 12$, this combination of n and p is close to the border of the values for which we are willing to use the approximation.

The distribution of the count of bad records in a sample of 15 is distinctly non-Normal, as Figure 5.8 showed. When we increase the sample size to 150, however, the shape of the binomial distribution becomes roughly Normal. Figure 5.12 displays the probability histogram of the binomial distribution with the density curve of the approximating Normal distribution superimposed. Both distributions have the same mean and standard deviation, and both the area under the histogram and the area under the curve are 1. The Normal curve fits the histogram reasonably well. Look closely: the histogram is slightly skewed to the right, a property that the symmetric Normal curve can't match.

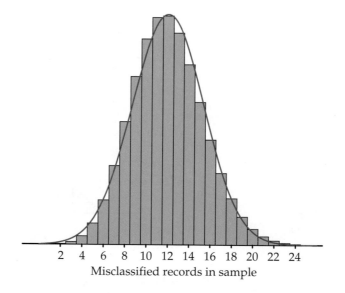

FIGURE 5.12 Probability histogram and Normal approximation for the binomial distribution with $n = 150$ and $p = 0.08$, for Example 5.22.

Misclassified records in sample

USE YOUR KNOWLEDGE

5.39 **Use the Normal approximation.** Suppose we toss a fair coin 100 times. Use the Normal approximation to find the probability that the sample proportion is

(a) between 0.3 and 0.7.

(b) between 0.35 and 0.65.

The continuity correction*

Figure 5.13 illustrates an idea that greatly improves the accuracy of the Normal approximation to binomial probabilities. The binomial probability $P(X \leq 10)$ is the area of the histogram bars for values 0 to 10. The bar for $X = 10$ actually extends from 9.5 to 10.5. Because the discrete binomial distribution puts probability only on whole numbers, the probabilities $P(X \leq 10)$ and $P(X \leq 10.5)$ are the same. The Normal distribution spreads probability continuously, so these two Normal probabilities are different. The Normal approximation is more accurate if we consider $X = 10$ to extend from 9.5 to 10.5, matching the bar in the probability histogram.

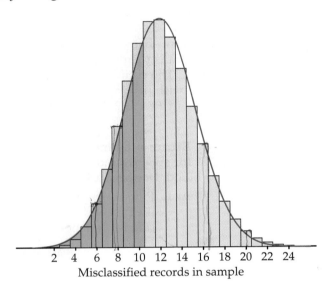

FIGURE 5.13 Area under the Normal approximation curve for the probability in Example 5.22.

Misclassified records in sample

The event $\{X \leq 10\}$ includes the outcome $X = 10$. Figure 5.13 shades the area under the Normal curve that matches all the histogram bars for outcomes 0 to 10, bounded on the right not by 10, but by 10.5. So $P(X \leq 10)$ is calculated as $P(X \leq 10.5)$. On the other hand, $P(X < 10)$ excludes the outcome $X = 10$, so we exclude the entire interval from 9.5 to 10.5 and calculate $P(X \leq 9.5)$ from the Normal table. Here is the result of the Normal calculation in Example 5.22 improved in this way:

$$P(X \leq 10) = P(X \leq 10.5)$$
$$= P\left(\frac{X - 12}{3.3226} \leq \frac{10.5 - 12}{3.3226}\right)$$
$$\doteq P(Z \leq -0.45) = 0.3264$$

The improved approximation misses the binomial probability by only 0.012. Acting as though a whole number occupies the interval from 0.5 below to **continuity correction**   0.5 above the number is called the **continuity correction** to the Normal approximation. If you need accurate values for binomial probabilities, try to use software to do exact calculations. If no software is available, use the continuity correction unless n is very large. Because most statistical purposes do not

---

*This material can be omitted if desired.

require extremely accurate probability calculations, we do not emphasize use of the continuity correction.

Binomial formula*

We can find a formula for the probability that a binomial random variable takes any value by adding probabilities for the different ways of getting exactly that many successes in n observations. Here is the example we will use to show the idea.

EXAMPLE

5.23 Blood types of children. Each child born to a particular set of parents has probability 0.25 of having blood type O. If these parents have 5 children, what is the probability that exactly 2 of them have type O blood?

The count of children with type O blood is a binomial random variable X with $n = 5$ tries and probability $p = 0.25$ of a success on each try. We want $P(X = 2)$.

Because the method doesn't depend on the specific example, we will use "S" for success and "F" for failure. In Example 5.23, "S" would stand for type O blood. Do the work in two steps.

Step 1. Find the probability that a specific 2 of the 5 tries give successes, say the first and the third. This is the outcome SFSFF. The multiplication rule for independent events tells us that

$$P(\text{SFSFF}) = P(S)P(F)P(S)P(F)P(F)$$
$$= (0.25)(0.75)(0.25)(0.75)(0.75)$$
$$= (0.25)^2(0.75)^3$$

Step 2. Observe that the probability of *any one* arrangement of 2 S's and 3 F's has this same probability. That's true because we multiply together 0.25 twice and 0.75 three times whenever we have 2 S's and 3 F's. The probability that $X = 2$ is the probability of getting 2 S's and 3 F's in any arrangement whatsoever. Here are all the possible arrangements:

SSFFF	SFSFF	SFFSF	SFFFS	FSSFF
FSFSF	FSFFS	FFSSF	FFSFS	FFFSS

There are 10 of them, all with the same probability. The overall probability of 2 successes is therefore

$$P(X = 2) = 10(0.25)^2(0.75)^3 = 0.2637$$

The pattern of this calculation works for any binomial probability. To use it, we need to be able to count the number of arrangements of k successes in n observations without actually listing them. We use the following fact to do the counting.

---

*The formula for binomial probabilities is useful in many settings, but we will not need it in our study of statistical inference. This section can therefore be omitted if desired.

BINOMIAL COEFFICIENT

The number of ways of arranging k successes among n observations is given by the **binomial coefficient**

$$\binom{n}{k} = \frac{n!}{k!\,(n-k)!}$$

for $k = 0, 1, 2, \ldots, n$.

---

factorial

   The formula for binomial coefficients uses the **factorial** notation. The factorial $n!$ for any positive whole number n is

$$n! = n \times (n-1) \times (n-2) \times \cdots \times 3 \times 2 \times 1$$

Also, $0! = 1$. Notice that the larger of the two factorials in the denominator of a binomial coefficient will cancel much of the $n!$ in the numerator. For example, the binomial coefficient we need for Example 5.23 is

$$\binom{5}{2} = \frac{5!}{2!\,3!}$$
$$= \frac{(5)(4)(3)(2)(1)}{(2)(1) \times (3)(2)(1)}$$
$$= \frac{(5)(4)}{(2)(1)} = \frac{20}{2} = 10$$

This agrees with our previous calculation.

   The notation $\binom{n}{k}$ *is not related to the fraction* $\frac{n}{k}$. A helpful way to remember its meaning is to read it as "binomial coefficient n choose k." Binomial coefficients have many uses in mathematics, but we are interested in them only as an aid to finding binomial probabilities. The binomial coefficient $\binom{n}{k}$ counts the number of ways in which k successes can be distributed among n observations. The binomial probability $P(X = k)$ is this count multiplied by the probability of any specific arrangement of the k successes. Here is the formula we seek.

BINOMIAL PROBABILITY

If X has the binomial distribution $B(n, p)$ with n observations and probability p of success on each observation, the possible values of X are $0, 1, 2, \ldots, n$. If k is any one of these values, the **binomial probability** is

$$P(X = k) = \binom{n}{k} p^k (1-p)^{n-k}$$

Here is an example of the use of the binomial probability formula.

EXAMPLE

5.24 Using the binomial probability formula. The number X of misclassified sales records in the auditor's sample in Example 5.16 has the $B(15, 0.08)$ distribution. The probability of finding no more than 1 misclassified record is

$$P(X \leq 1) = P(X = 0) + P(X = 1)$$

$$= \binom{15}{0}(0.08)^0(0.92)^{15} + \binom{15}{1}(0.08)^1(0.92)^{14}$$

$$= \frac{15!}{0! \, 15!}(1)(0.2863) + \frac{15!}{1! \, 14!}(0.08)(0.3112)$$

$$= (1)(1)(0.2863) + (15)(0.08)(0.3112)$$

$$= 0.2863 + 0.3734 = 0.6597$$

The calculation used the facts that $0! = 1$ and that $a^0 = 1$ for any number $a \neq 0$. The result agrees with that obtained from Table C in Example 5.16.

USE YOUR KNOWLEDGE

5.40 A coin is slightly bent, and as a result the probability of a head is 0.53. Suppose that you toss the coin four times.

(a) Use the binomial formula to find the probability of 3 or more heads.

(b) Compare your answer with the one that you would obtain if the coin were fair.

BEYOND THE BASICS

Weibull distributions

Our discussion of sampling distributions has concentrated on the Normal model for quantitative variables and the binomial model for count data. These models are important in statistical practice, but simplicity also contributes to their popularity. The parameters p in the binomial model and μ in the Normal model are easy to understand. To estimate them from data we use statistics $\hat{p}$ and $\bar{x}$ that are also easy to understand and that have simple sampling distributions.

There are many other probability distributions that are used to model data in various circumstances. The time that a product, such as a computer hard drive, lasts before failing rarely has a Normal distribution. Earlier we discussed the use of the exponential distribution to model time to failure. Another class of continuous distributions, the **Weibull distributions,** is more commonly used in these situations.

Weibull distributions

EXAMPLE

5.25 Weibull density curves. Figure 5.14 shows the density curves of three members of the Weibull family. Each describes a different type of distribution for the time to failure of a product.

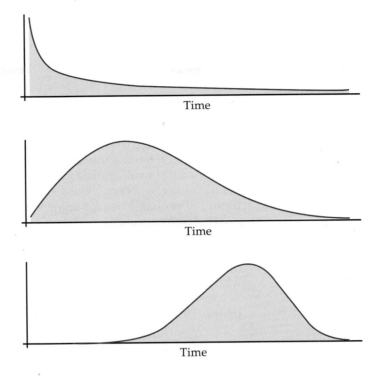

FIGURE 5.14 Density curves for three members of the Weibull family of distributions, for Example 5.25.

1. The top curve in Figure 5.14 is a model for *infant mortality*. This describes products that often fail immediately, prior to delivery to the customer. However, if the product does not fail right away, it will likely last a long time. For products like this, a manufacturer might test them and ship only the ones that do not fail immediately.

2. The middle curve in Figure 5.14 is a model for *early failure*. These products do not fail immediately, but many fail early in their lives after they are in the hands of customers. This is disastrous—the product or the process that makes it must be changed at once.

3. The bottom curve in Figure 5.14 is a model for *old-age wearout*. Most of these products fail only when they begin to wear out, and then many fail at about the same age.

A manufacturer certainly wants to know to which of these classes a new product belongs. To find out, engineers operate a random sample of products until they fail. From the failure time data we can estimate the parameter (called the "shape parameter") that distinguishes among the three Weibull distributions in Figure 5.14. The shape parameter has no simple definition like that of a population proportion or mean, and it cannot be estimated by a simple statistic such as $\hat{p}$ or $\bar{x}$.

Two things save the situation. First, statistical theory provides general approaches for finding good estimates of any parameter. These general methods not only tell us how to use $\hat{p}$ and $\bar{x}$ in the binomial and Normal settings but also tell us how to estimate the Weibull shape parameter. Second, modern software can calculate the estimate from data even though there is no algebraic formula that we can write for the estimate. Statistical practice often relies on both mathematical theory and methods of computation more elaborate than the ones we will meet in this book. Fortunately, big ideas such as sampling distributions carry over to more complicated situations.[10]

SECTION 5.2 Summary

A **count** X of successes has the **binomial distribution** $B(n, p)$ in the **binomial setting:** there are n trials, all independent, each resulting in a success or a failure, and each having the same probability p of a success.

Binomial probabilities are most easily found by software. There is an exact formula that is practical for calculations when n is small. Table C contains binomial probabilities for some values of n and p. For large n, you can use the Normal approximation.

The binomial distribution $B(n, p)$ is a good approximation to the **sampling distribution of the count of successes** in an SRS of size n from a large population containing proportion p of successes. We will use this approximation when the population is at least 20 times larger than the sample.

The mean and standard deviation of a **binomial count** X and a **sample proportion** of successes $\hat{p} = X/n$ are

$$\mu_X = np \qquad\qquad \mu_{\hat{p}} = p$$

$$\sigma_X = \sqrt{np(1-p)} \qquad\qquad \sigma_{\hat{p}} = \sqrt{\frac{p(1-p)}{n}}$$

The sample proportion $\hat{p}$ is therefore an unbiased estimator of the population proportion p.

The **Normal approximation** to the binomial distribution says that if X is a count having the $B(n, p)$ distribution, then when n is large,

$$X \text{ is approximately } N(np, \sqrt{np(1-p)})$$

$$\hat{p} \text{ is approximately } N\left(p, \sqrt{\frac{p(1-p)}{n}}\right)$$

We will use these approximations when $np \geq 10$ and $n(1-p) \geq 10$. The **continuity correction** improves the accuracy of the Normal approximations.

The exact **binomial probability formula** is

$$P(X = k) = \binom{n}{k} p^k (1-p)^{n-k}$$

where the possible values of X are $k = 0, 1, \ldots, n$. The binomial probability formula uses the **binomial coefficient**

$$\binom{n}{k} = \frac{n!}{k!\,(n-k)!}$$

Here the **factorial** $n!$ is

$$n! = n \times (n-1) \times (n-2) \times \cdots \times 3 \times 2 \times 1$$

for positive whole numbers n and $0! = 1$. The binomial coefficient counts the number of ways of distributing k successes among n trials.

SECTION 5.2 Exercises

For Exercises 5.32 and 5.33, see pages 313–314; for Exercises 5.34 and 5.35, see page 315; for Exercises 5.36 and 5.37, see page 319; for Exercise 5.38, see page 322; for Exercise 5.39, see page 326; and for Exercise 5.40, see page 330.

Most binomial probability calculations required in these exercises can be done by using Table C or the Normal approximation. Your instructor may request that you use the binomial probability formula or software. In exercises requiring the Normal approximation, you should use the continuity correction if you studied that topic.

5.41 What is wrong? Explain what is wrong in each of the following scenarios.

(a) If you toss a fair coin three times and heads appears each time, then the next toss is more likely to be a tail than a head.

(b) If you toss a fair coin three times and heads appears each time, then the next toss is more likely to be a head than a tail.

(c) $\hat{p}$ is one of the parameters for a binomial distribution.

5.42 What is wrong? Explain what is wrong in each of the following scenarios.

(a) In the binomial setting, X is a proportion.

(b) The variance for a binomial count is $\sqrt{p(1-p)/n}$.

(c) The Normal approximation to the binomial distribution is always accurate when n is greater than 1000.

5.43 Should you use the binomial distribution? In each of the following situations, is it reasonable to use a binomial distribution for the random variable X? Give reasons for your answer in each case. If a binomial distribution applies, give the values of n and p.

(a) A poll of 200 college students asks whether or not you are usually irritable in the morning. X is the number who reply that they are usually irritable in the morning.

(b) You toss a fair coin until a head appears. X is the count of the number of tosses that you make.

(c) Most calls made at random by sample surveys don't succeed in talking with a live person. Of calls to New York City, only one-twelfth succeed. A survey calls 500 randomly selected numbers in New York City. X is the number of times that a live person is reached.

(d) You deal 10 cards from a shuffled deck and count the number X of black cards.

5.44 Should you use the binomial distribution? In each of the following situations, is it reasonable to use a binomial distribution for the random variable X? Give reasons for your answer in each case.

(a) In a random sample of students in a fitness study, X is the mean systolic blood pressure of the sample.

(b) A manufacturer of running shoes picks a random sample of the production of shoes each day for a detailed inspection. Today's sample of 20 pairs of shoes includes one pair with a defect.

(c) A nutrition study chooses an SRS of college students. They are asked whether or not they usually eat at least five servings of fruits or vegetables per day. X is the number who say that they do.

5.45 Typographic errors. Typographic errors in a text are either nonword errors (as when "the" is typed as "teh") or word errors that result in a real but incorrect word. Spell-checking software will catch nonword errors but not word errors. Human proofreaders catch 70% of word errors. You ask a fellow student to proofread an essay in which you have deliberately made 10 word errors.

(a) If the student matches the usual 70% rate, what is the distribution of the number of errors caught? What is the distribution of the number of errors missed?

(b) Missing 4 or more out of 10 errors seems a poor performance. What is the probability that a proofreader who catches 70% of word errors misses 4 or more out of 10?

5.46 Streaming online music. A recent survey of 1000 United Kingdom music fans aged 14 to 64, revealed that roughly 30% of the teenage music fans are listening to streamed music on their computer everyday.[11] You decide to interview a random sample of 20 U.S. teenage music fans. For now assume they behave similarly to U.K. teenagers.

(a) What is the distribution of the number who listen to streamed music daily? Explain your answer.

(b) What is the probability that at least 8 of the 20 listen to streamed music daily?

5.47 Typographic errors. Return to the proofreading setting of Exercise 5.45.

(a) What is the mean number of errors caught? What is the mean number of errors missed? You see that these two means must add to 10, the total number of errors.

(b) What is the standard deviation σ of the number of errors caught?

(c) Suppose that a proofreader catches 90% of word errors, so that $p = 0.9$. What is σ in this case? What is σ if $p = 0.99$? What happens to the standard deviation of a binomial distribution as the probability of a success gets close to 1?

5.48 Streaming online music, continued. Recall Exercise 5.46. Suppose that only 25% of the U.S. teenage music fans listen to streamed music daily.

(a) If you interview 20 at random, what is the mean of the count X who listen to streamed music daily? What is the mean of the proportion $\hat{p}$ in your sample who listen to streamed music daily?

(b) Repeat the calculations in part (a) for samples of size 200 and 2000. What happens to the mean count of successes as the sample size increases? What happens to the mean proportion of successes?

5.49 🏅 **Typographic errors.** In the proofreading setting of Exercise 5.45, what is the smallest number of misses m with $P(X \geq m)$ no larger than 0.05? You might consider m or more misses as evidence that a proofreader actually catches fewer than 70% of word errors.

5.50 Attitudes toward drinking and behavior studies. Some of the methods in this section are approximations rather than exact probability results. We have given rules of thumb for safe use of these approximations.

(a) You are interested in attitudes toward drinking among the 75 members of a fraternity. You choose 30 members at random to interview. One question is, "Have you had five or more drinks at one time during the last week?" Suppose that in fact 30% of the 75 members would say "Yes." Explain why you *cannot* safely use the $B(30, 0.3)$ distribution for the count X in your sample who say "Yes."

(b) The National AIDS Behavioral Surveys found that 0.2% (that's 0.002 as a decimal fraction) of adult heterosexuals had both received a blood transfusion and had a sexual partner from a group at high risk of AIDS. Suppose that this national proportion holds for your region. Explain why you *cannot* safely use the Normal approximation for the sample proportion who fall in this group when you interview an SRS of 1000 adults.

5.51 Random digits. Each entry in a table of random digits like Table B has probability 0.1 of being a 0, and digits are independent of each other.

(a) What is the probability that a group of five digits from the table will contain at least one 5?

(b) What is the mean number of 5s in lines 40 digits long?

5.52 🅰 **Use the *Probability* applet.** The *Probability* applet simulates tosses of a coin. You can choose the number of tosses n and the probability p of a head. You can therefore use the applet to simulate binomial random variables.

The count of misclassified sales records in Example 5.15 (page 316) has the binomial distribution with $n = 15$ and $p = 0.08$. Set these values for the number of tosses and probability of heads in the applet. Table C shows that the probability of getting a sample with exactly 0 misclassified records is 0.2863. This is the long-run proportion of samples with no bad records. Click "Toss" and "Reset" repeatedly to simulate 25 samples. Record the number of bad records (the count of heads) in each of the 25 samples. What proportion of the 25 samples had exactly 0 bad records? Remember that probability tells us only what happens in the long run.

5.53 Inheritance of blood types. Children inherit their blood type from their parents, with probabilities that reflect the parents' genetic makeup. Children of Juan and Maria each have probability 1/4 of having blood type A and inherit independently of each other. Juan and Maria plan to have 4 children; let X be the number who have blood type A.

(a) What are n and p in the binomial distribution of X?

(b) Find the probability of each possible value of X, and draw a probability histogram for this distribution.

(c) Find the mean number of children with type A blood, and mark the location of the mean on your probability histogram.

5.54 The ideal number of children. "What do you think is the ideal number of children for a family to have?" A Gallup Poll asked this question of 1007 randomly chosen adults. Over half (52%) thought two children was ideal.[12] Suppose that $p = 0.52$ is exactly true for the population of all adults. Gallup announced a margin of error of ±3 percentage points for this poll. What is the probability that the sample proportion $\hat{p}$ for an SRS of size $n = 1007$ falls between 0.49 and 0.55? You see that it is likely, but not certain, that polls like this give results that are correct within their margin of error. We will say more about margins of error in Chapter 6.

5.55 Visiting a casino and betting on college sports. A Gallup Poll finds that 24% of adults visited a casino in the past 12 months, and that 4% bet on college sports.[13] These results come from a random sample of 1027 adults. For an SRS of size $n = 1027$:

(a) What is the probability that the sample proportion $\hat{p}$ is between 0.22 and 0.26 if the population proportion is $p = 0.24$?

EXAMPLE

6.7 How many applicants should we survey. Suppose that we are planning a credit card use survey similar to the one described in Example 6.4. If we want the margin of error to be $150 with 95% confidence, what sample size n do we need? For 95% confidence, Table D gives $z^* = 1.960$. For σ we will use the value from the previous study, $3500. If the margin of error is $150, we have

$$n = \left(\frac{z^*\sigma}{m}\right)^2 = \left(\frac{1.96 \times 3500}{150}\right)^2 = 2091.54$$

Because 2091 measurements will give a slightly wider interval than desired and 2092 measurements a slightly narrower interval, we could choose $n = 2092$. We need loan information from 2092 borrowers to determine an estimate of mean debt with the desired margin of error.

It is always safe to round *up* to the next higher whole number when finding n because this will give us a smaller margin of error. The purpose of this calculation is to determine a sample size that is sufficient to provide useful results, but the determination of what is useful is a matter of judgment. Would we need a much larger sample size to obtain a margin of error of $100? Here is the calculation:

$$n = \left(\frac{z^*\sigma}{m}\right)^2 = \left(\frac{1.96 \times 3500}{100}\right)^2 = 4705.96$$

A sample of $n = 4706$ is much larger, and the costs of such a large sample may be prohibitive.

Unfortunately, the actual number of usable observations is often less than what we plan at the beginning of a study. This is particularly true of data collected in surveys but is an important consideration in most studies. Careful study designers often assume a nonresponse rate or dropout rate that specifies what proportion of the originally planned sample will fail to provide data. We use this information to calculate the sample size to be used at the start of the study. For example, if in the preceding survey, we expect only 25% of those contacted to respond, we would need to start with a sample size of $4 \times 2092 = 8368$ to obtain usable information from 2092 borrowers.

USE YOUR KNOWLEDGE

6.7 Starting salaries. You are planning a survey of starting salaries for recent computer science majors. In the latest survey by the National Association of Colleges and Employers, the average starting salary was reported to be $61,467.[3] Assuming the standard deviation is $12,000, what sample size do you need to have a margin of error equal to $1000 with 95% confidence?

6.8 Changes in sample size. Suppose that in the setting of the previous exercise you have the resources to contact 1000 recent graduates. If all respond, will your margin of error be larger or smaller than $1000? What if only 50% respond? Verify your answers by performing the calculations.

Some cautions

We have already seen that small margins of error and high confidence can require large numbers of observations. You should also be keenly aware that *any formula for inference is correct only in specific circumstances.* If the government required statistical procedures to carry warning labels like those on drugs, most inference methods would have long labels. Our handy formula $\bar{x} \pm z^*\sigma/\sqrt{n}$ for estimating a population mean comes with the following list of warnings for the user:

- The data should be an SRS from the population. We are completely safe if we actually did a randomization and drew an SRS. We are not in great danger if the data can plausibly be thought of as independent observations from a population. That is the case in Examples 6.4 to 6.7, where we redefine our population to correspond to survey respondents.

- The formula is not correct for probability sampling designs more complex than an SRS. Correct methods for other designs are available. We will not discuss confidence intervals based on multistage or stratified samples. If you plan such samples, be sure that you (or your statistical consultant) know how to carry out the inference you desire.

- There is no correct method for inference from data haphazardly collected with bias of unknown size. Fancy formulas cannot rescue badly produced data.

- Because $\bar{x}$ is not a resistant measure, outliers can have a large effect on the confidence interval. You should search for outliers and try to correct them or justify their removal before computing the interval. If the outliers cannot be removed, ask your statistical consultant about procedures that are not sensitive to outliers.

  LOOK BACK
  resistant measure p. 45

- If the sample size is small and the population is not Normal, the true confidence level will be different from the value C used in computing the interval. Examine your data carefully for skewness and other signs of non-Normality. The interval relies only on the distribution of $\bar{x}$, which even for quite small sample sizes is much closer to Normal than that of the individual observations. When $n \geq 15$, the confidence level is not greatly disturbed by non-Normal populations unless extreme outliers or quite strong skewness are present. Our debt data in Example 6.4 are clearly skewed, but because of the large sample size, we are confident that the sample mean will be approximately Normal. We will discuss this issue in more detail in the next chapter.

- This interval $\bar{x} \pm z^*\sigma/\sqrt{n}$ assumes that the standard deviation σ of the population is known. This unrealistic requirement renders the interval of little use in statistical practice. We will learn in the next chapter what to do when σ is unknown. If, however, the sample is large, the sample standard deviation s will be close to the unknown σ. The interval $\bar{x} \pm z^*s/\sqrt{n}$ is then an approximate confidence interval for μ.

  LOOK BACK
  standard deviation p. 39

The most important caution concerning confidence intervals is a consequence of the first of these warnings. *The margin of error in a confidence interval covers only random sampling errors.* The margin of error is obtained from the sampling distribution and indicates how much error can be expected because of chance variation in randomized data production. *Practical difficulties such as undercoverage and nonresponse in a sample survey cause additional errors. These errors can be larger than the random sampling error.* This often happens when the sample size is large (so that $\sigma/\sqrt{n}$ is small). Remember this unpleasant fact when reading the results of an opinion poll or other sample survey. The practical conduct of the survey influences the trustworthiness of its results in ways that are not included in the announced margin of error.

Every inference procedure that we will meet has its own list of warnings. Because many of the warnings are similar to those we have mentioned, we will not print the full warning label each time. It is easy to state (from the mathematics of probability) conditions under which a method of inference is exactly correct. These conditions are *never fully met in practice.* For example, no population is exactly Normal. *Deciding when a statistical procedure should be used in practice often requires judgment assisted by exploratory analysis of the data.* Mathematical facts are therefore only a part of statistics. The difference between statistics and mathematics can be stated thus: mathematical theorems are true; statistical methods are often effective when used with skill.

Finally, you should understand what statistical confidence does not say. Based on our SRS in Example 6.3, we are 95% confident that the mean SATM score for the California students lies between 452 and 470. This says that this interval was calculated by a method that gives correct results in 95% of all possible samples. It does *not* say that the probability is 0.95 that the true mean falls between 452 and 470. No randomness remains after we draw a particular sample and compute the interval. The true mean either is or is not between 452 and 470. The probability calculations of standard statistical inference describe how often the *method,* not a particular sample, gives correct answers.

USE YOUR KNOWLEDGE

6.9 **Self-report survey.** As part of Sallie Mae's recent credit card usage study, a self-report survey was distributed to 5800 undergraduates. Of the 292 returned surveys, only 249 were complete and used in the analysis. Based on these responses, the report states[4]

> Ninety-two percent of undergraduate credit cardholders charged textbooks, school supplies, or other direct education expenses, up from 85 percent when the study was last conducted, in 2004.

The reported margin of error is 4%. Do you think that this small margin of error is a good measure of the accuracy of the survey's results? Explain your answer.

BEYOND THE BASICS

The Bootstrap

Confidence intervals are based on sampling distributions. In this section we have used the fact that the sampling distribution of $\bar{x}$ is $N(\mu, \sigma/\sqrt{n})$ when the data are an SRS from an $N(\mu, \sigma)$ population. If the data are not Normal, the central limit theorem tells us that this sampling distribution is still a reasonable approximation as long as the distribution of the data is not strongly skewed and there are no outliers. Even a fair amount of skewness can be tolerated when the sample size is large.

bootstrap

resample

What if the population does not appear to be Normal and we have only a small sample? Then we do not know what the sampling distribution of $\bar{x}$ looks like. The **bootstrap** is a procedure for approximating sampling distributions when theory cannot tell us their shape.[5]

The basic idea is to act as if our sample were the population. We take many samples from it. Each of these is called a **resample.** We calculate the mean $\bar{x}$ for each resample. We get different results from different resamples because we sample *with replacement.* An individual observation in the original sample can appear more than once in the resample.

For example, suppose we have four measurements of a student's daily time online last month (in minutes):

$$190.5 \quad 109.0 \quad 95.5 \quad 137.0$$

one resample could be

$$109.0 \quad 95.5 \quad 137.0 \quad 109.0$$

with $\bar{x} = 112.625$. Collect the $\bar{x}$'s from 1000 such resamples. Their distribution will be close to what we would get if we took 1000 samples from the entire population. We treat the distribution of $\bar{x}$'s from our 1000 resamples as if it were the sampling distribution. If we want a 95% confidence interval, for example, we use the middle 95% of this distribution.

The bootstrap is practical only when you can use a computer to take 1000 or more samples quickly. It is an example of how the use of fast and easy computing is changing the way we do statistics. More details about the bootstrap can be found in Chapter 16.

SECTION 6.1 Summary

The purpose of a **confidence interval** is to estimate an unknown parameter with an indication of how accurate the estimate is and of how confident we are that the result is correct.

Any confidence interval has two parts: an interval computed from the data and a confidence level. The interval often has the form

$$\text{estimate} \pm \text{margin of error}$$

The **confidence level** states the probability that the method will give a correct answer. That is, if you use 95% confidence intervals, in the long run 95% of your intervals will contain the true parameter value. When you apply the method once, you do not know if your interval gave a correct value (this happens 95% of the time) or not (this happens 5% of the time).

The **margin of error** for a level C confidence interval for the mean μ of a Normal population with known standard deviation σ, based on an SRS of size n, is given by

$$m = z^* \frac{\sigma}{\sqrt{n}}$$

Here z^* is obtained from the row labeled z^* at the bottom of Table D. The probability is C that a standard Normal random variable takes a value between $-z^*$ and z^*. The confidence interval is

$$\bar{x} \pm m$$

Other things being equal, the margin of error of a confidence interval decreases as

- the confidence level C decreases,
- the sample size n increases, and
- the population standard deviation σ decreases.

The sample size n required to obtain a confidence interval of specified margin of error m for a Normal mean is

$$n = \left(\frac{z^*\sigma}{m}\right)^2$$

where z^* is the critical point for the desired level of confidence.

A specific confidence interval recipe is correct only under specific conditions. The most important conditions concern the method used to produce the data. Other factors such as the form of the population distribution may also be important.

SECTION 6.1 Exercises

For Exercises 6.1 to 6.3, see page 346; for Exercise 6.4, see page 347; for Exercises 6.5 and 6.6, see page 351; for Exercises 6.7 and 6.8, see page 353; and for Exercise 6.9, see page 355.

6.10 Margin of error and the confidence interval. A study based on a sample of size 36 reported a mean of 87 with a margin of error of 10 for 95% confidence.

(a) Give the 95% confidence interval.

(b) If you wanted 99% confidence for the same study, would your margin of error be greater than, equal to, or less than 10? Explain your answer.

6.11 Changing the sample size. Suppose that the sample mean is 50 and the standard deviation is assumed to be 7. Make a diagram similar to Figure 6.5 (page 350) that illustrates the effect of sample size on the width of a 95% interval. Use the following sample sizes: 10, 20, 40, and 80. Summarize what the diagram shows.

6.12 Changing the confidence level. A study with 49 observations had a mean of 70. Assume that the standard deviation is 14. Make a diagram similar to Figure 6.6 (page 352) that illustrates the effect of the confidence level on the width of the interval. Use 80%, 90%, 95%, and 99%. Summarize what the diagram shows.

6.13 Confidence interval mistakes and misunderstandings. Suppose 400 randomly selected alumni of the University of Okoboji were asked to rate the university's counseling services on a 1 to 10 scale. The sample mean ($\bar{x}$) was found to be 8.6. Assume that the population standard deviation is known to be $\sigma = 2.0$.

(a) Ima Bitlost computes the 95% confidence interval for the average satisfaction score as $8.6 \pm 1.96(2.0)$. What is her mistake?

(b) After correcting her mistake in part (a), she states "I am 95% confident that the sample mean falls between 8.404 and 8.796." What is wrong with this statement?

(c) She quickly realizes her mistake in part (b) and instead states "The probability the true mean is between 8.404 and 8.796 is 0.95." What misinterpretation is she making now?

(d) Finally, in her defense for using the Normal distribution to determine the confidence coefficient she says "Because the sample size is quite large, the population of alumni ratings will be approximately Normal." Explain to Ima her misunderstanding and correct this statement.

6.14 More confidence interval mistakes and misunderstandings. Suppose 100 randomly selected members of MySpace Karaoke[6] were asked how much time they typically spend on the site during the week. The sample mean ($\bar{x}$) was found to be 4.2 hours. Assume that the population standard deviation is known to be $\sigma = 2.5$.

(a) Cary Oakey computes the 95% confidence interval for the average time on the site as $4.2 \pm 1.96(2.5/100)$. What is his mistake?

(b) He corrects this mistake and then states "95% of the members spend between 3.71 and 4.69 hours a week on the site." What is wrong with his interpretation of this interval?

(c) The margin of error is slightly less than a half hour. To reduce this down to 15 minutes, Gary says the sample size needs to be doubled to 200. What is wrong with this statement?

6.15 Importance of recreational sports. The National Intramural-Recreational Sports Association (NIRSA) performed a study to look at the value of recreational sports on college campuses.[7] A total of 2673 students were asked to indicate how important (on a 10-point scale) each of 21 factors was in terms of their college satisfaction and success. The factor "recreational sports and activities" resulted in a mean score of 7.5. Assume a standard deviation of 3.9.

(a) Give the margin of error and find the 95% confidence interval for this sample.

(b) Repeat these calculations for a 99% confidence interval. How do the results compare with those in part (a)?

6.16 Inference based on integer values. Refer to Exercise 6.15. The data for this study are integer values between 1 and 10. Explain why the confidence interval based on the Normal distribution should be a good approximation.

6.17 Mean serum TRAP in young women. For many important processes that occur in the body, direct measurement of characteristics of the process is not possible. In many cases, however, we can measure a *biomarker,* a biochemical substance that is relatively easy to measure and is associated with the process of interest. Bone turnover is the net effect of two processes: the breaking down of old bone, called resorption, and the building of new bone, called formation. One biochemical measure of bone resorption is tartrate resistant acid phosphatase (TRAP), which can be measured in blood. In a study of bone turnover in young women, serum TRAP was measured in 31 subjects.[8] The units are units per liter (U/l). The mean was 13.2 U/l. Assume that the standard deviation is known to be 6.5 U/l. Give the margin of error and find a 95% confidence interval for the mean for young women represented by this sample.

6.18 Mean OC in young women. Refer to the previous exercise. A biomarker for bone formation measured in the same study was osteocalcin (OC), measured in the blood. The units are nanograms per milliliter (ng/ml). For the 31 subjects in the study the mean was 33.4 ng/ml. Assume that the standard deviation is known to be 19.6 ng/ml. Report the 95% confidence interval.

6.19 Populations sampled and margins of error. Consider the following two scenarios. (A) Take a simple random sample of 100 sophomore students at your college

or university. (B) Take a simple random sample of 100 sophomore students in your major at your college or university. For each of these samples you will record the amount spent on textbooks used for classes during the fall semester. Which sample should have the smaller margin of error? Explain your answer.

6.20 🛡️ **Average starting salary.** The National Association of Colleges and Employers (NACE) Fall Salary Survey report shows that the current class of college graduates earned an average starting salary offer of $48,633.[9] Your institution collected an SRS ($n = 2500$) of its recent graduates and obtained a 95% confidence interval of ($45,330, $46,156). This interval is below the NACE average. Using this information, compute the 99% confidence interval of the *difference* between the average starting salary from recent graduates at your institution and the overall NACE mean.

6.21 Consumption of sugar-sweetened beverages. A recent study estimated the U.S. per capita consumption of sugar-sweetened beverages among adults 20 to 44 years of age to be 289 kcal/day with a standard deviation of the mean equal to 7 kcal/day.[10]

(a) The 68–95–99.7 rule says that the probability is about 0.95 that $\bar{x}$ is within _____ kcal/day of the population mean μ. Fill in the blank.

(b) About 95% of all samples will capture the true mean of kcals consumed per day in the interval $\bar{x}$ plus or minus _____ kcal/day. Fill in the blank.

6.22 Apartment rental rates. You want to rent an unfurnished one-bedroom apartment in Dallas next year. The mean monthly rent for a random sample of 10 apartments advertised in the local newspaper is $980. Assume the monthly rents in Dallas follow a Normal distribution with a standard deviation of $290. Find a 95% confidence interval for the mean monthly rent for unfurnished one-bedroom apartments available for rent in this community.

6.23 More on apartment rental rates. Refer to the previous exercise. Will the 95% confidence interval include approximately 95% of the rents of all unfurnished one-bedroom apartments in this area? Explain why or why not.

6.24 🛡️ **Inference based on skewed data.** The mean OC for the 31 subjects in Exercise 6.18 was 33.4 ng/ml. In our calculations, we assumed that the standard deviation was known to be 19.6 ng/ml. Use the 68–95–99.7 rule from Chapter 1 (page 56) to find the approximate bounds on the values of OC that would include these percents of the population. If the assumed standard deviation is correct, it would appear that this distribution may be highly skewed. Why? (*Hint:* The measured values for a variable such as this are all positive.) Do you think that this skewness will invalidate the use of the Normal confidence interval in this case? Explain your answer.

6.25 Average hours per week on the Internet. The *Student Monitor* surveys 1200 undergraduates from 100 colleges semiannually to understand trends among college students.[11] Recently, the *Student Monitor* reported that the average amount of time spent per week on the Internet was 19.0 hours. Assume that the standard deviation is 5.5 hours.

(a) Give a 95% confidence interval for the mean time spent per week on the Internet.

(b) Is it true that 95% of the students surveyed reported weekly times that lie in the interval you found in part (a)? Explain your answer.

6.26 Average minutes per week on the Internet. Refer to the previous exercise.

(a) Give the mean and standard deviation in minutes.

(b) Calculate the 95% confidence interval in minutes from your answer to part (a).

(c) Explain how you could have directly calculated this interval from the 95% interval that you calculated in the previous exercise.

6.27 Satisfied with your job? A Gallup Poll asked working adults about their job satisfaction.[12] One question was "How satisfied or dissatisfied are you with your job?" The possible answers were "completely satisfied," "somewhat satisfied," "somewhat dissatisfied," and "completely dissatisfied." Ninety percent responded that they were somewhat or completely satisfied. Material provided with the results of the poll noted:

> *Results are based on telephone interviews with 1,009 national adults, aged 18 and older, conducted Aug. 7–10, 2008. For results based on the total sample of national adults, one can say with 95% confidence that the maximum margin of sampling error is ±3 percentage points. For results based on the sample of 557 adults employed full or part-time, the maximum margin of sampling error is ±5 percentage points.*

The Gallup Poll uses a complex multistage sample design, but the sample percent has approximately a Normal sampling distribution.

(a) The announced poll result was 90% ± 5%. Can we be certain that the true population percent falls in this interval? Explain your answer.

(b) Explain to someone who knows no statistics what the announced result 90% ± 5% means.

(c) This confidence interval has the same form we have met earlier:

$$\text{estimate} \pm z^* \sigma_{\text{estimate}}$$

What is the standard deviation σ_{estimate} of the estimated percent?

(d) Does the announced margin of error include errors due to practical problems such as undercoverage and nonresponse?

6.28 Fuel efficiency. Computers in some vehicles calculate various quantities related to performance. One of these is the fuel efficiency, or gas mileage, usually expressed as miles per gallon (mpg). For one vehicle equipped in this way, the mpg were recorded each time the gas tank was filled, and the computer was then reset.[13] Here are the mpg values for a random sample of 20 of these records: ![icon] GASMILEAGE

| 41.5 | 50.7 | 36.6 | 37.3 | 34.2 | 45.0 | 48.0 | 43.2 | 47.7 | 42.2 |
| 43.2 | 44.6 | 48.4 | 46.4 | 46.8 | 39.2 | 37.3 | 43.5 | 44.3 | 43.3 |

Suppose that the standard deviation is known to be $\sigma = 3.5$ mpg.

(a) What is $\sigma_{\bar{x}}$, the standard deviation of $\bar{x}$?

(b) Examine the data for skewness and other signs of non-Normality. Show your plots and numerical summaries. Do you think it is reasonable to construct a confidence interval based on the Normal distribution? Explain your answer.

(c) Give a 95% confidence interval for μ, the mean mpg for this vehicle.

6.29 Fuel efficiency in metric units. In the previous exercise you found an estimate with a margin of error for the average miles per gallon. Convert your estimate and margin of error to the metric units kilometers per liter (kpl). To change mpg to kpl, use the fact that 1 mile = 1.609 kilometers and 1 gallon = 3.785 liters.

6.30 ![applet icon] **Percent coverage of 95% confidence interval.** The *Confidence Interval* applet lets you simulate large numbers of confidence intervals quickly. Select 95% confidence and then sample 50 intervals. Record the number of intervals that cover the true value (this appears in the "Hit" box in the applet). Press the reset button and repeat 30 times. Make a stemplot of the results and find the mean. Describe the results. If you repeated this experiment very many times, what would you expect the average number of hits to be?

6.31 Required sample size for specified margin of error. A new bone study is being planned that will measure the biomarker TRAP described in Exercise 6.17. Using the value of σ given there, 6.5 U/l, find the sample size required to provide an estimate of the mean TRAP with a margin of error of 1.5 U/l for 95% confidence.

6.32 ⚠ Adjusting required sample size for drop out. Refer to the previous exercise. In similar previous studies, about 20% of the subjects drop out before the study is completed. Adjust your sample size requirement to have enough subjects at the end of the study to meet the margin of error criterion.

6.33 Radio poll. A college radio station invites listeners to enter a dispute about a proposed "pay as you throw" waste collection program. The station asks listeners to call in and state how much each bag of trash should cost. A total of 633 listeners call in. The station calculates the 95% confidence interval for the average fee desired by city residents to be $0.83 to $1.28. Is this result trustworthy? Explain your answer.

6.34 Accuracy of a laboratory scale. To assess the accuracy of a laboratory scale, a standard weight known to weigh 10 grams is weighed repeatedly. The scale readings are Normally distributed with unknown mean (this mean is 10 grams if the scale has no bias). The standard deviation of the scale readings is known to be 0.0002 gram.

(a) The weight is measured five times. The mean result is 10.0023 grams. Give a 98% confidence interval for the mean of repeated measurements of the weight.

(b) How many measurements must be averaged to get a margin of error of ± 0.0001 with 98% confidence?

6.35 ⚠ More than one confidence interval. As we prepare to take a sample and compute a 95% confidence interval, we know that the probability that the interval we compute will cover the parameter is 0.95. That's the meaning of 95% confidence. If we use several such intervals, however, our confidence that *all* of them give correct results is less than 95%. Suppose we take independent samples each month for five months and report a 95% confidence interval for each set of data.

(a) What is the probability that all five intervals cover the true means? This probability (expressed as a percent) is our overall confidence level for the five simultaneous statements.

(b) What is the probability that at least four of the five intervals cover the true means?

6.2 Tests of Significance

The confidence interval is appropriate when our goal is to estimate population parameters. The second common type of inference is directed at a quite different goal: to assess the evidence provided by the data in favor of some claim about the population parameters.

The reasoning of significance tests

A significance test is a formal procedure for comparing observed data with a hypothesis whose truth we want to assess. The hypothesis is a statement about the population parameters. The results of a test are expressed in terms of a probability that measures how well the data and the hypothesis agree. We use the following examples to illustrate these concepts.

EXAMPLE

6.8 Credit card debt by U.S. region. One purpose of Sallie Mae's credit card usage studies described in Example 6.4 (page 349) is to compare the debt of different subgroups of student applicants. For example, the average debt of student applicants in the Midwest is $3260, while those student applicants in the West had an average debt of $3817. The difference of $557 is fairly large, but we know that these numbers are estimates of the true means. If we took different samples, we would get different estimates. Can we conclude from these data that the average credit card debts of student applicants in these two regions of the United States are different?

One way to answer this question is to compute the probability of obtaining a difference as large or larger than the observed $557 assuming that, in fact, there is no difference in the true means. This probability is 0.14. Because this probability is not particularly small, we conclude that observing a difference of $557 is not very surprising when the true means are equal. The data do not provide evidence for us to conclude that the credit card debts for student applicants in the Midwest and West are different.

Here is an example with a different conclusion.

EXAMPLE

6.9 Credit card debt by undergraduate grade level. This study also reports that the average debt among senior applicants is $4138, while it is $2912 among juniors. Is the average debt among senior applicants higher than the average debt among junior applicants? The observed difference is $1226 but as we learned in the previous example, an observed difference in means is not necessarily sufficient for us to conclude that the true means are different. Do the data provide evidence that there is a difference in debt? Again, we answer this question with a probability calculated under the assumption that there is *no difference in the true means*. The probability is 0.0000004 of observing a difference of mean debt that is $1226 or more when there really is no difference. Because this probability is so small, we have sufficient evidence in the data to conclude that senior applicants have a higher average credit card debt.

What are the key steps in these examples?

- We started each with a question about the difference between two mean debts. In Example 6.8, we compare student applicants in the West and Midwest. In Example 6.9, we compare junior applicants with senior applicants. In both cases, we ask whether or not the data are compatible with no difference, that is, a difference of $0.

- Next we compared the data, $557 in the first case and $1226 in the second, with the value that comes from the question, $0.

- The results of the comparisons are probabilities, 0.14 in the first case and 0.0000004 in the second.

The 0.14 probability is not particularly small, so we have limited evidence to question the possibility that the true difference is zero. In the second case, however, the probability is very small. Something that happens with probability 0.0000004 occurs only about 4 times out of 10,000,000. In this case we have two possible explanations:

1. We have observed something that is very unusual, or

2. The assumption that underlies the calculation, no difference in mean debt, is not true.

Because this probability is so small, we prefer the second conclusion: the mean debt of senior applicants is higher than the debt of junior applicants.

The probabilities in Examples 6.8 and 6.9 are measures of the compatibility of the data (a difference in means of $557 and $1226) with the *null hypothesis* that there is no difference in the true means. Figures 6.7 and 6.8 compare the two results graphically. For each a Normal curve centered at 0 is the sampling distribution. You can see that we are not particularly surprised to observe the difference $557 in Figure 6.7, but the difference $1226 in Figure 6.8 is clearly an unusual observation. We will now consider some of the formal aspects of significance testing.

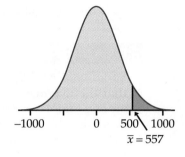

FIGURE 6.7 Comparison of the sample mean in Example 6.8 relative to the null hypothesized value 0.

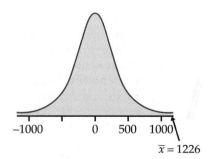

FIGURE 6.8 Comparison of the sample mean in Example 6.9 relative to the null hypothesized value 0.

Stating hypotheses

In Examples 6.8 and 6.9, we asked whether the difference in the observed means is reasonable if, in fact, there is no difference in the true means. To answer this, we begin by supposing that the statement following the "if" in the previous sentence is true. In other words, we suppose that the true difference is $0. We then ask whether the data provide evidence against the supposition we have made. If so, we have evidence in favor of an effect (the means are different) we are seeking. The first step in a test of significance is to state a claim that we will try to find evidence *against*.

NULL HYPOTHESIS

The statement being tested in a test of significance is called the **null hypothesis.** The test of significance is designed to assess the strength of the evidence against the null hypothesis. Usually the null hypothesis is a statement of "no effect" or "no difference in the true means."

We abbreviate "null hypothesis" as H_0. A null hypothesis is a statement about the population parameters. For example, our null hypothesis for Example 6.8 is

$$H_0: \text{there is no difference in the true means}$$

Note that the null hypothesis refers to the *true* means for all borrowers from either the Midwest or West regions, including those for whom we do not have data.

alternative hypothesis It is convenient also to give a name to the statement we hope or suspect is true instead of H_0. This is called the **alternative hypothesis** and is abbreviated as H_a. In Example 6.8, the alternative hypothesis states that the means are different. We write this as

$$H_a: \text{the true means are not the same}$$

Hypotheses always refer to some populations or a model, not to a particular outcome. For this reason, we must state H_0 and H_a in terms of population parameters.

Because H_a expresses the effect that we hope to find evidence *for,* we often begin with H_a and then set up H_0 as the statement that the hoped-for effect is not present. Stating H_a is often the more difficult task. It is not always clear,

one-sided or two-sided alternatives in particular, whether H_a should be **one-sided** or **two-sided,** which refers to whether a parameter differs from its null hypothesis value in a specific direction or in either direction.

The alternative hypothesis should express the hopes or suspicions we bring to the data. *It is cheating to first look at the data and then frame H_a to fit what the data show.* If you do not have a specific direction firmly in mind in advance, you must use a two-sided alternative. Moreover, some users of statistics argue that we should always use a two-sided alternative.

USE YOUR KNOWLEDGE

6.36 Food court survey. The food court closest to your dormitory has been redesigned. A survey is planned to determine whether or not students think that the new design is an improvement. Sampled students will respond on a seven-point scale with scores less than 4 favoring the old design and scores greater than 4 favoring the new design (to varying degrees). State the null and alternative hypotheses you would use for examining whether or not the new design is viewed as an improvement.

6.37 DXA scanners. A dual-energy X-ray absorptiometry (DXA) scanner is used to measure bone mineral density for people who may be at risk for osteoporosis. To ensure its accuracy, the company uses an object called a "phantom" that has known mineral density $\mu = 1.4$ grams per square centimeter. Once installed, the company scans the phantom 10 times and compares the sample mean reading $\bar{x}$ with the theoretical mean μ using a significance test. State the null and alternative hypotheses for this test.

Test statistics

We will learn the form of significance tests in a number of common situations. Here are some principles that apply to most tests and that help in understanding these tests:

- The test is based on a statistic that estimates the parameter that appears in the hypotheses. Usually this is the same estimate we would use in a confidence interval for the parameter. When H_0 is true, we expect the estimate to take a value near the parameter value specified by H_0.

- Values of the estimate far from the parameter value specified by H_0 give evidence against H_0. The alternative hypothesis determines which directions count against H_0.

- To assess how far the estimate is from the parameter, standardize the estimate. In many common situations the test statistic has the form

$$z = \frac{\text{estimate} - \text{hypothesized value}}{\text{standard deviation of the estimate}}$$

Let's return to our comparison of credit card debt among students in different regions of the United States.

EXAMPLE

6.10 Credit card debt of undergraduate applicants in two regions of the United States: the hypotheses. In Example 6.8, the hypotheses are stated in terms of the difference in debt between undergraduate applicants in the Midwest and West:

H_0: there is no difference in the true means

H_a: there is a difference in the true means

Because H_a is two-sided, large values of both positive and negative differences count as evidence against the null hypothesis.

---

test statistic   A **test statistic** measures compatibility between the null hypothesis and the data. We use it for the probability calculation that we need for our test of significance. It is a random variable with a distribution that we know.

EXAMPLE

6.11 Credit card debt of undergraduate applicants in two regions of the United States: the test statistic. In Example 6.8, we can state the null hypothesis as H_0: the true mean difference is 0. The estimate of the difference is $557. Using methods that we will discuss in detail later, we can determine that the standard deviation of the estimate is $374. For this problem the test statistic is

$$z = \frac{\text{estimate} - \text{hypothesized value}}{\text{standard deviation of the estimate}}$$

For our data,

$$z = \frac{557 - 0}{374} = 1.49$$

LOOK BACK
Normal distribution p. 54

We have observed a sample estimate that is about one and a half standard deviations away from the hypothesized value of the parameter. Because the sample sizes are sufficiently large for us to conclude that the distribution of the sample estimate is approximately Normal, the standardized test statistic z will have approximately the $N(0, 1)$ distribution.

We will use facts about the Normal distribution in what follows.

P-values

If all test statistics were Normal, we could base our conclusions on the value of the z test statistic. In fact, the Supreme Court of the United States has said that "two or three standard deviations" ($z = 2$ or 3) is its criterion for rejecting H_0 (see Exercise 6.42 on page 369), and this is the criterion used in most applications involving the law. Because not all test statistics are Normal, we translate the value of test statistics into a common language, the language of probability.

A test of significance finds the probability of getting an outcome *as extreme or more extreme than the actually observed outcome.* "Extreme" means "far from what we would expect if H_0 were true." The direction or directions that count as "far from what we would expect" are determined by H_a and H_0.

In Example 6.8 we want to know if the debt of undergraduate applicants in the Midwest is different from the debt of undergraduate applicants in the West. The difference we calculated based on our sample is \$557, which corresponds to 1.49 standard deviations away from zero—that is, $z = 1.49$. Because we are using a two-sided alternative for this problem, the evidence against H_0 is measured by the probability that we observe a value of Z as extreme or more extreme than 1.49. More formally, this probability is

$$P(Z \le -1.49 \text{ or } Z \ge 1.49)$$

where Z has the standard Normal distribution $N(0, 1)$.

P-VALUE

The probability, assuming H_0 is true, that the test statistic would take a value as extreme or more extreme than that actually observed is called the **P-value** of the test. The smaller the P-value, the stronger the evidence against H_0 provided by the data.

The key to calculating the P-value is the sampling distribution of the test statistic. For the problems we consider in this chapter, we need only the standard Normal distribution for the test statistic z.

EXAMPLE

6.12   Credit card debt of undergraduate applicants in two regions of the United States: the P-value. In Example 6.11 we found that the test statistic for testing

$$H_0\text{: the true mean difference is } 0$$

versus

$$H_a\text{: there is a difference in the true means}$$

is

$$z = \frac{557 - 0}{374} = 1.49$$

If H_0 is true, then z is a single observation from the standard Normal, $N(0, 1)$, distribution. Figure 6.9 illustrates this calculation. The P-value is the probability of observing a value of Z at least as extreme as the one that we observed, $z = 1.49$. From Table A, our table of standard Normal probabilities, we find

$$P(Z \geq 1.49) = 1 - 0.9319 = 0.0681$$

The probability for being extreme in the negative direction is the same:

$$P(Z \leq -1.49) = 0.0681$$

So the P-value is

$$P = 2P(Z \geq 1.49) = 2(0.0681) = 0.1362$$

This is the value that was reported on page 361. There is a 14% chance of observing a difference as extreme as the $557 in our sample if the true population difference is zero. The P-value tells us that our outcome is not particularly extreme, so we conclude that the data do not provide evidence that would cause us to doubt the validity of the null hypothesis.

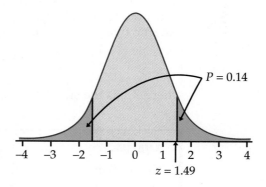

FIGURE 6.9 The P-value for Example 6.12. The P-value is the probability (when H_0 is true) that $\overline{x}$ takes a value as extreme or more extreme than the actual observed value, $z = 1.49$. Because this is a two-sided hypothesis, we use both tails of the distribution.

USE YOUR KNOWLEDGE

6.38 Normal curve and the P-value. A test statistic for a two-sided significance test for a population mean is $z = 2.42$. Sketch a standard Normal curve and mark this value of z on it. Find the P-value and shade the appropriate areas under the curve to illustrate your calculations.

6.39 More on the Normal curve and the *P*-value. A test statistic for a two-sided significance test for a population mean is $z = -1.63$. Sketch a standard Normal curve and mark this value of z on it. Find the *P*-value and shade the appropriate areas under the curve to illustrate your calculations.

Statistical significance

We started our discussion of the reasoning of significance tests with the statement of null and alternative hypotheses. We then learned that a test statistic is the tool used to examine the compatibility of the observed data with the null hypothesis. Finally, we translated the test statistic into a *P*-value to quantify the evidence against H_0. One important final step is needed: to state our conclusion.

We can compare the *P*-value we calculated with a fixed value that we regard as decisive. This amounts to announcing in advance how much evidence against H_0 we will require to reject H_0. The decisive value of P is called the **significance level.** It is commonly denoted by α. If we choose $\alpha = 0.05$, we are requiring that the data give evidence against H_0 so strong that it would happen no more than 5% of the time (1 time in 20) when H_0 is true. If we choose $\alpha = 0.01$, we are insisting on stronger evidence against H_0, evidence so strong that it would appear only 1% of the time (1 time in 100) if H_0 is in fact true.

significance level

STATISTICAL SIGNIFICANCE

If the *P*-value is as small or smaller than α, we say that the data are **statistically significant at level α.**

"Significant" in the statistical sense does not mean "important." The original meaning of the word is "signifying something." In statistics the term is used to indicate only that the evidence against the null hypothesis reached the standard set by α. Significance at level 0.01 is often expressed by the statement "The results were significant ($P < 0.01$)." Here P stands for the *P*-value. The *P*-value is more informative than a statement of significance because we can then assess significance at any level we choose. For example, a result with $P = 0.03$ is significant at the $\alpha = 0.05$ level but is not significant at the $\alpha = 0.01$ level.

A test of significance is a process for assessing the significance of the evidence provided by data against a null hypothesis. The four steps common to all tests of significance are as follows:

1. State the *null hypothesis* H_0 and the *alternative hypothesis* H_a. The test is designed to assess the strength of the evidence against H_0; H_a is the statement that we will accept if the evidence enables us to reject H_0.

2. Calculate the value of the *test statistic* on which the test will be based. This statistic usually measures how far the data are from H_0.

3. Find the *P-value* for the observed data. This is the probability, calculated assuming that H_0 is true, that the test statistic will weigh against H_0 at least as strongly as it does for these data.

4. State a conclusion. One way to do this is to choose a *significance level α*, how much evidence against H_0 you regard as decisive. If the P-value is less than or equal to α, you conclude that the alternative hypothesis is true; if it is greater than α, you conclude that the data do not provide sufficient evidence to reject the null hypothesis. Your conclusion is a sentence that summarizes what you have found by using a test of significance.

We will learn the details of many tests of significance in the following chapters. The proper test statistic is determined by the hypotheses and the data collection design. We use computer software or a calculator to find its numerical value and the P-value. The computer will not formulate your hypotheses for you, however. Nor will it decide if significance testing is appropriate or help you to interpret the P-value that it presents to you. The most difficult and important step is the last one: stating a conclusion.

EXAMPLE

6.13 Credit card debt of undergraduate applicants in two regions of the United States: significance. In Example 6.12 we found that the P-value is 0.1362. There is a 14% chance of observing a difference as extreme as the $557 in our sample if the true population difference is zero. The P-value tells us that our outcome is not particularly extreme. We could report the result as "the data do not provide evidence that would cause us to conclude that there is a difference in credit card debt between student applicants in the Midwest and West ($z = 1.49$, $P = 0.14$)."

If the P-value is small, we reject the null hypothesis. Here is an example.

EXAMPLE

6.14 Credit card debt by undergraduate grade level: significance. In Example 6.9 we found the difference in debt between senior and junior applicants was $1226. Since we would have a prior expectation that the debt would increase over grade levels because of the additional costs of a college education, it is appropriate to use a one-sided alternative in this situation. So, our hypotheses are

$$H_0: \text{the true mean difference is } 0$$

versus

$$H_a: \text{the difference between the mean debt of senior and junior}$$
$$\text{applicants is positive}$$

The standard deviation is $248 (again we defer details regarding this calculation), and the test statistic is

$$z = \frac{\text{estimate} - \text{hypothesized value}}{\text{standard deviation of the estimate}}$$
$$z = \frac{1226 - 0}{248}$$
$$= 4.94$$

Because only positive differences in debt count against the null hypothesis, the one-sided alternative leads to the calculation of the P-value using the upper tail of the Normal distribution. The P-value is

$$P = P(Z \geq 4.94)$$
$$= 0.0000004$$

The calculation is illustrated in Figure 6.10. There is about a 4 in 10,000,000 chance of observing a difference as large or larger than the $1226 in our sample if the true population difference is zero. This P-value tells us that our outcome is extremely rare. We conclude that the null hypothesis must be false. Here is one way to report the result: "The data clearly show that the mean credit card debt for senior applicants is larger than the average credit card debt for junior applicants ($z = 4.94$, $P < 0.001$)."

FIGURE 6.10 The P-value for Example 6.14. The P-value is the probability (when H_0 is true) that $\overline{x}$ takes a value as extreme or more extreme than the actual observed value, $z = 4.94$. We only look at the right tail because we are considering the one-sided (>) alternative.

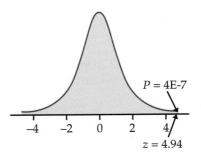

Note that the calculated P-value for this example is 0.0000004 but we reported the result as $P < 0.001$. The value 0.001, 1 in 1000, is sufficiently small to force a clear rejection of H_0. Standard practice is to report very small P-values as simply less than 0.001.

USE YOUR KNOWLEDGE

6.40 Finding significant z-scores. Consider a two-sided significance test for a population mean.

(a) Sketch a Normal curve similar to that shown in Figure 6.9 (page 366) but find the value z such that $P = 0.05$.

(b) Based on the figure from part (a), what values of the z statistic are statistically significant at the $\alpha = 0.05$ level?

6.41 More on finding significant z-scores. Consider a one-sided significance test for a population mean, where the alternative is greater than.

(a) Sketch a Normal curve similar to that shown in Figure 6.10 but find the value z such that $P = 0.05$.

(b) Based on the figure from part (a), what values of the z statistic are statistically significant at the $\alpha = 0.05$ level?

6.42 The Supreme Court speaks. The Supreme Court has said that z-scores beyond $z^* = 2$ or 3 are generally convincing statistical evidence. For a two-sided test, what significance level corresponds to $z^* = 2$? To $z^* = 3$?

Tests for a population mean

Our discussion has focused on the reasoning of statistical tests, and we have outlined the key ideas for one type of procedure. Here is a summary. We want to test the hypothesis that a parameter has a specified value. This is the null hypothesis. For a test of a population mean μ, the null hypothesis is

H_0: the true population mean is equal to μ_0

which often is expressed as

$$H_0: \mu = \mu_0$$

where μ_0 is the specified value of μ that we would like to examine.

The test is based on data summarized as an estimate of the parameter. For a population mean this is the sample mean $\overline{x}$. Our test statistic measures the difference between the sample estimate and the hypothesized parameter in terms of standard deviations of the test statistic:

$$z = \frac{\text{estimate} - \text{hypothesized value}}{\text{standard deviation of the estimate}}$$

LOOK BACK
distribution of sample mean
p. 302

Recall from Chapter 5 that the standard deviation of $\overline{x}$ is $\sigma/\sqrt{n}$. Therefore, the test statistic is

$$z = \frac{\overline{x} - \mu_0}{\sigma/\sqrt{n}}$$

LOOK BACK
central limit theorem p. 303

Again recall from Chapter 5 that, if the population is Normal, then $\overline{x}$ will be Normal and z will have the standard Normal distribution when H_0 is true. By the central limit theorem both distributions will be approximately Normal when the sample size is large even if the population is not Normal.

Suppose we have calculated a test statistic $z = 1.7$. If the alternative is one-sided on the high side, then the P-value is the probability that a standard Normal random variable Z takes a value as large or larger than the observed 1.7. That is,

$$P = P(Z \geq 1.7)$$
$$= 1 - P(Z < 1.7)$$
$$= 1 - 0.9554$$
$$= 0.0446$$

Similar reasoning applies when the alternative hypothesis states that the true μ lies below the hypothesized μ_0 (one-sided). When H_a states that μ is simply unequal to μ_0 (two-sided), values of z away from zero in either direction count against the null hypothesis. The P-value is the probability that a standard Normal Z is at least as far from zero as the observed z. Again, if the test statistic is $z = 1.7$, the two-sided P-value is the probability that $Z \leq -1.7$ or $Z \geq 1.7$. Because the standard Normal distribution is symmetric, we calculate this probability by finding $P(Z \geq 1.7)$ and *doubling* it:

$$P(Z \leq -1.7 \text{ or } Z \geq 1.7) = 2P(Z \geq 1.7)$$
$$= 2(1 - 0.9554) = 0.0892$$

We would make exactly the same calculation if we observed $z = -1.7$. It is the absolute value $|z|$ that matters, not whether z is positive or negative. Here is a statement of the test in general terms.

> **z TEST FOR A POPULATION MEAN**
>
> To test the hypothesis H_0: $\mu = \mu_0$ based on an SRS of size n from a population with unknown mean μ and known standard deviation σ, compute the test statistic
>
> $$z = \frac{\bar{x} - \mu_0}{\sigma/\sqrt{n}}$$
>
> In terms of a standard Normal random variable Z, the P-value for a test of H_0 against
>
> $$H_a: \mu > \mu_0 \text{ is } P(Z \geq z)$$
> $$H_a: \mu < \mu_0 \text{ is } P(Z \leq z)$$
> $$H_a: \mu \neq \mu_0 \text{ is } 2P(Z \geq |z|)$$
>
> These P-values are exact if the population distribution is Normal and are approximately correct for large n in other cases.

EXAMPLE

6.15 Cholesterol level of sedentary female undergraduates. Coronary heart disease (CHD) begins in young adulthood and is the fifth leading cause of death among adults aged 20 to 24 years.[14] Studies of serum cholesterol levels among college students, however, are very limited. A 1999 study looked at a large sample of students from a large southeastern university and reported that the mean serum cholesterol level among women is 168 mg/dl with a standard deviation of 27 mg/dl.[15] A more recent study at a southern university investigated the lipid levels in a cohort of sedentary university students.[16] The mean total cholesterol level among $n = 71$ females was $\bar{x} = 173.7$. Is there evidence that the mean cholesterol level among sedentary students differs from this average over all students?

The null hypothesis is "no difference" from the published mean $\mu_0 = 168$. The alternative is two-sided because the researcher did not have a particular direction in mind before examining the data. So the hypotheses about the unknown mean μ of the sedentary population are

$$H_0: \mu = 168$$
$$H_a: \mu \neq 168$$

As usual in this chapter, we make the unrealistic assumption that the population standard deviation is known. In this case we'll assume it is $\sigma = 27$. The z test requires that the 71 students in the sample are an SRS from the population of all sedentary female students. We check this assumption by asking how the data were produced. In this case, all participants were enrolled in a health class at this university so there may be some concerns about biases. We will press on for now.

We compute the test statistic:

$$z = \frac{\bar{x} - \mu_0}{\sigma/\sqrt{n}} = \frac{173.7 - 168}{27/\sqrt{71}}$$
$$= 1.78$$

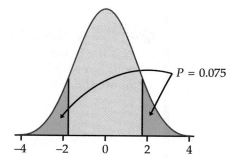

$P = 0.075$

FIGURE 6.11 Sketch of the
P-value calculation for the
two-sided test in Example 6.15.
The test statistic is $z = 1.78$.

Figure 6.11 illustrates the P-value, which is the probability that a standard Normal variable Z takes a value at least 1.78 away from zero. From Table A we find that this probability is

$$P = 2P(Z \geq 1.78) = 2(1 - 0.9625) = 0.075$$

That is, more than 7% of the time an SRS of size 71 from the general undergraduate female population would have a mean cholesterol level at least as far from 168 as that of the sedentary sample. The observed $\bar{x} = 173.7$ is therefore not strong evidence that the sedentary female undergraduate population differs from the general female undergraduate population.

The data in Example 6.15 do *not* establish that the mean cholesterol level μ for the sedentary population is 168. We sought evidence that μ differed from 168 and failed to find convincing evidence. That is all we can say. No doubt the mean cholesterol level of the entire sedentary population is not exactly equal to 168. A large enough sample would give evidence of the difference, even if it is very small. Tests of significance assess the evidence *against* H_0. If the evidence is strong, we can confidently reject H_0 in favor of the alternative. *Failing to find evidence against H_0 means only that the data are consistent with H_0, not that we have clear evidence that H_0 is true.*

EXAMPLE

6.16 Significance test of the mean SATM score. In a discussion of SAT Mathematics (SATM) scores, someone comments: "Because only a minority of California high school students take the test, the scores overestimate the ability of typical high school seniors. I think that if all seniors took the test, the mean score would be no more than 450." You decide to test this claim (H_0) using the SRS of 500 seniors from California from Example 6.3. These students had a mean SATM score of $\bar{x} = 461$. Is this good evidence against this claim? Because the claim states the mean is "no more than 450," the alternative hypothesis is one-sided. The hypotheses are

$$H_0: \mu = 450$$
$$H_a: \mu > 450$$

As we did in the discussion following Example 6.3, we assume that $\sigma = 100$. The z statistic is

$$z = \frac{\bar{x} - \mu_0}{\sigma/\sqrt{n}} = \frac{461 - 450}{100/\sqrt{500}} = 2.46$$

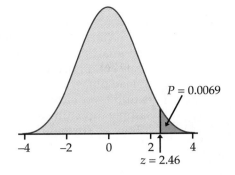

FIGURE 6.12 Sketch of the P-value calculation for the one-sided test in Example 6.16. The test statistic is $z = 2.46$ and the alternative is greater than.

Because H_a is one-sided on the high side, large values of z count against H_0. From Table A, we find that the P-value is

$$P = P(Z \geq 2.46) = 1 - 0.9931 = 0.0069$$

Figure 6.12 illustrates this P-value. A mean score as large as that observed would occur fewer than seven times in 1000 samples if the population mean were 450. This is convincing evidence that the mean SATM score for all California high school seniors is higher than 450.

USE YOUR KNOWLEDGE

6.43 Computing the test statistic and P-value. You will perform a significance test of H_0: $\mu = 25$ based on an SRS of $n = 36$. Assume $\sigma = 5$.

(a) If $\bar{x} = 27$, what is the test statistic z?

(b) What is the P-value if H_a: $\mu > 25$?

(c) What is the P-value if H_a: $\mu \neq 25$?

6.44 Testing a random number generator. Statistical software has a "random number generator" that is supposed to produce numbers uniformly distributed between 0 to 1. If this is true, the numbers generated come from a population with $\mu = 0.5$. A command to generate 100 random numbers gives outcomes with mean $\bar{x} = 0.514$ and $s = 0.314$. Because the sample is reasonably large, take the population standard deviation also to be $\sigma = 0.314$. Do we have evidence that the mean of all numbers produced by this software is not 0.5?

Two-sided significance tests and confidence intervals

Recall the basic idea of a confidence interval, discussed in the first section of this chapter. We constructed an interval that would include the true value of μ with a specified probability C. Suppose we use a 95% confidence interval ($C = 0.95$). Then the values of μ that are not in our interval would seem to be incompatible with the data. This sounds like a significance test with $\alpha = 0.05$ (or 5%) as our standard for drawing a conclusion. The following examples demonstrate that this is correct.

LEADTEST

EXAMPLE

6.17 Water quality testing. The Deely Laboratory is a drinking water testing and analysis service. One of the common contaminants it tests for is lead. Lead enters drinking water through corrosion of plumbing materials, such as lead pipes, fixtures, and solder. The service knows their analysis procedure is unbiased but not perfectly precise, so the laboratory analyzes each water sample three times and reports the mean result. The repeated measurements follow a Normal distribution quite closely. The standard deviation of this distribution is a property of the analytic procedure and is known to be $\sigma = 0.25$ parts per billion (ppb).

The service has been asked by the university to evaluate a claim that the drinking water in the Student Union has a concentration of 6 ppb. The university wants to demonstrate the level is well below the EPA action level of 15 ppb. Since the true concentration of the sample is the mean μ of the population of repeated analyses, the hypotheses are

$$H_0: \mu = 6$$
$$H_a: \mu \neq 6$$

The lab chooses the 1% level of significance, $\alpha = 0.01$.

Three analyses of one specimen give concentrations

$$6.79 \quad 6.13 \quad 7.17$$

The sample mean of these readings is

$$\bar{x} = \frac{6.79 + 6.13 + 7.17}{3} = 6.70$$

The test statistic is

$$z = \frac{\bar{x} - \mu_0}{\sigma/\sqrt{n}} = \frac{6.70 - 6.00}{0.25/\sqrt{3}} = 4.83 \text{ standard deviations}$$

Because the alternative is two-sided, the P-value is

$$P = 2P(Z \geq 4.83)$$

We cannot find this probability in Table A. The largest value of z in that table is 3.49. All that we can say from Table A is that P is less than $2P(X \geq 3.49) = 2(1 - 0.9998) = 0.0004$. If we use the bottom row of Table D, we find that the largest value of z^* is 3.291, corresponding to a P-value of $1 - 0.999 = 0.001$. Software or a calculator could be used to give an accurate value of the P-value. However, because the P-value is clearly less than the lab's standard of 1%, we reject H_0. It appears the level is higher than the university's claim.

We can compute a 99% confidence interval for the same data to get a likely region for this level μ.

EXAMPLE

6.18 99% confidence interval for the mean concentration. The 99% confidence interval for μ in Example 6.17 is

$$\bar{x} \pm z^* \frac{\sigma}{\sqrt{n}} = 6.70 \pm 2.576(0.25/\sqrt{3})$$

$$= 6.70 \pm 0.37 = (6.33, 7.07)$$

The hypothesized value $\mu_0 = 6.00$ in Example 6.17 falls outside the confidence interval we computed in Example 6.18. In other words, it is in the region we are 99% confident that μ is *not* in. Thus, we can reject

$$H_0: \mu = 6.0$$

at the 1% significance level. On the other hand, we cannot reject

$$H_0: \mu = 7.0$$

at the 1% level in favor of the two-sided alternative $H_a: \mu \neq 7.0$, because 7.0 lies inside the 99% confidence interval for μ. Figure 6.13 illustrates both cases.

FIGURE 6.13 The link between two-sided significance tests and confidence intervals. For the study described in Examples 6.17 and 6.18, values of μ falling outside a 99% confidence interval can be rejected at the 1% significance level; values falling inside the interval cannot be rejected. This holds for any signficance level α and 1-α confidence interval.

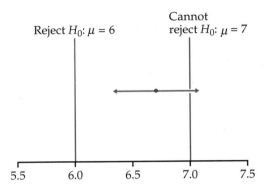

The calculation in Example 6.17 for a 1% significance test is very similar to the calculation for a 99% confidence interval. In fact, a two-sided test at significance level α can be carried out directly from a confidence interval with confidence level $C = 1 - \alpha$.

TWO-SIDED SIGNIFICANCE TESTS AND CONFIDENCE INTERVALS

A level α two-sided significance test rejects a hypothesis $H_0: \mu = \mu_0$ exactly when the value μ_0 falls outside a level $1 - \alpha$ confidence interval for μ.

USE YOUR KNOWLEDGE

6.45 Two-sided significance tests and confidence intervals. The P-value for a two-sided test of the null hypothesis $H_0: \mu = 30$ is 0.033.

(a) Does the 95% confidence interval include the value 30? Explain.

(b) Does the 99% confidence interval include the value 30? Explain.

6.46 More on two-sided tests and confidence intervals. A 95% confidence interval for a population mean is $(53, 62)$.

(a) Can you reject the null hypothesis that $\mu = 58$ at the 5% significance level against the two-sided alternative? Explain.

(b) Can you reject the null hypothesis that $\mu = 63$ at the 5% significance level against the two-sided alternative? Explain.

P-values versus fixed α

The observed result in Example 6.17 was $z = 4.83$. The conclusion that this result is significant at the 1% level does not tell the whole story. The observed z is far beyond the z corresponding to 1%, and the evidence against H_0 is far stronger than 1% significance suggests. The *P*-value

$$2P(Z \geq 4.83) = 0.0000014$$

gives a better sense of how strong the evidence is. *The P-value is the smallest level α at which the data are significant.* Knowing the *P*-value allows us to assess significance at any level.

EXAMPLE

6.19   Test of the mean SATM score: significance. In Example 6.16, we tested the hypotheses

$$H_0: \mu = 450$$
$$H_a: \mu > 450$$

concerning the mean SAT Mathematics score μ of California high school seniors. The test had the *P*-value $P = 0.0069$. This result is significant at the $\alpha = 0.01$ level because $0.0069 \leq 0.01$. It is not significant at the $\alpha = 0.005$ level, because the *P*-value is larger than 0.005. See Figure 6.14.

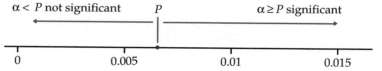

FIGURE 6.14   Link between the *P*-value and the significance level α. An outcome with *P*-value *P* is significant at all levels α at or above *P* and is not significant at smaller levels α.

A *P*-value is more informative than a reject-or-not finding at a fixed significance level. But assessing significance at a fixed level α is easier, because no probability calculation is required. You need only look up a number in a table. A value z^* with a specified area to its right under the standard Normal curve is **critical value**   called a **critical value** of the standard Normal distribution. Because the practice of statistics almost always employs computer software or a calculator that calculates *P*-values automatically, the use of tables of critical values is becoming outdated. We include the usual tables of critical values (such as Table D) at the end of the book for learning purposes and to rescue students without good computing facilities. The tables can be used directly to carry out fixed α tests. They also allow us to approximate *P*-values quickly without a probability calculation. The following example illustrates the use of Table D to find an approximate *P*-value.

EXAMPLE

6.20   Debt levels of undergraduate applicants in two regions of the United States: assessing significance. In Example 6.11 we found the test statistic $z = 1.49$ for testing the null hypothesis that there was no difference in the mean debt between undergraduates in the Midwest and West. The alternative was two-sided. Under the null hypothesis, z has a standard Normal distribution, and from the last row in Table D we can see that there is a 95% chance that z is between ± 1.96. Therefore, we reject H_0 in favor of H_a whenever z is outside this range. Since our calculated value is 1.49, we are within the range and we do not reject the null hypothesis at the 5% level of significance.

USE YOUR KNOWLEDGE

6.47 P-value and the significance level. The P-value for a significance test is 0.039.

(a) Do you reject the null hypothesis at level $\alpha = 0.05$?

(b) Do you reject the null hypothesis at level $\alpha = 0.01$?

(c) Explain how you determined your answers to parts (a) and (b).

6.48 More on the P-value and the significance level. The P-value for a significance test is 0.062.

(a) Do you reject the null hypothesis at level $\alpha = 0.05$?

(b) Do you reject the null hypothesis at level $\alpha = 0.01$?

(c) Explain how you determined your answers to parts (a) and (b).

6.49 One-sided and two-sided P-values. The P-value for a two-sided significance test is 0.044.

(a) State the P-values for the two one-sided tests.

(b) What additional information do you need to properly assign these P-values to the $>$ and $<$ (one-sided) alternatives?

SECTION 6.2 Summary

A **test of significance** is intended to assess the evidence provided by data against a **null hypothesis H_0** in favor of an **alternative hypothesis H_a**.

The hypotheses are stated in terms of population parameters. Usually H_0 is a statement that no effect or no difference is present, and H_a says that there is an effect or difference, in a specific direction (**one-sided alternative**) or in either direction (**two-sided alternative**).

The test is based on a **test statistic**. The **P-value** is the probability, computed assuming that H_0 is true, that the test statistic will take a value at least as extreme as that actually observed. Small P-values indicate strong evidence against H_0. Calculating P-values requires knowledge of the sampling distribution of the test statistic when H_0 is true.

If the P-value is as small or smaller than a specified value α, the data are **statistically significant** at significance level α.

Significance tests for the hypothesis $H_0: \mu = \mu_0$ concerning the unknown mean μ of a population are based on the **z statistic:**

$$z = \frac{\bar{x} - \mu_0}{\sigma/\sqrt{n}}$$

The z test assumes an SRS of size n, known population standard deviation σ, and either a Normal population or a large sample. *P*-values are computed from the Normal distribution (Table A). Fixed α tests use the table of **standard Normal critical values** (Table D).

SECTION 6.2 Exercises

For Exercises 6.36 and 6.37, see page 363; for Exercises 6.38 and 6.39, see pages 366–367; for Exercises 6.40 to 6.42, see page 369; for Exercises 6.43 and 6.44, see page 373; for Exercises 6.45 and 6.46, see page 375; and for Exercises 6.47 to 6.49, see page 377.

6.50 What's wrong? Here are several situations where there is an incorrect application of the ideas presented in this section. Write a short paragraph explaining what is wrong in each situation and why it is wrong.

(a) A researcher tests the following null hypothesis: $H_0: \bar{x} = 23$.

(b) A random sample of size 30 is taken from a population that is assumed to have a standard deviation of 5. The standard deviation of the sample mean is 5/30.

(c) A study with $\bar{x} = 45$ reports statistical significance for $H_a : \mu > 50$.

(d) A researcher tests the hypothesis $H_0: \mu = 350$ and concludes that the population mean is equal to 350.

6.51 What's wrong? Here are several situations where there is an incorrect application of the ideas presented in this section. Write a short paragraph explaining what is wrong in each situation and why it is wrong.

(a) A significance test rejected the null hypothesis that the sample mean is equal to 500.

(b) A test preparation company wants to test that the average score of their students on the ACT is better than the national average score of 21.2. They state their null hypothesis to be $H_0 : \mu > 21.2$.

(c) A study summary says that the results are statistically significant and the *P*-value is 0.98.

(d) The z statistic is equal to 0.018. Because this is less than $\alpha = 0.05$, the null hypothesis was rejected.

6.52 Determining hypotheses. State the appropriate null hypothesis H_0 and alternative hypothesis H_a in each of the following cases.

(a) A 2008 study reported that 88% of students owned a cell phone. You plan to take an SRS of students to see if the percentage has increased.

(b) The examinations in a large freshman chemistry class are scaled after grading so that the mean score is 75. The professor thinks that students who attend early morning recitation sections will have a higher mean score than the class as a whole. Her students this semester can be considered a sample from the population of all students she might teach, so she compares their mean score with 75.

(c) The student newspaper at your college recently changed the format of their opinion page. You take a random sample of students and select those who regularly read the newspaper. They are asked to indicate their opinions on the changes using a five-point scale: -2 if the new format is much worse than the old, -1 if the new format is somewhat worse than the old, 0 if the new format is the same as the old, $+1$ if the new format is somewhat better than the old, and $+2$ if the new format is much better than the old.

6.53 More on determining hypotheses. State the null hypothesis H_0 and the alternative hypothesis H_a in each case. Be sure to identify the parameters that you use to state the hypotheses.

(a) A university gives credit in first-year calculus to students who pass a placement test. The mathematics department wants to know if students who get credit in this way differ in their success with second-year calculus. Scores in second-year calculus are scaled so the average each year is equivalent to a 77. This year 21 students who took second-year calculus passed the placement test.

(b) Experiments on learning in animals sometimes measure how long it takes a mouse to find its way through a maze. The mean time is 20 seconds for one particular maze. A researcher thinks that playing rap music will cause the mice to complete the maze slower. She measures how long each of 12 mice takes with the rap music as a stimulus.

(c) The average square footage of one-bedroom apartments in a new student-housing development is advertised to be 880 square feet. A student group thinks that the apartments are smaller than advertised. They hire an engineer to measure a sample of apartments to test their suspicion.

6.54 Even more on determining hypotheses. In each of the following situations, state an appropriate null hypothesis H_0 and alternative hypothesis H_a. Be sure to identify the parameters that you use to state the hypotheses. (We have not yet learned how to test these hypotheses.)

(a) A sociologist asks a large sample of high school students which television channel they like best. She suspects that a higher percent of males than of females will name MTV as their favorite channel.

(b) An education researcher randomly divides sixth-grade students into two groups for physical education class. He teaches both groups basketball skills, using the same methods of instruction in both classes. He encourages Group A with compliments and other positive behavior but acts cool and neutral toward Group B. He hopes to show that positive teacher attitudes result in a higher mean score on a test of basketball skills than do neutral attitudes.

(c) An education researcher believes that among college students there is a negative correlation between time spent at social network sites and self-esteem. To test this, she gathers social networking information and self-esteem data from a sample of students at your college.

6.55 Translating research questions into hypotheses. Translate each of the following research questions into appropriate H_0 and H_a.

(a) Census Bureau data show that the mean household income in the area served by a shopping mall is $42,800 per year. A market research firm questions shoppers at the mall to find out whether the mean household income of mall shoppers is higher than that of the general population.

(b) Last year, your online registration technicians took an average of 0.4 hours to respond to trouble calls from students trying to register. Do this year's data show a different average response time?

6.56 Computing the P-value. A test of the null hypothesis H_0: $\mu = \mu_0$ gives test statistic $z = 1.63$.

(a) What is the P-value if the alternative is H_a: $\mu > \mu_0$?

(b) What is the P-value if the alternative is H_a: $\mu < \mu_0$?

(c) What is the P-value if the alternative is H_a: $\mu \neq \mu_0$?

6.57 More on computing the P-value. A test of the null hypothesis H_0: $\mu = \mu_0$ gives test statistic $z = -1.82$.

(a) What is the P-value if the alternative is H_a: $\mu > \mu_0$?

(b) What is the P-value if the alternative is H_a: $\mu < \mu_0$?

(c) What is the P-value if the alternative is H_a: $\mu \neq \mu_0$?

6.58 A two-sided test and the confidence interval. The P-value for a two-sided test of the null hypothesis H_0: $\mu = 30$ is 0.032.

(a) Does the 95% confidence interval include the value 30? Why?

(b) Does the 90% confidence interval include the value 30? Why?

6.59 More on a two-sided test and the confidence interval. A 90% confidence interval for a population mean is (25, 32).

(a) Can you reject the null hypothesis that $\mu = 24$ against the two-sided alternative at the 10% significance level? Why?

(b) Can you reject the null hypothesis that $\mu = 30$ against the two-sided alternative at the 10% significance level? Why?

6.60 Use of bed nets. A study found that the use of bed nets was associated with a lower prevalence of malarial infections in the Gambia.[17] A report of the study states that the significance is $P < 0.001$. Explain what this means in a way that could be understood by someone who has not studied statistics.

6.61 Peer pressure and choice of major. A recent study followed a cohort of students entering a business/economics program.[18] All students followed a common track during the first three semesters and then chose to specialize in either business or economics. Through a series of surveys, the researchers were able to classify roughly 50% of the students as either peer driven (ignored abilities and chose major to follow peers) or ability driven (ignored peers and chose major based on ability). When looking at entry wages after graduation, the researchers conclude that a peer-driven student can expect an average wage that is 13% less than that of an ability-driven student. The report states that the significance level is $P = 0.09$. Can you be confident of this statement regarding the wage decrease? Discuss.

6.62 Symbol of wealth in ancient China? Every society has its own symbols of wealth and prestige. In ancient China, it appears that owning pigs was such a symbol. Evidence comes from examining burial sites. If the skulls of sacrificed pigs tend to appear along with expensive ornaments, that suggests that the pigs, like the ornaments, signal the wealth and prestige of the person buried. A study of burials from around 3500 B.C. concluded that, "there are striking differences in grave goods between burials with pig skulls and burials without them.... A test indicates that the two samples of total artifacts are significantly different at the 0.01 level."[19] Explain clearly why "significantly different at the 0.01 level" gives good reason to think that there really is a systematic difference between burials that contain pig skulls and those that lack them.

6.63 Alcohol awareness among college students. A study of alcohol awareness among college students reported a higher awareness for students enrolled in a health and safety class than for those enrolled in a statistics class.[20] The difference is described as being statistically significant. Explain what this means in simple terms and offer an explanation for why the health and safety students had a higher mean score.

6.64 Change in eighth-grade average mathematics score. A report based on the 2009 National Assessment of Educational Progress (NAEP)[21] states that the average score on their mathematics test for eighth-grade students is significantly higher than in 2007. They also state that the average score in Minnesota was not significantly different from the average score in 2007. A footnote states that comparisons are determined by statistical tests with 0.05 as the level of significance. Explain what this means in language understandable to someone who knows no statistics. Do not use the word "significance" in your answer.

6.65 Sleep quality and elevated blood pressure. A recent study looked at $n = 238$ adolescents, all free of severe illness.[22] Subjects wore a wrist actigraph, which allowed the researchers to estimate sleep patterns. For those subjects classified as having low sleep efficiency, they are reported as having an average systolic blood pressure that is 4 mm Hg higher than other children with a standard deviation of the mean equal to 1.2 mm Hg. Based on these results, test whether this difference is significant at the 0.01 level.

6.66 🛡 **Are the pine trees randomly distributed north to south?** In Example 6.1 we looked at the distribution of longleaf pine trees in the Wade Tract. One way to formulate hypotheses about whether or not the trees are randomly distributed in the tract is to examine the average location in the north-south direction. The values range from 0 to 200, so if the trees are uniformly distributed in this direction, any difference from the middle value (100) should be due to chance variation. The sample mean for the 584 trees in the tract is 99.74. A theoretical calculation based on the assumption that the trees are uniformly distributed gives a standard deviation of 58. Carefully state the null and alternative hypotheses in terms of this variable. Note that this requires that you translate the research question about the random distribution of the trees into specific statements about the mean of a probability distribution. Test your hypotheses, report your results, and write a short summary of what you have found.

6.67 🛡 **Are the pine trees randomly distributed east to west?** Answer the questions in the previous exercise for the east-west direction, where the sample mean is 113.8.

6.68 Who is the author? Statistics can help decide the authorship of literary works. Sonnets by a certain Elizabethan poet are known to contain an average of $\mu = 8.9$ new words (words not used in the poet's other works). The standard deviation of the number of new words is $\sigma = 2.5$. Now a manuscript with six new sonnets has come to light, and scholars are debating whether it is the poet's work. The new sonnets contain an average of $\bar{x} = 10.2$ words not used in the poet's known works. We expect poems by another author to contain more new words, so to see if we have evidence that the new sonnets are not by our poet we test

$$H_0: \mu = 8.9$$
$$H_a: \mu > 8.9$$

Give the z test statistic and its P-value. What do you conclude about the authorship of the new poems?

6.69 Attitudes toward school. The Survey of Study Habits and Attitudes (SSHA) is a psychological test that measures the motivation, attitude toward school, and study habits of students. Scores range from 0 to 200. The mean score for U.S. college students is about 115, and the standard deviation is about 30. A teacher who suspects that older students have better attitudes toward school gives the SSHA to 25 students who are at least 30 years of age. Their mean score is $\bar{x} = 127.8$.

(a) Assuming that $\sigma = 30$ for the population of older students, carry out a test of

$$H_0: \mu = 115$$
$$H_a: \mu > 115$$

Report the P-value of your test, and state your conclusion clearly.

(b) Your test in part (a) required two important assumptions in addition to the assumption that the value of σ is known. What are they? Which of these assumptions is most important to the validity of your conclusion in part (a)?

6.70 Nutritional intake in Canadian high performance athletes. Since previous studies have reported that elite athletes are often deficient in their nutritional intake (e.g., total calories, carbohydrates, protein), a group of researchers decided to evaluate Canadian high performance athletes.[23] A total of $n = 324$ athletes from eight Canadian sports centers participated in the study. One reported finding was that the average caloric intake among the $n = 201$ women was

2403.7 kcal/day. The recommended amount is 2811.5 kcal/day. Is there evidence that female Canadian athletes are deficient in the caloric intake?

(a) State the appropriate H_0 and H_a to test this.

(b) Assuming a standard deviation of 880 kcal/day, carry out the test. Give the P-value, and then interpret the result in plain language.

6.71 Are the mpg measurements similar? Refer to Exercise 6.28 (page 359). In addition to the computer computing mpg, the driver also recorded the mpg by dividing the miles driven by the number of gallons at each fill-up. The following data are the differences between the computer's and the driver's calculations for that random sample of 20 records. The driver wants to determine if these calculations are different. Assume the standard deviation of a difference to be $\sigma = 3.0$. 🌀 MPGCOMPARISON

5.0	6.5	−0.6	1.7	3.7	4.5	8.0	2.2	4.9	3.0
4.4	0.1	3.0	1.1	1.1	5.0	2.1	3.7	−0.6	−4.2

(a) State the appropriate H_0 and H_a to test this suspicion.

(b) Carry out the test. Give the P-value, and then interpret the result in plain language.

6.72 Adjusting for the cost of living. In Example 6.8 (page 360), we compared the average credit card debt between undergraduate applicants in the Midwest and West. In computing the difference, we did not adjust for differing costs in the two regions. Assuming that $1 in the West is worth about $1.06 in the Midwest, redo the test based on Midwest dollars. For simplicity, assume the standard deviation is unchanged.

6.73 Level of nicotine in cigarettes. According to data from the Tobacco Institute Testing Laboratory, Camel Lights king size cigarettes contain an average of 0.9 milligrams of nicotine. An advocacy group commissions an independent test to see if the mean nicotine content is higher than the industry laboratory claims.

(a) What are H_0 and H_a?

(b) Suppose that the test statistic is $z = 1.83$. Is this result significant at the 5% level?

(c) Is the result significant at the 1% level?

6.74 🌀 **Changes of $\bar{x}$ on significance.** The *Statistical Significance* applet illustrates statistical tests with a fixed level of significance for Normally distributed data with known standard deviation. Open the applet and keep the default settings for the null ($\mu = 0$) and the alternative ($\mu > 0$) hypotheses, the sample size ($n = 10$), the standard deviation ($\sigma = 1$), and the significance level ($\alpha = 0.05$). In

the "I have data, and the observed $\bar{x}$ is $\bar{x} =$" box enter the value 1. Is the difference between $\bar{x}$ and μ_0 significant at the 5% level? Repeat for $\bar{x}$ equal to 0.1, 0.2, 0.3, 0.4, 0.5, 0.6, 0.7, 0.8, 0.9. Make a table giving $\bar{x}$ and the results of the significance tests. What do you conclude?

6.75 🌀 **Changes of α on significance.** Repeat the previous exercise with significance level $\alpha = 0.01$. How does the choice of α affect the values of $\bar{x}$ that are far enough away from μ_0 to be statistically significant?

6.76 🌀 **Changes of $\bar{x}$ on the P-value.** The *P-Value of a Test of Significance* applet illustrates P-values of significance tests for Normally distributed data with known standard deviation. Open the applet and keep the default settings for the null ($\mu = 0$) and the alternative ($\mu > 0$) hypotheses, the sample size ($n = 10$), the standard deviation ($\sigma = 1$), and the significance level ($\alpha = 0.05$). In the "I have data, and the observed $\bar{x}$ is $\bar{x} =$" box enter the value 1. What is the P-value? Repeat for $\bar{x}$ equal to 0.1, 0.2, 0.3, 0.4, 0.5, 0.6, 0.7, 0.8, 0.9. Make a table giving $\bar{x}$ and P-values. How does the P-value change as $\bar{x}$ moves farther away from μ_0?

6.77 Understanding levels of significance. Explain in plain language why a significance test that is significant at the 5% level must always be significant at the 10% level.

6.78 More on understanding levels of significance. You are told that a significance test is significant at the 5% level. From this information can you determine whether or not it is significant at the 1% level? Explain your answer.

6.79 Test statistic and levels of significance. Consider a significance test for a null hypothesis versus a two-sided alternative. Give a value of z that will give a result significant at the 1% level but not at the 0.5% level.

6.80 Using Table D to find a P-value. You have performed a two-sided test of significance and obtained a value of $z = 2.52$. Use Table D to find the approximate P-value for this test.

6.81 More on using Table D to find a P-value. You have performed a one-sided test of significance and obtained a value of $z = 0.63$. Use Table D to find the approximate P-value for this test.

6.82 Using Table A and Table D to find a P-value. Consider a significance test for a null hypothesis versus a two-sided alternative. Between what values from Table D does the P-value for an outcome $z = 1.92$ lie? Calculate the P-value using Table A, and verify that it lies between the values you found from Table D.

6.83 More on using Table A and Table D to find a P-value. Refer to the previous exercise. Find the P-value for $z = -1.92$.

6.3 Use and Abuse of Tests

Carrying out a test of significance is often quite simple, especially if the P-value is given effortlessly by a computer. Using tests wisely is not so simple. Each test is valid only in certain circumstances, with properly produced data being particularly important. The z test, for example, should bear the same warning label that was attached in Section 6.1 to the corresponding confidence interval (page 354). Similar warnings accompany the other tests that we will learn. There are additional caveats that concern tests more than confidence intervals, enough to warrant this separate section. Some hesitation about the unthinking use of significance tests is a sign of statistical maturity.

The reasoning of significance tests has appealed to researchers in many fields, so that tests are widely used to report research results. In this setting H_a is a "research hypothesis" asserting that some effect or difference is present. The null hypothesis H_0 says that there is no effect or no difference. A low P-value represents good evidence that the research hypothesis is true. Here are some comments on the use of significance tests, with emphasis on their use in reporting scientific research.

Choosing a level of significance

The spirit of a test of significance is to give a clear statement of the degree of evidence provided by the sample against the null hypothesis. The P-value does this. It is common practice to report P-values and to describe results as statistically significant whenever $P \leq 0.05$. *However, there is no sharp border between "significant" and "not significant," only increasingly strong evidence as the P-value decreases.* Having both the P-value and the statement that we reject or fail to reject H_0 allows us to draw better conclusions from our data.

EXAMPLE

6.21 Information provided by the P-value. Suppose the test statistic for a two-sided significance test for a population mean is $z = 1.95$. From Table A we can calculate the P-value. It is

$$P = 2[1 - P(Z \leq 1.95)] = 2(1 - 0.9744) = 0.0512$$

We have failed to meet the standard of evidence for $\alpha = 0.05$. However, with the information provided by the P-value, we can see that the result just barely missed the standard. If the effect in question is interesting and potentially important, we might want to design another study with a larger sample to investigate it further.

Here is another example where the P-value provides useful information beyond that provided by the statement that we reject or fail to reject the null hypothesis.

EXAMPLE

6.22 More on information provided by the P-value. We have a test statistic of $z = -4.66$ for a two-sided significance test on a population mean. Software tells us that the P-value is 0.000003. This means that there are 3 chances in

1,000,000 of observing a sample mean this far or farther away from the null hypothesized value of μ. This kind of event is virtually impossible if the null hypothesis is true. There is no ambiguity in the result; we can clearly reject the null hypothesis.

We frequently report small P-values such as that in the previous example as $P < 0.001$. This corresponds to a chance of 1 in 1000 and is sufficiently small to lead us to a clear rejection of the null hypothesis.

One reason for the common use of $\alpha = 0.05$ is the great influence of Sir R. A. Fisher, the inventor of formal statistical methods for analyzing experimental data. Here is his opinion on choosing a level of significance: "A scientific fact should be regarded as experimentally established only if a properly designed experiment *rarely fails* to give this level of significance."[24]

What statistical significance does not mean

When a null hypothesis ("no effect" or "no difference") can be rejected at the usual level $\alpha = 0.05$, there is good evidence that an effect is present. That effect, however, can be extremely small. *When large samples are available, even tiny deviations from the null hypothesis will be significant.*

EXAMPLE

6.23 It's significant but it is important? Suppose that we are testing the hypothesis of no correlation between two variables. With 400 observations, an observed correlation of only $r = 0.1$ is significant evidence at the $\alpha = 0.05$ level that the correlation in the population is not zero. The low significance level does *not* mean there is a strong association, only that there is strong evidence of some association. The proportion of the variability in one of the variables explained by the other is $r^2 = 0.01$, or 1%.

For practical purposes, we might well decide to ignore this association. *Statistical significance is not the same as practical significance.* Statistical significance rarely tells us about the importance of the experimental results. This depends on the context of the experiment.

The remedy for attaching too much importance to statistical significance is to pay attention to the actual experimental results as well as to the P-value. Plot your data and examine them carefully. Beware of outliers. *The foolish user of statistics who feeds the data to a computer without exploratory analysis will often be embarrassed.* It is usually wise to give a confidence interval for the parameter in which you are interested. Confidence intervals are not used as often as they should be, while tests of significance are perhaps overused.

USE YOUR KNOWLEDGE

6.84 Is it significant? More than 200,000 people worldwide take the GMAT examination each year as they apply for MBA programs. Their scores vary Normally with mean about $\mu = 525$ and standard deviation about $\sigma = 100$. One hundred students go through a rigorous training program designed to raise their GMAT scores. Test the following hypotheses about the training program

$$H_0: \mu = 525$$
$$H_a: \mu > 525$$

in each of the following situations:

(a) The students' average score is $\bar{x} = 541.4$. Is this result significant at the 5% level?

(b) Now suppose that the average score is $\bar{x} = 541.5$. Is this result significant at the 5% level?

(c) Explain how you would reconcile this difference in significance, especially if any increase greater than 15 points is considered a success.

Don't ignore lack of significance

There is a tendency to conclude that there is no effect whenever a P-value fails to attain the usual 5% standard. A provocative editorial in the *British Medical Journal* entitled "Absence of Evidence Is Not Evidence of Absence" deals with this issue.[25] Here is one of the examples they cite.

EXAMPLE

6.24 Interventions to reduce HIV-1 transmission. A randomized trial of interventions for reducing transmission of HIV-1 reported an incident rate ratio of 1.00, meaning that the intervention group and the control group both had the same rate of HIV-1 infection. The 95% confidence interval was reported as 0.63 to 1.58.[26] The editorial notes that a summary of these results that says the intervention has no effect on HIV-1 infection is misleading. The confidence interval indicates that the intervention may be capable of achieving a 37% decrease in infection; it might also be harmful and produce a 58% increase in infection. Clearly, more data are needed to distinguish between these possibilities.

The situation can be worse. Research in some fields has rarely been published unless significance at the 0.05 level is attained.

EXAMPLE

6.25 Journal survey of reported significance results. A survey of four journals published by the American Psychological Association showed that of 294 articles using statistical tests, only 8 reported results that did not attain the 5% significance level.[27] It is very unlikely that these were the only 8 studies of scientific merit that did not attain significance at the 0.05 level. Manuscripts describing other studies were likely rejected because of a lack of statistical significance or never submitted in the first place due to the expectation of rejection.

In some areas of research, small effects that are detectable only with large sample sizes can be of great practical significance. Data accumulated from a large number of patients taking a new drug may be needed before we can conclude that there are life-threatening consequences for a small number of people.

On the other hand, sometimes a meaningful result is not found significant.

EXAMPLE

6.26 A meaningful but statistically insignificant result. A sample of size 10 gave a correlation of $r = 0.5$ between two variables. The P-value is 0.102 for a two-sided significance test. In many situations, a correlation this large would be interesting and worthy of additional study. When it takes a lot of effort (say in terms of time or money) to obtain samples, researchers often use small studies like these as pilot projects to gain interest from various funding sources. With financial support, a larger, more powerful study can then be run.

Another important aspect of planning a study is to verify that the test you plan to use does have high probability of detecting an effect of the size you hope to find. This probability is the *power* of the test. Power calculations are discussed in Section 6.4.

LOOK BACK
design of experiments p. 170

Statistical inference is not valid for all sets of data

In Chapter 3, we learned that badly designed surveys or experiments often produce invalid results. *Formal statistical inference cannot correct basic flaws in the design.*

EXAMPLE

6.27 English vocabulary and studying a foreign language. There is no doubt that there is a significant difference in English vocabulary scores between high school seniors who have studied a foreign language and those who have not. But because the effect of actually studying a language is confounded with the differences between students who choose language study and those who do not, this statistical significance is hard to interpret. The most plausible explanation is that students who were already good at English chose to study another language. A randomized comparative experiment would isolate the actual effect of language study and so make significance meaningful. However, such an experiment probably could not be done.

Tests of significance and confidence intervals are based on the laws of probability. Randomization in sampling or experimentation ensures that these laws apply. *But we must often analyze data that do not arise from randomized samples or experiments. To apply statistical inference to such data, we must have confidence in a probability model for the data.* We can check a probability model by examining the data. If the Normal distribution model appears correct, we can apply the methods of this chapter to do inference about the mean μ.

USE YOUR KNOWLEDGE

6.85 Home security systems. A recent TV advertisement for home security systems said that homes without an alarm system are three times more likely to be broken into. Suppose this conclusion was obtained by examining an SRS of police records of break-ins and determining whether the percent of homes with alarm systems was significantly smaller than 50%. Explain why the significance of this study is suspect and propose an alternative study that would help clarify the importance of an alarm system.

Beware of searching for significance

Statistical significance is an outcome much desired by researchers. It means (or ought to mean) that you have found an effect that you were looking for. *The reasoning behind statistical significance works well if you decide what effect you are seeking, design an experiment or sample to search for it, and use a test of significance to weigh the evidence you get.* But because a successful search for a new scientific phenomenon often ends with statistical significance, it is all too tempting to make significance itself the object of the search. There are several ways to do this, none of them acceptable in polite scientific society.

EXAMPLE

6.28 Genomics studies. In genomic experiments, it is common to assess the differences in expression for tens of thousands of genes. If each of these genes was examined separately and statistical significance declared for all that had *P*-values that pass the 0.05 standard, we would have quite a mess. In the absence of any real biological effects, we would expect that, by chance alone, approximately 5% of these tests will show statistical significance. Much research in genomics is directed toward appropriate ways to deal with this situation.[28]

We do not mean that searching data for suggestive patterns is not proper scientific work. It certainly is. Many important discoveries have been made by accident rather than by design. Exploratory analysis of data is an essential part of statistics. We do mean that the usual reasoning of statistical inference does not apply when the search for a pattern is successful. *You cannot legitimately test a hypothesis on the same data that first suggested that hypothesis.* The remedy is clear. Once you have a hypothesis, design a study to search specifically for the effect you now think is there. If the result of this study is statistically significant, you have real evidence.

SECTION 6.3 Summary

P-values are more informative than the reject-or-not result of a fixed level α test. Beware of placing too much weight on traditional values of α, such as $\alpha = 0.05$.

Very small effects can be highly significant (small *P*), especially when a test is based on a large sample. A statistically significant effect need not be practically important. Plot the data to display the effect you are seeking, and use confidence intervals to estimate the actual values of parameters.

On the other hand, lack of significance does not imply that H_0 is true, especially when the test has low power.

Significance tests are not always valid. Faulty data collection, outliers in the data, and testing a hypothesis on the same data that suggested the hypothesis can invalidate a test. Many tests run at once will probably produce some significant results by chance alone, even if all the null hypotheses are true.

SECTION 6.3 Exercises

For Exercise 6.84, see page 383; and for Exercise 6.85, see page 386.

6.86 A role as a statistical consultant. You are the statistical expert for a graduate student planning her PhD research. After you carefully present the mechanics of significance testing, she suggests using $\alpha = 0.20$ for the study because she would be more likely to obtain statistically significant results and she *really* needs significant results to graduate. Explain in simple terms why this would not be a good use of statistical methods.

6.87 What do you know? A research report described two results that both achieved statistical significance at the 5% level. The *P*-value for the first is 0.048; for the second it is 0.0002. Do the *P*-values add any useful information beyond that conveyed by the statement that both results are statistically significant? Write a short paragraph explaining your views on this question.

6.88 Selective publication based on results. In addition to statistical significance, selective publication can also be due to the observed outcome. A recent review of 74 FDA-registered studies of antidepressant agents found 38 studies with positive results and 36 studies with negative or questionable results. All but 1 of the 38 positive studies were published. Of the remaining 36, 22 were not published with another 11 published in such a way as to convey a positive outcome.[29] Describe how this selective reporting can have adverse consequences on health care.

6.89 What a test of significance can answer. Explain whether a test of significance can answer each of the following questions.

(a) Is the sample or experiment properly designed?

(b) Is the observed effect compatible with the null hypothesis?

(c) Is the observed effect important?

6.90 Vitamin C and colds. In a study to investigate whether vitamin C will prevent colds, 400 subjects are assigned at random to one of two groups. The experimental group takes a vitamin C tablet daily, while the control group takes a placebo. At the end of the experiment, the researchers calculate the difference between the percents of subjects in the two groups who were free of colds. This difference is statistically significant ($P = 0.03$) in favor of the vitamin C group. Can we conclude that vitamin C has a strong effect in preventing colds? Explain your answer.

6.91 How far do rich parents take us? How much education children get is strongly associated with the wealth and social status of their parents, termed "socioeconomic status," or SES. The SES of parents, however, has little influence on whether children who have graduated from college continue their education. One study looked at whether college graduates took the graduate admissions tests for business, law, and other graduate programs. The effects of the parents' SES on taking the LSAT test for law school were "both statistically insignificant and small."

(a) What does "statistically insignificant" mean?

(b) Why is it important that the effects were small in size as well as insignificant?

6.92 Do you agree? State whether or not you agree with each of the following statements and provide a short summary of the reasons for your answers.

(a) If the *P*-value is larger than 0.05, the null hypothesis is true.

(b) Practical significance is not the same as statistical significance.

(c) We can perform a statistical analysis using any set of data.

(d) If you find an interesting pattern in a set of data, it is appropriate to then use a significance test to determine its significance.

6.93 Practical significance and sample size. Every user of statistics should understand the distinction between statistical significance and practical importance. A sufficiently large sample will declare very small effects statistically significant. Consider the study of elite female Canadian athletes in Exercise 6.70 (page 380). Female athletes were consuming an average of 2403.7 kcal/day with a standard deviation of 880 kcal/day. Suppose a nutritionist is brought in to implement a new health program for these athletes. This program should increase mean caloric intake but not change the standard deviation. Given the standard deviation and how caloric deficient these athletes are, a change in the mean of 50 kcal/day to 2453.7 is of little importance. However, with a large enough sample, this change can be significant. To see this, calculate the P-value for the test of

$$H_0: \mu = 2403.7$$
$$H_a: \mu > 2403.7$$

in each of the following situations:

(a) A sample of 100 athletes; their average caloric intake is $\bar{x} = 2453.7$.

(b) A sample of 500 athletes; their average caloric intake is $\bar{x} = 2453.7$.

(c) A sample of 2500 athletes; their average caloric intake is $\bar{x} = 2453.7$.

6.94 Statistical versus practical significance. A study with 7500 subjects reported a result that was statistically significant at the 5% level. Explain why this result might not be particularly important.

6.95 More on statistical versus practical significance. A study with 14 subjects reported a result that failed to achieve statistical significance at the 5% level. The P-value was 0.051. Write a short summary of how you would interpret these findings.

6.96 🛡 **Find journal articles.** Find two journal articles that report results with statistical analyses. For each article, summarize how the results are reported and write a critique of the presentation. Be sure to include details regarding use of significance testing at a particular level of significance, P-values, and confidence intervals.

6.97 Create example of your own. For each case, provide an example and an explanation as to why it is appropriate.

(a) A set of data or experiment for which statistical inference is not valid.

(b) A set of data or experiment for which statistical inference is valid.

6.98 🛡 **Predicting success of trainees.** What distinguishes managerial trainees who eventually become executives from those who, after expensive training, don't

succeed and leave the company? We have abundant data on past trainees—data on their personalities and goals, their college preparation and performance, even their family backgrounds and their hobbies. Statistical software makes it easy to perform dozens of significance tests on these dozens of variables to see which ones best predict later success. We find that future executives are significantly more likely than washouts to have an urban or suburban upbringing and an undergraduate degree in a technical field.

Explain clearly why using these "significant" variables to select future trainees is not wise. Then suggest a follow-up study using this year's trainees as subjects that should clarify the importance of the variables identified by the first study.

6.99 Searching for significance. Give an example of a situation where searching for significance would lead to misleading conclusions.

6.100 More on searching for significance. You perform 1000 significance tests using $\alpha = 0.05$. Assuming that all null hypotheses are true, about how many of the test results would you expect to be statistically significant? Explain how you obtained your answer.

6.101 Interpreting a very small P-value. Assume that you are performing a large number of significance tests. Let n be the number of these tests. How large would n need to be for you to expect about one P-value to be 0.00001 or smaller? Use this information to write an explanation of how to interpret a result that has $P = 0.00001$ in this setting.

6.102 🛡 **An adjustment for multiple tests.** One way to deal with the problem of misleading P-values when performing more than one significance test is to adjust the criterion you use for statistical significance. The **Bonferroni procedure** does this in a simple way. If you perform two tests and want to use the $\alpha = 5\%$ significance level, you would require a P-value of $0.05/2 = 0.025$ to declare either one of the tests significant. In general, if you perform k tests and want protection at level α, use α/k as your cutoff for statistical significance. You perform six tests and obtain individual P-values 0.083, 0.032, 0.246, 0.003, 0.010, and < 0.001. Which of these are statistically significant using the Bonferroni procedure with $\alpha = 0.05$?

6.103 🛡 **Significance using the Bonferroni procedure.** Refer to the previous problem. A researcher has performed 12 tests of significance and wants to apply the Bonferroni procedure with $\alpha = 0.05$. The calculated P-values are 0.041, 0.569, 0.050, 0.416, 0.002, 0.006, 0.286, 0.021, 0.888, 0.010, < 0.002, and 0.533. Which of these tests reject their null hypotheses with this procedure?

6.4 Power and Inference as a Decision*

Although we prefer to use P-values rather than the reject-or-not view of the fixed α significance test, the latter view is very important for planning studies and for understanding statistical decision theory. We will discuss these two topics in this section.

Power

Fixed level α significance tests are closely related to confidence intervals—in fact, we saw that a two-sided test can be carried out directly from a confidence interval. The significance level, like the confidence level, says how reliable the method is in repeated use. If we use 5% significance tests repeatedly when H_0 is in fact true, we will be wrong (the test will reject H_0) 5% of the time and right (the test will fail to reject H_0) 95% of the time.

The ability of a test to detect that H_0 is false is measured by the probability that the test will reject H_0 when an alternative is true. The higher this probability is, the more sensitive the test is.

> **POWER**
>
> The probability that a fixed level α significance test will reject H_0 when a particular alternative value of the parameter is true is called the **power** of the test to detect that alternative.

EXAMPLE

6.29 Power of TBBMC significance test. Can a six-month exercise program increase the total body bone mineral content (TBBMC) of young women? A team of researchers is planning a study to examine this question. Based on the results of a previous study, they are willing to assume that $\sigma = 2$ for the percent change in TBBMC over the six-month period. They also feel a change in TBBMC of 1% is important, so they would like to have a reasonable chance of detecting a change this large or larger. Is 25 subjects a large enough sample for this project?

We will answer this question by calculating the power of the significance test that will be used to evaluate the data to be collected. The calculation consists of three steps:

1. State H_0, H_a, the particular alternative we want to detect, and the significance level α.

2. Find the values of $\bar{x}$ that will lead us to reject H_0.

3. Calculate the probability of observing these values of $\bar{x}$ when the alternative is true.

---

*Although the topics in this section are important in planning and interpreting significance tests, they can be omitted without loss of continuity.

Step 1. The null hypothesis is that the exercise program has no effect on TBBMC. In other words, the mean percent change is zero. The alternative is that exercise is beneficial; that is, the mean change is positive. Formally, we have

$$H_0: \mu = 0$$
$$H_a: \mu > 0$$

The alternative of interest is $\mu = 1\%$ increase in TBBMC. A 5% test of significance will be used.

Step 2. The z test rejects H_0 at the $\alpha = 0.05$ level whenever

$$z = \frac{\bar{x} - \mu_0}{\sigma/\sqrt{n}} = \frac{\bar{x} - 0}{2/\sqrt{25}} \geq 1.645$$

Be sure you understand why we use 1.645. Rewrite this in terms of $\bar{x}$:

$$\bar{x} \geq 1.645 \frac{2}{\sqrt{25}}$$
$$\bar{x} \geq 0.658$$

Because the significance level is $\alpha = 0.05$, this event has probability 0.05 of occurring *when the population mean μ is 0.*

Step 3. The power against the alternative $\mu = 1\%$ is the probability that H_0 will be rejected *when in fact $\mu = 1\%$.* We calculate this probability by standardizing $\bar{x}$, using the value $\mu = 1$, the population standard deviation $\sigma = 2$, and the sample size $n = 25$. The power is

$$P(\bar{x} \geq 0.658 \quad \text{when} \quad \mu = 1) = P\left(\frac{\bar{x} - \mu}{\sigma/\sqrt{n}} \geq \frac{0.658 - 1}{2/\sqrt{25}}\right)$$
$$= P(Z \geq -0.855) = 0.80$$

Figure 6.15 illustrates the power with the sampling distribution of $\bar{x}$ when $\mu = 1$. This significance test rejects the null hypothesis that exercise has no effect on TBBMC 80% of the time if the true effect of exercise is a 1% increase in TBBMC. If the true effect of exercise is a greater percent increase, the test will have greater power; it will reject with a higher probability.

Here is another example of a power calculation, this time for a two-sided z test.

EXAMPLE

6.30 Power of the lead concentration test. Example 6.17 (page 374) presented a test of

$$H_0: \mu = 6.0$$
$$H_a: \mu \neq 6.0$$

at the 1% level of significance. What is the power of this test against the specific alternative $\mu = 6.5$?

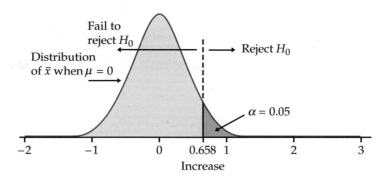

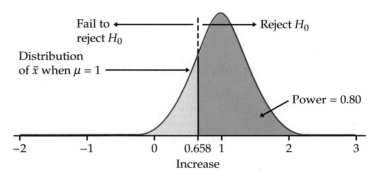

FIGURE 6.15   The sampling distributions of $\bar{x}$ when $\mu = 0$ and when $\mu = 1$. The power is the probability that the test rejects H_0 when the alternative is true.

The test rejects H_0 when $|z| \geq 2.576$. The test statistic is

$$z = \frac{\bar{x} - 6.0}{0.25/\sqrt{3}}$$

Some arithmetic shows that the test rejects when either of the following is true:

$$z \geq 2.576 \quad \text{(in other words, } \bar{x} \geq 6.37)$$
$$z \leq -2.576 \quad \text{(in other words, } \bar{x} \leq 5.63)$$

These are disjoint events, so the power is the sum of their probabilities, computed assuming that the alternative $\mu = 6.5$ is true. We find that

$$P(\bar{x} \geq 6.37) = P\left(\frac{\bar{x} - \mu}{\sigma/\sqrt{n}} \geq \frac{6.37 - 6.50}{0.25/\sqrt{3}}\right)$$
$$= P(Z \geq -0.90) = 0.8159$$
$$P(\bar{x} \leq 5.63) = P\left(\frac{\bar{x} - \mu}{\sigma/\sqrt{n}} \leq \frac{5.63 - 6.50}{0.25/\sqrt{3}}\right)$$
$$= P(Z \leq -6.03) \doteq 0$$

Figure 6.16 illustrates this calculation. Because the power is about 0.82, we are quite confident that the test will reject H_0 when this alternative is true.

High power is desirable. Along with 95% confidence intervals and 5% significance tests, 80% power is becoming a standard. Many U.S. government agencies that provide research funds require that the sample size for the funded studies be sufficient to detect important results 80% of the time using a 5% test of significance.

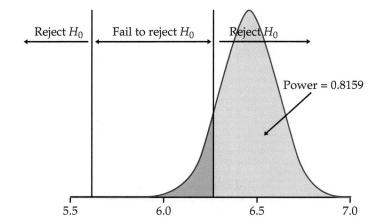

Reject H_0 | Fail to reject H_0 | Reject H_0

Power = 0.8159

5.5   6.0   6.5   7.0

FIGURE 6.16 The power for Example 6.30. Unlike Figure 6.15, only the sampling distribution under the alterntive is shown.

Increasing the power

Suppose you have performed a power calculation and found that the power is too small. What can you do to increase it? Here are four ways:

- **Increase α.** A 5% test of significance will have a greater chance of rejecting the alternative than a 1% test because the strength of evidence required for rejection is less.

- **Consider a particular alternative that is farther away from μ_0.** Values of μ that are in H_a but lie close to the hypothesized value μ_0 are harder to detect (lower power) than values of μ that are far from μ_0.

- **Increase the sample size.** More data will provide more information about $\bar{x}$ so we have a better chance of distinguishing values of μ.

- **Decrease σ.** This has the same effect as increasing the sample size: more information about μ. Improving the measurement process and restricting attention to a subpopulation are two common ways to decrease σ.

Power calculations are important in planning studies. Using a significance test with low power makes it unlikely that you will find a significant effect even if the truth is far from the null hypothesis. A null hypothesis that is in fact false can become widely believed if repeated attempts to find evidence against it fail because of low power. The following example illustrates this point.

EXAMPLE

6.31 Are stock markets efficient? The "efficient market hypothesis" for the time series of stock prices says that future stock prices (when adjusted for inflation) show only random variation. No information available now will help us predict stock prices in the future, because the efficient working of the market has already incorporated all available information in the present price. Many studies have tested the claim that one or another kind of information is helpful. In these studies, the efficient market hypothesis is H_0, and the claim that prediction is possible is H_a. Almost all the studies have failed to find good evidence against H_0. As a result, the efficient market hypothesis is quite popular. But an examination of the significance tests employed finds that the power is generally low. Failure to reject H_0 when using tests of low power is not evidence that H_0 is true. As one expert says, "The widespread impression that there is strong evidence for market efficiency may be due just to a lack of appreciation of the low power of many statistical tests."[30]

Inference as decision*

We have presented tests of significance as methods for assessing the strength of evidence against the null hypothesis. This assessment is made by the P-value, which is a probability computed under the assumption that H_0 is true. The alternative hypothesis (the statement we seek evidence for) enters the test only to help us see what outcomes count against the null hypothesis.

There is another way to think about these issues. Sometimes we are really concerned about making a decision or choosing an action based on our evaluation of the data. **Acceptance sampling** is one such circumstance. A producer of bearings and a skateboard manufacturer agree that each carload lot of bearings shall meet certain quality standards. When a carload arrives, the manufacturer chooses a sample of bearings to be inspected. On the basis of the sample outcome, the manufacturer will either accept or reject the carload. Let's examine how the idea of inference as a decision changes the reasoning used in tests of significance.

acceptance sampling

Two types of error

Tests of significance concentrate on H_0, the null hypothesis. If a decision is called for, however, there is no reason to single out H_0. There are simply two hypotheses, and we must accept one and reject the other. It is convenient to call the two hypotheses H_0 and H_a, but H_0 no longer has the special status (the statement we try to find evidence against) that it had in tests of significance. In the acceptance sampling problem, we must decide between

H_0: the lot of bearings meets standards
H_a: the lot does not meet standards

on the basis of a sample of bearings.

We hope that our decision will be correct, but sometimes it will be wrong. There are two types of incorrect decisions. We can accept a bad lot of bearings, or we can reject a good lot. Accepting a bad lot injures the consumer, while rejecting a good lot hurts the producer. To help distinguish these two types of error, we give them specific names.

TYPE I AND TYPE II ERRORS

If we reject H_0 (accept H_a) when in fact H_0 is true, this is a **Type I error.**
If we accept H_0 (reject H_a) when in fact H_a is true, this is a **Type II error.**

---

*The purpose of this discussion is to clarify the reasoning of significance tests by contrast with a related type of reasoning. It can be omitted without loss of continuity.

The possibilities are summed up in Figure 6.17. If H_0 is true, our decision either is correct (if we accept H_0) or is a Type I error. If H_a is true, our decision either is correct or is a Type II error. Only one error is possible at one time. Figure 6.18 applies these ideas to the acceptance sampling example.

		Truth about the population	
		H_0 true	H_a true
Decision based on sample	Reject H_0	Type I error	Correct decision
	Accept H_0	Correct decision	Type II error

FIGURE 6.17 The two types of error in testing hypotheses.

		Truth about the lot	
		Does meet standards	Does not meet standards
Decision based on sample	Reject the lot	Type I error	Correct decision
	Accept the lot	Correct decision	Type II error

FIGURE 6.18 The two types of error in the acceptance sampling setting.

Error probabilities

Any rule for making decisions is assessed in terms of the probabilities of the two types of error. This is in keeping with the idea that statistical inference is based on probability. We cannot (short of inspecting the whole lot) guarantee that good lots of bearings will never be rejected and bad lots never be accepted. But by random sampling and the laws of probability, we can say what the probabilities of both kinds of error are.

Significance tests with fixed level α give a rule for making decisions because the test either rejects H_0 or fails to reject it. If we adopt the decision-making way of thought, failing to reject H_0 means deciding that H_0 is true. We can then describe the performance of a test by the probabilities of Type I and Type II errors.

EXAMPLE

6.32 Outer diameter of a skateboard bearing. The mean outer diameter of a skateboard bearing is supposed to be 22.000 millimeters (mm). The outer diameters vary Normally with standard deviation $\sigma = 0.010$ mm. When a lot of the bearings arrives, the skateboard manufacturer takes an SRS of

5 bearings from the lot and measures their outer diameters. The manufacturer rejects the bearings if the sample mean diameter is significantly different from 22 at the 5% significance level.

This is a test of the hypotheses

$$H_0: \mu = 22$$
$$H_a: \mu \neq 22$$

To carry out the test, the manufacturer computes the z statistic:

$$z = \frac{\bar{x} - 22}{0.01/\sqrt{5}}$$

and rejects H_0 if

$$z < -1.96 \quad \text{or} \quad z > 1.96$$

A Type I error is to reject H_0 when in fact $\mu = 22$.

What about Type II errors? Because there are many values of μ in H_a, we will concentrate on one value. The producer and the manufacturer agree that a lot of bearings with mean 0.015 cm away from the desired mean 22.000 should be rejected. So a particular Type II error is to accept H_0 when in fact $\mu = 22.015$.

Figure 6.19 shows how the two probabilities of error are obtained from the two sampling distributions of $\bar{x}$, for $\mu = 22$ and for $\mu = 22.015$. When $\mu = 22$, H_0 is true and to reject H_0 is a Type I error. When $\mu = 22.015$, accepting H_0 is a Type II error. We will now calculate these error probabilities.

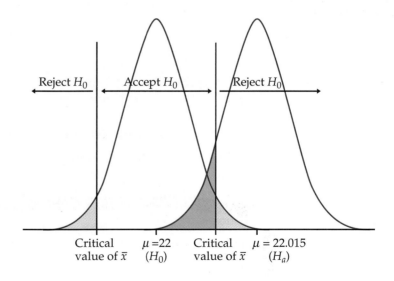

FIGURE 6.19 The two error probabilities for Example 6.32. The probability of a Type I error (yellow area) is the probability of rejecting $H_0: \mu = 22$ when in fact $\mu = 22$. The probability of a Type II error (blue area) is the probability of accepting H_0 when in fact $\mu = 22.015$.

The probability of a Type I error is the probability of rejecting H_0 when it is really true. In Example 6.32, this is the probability that $|z| \geq 1.96$ when $\mu = 22$. But this is exactly the significance level of the test. The critical value 1.96 was chosen to make this probability 0.05, so we do not have to compute it again. The definition of "significant at level 0.05" is that sample outcomes this extreme will occur with probability 0.05 when H_0 is true.

SIGNIFICANCE AND TYPE I ERROR

The significance level α of any fixed level test is the probability of a Type I error. That is, α is the probability that the test will reject the null hypothesis H_0 when H_0 is in fact true.

The probability of a Type II error for the particular alternative $\mu = 22.015$ in Example 6.32 is the probability that the test will fail to reject H_0 when μ has this alternative value. The *power* of the test against the alternative $\mu = 22.015$ is just the probability that the test *does* reject H_0. By following the method of Example 6.30, we can calculate that the power is about 0.92. The probability of a Type II error is therefore $1 - 0.92$, or 0.08.

POWER AND TYPE II ERROR

The power of a fixed level test to detect a particular alternative is 1 minus the probability of a Type II error for that alternative.

The two types of error and their probabilities give another interpretation of the significance level and power of a test. The distinction between tests of significance and tests as rules for deciding between two hypotheses does not lie in the calculations *but in the reasoning that motivates the calculations.* In a test of significance we focus on a single hypothesis (H_0) and a single probability (the P-value). The goal is to measure the strength of the sample evidence against H_0. Calculations of power are done to check the sensitivity of the test. If we cannot reject H_0, we conclude only that there is not sufficient evidence against H_0, not that H_0 is actually true. If the same inference problem is thought of as a decision problem, we focus on two hypotheses and give a rule for deciding between them based on the sample evidence. We therefore must focus equally on two probabilities, the probabilities of the two types of error. We must choose one hypothesis and cannot abstain on grounds of insufficient evidence.

The common practice of testing hypotheses

Such a clear distinction between the two ways of thinking is helpful for understanding. In practice, the two approaches often merge. We continued to call one of the hypotheses in a decision problem H_0. The common practice of *testing hypotheses* mixes the reasoning of significance tests and decision rules as follows:

1. State H_0 and H_a just as in a test of significance.

2. Think of the problem as a decision problem, so that the probabilities of Type I and Type II errors are relevant.

3. Because of Step 1, Type I errors are more serious. So choose an α (significance level) and consider only tests with probability of Type I error no greater than α.

4. Among these tests, select one that makes the probability of a Type II error as small as possible (that is, power as large as possible). If this probability

is too large, you will have to take a larger sample to reduce the chance of an error.

Testing hypotheses may seem to be a hybrid approach. It was, historically, the effective beginning of decision-oriented ideas in statistics. An impressive mathematical theory of hypothesis testing was developed between 1928 and 1938 by Jerzy Neyman and Egon Pearson. The decision-making approach came later (1940s). Because decision theory in its pure form leaves you with two error probabilities and no simple rule on how to balance them, it has been used less often than either tests of significance or tests of hypotheses. Decision ideas have been applied in testing problems mainly by way of the Neyman-Pearson hypothesis-testing theory. That theory asks you first to choose α, and the influence of Fisher has often led users of hypothesis testing comfortably back to $\alpha = 0.05$ or $\alpha = 0.01$. Fisher, who was exceedingly argumentative, violently attacked the Neyman-Pearson decision-oriented ideas, and the argument still continues.

SECTION 6.4 Summary

The **power** of a significance test measures its ability to detect an alternative hypothesis. The power against a specific alternative is calculated as the probability that the test will reject H_0 when that alternative is true. This calculation requires knowledge of the sampling distribution of the test statistic under the alternative hypothesis. Increasing the size of the sample increases the power when the significance level remains fixed.

An alternative to significance testing regards H_0 and H_a as two statements of equal status that we must decide between. This **decision theory** point of view regards statistical inference in general as giving rules for making decisions in the presence of uncertainty.

In the case of testing H_0 versus H_a, decision analysis chooses a decision rule on the basis of the probabilities of two types of error. A **Type I error** occurs if H_0 is rejected when it is in fact true. A **Type II error** occurs if H_0 is accepted when in fact H_a is true.

In a fixed level α significance test, the significance level α is the probability of a Type I error, and the power against a specific alternative is 1 minus the probability of a Type II error for that alternative.

SECTION 6.4 Exercises

6.104 Make a recommendation. Your manager has asked you to review a research proposal that includes a section on sample size justification. A careful reading of this section indicates that the power is 35% for detecting an effect that you would consider important. Write a short report for your manager explaining what this means and make a recommendation on whether or not this study should be run.

6.105 Explain power and sample size. Two studies are identical in all respects except for the sample sizes. Consider the power versus a particular sample size. Will the study with the larger sample size have more power or less power than the one with the smaller sample size? Explain your answer in terms that could be understood by someone with very little knowledge of statistics.

6.106 Power for a different alternative. The power for a two-sided test of the null hypothesis $\mu = 0$ versus the alternative $\mu = 3$ is 0.82. What is the power versus the alternative $\mu = -3$? Explain your answer.

6.107 More on the power for a different alternative. A one-sided test of the null hypothesis $\mu = 20$ versus the alternative $\mu = 30$ has power equal to 0.6. Will the power for the alternative $\mu = 40$ be higher or lower than 0.6? Draw a picture and use this to explain your answer.

6.108 **Power of the random north-south distribution of trees test.** In Exercise 6.66 (page 380) you performed a two-sided significance test of the null hypothesis that the average north-south location of the longleaf pine trees sampled in the Wade Tract was $\mu = 100$. There were 584 trees in the sample and the standard deviation was assumed to be 58. The sample mean in that analysis was $\bar{x} = 99.74$. Use the *Power* applet to compute the power for the alternative $\mu = 99$ using a two-sided test at the 5% level of significance.

6.109 **Power of the random east-west distribution of trees test.** Refer to the previous exercise. Note that in the east-west direction, the average location was 113.8. Use the *Power* applet to find the power for the alternative $\mu = 110$.

6.110 Planning another test to compare cholesterol levels. Example 6.15 (page 371) gives a test of a hypothesis about the cholesterol level for female sedentary undergraduates based on a sample of size $n = 71$. The hypotheses are

$$H_0: \mu = 168$$
$$H_a: \mu \neq 168$$

While the result was not statistically significant, it did provide some evidence that the mean was larger than 168. Thus, the researcher plans to recruit another sample of sedentary females but this time using a one-sided alternative. He plans to obtain $n = 70$ subjects and wonders if this sample size gives him adequate power to detect an increase of 5 mg/dl to $\mu = 173$.

(a) Given $\alpha = 0.05$, for what values of z will he reject the null hypothesis?

(b) Using $\sigma = 27$ and $\mu = 168$, for what values of $\bar{x}$ will he reject H_0?

(c) Using $\sigma = 27$ and $\mu = 173$, what is the probability $\bar{x}$ will fall in the region defined in part (b)?

(d) Does it appear he has adequate power for a sample size of $n = 70$? Or does he need to find ways to increase it? Explain your answer.

6.111 Power of the mean SAT score test. Example 6.16 (page 372) gives a test of a hypothesis about the SAT scores of California high school students based

on an SRS of 500 students. The hypotheses are

$$H_0: \mu = 450$$
$$H_a: \mu > 450$$

Assume that the population standard deviation is $\sigma = 100$. The test rejects H_0 at the 1% level of significance when $z \geq 2.326$, where

$$z = \frac{\bar{x} - 450}{100/\sqrt{500}}$$

Is this test sufficiently sensitive to usually detect an increase of 10 points in the population mean SAT score? Answer this question by calculating the power of the test against the alternative $\mu = 460$.

6.112 **Choose the appropriate distribution.** You must decide which of two discrete distributions a random variable X has. We will call the distributions p_0 and p_1. Here are the probabilities they assign to the values x of X:

x	0	1	2	3	4	5	6
p_0	0.1	0.1	0.2	0.1	0.1	0.1	0.3
p_1	0.2	0.2	0.2	0.1	0.1	0.1	0.1

You have a single observation on X and wish to test

$$H_0: p_0 \text{ is correct}$$
$$H_a: p_1 \text{ is correct}$$

One possible decision procedure is to reject H_0 only if $X \leq 2$.

(a) Find the probability of a Type I error, that is, the probability that you reject H_0 when p_0 is the correct distribution.

(b) Find the probability of a Type II error.

6.113 Computer-assisted career guidance systems. A wide variety of computer-assisted career guidance systems have been developed over the last decade. These programs use factors such as student interests, aptitude, skills, personality, and family history to recommend a career path. For simplicity, suppose a program either recommends a high school graduate either go to college or join the workforce.

(a) What are the two hypotheses and the two types of error that the program can make?

(b) The program can be adjusted to decrease one error probability at the cost of an increase in the other error probability. Which error probability would you choose to make smaller, and why? (This is a matter of judgment. There is no single correct answer.)

CHAPTER 6 Exercises

6.114 Telemarketing wages. An advertisement in the student newspaper asks you to consider working for a telemarketing company. The ad states, "Earn between $500 and $1000 per week." Do you think that the ad is describing a confidence interval? Explain your answer.

6.115 Exercise and statistics exams. A study examined whether exercise affects how students perform on their final exam in statistics. The P-value was given as 0.38.

(a) State null and alternative hypotheses that could be used for this study. (Note: There is more than one correct answer.)

(b) Do you reject the null hypothesis? State your conclusion in plain language.

(c) What other facts about the study would you like to know for a proper interpretation of the results?

6.116 🔺 **Stress by occupation.** As part of a study on the impact of job stress on smoking, researchers used data from the Health and Retirement Study (HRS) to collect information on 3825 ever-smoker individuals who were 50 to 64 years of age.[31] An ever-smoker is someone that was a smoker at some time in his or her life. The HRS is a biennial survey, thus providing the researchers with 17,043 person-year observations. One of the questions on the survey asked a participant how much he or she agrees or disagrees with the statement "My job involves a lot of stress." The answers were coded as a 1 if a participant "strongly agreed" and 0 otherwise. The following table summarizes these responses by occupation.

Occupation	$\hat{p}$	n
Professional	0.23	2447
Managerial	0.22	2552
Administrative	0.17	2309
Sales	0.15	1811
Mechanical	0.12	1979
Service	0.13	2592
Operator	0.12	2782
Farm	0.08	571

(a) Because these responses are binary, use the formula for the standard deviation of a sample proportion (page 322) and construct 95% confidence intervals for each occupation.

(b) Summarize the results. Do there appear to be certain groups of occupations with similar stress levels?

(c) A friend questions the use of the standard deviation formula in part (a). Refer back to the binomial setting (page 314). What might your friend be concerned with?

6.117 🔺 **Workers' perceptions about safety.** The Safety Climate Index (SCI) measures workers' perceptions about the safety of their work environment. A study of safe work practices of industrial workers reported mean SCI scores for workers classified by workplace size.[32] Here are the means:

Workplace size	Fewer than 50 workers	50 to 200 workers	More than 200 workers
Mean SCI	67.23	70.37	74.83

Assume that the standard deviation is 19 and the sample sizes are all 180.

(a) Calculate the 95% confidence interval for each mean.

(b) Plot the means versus workplace size. Draw a vertical line through the first mean extending up to the upper confidence limit and down to the lower limit. At the ends of the line, draw a short dash. Do the same for each of the other means.

(c) One way to adjust for the fact that we are reporting three confidence intervals is a procedure that uses a larger value of z^* in the calculation of the margin of error. For this problem one recommendation would be to use $z^* = 2.40$. Repeat parts (a) and (b) making this adjustment.

(d) Summarize your results. Be sure to include comments on the effects of the adjustment on your results.

6.118 🔺 🔺 **Coverage percent of 95% confidence interval.** For this exercise you will use the *Confidence Interval* applet. Set the confidence level at 95% and click the "Sample" button 10 times to simulate 10 confidence intervals. Record the percent hit. Simulate another 10 intervals by clicking another 10 times (do not click the "Reset" button). Record the percent hit for your 20 intervals. Repeat the process of simulating 10 additional intervals and recording the results until you have a total of 200 intervals. Plot your results and write a summary of what you have found.

6.119 🔺 🔺 **Coverage percent of 90% confidence interval.** Refer to the previous exercise. Do the simulations and report the results for 90% confidence.

6.120 🔺 **Effect of sample size on significance.** You are testing the null hypothesis that $\mu = 0$ versus the alternative $\mu > 0$ using $\alpha = 0.05$. Assume $\sigma = 14$. Suppose $\bar{x} = 4$ and $n = 10$. Calculate the test statistic and its P-value. Repeat assuming the same value of $\bar{x}$ but with

$n = 20$. Do the same for sample sizes of 30, 40, and 50. Plot the values of the test statistic versus the sample size. Do the same for the P-values. Summarize what this demonstration shows about the effect of the sample size on significance testing.

6.121 ⚠ **Blood phosphorus level in dialysis patients.** Patients with chronic kidney failure may be treated by dialysis, using a machine that removes toxic wastes from the blood, a function normally performed by the kidneys. Kidney failure and dialysis can cause other changes, such as retention of phosphorus, that must be corrected by changes in diet. A study of the nutrition of dialysis patients measured the level of phosphorus in the blood of several patients on six occasions. Here are the data for one patient (in milligrams of phosphorus per deciliter of blood):[33]

5.4 5.2 4.5 4.9 5.7 6.3

The measurements are separated in time and can be considered an SRS of the patient's blood phosphorus level. Assume that this level varies Normally with $\sigma = 0.9$ mg/dl. 🔵 PHOSPHORUS

(a) Give a 95% confidence interval for the mean blood phosphorus level.

(b) The normal range of phosphorus in the blood is considered to be 2.6 to 4.8 mg/dl. Is there strong evidence that this patient has a mean phosphorus level that exceeds 4.8?

6.122 Cellulose content in alfalfa hay. An agronomist examines the cellulose content of a variety of alfalfa hay. Suppose that the cellulose content in the population has standard deviation $\sigma = 8$ milligrams per gram (mg/g). A sample of 15 cuttings has mean cellulose content $\bar{x} = 145$ mg/g.

(a) Give a 90% confidence interval for the mean cellulose content in the population.

(b) A previous study claimed that the mean cellulose content was $\mu = 140$ mg/g, but the agronomist believes that the mean is higher than that figure. State H_0 and H_a and carry out a significance test to see if the new data support this belief.

(c) The statistical procedures used in parts (a) and (b) are valid when several assumptions are met. What are these assumptions?

6.123 Odor threshold of future wine experts. Many food products contain small quantities of substances that would give an undesirable taste or smell if they are present in large amounts. An example is the "off-odors" caused by sulfur compounds in wine. Oenologists (wine experts) have determined the odor threshold, the lowest concentration of a compound that the human nose can detect. For example, the odor threshold for dimethyl sulfide (DMS) is given in the oenology literature as 25 micrograms per liter of wine (μg/l). Untrained noses may be less sensitive, however. Here are the DMS odor thresholds for 10 beginning students of oenology:

31 31 43 36 23 34 32 30 20 24

Assume (this is not realistic) that the standard deviation of the odor threshold for untrained noses is known to be $\sigma = 7 \; \mu$g/l. 🔵 WINEODOR

(a) Make a stemplot to verify that the distribution is roughly symmetric with no outliers. (A Normal quantile plot confirms that there are no systematic departures from Normality.)

(b) Give a 95% confidence interval for the mean DMS odor threshold among all beginning oenology students.

(c) Are you convinced that the mean odor threshold for beginning students is higher than the published threshold, 25 μg/l? Carry out a significance test to justify your answer.

6.124 ⚠ **Where do you buy?** Consumers can purchase nonprescription medications at food stores, mass merchandise stores such as Kmart and Wal-Mart, or pharmacies. About 45% of consumers make such purchases at pharmacies. What accounts for the popularity of pharmacies, which often charge higher prices?

A study examined consumers' perceptions of overall performance of the three types of stores, using a long questionnaire that asked about such things as "neat and attractive store," "knowledgeable staff," and "assistance in choosing among various types of nonprescription medication." A performance score was based on 27 such questions. The subjects were 201 people chosen at random from the Indianapolis telephone directory. Here are the means and standard deviations of the performance scores for the sample:[34]

Store type	$\bar{x}$	s
Food stores	18.67	24.95
Mass merchandisers	32.38	33.37
Pharmacies	48.60	35.62

We do not know the population standard deviations, but a sample standard deviation s from so large a sample is usually close to σ. Use s in place of the unknown σ in this exercise.

(a) What population do you think the authors of the study want to draw conclusions about? What population are you certain they can draw conclusions about?

(b) Give 95% confidence intervals for the mean performance for each type of store.

(c) Based on these confidence intervals, are you convinced that consumers think that pharmacies offer higher performance than the other types of stores? (In Chapter 12, we will study a statistical method for comparing means of several groups.)

6.125 CEO pay. A study of the pay of corporate chief executive officers (CEOs) examined the increase in cash compensation of the CEOs of 104 companies, adjusted for inflation, in a recent year. The mean increase in real compensation was $\bar{x} = 6.9\%$, and the standard deviation of the increases was $s = 55\%$. Is this good evidence that the mean real compensation μ of all CEOs increased that year? The hypotheses are

$$H_0: \mu = 0 \quad \text{(no increase)}$$
$$H_a: \mu > 0 \quad \text{(an increase)}$$

Because the sample size is large, the sample s is close to the population σ, so take $\sigma = 55\%$.

(a) Sketch the Normal curve for the sampling distribution of $\bar{x}$ when H_0 is true. Shade the area that represents the P-value for the observed outcome $\bar{x} = 6.9\%$.

(b) Calculate the P-value.

(c) Is the result significant at the $\alpha = 0.05$ level? Do you think the study gives strong evidence that the mean compensation of all CEOs went up?

6.126 Meaning of "statistically significant." When asked to explain the meaning of "statistically significant at the $\alpha = 0.01$ level," a student says, "This means there is only probability 0.01 that the null hypothesis is true." Is this an essentially correct explanation of statistical significance? Explain your answer.

6.127 More on the meaning of "statistically significant." Another student, when asked why statistical significance appears so often in research reports, says, "Because saying that results are significant tells us that they cannot easily be explained by chance variation alone." Do you think that this statement is essentially correct? Explain your answer.

6.128 Roulette. A roulette wheel has 18 red slots among its 38 slots. You observe many spins and record the number of times that red occurs. Now you want to use these data to test whether the probability of a red has the value that is correct for a fair roulette wheel. State the hypotheses H_0 and H_a that you will test.

6.129 Simulation study of the confidence interval. Use a computer to generate $n = 12$ observations from a Normal distribution with mean 25 and standard deviation 4: $N(25, 4)$. Find the 95% confidence interval for μ. Repeat this process 100 times and then count the number of times that the confidence interval includes the value $\mu = 25$. Explain your results.

6.130 Simulation study of a test of significance. Use a computer to generate $n = 12$ observations from a Normal distribution with mean 25 and standard deviation 4: $N(25, 4)$. Test the null hypothesis that $\mu = 25$ using a two-sided significance test. Repeat this process 100 times and then count the number of times that you reject H_0. Explain your results.

6.131 Another simulation study of a test of significance. Use the same procedure for generating data as in the previous exercise. Now test the null hypothesis that $\mu = 23$. Explain your results.

6.132 The handshake during employment interviews. Nonverbal clues, such as eye contact and smiling, have been shown to positively influence the assessment of an interview. Because a firm handshake is often viewed as a sign of confidence and strength, it is thought that it may also influence the assessment. To investigate this, some researchers recruited 98 undergraduate students enrolled in an elective, one-credit career preparation course and had them participate in a mock interview.[35] The following table of means, broken down by gender, summarizes the interviewer's impression of a series of characteristics associated with the interview. These impressions were all rated on a 1 to 5 scale with 1 representing "weak" and 5 representing "strong." There were 50 females and 48 males in the study.

Characteristic	Men	Women
Conscientiousness	3.80	3.88
Extraversion	3.88	3.79
Agreeableness	3.72	3.94
Emotional stability	3.58	3.43
Openness to experience	3.50	3.45
Overall handshake	3.70	3.47
Handshake strength	3.64	3.11
Handshake vigor	3.42	3.25
Handshake grip	3.89	3.51
Handshake duration	3.65	3.50
Eye contact	3.90	3.96
Professional dress	4.33	4.53
Interviewer assessment	3.72	3.83

For each characteristic compute the z statistic and the associated P-value for the comparison between the two groups. For all characteristics but the overall interviewer assessment, assume that the standard deviation of the difference is 0.10, so z is simply the difference in the means divided by 0.10. For interviewer assessment, assume the standard deviation of the difference is 0.19. Note that you are performing 13 significance tests in this exercise. Keep this in mind when you interpret your results. Write a report summarizing your work.

6.133 Find published studies with confidence intervals. Search the Internet or some journals that report research in your field and find two reports that provide an estimate with a margin of error or a confidence interval. For each report,

(a) describe the method used to collect the data.

(b) describe the variable being studied.

(c) give the estimate and the confidence interval.

(d) describe any practical difficulties that may have led to errors in addition to the sampling errors quantified by the margin of error.

Inference for Distributions

Introduction

We began our study of data analysis in Chapter 1 by learning graphical and numerical tools for describing the distribution of a single variable and for comparing several distributions. Our study of the practice of statistical inference begins in the same way, with inference about a single distribution and comparison of two distributions. Comparing more than two distributions requires more elaborate methods, which are presented in Chapters 12 and 13.

Two important aspects of any distribution are its center and spread. If the distribution is Normal, we describe its center by the mean μ and its spread by the standard deviation σ. In this chapter, we will meet confidence intervals and significance tests for inference about a population mean μ and the difference between two population means $\mu_1 - \mu_2$. The previous chapter emphasized the reasoning of significance tests and confidence intervals; now we emphasize statistical practice, so we no longer assume that population standard deviations are known. The t procedures for inference about means are among the most commonly used statistical methods.

The methods in this chapter will allow us to address questions such as,

- How many hours per month, on average, does a college student spend watching streaming videos on a mobile phone?

- Do male and female college students differ in terms of "social insight," the ability to appraise other people?

- Does the daily number of disruptive behaviors in dementia patients change when there is a full moon?

7.1 Inference for the Mean of a Population

LOOK BACK
sampling distribution of $\bar{x}$
p. 302

Both confidence intervals and tests of significance for the mean μ of a Normal population are based on the sample mean $\bar{x}$, which estimates the unknown μ. The sampling distribution of $\bar{x}$ depends on σ. This fact causes no difficulty when σ is known. When σ is unknown, however, we must estimate σ even though we are primarily interested in μ. The sample standard deviation s is used to estimate the population standard deviation σ.

The *t* distributions

Suppose that we have a simple random sample (SRS) of size n from a Normally distributed population with mean μ and standard deviation σ. The sample mean $\bar{x}$ is then Normally distributed with mean μ and standard deviation $\sigma/\sqrt{n}$. When σ is not known, we estimate it with the sample standard deviation s, and then we estimate the standard deviation of $\bar{x}$ by $s/\sqrt{n}$. This quantity is called the *standard error* of the sample mean $\bar{x}$ and we denote it by $\text{SE}_{\bar{x}}$.

STANDARD ERROR

When the standard deviation of a statistic is estimated from the data, the result is called the **standard error** of the statistic. The standard error of the sample mean is

$$\text{SE}_{\bar{x}} = \frac{s}{\sqrt{n}}$$

The term "standard error" is sometimes used for the actual standard deviation of a statistic. The estimated value is then called the "estimated standard error." In this book we will use the term "standard error" only when the standard deviation of a statistic is estimated from the data. The term has this meaning in the output of many statistical computer packages and in research reports that apply statistical methods.

The standardized sample mean, or one-sample z statistic,

$$z = \frac{\bar{x} - \mu}{\sigma/\sqrt{n}}$$

is the basis of the z procedures for inference about μ when σ is known. This statistic has the standard Normal distribution $N(0, 1)$. When we substitute the standard error $s/\sqrt{n}$ for the standard deviation $\sigma/\sqrt{n}$ of $\bar{x}$, the statistic does *not* have a Normal distribution. It has a distribution that is new to us, called a *t distribution*.

THE *t* DISTRIBUTIONS

Suppose that an SRS of size n is drawn from an $N(\mu, \sigma)$ population. Then the **one-sample *t* statistic**

$$t = \frac{\overline{x} - \mu}{s/\sqrt{n}}$$

has the ***t* distribution** with $n - 1$ **degrees of freedom.**

LOOK BACK
degrees of freedom
p. 41

A particular t distribution is specified by giving the *degrees of freedom*. We use $t(k)$ to stand for the t distribution with k degrees of freedom. The degrees of freedom for this t statistic come from the sample standard deviation s in the denominator of t. We showed earlier that s has $n - 1$ degrees of freedom. Thus, there is a different t distribution for each sample size. There are also other t statistics with different degrees of freedom, some of which we will meet later in this chapter.

The t distributions were discovered in 1908 by William S. Gosset. Gosset was a statistician employed by the Guinness brewing company, which prohibited its employees from publishing their discoveries that were brewing related. In this case, the company let him publish under the pen name "Student" using an example that did not involve brewing. The t distribution is often called "Student's t" in his honor.

The density curves of the $t(k)$ distributions are similar in shape to the standard Normal curve. That is, they are symmetric about 0 and are bell shaped. Figure 7.1 compares the density curves of the standard Normal distribution and the t distributions with 5 and 10 degrees of freedom. The similarity in shape is apparent, as is the fact that the t distributions have more probability in the tails and less in the center. This greater spread is due to the extra variability

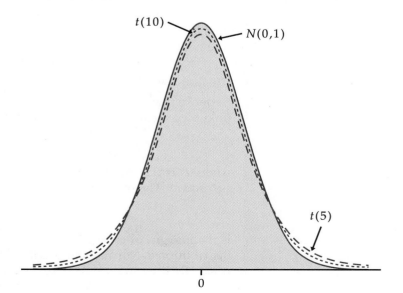

FIGURE 7.1 Density curves for the standard Normal, $t(10)$, and $t(5)$ distributions. All are symmetric with center 0. The t distributions have more probability in the tails than the standard Normal distribution.

caused by substituting the random variable s for the fixed parameter σ. Figure 7.1 also shows that as the degrees of freedom k increase, the $t(k)$ density curve gets closer to the $N(0, 1)$ curve. This reflects the fact that s will be closer to σ as the sample size increases.

Table D in the back of the book gives critical values t^* for the t distributions. For convenience, we have labeled the table entries both by the value of p needed for significance tests and by the confidence level C (in percent) required for confidence intervals. The standard Normal critical values are in the bottom row of entries and labeled z^*. As in the case of the Normal table (Table A), computer software often makes Table D unnecessary.

USE YOUR KNOWLEDGE

7.1   Apartment rates. You randomly choose 16 unfurnished one-bedroom apartments from a large number of advertisements in your local newspaper. You calculate that their mean monthly rent is \$613 and their standard deviation is \$96.

   (a) What is the standard error of the mean?

   (b) What are the degrees of freedom for a one-sample t statistic?

7.2   Finding critical t^* values. What critical value t^* from Table D should be used to construct

   (a) a 95% confidence interval when $n = 11$?

   (b) a 99% confidence interval when $n = 33$?

   (c) a 90% confidence interval when $n = 250$?

The one-sample t confidence interval

LOOK BACK
z confidence interval
p. 349

With the t distributions to help us, we can now analyze a sample from a Normal population with unknown σ. The one-sample t confidence interval is similar in both reasoning and computational detail to the z confidence interval of Chapter 6. There, the margin of error for the population mean was $z^*\sigma/\sqrt{n}$. Here, we replace σ by its estimate s and z^* by t^*. This means that the margin of error for the population mean when we use the data to estimate σ is $t^*s/\sqrt{n}$.

THE ONE-SAMPLE t CONFIDENCE INTERVAL

Suppose that an SRS of size n is drawn from a population having unknown mean μ. A level C confidence interval for μ is

$$\bar{x} \pm t^*\frac{s}{\sqrt{n}}$$

where t^* is the value for the $t(n - 1)$ density curve with area C between $-t^*$ and t^*. The quantity

$$t^*\frac{s}{\sqrt{n}}$$

is the **margin of error.** This interval is exact when the population distribution is Normal and is approximately correct for large n in other cases.

EXAMPLE

7.1 Watching videos on a mobile phone. The Nielsen Company is a global information and media company and one of the leading suppliers of media information. Recently, they reported that U.S. mobile phone subscribers average 3.7 hours per month watching videos on their phone.[1] Suppose we want to determine a 95% confidence interval for the average among U.S. college students and draw the following SRS of size 8 from this population:

$$7\ \ 9\ \ 1\ \ 6\ \ 13\ \ 10\ \ 3\ \ 5$$

The sample mean is

$$\bar{x} = \frac{7 + 9 + \cdots + 5}{8} = 6.75$$

and the standard deviation is

$$s = \sqrt{\frac{(7 - 6.75)^2 + (9 - 6.75)^2 + \cdots + (5 - 6.75)^2}{8 - 1}} = 3.88$$

with degrees of freedom $n - 1 = 7$. The standard error is

$$SE_{\bar{x}} = s/\sqrt{n} = 3.88/\sqrt{8} = 1.37$$

From Table D we find $t^* = 2.365$. The 95% confidence interval is

$$\bar{x} \pm t^* \frac{s}{\sqrt{n}} = 6.75 \pm 2.365 \frac{3.88}{\sqrt{8}}$$

$$= 6.75 \pm (2.365)(1.37)$$

$$= 6.75 \pm 3.25$$

$$= (3.5,\ 10.0)$$

We are 95% confident that among U.S. college students the average time watching videos on a mobile phone is between 3.5 and 10.0 hours per month.

In this example we have given the actual interval (3.5, 10.0) as our answer. Sometimes we prefer to report the mean and margin of error: the mean time is 6.75 hours per month with a margin of error of 3.25 hours.

The use of the *t* confidence interval in Example 7.1 rests on assumptions that appear reasonable here. First, we assume our random sample is an SRS from the U.S. population of college student mobile phone users. Second, we assume the distribution of watching times is Normal. With only 8 observations, this assumption cannot be effectively checked. In fact, because the watching time cannot be negative, we might expect this distribution to be skewed to the right. With these data, however, there are no extreme outliers to suggest a severe departure from Normality.

USE YOUR KNOWLEDGE

7.3 More on apartment rents. Recall Exercise 7.1 (page 406). Construct a 95% confidence interval for the mean monthly rent of all advertised one-bedroom apartments.

7.4 90% versus 95% confidence interval. If you were to use 90% confidence, rather than 95% confidence, would the margin of error be larger or smaller? Explain your answer.

The one-sample *t* test

LOOK BACK
z significance test
p. 371

Significance tests using the standard error are also very similar to the *z* test that we studied in the last chapter.

THE ONE-SAMPLE *t* TEST

Suppose that an SRS of size n is drawn from a population having unknown mean μ. To test the hypothesis $H_0: \mu = \mu_0$ based on an SRS of size n, compute the one-sample *t* statistic

$$t = \frac{\bar{x} - \mu_0}{s/\sqrt{n}}$$

In terms of a random variable T having the $t(n-1)$ distribution, the *P*-value for a test of H_0 against

$H_a: \mu > \mu_0$ is $P(T \geq t)$

$H_a: \mu < \mu_0$ is $P(T \leq t)$

$H_a: \mu \neq \mu_0$ is $2P(T \geq |t|)$

These *P*-values are exact if the population distribution is Normal and are approximately correct for large n in other cases.

EXAMPLE

7.2 Significance test for mobile phone use. Suppose we want to test whether the U.S. college student average is different from the reported overall U.S. average at the 0.05 significance level. Specifically, we want to test

$$H_0: \mu = 3.7$$
$$H_a: \mu \neq 3.7$$

Recall that $n = 8$, $\bar{x} = 6.75$, and $s = 3.88$. The *t* test statistic is

$$t = \frac{\bar{x} - \mu_0}{s/\sqrt{n}} = \frac{6.75 - 3.7}{3.88/\sqrt{8}}$$
$$= 2.22$$

This means that the sample mean $\bar{x} = 6.75$ is slightly less than 2.25 standard deviations away from the null hypothesized value $\mu = 3.7$. Because the degrees of freedom are $n - 1 = 7$, this *t* statistic has the $t(7)$ distribution. Figure 7.2 shows that the *P*-value is $2P(T \geq 2.22)$, where T has the $t(7)$ distribution. From Table D we see that $P(T \geq 1.895) = 0.05$ and $P(T \geq 2.365) = 0.025$.

VIDEOPHONE

df = 7		
p	0.05	0.025
t*	1.895	2.365

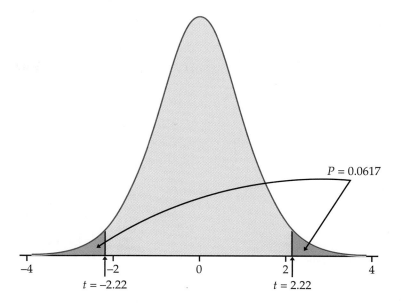

$P = 0.0617$

-4 -2 0 2 4

$t = -2.22$ $t = 2.22$

FIGURE 7.2 Sketch of the *P*-value calculation for Example 7.2.

Therefore, we conclude that the *P*-value is between $2 \times 0.025 = 0.05$ and $2 \times 0.05 = 0.10$. Software gives the exact value as $P = 0.0617$. These data are compatible with a mean of 3.7 hours per month at the $\alpha = 0.05$ level. In other words, there is not enough evidence to reject the null hypothesis.

In this example we tested the null hypothesis $\mu = 3.7$ hours per month against the two-sided alternative $\mu \neq 3.7$ hours per month because we had no prior suspicion that the average among college students would be larger or smaller. If we had suspected that the average would be larger, we would have used a one-sided test.

EXAMPLE

7.3 One-sided test for mobile phone use. For the mobile phone problem described in the previous example, we want to test whether the U.S. college student average is larger than the overall U.S. population average. Here we test

$$H_0: \mu = 3.7$$

versus

$$H_a: \mu > 3.7$$

The *t* test statistic does not change: $t = 2.22$. As Figure 7.3 illustrates, however, the *P*-value is now $P(T \geq 2.22)$, half of the value in the previous example. From Table D we can determine that $0.025 < P < 0.05$; software gives the exact value as $P = 0.0308$. At the $\alpha = 0.05$ level, we reject the null hypothesis and conclude that the U.S. college student average is larger than the U.S. average.

For the mobile phone problem, our conclusion depends on the choice of a one-sided or two-sided test. This choice needs to be done prior to analysis. If in

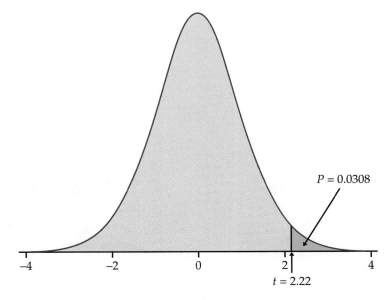

FIGURE 7.3 Sketch of the *P*-value calculation for Example 7.3.

doubt, always use a two-sided test. *It is wrong to examine the data first and then decide to do a one-sided test in the direction indicated by the data.* In the present circumstance we could use our results from Example 7.2 to justify a one-sided test for *another* sample from the same population.

For small data sets, such as the one in Example 7.1, it is easy to perform the computations for confidence intervals and significance tests with an ordinary calculator. For larger data sets, however, we prefer to use software or a statistical calculator.

DIVERSIFY

> ### EXAMPLE
>
> **7.4 Stock portfolio diversification?** An investor with a stock portfolio worth several hundred thousand dollars sued his broker and brokerage firm because lack of diversification in his portfolio led to poor performance. Table 7.1 gives the rates of return for the 39 months that the account was managed by the broker.[2] Figure 7.4 gives a histogram for these data and Figure 7.5 gives the Normal quantile plot. There are no outliers and the distribution shows no strong skewness. We are reasonably confident that the distribution of $\bar{x}$ is approximately Normal, and we proceed with our inference based on Normal theory.

> ### TABLE 7.1

Monthly rates of return on a portfolio (%)

−8.36	1.63	−2.27	−2.93	−2.70	−2.93	−9.14	−2.64
6.82	−2.35	−3.58	6.13	7.00	−15.25	−8.66	−1.03
−9.16	−1.25	−1.22	−10.27	−5.11	−0.80	−1.44	1.28
−0.65	4.34	12.22	−7.21	−0.09	7.34	5.04	−7.24
−2.14	−1.01	−1.41	12.03	−2.56	4.33	2.35	

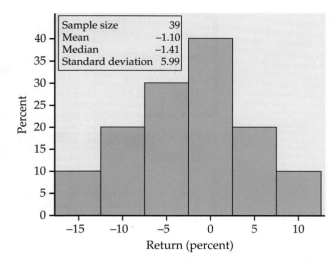

FIGURE 7.4 Histogram for Example 7.4.

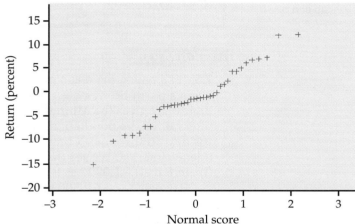

FIGURE 7.5 Normal quantile plot for Example 7.4.

The arbitration panel compared these returns with the average of the Standard and Poor's 500 stock index for the same period. Consider the 39 monthly returns as a random sample from the population of monthly returns the brokerage would generate if it managed the account forever. Are these returns compatible with a population mean of $\mu = 0.95\%$, the S&P 500 average? Our hypotheses are

$$H_0: \mu = 0.95$$
$$H_a: \mu \neq 0.95$$

Minitab and SPSS outputs appear in Figure 7.6. Output from other software will look similar.

Here is one way to report the conclusion: the mean monthly return on investment for this client's account was $\bar{x} = -1.1\%$. This differs significantly from the performance of the S&P 500 stock index for the same period ($t = -2.14$, df = 38, $P = 0.039$).

The hypothesis test in Example 7.4 leads us to conclude that the mean return on the client's account differs from that of the S&P 500 stock index. Now let's assess the return on the client's account with a confidence interval.

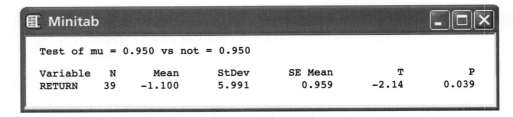

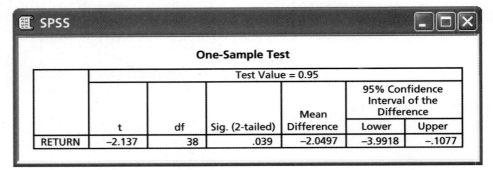

FIGURE 7.6 Minitab and SPSS output for Example 7.4.

EXAMPLE

7.5 Estimating the mean monthly return. The mean monthly return on the client's portfolio was $\bar{x} = -1.1\%$ and the standard deviation was $s = 5.99\%$. Figure 7.7 gives the Minitab, SPSS, and Excel outputs for a 95% confidence interval for the population mean μ. Note that Excel gives the margin of error next to the label "Confidence Level (95.0%)" rather than the actual confidence interval. We see that the 95% confidence interval is $(-3.04, 0.84)$, or (from Excel) -1.0997 ± 1.9420.

Because the S&P 500 return, 0.95%, falls outside this interval, we know that μ differs significantly from 0.95% at the $\alpha = 0.05$ level. Example 7.4 gave the actual P-value as $P = 0.039$.

The confidence interval suggests that the broker's management of this account had a long-term mean somewhere between a loss of 3.04% and a gain of 0.84% per month. We are interested not in the actual mean, but in the difference between the performance of the client's portfolio and that of the diversified S&P 500 stock index.

EXAMPLE

7.6 Estimating the difference from a standard. Following the analysis accepted by the arbitration panel, we are considering the S&P 500 monthly average return as a constant standard. (It is easy to envision scenarios where we would want to treat this type of quantity as random.) The difference between the mean of the investor's account and the S&P 500 is $\bar{x} - \mu = -1.10 - 0.95 = -2.05\%$. In Example 7.5 we found that the 95% confidence interval for the investor's account was $(-3.04, 0.84)$. To obtain the corresponding interval for the difference, subtract 0.95 from each of the endpoints. The resulting interval is $(-3.04-0.95, 0.84-0.95)$, or $(-3.99, -0.11)$. We conclude with 95% confidence that the underperformance was between -3.99% and -0.11%. This interval is presented in the SPSS output of Figure 7.6. This estimate helps to set the compensation owed the investor.

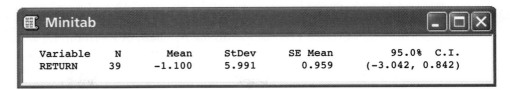

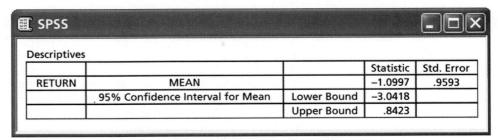

FIGURE 7.7 Minitab, SPSS, and Excel output for Example 7.5.

The assumption that these 39 monthly returns represent an SRS from the population of monthly returns is certainly questionable. If the monthly S&P 500 returns were available, an alternative analysis would be to compare the average difference between the monthly returns for this account and for the S&P 500. This method of analysis is discussed next.

USE YOUR KNOWLEDGE

7.5 Significance test using the t distribution. A test of a null hypothesis versus a two-sided alternative gives $t = 2.18$.

 (a) The sample size is 18. Is the test result significant at the 5% level? Explain how you obtained your answer.

 (b) The sample size is 10. Is the test result significant at the 5% level? Explain how you obtained your answer.

 (c) Sketch the two t distributions to illustrate your answers.

7.6 Significance test for apartment rents. Recall Exercise 7.1 (page 406). Does this SRS give good reason to believe that the mean rent of all advertised one-bedroom apartments is greater than $550? State the hypotheses, find the t statistic and its P-value, and state your conclusion.

7.7 Using software. In Example 7.1 (page 407) we calculated the 95% confidence interval for the U.S. college student average of hours per month spent watching videos on a mobile phone. Use software to compute this interval and verify that you obtain the same interval.

Matched pairs *t* procedures

The mobile phone problem of Example 7.1 concerns only a single population. We know that comparative studies are usually preferred to single-sample investigations because of the protection they offer against confounding. For that reason, inference about a parameter of a single distribution is less common than comparative inference.

LOOK BACK
confounding p. 150

LOOK BACK
matched pairs design
p. 181

One common comparative design, however, makes use of single-sample procedures. In a matched pairs study, subjects are matched in pairs and their outcomes are compared within each matched pair. For example, an experiment to compare two cell phone packages might use pairs of subjects that are the same age, sex, and income level. The experimenter could toss a coin to assign the two packages to the two subjects in each pair. The idea is that matched subjects are more similar than unmatched subjects, so comparing outcomes within each pair is more efficient. Matched pairs are also common when randomization is not possible. One situation calling for matched pairs is when observations are taken on the same subjects, under two different conditions, or before and after some intervention. Here is an example.

EXAMPLE

7.7 Does a full moon affect behavior? Many people believe that the moon influences the actions of some individuals. A study of dementia patients in nursing homes recorded various types of disruptive behaviors every day for 12 weeks. Days were classified as moon days if they were in a three-day period centered at the day of the full moon. For each patient the average number of disruptive behaviors was computed for moon days and for all other days. The data for the 15 subjects whose behaviors were classified as aggressive are presented in Table 7.2.[3] The patients in this study are not a random sample of dementia patients. However, we examine their data in the hope that what we find is not unique to this particular group of individuals and applies to other patients who have similar characteristics.

To analyze these paired data, we first subtract the disruptive behaviors for moon days from the disruptive behaviors for other days. These 15 differences form a single sample. They appear in the "Difference" columns in Table 7.2. The first patient, for example, averaged 3.33 aggressive behaviors

TABLE 7.2

Aggressive behaviors of dementia patients

Patient	Moon days	Other days	Difference	Patient	Moon days	Other days	Difference
1	3.33	0.27	3.06	9	6.00	1.59	4.41
2	3.67	0.59	3.08	10	4.33	0.60	3.73
3	2.67	0.32	2.35	11	3.33	0.65	2.68
4	3.33	0.19	3.14	12	0.67	0.69	−0.02
5	3.33	1.26	2.07	13	1.33	1.26	0.07
6	3.67	0.11	3.56	14	0.33	0.23	0.10
7	4.67	0.30	4.37	15	2.00	0.38	1.62
8	2.67	0.40	2.27				

```
4 | 4 4
3 | 1 1 1 6 7
2 | 1 3 4 7
1 | 6
0 | 0 1 1
```

FIGURE 7.8 Stemplot of differences in aggressive behaviors for Examples 7.7 and 7.8.

on moon days but only 0.27 aggressive behaviors on other days. The difference $3.33 - 0.27 = 3.06$ is what we will use in our analysis.

Next, we examine the distribution of these differences. Figure 7.8 gives a stemplot of the differences. This plot indicates that there are three patients with very small differences, but there are no indications of extreme outliers or strong skewness. We will proceed with our analysis using the Normality-based methods of this section.

To assess whether there is a difference in aggressive behaviors on moon days versus other days, we test

$$H_0: \mu = 0$$
$$H_a: \mu \neq 0$$

Here μ is the mean difference in aggressive behaviors, moon versus other days, for patients of this type. The null hypothesis says that aggressive behaviors occur at the same frequency for both types of days, and H_a says that the behaviors on moon days are not the same as on other days. The 15 differences have

$$\bar{x} = 2.433 \quad \text{and} \quad s = 1.460$$

The one-sample t statistic is therefore

$$t = \frac{\bar{x} - 0}{s/\sqrt{n}} = \frac{2.433}{1.460/\sqrt{15}}$$
$$= 6.45$$

df = 14		
p	0.001	0.0005
t^*	3.787	4.140

The P-value is found from the $t(14)$ distribution (remember that the degrees of freedom are 1 less than the sample size).

Table D shows that 6.45 lies beyond the upper 0.0005 critical value of the $t(14)$ distribution. Since we are using a two-sided alternative, we know that the P-value is less than two times this value, or 0.0010. Software gives a value that is much smaller, $P = 0.000015$. In practice, there is little difference between these two P-values; the data provide clear evidence in favor of the alternative hypothesis. A difference this large is very unlikely to occur by chance if there is, in fact, no effect of the moon on aggressive behaviors. In scholarly publications, the details of routine statistical procedures are omitted; our test would be reported in the form: "There was more aggressive behavior on moon days than on other days ($t = 6.45$, df = 14, $P < 0.001$)."

Note that we could have justified a one-sided alternative in this example. Based on previous research, we expect more aggressive behaviors on moon days, and the alternative $H_a: \mu > 0$ is reasonable in this setting. The choice of the alternative here, however, has no effect on the conclusion: from Table D we determine that P is less than 0.0005; from software it is 0.000008. These are very small values and we would still report $P < 0.001$. *In most circumstances we cannot be absolutely certain about the direction and the safest strategy is to use the two-sided alternative.*

The results of the significance test allow us to conclude that dementia patients exhibit more aggressive behaviors in the days around a full moon. What are the implications of the study for the administrators who run the facilities where these patients live? For example, should they increase staff on these days? To make these kinds of decisions, an estimate of the magnitude of the problem, with a margin of error, would be helpful.

EXAMPLE

7.8 95% confidence interval for the full-moon study. A 95% confidence interval for the mean difference in aggressive behaviors per day requires the critical value $t^* = 2.145$ from Table D. The margin of error is

$$t^* \frac{s}{\sqrt{n}} = 2.145 \frac{1.460}{\sqrt{15}}$$
$$= 0.81$$

and the confidence interval is

$$\bar{x} \pm t^* \frac{s}{\sqrt{n}} = 2.43 \pm 0.81$$
$$= (1.62, \ 3.24)$$

The estimated average difference is 2.43 aggressive behaviors per day, with margin of error 0.81 for 95% confidence. The increase needs to be interpreted in terms of the baseline values. The average number of aggressive behaviors per day on other days is 0.59; on moon days it is 3.02. This is approximately a 400% increase. If aggressive behaviors require a substantial amount of attention by staff, then administrators should be aware of the increased level of these activities during the full-moon period. Additional staff may be needed.

The following are key points to remember concerning matched pairs:

1. A matched pairs analysis is called for when subjects are matched in pairs or there are two measurements or observations on each individual and we want to examine the difference.

2. For each pair or individual, use the difference between the two measurements as the data for your analysis.

3. Use the one-sample confidence interval and significance-testing procedures that we learned in this section.

Use of the t procedures in Examples 7.7 and 7.8 faces several issues. First, no randomization is possible in a study like this. Our inference procedures assume that there is a process that generates these aggressive behaviors and that the process produces them at possibly different rates during the days near the full moon. Second, many of the patients in these nursing homes did not exhibit any disruptive behaviors. These were not included in our analysis, so our inference is restricted to patients who do exhibit disruptive behaviors.

A final difficulty is that the data show departures from Normality. In a matched pairs analysis, the t procedures are applied to the differences, so we are assuming that the differences are Normally distributed. Figure 7.8 gives a stemplot of the differences. There are 3 patients with very small differences in aggressive behaviors while the other 12 have a large increase. We have a dilemma here similar to that in Example 7.1. *The data may not be Normal, and our sample size is very small.* We can try an alternative procedure that does not require the Normality assumption—but there is a price to pay. The alternative procedures have less power to detect differences. Despite these caveats, for Example 7.7 the P-value is so small that we are very confident that we have found an effect of the moon phase on behavior.

USE YOUR KNOWLEDGE

7.8 **Comparison of two energy drinks.** Consider the following study to compare two popular energy drinks. For each subject, a coin was flipped to determine which drink to rate first. Each drink was rated on a 0 to 100 scale, with 100 being the highest rating.

Drink	Subject				
	1	2	3	4	5
A	43	83	66	89	78
B	45	78	64	79	71

Is there a difference in preference? State appropriate hypotheses and carry out a matched pairs *t* test for these data.

7.9 **95% confidence interval for the difference in energy drinks.** For the companies producing these drinks, the real question is how much difference there is between the two preferences. Use the data above to give a 95% confidence interval for the difference in preference between Drink A and Drink B.

Robustness of the *t* procedures

The results of one-sample *t* procedures are exactly correct only when the population is Normal. Real populations are never exactly Normal. The usefulness of the *t* procedures in practice therefore depends on how strongly they are affected by non-Normality. Procedures that are not strongly affected are called *robust*.

ROBUST PROCEDURES

A statistical inference procedure is called **robust** if the required probability calculations are insensitive to violations of the assumptions made.

LOOK BACK
resistant measure p. 31

The assumption that the population is Normal rules out outliers, so the presence of outliers shows that this assumption is not valid. The *t* procedures are not robust against outliers, because $\bar{x}$ and *s* are not resistant to outliers.

In Example 7.7, there are three patients with fairly low values of the difference. Whether or not these are outliers is a matter of judgment. If we rerun the analysis without these three patients, the *t* statistic would increase to 11.89 and the *P*-value would be much lower. Careful inspection of the records may reveal some characteristic of these patients which distinguishes them from the others in the study. Without such information, it is difficult to justify excluding them from the analysis. *In general, we should be very cautious about discarding suspected outliers, particularly when they make up a substantial proportion of the data, as they do in this example.*

Fortunately, the t procedures are quite robust against non-Normality of the population except in the case of outliers or strong skewness. Larger samples improve the accuracy of P-values and critical values from the t distributions when the population is not Normal. This is true for two reasons:

LOOK BACK
central limit theorem
p. 303

1. The sampling distribution of the sample mean $\bar{x}$ from a large sample is close to Normal (that's the central limit theorem). Normality of the individual observations is of little concern when the sample is large.

LOOK BACK
law of large numbers
p. 263

2. As the sample size n grows, the sample standard deviation s will be an accurate estimate of σ whether or not the population has a Normal distribution. This fact is closely related to the law of large numbers.

Constructing a Normal quantile plot, stemplot, or boxplot to check for skewness and outliers is an important preliminary to the use of t procedures for small samples. For most purposes, the one-sample t procedures can be safely used when $n \geq 15$ unless an outlier or clearly marked skewness is present. *Except in the case of small samples, the assumption that the data are an SRS from the population of interest is more crucial than the assumption that the population distribution is Normal.* Here are practical guidelines for inference on a single mean:[4]

- *Sample size less than 15:* Use t procedures if the data are close to Normal. If the data are clearly non-Normal or if outliers are present, do not use t.

- *Sample size at least 15:* The t procedures can be used except in the presence of outliers or strong skewness.

- *Large samples:* The t procedures can be used even for clearly skewed distributions when the sample is large, roughly $n \geq 40$.

Consider, for example, some of the data we studied in Chapter 1. The service center call lengths in Figure 1.30 (page 66) are strongly skewed to the right. Since there are 80 observations, we could use the t procedures here. On the other hand, many would prefer to use a transformation to make these data more nearly Normal. (See the material on inference for non-Normal populations on page 420 and in Chapter 16.) The time to start a business data in Figure 1.31 (page 66) contain one outlier in a sample of size 24, which makes the use of t procedures more risky. Figure 1.32 (page 66) gives the Normal quantile plot for 105 acidity measurements of rainwater. These data appear to be Normal and we would apply the t procedures in this case.

USE YOUR KNOWLEDGE

7.10 t procedures for CO_2 emissions? Consider the CO_2 emissions data presented in Figure 1.36 (page 73). Would you feel comfortable applying the t procedures in this case? Explain your answer.

7.11 t procedures for ticket prices? Consider the data on StubHub! tickets prices presented in Figure 1.33 (page 68). Would you feel comfortable applying the t procedures in this case? In explaining your answer, recall that these t procedures focus on the mean μ.

The power of the *t* test*

The power of a statistical test measures its ability to detect deviations from the null hypothesis. In practice, we carry out the test in the hope of showing that the null hypothesis is false, so high power is important. The power of the one-sample *t* test for a specific alternative value of the population mean μ is the probability that the test will reject the null hypothesis when the alternative value of the mean is true. To calculate the power, we assume a fixed level of significance, often $\alpha = 0.05$.

Calculation of the exact power of the *t* test takes into account the estimation of σ by s and is a bit complex. But an approximate calculation that acts as if σ were known is almost always adequate for planning a study. This calculation is very much like that for the *z* test:

LOOK BACK
power of the *z* test
p. 389

1. Decide on a standard deviation, significance level, whether the test is one-sided or two-sided, and an alternative value of μ to detect.

2. Write the event that the test rejects H_0 in terms of $\bar{x}$.

3. Find the probability of this event when the population mean has this alternative value.

Consider Example 7.7, where we examined the effect of the moon on the aggressive behavior of dementia patients in nursing homes. Suppose that we wanted to perform a similar study in a different setting. How many patients should we include in our new study? To answer this question, we do a power calculation.

In Example 7.7, we found $\bar{x} = 2.433$ and $s = 1.460$. Let's use $s = 1.5$ for our calculations. *It is always better to use a value of the standard deviation that is a little larger than what we expect than one that is smaller.* This may give a sample size that is a little larger than we need. We want to avoid a situation where we fail to find the effect that we are looking for because we did not have enough data. Let's use $\mu = 1.0$ as the alternative value to detect. We are very confident that the effect was larger than this in our previous study, and this amount of an increase in aggressive behavior would still be important to those who work in these facilities. Finally, based on the previous study, we can justify using a one-sided alternative; we expect the moon days to be associated with an increase in aggressive behavior.

EXAMPLE

7.9 Computing the power of a *t* test. Let's compute the power of the *t* test for

$$H_0: \mu = 0$$
$$H_a: \mu > 0$$

when the alternative $\mu = 1.0$. We will use a 5% level of significance. The *t* test with n observations rejects H_0 at the 5% significance level if the *t* statistic

$$t = \frac{\bar{x} - 0}{s/\sqrt{n}}$$

---

*This section can be omitted without loss of continuity.

exceeds the upper 5% point of $t(n-1)$. Taking $n = 20$ and $s = 1.5$, the upper 5% point of $t(19)$ is 1.729. The event that the test rejects H_0 is therefore

$$t = \frac{\bar{x}}{1.5/\sqrt{20}} \geq 1.729$$

$$\bar{x} \geq 1.729 \frac{1.5}{\sqrt{20}}$$

$$\bar{x} \geq 0.580$$

The power is the probability that $\bar{x} \geq 0.580$ when $\mu = 1.0$. Taking $\sigma = 1.5$, this probability is found by standardizing $\bar{x}$:

$$P(\bar{x} \geq 0.580 \text{ when } \mu = 1.0) = P\left(\frac{\bar{x} - 1.0}{1.5/\sqrt{20}} \geq \frac{0.580 - 1.0}{1.5/\sqrt{20}}\right)$$

$$= P(Z \geq -1.25)$$

$$= 1 - 0.1056 = 0.89$$

The power is 89% that we will detect an increase of 1.0 aggressive behavior per day during moon days. This is sufficient power for most situations. For many studies, 80% is considered the standard value for desirable power. We could repeat the calculations for some smaller values of n to determine the smallest value that would meet the 80% criterion.

Power calculations are used in planning studies to ensure that we have a reasonable chance of detecting effects of interest. They give us some guidance in selecting a sample size. In making these calculations, we need assumptions about the standard deviation and the alternative of interest. In our example we assumed that the standard deviation would be 1.5, but in practice we are hoping that the value will be somewhere around this value. Similarly, we have used a somewhat arbitrary alternative of 1.0. This is a guess based on the results of the previous study. *Beware of putting too much trust in fine details of the results of these calculations.* They serve as a guide, not a mandate.

USE YOUR KNOWLEDGE

7.12 Power and the alternative mean μ. If you were to repeat the power calculation in Example 7.9 for a value of μ that is greater than 1, would you expect the power to be higher or lower than 89%? Why?

7.13 More on power and the alternative mean μ. Verify your answer to the previous question by doing the calculation for the alternative $\mu = 1.10$.

Inference for non-Normal populations*

We have not discussed how to do inference about the mean of a clearly non-Normal distribution based on a small sample. If you face this problem, you should consult an expert. Three general strategies are available:

1. In some cases a distribution other than a Normal distribution will describe the data well. There are many non-Normal models for data, and inference procedures for these models are available.

---

*This section can be omitted without loss of continuity.

2. Because skewness is the chief barrier to the use of t procedures on data without outliers, you can attempt to transform skewed data so that the distribution is symmetric and as close to Normal as possible. Confidence levels and P-values from the t procedures applied to the transformed data will be quite accurate for even moderate sample sizes.

distribution-free procedures

nonparametric procedures

3. Use a **distribution-free** inference procedure. Such procedures do not assume that the population distribution has any specific form, such as Normal. Distribution-free procedures are often called **nonparametric procedures.** Chapter 15 discusses several of these procedures.

Each of these strategies can be effective, but each quickly carries us beyond the basic practice of statistics. We emphasize procedures based on Normal distributions because they are the most common in practice, because their robustness makes them widely useful, and (most important) because we are first of all concerned with understanding the principles of inference. We will therefore not discuss procedures for non-Normal continuous distributions. We will be content with illustrating by example the use of a transformation and of a simple distribution-free procedure.

Transforming data   When the distribution of a variable is skewed, it often happens that a simple transformation results in a variable whose distribution is symmetric and even close to Normal. The most common transformation is the logarithm, or log. The logarithm tends to pull in the right tail of a distribution. For example, the data 2, 3, 4, 20 show an outlier in the right tail. Their logarithms 0.30, 0.48, 0.60, 1.30 are much less skewed. Taking logarithms is a possible remedy for right-skewness. Instead of analyzing values of the original variable X, we first compute their logarithms and analyze the values of log X. Here is an example of this approach.

LOOK BACK
log transformation p. 90

EXAMPLE

SONGLENGTH

7.10 Length of audio files on an iPod.   Table 7.3 presents data on the length (in seconds) of audio files found on an iPod. There was a total of 10,003 audio files and 50 files were randomly selected using the "shuffle songs" command.[5] We would like to give a confidence interval for the average audio file length μ for this iPod.

A Normal quantile plot of the audio data from Table 7.3 (Figure 7.9) shows that the distribution is skewed to the right. Because there are no

TABLE 7.3

Length (in seconds) of audio files sampled from an iPod

240	316	259	46	871	411	1366
233	520	239	259	535	213	492
315	696	181	357	130	373	245
305	188	398	140	252	331	47
309	245	69	293	160	245	184
326	612	474	171	498	484	271
207	169	171	180	269	297	266
1847						

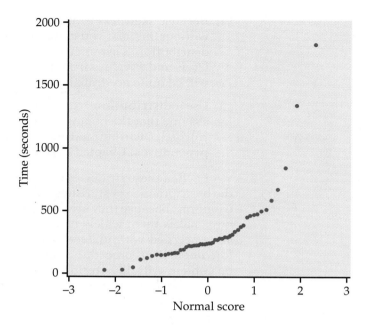

FIGURE 7.9   Normal quantile plot of audio file length, for Example 7.10. This sort of pattern occurs when a distribution is skewed to the right.

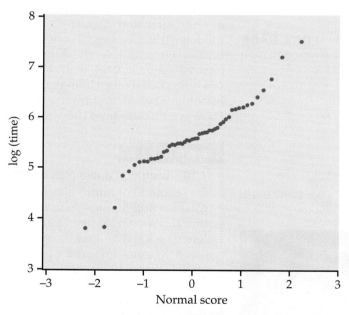

FIGURE 7.10   Normal quantile plot of the logarithms of the audio file lengths, for Example 7.10. This distribution appears approximately Normal.

extreme outliers, the sample mean of the 50 observations will nonetheless have an approximately Normal sampling distribution. The t procedures could be used for approximate inference. For more exact inference, we will seek to transform the data so that the distribution is more nearly Normal. Figure 7.10 is a Normal quantile plot of the natural logarithms of the time measurements. The transformed data are very close to Normal, so t procedures will give quite exact results.

The application of the t procedures to the transformed data is straightforward. Call the original length values from Table 7.3 the variable X. The transformed data are values of $X_{new} = \log X$. In most software packages, it is an easy task to transform data in this way and then analyze the new variable.

EXAMPLE

7.11 Software output of audio length data. Analysis of the natural log of the length values in Minitab produces the following output:

```
    N     MEAN     STDEV     SE MEAN    95.0 PERCENT C.I.
   50    5.6315    0.6840     0.0967    ( 5.4371, 5.8259)
```

For comparison, the 95% t confidence interval for the original mean μ is found from the original data as follows:

```
    N     MEAN     STDEV     SE MEAN    95.0 PERCENT C.I.
   50    354.1    307.9       43.6      (266.6, 441.6)
```

The advantage of analyzing transformed data is that use of procedures based on the Normal distributions is better justified and the results are more exact. The disadvantage is that a confidence interval for the mean μ in the original scale (in our example, seconds) cannot be easily recovered from the confidence interval for the mean of the logs. One approach based on the log-normal distribution[6] results in an interval of (290.33, 428.30), which is much narrower than the t interval.

The sign test Perhaps the most straightforward way to cope with non-Normal data is to use a *distribution-free*, or *nonparametric*, procedure. As the name indicates, these procedures do not require the population distribution to have any specific form, such as Normal. Distribution-free significance tests are quite simple and are available in most statistical software packages. Distribution-free tests have two drawbacks. First, they are generally less powerful than tests designed for use with a specific distribution, such as the t test. Second, we must often modify the statement of the hypotheses in order to use a distribution-free test. A distribution-free test concerning the center of a distribution, for example, is usually stated in terms of the median rather than the mean. This is sensible when the distribution may be skewed. But the distribution-free test does not ask the same question (Has the mean changed?) that the t test does. The simplest distribution-free test, and one of the most useful, is the **sign test.**

sign test

Let's examine again the aggressive-behavior data of Example 7.7 (page 414). In that example we concluded that there was more aggressive behavior on moon days than on other days. The stemplot given in Figure 7.8 was not very reassuring concerning the assumption that the data are Normal. There were 3 patients with low values that seemed to be somewhat different from the observations on the other 12 patients. How does the sign test deal with these data?

DATA FILE
MOONEFFECT

EXAMPLE

7.12 Sign test for the full-moon effect. The sign test is based on the following simple observation: of the 15 patients in our sample, 14 had more aggressive behaviors on moon days than on other days. This sounds like convincing evidence in favor of a moon effect on behavior, but we need to do some calculations to confirm this.

Let p be the probability that a randomly chosen dementia patient will have more aggressive behaviors on moon days than on other days. The null hypothesis of "no moon effect" says that the moon days are no different from other days, so a patient is equally likely to have more aggressive behaviors on moon days as on other days. We therefore want to test

$$H_0: p = 1/2$$
$$H_a: p > 1/2$$

There are 15 patients in the study, so the number who have more aggressive behaviors on moon days has the binomial distribution $B(15, 1/2)$ if H_0 is true. The P-value for the observed count 14 is therefore $P(X \geq 14)$, where X has the $B(15, 1/2)$ distribution. You can compute this probability with software or from the binomial probability formula:

$$P(X \geq 14) = P(X = 14) + P(X = 15)$$

$$= \binom{15}{14}\left(\frac{1}{2}\right)^{14}\left(\frac{1}{2}\right)^{1} + \binom{15}{15}\left(\frac{1}{2}\right)^{15}\left(\frac{1}{2}\right)^{0}$$

$$= (15)\left(\frac{1}{2}\right)^{15} + \left(\frac{1}{2}\right)^{15}$$

$$= 0.000488$$

LOOK BACK
binomial probability formula
p. 329

Using Table C we would approximate this value as 0.0005. As in Example 7.7, there is very strong evidence in favor of an increase in aggressive behavior on moon days.

There are several varieties of sign test, all based on counts and the binomial distribution. The sign test for matched pairs (Example 7.12) is the most useful. The null hypothesis of "no effect" is then always $H_0: p = 1/2$. The alternative can be one-sided in either direction or two-sided, depending on the type of change we are looking for. The test gets its name from the fact that we look only at the signs of the differences, not their actual values.

THE SIGN TEST FOR MATCHED PAIRS

Ignore pairs with difference 0; the number of trials n is the count of the remaining pairs. The test statistic is the count X of pairs with a positive difference. P-values for X are based on the binomial $B(n, 1/2)$ distribution.

The matched pairs t test in Example 7.7 tested the hypothesis that the mean of the distribution of differences (moon days minus other days) is 0. The sign test in Example 7.12 is in fact testing the hypothesis that the *median* of the differences is 0. If p is the probability that a difference is positive, then $p = 1/2$ when the median is 0. This is true because the median of the distribution is the point with probability 1/2 lying to its right. As Figure 7.11 illustrates, $p > 1/2$ when the median is greater than 0, again because the probability to the right of the median is always 1/2. The sign test of $H_0: p = 1/2$ against $H_a: p > 1/2$ is a test of

$$H_0: \text{population median} = 0$$
$$H_a: \text{population median} > 0$$

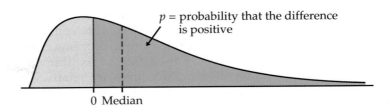

FIGURE 7.11 Why the sign test tests the median difference: when the median is greater than 0, the probability p of a positive difference is greater than 1/2, and vice versa.

The sign test in Example 7.12 makes no use of the actual differences—it just counts how many patients had more aggressive behaviors on moon days than on other days. Because the sign test uses so little of the available information, it is much less powerful than the t test when the population is close to Normal. *It is better to use a test that is powerful when we believe our assumptions are approximately satisfied than a less powerful test with fewer assumptions.* There are other distribution-free tests that are more powerful than the sign test.[7]

USE YOUR KNOWLEDGE

7.14 Sign test for energy drink comparison. Exercise 7.8 (page 417) gives data on the appeal of two popular energy drinks. Is there evidence that the medians are different? State the hypotheses, carry out the sign test, and report your conclusion.

SECTION 7.1 Summary

Significance tests and confidence intervals for the mean μ of a Normal population are based on the sample mean $\bar{x}$ of an SRS. Because of the central limit theorem, the resulting procedures are approximately correct for other population distributions when the sample is large.

The standardized sample mean, or **one-sample z statistic,**

$$z = \frac{\bar{x} - \mu}{\sigma/\sqrt{n}}$$

has the $N(0, 1)$ distribution. If the standard deviation $\sigma/\sqrt{n}$ of $\bar{x}$ is replaced by the **standard error** $s/\sqrt{n}$, the **one-sample t statistic**

$$t = \frac{\bar{x} - \mu}{s/\sqrt{n}}$$

has the t **distribution** with $n - 1$ degrees of freedom.

There is a t distribution for every positive **degrees of freedom k.** All are symmetric distributions similar in shape to Normal distributions. The $t(k)$ distribution approaches the $N(0, 1)$ distribution as k increases.

A level C **confidence interval for the mean** μ of a Normal population is

$$\bar{x} \pm t^* \frac{s}{\sqrt{n}}$$

where t^* is the value for the $t(n - 1)$ density curve with area C between $-t^*$ and t^*. The quantity

$$t^* \frac{s}{\sqrt{n}}$$

is the **margin of error.**

Significance tests for H_0: $\mu = \mu_0$ are based on the t statistic. P-values or fixed significance levels are computed from the $t(n-1)$ distribution.

These one-sample procedures are used to analyze **matched pairs** data by first taking the differences within the matched pairs to produce a single sample.

The t procedures are relatively **robust** against non-Normal populations. The t procedures are useful for non-Normal data when $15 \leq n < 40$ unless the data show outliers or strong skewness. When $n \geq 40$, the t procedures can be used even for clearly skewed distributions.

The **power** of the t test is calculated like that of the z test, using an approximate value for both σ and s.

Small samples from skewed populations can sometimes be analyzed by first applying a transformation (such as the logarithm) to obtain an approximately Normally distributed variable. The t procedures then apply to the transformed data.

The **sign test** is a **distribution-free test** because it uses probability calculations that are correct for a wide range of population distributions.

The sign test for "no treatment effect" in matched pairs counts the number of positive differences. The P-value is computed from the $B(n, 1/2)$ distribution, where n is the number of non-0 differences. The sign test is less powerful than the t test in cases where use of the t test is justified.

SECTION 7.1 Exercises

For Exercises 7.1 and 7.2, see page 406; for Exercises 7.3 and 7.4, see page 407; for Exercises 7.5 to 7.7, see page 413; for Exercises 7.8 and 7.9, see page 417; for Exercises 7.10 and 7.11, see page 418; for Exercises 7.12 and 7.13, see page 420; and for Exercise 7.14, see page 425.

7.15 Finding the critical value t^*. What critical value t^* from Table D should be used to calculate the margin of error for a confidence interval for the mean of the population in each of the following situations?

(a) A 95% confidence interval based on $n = 11$ observations.

(b) A 95% confidence interval from an SRS of 22 observations.

(c) A 90% confidence interval from a sample of size 22.

(d) These cases illustrate how the size of the margin of error depends upon the confidence level and the sample size. Summarize these relationships.

7.16 Distribution of the t statistic. Assume a sample size of $n = 18$. Draw a picture of the distribution of the t statistic under the null hypothesis. Use Table D and your picture to illustrate the values of the test statistic that would lead to rejection of the null hypothesis at the 5% level for a two-sided alternative.

7.17 More on the distribution of the t statistic. Repeat the previous exercise for the two situations where the alternative is one-sided.

7.18 One-sided versus two-sided P-values. Computer software reports $\bar{x} = 15.3$ and $P = 0.074$ for a t test of H_0: $\mu = 0$ versus H_a: $\mu \neq 0$. Based on prior knowledge, you can justify testing the alternative H_a: $\mu > 0$. What is the P-value for your significance test?

7.19 More on one-sided versus two-sided P-values. Suppose that $\bar{x} = -15.3$ in the setting of the previous exercise. Would this change your P-value? Use a sketch of the distribution of the test statistic under the null hypothesis to illustrate and explain your answer.

7.20 A one-sample t test. The one-sample t statistic for testing

$$H_0: \mu = 8$$
$$H_a: \mu > 8$$

from a sample of $n = 16$ observations has the value $t = 2.10$.

(a) What are the degrees of freedom for this statistic?

(b) Give the two critical values t^* from Table D that bracket t.

(c) Between what two values does the *P*-value of the test fall?

(d) Is the value $t = 2.10$ significant at the 5% level? Is it significant at the 1% level?

(e) If you have software available, find the exact *P*-value.

7.21 Another one-sample *t* test. The one-sample *t* statistic for testing

$$H_0: \mu = 40$$
$$H_a: \mu \neq 40$$

from a sample of $n = 28$ observations has the value $t = 2.01$.

(a) What are the degrees of freedom for *t*?

(b) Locate the two critical values t^* from Table D that bracket *t*.

(c) Between what two values does the *P*-value of the test fall?

(d) Is the value $t = 2.01$ statistically significant at the 5% level? At the 1% level?

(e) If you have software available, find the exact *P*-value.

7.22 A final one-sample *t* test. The one-sample *t* statistic for testing

$$H_0: \mu = 20$$
$$H_a: \mu < 20$$

based on $n = 14$ observations has the value $t = -2.55$.

(a) What are the degrees of freedom for this statistic?

(b) Between what two values does the *P*-value of the test fall?

(c) If you have software available, find the exact *P*-value.

7.23 Two-sided to one-sided *P*-value. Most software gives *P*-values for two-sided alternatives. Explain why you cannot always divide these *P*-values by 2 to obtain *P*-values for one-sided alternatives.

7.24 Number of friends on Facebook. Facebook provides a variety of statistics on their Web site that detail the growth and popularity of the site. One such statistic is that the average user has 130 friends.[8] Consider the following SRS of $n = 30$ Facebook users from a large university. FACEBOOKFRIENDS

99	148	158	126	118	112	103	111	154	85
120	127	137	74	85	104	106	72	119	160
83	110	97	193	96	152	105	119	171	128

(a) Do you think these data are Normally distributed? Use graphical methods to examine the distribution. Write a short summary of your findings.

(b) Is it appropriate to use the *t* methods of this section to compute a 95% confidence interval for the mean number of Facebook users at this large university? Explain why or why not.

(c) Find the mean, standard deviation, standard error, and margin of error for 95% confidence.

(d) Report the 95% confidence interval for μ, the average number of friends for Facebook users at this large university.

7.25 Rudeness and its effect on onlookers. Many believe that an uncivil environment has a negative effect on people. A pair of researchers recently performed a series of experiments to test whether witnessing rudeness and disrespect affects task performance.[9] In one study, 34 participants met in small groups and witnessed the group organizer being rude to a "participant" who showed up late for the group meeting. After the exchange, each participant went through an individual brainstorming task in which they were asked to produce as many uses for a brick as possible in 5 minutes. The mean number of uses was 7.88 with a standard deviation of 2.35.

(a) Suppose prior research has shown that the average number of uses a person can produce in 5 minutes under normal conditions is 10. Given that the researchers hypothesize witnessing this rudeness will decrease performance, state the appropriate null and alternative hypotheses.

(b) Carry out the significance test using a significance level of 0.05. Give the *P*-value and state your conclusion.

7.26 Fuel efficiency *t* test. Computers in some vehicles calculate various quantities related to performance. One of these is the fuel efficiency, or gas mileage, usually expressed as miles per gallon (mpg). For one vehicle equipped in this way, the mpg were recorded each time the gas tank was filled, and the computer was then reset.[10] Here are the mpg values for a random sample of 20 of these records: GASMILEAGE

| 41.5 | 50.7 | 36.6 | 37.3 | 34.2 | 45.0 | 48.0 | 43.2 | 47.7 | 42.2 |
| 43.2 | 44.6 | 48.4 | 46.4 | 46.8 | 39.2 | 37.3 | 43.5 | 44.3 | 43.3 |

(a) Describe the distribution using graphical methods. Is it appropriate to analyze these data using methods based on Normal distributions? Explain why or why not.

(b) Find the mean, standard deviation, standard error, and margin of error for 95% confidence.

(c) Report the 95% confidence interval for μ, the mean mpg for this vehicle based on these data.

7.27 Random distribution of trees t test A study of 584 longleaf pine trees in the Wade Tract in Thomas County, Georgia, is described in Example 6.1 (page 342). For each tree in the tract, the researchers measured the diameter at breast height (DBH). This is the diameter of the tree at 4.5 feet and the units are centimeters (cm). Only trees with DBH greater than 1.5 cm were sampled. Here are the diameters of a random sample of 40 of these trees: 🌲 TREEDIAMETER

10.5	13.3	26.0	18.3	52.2	9.2	26.1	17.6	40.5	31.8
47.2	11.4	2.7	69.3	44.4	16.9	35.7	5.4	44.2	2.2
4.3	7.8	38.1	2.2	11.4	51.5	4.9	39.7	32.6	51.8
43.6	2.3	44.6	31.5	40.3	22.3	43.3	37.5	29.1	27.9

(a) Use a histogram or stemplot and a boxplot to examine the distribution of DBHs. Include a Normal quantile plot if you have the necessary software. Write a careful description of the distribution.

(b) Is it appropriate to use the methods of this section to find a 95% confidence interval for the mean DBH of all trees in the Wade Tract? Explain why or why not.

(c) Report the mean with the margin of error and the confidence interval. Write a short summary describing the meaning of the confidence interval.

(d) Do you think these results would apply to other similar trees in the same area? Give reasons for your answer.

7.28 Nutritional intake in Canadian high performance male athletes. Recall Exercise 6.70 (page 380). A total of $n = 114$ male athletes from eight Canadian sports centers were surveyed. The average caloric intake was 3077.0 kcal/day with a standard deviation of 987.0. The recommended amount is 3421.7. Is there evidence that Canadian high performance male athletes are deficient in their caloric intake?

(a) State the appropriate H_0 and H_a to test this.

(b) Carry out the test, give the P-value, and state your conclusion.

(c) Construct a 95% confidence interval for the average deficiency in caloric intake.

7.29 Do you feel lucky? Children in a psychology study were asked to solve some puzzles and were then given feedback on their performance. Then they were asked to rate how luck played a role in determining their scores.[11] This variable was recorded on a 1 to 10 scale with 1

corresponding to very lucky and 10 corresponding to very unlucky. Here are the scores for 60 children: 🍀 FEELLUCKY

1	10	1	10	1	1	10	5	1	1	8	1	10	2	1
9	5	2	1	8	10	5	9	10	10	9	6	10	1	5
1	9	2	1	7	10	9	5	10	10	10	1	8	1	6
10	1	6	10	10	8	10	3	10	8	1	8	10	4	2

(a) Use graphical methods to display the distribution. Describe any unusual characteristics. Do you think that these would lead you to hesitate before using the Normality-based methods of this section?

(b) Give a 95% confidence interval for the mean luck score.

(c) The children in this study were volunteers whose parents agreed to have them participate in the study. To what extent do you think your results would apply to all similar children in this community?

7.30 Perceived organizational skills. In a study of children with attention deficit hyperactivity disorder (ADHD), parents were asked to rate their child on a variety of items related to how well their child performs different tasks.[12] One item was "Has difficulty organizing work," rated on a five-point scale of 0 to 4 with 0 corresponding to "not at all" and 4 corresponding to "very much." The mean rating for 282 boys with ADHD was reported as 2.22 with a standard deviation of 1.03.

(a) Do you think that these data are Normally distributed? Explain why or why not.

(b) Is it appropriate to use the methods of this section to compute a 99% confidence interval? Explain why or why not.

(c) Find the 99% margin of error and the corresponding confidence interval. Write a sentence explaining the interval and the meaning of the 99% confidence level.

(d) The boys in this study were all evaluated at the Western Psychiatric Institute and Clinic at the University of Pittsburgh. To what extent do you think the results could be generalized to boys with ADHD in other locations?

7.31 Confidence level and interval width. Refer to the previous exercise. Compute the 90% and the 95% confidence intervals. Display the three intervals graphically and write a short explanation of the effect of the confidence level on the width of the interval using your display as an example.

7.32 🔺 **Food intake and weight gain.** If we increase our food intake, we generally gain weight. Nutrition scientists can calculate the amount of weight gain that would be associated with a given increase in calories. In one study, 16 nonobese adults, aged 25 to 36 years, were fed 1000 calories per day in excess of the calories needed to maintain a stable body weight. The subjects maintained this diet for 8 weeks, so they consumed a total of 56,000 extra calories.[13] According to theory, 3500 extra calories will translate into a weight gain of 1 pound. Therefore, we expect each of these subjects to gain 56,000/3500 = 16 pounds (lb). Here are the weights before and after the 8-week period expressed in kilograms (kg): 🌀 WEIGHTGAIN

Subject	1	2	3	4	5	6	7	8
Weight before	55.7	54.9	59.6	62.3	74.2	75.6	70.7	53.3
Weight after	61.7	58.8	66.0	66.2	79.0	82.3	74.3	59.3

Subject	9	10	11	12	13	14	15	16
Weight before	73.3	63.4	68.1	73.7	91.7	55.9	61.7	57.8
Weight after	79.1	66.0	73.4	76.9	93.1	63.0	68.2	60.3

(a) For each subject, subtract the weight before from the weight after to determine the weight change.

(b) Find the mean and the standard deviation for the weight change.

(c) Calculate the standard error and the margin of error for 95% confidence. Report the 95% confidence interval in a sentence that explains the meaning of the 95%.

(d) Convert the mean weight gain in kilograms to mean weight gain in pounds. Because there are 2.2 kg per pound, multiply the value in kilograms by 2.2 to obtain pounds. Do the same for the standard deviation and the confidence interval.

(e) Test the null hypothesis that the mean weight gain is 16 lb. Be sure to specify the null and alternative hypotheses, the test statistic with degrees of freedom, and the P-value. What do you conclude?

(f) Write a short paragraph explaining your results.

7.33 Food intake and NEAT. Nonexercise activity thermogenesis (NEAT) provides a partial explanation for the results you found in the previous analysis. NEAT is energy burned by fidgeting, maintenance of posture, spontaneous muscle contraction, and other activities of daily living. In the study of the previous exercise, the 16 subjects increased their NEAT by 328 calories per day, on average, in response to the additional food intake. The standard deviation was 256.

(a) Test the null hypothesis that there was no change in NEAT versus the two-sided alternative. Summarize the results of the test and give your conclusion.

(b) Find a 95% confidence interval for the change in NEAT. Discuss the additional information provided by the

confidence interval that is not evident from the results of the significance test.

7.34 Potential insurance fraud? Insurance adjusters are concerned about the high estimates they are receiving from Jocko's Garage. To see if the estimates are unreasonably high, each of 10 damaged cars was taken to Jocko's and to another garage and the estimates recorded. Here are the results: 🌀 JOCKOGARAGE

Car	1	2	3	4	5
Jocko's	1375	1550	1250	1300	900
Other	1250	1300	1250	1200	950

Car	6	7	8	9	10
Jocko's	1500	1750	3600	2250	2800
Other	1575	1600	3300	2125	2600

(a) For each car, subtract the estimate of the other garage from Jocko's estimate. Find the mean and the standard deviation for this difference.

(b) Test the null hypothesis that there is no difference between the estimates of the two garages. Be sure to specify the null and alternative hypotheses, the test statistic with degrees of freedom, and the P-value. What do you conclude using the 0.05 significance level?

(c) Construct a 95% confidence interval for the difference in estimates.

(d) The insurance company is considering seeking repayment from 1000 claims filed with Jocko's last year. Using your answer to part (c), what repayment would you recommend the insurance company seek? Explain your answer.

7.35 Fuel efficiency comparison t test. Refer to Exercise 7.26. In addition to the computer calculating mpg, the driver also recorded the mpg by dividing the miles driven by the amount of gallons at fill-up. The driver wants to determine if these calculations are different. 🌀 MPGCOMPARISON

Fill-up	1	2	3	4	5	6	7	8	9	10
Computer	41.5	50.7	36.6	37.3	34.2	45.0	48.0	43.2	47.7	42.2
Driver	36.5	44.2	37.2	35.6	30.5	40.5	40.0	41.0	42.8	39.2

Fill-up	11	12	13	14	15	16	17	18	19	20
Computer	43.2	44.6	48.4	46.4	46.8	39.2	37.3	43.5	44.3	43.3
Driver	38.8	44.5	45.4	45.3	45.7	34.2	35.2	39.8	44.9	47.5

(a) State the appropriate H_0 and H_a.

(b) Carry out the test using a significance level of 0.05. Give the P-value, and then interpret the result.

7.36 Counts of picks in a 1-pound bag. A guitar supply company must maintain strict oversight on the number of picks they package for sale to customers. Their current advertisement specifies between 900 and 1000 picks in every bag. An SRS of thirty-six 1-pound bags of picks were collected as part of a Six Sigma Quality Improvement effort within the company. The number of picks in each bag are shown in the following table. 🌀 PICKCOUNT

924	925	967	909	959	937	970	936	952
919	965	921	913	886	956	962	916	945
957	912	961	950	923	935	969	916	952
917	977	940	924	957	920	986	895	923

(a) Create a histogram or stemplot, boxplot, and a Normal quantile plot of these counts. Write a careful description of the distribution. Make sure to note any outliers and comment on the skewness and Normality of the data.

(b) Based on your observations in part (a), is it appropriate to analyze these data using the t procedures? Briefly explain your response.

(c) Find the mean, the standard deviation, and the standard error of the mean for this sample.

(d) Calculate the 90% confidence interval for the mean number of picks in a 1-pound bag.

7.37 Significance test for the average number of picks. Refer to the previous exercise.

(a) Do these data provide evidence that the average number of picks in a 1-pound bag is greater than 925? Using a significance level of 5%, state your hypotheses, the P-value, and your conclusions.

(b) Do these data provide evidence that the average number of picks in a 1-pound bag is greater than 935? Using a significance level of 5%, state your hypotheses, the P-value, and your conclusion.

(c) Explain the relationship between your conclusions to parts (a) and (b) and the 90% confidence interval calculated in the previous problem.

7.38 A customer satisfaction survey. Many organizations are doing surveys to determine the satisfaction of their customers. Attitudes toward various aspects of campus life were the subject of one such study conducted at Purdue University. Each item was rated on a 1 to 5 scale, with 5 being the highest rating. The average response of 1783 first-year students to "Feeling welcomed at Purdue" was 3.8 with a standard deviation of 1.02. Assuming that the respondents are an SRS, give a 90% confidence interval for the mean of all first-year students.

7.39 Comparing operators of a DXA machine. Dual-energy X-ray absorptiometry (DXA) is a technique

for measuring bone health. One of the most common measures is total body bone mineral content (TBBMC). A highly skilled operator is required to take the measurements. Recently, a new DXA machine was purchased by a research lab and two operators were trained to take the measurements. TBBMC for eight subjects was measured by both operators.[14] The units are grams (g). A comparison of the means for the two operators provides a check on the training they received and allows us to determine if one of the operators is producing measurements that are consistently higher than the other. Here are the data: 🌀 TBBMC

	Subject							
Operator	1	2	3	4	5	6	7	8
1	1.328	1.342	1.075	1.228	0.939	1.004	1.178	1.286
2	1.323	1.322	1.073	1.233	0.934	1.019	1.184	1.304

(a) Take the difference between the TBBMC recorded for Operator 1 and the TBBMC for Operator 2. Describe the distribution of these differences.

(b) Use a significance test to examine the null hypothesis that the two operators have the same mean. Be sure to give the test statistic with its degrees of freedom, the P-value, and your conclusion.

(c) The sample here is rather small, so we may not have much power to detect differences of interest. Use a 95% confidence interval to provide a range of differences that are compatible with these data.

(d) The eight subjects used for this comparison were not a random sample. In fact, they were friends of the researchers whose ages and weights were similar to the types of people who would be measured with this DXA. Comment on the appropriateness of this procedure for selecting a sample, and discuss any consequences regarding the interpretation of the significance testing and confidence interval results.

7.40 Another comparison of DXA machine operators. Refer to the previous exercise. TBBMC measures the total amount of mineral in the bones. Another important variable is total body bone mineral density (TBBMD). This variable is calculated by dividing TBBMC by the area corresponding to bone in the DXA scan. The units are grams per squared centimeter (g/cm^2). Here are the TBBMD values for the same subjects: 🌀 TBBMD

	Subject							
Operator	1	2	3	4	5	6	7	8
1	4042	3703	2626	2673	1724	2136	2808	3322
2	4041	3697	2613	2628	1755	2140	2836	3287

Analyze these data using the questions in the previous exercise as a guide.

7.41 🏔 **Assessment of a foreign-language institute.** The National Endowment for the Humanities sponsors summer institutes to improve the skills of high school teachers of foreign languages. One such institute hosted 20 French teachers for 4 weeks. At the beginning of the period, the teachers were given the Modern Language Association's listening test of understanding of spoken French. After 4 weeks of immersion in French in and out of class, the listening test was given again. (The actual French spoken in the two tests was different, so that simply taking the first test should not improve the score on the second test.) The maximum possible score on the test is 36.[15] Here are the data: 🎯 FRENCHTEST

Teacher	Pretest	Posttest	Gain	Teacher	Pretest	Posttest	Gain
1	32	34	2	11	30	36	6
2	31	31	0	12	20	26	6
3	29	35	6	13	24	27	3
4	10	16	6	14	24	24	0
5	30	33	3	15	31	32	1
6	33	36	3	16	30	31	1
7	22	24	2	17	15	15	0
8	25	28	3	18	32	34	2
9	32	26	−6	19	23	26	3
10	20	26	6	20	23	26	3

To analyze these data, we first subtract the pretest score from the posttest score to obtain the improvement for each teacher. These 20 differences form a single sample. They appear in the "Gain" columns. The first teacher, for example, improved from 32 to 34, so the gain is $34 - 32 = 2$.

(a) State appropriate null and alternative hypotheses for examining the question of whether or not the course improves French spoken-language skills.

(b) Describe the gain data. Use numerical and graphical summaries.

(c) Perform the significance test. Give the test statistic, the degrees of freedom, and the P-value. Summarize your conclusion.

(d) Give a 95% confidence interval for the mean improvement.

7.42 Length of calls to customer service center. Refer to the lengths of calls to a customer service center in Table 1.2 (page 15). Give graphical and numerical summaries for these data. Compute a 95% confidence interval for the mean call length. Comment on the validity of your interval. 🎯 CALLCENTER80

7.43 IQ test scores. Refer to the IQ test scores for fifth-grade students in Table 1.1 (page 13). Give numerical and graphical summaries of the data and compute a 95% confidence interval. Comment on the validity of the interval. 🎯 IQ

7.44 Carbon dioxide emissions. Table 1.3 (page 26) gives the carbon dioxide emissions per person for countries with population at least 20 million. Does it make sense to use the t procedures on these 48 values to obtain a 95% confidence interval for the average carbon dioxide emissions per person? Explain your answer. 🎯 CO_2

The following exercises concern the optional material in the sections on the power of the t test and on non-Normal populations.

7.45 Sign test for potential insurance fraud. The differences in the repair estimates in Exercise 7.34 can also be analyzed using a sign test. Set up the appropriate null and alternative hypotheses, carry out the test, and summarize the results. How do these results compare with those that you obtained in Exercise 7.34? 🎯 JOCKOGARAGE

7.46 Sign test for the comparison of operators. The differences in the TBBMC measures in Exercise 7.39 can also be analyzed using a sign test. Set up the appropriate null and alternative hypotheses, carry out the test, and summarize the results. How do these results compare with those that you obtained in Exercise 7.39? 🎯 TBBMC

7.47 Another sign test for the comparison of operators. TBBMD values for the same subjects that you studied in the previous exercise are given in Exercise 7.40. Answer the questions given in the previous exercise for TBBMD. 🎯 TBBMD

7.48 Sign test for assessment of foreign-language institute. Use the sign test to assess whether the summer institute of Exercise 7.41 improves French listening skills. State the hypotheses, give the P-value using the binomial table (Table C), and report your conclusion. 🎯 FRENCHTEST

7.49 Sign test for fuel efficiency comparison. Use the sign test to assess whether the computer calculates a higher mpg than the driver in Exercise 7.35. State the hypotheses, give the P-value using the binomial table (Table C), and report your conclusion. 🎯 MPGCOMPARISON

7.50 Insulation study. A manufacturer of electric motors tests insulation at a high temperature (250°C) and records the number of hours until the insulation fails.[16] The data for 5 specimens are

$$446 \quad 326 \quad 372 \quad 377 \quad 310$$

The small sample size makes judgment from the data difficult, but engineering experience suggests that the logarithm of the failure time will have a Normal distribution. Take the logarithms of the 5 observations, and use t procedures to give a 90% confidence interval for the mean of the log failure time for insulation of this type. INSULATION

7.51 Power of the comparison of DXA machine operators. Suppose that the bone researchers in Exercise 7.39 wanted to be able to detect an alternative mean difference of 0.002. Find the power for this alternative for a sample size of 15. Use the standard deviation that you found in Exercise 7.39 for these calculations.

7.52 Sample size calculations. You are designing a study to test the null hypothesis that $\mu = 0$ versus the alternative that μ is positive. Assume that σ is 15. Suppose that it would be important to be able to detect the alternative $\mu = 2$. Perform power calculations for a variety of sample sizes and determine how large a sample you would need to detect this alternative with power of at least 0.80.

7.53 Determining the sample size. Consider Example 7.9 (page 419). What is the minimum sample size needed for the power to be greater than 80% when $\mu = 0.75$?

7.2 Comparing Two Means

A psychologist wants to compare male and female college students' impressions of personality based on selected MySpace pages. A nutritionist is interested in the effect of increased calcium on blood pressure. A bank wants to know which of two incentive plans will most increase the use of its debit cards. Two-sample problems such as these are among the most common situations encountered in statistical practice.

> **TWO-SAMPLE PROBLEMS**
>
> - The goal of inference is to compare the responses in two groups.
> - Each group is considered to be a sample from a distinct population.
> - The responses in each group are independent of those in the other group.

LOOK BACK
randomized comparative
experiment p. 175

A two-sample problem can arise from a randomized comparative experiment that randomly divides the subjects into two groups and exposes each group to a different treatment. Comparing random samples separately selected from two populations is also a two-sample problem. Unlike the matched pairs designs studied earlier, there is no matching of the units in the two samples, and the two samples may be of different sizes. Inference procedures for two-sample data differ from those for matched pairs.

We can present two-sample data graphically by a back-to-back stemplot (for small samples) or by side-by-side boxplots (for larger samples). Now we will apply the ideas of formal inference in this setting. When both population distributions are symmetric, and especially when they are at least approximately Normal, a comparison of the mean responses in the two populations is most often the goal of inference.

We have two independent samples, from two distinct populations (such as subjects given a treatment and those given a placebo). The same variable is measured for both samples. We will call the variable x_1 in the first population and x_2 in the second because the variable may have different distributions in the two populations. Here is the notation that we will use to describe the two populations:

Population	Variable	Mean	Standard deviation
1	x_1	μ_1	σ_1
2	x_2	μ_2	σ_2

We want to compare the two population means, either by giving a confidence interval for $\mu_1 - \mu_2$ or by testing the hypothesis of no difference, $H_0: \mu_1 = \mu_2$.

Inference is based on two independent SRSs, one from each population. Here is the notation that describes the samples:

Population	Sample size	Sample mean	Sample standard deviation
1	n_1	$\overline{x}_1$	s_1
2	n_2	$\overline{x}_2$	s_2

Throughout this section, the subscripts 1 and 2 show the population to which a parameter or a sample statistic refers.

The two-sample *z* statistic

The natural estimator of the difference $\mu_1 - \mu_2$ is the difference between the sample means, $\overline{x}_1 - \overline{x}_2$. If we are to base inference on this statistic, we must know its sampling distribution. First, the mean of the difference $\overline{x}_1 - \overline{x}_2$ is the difference of the means $\mu_1 - \mu_2$. This follows from the addition rule for means and the fact that the mean of any $\overline{x}$ is the mean of the population. To compute the variance, we use addition rule for variances. Because the samples are independent, their sample means $\overline{x}_1$ and $\overline{x}_2$ are independent random variables. Thus the variance of the difference $\overline{x}_1 - \overline{x}_2$ is the sum of their variances, which is

LOOK BACK
addition rule for means
p. 267

LOOK BACK
addition rule for variances
p. 271

$$\frac{\sigma_1^2}{n_1} + \frac{\sigma_2^2}{n_2}$$

We now know the mean and variance of the distribution of $\overline{x}_1 - \overline{x}_2$ in terms of the parameters of the two populations. If the two population distributions are both Normal, then the distribution of $\overline{x}_1 - \overline{x}_2$ is also Normal. This is true because each sample mean alone is Normally distributed and because a difference of independent Normal random variables is also Normal.

EXAMPLE

7.13 Heights of 10-year-old girls and boys. A fourth-grade class has 12 girls and 8 boys. The children's heights are recorded on their tenth birthdays. What is the chance that the girls are taller than the boys? Of course, it is very unlikely that all the girls are taller than all the boys. We translate the

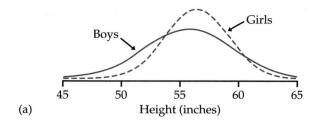

(a)

FIGURE 7.12   Distributions for Example 7.13. (a) Distributions of heights of 10-year-old boys and girls. (b) Distribution of the difference between mean heights of 12 girls and 8 boys.

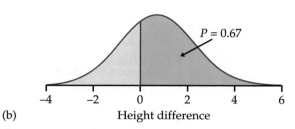

(b)

question into the following: what is the probability that the mean height of the girls is greater than the mean height of the boys?

Based on information from the National Health and Nutrition Examination Survey,[17] we assume that the heights (in inches) of 10-year-old girls are $N(56.4, 2.7)$ and the heights of 10-year-old boys are $N(55.7, 3.8)$. The heights of the students in our class are assumed to be random samples from these populations. The two distributions are shown in Figure 7.12(a).

The difference $\bar{x}_1 - \bar{x}_2$ between the female and male mean heights varies in different random samples. The sampling distribution has mean

$$\mu_1 - \mu_2 = 56.4 - 55.7 = 0.7 \text{ inch}$$

and variance

$$\frac{\sigma_1^2}{n_1} + \frac{\sigma_2^2}{n_2} = \frac{2.7^2}{12} + \frac{3.8^2}{8}$$

$$= 2.41$$

The standard deviation of the difference in sample means is therefore $\sqrt{2.41} = 1.55$ inches.

If the heights vary Normally, the difference in sample means is also Normally distributed. The distribution of the difference in heights is shown in Figure 7.12(b). We standardize $\bar{x}_1 - \bar{x}_2$ by subtracting its mean (0.7) and dividing by its standard deviation (1.55). Therefore, the probability that the girls are taller than the boys is

$$P(\bar{x}_1 - \bar{x}_2 > 0) = P\left(\frac{(\bar{x}_1 - \bar{x}_2) - 0.7}{1.55} > \frac{0 - 0.7}{1.55}\right)$$

$$= P(Z > -0.45) = 0.67$$

Even though the population mean height of 10-year-old girls is greater than the population mean height of 10-year-old boys, the probability that the sample mean of the girls is greater than the sample mean of the boys in our class is only 67%. *Large samples are needed to see the effects of small differences.*

As Example 7.13 reminds us, any Normal random variable has the $N(0, 1)$ distribution when standardized. We have arrived at a new z statistic.

TWO-SAMPLE z STATISTIC

Suppose that $\bar{x}_1$ is the mean of an SRS of size n_1 drawn from an $N(\mu_1, \sigma_1)$ population and that $\bar{x}_2$ is the mean of an independent SRS of size n_2 drawn from an $N(\mu_2, \sigma_2)$ population. Then the **two-sample z statistic**

$$z = \frac{(\bar{x}_1 - \bar{x}_2) - (\mu_1 - \mu_2)}{\sqrt{\dfrac{\sigma_1^2}{n_1} + \dfrac{\sigma_2^2}{n_2}}}$$

has the standard Normal $N(0, 1)$ sampling distribution.

In the unlikely event that both population standard deviations are known, the two-sample z statistic is the basis for inference about $\mu_1 - \mu_2$. Exact z procedures are seldom used, however, because σ_1 and σ_2 are rarely known. In Chapter 6, we discussed the one-sample z procedures in order to introduce the ideas of inference. Here we move directly to the more useful t procedures.

The two-sample t procedures

Suppose now that the population standard deviations σ_1 and σ_2 are not known. We estimate them by the sample standard deviations s_1 and s_2 from our two samples. Following the pattern of the one-sample case, we substitute the standard errors for the standard deviations used in the two-sample z statistic. The result is the *two-sample t statistic:*

$$t = \frac{(\bar{x}_1 - \bar{x}_2) - (\mu_1 - \mu_2)}{\sqrt{\dfrac{s_1^2}{n_1} + \dfrac{s_2^2}{n_2}}}$$

Unfortunately, this statistic does *not* have a t distribution. A t distribution replaces the $N(0, 1)$ distribution only when a single standard deviation (σ) in a z statistic is replaced by its sample standard deviation (s). In this case, we replace two standard deviations (σ_1 and σ_2) by their estimates (s_1 and s_2), which does not produce a statistic having a t distribution.

Nonetheless, we can approximate the distribution of the two-sample t statis-

approximations for the degrees of freedom

tic by using the $t(k)$ distribution with an **approximation for the degrees of freedom k.** We use these approximations to find approximate values of t^* for confidence intervals and to find approximate P-values for significance tests. Here are two approximations:

1. Use a value of k that is calculated from the data. In general, it will not be a whole number.

2. Use k equal to the smaller of $n_1 - 1$ and $n_2 - 1$.

In practice, the choice of approximation rarely makes a difference in our conclusion. Most statistical software uses the first option to approximate the $t(k)$ distribution for two-sample problems unless the user requests another method. Use of this approximation without software is a bit complicated; we will give the details later in this section. If you are not using software, the second approximation is preferred. This approximation is appealing because it is conservative.[18] Margins of error for the level C confidence intervals are a bit larger than they need to be, so the true confidence level is larger than C. For significance testing, the true P-values are a bit smaller than those we obtain from this approximation; thus for tests at a fixed significance level, we are a little less likely to reject H_0 when it is true.

The two-sample t significance test

> ### THE TWO-SAMPLE t SIGNIFICANCE TEST
>
> Suppose that an SRS of size n_1 is drawn from a Normal population with unknown mean μ_1 and that an independent SRS of size n_2 is drawn from another Normal population with unknown mean μ_2. To test the hypothesis $H_0: \mu_1 = \mu_2$, compute the **two-sample t statistic**
>
> $$t = \frac{\overline{x}_1 - \overline{x}_2}{\sqrt{\dfrac{s_1^2}{n_1} + \dfrac{s_2^2}{n_2}}}$$
>
> and use P-values or critical values for the $t(k)$ distribution, where the degrees of freedom k are either approximated by software or are the smaller of $n_1 - 1$ and $n_2 - 1$.

DRP

EXAMPLE

7.14 Directed reading activities assessment. An educator believes that new directed reading activities in the classroom will help elementary school pupils improve some aspects of their reading ability. She arranges for a third-grade class of 21 students to take part in these activities for an eight-week period. A control classroom of 23 third-graders follows the same curriculum without the activities. At the end of the eight weeks, all students are given a Degree of Reading Power (DRP) test, which measures the aspects of reading ability that the treatment is designed to improve. The data appear in Table 7.4.[19]

TABLE 7.4

DRP scores for third-graders

Treatment Group				Control Group			
24	61	59	46	42	33	46	37
43	44	52	43	43	41	10	42
58	67	62	57	55	19	17	55
71	49	54		26	54	60	28
43	53	57		62	20	53	48
49	56	33		37	85	42	

First examine the data:

```
        Control    Treatment
          970 │ 1 │
          860 │ 2 │ 4
          773 │ 3 │ 3
      8632221 │ 4 │ 3334699
         5543 │ 5 │ 23467789
           20 │ 6 │ 127
              │ 7 │ 1
            5 │ 8 │
```

A back-to-back stemplot suggests that there is a mild outlier in the control group but no deviation from Normality serious enough to forbid use of t procedures. Separate Normal quantile plots for both groups (Figure 7.13) confirm that both are approximately Normal. The scores of the treatment group appear to be somewhat higher than those of the control group. The summary statistics are

Group	n	$\bar{x}$	s
Treatment	21	51.48	11.01
Control	23	41.52	17.15

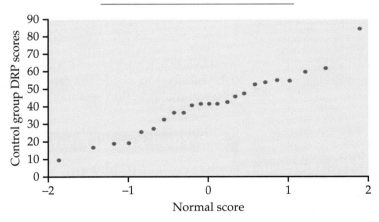

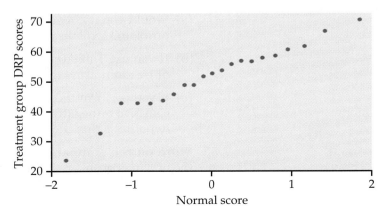

FIGURE 7.13 Normal quantile plots of the DRP scores in Table 7.4.

Because we hope to show that the treatment (Group 1) is better than the control (Group 2), the hypotheses are

$$H_0: \mu_1 = \mu_2$$
$$H_a: \mu_1 > \mu_2$$

The two-sample t test statistic is

$$t = \frac{\bar{x}_1 - \bar{x}_2}{\sqrt{\dfrac{s_1^2}{n_1} + \dfrac{s_2^2}{n_2}}}$$

$$= \frac{51.48 - 41.52}{\sqrt{\dfrac{11.01^2}{21} + \dfrac{17.15^2}{23}}}$$

$$= 2.31$$

The P-value for the one-sided test is $P(T \geq 2.31)$. Software gives the approximate P-value as 0.0132 and uses 37.9 as the degrees of freedom. For the second approximation, the degrees of freedom k are equal to the smaller of

$$n_1 - 1 = 21 - 1 = 20 \quad \text{and} \quad n_2 - 1 = 23 - 1 = 22$$

df = 20		
p	0.02	0.01
t^*	2.197	2.528

Comparing 2.31 with the entries in Table D for 20 degrees of freedom, we see that P lies between 0.01 and 0.02. The data strongly suggest that directed reading activity improves the DRP score ($t = 2.31$, df $= 20$, $0.01 < P < 0.02$).

If your software gives P-values for only the two-sided alternative, $2P(T \geq |t|)$, you need to divide the reported value by 2 after checking that the means differ in the direction specified by the alternative hypothesis.

USE YOUR KNOWLEDGE

7.54 Comparison of two MySpace page designs. You want to compare the daily number of hits for two different MySpace page designs that advertise your indie rock band. You assign the next 30 days to either Design A or Design B, 15 days to each.

  (a) Would you use a one-sided or two-sided significance test for this problem? Explain your choice.

  (b) If you use Table D to find the critical value, what are the degrees of freedom using the second approximation?

  (c) If you perform the significance test using $\alpha = 0.05$, how large (positive or negative) must the t statistic be to reject the null hypothesis that the two designs result in the same average hits?

7.55 More on the comparison of two MySpace page designs. Consider the previous problem. If the t statistic for comparing the mean hits was 2.45, what P-value would you report? What would you conclude using $\alpha = 0.05$?

The two-sample t confidence interval

The same ideas that we used for the two-sample t significance tests also apply to give us *two-sample t confidence intervals*. We can use either software or the conservative approach with Table D to approximate the value of t^*.

THE TWO-SAMPLE t CONFIDENCE INTERVAL

Suppose that an SRS of size n_1 is drawn from a Normal population with unknown mean μ_1 and that an independent SRS of size n_2 is drawn from another Normal population with unknown mean μ_2. The **confidence interval for $\mu_1 - \mu_2$** given by

$$(\bar{x}_1 - \bar{x}_2) \pm t^* \sqrt{\frac{s_1^2}{n_1} + \frac{s_2^2}{n_2}}$$

has confidence level at least C no matter what the population standard deviations may be. Here, t^* is the value for the $t(k)$ density curve with area C between $-t^*$ and t^*. The value of the degrees of freedom k is approximated by software or we use the smaller of $n_1 - 1$ and $n_2 - 1$.

To complete the analysis of the DRP scores we examined in Example 7.14, we need to describe the size of the treatment effect. We do this with a confidence interval for the difference between the treatment group and the control group means.

EXAMPLE

7.15 How much improvement? We will find a 95% confidence interval for the mean improvement in the entire population of third-graders. The interval is

$$(\bar{x}_1 - \bar{x}_2) \pm t^* \sqrt{\frac{s_1^2}{n_1} + \frac{s_2^2}{n_2}} = (51.48 - 41.52) \pm t^* \sqrt{\frac{11.01^2}{21} + \frac{17.15^2}{23}}$$

$$= 9.96 \pm 4.31 t^*$$

Using software, the degrees of freedom are 37.9 and $t^* = 2.025$. This approximation gives

$$9.96 \pm (4.31 \times 2.025) = 9.96 \pm 8.72 = (1.2, \ 18.7)$$

The conservative approach uses the $t(20)$ distribution. Table D gives $t^* = 2.086$. With this approximation we have

$$9.96 \pm (4.31 \times 2.086) = 9.96 \pm 8.99 = (1.0, \ 18.9)$$

We can see that the conservative approach does, in fact, give a wider interval than the more accurate approximation used by software. However, the difference is pretty small.

We estimate the mean improvement to be about 10 points, but with a margin of error of almost 9 points with either method. Although we have good evidence of some improvement, the data do not allow a very precise estimate of the size of the average improvement.

The design of the study in Example 7.14 is not ideal. Random assignment of students was not possible in a school environment, so existing third-grade classes were used. The effect of the reading programs is therefore confounded with any other differences between the two classes. The classes were chosen to be as similar as possible—for example, in terms of the social and economic status of the students. Extensive pretesting showed that the two classes were on the average quite similar in reading ability at the beginning of the experiment. To avoid the effect of two different teachers, the researcher herself taught reading in both classes during the eight-week period of the experiment. We can therefore be somewhat confident that the two-sample test is detecting the effect of the treatment and not some other difference between the classes. This example is typical of many situations in which an experiment is carried out but randomization is not possible.

USE YOUR KNOWLEDGE

7.56 Two-sample t confidence interval. Assume $\bar{x}_1 = 110$, $\bar{x}_2 = 120$, $s_1 = 8$, $s_2 = 12$, $n_1 = 50$, and $n_2 = 50$. Find a 95% confidence interval for the difference in the corresponding values of μ using the second approximation for degrees of freedom. Does this interval include more or fewer values than a 99% confidence interval? Explain your answer.

7.57 Another two-sample t confidence interval. Assume $\bar{x}_1 = 110$, $\bar{x}_2 = 120$, $s_1 = 8$, $s_2 = 12$, $n_1 = 10$, and $n_2 = 10$. Find a 95% confidence interval for the difference in the corresponding values of μ using the second approximation for degrees of freedom. Would you reject the null hypothesis that the population means are equal in favor of the two-sided alternative at significance level 0.05? Explain.

Robustness of the two-sample procedures

The two-sample t procedures are more robust than the one-sample t methods. When the sizes of the two samples are equal and the distributions of the two populations being compared have similar shapes, probability values from the t table are quite accurate for a broad range of distributions when the sample sizes are as small as $n_1 = n_2 = 5$.[20] When the two population distributions have different shapes, larger samples are needed. The guidelines for the use of one-sample t procedures can be adapted to two-sample procedures by replacing "sample size" with the "sum of the sample sizes" $n_1 + n_2$. These guidelines are rather conservative, especially when the two samples are of equal size. *In planning a two-sample study, choose equal sample sizes if you can.* The two-sample t procedures are most robust against non-Normality in this case, and the conservative probability values are most accurate.

Here is an example with reasonably large sample sizes that are not equal. Even if the distributions are not Normal, we are confident that the sample means will be approximately Normal. The two-sample t test is very robust in this case.

EXAMPLE

7.16 Sleep and blood pressure in adolescents. Hypertension is an increasingly common health problem in both adults and adolescents. Childhood hypertension is associated with hypertension in adulthood, a risk factor for cardiovascular disease and death. While several studies have implicated insufficient sleep as a risk factor for hypertension in adults, only a few studies have looked at the relationship between sleep and hypertension in children. One study examined 238 adolescents between the ages of 13 and 16.[21] Based on in-home monitoring and overnight observance, each child was classified as having either high or low sleep efficiency. Here are the summary statistics of their systolic blood pressures in mm Hg:

Sleep efficiency	n	$\bar{x}$	s
Low	61	118.4	9.9
High	177	112.6	7.5

The low sleep efficiency children have higher pressures on the average. Can we conclude that the systolic blood pressure of children with low and high sleep efficiency are not the same? Or is this observed difference merely what we could expect to see given the variation among children?

Even though prior evidence suggested that the blood pressure would be elevated in low sleep efficiency children, the researchers did not specify a direction for the difference. Thus, the hypotheses are

$$H_0: \mu_1 = \mu_2$$
$$H_a: \mu_1 \neq \mu_2$$

Because the samples are relatively large, we can confidently use the t procedures even though we lack the detailed data and so cannot verify the Normality condition.

The two-sample t statistic is

$$t = \frac{\bar{x}_1 - \bar{x}_2}{\sqrt{\frac{s_1^2}{n_1} + \frac{s_2^2}{n_2}}}$$

$$= \frac{118.4 - 112.6}{\sqrt{\frac{9.9^2}{61} + \frac{7.5^2}{177}}}$$

$$= 4.18$$

The conservative approach finds the P-value by comparing 4.18 to critical values for the $t(60)$ distribution because the smaller sample has 61 observations. We must double the table tail area p because the alternative is two-sided.

Our calculated value of t is larger than the $p = 0.0005$ entry in the table. Doubling 0.0005, we conclude that the P-value is less than 0.001. The data give conclusive evidence that the mean systolic blood pressure is higher in low sleep efficiency children ($t = 4.18$, df $= 60$, $P < 0.001$).

df = 60	
p	0.0005
t^*	3.460

In this example the exact P-value is very small because $t = 4.18$ says that the observed difference in means is over 4 standard errors above the hypothesized difference of zero ($\mu_1 = \mu_2$). The difference of 5.8 mm Hg may not appear that large but in terms of having an elevated systolic blood pressure, only 6.2% of the high sleep efficiency children had elevated pressure while 26.2% of the low sleep efficiency children had elevated pressure.

In this and other examples, we can choose which population to label 1 and which to label 2. After inspecting the data, we chose low sleep efficiency children as Population 1 because this choice makes the *t* statistic a positive number. This avoids any possible confusion from reporting a negative value for *t*. *Choosing the population labels is* **not** *the same as choosing a one-sided alternative after looking at the data.* Choosing hypotheses after seeing a result in the data is a violation of sound statistical practice.

Inference for small samples

Small samples require special care. We do not have enough observations to examine the distribution shapes, and only extreme outliers stand out. The power of significance tests tends to be low, and the margins of error of confidence intervals tend to be large. Despite these difficulties, we can often draw important conclusions from studies with small sample sizes. If the size of an effect is very large, it should still be evident even if the *n*'s are small.

> **EXAMPLE**
>
> **7.17 Sleep efficiency and blood pressure.** In the setting of Example 7.16, let's consider a much smaller study that collects systolic blood pressures from only 5 children in each sleep efficiency group. Also, given the results of this past example, we choose the one-sided alternative. The data are

Sleep efficiency	Systolic blood pressure (mm Hg)				
Low	110	118	128	126	119
High	113	120	102	108	114

First, examine the distributions with a back-to-back stemplot.

Low		High
	100	2
	100	8
0	110	34
98	110	
	120	0
86	120	

While there is variation among pressures within each group, there is also a noticeable separation. The high sleep efficiency group contains 4 of the 5 lowest pressures and the low efficiency group contains the 4 of the 5 highest blood pressures. A significance test can confirm whether this pattern can arise just by chance or if the low sleep efficiency group has a higher mean. We test

$$H_0: \mu_1 = \mu_2$$

$$H_a: \mu_1 > \mu_2$$

The blood pressure is higher in the low efficiency group ($t = 2.00$, df $= 7.98$, $P = 0.0404$). The difference in sample means is 8.8 mm Hg.

Figure 7.14 gives outputs for this analysis from several software packages. Although the formats differ, the basic information is the same. All report the sample sizes, the sample means and standard deviations (or variances), the t statistic, and its P-value. All agree that the P-value is small, though some give

SAS

The TTEST Procedure

Variable: Pressure

Group	N	Mean	Std Dev	Std Err
High	5	111.4	6.7676	3.0265
Low	5	120.2	7.1554	3.2000
Diff (1-2)		-8.8000	6.9642	4.4045

Group	Method	Mean	95% CL Mean	
High		111.4	103.0	119.8
Low		120.2	111.3	129.1
Diff (1-2)	Pooled	-8.8000	-18.9569	1.3569
Diff (1-2)	Satterthwaite	-8.8000	-18.9624	1.3624

Method	Variances	DF	t Value	Pr > |t|
Pooled	Equal	8	-2.00	0.0808
Satterthwaite	Unequal	7.9753	-2.00	0.0809

Excel

	A	B	C
1	t-Test: Two-Sample Assuming Unequal Variances		
2			
3		Low	High
4	Mean	120.2	111.4
5	Variance	51.2	45.8
6	Observations	5	5
7	Hypothesized Mean Difference	0	
8	df	8	
9	t Stat	1.99793708	
10	P(T<=t) one-tail	0.040386978	
11	t Critical one-tail	1.859548033	
12	P(T<=t) two-tail	0.080773957	
13	t Critical two-tail	2.306004133	

FIGURE 7.14 SAS, Excel, JMP, and SPSS output for Example 7.17.

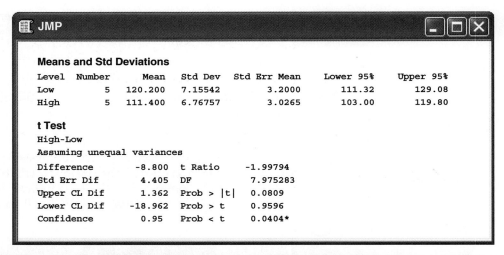

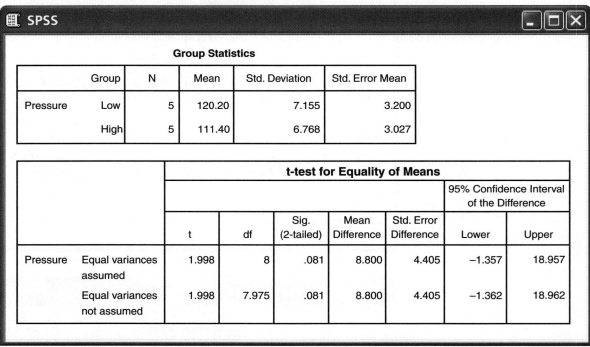

FIGURE 7.14 *(Continued)*

more detail than others. Software often labels the groups in alphabetical order. In this example, High is then the first population and $t = -2.00$, the negative of our result. Always check the means first and report the statistic (you may need to change the sign) in an appropriate way. Be sure to also mention the size of the effect you observed, such as "The mean systolic blood pressure for low sleep efficiency children was 8.8 mm Hg higher than in the high sleep efficiency group."

There are two other things to notice in the outputs. First, SAS and SPSS only give results for the two-sided alternative. To get the P-value for the one-sided alternative, we must first check the mean difference to make sure it is in the proper direction. If it is, we divide the given P-value by 2. Also, SAS and SPSS report the results of *two t* procedures: a special procedure that assumes that the two population variances are equal and the general two-sample procedure that we have just studied. We don't recommend the "equal-variances" procedures, but we describe them later, in the section on pooled two-sample t procedures.

Software approximation for the degrees of freedom*

We noted earlier that the two-sample t statistic does not have an exact t distribution. Moreover, the exact distribution changes as the unknown population standard deviations σ_1 and σ_2 change. However, the distribution can be approximated by a t distribution with degrees of freedom given by

$$df = \frac{\left(\dfrac{s_1^2}{n_1} + \dfrac{s_2^2}{n_2} \right)^2}{\dfrac{1}{n_1 - 1}\left(\dfrac{s_1^2}{n_1} \right)^2 + \dfrac{1}{n_2 - 1}\left(\dfrac{s_2^2}{n_2} \right)^2}$$

This is the approximation used by most statistical software. It is quite accurate when both sample sizes n_1 and n_2 are 5 or larger.

> ### EXAMPLE
>
> **7.18   Degrees of freedom for directed reading assessment.**   For the DRP study of Example 7.14, the following table summarizes the data:
>
> | Group | n | $\bar{x}$ | s |
> |-------|-----|-----------|------|
> | 1 | 21 | 51.48 | 11.01 |
> | 2 | 23 | 41.52 | 17.15 |
>
> For greatest accuracy, we will use critical points from the t distribution with degrees of freedom given by the equation above:
>
> $$df = \frac{\left(\dfrac{11.01^2}{21} + \dfrac{17.15^2}{23} \right)^2}{\dfrac{1}{20}\left(\dfrac{11.01^2}{21} \right)^2 + \dfrac{1}{22}\left(\dfrac{17.15^2}{23} \right)^2}$$
>
> $$= \frac{344.486}{9.099} = 37.86$$
>
> This is the value that we reported in Examples 7.14 and 7.15, where we gave the results produced by software.

*This material can be omitted unless you are using statistical software and wish to understand what the software does.

The number df given by the preceding approximation is always at least as large as the smaller of $n_1 - 1$ and $n_2 - 1$. On the other hand, df is never larger than the sum $n_1 + n_2 - 2$ of the two individual degrees of freedom. The number of degrees of freedom is generally not a whole number. There is a t distribution with any positive degrees of freedom, even though Table D contains entries only for whole-number degrees of freedom. When df is small and is not a whole number, interpolation between entries in Table D may be needed to obtain an accurate critical value or P-value. Because of this and the need to calculate df, we do not recommend regular use of this approximation if a computer is not doing the arithmetic. With a computer, however, the more accurate procedures are painless.

USE YOUR KNOWLEDGE

7.58   Calculating the degrees of freedom. Assume $s_1 = 12$, $s_2 = 9$, $n_1 = 30$, and $n_2 = 25$. Find the software approximate degrees of freedom.

The pooled two-sample t procedures*

There is one situation in which a t statistic for comparing two means has exactly a t distribution. This is when the two Normal population distributions have the *same* standard deviation. As we've done with other t statistics, we will first develop the z statistic, and from it the t statistic. In this case, notice that we need to substitute only a single standard error going from the z to t statistic. This is why the resulting t statistic has a t distribution.

Call the common—and still unknown—standard deviation of both populations σ. Both sample variances s_1^2 and s_2^2 estimate σ^2. The best way to combine these two estimates is to average them with weights equal to their degrees of freedom. This gives more weight to the sample variance from the larger sample, which is reasonable. The resulting estimator of σ^2 is

$$s_p^2 = \frac{(n_1 - 1)s_1^2 + (n_2 - 1)s_2^2}{n_1 + n_2 - 2}$$

pooled estimator of σ^2 This is called the **pooled estimator of σ^2** because it combines the information in both samples.

When both populations have variance σ^2, the addition rule for variances says that $\bar{x}_1 - \bar{x}_2$ has variance equal to the *sum* of the individual variances, which is

$$\frac{\sigma^2}{n_1} + \frac{\sigma^2}{n_2} = \sigma^2 \left(\frac{1}{n_1} + \frac{1}{n_2} \right)$$

The standardized difference of means in this equal-variance case is therefore

$$z = \frac{(\bar{x}_1 - \bar{x}_2) - (\mu_1 - \mu_2)}{\sigma \sqrt{\dfrac{1}{n_1} + \dfrac{1}{n_2}}}$$

---

*This section can be omitted if desired, but it should be read if you plan to read Chapters 12 and 13.

This is a special two-sample z statistic for the case in which the populations have the same σ. Replacing the unknown σ by the estimate s_p gives a t statistic. The degrees of freedom are $n_1 + n_2 - 2$, the sum of the degrees of freedom of the two sample variances. This statistic is the basis of the pooled two-sample t inference procedures.

THE POOLED TWO-SAMPLE t PROCEDURES

Suppose that an SRS of size n_1 is drawn from a Normal population with unknown mean μ_1 and that an independent SRS of size n_2 is drawn from another Normal population with unknown mean μ_2. Suppose also that the two populations have the same standard deviation. A level C confidence interval for $\mu_1 - \mu_2$ is

$$(\bar{x}_1 - \bar{x}_2) \pm t^* s_p \sqrt{\frac{1}{n_1} + \frac{1}{n_2}}$$

Here t^* is the value for the $t(n_1 + n_2 - 2)$ density curve with area C between $-t^*$ and t^*.

To test the hypothesis H_0: $\mu_1 = \mu_2$, compute the pooled two-sample t statistic

$$t = \frac{\bar{x}_1 - \bar{x}_2}{s_p \sqrt{\frac{1}{n_1} + \frac{1}{n_2}}}$$

In terms of a random variable T having the $t(n_1 + n_2 - 2)$ distribution, the P-value for a test of H_0 against

$$H_a: \mu_1 > \mu_2 \quad \text{is} \quad P(T \geq t)$$

$$H_a: \mu_1 < \mu_2 \quad \text{is} \quad P(T \leq t)$$

$$H_a: \mu_1 \neq \mu_2 \quad \text{is} \quad 2P(T \geq |t|)$$

EXAMPLE

BPANDCALCIUM

7.19 Calcium and blood pressure.   Does increasing the amount of calcium in our diet reduce blood pressure? Examination of a large sample of people revealed a relationship between calcium intake and blood pressure, but such observational studies do not establish causation. Animal experiments, however, showed that calcium supplements do reduce blood pressure in rats, justifying an experiment with human subjects. A randomized comparative experiment gave one group of 10 black men a calcium supplement for 12 weeks. The control group of 11 black men received a placebo that appeared identical. (In fact, a block design with black and white men as the blocks was used. We will look only at the results for blacks, because the earlier survey suggested that calcium is more effective for blacks.) The experiment was double-blind. Table 7.5 gives the seated systolic (heart contracted) blood pressure for all subjects at the beginning and end of the 12-week period, in millimeters (mm) of mercury. Because the researchers were interested in decreasing blood pressure, Table 7.5 also shows the decrease for each subject. An increase appears as a negative entry.[22]

Seated systolic blood pressure

Calcium Group			Placebo Group		
Begin	**End**	**Decrease**	**Begin**	**End**	**Decrease**
107	100	7	123	124	−1
110	114	−4	109	97	12
123	105	18	112	113	−1
129	112	17	102	105	−3
112	115	−3	98	95	3
111	116	−5	114	119	−5
107	106	1	119	114	5
112	102	10	114	112	2
136	125	11	110	121	−11
102	104	−2	117	118	−1
			130	133	−3

As usual, we first examine the data. To compare the effects of the two treatments, take the response variable to be the amount of the decrease in blood pressure. Inspection of the data reveals that there are no outliers. Side-by-side boxplots and normal quantile plots (Figures 7.15 and 7.16) give a more detailed picture. The calcium group has a somewhat short left tail, but there are no severe departures from Normality that will prevent use of t procedures. To examine the question of the researchers who collected these data, we perform a significance test.

EXAMPLE

7.20 Does increased calcium reduce blood pressure? Take Group 1 to be the calcium group and Group 2 to be the placebo group. The evidence that calcium lowers blood pressure more than a placebo is assessed by testing

$$H_0: \mu_1 = \mu_2$$
$$H_a: \mu_1 > \mu_2$$

Here are the summary statistics for the decrease in blood pressure:

Group	Treatment	n	$\bar{x}$	s
1	Calcium	10	5.000	8.743
2	Placebo	11	−0.273	5.901

The calcium group shows a drop in blood pressure, and the placebo group has a small increase. The sample standard deviations do not rule out equal population standard deviations. A difference this large will often arise by chance in samples this small. We are willing to assume equal population

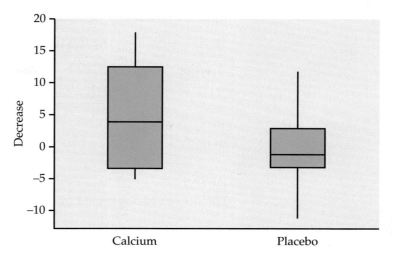

FIGURE 7.15 Side-by-side boxplots of the decrease in blood pressure from Table 7.5.

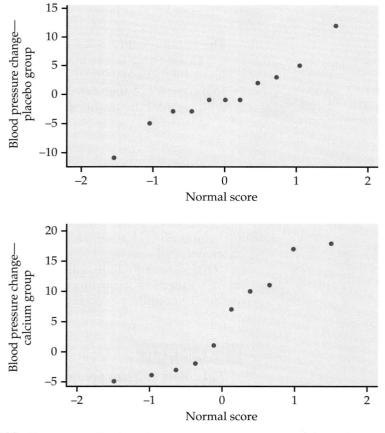

FIGURE 7.16 Normal quantile plots of the change in blood pressure from Table 7.5.

standard deviations. The pooled sample variance is

$$s_p^2 = \frac{(n_1 - 1)s_1^2 + (n_2 - 1)s_2^2}{n_1 + n_2 - 2}$$

$$= \frac{(10 - 1)8.743^2 + (11 - 1)5.901^2}{10 + 11 - 2} = 54.536$$

so that

$$s_p = \sqrt{54.536} = 7.385$$

The pooled two-sample t statistic is

$$t = \frac{\bar{x}_1 - \bar{x}_2}{s_p\sqrt{\dfrac{1}{n_1} + \dfrac{1}{n_2}}}$$

$$= \frac{5.000 - (-0.273)}{7.385\sqrt{\dfrac{1}{10} + \dfrac{1}{11}}}$$

$$= \frac{5.273}{3.227} = 1.634$$

The P-value is $P(T \geq 1.634)$, where T has the $t(19)$ distribution.

From Table D we can see that P falls between the $\alpha = 0.10$ and $\alpha = 0.05$ levels. Statistical software gives the exact value $P = 0.059$. The experiment found evidence that calcium reduces blood pressure, but the evidence falls a bit short of the traditional 5% and 1% levels.

df = 19		
p	0.10	0.05
t^*	1.328	1.729

Sample size strongly influences the P-value of a test. An effect that fails to be significant at a specified level α in a small sample can be significant in a larger sample. In the light of the rather small samples in Example 7.20, the evidence for some effect of calcium on blood pressure is rather good. The published account of the study combined these results for blacks with the results for whites and adjusted for pretest differences among the subjects. Using this more detailed analysis, the researchers were able to report a P-value of 0.008.

Of course, a P-value is almost never the last part of a statistical analysis. To make a judgment regarding the size of the effect of calcium on blood pressure, we need a confidence interval.

EXAMPLE

7.21 How different are the calcium and placebo groups? We estimate that the effect of calcium supplementation is the difference between the sample means of the calcium and the placebo groups, $\bar{x}_1 - \bar{x}_2 = 5.273$ mm Hg. A 90% confidence interval for $\mu_1 - \mu_2$ uses the critical value $t^* = 1.729$ from the $t(19)$ distribution. The interval is

$$(\bar{x}_1 - \bar{x}_2) \pm t^* s_p \sqrt{\frac{1}{n_1} + \frac{1}{n_2}} = [5.000 - (-0.273)] \pm (1.729)(7.385)\sqrt{\frac{1}{10} + \frac{1}{11}}$$

$$= 5.273 \pm 5.579$$

We are 90% confident that the difference in means is in the interval $(-0.306, 10.852)$. The calcium treatment reduced blood pressure by about 5.3 mm Hg more than a placebo on the average, but the margin of error for this estimate is 5.6 mm Hg.

The pooled two-sample t procedures are anchored in statistical theory and so have long been the standard version of the two-sample t in textbooks. *But they require the assumption that the two unknown population standard deviations are equal.* As we shall see in Section 7.3, this assumption is hard to verify. The pooled t procedures are therefore a bit risky. They are reasonably robust against both non-Normality and unequal standard deviations when the sample sizes are nearly the same. When the samples are quite different in size, the pooled t procedures become sensitive to unequal standard deviations and should be used with caution unless the samples are large. Unequal standard deviations are quite common. In particular, it is not unusual for the spread of data to increase when the center gets larger. Statistical software often calculates both the pooled and the unpooled t statistics, as in Figure 7.14.

USE YOUR KNOWLEDGE

7.59 Sleep efficiency revisited. Figure 7.14 (page 443) gives the outputs from four software packages for comparing the systolic blood pressures across two sleep efficiency groups. Some of the software reports both pooled and unpooled analyses. Which outputs give the pooled results? What are the pooled t and its P-value?

7.60 Equal sample sizes. The software outputs in Figure 7.14 give the *same value* for the pooled and unpooled t statistics. Do some simple algebra to show that this is always true when the two sample sizes n_1 and n_2 are the same. In other cases, the two t statistics usually differ.

SECTION 7.2 Summary

Significance tests and confidence intervals for the difference of the means μ_1 and μ_2 of two Normal populations are based on the difference $\bar{x}_1 - \bar{x}_2$ of the sample means from two independent SRSs. Because of the central limit theorem, the resulting procedures are approximately correct for other population distributions when the sample sizes are large.

When independent SRSs of sizes n_1 and n_2 are drawn from two Normal populations with parameters μ_1, σ_1 and μ_2, σ_2 the **two-sample z statistic**

$$z = \frac{(\bar{x}_1 - \bar{x}_2) - (\mu_1 - \mu_2)}{\sqrt{\dfrac{\sigma_1^2}{n_1} + \dfrac{\sigma_2^2}{n_2}}}$$

has the $N(0, 1)$ distribution.

The **two-sample t statistic**

$$t = \frac{(\bar{x}_1 - \bar{x}_2) - (\mu_1 - \mu_2)}{\sqrt{\dfrac{s_1^2}{n_1} + \dfrac{s_2^2}{n_2}}}$$

does *not* have a t distribution. However, good approximations are available.

Conservative inference procedures for comparing μ_1 and μ_2 are obtained from the two-sample t statistic by using the $t(k)$ distribution with degrees of freedom k equal to the smaller of $n_1 - 1$ and $n_2 - 1$.

More accurate probability values can be obtained by estimating the degrees of freedom from the data. This is the usual procedure for statistical software.

An *approximate* level C **confidence interval** for $\mu_1 - \mu_2$ is given by

$$(\bar{x}_1 - \bar{x}_2) \pm t^* \sqrt{\frac{s_1^2}{n_1} + \frac{s_2^2}{n_2}}$$

Here, t^* is the value for the $t(k)$ density curve with area C between $-t^*$ and t^*, where k is computed from the data by software or is the smaller of $n_1 - 1$ and $n_2 - 1$. The quantity

$$t^* \sqrt{\frac{s_1^2}{n_1} + \frac{s_2^2}{n_2}}$$

is the **margin of error.**

Significance tests for H_0: $\mu_1 = \mu_2$ use the **two-sample t statistic**

$$t = \frac{\bar{x}_1 - \bar{x}_2}{\sqrt{\dfrac{s_1^2}{n_1} + \dfrac{s_2^2}{n_2}}}$$

The P-value is approximated using the $t(k)$ distribution where k is estimated from the data using software or is the smaller of $n_1 - 1$ and $n_2 - 1$.

The guidelines for practical use of two-sample t procedures are similar to those for one-sample t procedures. Equal sample sizes are recommended.

If we can assume that the two populations have equal variances, **pooled two-sample t procedures** can be used. These are based on the **pooled estimator**

$$s_p^2 = \frac{(n_1 - 1)s_1^2 + (n_2 - 1)s_2^2}{n_1 + n_2 - 2}$$

of the unknown common variance and the $t(n_1 + n_2 - 2)$ distribution.

SECTION 7.2 Exercises

For Exercises 7.54 and 7.55, see page 438; for Exercises 7.56 and 7.57, see page 440; for Exercise 7.58, see page 446; and for Exercises 7.59 and 7.60, see page 451.

In exercises that call for two-sample t procedures, you may use either of the two approximations for the degrees of freedom that we have discussed: the value given by your software or the smaller of $n_1 - 1$ and $n_2 - 1$. Be sure to state clearly which approximation you have used.

7.61 What is wrong? In each of the following situations explain what is wrong and why.

(a) A researcher wants to test $H_0: \bar{x}_1 = \bar{x}_2$ versus the two-sided alternative $H_a: \bar{x}_1 \neq \bar{x}_2$.

(b) A study recorded the IQ scores of 100 college freshmen. The scores of the 56 males in the study were compared with the scores of all 100 freshmen using the two-sample methods of this section.

(c) A two-sample t statistic gave a P-value of 0.94. From this we can reject the null hypothesis with 90% confidence.

(d) A researcher is interested in testing the one-sided alternative $H_a: \mu_1 < \mu_2$. The significance test gave $t = 2.15$. Since the P-value for the two-sided alternative is 0.036, he concluded that his P-value was 0.018.

7.62 Basic concepts. For each of the following, answer the question and give a short explanation of your reasoning.

(a) A 95% confidence interval for the difference between two means is reported as (0.8, 2.3). What can you conclude about the results of a significance test of the null hypothesis that the population means are equal versus the two-sided alternative?

(b) Will larger samples generally give a larger or smaller margin of error for the difference between two sample means?

7.63 More basic concepts. For each of the following, answer the question and give a short explanation of your reasoning.

(a) A significance test for comparing two means gave $t = -2.18$ with 10 degrees of freedom. Can you reject the null hypothesis that the μ's are equal versus the two-sided alternative at the 5% significance level?

(b) Answer part (a) for the one-sided alternative that the difference in means is negative.

7.64 Effect of the confidence level. Assume $\bar{x}_1 = 100$, $\bar{x}_2 = 110$, $s_1 = 18$, $s_2 = 15$, $n_1 = 50$, and $n_2 = 40$. Find a 95% confidence interval for the difference in the corresponding values of μ. Does this interval include more or fewer values than a 99% confidence interval? Explain your answer.

7.65 Sadness and spending. The "misery is not miserly" phenomenon refers to a person's spending judgment going haywire when sad. In a recent study, 31 young adults were given $10 and randomly assigned to either a sad or neutral group. The participants in the sad group watched a video about the death of a boy's mentor (from *The Champ*) and those in the neutral group watched a video on the Great Barrier Reef. After the video, each participant was offered the chance to trade $0.50 increments of the $10 for an insulated water bottle.[23] Consider the data as follows: 🌊 SADNESS

Group	Purchase Price								
Neutral	0.00	2.00	0.00	1.00	0.50	0.00	0.50		
	2.00	1.00	0.00	0.00	0.00	0.00	1.00		
Sad	3.00	4.00	0.50	1.00	2.50	2.00	1.50	0.00	1.00
	1.50	1.50	2.50	4.00	3.00	3.50	1.00	3.50	

(a) Examine each group's prices graphically. Is use of the t procedures appropriate for these data? Carefully explain your answer.

(b) Make a table with the sample size, mean, and standard deviation for each of the two groups.

(c) State appropriate null and alternative hypotheses for comparing these two groups.

(d) Perform the significance test at the $\alpha = 0.05$ level, making sure to report the test statistic, degrees of freedom, and P-value. What is your conclusion?

(e) Construct a 95% confidence interval for the mean difference in purchase price between the two groups.

7.66 Wine labels with animals? Traditional brand research argues that successful logos are ones that are highly relevant to the product they represent. However, a market research firm recently reported that nearly 20% of all table wine brands introduced in the last three years feature an animal on the label. Since animals have little to do with the product, why are marketers using this tactic? Some researchers have proposed that consumers who are "primed"—in other words, they've thought about the image earlier in an unrelated context—process visual information easier.[24] To demonstrate this, the researchers randomly assigned participants to either a primed or non-primed group. Each participant was asked to indicate their attitude toward a product on a seven-point scale (from 1 = dislike very much to 7 = like very much). A bottle of MagicCoat pet shampoo, with a picture of a collie on the label, was the product. Prior to giving this score, however, participants were asked to do a word find where four of the words were common across groups (pet, grooming, bottle, label) and four were either related to the image (dog, collie, puppy, woof) or image conflicting (cat, feline, kitten, meow). The following table contains the responses listed from smallest to largest.

BRANDPREFERENCE

Group	Brand attitude
Primed	2 2 3 3 3 4 4 4 4 4 4 4 4 4 4 5 5 5 5 5 5 5
Non-primed	1 1 2 2 3 3 3 3 3 3 3 3 3 3 3 3 4 4 4 5

(a) Examine the scores of each group graphically. Is it appropriate to use the two-sample t procedures? Explain your answer.

(b) Test whether these two groups show the same preference for this product. Use a two-sided alternative hypothesis and a significance level of 5%.

(c) Construct a 95% confidence interval for the difference in average preference.

(d) Write a short summary of your conclusions.

7.67 Drive-thru speaker clarity. *QSRMagazine.com* surveyed 689 adults on their drive-thru window experiences at quick service restaurants.[25] One question was, "Thinking about your most recent drive-thru experience, please rate how satisfied you were with the clarity of communication through the speaker." Responses ranged from "Very Dissatisfied (1)" to "Very Satisfied (5)." The following table breaks down the responses according to gender. SPEAKERCLARITY

			Rating		
Gender	1	2	3	4	5
Female	5	44	48	183	188
Male	5	29	30	91	66

(a) Report the means and standard deviations of the rating for the male and female participants separately.

(b) Comment on the appropriateness of t procedures for these data.

(c) Test whether males and females are, on average, equally satisfied with speaker clarity. Use a two-sided alternative hypothesis and a significance level of 5%.

(d) Construct a 95% confidence interval for the difference in average satisfaction.

(e) Given the coarseness of the rating, the owner of the "Sir Beef-a-lot" chain only considers a difference in the means of at least 0.25 units as meaningful. Based on your results in parts (c) and (d), what would you tell this owner?

7.68 Diet and mood. Researchers were interested in comparing the long-term psychological effects of dieters on a high-carbohydrate, low-fat (LF) diet with those on a high-fat, low-carbohydrate (LC) diet.[26] A total of 106 overweight and obese participants were randomly assigned to one of these two energy-restricted diets. At 52 weeks a total of 32 LC dieters and 33 LF dieters remained. Mood was assessed using a total mood disturbance score (TMDS), where a lower score is associated with a less negative mood. A summary of these results follows:

Group	n	$\bar{x}$	s
LC	32	47.3	28.3
LF	33	19.3	25.8

(a) Is there a difference in the TMDS at week 52? Test the null hypothesis that the dieters' average mood in the two groups is the same. Use a significance level of 0.05.

(b) Critics of this study focus on the specific LC diet (i.e., the science) and the dropout rate. Explain why the dropout rate is important to consider when drawing conclusions from this study.

7.69 Comparison of blood lipid levels in males and females. Recall Example 6.15 (page 371). That study included both female and male sedentary students. A total of 108 students volunteered for the study and met the eligibility criteria. The following table summarizes the blood lipid levels, in milligrams per deciliter (mg/dl), of the participants broken down by gender:

	Females ($n = 71$)		Males ($n = 37$)	
	$\bar{x}$	s	$\bar{x}$	s
Total Cholesterol	173.70	34.79	171.81	33.24
LDL	96.38	29.78	109.44	31.05
HDL	61.62	13.75	46.47	7.94

(a) Is it appropriate to use the two-sample t procedures that we studied in this section to analyze these data for gender differences? Give reasons for your answer.

(b) Describe appropriate null and alternative hypotheses for comparing male and female total cholesterol levels.

(c) Carry out the significance test using $\alpha = 0.05$. Report the test statistic with the degrees of freedom and the P-value. Write a short summary of your conclusion.

(d) Find a 95% confidence interval for the difference between the two means. Compare the information given by the interval with the information given by the significance test.

(e) The participants in this study were all taking an introductory health class. To what extent do you think the results can be generalized to other populations?

7.70 More on blood lipid levels. Refer to the previous exercise. LDL is also known as "bad" cholesterol. Suppose the researchers wanted to test the hypothesis that LDL levels are higher in sedentary males than in sedentary females. Describe appropriate null and alternative hypotheses and carry out the significance test using $\alpha = 0.05$. Report the test statistic with the degrees of freedom and the P-value. Write a short summary of your conclusion.

7.71 Evaluating a multimedia program. A multimedia program designed to improve dietary behavior among low-income women was evaluated by comparing women who were randomly assigned to intervention and control groups. The intervention was a 30-minute session in a computer kiosk in the Food Stamp office.[27] One of the outcomes was the score on a knowledge test taken about two months after the program. Here is a summary of the data:

Group	n	$\bar{x}$	s
Intervention	165	5.08	1.15
Control	212	4.33	1.16

(a) The test had six multiple-choice items that were scored as correct or incorrect, so the total score was an integer between 0 and 6. Do you think that these data are Normally distributed? Explain why or why not.

(b) Is it appropriate to use the two-sample t procedures that we studied in this section to analyze these data? Give reasons for your answer.

(c) Describe appropriate null and alternative hypotheses for evaluating the intervention. Some people would prefer a two-sided alternative in this situation while others would use a one-sided significance test. Give reasons for each point of view.

(d) Carry out the significance test. Report the test statistic with the degrees of freedom and the P-value. Write a short summary of your conclusion.

(e) Find a 95% confidence interval for the difference between the two means. Compare the information given by the interval with the information given by the significance test.

(f) The women in this study were all residents of Durham, North Carolina. To what extent do you think the results can be generalized to other populations?

7.72 Self-control and food. Self-efficacy is a general concept that measures how well we think we can control different situations. In the study described in the previous exercise, the participants were asked, "How sure are you that you can eat foods low in fat over the next month?" The response was measured on a five-point scale with 1 corresponding to "not sure at all" and 5 corresponding to "very sure." Here is a summary of the self-efficacy scores obtained about two months after the intervention:

Group	n	$\bar{x}$	s
Intervention	165	4.10	1.19
Control	212	3.67	1.12

Analyze the data using the questions in the previous exercise as a guide.

7.73 Dust exposure at work. Exposure to dust at work can lead to lung disease later in life. One study measured the workplace exposure of tunnel construction workers.[28] Part of the study compared 115 drill and blast workers with 220 outdoor concrete workers. Total dust exposure was measured in milligram years per cubic meter ($mg.y/m^3$). The mean exposure for the drill and blast workers was 18.0 $mg.y/m^3$ with a standard deviation of 7.8 $mg.y/m^3$. For the outdoor concrete workers, the corresponding values were 6.5 $mg.y/m^3$ and 3.4 $mg.y/m^3$.

(a) The sample included all workers for a tunnel construction company who received medical examinations as part of routine health checkups. Discuss the extent to which you think these results apply to other similar types of workers.

(b) Use a 95% confidence interval to describe the difference in the exposures. Write a sentence that gives the interval and provides the meaning of 95% confidence.

(c) Test the null hypothesis that the exposures for these two types of workers are the same. Justify your choice of a one-sided or two-sided alternative. Report the test statistic, the degrees of freedom, and the P-value. Give a short summary of your conclusion.

(d) The authors of the article describing these results note that the distributions are somewhat skewed. Do you think that this fact makes your analysis invalid? Give reasons for your answer.

7.74 Not all dust is the same. Not all dust particles that are in the air around us cause problems for our lungs.

Some particles are too large and stick to other areas of our body before they can get to our lungs. Others are so small that we can breathe them in and out and they will not deposit on our lungs. The researchers in the study described in the previous exercise also measured respirable dust. This is dust that deposits in our lungs when we breathe it. For the drill and blast workers, the mean exposure to respirable dust was 6.3 mg.y/m^3 with a standard deviation of 2.8 mg.y/m^3. The corresponding values for the outdoor concrete workers were 1.4 mg.y/m^3 and 0.7 mg.y/m^3. Analyze these data using the questions in the previous exercise as a guide.

7.75 Change in portion size. A recent study of food portion sizes reported that over a 17-year period, the average size of a soft drink consumed by Americans aged 2 years and older increased from 13.1 ounces (oz) to 19.9 oz. The authors state that the difference is statistically significant with $P < 0.01$.[29] Explain what additional information you would need to compute a confidence interval for the increase, and outline the procedure that you would use for the computations. Do you think that a confidence interval would provide useful additional information? Explain why or why not.

7.76 Beverage consumption. The results in the previous exercise were based on two national surveys with a very large number of individuals. Here is a study that also looked at beverage consumption, but the sample sizes are much smaller. One part of this study compared 20 children who were 7 to 10 years old with 5 children who were 11 to 13.[30] The younger children consumed an average of 8.2 oz of sweetened drinks per day while the older ones averaged 14.5 oz. The standard deviations were 10.7 oz and 8.2 oz, respectively.

(a) Do you think that it is reasonable to assume that these data are Normally distributed? Explain why or why not. (*Hint:* Think about the 68–95–99.7 rule.)

(b) Using the methods in this section, test the null hypothesis that the two groups of children consume equal amounts of sweetened drinks versus the two-sided alternative. Report all details of the significance-testing procedure with your conclusion.

(c) Give a 95% confidence interval for the difference in means.

(d) Do you think that the analyses performed in parts (b) and (c) are appropriate for these data? Explain why or why not.

(e) The children in this study were all participants in an intervention study at the Cornell Summer Day Camp at Cornell University. To what extent do you think that these results apply to other groups of children?

7.77 Study design is important! Recall Exercise 7.54 (page 438). You are concerned that day of the week may affect online sales. So to compare the two MySpace page designs, you choose two successive weeks in the middle of a month. You flip a coin to assign one Monday to the first design and the other Monday to the second. You repeat this for each of the seven days of the week. You now have 7 hit amounts for each design. It is *incorrect* to use the two-sample t test to see if the mean hits differ for the two designs. Carefully explain why.

7.78 New computer monitors? The purchasing department has suggested that all new computer monitors for your company should be flat screens. You want data to assure you that employees will like the new screens. The next 20 employees needing a new computer are the subjects for an experiment.

(a) Label the employees 01 to 20. Randomly choose 10 to receive flat screens. The remaining 10 get standard monitors.

(b) After a month of use, employees express their satisfaction with their new monitors by responding to the statement "I like my new monitor" on a scale from 1 to 5, where 1 represents "strongly disagree," 2 is "disagree," 3 is "neutral," 4 is "agree," and 5 stands for "strongly agree." The employees with the flat screens have average satisfaction 4.8 with standard deviation 0.7. The employees with the standard monitors have average 3.0 with standard deviation 1.5. Give a 95% confidence interval for the difference in the mean satisfaction scores for all employees.

(c) Would you reject the null hypothesis that the mean satisfaction for the two types of monitors is the same versus the two-sided alternative at significance level 0.05? Use your confidence interval to answer this question. Explain why you do not need to calculate the test statistic.

7.79 Why randomize? Refer to the previous exercise. A coworker suggested that you give the flat screens to the next 10 employees who need new screens and the standard monitor to the following 10. Explain why your randomized design is better.

7.80 Does ad placement matter? Corporate advertising tries to enhance the image of the corporation. A study compared two ads from two sources, the *Wall Street Journal* and the *National Enquirer*. Subjects were asked to pretend that their company was considering a major investment in Performax, the fictitious sportswear firm in the ads. Each subject was asked to respond to the question, "How trustworthy was the source in the sportswear company ad for Performax?" on a 7-point scale. Higher values indicated more trustworthiness.[31]

Here is a summary of the results:

Ad source	n	$\bar{x}$	s
Wall Street Journal	66	4.77	1.50
National Enquirer	61	2.43	1.64

(a) Compare the two sources of ads using a t test. Be sure to state your null and alternative hypotheses, the test statistic with degrees of freedom, the P-value, and your conclusion.

(b) Give a 95% confidence interval for the difference.

(c) Write a short paragraph summarizing the results of your analyses.

7.81 ⚔ **Size of trees in the northern and southern halves.** The study of 584 longleaf pine trees in the Wade Tract in Thomas County, Georgia, had several purposes. Are trees in one part of the tract more or less like trees in any other part of the tract or are there differences? In Example 6.1 (page 342) we examined how the trees were distributed in the tract and found that the pattern was not random. In this exercise we will examine the sizes of the trees. In Exercise 7.25 we analyzed the sizes, measured as diameter at breast height (DBH), for a random sample of 40 trees. Here we divide the tract into northern and southern halves and take random samples of 30 trees from each half. Here are the diameters in centimeters (cm) of the sampled trees: 🌑 NSTREEDIAMETER

North	27.8	14.5	39.1	3.2	58.8	55.5	25.0	5.4	19.0	30.6
	15.1	3.6	28.4	15.0	2.2	14.2	44.2	25.7	11.2	46.8
	36.9	54.1	10.2	2.5	13.8	43.5	13.8	39.7	6.4	4.8

South	44.4	26.1	50.4	23.3	39.5	51.0	48.1	47.2	40.3	37.4
	36.8	21.7	35.7	32.0	40.4	12.8	5.6	44.3	52.9	38.0
	2.6	44.6	45.5	29.1	18.7	7.0	43.8	28.3	36.9	51.6

(a) Use a back-to-back stemplot and side-by-side boxplots to examine the data graphically. Describe the patterns in the data.

(b) Is it appropriate to use the methods of this section to compare the mean DBH of the trees in the north half of the tract with the mean DBH of trees in the south half? Give reasons for your answer.

(c) What are appropriate null and alternative hypotheses for comparing the two samples of tree DBHs? Give reasons for your choices.

(d) Perform the significance test. Report the test statistic, the degrees of freedom, and the P-value. Summarize your conclusion.

(e) Find a 95% confidence interval for the difference in mean DBHs. Explain how this interval provides additional information about this problem.

7.82 ⚔ **Size of trees in the eastern and western halves.** The Wade Tract can also be divided into eastern and western halves. Here are the DBHs of 30 randomly selected longleaf pine trees from each half: 🌑 EWTREEDIAMETER

East	23.5	43.5	6.6	11.5	17.2	38.7	2.3	31.5	10.5	23.7
	13.8	5.2	31.5	22.1	6.7	2.6	6.3	51.1	5.4	9.0
	43.0	8.7	22.8	2.9	22.3	43.8	48.1	46.5	39.8	10.9

West	17.2	44.6	44.1	35.5	51.0	21.6	44.1	11.2	36.0	42.1
	3.2	25.5	36.5	39.0	25.9	20.8	3.2	57.7	43.3	58.0
	21.7	35.6	30.9	40.6	30.7	35.6	18.2	2.9	20.4	11.4

Using the questions in the previous exercise, analyze these data.

7.83 Sales of a small appliance across months. A market research firm supplies manufacturers with estimates of the retail sales of their products from samples of retail stores. Marketing managers are prone to look at the estimate and ignore sampling error. Suppose that an SRS of 70 stores this month shows mean sales of 53 units of a small appliance, with standard deviation 12 units. During the same month last year, an SRS of 55 stores gave mean sales of 50 units, with standard deviation 10 units. An increase from 50 to 53 is a rise of 6%. The marketing manager is happy because sales are up 6%.

(a) Use the two-sample t procedure to give a 95% confidence interval for the difference in mean number of units sold at all retail stores.

(b) Explain in language that the manager can understand why he cannot be certain that sales rose by 6%, and that in fact sales may even have dropped.

7.84 An improper significance test. A friend has performed a significance test of the null hypothesis that two means are equal. His report states that the null hypothesis is rejected in favor of the alternative that the first mean is larger than the second. In a presentation on his work, he notes that the first sample mean was larger than the second mean and this is why he chose this particular one-sided alternative.

(a) Explain what is wrong with your friend's procedure and why.

(b) Suppose he reported $t = 1.70$ with a P-value of 0.06. What is the correct P-value that he should report?

7.85 Breast-feeding versus baby formula. A study of iron deficiency among infants compared samples of infants following different feeding regimens. One group contained breast-fed infants, while the children in another group were fed a standard baby formula without any iron supplements. Here are summary results on blood hemoglobin levels at 12 months of age:[32]

Group	n	$\bar{x}$	s
Breast-fed	23	13.3	1.7
Formula	19	12.4	1.8

(a) Is there significant evidence that the mean hemoglobin level is higher among breast-fed babies? State H_0 and H_a and carry out a t test. Give the P-value. What is your conclusion?

(b) Give a 95% confidence interval for the mean difference in hemoglobin level between the two populations of infants.

(c) State the assumptions that your procedures in parts (a) and (b) require in order to be valid.

The following exercises concern optional material on the pooled two-sample t procedures and on the power of tests.

7.86 Revisiting the sadness and spending study. In Exercise 7.65 (page 453), the purchase price of a water bottle was compared using the two-sample t procedures that do not assume equal standard deviations. Compare the means using a significance test and find the 95% confidence interval for the difference using the pooled methods. How do the results compare with those you obtained in Exercise 7.65?

7.87 Revisiting wine labels with animals. In Exercise 7.66 (page 453), the attitude toward a product was compared using the two-sample t procedures that do not assume equal standard deviations. Compare the means using a significance test and find the 95% confidence interval for the difference using the pooled methods. How do the results compare with those you obtained in Exercise 7.66? BRANDPREFERENCE

7.88 Revisiting drive-thru speaker clarity. In Exercise 7.67 (page 454), the satisfaction with drive-thru speaker clarity was compared using the two-sample t procedures that do not assume equal standard deviations. Examine the standard deviations for the two groups and verify that it is appropriate to use the pooled procedures for these data. Compare the means using a significance test and find the 95% confidence interval for the difference using the pooled methods. How do the results compare with those you obtained in Exercise 7.67? SPEAKERCLARITY

7.89 Revisiting the size of trees. Refer to the Wade Tract DBH data in Exercise 7.81 (page 457) where we compared a sample of trees from the northern half of the tract with a sample from the southern half. Because the standard deviations for the two samples are quite close, it is reasonable to analyze these data using the pooled procedures. Perform the significance test and find the 95% confidence interval for the difference in means using these methods. Summarize your results and compare them with what you found in Exercise 7.81. NSTREEDIAMETER

7.90 Revisiting the sleep efficiency study. Example 7.16 (page 440) gives summary statistics for systolic blood pressure in high sleep efficiency and low sleep efficiency children. The two sample standard deviations are somewhat similar, so we may be willing to assume equal population standard deviations. Calculate the pooled t test statistic and its degrees of freedom from the summary statistics. Use Table D to assess significance. How do your results compare with the unpooled analysis in the example?

7.91 Computing the degrees of freedom. Use the Wade Tract data in Exercise 7.81 to calculate the software approximation to the degrees of freedom using the formula on page 445. Verify your calculation with software.

7.92 Again computing the degrees of freedom. Use the Wade Tract data in Exercise 7.82 to calculate the software approximation to the degrees of freedom using the formula on page 445. Verify your calculation with software.

7.93 Revisiting the dust exposure study. The data on occupational exposure to dust that we analyzed in Exercise 7.73 (page 455) come from two groups of workers that are quite different in size. This complicates the issue regarding pooling because the sample that is larger will dominate the calculations.

(a) Calculate the degrees of freedom approximation using the formula for the degrees of freedom given on page 445. Then verify your calculations with software.

(b) Find the pooled estimate of the standard deviation. Write a short summary comparing it with the estimates of the standard deviations that come from each group.

(c) Find the standard error of the difference in sample means that you would use for the method that does not assume equal variances. Do the same for the pooled approach. Compare these two estimates with each other.

(d) Perform the significance test and find the 95% confidence interval using the pooled methods. How do these results compare with those you found in Exercise 7.73 (page 455)?

(e) Exercise 7.74 has data for the same workers but for respirable dust. Here the standard deviations differ more than those in Exercise 7.73 do. Answer parts (a) through (d) for these data. Write a summary of what you have found in this exercise.

7.94 **Revisiting the small sample exercise.** Recall Example 7.17 (page 442). This is a case where the sample sizes are quite small. With only 5 observations per group, we have very little information to make a judgment about whether the population standard deviations are equal. The potential gain from pooling is large when the sample sizes are small. Assume that we will perform a two-sided test using the 5% significance level. 🌀 BPANDSLEEP

(a) Find the critical value for the unpooled t test statistic that does not assume equal variances. Use the minimum of $n_1 - 1$ and $n_2 - 1$ for the degrees of freedom.

(b) Find the critical value for the pooled t test statistic.

(c) How does comparing these critical values show an advantage of the pooled test?

7.3 Optional Topics in Comparing Distributions*

In this section we discuss three topics that are related to the material that we have already covered in this chapter. If we can do inference for means, it is natural to ask if we can do something similar for spread. The answer is yes, but there are many cautions. We also discuss robustness and show how to find the power for the two-sample t test. If you plan to design studies, you should become familiar with this last topic.

Inference for population spread

The two most basic descriptive features of a distribution are its center and spread. In a Normal population, these aspects are measured by the mean and the standard deviation. We have described procedures for inference about population means for Normal populations and found that these procedures are often useful for non-Normal populations as well. It is natural to turn next to inference about the standard deviations of Normal populations. Our recommendation here is short and clear: don't do it without expert advice.

CAUTION

We will describe the F test for comparing the spread of two Normal populations. *Unlike the t procedures for means, the F test and other procedures for standard deviations are extremely sensitive to non-Normal distributions.* This lack of robustness does not improve in large samples. It is difficult in practice to tell whether a significant F-value is evidence of unequal population spreads or simply evidence that the populations are not Normal. Consequently, we do not recommend use of inference about population standard deviations in basic statistical practice.[33]

It was once common to test equality of standard deviations as a preliminary to performing the pooled two-sample t test for equality of two population means. It is better practice to check the distributions graphically, with special attention to skewness and outliers, and to use the software-based two-sample t that does not require equal standard deviations. In the words of one distinguished statistician, "To make a preliminary test on variances is rather like putting to sea in a rowing boat to find out whether conditions are sufficiently calm for an ocean liner to leave port!"[34]

---

*This section can be omitted without loss of continuity.

The *F* test for equality of spread

Because of the limited usefulness of procedures for inference about the standard deviations of Normal distributions, we will present only one such procedure. Suppose that we have independent SRSs from two Normal populations, a sample of size n_1 from $N(\mu_1, \sigma_1)$ and a sample of size n_2 from $N(\mu_2, \sigma_2)$. The population means and standard deviations are all unknown. The hypothesis of equal spread

$$H_0: \sigma_1 = \sigma_2$$

is tested against

$$H_a: \sigma_1 \neq \sigma_2$$

by a simple statistic, the ratio of the sample variances.

THE *F* STATISTIC AND *F* DISTRIBUTIONS

When s_1^2 and s_2^2 are sample variances from independent SRSs of sizes n_1 and n_2 drawn from Normal populations, the *F* statistic

$$F = \frac{s_1^2}{s_2^2}$$

has the *F* distribution with $n_1 - 1$ and $n_2 - 1$ degrees of freedom when $H_0: \sigma_1 = \sigma_2$ is true.

F distributions

    The **F distributions** are a family of distributions with two parameters: the degrees of freedom of the sample variances in the numerator and denominator of the *F* statistic. The *F* distributions are another of R. A. Fisher's contributions to statistics and are called *F* in his honor. Fisher introduced *F* statistics for comparing several means. We will meet these useful statistics in later chapters. The numerator degrees of freedom are always mentioned first. Interchanging the degrees of freedom changes the distribution, so the order is important. Our brief notation will be $F(j, k)$ for the *F* distribution with j degrees of freedom in the numerator and k degrees of freedom in the denominator. The *F* distributions are not symmetric but are right-skewed. The density curve in Figure 7.17 illustrates the shape. Because sample variances cannot be negative,

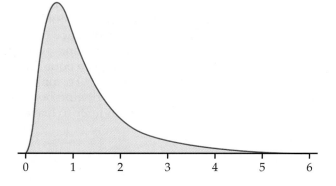

FIGURE 7.17 The density curve for the $F(9, 10)$ distribution. The *F* distributions are skewed to the right.

the F statistic takes only positive values and the F distribution has no probability below 0. The peak of the F density curve is near 1; values far from 1 in either direction provide evidence against the hypothesis of equal standard deviations.

Tables of F critical values are awkward because a separate table is needed for every pair of degrees of freedom j and k. Table E in the back of the book gives upper p critical values of the F distributions for $p = 0.10, 0.05, 0.025, 0.01,$ and 0.001. For example, these critical values for the $F(9, 10)$ distribution shown in Figure 7.17 are

p	0.10	0.05	0.025	0.01	0.001
F^*	2.35	3.02	3.78	4.94	8.96

The skewness of F distributions causes additional complications. In the symmetric Normal and t distributions, the point with probability 0.05 below it is just the negative of the point with probability 0.05 above it. This is not true for F distributions. We therefore require either tables of both the upper and lower tails or means of eliminating the need for lower-tail critical values. Statistical software that eliminates the need for tables is plainly very convenient. If you do not use statistical software, arrange the F test as follows:

1. Take the test statistic to be

$$F = \frac{\text{larger } s^2}{\text{smaller } s^2}$$

This amounts to naming the populations so that s_1^2 is the larger of the observed sample variances. The resulting F is always 1 or greater.

2. Compare the value of F with the critical values from Table E. Then *double* the probabilities obtained from the table to get the significance level for the two-sided F test.

The idea is that we calculate the probability in the upper tail and double to obtain the probability of all ratios on either side of 1 that are at least as improbable as that observed. Remember that the order of the degrees of freedom is important in using Table E.

BPANDCALCIUM

EXAMPLE

7.22 Comparing calcium and placebo groups. Example 7.19 (page 447) recounts a medical experiment comparing the effects of calcium and a placebo on the blood pressure of black men. The analysis (Example 7.20) employed the pooled two-sample t procedures. Because these procedures require equal population standard deviations, it is tempting to first test

$$H_0: \sigma_1 = \sigma_2$$
$$H_a: \sigma_1 \neq \sigma_2$$

The larger of the two sample standard deviations is $s = 8.743$ from 10 observations. The other is $s = 5.901$ from 11 observations. The two-sided test statistic is therefore

$$F = \frac{\text{larger } s^2}{\text{smaller } s^2} = \frac{8.743^2}{5.901^2} = 2.20$$

We compare the calculated value $F = 2.20$ with critical points for the $F(9, 10)$ distribution. Table E shows that 2.20 is *less* than the 0.10 critical value of the $F(9, 10)$ distribution, which is $F^* = 2.35$. Doubling 0.10, we know that the observed F falls short of the 0.20 significance level. The results are not significant at the 20% level (or any lower level). Statistical software shows that the exact upper-tail probability is 0.118, and hence $P = 0.236$. *If* the populations were Normal, the observed standard deviations would give little reason to suspect unequal population standard deviations. Because one of the populations shows some non-Normality, however, we cannot be fully confident of this conclusion.

USE YOUR KNOWLEDGE

7.95 The F statistic. The F statistic $F = s_1^2/s_2^2$ is calculated from samples of size $n_1 = 16$ and $n_2 = 23$.

(a) What is the upper 5% critical value for this F?

(b) In a test of equality of standard deviations against the two-sided alternative, this statistic has the value $F = 2.45$. Is this value significant at the 10% level? Is it significant at the 5% level?

Robustness of Normal inference procedures

We have claimed that

- The t procedures for inference about means are quite robust against non-Normal population distributions. These procedures are particularly robust when the population distributions are symmetric and (for the two-sample case) when the two sample sizes are equal.

- The F test and other procedures for inference about variances are so lacking in robustness as to be of little use in practice.

Simulations with a large variety of non-Normal distributions support these claims. One set of simulations was carried out with samples of size 25 and used significance tests with fixed level $\alpha = 0.05$. The three types of tests studied were the one-sample and pooled two-sample t tests and the F test for comparing two variances.

The robustness of the one-sample and two-sample t procedures is remarkable. The true significance level remains between about 4% and 6% for a large range of populations. The t test and the corresponding confidence intervals are among the most reliable tools that statisticians use. Remember, however, that outliers can greatly disturb the t procedures. Also, two-sample procedures are less robust when the sample sizes are not similar.

The lack of robustness of the tests for variances is equally remarkable. The true significance levels depart rapidly from the target 5% as the population distribution departs from Normality. The two-sided *F* test carried out with 5% critical values can have a true level of less than 1% or greater than 11% even in symmetric populations with no outliers. Results such as these are the basis for our recommendation that these procedures not be used.

The power of the two-sample *t* test

The two-sample *t* test is one of the most used statistical procedures. Unfortunately, because of inadequate planning, users frequently fail to find evidence for the effects that they believe to be true. Power calculations should be part of the planning of any statistical study. Information from a pilot study or previous research is needed.

In Section 7.1 (optional material), we learned how to find an approximation for the power of the one-sample *t* test. The basic concepts for the two-sample case are the same. Here, we give the exact method, which involves a new distribution, the **noncentral *t* distribution.** To perform the calculations, we simply need software to calculate probabilities for this distribution.

noncentral t *distribution*

We first present the method for the pooled two-sample *t* test, where the parameters are μ_1, μ_2, and the common standard deviation σ. Modifications to get approximate results when we do not pool are then described.

To find the power for the pooled two-sample *t* test, use the following steps. We consider only the case where the null hypothesis is $\mu_1 - \mu_2 = 0$.

1. Specify
   (a) an alternative value for $\mu_1 - \mu_2$ that you consider important to detect;
   (b) the sample sizes, n_1 and n_2;
   (c) a fixed significance level, α;
   (d) a guess at the standard deviation, σ.

2. Find the degrees of freedom df $= n_1 + n_2 - 2$ and the value of t^* that will lead to rejection of H_0.

noncentrality parameter

3. Calculate the **noncentrality parameter**

$$\delta = \frac{|\mu_1 - \mu_2|}{\sigma\sqrt{\dfrac{1}{n_1} + \dfrac{1}{n_2}}}$$

4. Find the power as the probability that a noncentral *t* random variable with degrees of freedom df and noncentrality parameter δ will be greater than t^*. In SAS the command is 1-PROBT(tstar,df,delta). If you do not have software that can perform this calculation, you can approximate the power as the probability that a standard Normal random variable is greater than $t^* - \delta$, that is, $P(z > t^* - \delta)$, and use Table A.

Note that the denominator in the noncentrality parameter,

$$\sigma\sqrt{\frac{1}{n_1} + \frac{1}{n_2}}$$

is our guess at the standard error for the difference in the sample means. Therefore, if we wanted to assess a possible study in terms of the margin of error for the estimated difference, we would examine t^* times this quantity.

If we do not assume that the standard deviations are equal, we need to guess both standard deviations and then combine these for our guess at the standard error:

$$\sqrt{\frac{\sigma_1^2}{n_1} + \frac{\sigma_2^2}{n_2}}$$

This guess is then used in the denominator of the noncentrality parameter. For the degrees of freedom, the conservative approximation is appropriate.

EXAMPLE

7.23   Planning a new study of calcium versus placebo groups.   In Example 7.20 we examined the effect of calcium on blood pressure by comparing the means of a treatment group and a placebo group using a pooled two-sample t test. The P-value was 0.059, failing to achieve the usual standard of 0.05 for statistical significance. Suppose that we wanted to plan a new study that would provide convincing evidence, say at the 0.01 level, with high probability. Let's examine a study design with 45 subjects in each group ($n_1 = n_2 = 45$). Based on our previous results we choose $\mu_1 - \mu_2 = 5$ as an alternative that we would like to be able to detect with $\alpha = 0.01$. For σ we use 7.4, our pooled estimate from Example 7.20. The degrees of freedom are $n_1 + n_2 - 2 = 88$ and $t^* = 2.37$ for the significance test. The noncentrality parameter is

$$\delta = \frac{5}{7.4\sqrt{\frac{1}{45} + \frac{1}{45}}} = \frac{5}{1.56} = 3.21$$

Software gives the power as 0.7965, or 80%. The Normal approximation gives 0.7983, a very accurate result. With this choice of sample sizes we would expect the margin of error for a 95% confidence interval ($t^* = 1.99$) for the difference in means to be

$$t^* \times 7.4\sqrt{\frac{1}{45} + \frac{1}{45}} = 1.99 \times 1.56 = 3.1$$

With software it is very easy to examine the effects of variations on a study design. In the preceding example, we might want to examine the power for $\alpha = 0.05$ and the effects of reducing the sample sizes.

USE YOUR KNOWLEDGE

7.96   Power and $\mu_1 - \mu_2$. If you repeat the calculation in Example 7.23 for other values of $\mu_1 - \mu_2$ that are larger than 5, would you expect the power to be higher or lower than 0.7965? Why?

7.97   Power and the standard deviation. If the true population standard deviation were 7.8 instead of the 7.4 hypothesized in Example 7.23, would the power for this new experiment be greater or smaller than 0.7965? Explain.

SECTION 7.3 Summary

Inference procedures for comparing the standard deviations of two Normal populations are based on the **F statistic,** which is the ratio of sample variances:

$$F = \frac{s_1^2}{s_2^2}$$

If an SRS of size n_1 is drawn from the x_1 population and an independent SRS of size n_2 is drawn from the x_2 population, the F statistic has the **F distribution** $F(n_1 - 1, n_2 - 1)$ if the two population standard deviations σ_1 and σ_2 are in fact equal.

The **F test for equality of standard deviations** tests $H_0: \sigma_1 = \sigma_2$ versus $H_a: \sigma_1 \neq \sigma_2$ using the statistic

$$F = \frac{\text{larger } s^2}{\text{smaller } s^2}$$

and doubles the upper-tail probability to obtain the P-value.

The t procedures are quite **robust** when the distributions are not Normal. The F tests and other procedures for inference about the spread of one or more Normal distributions are so strongly affected by non-Normality that we do not recommend them for regular use.

The **power** of the pooled two-sample t test is found by first computing the critical value for the significance test, the degrees of freedom, and the **noncentrality parameter** for the alternative of interest. These are used to find the power from the **noncentral t distribution.** A Normal approximation works quite well. Calculating margins of error for various study designs and assumptions is an alternative procedure for evaluating designs.

SECTION 7.3 Exercises

For Exercise 7.95, see page 462; and for Exercises 7.96 and 7.97, see page 464.

In all exercises calling for use of the F test, assume that both population distributions are very close to Normal. The actual data are not always sufficiently Normal to justify use of the F test.

7.98 Comparison of standard deviations. Here are some summary statistics from two independent samples from Normal distributions:

Sample	n	s^2
1	9	3.5
2	18	9.1

You want to test the null hypothesis that the two population standard deviations are equal versus the two-sided alternative at the 5% significance level.

(a) Calculate the test statistic.

(b) Find the appropriate value from Table E that you need to perform the significance test.

(c) What do you conclude?

7.99 Revisiting the sleep efficiency comparison. Compare the standard deviations of blood pressure in Example 7.16 (page 440). Give the test statistic, the degrees of freedom, and the P-value. Write a short summary of your analysis, including comments on the assumptions for the test.

7.100 An HDL comparison. HDL is also known as "good" cholesterol. Compare the standard deviations of HDL in Exercise 7.69 (page 454). Give the test statistic, the degrees of freedom, and the P-value. Write a short summary of your analysis, including comments on the assumptions for the test.

7.101 Revisiting the multimedia evaluation study. Mean scores on a knowledge test are compared for two groups of women in Exercise 7.71 (page 455). Compare the standard deviations using an F test. What do you conclude? Comment on the Normal assumption for these data. These standard deviations are so close that we are not particularly surprised at the result of the significance test. Assume that the sample standard deviation in the intervention is the value given in

Exercise 7.71 (1.15). How large would the standard deviation in the control group need to be to reject the null hypothesis of equal standard deviations at the 5% level?

7.102 Revisiting the self-control and food study. Compare the standard deviations of the self-efficacy scores in Exercise 7.72 (page 455). Give the test statistic, the degrees of freedom, and the P-value. Write a short summary of your analysis, including comments on the assumptions for the test.

7.103 Revisiting the dust exposure study. The two-sample problem in Exercise 7.73 (page 455) compares drill and blast workers with outdoor concrete workers with respect to the total dust that they are exposed to in the workplace. Here it may be useful to know whether or not the standard deviations differ in the two groups. Perform the F test and summarize the results. Are you concerned about the assumptions here? Explain why or why not.

7.104 More on the dust exposure study. Exercise 7.74 (page 455) is similar to Exercise 7.73, but the response variable here is exposure to dust particles that can enter and stay in the lungs. Compare the standard deviations with a significance test and summarize the results. Be sure to comment on the assumptions.

7.105 Revisiting the size of trees in the north and south. The diameters of trees in the Wade Tract for random samples selected from the north and south portions of the tract are compared in Exercise 7.81 (page 457). Are there statistically significant differences in the standard deviations for these two parts of the tract? Perform the significance test and summarize the results. Does the Normal assumption appear reasonable for these data? NSTREEDIAMETER

7.106 Revisiting the size of trees in the east and west. Tree diameters for the east and west halves of the Wade Tract are compared in Exercise 7.82 (page 457). Using the questions in the previous exercise as a guide, analyze these data. EWTREEDIAMETER

7.107 Revisiting the small sample size example. In Example 7.17 (page 442), we addressed a study with only five observations per group. BPANDSLEEP

(a) Are the statistically significant differences in the standard deviations for these two groups? Perform the test using a significance level of 0.05 and state your conclusion.

(b) Using Table E, state the value that the ratio of variances would have to exceed for us to reject the null hypothesis (at the 5% level) that the standard deviations

are equal. Also, report this value for group sample sizes of $n = 4$, 3 and 2. What does this suggest about the power of this test when sample sizes are small?

7.108 Planning a study to compare tree size. In Exercise 7.81 (page 457) DBH data for longleaf pine trees in two parts of the Wade Tract are compared. Suppose that you are planning a similar study where you will measure the diameters of longleaf pine trees. Based on Exercise 7.81, you are willing to assume that the standard deviation is 20 cm. Suppose that a difference in mean DBH of 10 cm or more would be important to detect. You will use a t statistic and a two-sided alternative for the comparison.

(a) Find the power if you randomly sample 20 trees from each area to be compared.

(b) Repeat the calculations for 60 trees in each sample.

(c) If you had to choose between the 20 and 60 trees per sample, which would be acceptable? Give reasons for your answer.

7.109 More on planning a study to compare tree size. Refer to the previous exercise. Find the two standard deviations from Exercise 7.81. Do the same for the data in Exercise 7.82, which is a similar setting. These are somewhat smaller than the assumed value that you used in the previous exercise. Explain why it is generally a better idea to assume a standard deviation that is larger than you expect than one that is smaller. Repeat the power calculations for some other reasonable values of σ and comment on the impact of the size of σ for planning the new study.

7.110 Planning a study to compare ad placement. Refer to Exercise 7.80 (page 456), where we compared trustworthy ratings for ads from two different publications. Suppose that you are planning a similar study using two different publications that are not expected to show the differences seen when comparing the *Wall Street Journal* with the *National Enquirer*. You would like to detect a difference of 1.5 points using a two-sided significance test with a 5% level of significance. Based on Exercise 7.80, it is reasonable to use 1.6 as the value of the standard deviation for planning purposes.

(a) What is the power if you use sample sizes similar to those used in the previous study, for example, 65 for each publication?

(b) Repeat the calculations for 100 in each group.

(c) What sample size would you recommend for the new study?

CHAPTER 7 Exercises

7.111 LSAT scores. The scores of four senior roommates on the Law School Admission Test (LSAT) are

$$158, \ 138, \ 141, \ 125$$

Find the mean, the standard deviation, and the standard error of the mean. Is it appropriate to calculate a confidence interval based on these data? Explain why or why not. 🎯 LSAT

7.112 Converting a two-sided *P*-value. You use statistical software to perform a significance test of the null hypothesis that two means are equal. The software reports *P*-values for the two-sided alternative. Your alternative is that the first mean is greater than the second mean.

(a) The software reports $t = 2.08$ with a *P*-value of 0.08. Would you reject H_0 with $\alpha = 0.05$? Explain your answer.

(b) The software reports $t = -2.08$ with a *P*-value of 0.08. Would you reject H_0 with $\alpha = 0.05$? Explain your answer.

7.113 Degrees of freedom and confidence interval width. As the degrees of freedom increase, the t distributions get closer and closer to the z ($N(0, 1)$) distribution. One way to see this is to look at how the value of t^* for a 95% confidence interval changes with the degrees of freedom. Make a plot with degrees of freedom from 2 to 100 on the x axis and t^* on the y axis. Draw a horizontal line on the plot corresponding to the value of $z^* = 1.96$. Summarize the main features of the plot.

7.114 Degrees of freedom and t^*. Refer to the previous exercise. Make a similar plot and summarize its features for the value of t^* for a 90% confidence interval.

7.115 Sample size and margin of error. The margin of error for a confidence interval depends on the confidence level, the standard deviation, and the sample size. Fix the confidence level at 95% and the standard deviation at 1 to examine the effect of the sample size. Find the margin of error for sample sizes of 5 to 100 by 5's—that is, let $n = 5, 10, 15, \ldots, 100$. Plot the margins of error versus the sample size and summarize the relationship.

7.116 More on sample size and margin of error. Refer to the previous exercise. Make a similar plot and summarize its features for a 99% confidence interval.

7.117 Which design? The following situations all require inference about a mean or means. Identify each as (1) a single sample, (2) matched pairs, or (3) two independent samples. Explain your answers.

(a) Your customers are college students. You are interested in comparing the interest in a new product that you are developing between those students that live in the dorms and those that live elsewhere.

(b) Your customers are college students. You are interested in comparing which of two new product labels is more appealing.

(c) Your customers are college students. You are interested in assessing their interest in a new product.

7.118 Which design? The following situations all require inference about a mean or means. Identify each as (1) a single sample, (2) matched pairs, or (3) two independent samples. Explain your answers.

(a) You want to estimate the average age of your store's customers.

(b) You do an SRS survey of your customers every year. One of the questions in the survey asks about customer satisfaction on a seven-point scale with the response 1 indicating "very dissatisfied" and 7 indicating "very satisfied." You want to see if the mean customer satisfaction has improved from last year.

(c) You ask an SRS of customers their opinions on each of two new floor plans for your store.

7.119 Number of critical food violations. The results of a major city's restaurant inspections are available through its online newspaper.[35] Critical food violations are those that put patrons at risk of getting sick and must immediately be corrected by the restaurant. An SRS of $n = 200$ inspections from the more than 4400 inspections since January 2008 were collected resulting in $\bar{x} = 1.20$ violations and $s = 1.81$ violations.

(a) Test the hypothesis that the average number of critical violations is less than 1.5 using a significance level of 0.05. State the two hypotheses, the test statistic, and *P*-value.

(b) Construct a 95% confidence interval for the average number of critical violations and summarize your result.

(c) Which of the two summaries (significance test versus confidence interval) do you find more helpful in this case? Explain your answer.

(d) These data are integers ranging from 0 to 9. The data are also skewed to the right with 70% of the values either a 0 or 1. Given this information, do you feel use of the t procedures is appropriate? Explain your answer.

7.120 Brain training. The assessment of computerized brain-training programs is a rapidly growing area of research. Researchers are now focusing on whom does this training benefit most, what brain functions can be best improved, and which products are most effective. A recent study looked at 487 community-dwelling adults aged 65 and older, each randomly assigned to one of two training groups. In one group, the participants used a computerized program 1 hour per day. In the other, DVD-based educational programs were shown with quizzes following each video. The training period lasted 8 weeks. The response was the improvement in a composite score obtained from an auditory memory/attention survey given before and after the 8 weeks.[36] The results are summarized in the following table.

Group	n	$\bar{x}$	s
Computer Program	242	3.9	8.28
DVD Program	245	1.8	8.33

(a) Given that there are other studies showing a benefit of computerized brain training, state the null and alternative hypotheses.

(b) Report the test statistic, its degrees of freedom, and the P-value. What is your conclusion using significance level $\alpha = 0.05$?

(c) Can you conclude that this computerized brain training always improves a person's auditory memory better than the DVD program? If not, explain why.

7.121 ⚠ **Alcohol consumption and body composition.** Individuals who consume large amounts of alcohol do not use the calories from this source as efficiently as calories from other sources. One study examined the effects of moderate alcohol consumption on body composition and the intake of other foods. Fourteen subjects participated in a crossover design where they either drank wine for the first 6 weeks and then abstained for the next 6 weeks or vice versa.[37] During the period when they drank wine, the subjects, on average, lost 0.4 kilograms (kg) of body weight; when they did not drink wine, they lost an average of 1.1 kg. The standard deviation of the difference between the weight lost under these two conditions is 8.6 kg. During the wine period, they consumed an average of 2589 calories; with no wine, the mean consumption was 2575. The standard deviation of the difference was 210.

(a) Compute the differences in means and the standard errors for comparing body weight and caloric intake under the two experimental conditions.

(b) A report of the study indicated that there were no significant differences in these two outcome measures.

Verify this result for each measure, giving the test statistic, degrees of freedom, and the P-value.

(c) One concern with studies such as this, with a small number of subjects, is that there may not be sufficient power to detect differences that are potentially important. Address this question by computing 95% confidence intervals for the two measures and discuss the information provided by the intervals.

(d) Here are some other characteristics of the study. The study periods lasted for 6 weeks. All subjects were males between the ages of 21 and 50 years who weighed between 68 and 91 kilograms (kg). They were all from the same city. During the wine period, subjects were told to consume two 135 milliliter (ml) servings of red wine and no other alcohol. The entire 6-week supply was given to each subject at the beginning of the period. During the other period, subjects were instructed to refrain from any use of alcohol. All subjects reported that they complied with these instructions except for three subjects, who said that they drank no more than three to four 12-ounce bottles of beer during the no-alcohol period. Discuss how these factors could influence the interpretation of the results.

7.122 The wine makes the meal? In a recent study, 39 diners were given a free glass of Cabernet Sauvignon to accompany a French meal.[38] Although the wine was identical, half the bottle labels claimed the wine was from California and the other half from North Dakota. The following table summarizes the grams of entrée and wine consumed during the meal.

	Wine label	n	Mean	St. Dev
Entrée	California	24	499.8	87.2
	North Dakota	15	439.0	89.2
Wine	California	24	100.8	23.3
	North Dakota	15	110.4	9.0

Did those patrons, thinking the wine was from California, consume more? Analyze the data and write a report summarizing your work. Be sure to include details regarding the statistical methods you used, your assumptions, and your conclusions.

7.123 Study design information. In the previous study, diners were seated alone or in groups of two, three, four, and in one case, nine (for a total of $n = 16$ tables). Also, each table, not each patron, was randomly assigned a particular wine label. Does this information alter how you might do the analysis in the previous problem? Explain your answer.

7.124 ⚠ **Analysis of tree size using the complete data set.** The data used in Exercises 7.27 (page 428),

7.81, and 7.82 (page 457) were obtained by taking simple random samples from the 584 longleaf pine trees that were measured in the Wade Tract. The entire data set is given in the LONGLEAF1 data set. Find the 95% confidence interval for the mean DBH using the entire data set, and compare this interval with the one that you calculated in Exercise 7.25. Write a report about these data. Include comments on the effect of the sample size on the margin of error, the distribution of the data, the appropriateness of the Normality-based methods for this problem, and the generalizability of the results to other similar stands of longleaf pine or other kinds of trees in this area of the United States and other areas. 🌐 LONGLEAF1

7.125 🔺 **More on the complete tree size data set.** Use the LONGLEAF1 data set to repeat the calculations that you performed in Exercises 7.81 and 7.82. Discuss the effect of the sample size on the results. 🌐 LONGLEAF1

7.126 🔺 **Even more on the complete tree size data set.** The DBH measures in the LONGLEAF1 data set do not appear to be Normally distributed. Make a histogram of the data and a Normal quantile plot if you have the software available. Mark the mean and the median on the histogram. Now, transform the data using a logarithm. Does this make the distribution appear to be Normal? Use the same graphical summaries with the mean and the median marked on the histogram. Write a summary of your conclusions, paying particular attention to the use of data such as these for inference using the methods based on Normal distributions. 🌐 LONGLEAF1

7.127 🔺 **Can mockingbirds learn to identify specific humans?** A central question in urban ecology is why some animals adapt well to the presence of humans and others do not. The following results summarize part of a study of the Northern Mockingbird (*Mimus polyglottos*) that took place on a campus of a large university.[39] For 4 consecutive days, the same human approached a nest and stood 1 meter away for 30 seconds, placing his/her hand on the rim of the nest. On the 5th day, a new person did the same thing. Each day, the distance from the nest when the bird flushed was recorded. This was repeated for 24 nests. The human intruder varied his or her appearance (i.e., wore different clothes) over the 4 days. We only report results for days 1, 4, and 5 here.

	Flush dist (m)	
Day	Mean	s
1	6.1	4.9
4	15.1	7.3
5	4.9	5.3

(a) Explain why this should be treated as a matched design.

(b) Unfortunately the research article does not provide the standard error of the difference, only the standard error of the mean flush distance for each day. However, we can use the general addition rule for variances (page 271) to approximate it. Assuming the correlation between the flush distance at day 1 and day 4 in each nest is $\rho = 0.40$, what is the standard deviation for the difference in distance?

(c) Using your result in part (b), test the hypothesis that there is no difference in the flush distance across these two days. Use a significance level of 0.05.

(d) Repeat parts (b) and (c) but now compare day 1 and day 5.

(e) Write a brief summary of your conclusions.

7.128 A comparison of female high school students. A recent study was performed to determine the prevalence of the female athlete triad (low energy availability, menstrual dysfunction, and low bone mineral density) in high school students.[40] A total of 80 high school athletes and 80 sedentary students were assessed. The following table summarizes several measured characteristics:

	Athletes		Sedentary	
Characteristic	$\bar{x}$	s	$\bar{x}$	s
Body fat (%)	25.61	5.54	32.51	8.05
Body mass index	21.60	2.46	26.41	2.73
Calcium deficit (mg)	297.13	516.63	580.54	372.77
Glasses of milk/day	2.21	1.46	1.82	1.24

(a) For each of the characteristics, test the hypothesis that the means are the same in the two groups. Use a significance level of 0.05 for each test.

(b) Write a short report summarizing your result.

7.129 Competitive prices? A retailer entered into an exclusive agreement with a supplier who guaranteed to provide all products at competitive prices. The retailer eventually began to purchase supplies from other vendors who offered better prices. The original supplier filed a legal action claiming violation of the agreement. In defense, the retailer had an audit performed on a random sample of invoices. For each audited invoice, all purchases made from other suppliers were examined and the prices were compared with those offered by the original supplier. For each invoice, the percent of purchases for which the alternate supplier offered a lower price than the original supplier was recorded.[41] Here are the data:

| 0 | 100 | 0 | 100 | 33 | 34 | 100 | 48 | 78 | 100 | 77 | 100 | 38 |
| 68 | 100 | 79 | 100 | 100 | 100 | 100 | 100 | 100 | 89 | 100 | 100 | |

Report the average of the percents with a 95% margin of error. Do the sample invoices suggest that the original supplier's prices are not competitive on the average? COMPPRICE

7.130 Weight-loss programs. In a study of the effectiveness of weight-loss programs, 47 subjects who were at least 20% overweight took part in a group support program for 10 weeks. Private weighings determined each subject's weight at the beginning of the program and 6 months after the program's end. The matched pairs t test was used to assess the significance of the average weight loss. The paper reporting the study said, "The subjects lost a significant amount of weight over time, $t(46) = 4.68$, $p < 0.01$." It is common to report the results of statistical tests in this abbreviated style.[42]

(a) Why was the matched pairs statistic appropriate?

(b) Explain to someone who knows no statistics but is interested in weight-loss programs what the practical conclusion is.

(c) The paper follows the tradition of reporting significance only at fixed levels such as $\alpha = 0.01$. In fact, the results are more significant than "$p < 0.01$" suggests. What can you say about the P-value of the t test?

7.131 Do women perform better in school? Some research suggests that women perform better than men in school, but men score higher on standardized tests. Table 1.4 (page 28) presents data on a measure of school performance, grade point average (GPA), and a standardized test, IQ, for 78 seventh-grade students. Do these data lend further support to the previously found gender differences? Give graphical displays of the data and describe the distributions. Use significance tests and confidence intervals to examine this question, and prepare a short report summarizing your findings. SEVENTHGRADE

7.132 Self-concept and school performance. Refer to the previous exercise. Although self-concept in this study was measured on a scale with values in the data set ranging from 20 to 80, many prefer to think of this kind of variable as having only two possible values: low self-concept or high self-concept. Find the median of the self-concept scores in Table 1.4 (page 28) and define those students with scores at or below the median to be low-self-concept students and those with scores above the median to be high-self-concept students. Do high-self-concept students have grade point averages that are different from low-self-concept students? What about IQ? Prepare a report addressing these questions. Be sure

to include graphical and numerical summaries and confidence intervals, and state clearly the details of significance tests. SEVENTHGRADE

7.133 Behavior of pet owners. On the morning of March 5, 1996, a train with 14 tankers of propane derailed near the center of the small Wisconsin town of Weyauwega. Six of the tankers were ruptured and burning when the 1700 residents were ordered to evacuate the town. Researchers study disasters like this so that effective relief efforts can be designed for future disasters. About half the households with pets did not evacuate all of their pets. A study conducted after the derailment focused on problems associated with retrieval of the pets after the evacuation and characteristics of the pet owners. One of the scales measured "commitment to adult animals," and the people who evacuated all or some of their pets were compared with those who did not evacuate any of their pets. Higher scores indicate that the pet owner is more likely to take actions that benefit the pet.[43] Here are the data summaries:

Group	n	$\bar{x}$	s
Evacuated all or some pets	116	7.95	3.62
Did not evacuate any pets	125	6.26	3.56

Analyze the data and prepare a short report describing the results.

7.134 Occupation and diet. Do various occupational groups differ in their diets? A British study of this question compared 98 drivers and 83 conductors of London double-decker buses.[44] The conductors' jobs require more physical activity. The article reporting the study gives the data as "Mean daily consumption ($\pm$ se)." Some of the study results follow:

	Drivers	Conductors
Total calories	2821 ± 44	2844 ± 48
Alcohol (grams)	0.24 ± 0.06	0.39 ± 0.11

(a) What does "se" stand for? Give $\bar{x}$ and s for each of the four sets of measurements.

(b) Is there significant evidence at the 5% level that conductors consume more calories per day than do drivers? Use the two-sample t method to give a P-value, and then assess significance.

(c) How significant is the observed difference in mean alcohol consumption? Use two-sample t methods to obtain the P-value.

(d) Give a 95% confidence interval for the mean daily alcohol consumption of London double-decker bus conductors.

(e) Give a 99% confidence interval for the difference in mean daily alcohol consumption between drivers and conductors.

7.135 Occupation and diet, continued (optional). Use of the pooled two-sample t test is justified in part (b) of the previous exercise. Explain why. Find the P-value for the pooled t statistic, and compare it with your result in the previous exercise.

7.136 Conditions for inference. The report cited in Exercise 7.134 says that the distribution of alcohol consumption among the individuals studied is "grossly skew."

(a) Do you think that this skewness prevents the use of the two-sample t test for equality of means? Explain your answer.

(b) (Optional) Do you think that the skewness of the distributions prevents the use of the F test for equality of standard deviations? Explain your answer.

7.137 More on conditions for inference. Suppose your state contains 85 school corporations and each corporation reports its expenditures per pupil. Is it proper to apply the one-sample t method to these data to give a 95% confidence interval for the average expenditure per pupil? Explain your answer.

7.138 Male and female CS students (optional). Is there a difference between the average SAT scores of males and females? The CSDATA data set gives the Math (SATM) and Verbal (SATV) scores for a group of 224 computer science majors. The variable SEX indicates whether each individual is male or female. CSDATA

(a) Compare the two distributions graphically, and then use the two-sample t test to compare the average SATM scores of males and females. Is it appropriate to use the pooled t test for this comparison? Write a brief summary of your results and conclusions that refers to both versions of the t test and to the F test for equality of standard deviations. Also give a 95% confidence interval for the difference in the means.

(b) Answer part (a) for the SATV scores.

(c) The students in the CSDATA data set were all computer science majors who began college during a particular year. To what extent do you think that your results would generalize to (i) computer science students

entering in different years, (ii) computer science majors at other colleges and universities, and (iii) college students in general?

7.139 Different methods of teaching reading. In the READING data set, the response variable Post3 is to be compared for three methods of teaching reading. The Basal method is the standard, or control, method, and the two new methods are DRTA and Strat. We can use the methods of this chapter to compare Basal with DRTA and Basal with Strat. Note that to make comparisons among three treatments it is more appropriate to use the procedures that we will learn in Chapter 12. READING

(a) Is the mean reading score with the DRTA method higher than that for the Basal method? Perform an analysis to answer this question, and summarize your results.

(b) Answer part (a) for the Strat method in place of DRTA.

7.140 Sample size calculation (optional). Example 7.13 (page 433) tells us that the mean height of 10-year-old girls is $N(56.4, 2.7)$ and for boys it is $N(55.7, 3.8)$. The null hypothesis that the mean heights of 10-year-old boys and girls are equal is clearly false. The difference in mean heights is $56.4 - 55.7 = 0.7$ inch. Small differences such as this can require large sample sizes to detect. To simplify our calculations, let's assume that the standard deviations are the same, say $\sigma = 3.2$, and that we will measure the heights of an equal number of girls and boys. How many would we need to measure to have a 90% chance of detecting the (true) alternative hypothesis?

7.141 House prices. How much more would you expect to pay for a home that has four bedrooms than for a home that has three? Here are some data for West Lafayette, Indiana.[45] These are the asking prices (in dollars) that the owners have set for their homes. HOUSEPRICE

Four-bedroom homes

149,900	169,900	175,000	189,000	206,900	225,000
249,900	289,900	320,000	339,900	399,900	429,900
320,000	269,900				

Three-bedroom homes

79,500	82,000	89,999	90,000	99,900	100,000
106,900	113,900	115,000	117,500	122,900	129,900
139,900	145,000	149,000	150,000	157,900	164,900
189,900	219,900	260,000	274,900	295,000	

(a) Plot the asking prices for the two sets of homes and describe the two distributions.

(b) Test the null hypothesis that the mean asking prices for the two sets of homes are equal versus the two-sided alternative. Give the test statistic with degrees of freedom, the P-value, and your conclusion.

(c) Would you consider using a one-sided alternative for this analysis? Explain why or why not.

(d) Give a 95% confidence interval for the difference in mean asking prices.

(e) These data are not SRSs from a population. Give a justification for use of the two-sample t procedures in this case.

7.142 **More on house prices.** Go to the Web site www.realtor.com and select two geographical areas of interest to you. You will compare the prices of similar types of homes in these two areas. State clearly how you define the areas and the type of homes. For example, you can use city names or zip codes to define the area and you can select single-family homes or condominiums. We view these homes as representative of the asking prices of homes for these areas at the time of your search. If the search gives a large number of homes, select a random sample. Be sure to explain exactly how you do this. Use the methods you have learned in this chapter to compare the asking prices. Be sure to include a graphical summary.

Inference for Proportions

Introduction

We frequently collect data on *categorical variables*, such as whether or not a person is employed, the brand name of a cell phone, or the country where a college student studies abroad. When we record categorical variables, our data consist of *counts* or of *percents* obtained from counts.

In these settings, our goal is to say something about the corresponding *population proportions*. Just as in the case of inference about population means, we may be concerned with a single population or with comparing two populations. Inference about one or two proportions is very similar to inference about means, which we discussed in Chapter 7. In particular, inference for both means and proportions is based on sampling distributions that are approximately Normal.

We begin in Section 8.1 with inference about a single population proportion. Section 8.2 concerns methods for comparing two proportions.

8.1 Inference for a Single Proportion

We want to estimate the proportion p of some characteristic in a large population. For example, the proportion of likely voters who approve of the president's conduct in office. We select a simple random sample (SRS) of size n from the population and record the count X of "successes" (such as "Yes" answers to a question about the president). We will use "success" to represent the characteristic of interest. The sample proportion of successes $\hat{p} = X/n$ estimates the unknown population proportion p. If the population is much larger than the sample (say, at least 20 times as large), the count X has approximately the binomial distribution $B(n, p)$.[1] In statistical terms, we are concerned with inference about the probability p of a success in the binomial setting.

EXAMPLE

8.1 Adults and video games. A PriceWaterhouseCooper report estimates that the U.S. video game market was approximately $8.6 billion in 2007 and is expected to increase at an annual rate of 6.3% through 2012.[2] Who plays video games? A Pew survey, conducted by Princeton Survey Research International, reports that over half of American adults aged 18 and over play video games.[3] The Pew survey was conducted by Princeton Survey Research International and used a nationally representative sample of 2054 adults. Of the total, 1063 adults said that they played video games. Here, p is the proportion of adults in the population who play video games and

$$\hat{p} = \frac{X}{n} = \frac{1063}{2054} = 0.51753$$

is the sample proportion. We estimate that 52% of adults play video games.

USE YOUR KNOWLEDGE

8.1 Bank acquisitions. The American Bankers Association Community Bank Competitiveness Survey for 2008 had responses from 760 community banks. Of these, 283 reported that they expected to acquire another bank within five years.[4]

(a) What is the sample size n for this survey?

(b) What is the count X? Describe the count in a short sentence.

(c) Find the sample proportion $\hat{p}$.

8.2 How often do they play? In the Pew survey described in Example 8.1, those who played video games were asked how often they played. In this subpopulation, 223 adults said that they played every day or almost every day.

(a) What is the sample size n for the subpopulation of U.S. adults who play video games? (*Hint:* Look at Example 8.1.)

(b) What is the count X of those who said that they played everyday or almost everyday?

(c) Find the sample proportion $\hat{p}$.

If the sample size n is very small, we must base tests and confidence intervals for p on the binomial distributions. These are awkward to work with because of the discreteness of the binomial distributions.[5] But we know that when the sample is large, both the count X and the sample proportion $\hat{p}$ are approximately Normal. We will consider only inference procedures based on the Normal approximation. These procedures are similar to those for inference about the mean of a Normal distribution.

LOOK BACK
Normal approximation for
counts p. 324

Large-sample confidence interval for a single proportion

The unknown population proportion p is estimated by the sample proportion $\hat{p} = X/n$. If the sample size n is sufficiently large, $\hat{p}$ has approximately the Normal distribution, with mean $\mu_{\hat{p}} = p$ and standard deviation $\sigma_{\hat{p}} = \sqrt{p(1-p)/n}$. This means that approximately 95% of the time $\hat{p}$ will be within $2\sqrt{p(1-p)/n}$ of the unknown population proportion p.

LOOK BACK
Normal approximation for
proportions p. 324

Note that the standard deviation $\sigma_{\hat{p}}$ depends upon the unknown parameter p. To estimate this standard deviation using the data, we replace p in the formula by the sample proportion $\hat{p}$. As we did in Chapter 7, we use the term *standard error* for the standard deviation of a statistic that is estimated from data. Here is a summary of the procedure.

LOOK BACK
standard error p. 404

LARGE-SAMPLE CONFIDENCE INTERVAL FOR A POPULATION PROPORTION

Choose an SRS of size n from a large population with an unknown proportion p of successes. The **sample proportion** is

$$\hat{p} = \frac{X}{n}$$

where X is the number of successes. The **standard error of $\hat{p}$** is

$$SE_{\hat{p}} = \sqrt{\frac{\hat{p}(1-\hat{p})}{n}}$$

and the **margin of error** for confidence level C is

$$m = z^* SE_{\hat{p}}$$

where the critical value z^* is the value for the standard Normal density curve with area C between $-z^*$ and z^*. An **approximate level C confidence interval** for p is

$$\hat{p} \pm m$$

Use this interval for 90%, 95%, or 99% confidence when the number of successes and the number of failures are both at least 15.

Table D includes a line at the bottom with values of z^* for selected values of C. Use Table A for other values of C.

> ### EXAMPLE
>
> **8.2 Inference for adults and video games.** The sample survey in Example 8.1 found that 1063 of a sample of 2054 adults reported that they played video games. In that example we calculated $\hat{p} = 0.5175$. The standard error is
>
> $$SE_{\hat{p}} = \sqrt{\frac{\hat{p}(1 - \hat{p})}{n}} = \sqrt{\frac{0.5175(1 - 0.5175)}{2054}} = 0.011026$$
>
> The z critical value for 95% confidence is $z^* = 1.96$, so the margin of error is
>
> $$m = 1.96 SE_{\hat{p}} = (1.96)(0.011026) = 0.021610$$
>
> The confidence interval is
>
> $$\hat{p} \pm m = 0.52 \pm 0.02$$
>
> We are 95% confident that between 50% and 54% of adults play video games.

In performing these calculations we have kept a large number of digits for our intermediate calculations. However, when reporting the results we prefer to use rounded values. For example, 52% with a margin of error of 2%. In this way we focus attention on the important parts that we have found. There is no additional information to be gained by reporting 0.51753 with a margin of error of 0.021610.

Remember that the margin of error in any confidence interval includes only random sampling error. If people do not respond honestly to the questions asked, for example, your estimate is likely to miss by more than the margin of error.

Because the calculations for statistical inference for a single proportion are relatively straightforward, we often do them with a calculator or in a spreadsheet. Figure 8.1 gives output from Minitab and SAS for the data in Example 8.1. As usual, the output reports more digits than are useful. When you use software, be sure to think about how many digits are meaningful for your purposes. Do not clutter your report with information that is not meaningful. SAS

gives the standard error next to the label ASE, which stands for asymptotic standard error. The SAS output also includes an alternative interval based on an "exact" method.

We recommend the large-sample confidence interval for 90%, 95%, and 99% confidence whenever the number of successes and the number of failures are both at least 15. For smaller sample sizes, we recommend exact methods that use the binomial distribution. These are available as the default or as options in many statistical software packages and we do not cover them here. There is also an intermediate case between large samples and very small samples where a slight modification of the large-sample approach works quite well.[6] This method is called the "plus four" procedure and is described later.

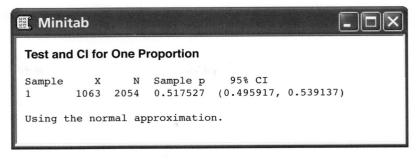

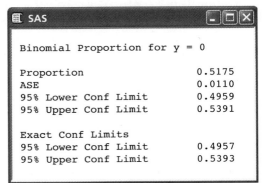

FIGURE 8.1 Minitab and SAS output for the confidence interval in Example 8.1.

8.3 Bank acquisitions. Refer to Exercise 8.1 (page 474).

(a) Find $SE_{\hat{p}}$, the standard error of $\hat{p}$.

(b) Give the 95% confidence interval for p in the form of estimate plus or minus the margin of error.

(c) Give the confidence interval as an interval of percents.

8.4 How often do they play? Refer to Exercise 8.2 (page 474).

(a) Find $SE_{\hat{p}}$, the standard error of $\hat{p}$.

(b) Give the 95% confidence interval for p in the form of estimate plus or minus the margin of error.

(c) Give the confidence interval as an interval of percents.

The plus four confidence interval for a single proportion

Computer studies reveal that confidence intervals based on the large-sample approach can be quite inaccurate when the number of successes and the number of failures are not at least 15. When this occurs, a simple adjustment to the confidence interval works very well in practice. The adjustment is based on assuming that the sample contains 4 additional observations, 2 of which are successes and 2 of which are failures. The estimator of the population proportion based on this *plus four* rule is

$$\tilde{p} = \frac{X + 2}{n + 4}$$

plus four estimate

This estimate was first suggested by Edwin Bidwell Wilson in 1927 and we call it the **plus four estimate.** The confidence interval is based on the z statistic obtained by standardizing the plus four estimate $\tilde{p}$. Because $\tilde{p}$ is the sample proportion for our modified sample of size $n + 4$, it isn't surprising that the distribution of $\tilde{p}$ is close to the Normal distribution with mean p and standard deviation $\sqrt{p(1 - p)/(n + 4)}$. To get a confidence interval, we estimate p by $\tilde{p}$ in this standard deviation to get the standard error of $\tilde{p}$. Here is an example.

EXAMPLE

8.3   Percent of equol producers.   Research has shown that there are many health benefits associated with a diet that contains soy foods. Substances in soy called isoflavones are known to be responsible for these benefits. When soy foods are consumed, some subjects produce a chemical called equol, and it is thought that production of equol is a key factor in the health benefits of a soy diet. Unfortunately, not all people are equol producers; there appear to be two distinct subpopulations: equol producers and equol nonproducers.

A nutrition researcher planning some bone health experiments would like to include some equol producers and some nonproducers among her subjects. A preliminary sample of 12 female subjects were measured, and 4 were found to be equol producers. We would like to estimate the proportion of equol producers in the population from which this researcher will draw her subjects.

The plus four estimate of the proportion of equol producers is

$$\tilde{p} = \frac{4 + 2}{12 + 4} = \frac{6}{16} = 0.375$$

For a 95% confidence interval, we use Table D to find $z^* = 1.96$. We first compute the standard error

$$SE_{\tilde{p}} = \sqrt{\frac{\tilde{p}(1 - \tilde{p})}{n + 4}}$$

$$= \sqrt{\frac{(0.375)(1 - 0.375)}{16}}$$

$$= 0.12103$$

and then the margin of error

$$m = z^* SE_{\tilde{p}}$$

$$= (1.96)(0.12103)$$

$$= 0.237$$

So the confidence interval is

$$\tilde{p} \pm m = 0.375 \pm 0.237$$

$$= (0.138, 0.612)$$

We estimate with 95% confidence that between 14% and 61% of women from this population are equol producers.

If the true proportion of equol users is near 14%, the lower limit of this interval, there may not be a sufficient number of equol producers in the study if subjects are tested only after they are enrolled in the experiment. It may be necessary to determine whether or not a potential subject is an equol producer. The study could then be designed to have the same number of equol producers and nonproducers.

Significance test for a single proportion

LOOK BACK
Normal approximation for proportions p. 324

Recall that the sample proportion $\hat{p} = X/n$ is approximately Normal, with mean $\mu_{\hat{p}} = p$ and standard deviation $\sigma_{\hat{p}} = \sqrt{p(1-p)/n}$. For confidence intervals, we substitute $\hat{p}$ for p in the last expression to obtain the standard error. When performing a significance test, however, the null hypothesis specifies a value for p, and we assume that this is the true value when calculating the P-value. Therefore, when we test $H_0: p = p_0$, we substitute p_0 into the expression for $\sigma_{\hat{p}}$ and then standardize $\hat{p}$. Here are the details.

LARGE-SAMPLE SIGNIFICANCE TEST FOR A POPULATION PROPORTION

Draw an SRS of size n from a large population with an unknown proportion p of successes. To test the hypothesis $H_0: p = p_0$, compute the **z statistic**

$$z = \frac{\hat{p} - p_0}{\sqrt{\dfrac{p_0(1-p_0)}{n}}}$$

In terms of a standard Normal random variable Z, the approximate P-value for a test of H_0 against

$H_a: p > p_0$ is $P(Z \geq z)$

$H_a: p < p_0$ is $P(Z \leq z)$

$H_a: p \neq p_0$ is $2P(Z \geq |z|)$

LOOK BACK
sign test for matched pairs p. 423

We recommend the large-sample z significance test as long as the expected number of successes, np_0, and the expected number of failures, $n(1-p_0)$, are both greater than 10. If this rule of thumb is not met, or if the population is less than 20 times as large as the sample, other procedures should be used. One such approach is to use the binomial distribution as we did with the sign test. Here is a large-sample example.

EXAMPLE

8.4 Comparing two sun block lotions. Your company produces a sun block lotion designed to protect the skin from both UVA and UVB exposure to the sun. You hire a company to compare your product with the product sold by your major competitor. The testing company exposes skin on the back of a sample of 20 people to UVA and UVB rays and measures the protection provided by each product. For 13 of the subjects, your product provided better protection while for the other 7 subjects, your competitor's product provided better protection. Do you have evidence to support a commercial claiming that your product provides superior UVA and UVB protection? For the data we have $n = 20$ subjects and $X = 13$ successes. The parameter p is the proportion of people who would receive superior UVA and UVB protection from your product. To answer the claim question, we test

$$H_0: p = 0.5$$
$$H_a: p \neq 0.5$$

The expected numbers of successes (your product provides better protection) and failures (your competitor's product provides better protection) are $20 \times 0.5 = 10$ and $20 \times 0.5 = 10$. Both are at least 10, so we can use the z test. The sample proportion is

$$\hat{p} = \frac{X}{n} = \frac{13}{20} = 0.65$$

The test statistic is

$$z = \frac{\hat{p} - p_0}{\sqrt{\dfrac{p_0(1 - p_0)}{n}}} = \frac{0.65 - 0.5}{\sqrt{\dfrac{(0.5)(0.5)}{20}}} = 1.34$$

From Table A we find $P(Z < 1.34) = 0.9099$, so the probability in the upper tail is $1 - 0.9099 = 0.0901$. The P-value is the area in both tails, $P = 2 \times 0.0901 = 0.1802$. Minitab and SAS outputs for the analysis in Example 8.4 appear in Figure 8.2. We conclude that the sun block testing data are compatible with the hypothesis of no difference between your product and your competitor's ($\hat{p} = 0.65$, $z = 1.34$, $P = 0.18$). The data do not provide you with a basis to support your advertising claim.

Note that we used a two-sided hypothesis test when we compared the two sum block lotions in Example 8.4. In settings like this, we must start with the view that either product could be better if we want to prove a claim of superiority. Thinking or hoping that your product is superior cannot be used to justify a one-sided test.

USE YOUR KNOWLEDGE

8.5 Draw a picture. Draw a picture of a standard Normal curve and shade the tail areas to illustrate the calculation of the P-value for Example 8.4.

8.6 What does the confidence interval tell us? Inspect the outputs in Figure 8.2 and report the confidence interval for the percent of people who would get better sun protection from your product than from your

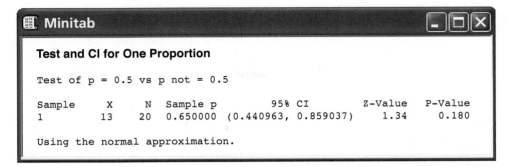

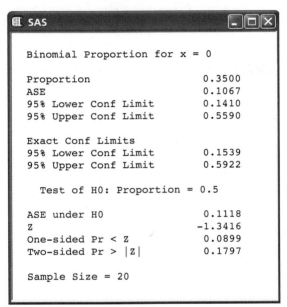

FIGURE 8.2 Minitab and SAS output for the significance test in Example 8.4.

competitor's. Be sure to convert from proportions to percents and to round appropriately. Interpret the confidence interval and compare this way of analyzing data with the significance test.

8.7 The effect of X. In Example 8.4, suppose that your product provided better UVA and UVB protection for 15 of the 20 subjects. Perform the significance test and summarize the results. Compare your results to those obtained in Example 8.4.

8.8 The effect of n. In Example 8.4, consider what would have happened if you had paid for twice as many subjects to be tested. Assume that the results would be similar to those in your test, that is 65% of the subjects had better UVA and UVB protection with your product. Perform the significance test and summarize the results. Compare your results to those obtained in Example 8.4.

In Example 8.4, we treated an outcome as a success whenever your product provided better sun protection. Would we get the same results if we defined success as an outcome where your competitor's product was superior? In this setting the null hypothesis is still H_0: $p = 0.5$. You will find that the z test statistic is unchanged except for its sign and that the P-value remains the same.

USE YOUR KNOWLEDGE

8.9 Yes or no? In Example 8.4 we performed a significance test to compare your product with your competitor's. Success was defined as the outcome where your product provided better protection. Now, take the viewpoint of your competitor where success is defined to be the outcome where your competitor's product provides better protection. In other words, n remains the same (20) but X is now 7.

(a) Perform the two-sided significance test and report the results. How do these compare with what we did in Example 8.4?

(b) Find the 95% confidence interval for this setting and compare it with the interval calculated where success is defined as the outcome where your product provides better protection.

We do not often use significance tests for a single proportion, because it is uncommon to have a situation where there is a precise p_0 that we want to test. For physical experiments such as coin tossing or drawing cards from a well-shuffled deck, probability arguments lead to an ideal p_0. Even here, however, it can be argued, for example, that no real coin has a probability of heads *exactly* equal to 0.5. Data from past large samples can sometimes provide a p_0 for the null hypothesis of a significance test. In some types of epidemiology research, for example, "historical controls" from past studies serve as the benchmark for evaluating new treatments. Medical researchers argue about the validity of these approaches, because the past never quite resembles the present. In general, we prefer comparative studies whenever possible.

Choosing a sample size

LOOK BACK
choosing sample size p. 352

In Chapter 6, we showed how to choose the sample size n to obtain a confidence interval with specified margin of error m for the mean of a Normal distribution. Because we are using a Normal approximation for inference about a population proportion, sample size selection proceeds in much the same way.

Recall that the margin of error for the large-sample confidence interval for a population proportion is

$$m = z^* \text{SE}_{\hat{p}} = z^* \sqrt{\frac{\hat{p}(1 - \hat{p})}{n}}$$

Choosing a confidence level C fixes the critical value z^*. The margin of error also depends on the value of $\hat{p}$ and the sample size n. Because we don't know the value of $\hat{p}$ until we gather the data, we must guess a value to use in the calculations. We will call the guessed value p^*. There are two common ways to get p^*:

1. Use the sample estimate from a pilot study or from similar studies done earlier.

2. Use $p^* = 0.5$. Because the margin of error is largest when $\hat{p} = 0.5$, this choice gives a sample size that is somewhat larger than we really need for the confidence level we choose. It is a safe choice no matter what the data later show.

Once we have chosen p^* and the margin of error m that we want, we can find the n we need to achieve this margin of error. Here is the result.

SAMPLE SIZE FOR DESIRED MARGIN OF ERROR

The level C confidence interval for a proportion p will have a margin of error approximately equal to a specified value m when the sample size satisfies

$$n = \left(\frac{z^*}{m}\right)^2 p^*(1 - p^*)$$

Here z^* is the critical value for confidence C, and p^* is a guessed value for the proportion of successes in the future sample.

The margin of error will be less than or equal to m if p^* is chosen to be 0.5. The sample size required when $p^* = 0.5$ is

$$n = \frac{1}{4}\left(\frac{z^*}{m}\right)^2$$

The value of n obtained by this method is not particularly sensitive to the choice of p^* when p^* is fairly close to 0.5. However, if the value of p is likely to be smaller than about 0.3 or larger than about 0.7, use of $p^* = 0.5$ may result in a sample size that is much larger than needed.

EXAMPLE

8.5 Planning a survey of students. A large university is interested in assessing student satisfaction with the overall campus environment. The plan is to distribute a questionnaire to an SRS of students, but before proceeding, the university wants to determine how many students to sample. The questionnaire asks about a student's degree of satisfaction with various student services, each measured on a five-point scale. The university is interested in the proportion p of students who are satisfied (that is, who choose either "satisfied" or "very satisfied," the two highest levels on the five-point scale).

The university wants to estimate p with 95% confidence and a margin of error less than or equal to 3%, or 0.03. For planning purposes, they are willing to use $p^* = 0.5$. To find the sample size required,

$$n = \frac{1}{4}\left(\frac{z^*}{m}\right)^2 = \frac{1}{4}\left[\frac{1.96}{0.03}\right]^2 = 1067.1$$

Round up to get $n = 1068$. (Always round up. Rounding down would give a margin of error slightly greater than 0.03.)

Similarly, for a 2.5% margin of error we have (after rounding up)

$$n = \frac{1}{4}\left[\frac{1.96}{0.025}\right]^2 = 1537$$

and for a 2% margin of error,

$$n = \frac{1}{4}\left[\frac{1.96}{0.02}\right]^2 = 2401$$

News reports frequently describe the results of surveys with sample sizes between 1000 and 1500 and a margin of error of about 3%. These surveys generally use sampling procedures more complicated than simple random sampling, so the calculation of confidence intervals is more involved than what we have studied in this section. The calculations in Example 8.6 nonetheless show in principle how such surveys are planned.

In practice, many factors influence the choice of a sample size. The following example illustrates one set of factors.

EXAMPLE

8.6   Assessing interest in Pilates classes. The Division of Recreational Sports (Rec Sports) at a major university is responsible for offering comprehensive recreational programs, services, and facilities to the students. Rec Sports is continually examining its programs to determine how well it is meeting the needs of the students. Rec Sports is considering adding some new programs and would like to know how much interest there is in a new exercise program based on the Pilates method.[7] They will take a survey of undergraduate students. In the past, they emailed short surveys to all undergraduate students. The response rate obtained in this way was about 5%. This time they will send emails to a simple random sample of the students and will follow up with additional emails and eventually a phone call to get a higher response rate. Because of limited staff and the work involved with the follow-up, they would like to use a sample size of about 200. One of the questions they will ask is "Have you ever heard about the Pilates method of exercise?"

The primary purpose of the survey is to estimate various sample proportions for undergraduate students. Will the proposed sample size of $n = 200$ be adequate to provide Rec Sports with the needed information? To address this question, we calculate the margins of error of 95% confidence intervals for various values of $\hat{p}$.

EXAMPLE

8.7   Margins of error. In the Rec Sports survey, the margin of error of a 95% confidence interval for any value of $\hat{p}$ and $n = 200$ is

$$m = z^* \text{SE}_{\hat{p}}$$

$$= 1.96\sqrt{\frac{\hat{p}(1 - \hat{p})}{200}} = 0.139\sqrt{\hat{p}(1 - \hat{p})}$$

The results for various values of $\hat{p}$ are

$\hat{p}$	m	$\hat{p}$	m
0.05	0.030	0.60	0.068
0.10	0.042	0.70	0.064
0.20	0.056	0.80	0.056
0.30	0.064	0.90	0.042
0.40	0.068	0.95	0.030
0.50	0.070		

Rec Sports judged these margins of error to be acceptable, and they used a sample size of 200 in their survey.

The table in Example 8.7 illustrates two points. First, the margins of error for $\hat{p} = 0.05$ and $\hat{p} = 0.95$ are the same. The margins of error will always be the same for $\hat{p}$ and $1 - \hat{p}$. This is a direct consequence of the form of the confidence interval. Second, the margin of error varies only between 0.064 and 0.070 as $\hat{p}$ varies from 0.3 to 0.7, and the margin of error is greatest when $\hat{p} = 0.5$, as we claimed earlier. It is true in general that the margin of error will vary relatively little for values of $\hat{p}$ between 0.3 and 0.7. Therefore, when planning a study, it is not necessary to have a very precise guess for p. If $p^* = 0.5$ is used and the observed $\hat{p}$ is between 0.3 and 0.7, the actual interval will be a little shorter than needed, but the difference will be small.

Again it is important to emphasize that these calculations consider only the effects of sampling variability that are quantified in the margin of error. Other sources of error, such as **nonresponse** and possible **misinterpretation** of questions, are not included in the table of margins of error for Example 8.7. Rec Sports is trying to minimize these kinds of errors. They did a pilot study using a small group of current users of their facilities to check the wording of the questions, and they devised a careful plan to follow up with the students who did not respond to the initial email.

USE YOUR KNOWLEDGE

8.10 Confidence level and sample size. Refer to Example 8.5 (page 483). Suppose the university was interested in a 90% confidence interval with margin of error 0.03. Would the required sample size be smaller or larger than 1068 students? Verify this by performing the calculation.

8.11 Make a plot. Use the values for $\hat{p}$ and m given in Example 8.7 to draw a plot of the sample proportion versus the margin of error. Summarize the major features of your plot.

SECTION 8.1 Summary

Inference about a population proportion p from an SRS of size n is based on the **sample proportion** $\hat{p} = X/n$. When n is large, $\hat{p}$ has approximately the Normal distribution with mean p and standard deviation $\sqrt{p(1-p)/n}$.

For large samples, the **margin of error for confidence level C** is

$$m = z^* \text{SE}_{\hat{p}}$$

where the critical value z^* is the value for the standard Normal density curve with area C between $-z^*$ and z^*, and the **standard error of $\hat{p}$** is

$$\text{SE}_{\hat{p}} = \sqrt{\frac{\hat{p}(1 - \hat{p})}{n}}$$

The **level C large-sample confidence interval** is

$$\hat{p} \pm m$$

We recommend using this interval for 90%, 95%, and 99% confidence whenever the number of successes and the number of failures are both at least 15. When sample sizes are smaller, alternative procedures such as the **plus four estimate of the population proportion** are recommended.

The **sample size** required to obtain a confidence interval of approximate margin of error m for a proportion is found from

$$n = \left(\frac{z^*}{m}\right)^2 p^*(1 - p^*)$$

where p^* is a guessed value for the proportion, and z^* is the standard Normal critical value for the desired level of confidence. To ensure that the margin of error of the interval is less than or equal to m no matter what $\hat{p}$ may be, use

$$n = \left(\frac{z^*}{2m}\right)^2$$

Tests of H_0: $p = p_0$ are based on the **z statistic**

$$z = \frac{\hat{p} - p_0}{\sqrt{\dfrac{p_0(1 - p_0)}{n}}}$$

with P-values calculated from the $N(0, 1)$ distribution. Use this procedure when the expected number of successes, np_0, and the expected number of failures, $n(1 - p_0)$, are both greater than 10.

SECTION 8.1   Exercises

For Exercises 8.1 and 8.2, see page 474; for Exercises 8.3 and 8.4, see page 477; for Exercises 8.5 to 8.8, see pages 480–481; for Exercise 8.9, see page 482; and for Exercises 8.10 and 8.11, see page 485.

8.12 What's wrong? Explain what is wrong with each of the following:

(a) You can use a significance test to evaluate the hypothesis H_0: $\hat{p} = 0.6$ versus the two-sided alternative.

(b) The large-sample significance test for a population proportion is based on a t statistic.

(c) An approximate 95% confidence interval for an unknown proportion p is $\hat{p}$ plus or minus its standard error.

8.13 What's wrong? Explain what is wrong with each of the following:

(a) The margin of error for a confidence interval used for an opinion poll takes into account the fact that people who did not answer the poll questions may have had different responses from those who did answer the questions.

(b) If the P-value for a significance test is 0.35, we can conclude that the null hypothesis has a 35% chance of being true.

(c) A student project used a confidence interval to describe the results in a final report. The confidence level was 110%.

8.14 Draw some pictures. Consider the binomial setting with $n = 50$ and $p = 0.4$.

(a) The sample proportion $\hat{p}$ will have a distribution that is approximately Normal. Give the mean and the standard deviation of this Normal distribution.

(b) Draw a sketch of this Normal distribution. Mark the location of the mean.

(c) Find a value p^* for which the probability is 95% that $\hat{p}$ will be between $\pm p^*$. Mark these two values on your sketch.

8.15 Country food and Inuits. Country food for Inuits includes seal, caribou, whale, ducks, fish, and berries and is an important part of the diet of the aboriginal people called Inuits who inhabit Inuit Nunangat, the northern region of what is now called Canada. A survey of Inuits in Inuit Nunangat reported that 3274 out of 5000 respondents said that at least half of the meat and fish that they eat is country food.[8] Find the sample proportion and a 95% confidence interval for the population proportion of Inuits whose meat and fish consumption consists of at least half country food.

8.16 Most desirable mates. A poll of 5000 residents in Brazil, Canada, China, France, Malaysia, South Africa, and the United States asked about what profession they would prefer a marriage partner to have. The choice receiving the highest percent, 16% of the responses, was doctors, nurses, and other health care professionals.[9]

(a) Find the sample proportion with a 95% confidence interval for the proportion of people who would prefer a doctor, nurse, or other health care professional as a marriage partner.

(b) Convert the estimate and the confidence interval to percents.

8.17 Guitar Hero and Rock Band. An electronic survey of 7,061 players of Guitar Hero and Rock Band reported that 67% of those who do not currently play a musical instrument said that they are likely to begin playing a real musical instrument in the next two years.[10] The reports describing the survey do not give the number of respondents who do not currently play a musical instrument.

(a) Explain why it is important to know the number of respondents who do not currently play a musical instrument.

(b) Assume that half of the respondents do not currently play a musical instrument. Find the count of players who said that they are likely to begin playing a real musical instrument in the next two years.

(c) Give a 99% confidence interval for the population proportion who would say that they are likely to begin playing a real musical instrument in the next two years.

(d) The survey collected data from two separate consumer panels. There were 3300 respondents from the LightSpeed consumer panel and the others were from Guitar Center's proprietary consumer panel. Comment on the sampling procedure used for this survey and how it would influence your interpretation of the findings.

8.18 Guitar Hero and Rock Band. Refer to the previous exercise.

(a) How would the result that you reported in part (c) change if only 25% of the respondents said that they did not currently play a musical instrument?

(b) Do the same calculations if the percent was 75%.

(c) The main conclusion of the survey that appeared in many news stories was that 67% of players of Guitar Hero and Rock Band who do not currently play a musical instrument said that they are likely to begin playing a real musical instrument in the next two years. What can you conclude about the effect of the three scenarios [part(b) in the previous exercise and parts (a) and (b) in this exercise] on the margin of error for the main result?

8.19 $\hat{p}$ and the Normal distribution. Consider the binomial setting with $n = 60$. You are testing the null hypothesis that $p = 0.4$ versus the two-sided alternative with a 5% chance or rejecting the null hypothesis when it is true.

(a) Find the values of the sample proportion $\hat{p}$ that will lead to rejection of the null hypothesis.

(b) Repeat part (a) assuming a sample size of $n = 100$.

(c) Make a sketch illustrating what you have found in parts (a) and (b). What does your sketch show about the effect of the sample size in this setting?

8.20 Students doing community service. In a sample of 159,949 first-year college students, the National Survey of Student Engagement reported that 39% participated in community service or volunteer work.[11]

(a) Find the margin of error for 99% confidence.

(b) Here are some facts from the report that summarizes the survey. The students were from 617 four-year colleges and universities. The response rate was 36%. Institutions paid a participation fee of between $1800 and $7800 based on the size of their undergraduate enrollment. Discuss these facts as possible sources of error in this study. How do you think these errors would compare with the error that you calculated in part (a)?

8.21 Plans to study abroad. The survey described in the previous exercise also asked about items related to academics. In response to one of these questions, 42% of first-year students reported that they plan to study abroad.

(a) Based on the information available, how many students plan to study abroad?

(b) Give a 99% confidence interval for the population proportion of first-year college students who plan to study abroad.

8.22 Student credit cards. In a survey of 1430 undergraduate students, 1087 reported that they had one or more credit cards.[12] Give a 95% confidence interval for the proportion of all college students who have at least one credit card.

8.23 How many credit cards? The summary of the survey described in the previous exercise reported that 43% of undergraduates had four or more credit cards. Give a 95% confidence interval for the proportion of all college students who have four or more credit cards.

8.24 How would the confidence interval change? Refer to Exercise 8.23.

(a) Would a 99% confidence interval be wider or narrower than the one that you found in that exercise? Verify your results by computing the interval.

(b) Would a 90% confidence interval be wider or narrower than the one that you found in Exercise 8.23? Verify your results by computing the interval.

8.25 Do students report Internet sources? The National Survey of Student Engagement found that 87% of students report that their peers at least "sometimes" copy information from the Internet in their papers without reporting the source.[13] Assume that the sample size is 430,000.

(a) Find the margin of error for 99% confidence.

(b) Here are some items from the report that summarizes the survey. More than 430,000 students from 730 four-year colleges and universities participated. The average response rate was 43% and ranged from 15% to 89%. Institutions pay a participation fee of between $3000 and $7500 based on the size of their undergraduate enrollment. Discuss these as sources of error in this study. How do you think these errors would compare with the error that you calculated in part (a)?

8.26 Do you enjoy driving your car? The Pew Research Center recently polled $n = 1048$ U.S. drivers and found that 69% enjoyed driving their automobiles.[14]

(a) Construct a 95% confidence interval for the proportion of U.S. drivers who enjoy driving their automobiles.

(b) In 1991, a Gallup Poll reported this percent to be 79%. Using the data from this poll, test the claim that the percent of drivers who enjoy driving their cars has declined since 1991. Report the large-sample z statistic and its P-value.

8.27 Getting angry at other drivers. Refer to Exercise 8.26. The same Pew Poll found that 38% of the respondents "shouted, cursed or made gestures to other drivers" in the last year.

(a) Construct a 95% confidence interval for the true proportion of U.S. drivers who did these actions in the last year.

(b) Does the fact that the respondent is self-reporting these actions affect the way that you interpret the results? Write a short paragraph explaining your answer.

8.28 Cheating during a test. A national survey of high school students conducted by the Josephson Institute of Ethics was sent to 37,328 students, and 24,142 were returned. One question asked students if they had cheated during a test in the last school year.[15] Of those who returned the survey, 9054 responded that they had cheated at least two times in the last year.

(a) What is the sample proportion of respondents who cheated at least twice?

(b) Compute the 95% confidence interval for the true proportion of students who have cheated on at least two tests in the last year.

(c) Compute the nonresponse rate for this study. Does this influence how you interpret these results? Write a short discussion of this issue.

8.29 ⚠ **Long sermons.** The National Congregations Study collected data in a one-hour interview with a key informant—that is, a minister, priest, rabbi, or other staff person or leader.[16] One question asked concerned the length of the typical sermon. For this question 390 out of 1191 congregations reported that the typical sermon lasted more than 30 minutes.

(a) Use the large-sample inference procedures to estimate the true proportion for this question with a 95% confidence interval.

(b) The respondents to this question were not asked to use a stopwatch to record the lengths of a random sample of sermons at their congregations. They responded based on their impressions of the sermons. Do you think that ministers, priests, rabbis, or other staff persons or leaders might perceive sermon lengths differently from the people listening to the sermons? Discuss how your ideas would influence your interpretation of the results of this study.

8.30 Confidence level and interval width. Refer to Exercise 8.17. Would a 90% confidence interval be wider or narrower than the one that you found in that exercise? Verify your results by computing the interval.

8.31 Can we use the z test? In each of the following cases state whether or not the Normal approximation to the binomial should be used for a significance test on the population proportion p.

(a) $n = 30$ and H_0: $p = 0.2$.

(b) $n = 30$ and H_0: $p = 0.6$.

(c) $n = 100$ and H_0: $p = 0.5$.

(d) $n = 200$ and H_0: $p = 0.01$.

8.32 Instant versus fresh-brewed coffee. A matched pairs experiment compares the taste of instant versus fresh-brewed coffee. Each subject tastes two unmarked cups of coffee, one of each type, in random order and states which he or she prefers. Of the 40 subjects who participate in the study, 12 prefer the instant coffee. Let p be the probability that a randomly chosen subject prefers fresh-brewed coffee to instant coffee. (In practical terms, p is the proportion of the population who prefer fresh-brewed coffee.)

(a) Test the claim that a majority of people prefer the taste of fresh-brewed coffee. Report the large-sample z statistic and its P-value.

(b) Draw a sketch of a standard Normal curve and mark the location of your z statistic. Shade the appropriate area that corresponds to the P-value.

(c) Is your result significant at the 5% level? What is your practical conclusion?

8.33 College students and dieting. For a study of unhealthy eating behaviors, 267 college women aged 18 to 25 years were surveyed.[17] Of these, 69% reported that they had been on a diet sometime during the past year. Give a 95% confidence interval for the true proportion of college women aged 18 to 25 years in this population who dieted last year.

8.34 High school students and dieting. In the study described in the previous exercise, the researchers also surveyed 266 high school students who were 18 years old. In this sample 58.3% reported that they had dieted sometime in the past year. Give a 95% confidence interval for the true proportion of 18-year-old high school students in this population who were on a diet sometime during the past year.

8.35 Pet ownership among older adults. In a study of the relationship between pet ownership and physical activity in older adults,[18] 594 subjects reported that they owned a pet, while 1939 reported that they did not. Give a 95% confidence interval for the proportion of older adults in this population who are pet owners.

8.36 ⚠ **Annual income of older adults.** In the study described in the previous exercise, 1434 subjects out of a total of 2533 reported that their annual income was $25,000 or more.

(a) Give a 95% confidence interval for the true proportion of subjects in this population with incomes of at least $25,000.

(b) Do you think that some respondents might not give truthful answers to a question about their income? Discuss the possible effects on your estimate and confidence interval.

(c) In the previous exercise, the question analyzed concerned pet ownership. Compare this question with the income question with respect to the possibility that the respondents were not truthful.

8.37 Dogs sniffing out cancer. A 2005 study by researchers set out to determine whether dogs could be trained to detect lung and breast cancer by sniffing exhaled breath samples.[19] For the breast cancer portion, breath samples from 6 cancer patients and 17 cancer-free volunteers were used. Each dog had to sniff five breath samples. For 125 trials, there were four control samples and one cancer sample. A correct response involved lying down next to the sample from the breast cancer patient. Collectively, the dogs correctly identified the cancer sample in 110 of these trials. Construct a 95% confidence interval for the true proportion of times these dogs will correctly identify a breast cancer sample.

8.38 ⚠ **Bicycle accidents and alcohol.** In the United States approximately 900 people die in bicycle accidents each year. One study examined the records of 1711 bicyclists aged 15 or older who were fatally injured in bicycle accidents between 1987 and 1991 and were tested for alcohol. Of these, 542 tested positive for alcohol (blood alcohol concentration of 0.01% or higher).[20]

(a) Summarize the data with appropriate descriptive statistics.

(b) To do statistical inference for these data, we think in terms of a model where p is a parameter that represents the probability that a tested bicycle rider is positive for alcohol. Find a 99% confidence interval for p.

(c) Can you conclude from your analysis of this study that alcohol causes fatal bicycle accidents? Explain.

(d) In this study 386 bicyclists had blood alcohol levels above 0.10%, a level defining legally drunk in many states at the time. Give a 99% confidence interval for the proportion who were legally drunk according to this criterion.

8.39 Tossing a coin 10,000 times! The South African mathematician John Kerrich, while a prisoner of war during World War II, tossed a coin 10,000 times and obtained 5067 heads.

(a) Is this significant evidence at the 5% level that the probability that Kerrich's coin comes up heads is not 0.5? Use a sketch of the standard Normal distribution to illustrate the P-value.

(b) Use a 95% confidence interval to find the range of probabilities of heads that would not be rejected at the 5% level.

8.40 Is there interest in a new product? One of your employees has suggested that your company develop a new product. You decide to take a random sample of your customers and ask whether or not there is interest in the new product. The response is on a 1 to 5 scale with 1 indicating "definitely would not purchase"; 2, "probably would not purchase"; 3, "not sure"; 4, "probably would purchase"; and 5, "definitely would purchase." For an initial analysis, you will record the responses 1, 2, and 3 as "No" and 4 and 5 as "Yes." What sample size would you use if you wanted the 95% margin of error to be 0.15 or less?

8.41 More information is needed. Refer to the previous exercise. Suppose that after reviewing the results of the previous survey, you proceeded with preliminary development of the product. Now you are at the stage where you need to decide whether or not to make a major investment to produce and market it. You will use another random sample of your customers, but now you want the margin of error to be smaller. What sample size would you use if you wanted the 95% margin of error to be 0.075 or less?

8.42 Sample size needed for an evaluation. You are planning an evaluation of a semester-long alcohol awareness campaign at your college. Previous evaluations indicate that about 25% of the students surveyed will respond "Yes" to the question "Did the campaign alter your behavior toward alcohol consumption?" How large a sample of students should you take if you want the margin of error for 95% confidence to be about 0.1?

8.43 🏆 **Sample size needed for an evaluation, continued.** The evaluation in the previous exercise will also have questions that have not been asked before, so you do not have previous information about the possible value of p. Repeat the preceding calculation for the following values of p^*: 0.1, 0.2, 0.3, 0.4, 0.5, 0.6, 0.7, 0.8, and 0.9. Summarize the results in a table and graphically. What sample size will you use?

8.44 Are the customers dissatisfied? An automobile manufacturer would like to know what proportion of its customers are dissatisfied with the service received from their local dealer. The customer relations department will survey a random sample of customers and compute a 95% confidence interval for the proportion that are dissatisfied. From past studies, they believe that this proportion will be about 0.15. Find the sample size needed if the margin of error of the confidence interval is to be no more than 0.02.

8.2 Comparing Two Proportions

Because comparative studies are so common, we often want to compare the proportions of two groups (such as men and women) that have some characteristic. In the previous section, we learned how to estimate a single proportion. Our problem now concerns the comparison of two proportions.

We call the two groups being compared Population 1 and Population 2, and the two population proportions of "successes" p_1 and p_2. The data consist of two independent SRSs, of size n_1 from Population 1 and size n_2 from Population 2. The proportion of successes in each sample estimates the corresponding population proportion. Here is the notation we will use in this section:

Population	Population proportion	Sample size	Count of successes	Sample proportion
1	p_1	n_1	X_1	$\hat{p}_1 = X_1/n_1$
2	p_2	n_2	X_2	$\hat{p}_2 = X_2/n_2$

To compare the two populations, we use the difference between the two sample proportions:

$$D = \hat{p}_1 - \hat{p}_2$$

When both sample sizes are sufficiently large, the sampling distribution of the difference D is approximately Normal.

Inference procedures for comparing proportions are z procedures based on the Normal approximation and on standardizing the difference D. The first

◄ LOOK BACK
addition rule for means p. 267

step is to obtain the mean and standard deviation of D. By the addition rule for means, the mean of D is the difference of the means:

$$\mu_D = \mu_{\hat{p}_1} - \mu_{\hat{p}_2} = p_1 - p_2$$

That is, the difference $D = \hat{p}_1 - \hat{p}_2$ between the sample proportions is an unbiased estimator of the population difference $p_1 - p_2$. Similarly, the addition rule for variances tells us that the variance of D is the *sum* of the variances:

◄ LOOK BACK
addition rule for variances
p. 271

$$\sigma_D^2 = \sigma_{\hat{p}_1}^2 + \sigma_{\hat{p}_2}^2$$
$$= \frac{p_1(1 - p_1)}{n_1} + \frac{p_2(1 - p_2)}{n_2}$$

Therefore, when n_1 and n_2 are large, D is approximately Normal with mean $\mu_D = p_1 - p_2$ and standard deviation

$$\sigma_D = \sqrt{\frac{p_1(1 - p_1)}{n_1} + \frac{p_2(1 - p_2)}{n_2}}$$

USE YOUR KNOWLEDGE

8.45 Rules for means and variances. Suppose $p_1 = 0.4$, $n_1 = 25$, $p_2 = 0.5$, $n_2 = 30$. Find the mean and the standard deviation of the sampling distribution of $\hat{p}_1 - \hat{p}_2$.

8.46 Effect of the sample sizes. Suppose $p_1 = 0.4$, $n_1 = 100$, $p_2 = 0.5$, $n_2 = 120$.

(a) Find the mean and the standard deviation of the sampling distribution of $\hat{p}_1 - \hat{p}_2$.

(b) The sample sizes here are four times as large as those in the previous exercise while the population proportions are the same. Compare the results for this exercise with those that you found in the previous exercise. What is the effect of multiplying the sample sizes by 4?

8.47 Rules for means and variances. It is quite easy to verify the formulas for the mean and standard deviation of the difference D.

(a) What are the means and standard deviations of the two sample proportions $\hat{p}_1$ and $\hat{p}_2$?

(b) Use the addition rule for means of random variables: what is the mean of $D = \hat{p}_1 - \hat{p}_2$?

(c) The two samples are independent. Use the addition rule for variances of random variables: what is the variance of D?

Large-sample confidence interval for a difference in proportions

To obtain a confidence interval for $p_1 - p_2$, we once again replace the unknown parameters in the standard deviation by estimates to obtain an estimated standard deviation, or standard error. Here is the confidence interval we want.

LARGE-SAMPLE CONFIDENCE INTERVAL FOR COMPARING TWO PROPORTIONS

Choose an SRS of size n_1 from a large population having proportion p_1 of successes and an independent SRS of size n_2 from another population having proportion p_2 of successes. The estimate of the difference in the population proportions is

$$D = \hat{p}_1 - \hat{p}_2$$

The **standard error of D** is

$$SE_D = \sqrt{\frac{\hat{p}_1(1 - \hat{p}_1)}{n_1} + \frac{\hat{p}_2(1 - \hat{p}_2)}{n_2}}$$

and the **margin of error** for confidence level C is

$$m = z^* SE_D$$

where the critical value z^* is the value for the standard Normal density curve with area C between $-z^*$ and z^*. An **approximate level C confidence interval** for $p_1 - p_2$ is

$$D \pm m$$

Use this method for 90%, 95%, or 99% confidence when the number of successes and the number of failures in each sample are at least 10.

EXAMPLE

8.8 Gender and the proportion of frequent binge drinkers. Many studies have documented binge drinking as a major problem among college students.[21] Here are some data that let us compare men and women:

Population	n	X	$\hat{p} = X/n$
1 (men)	5,348	1,392	0.260
2 (women)	8,471	1,748	0.206
Total	13,819	3,140	0.227

In this table the $\hat{p}$ column gives the sample proportions of frequent binge drinkers.

Let's find a 95% confidence interval for the difference between the proportions of men and of women who are frequent binge drinkers. Output from Minitab and CrunchIt! is given in Figure 8.3. To perform the computations using our formulas, we first find the difference in the proportions:

$$D = \hat{p}_1 - \hat{p}_2$$
$$= 0.260 - 0.206$$
$$= 0.054$$

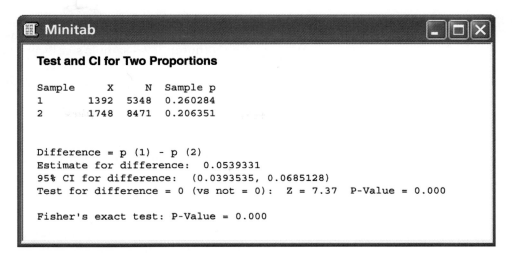

FIGURE 8.3 Minitab and Crunch It! output for Example 8.8.

The Minitab output shows:

Test and CI for Two Proportions

```
Sample    X      N   Sample p
1       1392   5348  0.260284
2       1748   8471  0.206351


Difference = p (1) - p (2)
Estimate for difference:  0.0539331
95% CI for difference:  (0.0393535, 0.0685128)
Test for difference = 0 (vs not = 0):  Z = 7.37   P-Value = 0.000

Fisher's exact test: P-Value = 0.000
```

CrunchIt!

Proportion Two-Sample with Summary

Alternative Hypothesis Two-Sided

Confidence Level 0.9500

Statistic	Sample 1 Size	Sample 1 Successes	Sample 1 Proportion	Sample 2 Size	Sample 2 Successes	Sample 2 Proportion	Difference of Proportions	Z statistic	p Value	CI Upper Bound	CI Lower Bound
Result	5348	1392	0.2603	8471	1748	0.2064	0.0539	7.3693	1.7141e-13	0.0685	0.0394

Then we calculate the standard error of D:

$$SE_D = \sqrt{\frac{\hat{p}_1(1 - \hat{p}_1)}{n_1} + \frac{\hat{p}_2(1 - \hat{p}_2)}{n_2}}$$

$$= \sqrt{\frac{(0.260)(0.740)}{5348} + \frac{(0.206)(0.794)}{8471}}$$

$$= 0.00744$$

For 95% confidence, we have $z^* = 1.96$, so the margin of error is

$$m = z^*SE_D = (1.96)(0.00744)$$

$$= 0.015$$

The 95% confidence interval is

$$D \pm m = 0.054 \pm 0.015$$

$$= (0.039, 0.069)$$

With 95% confidence we can say that the difference in the proportions is between 0.039 and 0.069. Alternatively, we can report that the difference in the percent of men who are frequent binge drinkers and the percent of women who are frequent binge drinkers is 5.4%, with a 95% margin of error of 1.5%.

In this example men and women were not sampled separately. The sample sizes are in fact random and reflect the gender distributions of the colleges that were randomly chosen. Two-sample significance tests and confidence intervals are still approximately correct in this situation. The authors of the report note that women are somewhat overrepresented partly because 6 of the 140 colleges in the study were all-women institutions.

In the example above we chose men to be the first population. Had we chosen women to be the first population, the estimate of the difference would be negative (-0.054). Because it is easier to discuss positive numbers, we generally choose the first population to be the one with the higher proportion.

USE YOUR KNOWLEDGE

8.48 Gender and commercial preference. A study was designed to compare two energy drink commercials. Each participant was shown the commercials in random order and asked to select the better one. Commercial A was selected by 44 out of 100 women and 79 out of 140 men. Give an estimate of the difference in gender proportions that favored Commercial A. Also construct a large-sample 95% confidence interval for this difference.

8.49 Gender and commercial preference, revisited. Refer to Exercise 8.48. Construct a 95% confidence interval for the difference in proportions that favor Commercial B. Explain how you could have obtained these results from the calculations you did in Exercise 8.48.

BEYOND THE BASICS

Plus four confidence interval for a difference in proportions

Just as in the case of estimating a single proportion, a small modification of the sample proportions can greatly improve the accuracy of confidence intervals.[22] As before, we add 2 successes and 2 failures to the actual data, but now we divide them equally between the two samples. That is, we *add 1 success and 1 failure to each sample*. We will again call the estimates produced by adding hypothetical observations plus four estimates. The plus four estimates of the two population proportions are

$$\tilde{p}_1 = \frac{X_1 + 1}{n_1 + 2} \qquad \text{and} \qquad \tilde{p}_2 = \frac{X_2 + 1}{n_2 + 2}$$

The estimated difference between the populations is

$$\tilde{D} = \tilde{p}_1 - \tilde{p}_2$$

and the standard deviation of $\tilde{D}$ is approximately

$$\sigma_{\tilde{D}} = \sqrt{\frac{p_1(1 - p_1)}{n_1 + 2} + \frac{p_2(1 - p_2)}{n_2 + 2}}$$

This is similar to the formula for σ_D, adjusted for the sizes of the modified samples.

To obtain a confidence interval for $p_1 - p_2$, we once again replace the unknown parameters in the standard deviation by estimates to obtain an estimated standard deviation, or standard error. Here is the confidence interval we want.

PLUS FOUR CONFIDENCE INTERVAL FOR COMPARING TWO PROPORTIONS

Choose an SRS of size n_1 from a large population having proportion p_1 of successes and an independent SRS of size n_2 from another population having proportion p_2 of successes. The **plus four estimate of the difference in proportions** is

$$\tilde{D} = \tilde{p}_1 - \tilde{p}_2$$

where

$$\tilde{p}_1 = \frac{X_1 + 1}{n_1 + 2} \qquad \tilde{p}_2 = \frac{X_2 + 1}{n_2 + 2}$$

The **standard error of $\tilde{D}$** is

$$\mathrm{SE}_{\tilde{D}} = \sqrt{\frac{\tilde{p}_1(1 - \tilde{p}_1)}{n_1 + 2} + \frac{\tilde{p}_2(1 - \tilde{p}_2)}{n_2 + 2}}$$

and the **margin of error** for confidence level C is

$$m = z^* \mathrm{SE}_{\tilde{D}}$$

where z^* is the value for the standard Normal density curve with area C between $-z^*$ and z^*. An **approximate level C confidence interval** for $p_1 - p_2$ is

$$\tilde{D} \pm m$$

Use this method for 90%, 95%, or 99% confidence when both sample sizes are at least 5.

EXAMPLE

8.9 Gender and sexual maturity. In studies that look for a difference between genders, a major concern is whether or not apparent differences are due to other variables that are associated with gender. Because boys mature more slowly than girls, a study of adolescents that compares boys and girls of the same age may confuse a gender effect with an effect of sexual maturity. The "Tanner score" is a commonly used measure of sexual maturity.[23] Subjects are asked to determine their score by placing a mark next to a rough drawing of an individual at their level of sexual maturity. There are five different drawings, so the score is an integer between 1 and 5.

A pilot study included 12 girls and 12 boys from a population that will be used for a large experiment. Four of the boys and three of the girls had Tanner scores of 4 or 5, a high level of sexual maturity. Let's find a 95% confidence interval for the difference between the proportions of boys and girls who have high (4 or 5) Tanner scores in this population. The numbers of successes and failures in both groups are not all at least 10, so the large-sample approach is not recommended. On the other hand, the sample sizes are both at least 5, so the plus four method is appropriate.

The plus four estimate of the population proportion for boys is

$$\tilde{p}_1 = \frac{X_1 + 1}{n_1 + 2} = \frac{4 + 1}{12 + 2} = 0.3571$$

For girls, the estimate is

$$\tilde{p}_2 = \frac{X_2 + 1}{n_2 + 2} = \frac{3 + 1}{12 + 2} = 0.2857$$

Therefore, the estimate of the difference is

$$\tilde{D} = \tilde{p}_1 - \tilde{p}_2 = 0.3571 - 0.2857 = 0.071$$

The standard error of $\tilde{D}$ is

$$\text{SE}_{\tilde{D}} = \sqrt{\frac{\tilde{p}_1(1 - \tilde{p}_1)}{n_1 + 2} + \frac{\tilde{p}_2(1 - \tilde{p}_2)}{n_2 + 2}}$$

$$= \sqrt{\frac{(0.3571)(1 - 0.3571)}{12 + 2} + \frac{(0.2857)(1 - 0.2857)}{12 + 2}}$$

$$= 0.1760$$

For 95% confidence, $z^* = 1.96$ and the margin of error is

$$m = z^* \text{SE}_{\tilde{D}} = (1.96)(0.1760) = 0.345$$

The confidence interval is

$$\tilde{D} \pm m = 0.071 \pm 0.345$$
$$= (-0.274,\ 0.416)$$

With 95% confidence we can say that the difference in the proportions is between -0.274 and 0.416. Alternatively, we can report that the difference in the proportions of boys and girls with high Tanner scores in this population is 7.1% with a 95% margin of error of 34.5%.

The very large margin of error in this example indicates that either boys or girls could be more sexually mature in this population and that the difference could be quite large. *Although the interval includes the possibility that there is no difference, corresponding to $p_1 = p_2$ or $p_1 - p_2 = 0$, we should not conclude that there is **no** difference in the proportions.* With small sample sizes such as these, the data do not provide us with a lot of information for our inference. This fact is expressed quantitatively through the very large margin of error.

Significance test for a difference in proportions

Although we prefer to compare two proportions by giving a confidence interval for the difference between the two population proportions, it is sometimes useful to test the null hypothesis that the two population proportions are the same.

We standardize $D = \hat{p}_1 - \hat{p}_2$ by subtracting its mean $p_1 - p_2$ and then dividing by its standard deviation

$$\sigma_D = \sqrt{\frac{p_1(1 - p_1)}{n_1} + \frac{p_2(1 - p_2)}{n_2}}$$

If n_1 and n_2 are large, the standardized difference is approximately $N(0, 1)$. For the large-sample confidence interval we used sample estimates in place of the unknown population values in the expression for σ_D. Although this approach would lead to a valid significance test, we instead adopt the more common practice of replacing the unknown σ_D with an estimate that takes into account our null hypothesis H_0: $p_1 = p_2$. If these two proportions are equal, then we can view all the data as coming from a single population. Let p denote the common value of p_1 and p_2; then the standard deviation of $D = \hat{p}_1 - \hat{p}_2$ is

$$\sigma_D = \sqrt{\frac{p(1 - p)}{n_1} + \frac{p(1 - p)}{n_2}}$$

$$= \sqrt{p(1 - p)\left(\frac{1}{n_1} + \frac{1}{n_2}\right)}$$

We estimate the common value of p by the overall proportion of successes in the two samples:

$$\hat{p} = \frac{\text{number of successes in both samples}}{\text{number of observations in both samples}} = \frac{X_1 + X_2}{n_1 + n_2}$$

pooled estimate of p   This estimate of p is called the **pooled estimate** because it combines, or pools, the information from both samples.

To estimate σ_D under the null hypothesis, we substitute $\hat{p}$ for p in the expression for σ_D. The result is a standard error for D that assumes H_0: $p_1 = p_2$:

$$\text{SE}_{Dp} = \sqrt{\hat{p}(1 - \hat{p})\left(\frac{1}{n_1} + \frac{1}{n_2}\right)}$$

The subscript on SE_{Dp} reminds us that we pooled data from the two samples to construct the estimate.

SIGNIFICANCE TEST FOR COMPARING TWO PROPORTIONS

To test the hypothesis

$$H_0: p_1 = p_2$$

compute the **z statistic**

$$z = \frac{\hat{p}_1 - \hat{p}_2}{\text{SE}_{Dp}}$$

where the **pooled standard error** is

$$\text{SE}_{Dp} = \sqrt{\hat{p}(1 - \hat{p})\left(\frac{1}{n_1} + \frac{1}{n_2}\right)}$$

and where

$$\hat{p} = \frac{X_1 + X_2}{n_1 + n_2}$$

In terms of a standard Normal random variable Z, the P-value for a test of H_0 against

H_a: $p_1 > p_2$ is $P(Z \geq z)$

H_a: $p_1 < p_2$ is $P(Z \leq z)$

H_a: $p_1 \neq p_2$ is $2P(Z \geq |z|)$

This z test is based on the Normal approximation to the binomial distribution. As a general rule, we will use it when the number of successes and the number of failures in each of the samples are at least 5.

EXAMPLE

8.10 Gender and the proportion of frequent binge drinkers: the z test. Are men and women college students equally likely to be frequent binge drinkers? We examine the survey data in Example 8.8 (page 492) to answer this question. Here is the data summary:

Population	n	X	$\hat{p} = X/n$
1 (men)	5,348	1,392	0.260
2 (women)	8,471	1,748	0.206
Total	13,819	3,140	0.227

The sample proportions are certainly quite different, but we will perform a significance test to see if the difference is large enough to lead us to believe that the population proportions are not equal. Formally, we test the hypotheses

$$H_0: p_1 = p_2$$
$$H_a: p_1 \neq p_2$$

The pooled estimate of the common value of p is

$$\hat{p} = \frac{1392 + 1748}{5348 + 8471} = \frac{3140}{13,819} = 0.227$$

Note that this is the estimate on the bottom line of the preceding data summary.

The test statistic is calculated as follows:

$$SE_{Dp} = \sqrt{(0.227)(0.773)\left(\frac{1}{5348} + \frac{1}{8471}\right)} = 0.007316$$

$$z = \frac{\hat{p}_1 - \hat{p}_2}{SE_{Dp}} = \frac{0.260 - 0.206}{0.007316}$$

$$= 7.37$$

The P-value is $2P(Z \geq 7.37)$. The largest value of z in Table A is 3.49, so from this table we can conclude $P < 2 \times 0.0002 = 0.0004$. Most software reports this result as simply 0 or a very small number. Output from Minitab and CrunchIt! is given in Figure 8.4. Minitab reports the P-value as 0.000. This means that the calculated value is less than 0.0005; this is certainly a very small number. CrunchIt! gives < 0.0001. The exact value is not particularly important; it is clear that we should reject the null hypothesis. For most situations 0.001 (1 chance in 1000) is sufficiently small. We report: among college students in the study, 26.0% of the men and 20.6% of the women were frequent binge drinkers; the difference is statistically significant ($z = 7.37$, $P < 0.001$).

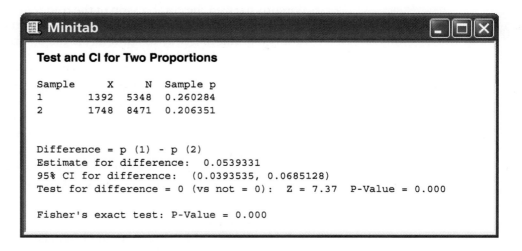

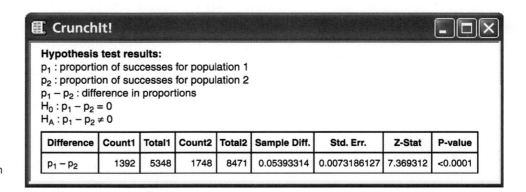

FIGURE 8.4 Minitab and Crunch It! output for Example 8.10.

We could have argued that we expect the proportion to be higher for men than for women in this example. This would justify using the one-sided alternative $H_a: p_1 > p_2$. The P-value would be half of the value obtained for the two-sided test. Because the z statistic is so large, this distinction is of no practical importance.

USE YOUR KNOWLEDGE

8.50 Gender and commercial preference: the z test. Refer to Exercise 8.48 (page 494). Test that the proportions of women and men who liked Commercial A are the same versus the two-sided alternative at the 5% level.

8.51 Changing the alternative hypothesis. Refer to the previous exercise. Does your conclusion change if you test whether the proportion of men who favor Commercial A is larger than the proportion of females? Explain.

BEYOND THE BASICS

Relative risk

risk

relative risk

We summarized the comparison of the frequent binge-drinking proportions for men and for women by reporting a confidence interval for the difference in Example 8.8. Another way to summarize the comparison is to view each sample proportion as the **risk** that a college student of that gender is a frequent binge drinker. We then compare these two risks with the ratio of the two proportions, which is called the **relative risk** (RR) in many applications. Note that a relative risk of 1 means that the two proportions, $\hat{p}_1$ and $\hat{p}_2$, are equal. The procedure for calculating confidence intervals for relative risk is based on the same kind of principles that we have studied, but the details are somewhat more complicated. Fortunately, we can leave the details to software and concentrate on interpretation and communication of the results.

EXAMPLE

8.11 Gender and the proportion of frequent binge drinkers: the relative risk. On page 498 we summarized the data on the proportions of men and women who are frequent binge drinkers with the following table:

Population	n	X	$\hat{p} = X/n$
1 (men)	5,348	1,392	0.260
2 (women)	8,471	1,748	0.206
Total	13,819	3,140	0.227

The relative risk is

$$\text{RR} = \frac{\hat{p}_1}{\hat{p}_2} = \frac{0.260}{0.206} = 1.26$$

> Software gives the 95% confidence interval as 1.19 to 1.34. Men are 1.26 times as likely as women to be frequent binge drinkers; the 95% confidence interval is (1.19, 1.34).

In this example the confidence interval appears to be symmetric about the estimate. If we reported the results with more accuracy (RR = 1.261, 95% confidence interval = 1.186 to 1.341), we would see that the interval is *not* symmetric, and this is true in general.

SECTION 8.2 Summary

The **large-sample estimate of the difference in two population proportions** is

$$D = \hat{p}_1 - \hat{p}_2$$

where $\hat{p}_1$ and $\hat{p}_2$ are the sample proportions

$$\hat{p}_1 = \frac{X_1}{n_1} \quad \text{and} \quad \hat{p}_2 = \frac{X_2}{n_2}$$

The **standard error of the difference D** is

$$\text{SE}_D = \sqrt{\frac{\hat{p}_1(1 - \hat{p}_1)}{n_1} + \frac{\hat{p}_2(1 - \hat{p}_2)}{n_2}}$$

The **margin of error for confidence level C** is

$$m = z^* \text{SE}_D$$

where z^* is the value for the standard Normal density curve with area C between $-z^*$ and z^*. The **large-sample level C confidence interval** is

$$D \pm m$$

We recommend using this interval for 90%, 95%, or 99% confidence when the number of successes and the number of failures in both samples are all at least 10. When sample sizes are smaller, alternative procedures such as the **plus four estimate of the difference in two population proportions** are recommended.

Significance tests of H_0: $p_1 = p_2$ use the z **statistic**

$$z = \frac{\hat{p}_1 - \hat{p}_2}{\text{SE}_{Dp}}$$

with P-values from the $N(0, 1)$ distribution. In this statistic,

$$\text{SE}_{Dp} = \sqrt{\hat{p}(1 - \hat{p})\left(\frac{1}{n_1} + \frac{1}{n_2}\right)}$$

and $\hat{p}$ is the **pooled estimate** of the common value of p_1 and p_2:

$$\hat{p} = \frac{X_1 + X_2}{n_1 + n_2}$$

Use this test when the number of successes and the number of failures in each of the samples are at least 5.

Relative risk is the ratio of two sample proportions:

$$RR = \frac{\hat{p}_1}{\hat{p}_2}$$

Confidence intervals for relative risk are often used to summarize the comparison of two proportions.

SECTION 8.2 Exercises

For Exercises 8.45 to 8.47, see page 491; for Exercises 8.48 and 8.49, see page 494; and for Exercises 8.50 and 8.51, see page 500.

8.52 Podcast downloading. The Podcast Alley Web site recently reported that they have 53,501 podcasts available for downloading with 3,447,545 episodes.[24] A Pew survey of Internet users described the results of two surveys about podcast downloading. The first was conducted between February and April 2006 and surveyed 2822 Internet users. They found that 198 of these said that they had downloaded a podcast to listen to it or view it later at least once. In a more recent survey, conducted in May 2008, there were 1553 Internet users. Of this total, 295 said that they had downloaded a podcast to listen to it or view it later.[25]

(a) Refer to the table that appears at the beginning of this section (page 490). Fill in the numerical values of all quantities that are known.

(b) Find the estimate of the difference between the proportion of Internet users who had ever downloaded podcasts as of February to April 2006 and the proportion as of May 2008.

(c) Is the large-sample confidence interval for the difference in two proportions appropriate to use in this setting? Explain your answer.

(d) Find the 95% confidence interval for difference.

(e) Convert your estimated difference and confidence interval to percents.

(f) One of the surveys was conducted between February and April whereas the other was conducted in May. Do you think that this difference should have any effect on the interpretation of the results? Be sure to explain your answer.

8.53 Significance test for podcast downloading. Refer to the previous exercise. Test the null hypothesis that the two proportions are equal. Report the test statistic with the *P*-value and summarize your conclusion.

8.54 Are more Internet users downloading podcasts? Refer to the previous two exercises. The ratio of the proportion in the 2008 sample to the proportion in the 2006 sample is about 2.7.

(a) Can you conclude that 2.7 times as many people are downloading podcasts? Explain why or why not.

(b) Can you conclude from the data available that there has been an increase from 2006 to 2008 in the number of people who downloaded podcasts? If your answer is no, explain what additional data you would need or what additional assumptions you would need to make to be able to draw this conclusion.

8.55 Adult gamers versus teen gamers. A Pew Internet Project Data Memo presented data comparing adult gamers with teen gamers with respect to the devices on which they play. The data are from two surveys. The adult survey had 1063 gamers while the teen survey had 1064 gamers. The memo reports that 54% of adult gamers played on game consoles (Xbox, PlayStation, Wii, etc.) while 89% of teen gamers played on game consoles.[26]

(a) Refer to the table that appears at the beginning of this section (page 490). Fill in the numerical values of all quantities that are known.

(b) Find the estimate of the difference between the proportion of teen gamers who played on game consoles and the proportion of adults who played on these devices.

(c) Is the large sample confidence interval for the difference in two proportions appropriate to use in this setting? Explain your answer.

(d) Find the 95% confidence interval for difference.

(e) Convert your estimated difference and confidence interval to percents.

(f) The adult survey was conducted between October and December 2008 whereas the teen survey was conducted between November 2007 and February 2008. Do you think that this difference should have any effect on the interpretation of the results? Be sure to explain your answer.

8.56 Significance test for gaming on consoles. Refer to the previous exercise. Test the null hypothesis that the two proportions are equal. Report the test statistic with the P-value and summarize your conclusion.

8.57 Gamers on computers. The report described in Exercise 8.55 also presented data from the same surveys for gaming on computers (desktops or laptops). These devices were used by 73% of adult gamers and by 76% of teen gamers. Answer the questions given in Exercise 8.55 for gaming on computers.

8.58 Significance test for gaming on consoles. Refer to the previous exercise. Test the null hypothesis that the two proportions are equal. Report the test statistic with the P-value and summarize your conclusion.

8.59 Can we compare gaming on consoles with gaming on computers? Refer to the previous four exercises. Do you think that you can use the large-sample confidence intervals for a difference in proportions to compare teens' use of computers with teens' use of consoles? Write a short paragraph giving the reason for your answer. (*Hint:* Look carefully in the box giving the assumptions needed for this procedure.)

8.60 Draw a picture. Suppose there are two binomial populations. For the first, the true proportion of successes is 0.4; for the second, it is 0.5. Consider taking independent samples from these populations, 50 from the first and 60 from the second.

(a) Find the mean and the standard deviation of the distribution of $\hat{p}_1 - \hat{p}_2$.

(b) This distribution is approximately Normal. Sketch this Normal distribution and mark the location of the mean.

(c) Find a value d for which the probability is 0.95 that the difference in sample proportions is within $\pm d$. Mark these values on your sketch.

8.61 What's wrong? For each of the following, explain what is wrong and why.

(a) A z statistic is used to test the null hypothesis that $\hat{p}_1 = \hat{p}_2$.

(b) If two sample proportions are equal, then the sample counts are equal.

(c) A 95% confidence interval for the difference in two proportions includes errors due to nonresponse.

8.62 $\hat{p}_1 - \hat{p}_2$ **and the Normal distribution.** Refer to Exercise 8.60. Assume all the conditions for that

exercise remain the same, with the exception that $n_2 = 1000$.

(a) Find the mean and standard deviation of $\hat{p}_1 - \hat{p}_2$.

(b) Find the mean and standard deviation of $\hat{p}_2 - 0.5$.

(c) Because n_2 is very large, we expect $\hat{p}_2$ to be very close to 0.5. How close?

(d) Summarize what you have found in parts (a), (b), and (c) of this exercise. Interpret your results in terms of inference for comparing two proportions when the sample size in one of the samples is much larger than the sample size in the other.

8.63 Pet ownership and gender. In the Health ABC Study, 595 subjects owned a pet and 1939 subjects did not.[27] Among the pet owners, there were 285 women; 1024 of the non–pet owners were women. Find the proportion of pet owners who were women. Do the same for the non–pet owners. Give a 95% confidence interval for the difference in the two proportions. (Be sure to let Population 1 correspond to the group with the higher proportion so that the difference will be positive.)

8.64 Effects of reducing air pollution. A study that evaluated the effects of a reduction in exposure to traffic-related air pollutants compared respiratory symptoms of 283 residents of an area with congested streets with 165 residents in a similar area where the congestion was removed because a bypass was constructed. The symptoms of the residents of both areas were evaluated at baseline and again a year after the bypass was completed.[28] For the residents of the congested streets, 17 reported that their symptoms of wheezing improved between baseline and one year later, while 35 of the residents of the bypass streets reported improvement.

(a) Find the two sample proportions.

(b) Report the difference in the proportions and the standard error of the difference.

(c) What are the appropriate null and alternative hypotheses for examining the question of interest? Be sure to explain your choice of the alternative hypothesis.

(d) Find the test statistic. Construct a sketch of the distribution of the test statistic under the assumption that the null hypothesis is true. Find the P-value and use your sketch to explain its meaning.

(e) Is no evidence of an effect the same as evidence that there is no effect? Use a 95% confidence interval to answer this question. Summarize your ideas in a way that could be understood by someone who has very little experience with statistics.

(f) The study was done in the United Kingdom. To what extent do you think that the results can be generalized to other circumstances?

8.65 Downloading music from the Internet. A 2005 survey of Internet users reported that 22% downloaded music onto their computers. The filing of lawsuits by the recording industry may be a reason why this percent has decreased from the estimate of 29% from a survey taken two years before.[29] Assume that the sample sizes are both 1421. Using a significance test, evaluate whether or not there has been a change in the percent of Internet users who download music. Provide all details for the test and summarize your conclusion. Also report a 95% confidence interval for the difference in proportions and explain what information is provided in the interval that is not in the significance test results.

8.66 More on downloading music from the Internet. Refer to the previous exercise. Suppose we are not exactly sure about the sizes of the samples. Redo the calculations for the significance test and the confidence interval under the following assumptions: (*i*) both sample sizes are 1000, (*ii*) both sample sizes are 1600, (*iii*) the first sample size is 1000 and the second is 1600. Summarize the effects of the sample sizes on the results.

8.67 Who gets stock options? Different kinds of companies compensate their key employees in different ways. Established companies may pay higher salaries, while new companies may offer stock options that will be valuable if the company succeeds. Do high-tech companies tend to offer stock options more often than other companies? One study looked at a random sample of 200 companies. Of these, 91 were listed in the *Directory of Public High Technology Corporations* and 109 were not listed. Treat these two groups as SRSs of high-tech and non-high-tech companies. Seventy-three of the high-tech companies and 75 of the non-high-tech companies offered incentive stock options to key employees.[30]

(a) Give a 95% confidence interval for the difference in the proportions of the two types of companies that offer stock options.

(b) Compare the two groups of companies with a significance test.

(c) Summarize your analysis and conclusions.

8.68 Cheating during a test: 2002 versus 2004. In Exercise 8.28, you examined the proportion of high school students who cheated on tests at least twice during the past year. Included in that study were the results for both 2002 and 2004. A reported 9054 out of 24,142 students said they cheated at least twice in 2004. A reported 5794 out of 12,121 students said they cheated at least twice in

2002. Give an estimate of the difference between these two proportions with a 90% confidence interval.

8.69 Gender bias in textbooks. To what extent do syntax textbooks, which analyze the structure of sentences, illustrate gender bias? A study of this question sampled sentences from 10 texts.[31] One part of the study examined the use of the words "girl," "boy," "man," and "woman." We will call the first two words *juvenile* and the last two *adult*. Is the proportion of female references that are juvenile (girl) equal to the proportion of male references that are juvenile (boy)? Here are data from one of the texts:

Gender	n	X (juvenile)
Female	60	48
Male	132	52

(a) Find the proportion of juvenile references for females and its standard error. Do the same for the males.

(b) Give a 90% confidence interval for the difference and briefly summarize what the data show.

(c) Use a test of significance to examine whether the two proportions are equal.

8.70 ⚔ **Bicycle accidents, alcohol, and gender.** In Exercise 8.38 (page 489) we examined the percent of fatally injured bicyclists tested for alcohol who tested positive. Here we examine the same data with respect to gender.

Gender	n	X (tested positive)
Female	191	27
Male	1520	515

(a) Summarize the data by giving the estimates of the two population proportions and a 95% confidence interval for their difference.

(b) The standard error SE_D contains a contribution from each sample, $\hat{p}_1(1 - \hat{p}_1)/n_1$ and $\hat{p}_2(1 - \hat{p}_2)/n_2$. Which of these contributes the larger amount to the standard error of the difference? Explain why.

(c) Use a test of significance to examine whether the two proportions are equal.

8.71 Pet ownership and gender: the significance test. In Exercise 8.63 (page 503) we compared the proportion of pet owners who were women with the proportion of non–pet owners who were women in the Health ABC Study. Use a significance test to make the comparison and summarize the results of your analysis.

CHAPTER 8   Exercises

8.72 Video game genres. U.S. computer and video game software sales were $9.5 billion in 2007.[32] A survey of 1102 teens collected data about video game use by teens. According to the survey, the most popular game genres are[33]

Genre	Examples	Percent who play
Racing	NASCAR, Mario Kart, Burnout	74
Puzzle	Bejeweled, Tetris, Solitaire	72
Sports	Madden, FIFA, Tony Hawk	68
Action	Grand Theft Auto, Devil May Cry, Ratchet and Clank	67
Adventure	Legend of Zelda, Tomb Raider	66
Rhythm	Guitar Hero, Dance Dance Revolution, Lumines	61

Give a 95% confidence interval for the proportion who play games in each of these six genres.

8.73 Too many errors. Refer to the previous exercise. The chance that each of the six intervals that you calculated includes the true proportion for that genre is approximately 95%. In other words, the chance that you make an error and your interval misses the true value is approximately 5%.

(a) Explain why your chance that at least one of your intervals does not contain the true value of the parameter is greater than 5%.

(b) One way to deal with this problem is to adjust the confidence level for each interval so that the overall probability of at least one miss is 5%. One simple way to do this is to use a **Bonferroni procedure.** Here is the basic idea: You have an error budget of 5% and you choose to spend it equally on six intervals. Each interval has a budget of $0.05/6 = 0.008$. So, each confidence interval should have a 0.8% chance of missing the true value. In other words, the confidence level for each interval should be $1 - 0.008 = 0.992$. Use Table A to find the value of z^* for a large sample confidence interval for a single proportion corresponding to 99.2% confidence.

(c) Calculate the six confidence intervals using the Bonferroni procedure.

8.74 Wireless only. Are customers giving up their landlines and relying on wireless for all their phone needs? Surveys have collected data to answer this question.[34] In December 2003, 4.2% of households were wireless only. Assume that this survey is based on sampling 15,000 households.

(a) Convert the percent to a proportion. Then use the proportion and the sample size to find the count of households who were wireless only.

(b) Find a 95% confidence interval for the proportion of households that were wireless only in December 2003.

8.75 Change in wireless only. Refer to the previous exercise. The percentage increased to 16.4% in December 2007. Assume the same sample size for this sample.

(a) Find the proportion and the count for this sample.

(b) Compute the 95% confidence interval for the proportion.

(c) Convert the estimate and confidence interval in terms of proportions to an estimate and confidence in terms of percents.

(d) Find the estimate of the difference in the proportions of households that are wireless only in December 2007 and the households that are wireless only in December 2003.

(e) Give the margin of error for 95% confidence for the difference in proportions.

8.76 Do the change in terms of relative risk. Refer to the previous two exercises.

(a) Summarize the change data in terms of relative risk. The term "relative risk" is a poor description of the ratio that you are using for this exercise. Give a better term for this ratio.

(b) Analyze the data in terms of relative risk and write a summary of your results.

(c) Compare your results in part (b) with your findings in terms of a difference in proportions from the previous exercise.

(d) Which approach do you prefer, difference in proportions or relative risk? Give reasons for your answer.

8.77 Changes in credit card usage by undergraduates. In Exercise 8.22 (page 487) we looked at data from a survey of 1430 undergraduate students and their credit card use. These students were surveyed in 2004. In the sample, 43% said that they had four or more credit cards. A similar study performed in 2000 by the same organization reported that 32% of the sample said that they had four or more credit cards.[35] Assume that the sample sizes for the two studies are the same. Find a 95%

confidence interval for the change from 2000 to 2004 in the percent of undergraduates who report having four or more credit cards.

8.78 Do the significance test for the change. Refer to the previous exercise. Perform the significance test for comparing the two proportions. Report your test statistic, the *P*-value, and summarize your conclusion.

8.79 We did not know the sample size. Refer to the previous two exercises. We did not report the sample size for the 2000 study, but it is reasonable to assume that it is reasonably close to the sample size for the 2004 study.

(a) Suppose that the sample size for the 2000 study was only 1000. Redo the confidence interval and significance testing calculations for this scenario.

(b) Suppose that the sample size for the 2000 study was 2000. Redo the confidence interval and significance testing calculations for this scenario.

(c) Compare your results for parts (a) and (b) of this exercise with the results that you found in the previous two exercises. Write a short paragraph about the effects of assuming a value for the sample size on your conclusions.

8.80 Student employment during the school year. A study of 1430 undergraduate students reported that 994 work 10 or more hours a week during the school year. Give a 95% confidence interval for the proportion of all undergraduate students who work 10 or more hours a week during the school year.

8.81 Examine the effect of the sample size. Refer to the previous exercise. Assume a variety of different scenarios where the sample size changes, but the proportion in the sample who work 10 or more hours a week during the school year remains the same. Write a short report summarizing your results and conclusions. Be sure to include numerical and graphical summaries of what you have found.

8.82 Using a handheld phone while driving. Refer to Exercise 8.26 (page 488). This same poll found that 58% of the respondents talked on a handheld phone while driving in the last year. Construct a 90% confidence interval for the proportion of U.S. drivers who talked on a handheld phone while driving in the last year.

8.83 Gender and using a handheld phone while driving. Refer to the previous exercise. In this same report, this percent was broken down into 59% for men and 56% for women. Assuming that among the 1048 respondents, there were an equal number of men and women, construct a 95% confidence interval for the difference in these proportions.

8.84 ⚠ **Even more on downloading music from the Internet.** The following quotation is from a survey of Internet users. The sample size for the survey was 1371. Since 18% of those surveyed said they download music, the sample size for this subsample is 247.

> Among current music down loaders, 38% say they are downloading less because of the RIAA suits. . . . About a third of current music down loaders say they use peer-to-peer networks. . . . 24% of them say they swap files using email and instant messaging; 20% download files from music-related Web sites like those run by music magazines or musician homepages. And while online music services like iTunes are far from trumping the popularity of file-sharing networks, 17% of current music downloaders say they are using these paid services. Overall, 7% of Internet users say they have bought music at these new services at one time or another, including 3% who currently use paid services.[36]

(a) For each percent quoted, give the margin of error. You should express these in percents, as given in the quote.

(b) Rewrite the paragraph more concisely and include the margins of error.

(c) Pick either side A or side B below and give arguments in favor of the view that you select.

(A) The margins of error should be included because they are necessary for the reader to properly interpret the results. (B) The margins of error interfere with the flow of the important ideas. It would be better to just report one margin of error and say that all of the others are no greater than this number.

If you choose view B, be sure to give the value of the margin of error that you report.

8.85 ⚠ **More on the effects of reducing air pollution.** In Exercise 8.64 the effects of a reduction in air pollution on wheezing was examined by comparing the one-year change in symptoms in a group of residents who lived on congested streets with a group who lived in an area that had been congested but from which the congestion was removed when a bypass was built. The effect of the reduction in air pollution was assessed by comparing the proportions of residents in the two groups who reported that their wheezing symptoms improved. Here are some additional data from the same study: 🔵 AIRPOLLUTION

Symptom	Bypass		Congested	
	n	Improved	n	Improved
Number of wheezing attacks	282	45	163	21
Wheezing disturbs sleep	282	45	164	12
Wheezing limits speech	282	12	164	4
Wheezing affects activities	281	26	165	13
Winter cough	261	15	156	14
Winter phlegm	253	12	144	10
Consulted doctor	247	29	140	18

The table gives the number of subjects in each group and the number reporting improvement. So, for example, the proportion who reported improvement in the number of wheezing attacks was 21/163 in the congested group.

(a) The reported sample sizes vary from symptom to symptom. Give possible reasons for this and discuss the possible impact on the results.

(b) Calculate the difference in the proportions for each symptom. Make a table of symptoms ordered from highest to lowest based on these differences. Include the estimates of the differences and the 95% confidence intervals in the table. Summarize your conclusions.

(c) Can you justify a one-sided alternative in this situation? Give reasons for your answer.

(d) Perform a significance test to compare the two groups for each of the symptoms. Summarize the results.

(e) Reanalyze the data using only the data from the bypass group. Give confidence intervals for the proportions that reported improved symptoms. Compare the conclusions that someone might make from these results with those you presented in part (b).

(f) Use your analyses of the data in this exercise to discuss the importance of a control group in studies such as this.

8.86 Improving the time to repair golf clubs. The Ping Company makes custom-built golf clubs and competes in the $4 billion golf equipment industry. To improve its business processes, Ping decided to seek ISO 9001 certification.[37] As part of this process, a study of the time it took to repair golf clubs that were sent to the company by mail determined that 16% of orders were sent back to the customers in 5 days or less. Ping examined the processing of repair orders and made changes. Following the changes, 90% of orders were completed within 5 days. Assume that each of the estimated percents is based on a random sample of 200 orders.

(a) How many orders were completed in 5 days or less before the changes? Give a 95% confidence interval for the proportion of orders completed in this time.

(b) Do the same for orders after the changes.

(c) Give a 95% confidence interval for the improvement. Express this both for a difference in proportions and for a difference in percents.

8.87 Parental pressure to succeed in school. A Pew Research Center Poll used telephone interviews to ask American adults if parents are pushing their kids too hard to succeed in school. Of those responding, 56% said parents are placing too little pressure on their children.[38] Assuming that this is an SRS of 1200 U.S. residents over the age of 18, give the 95% margin of error for this estimate.

8.88 Brand loyalty and the Chicago Cubs. According to literature on brand loyalty, consumers who are loyal to a brand are likely to consistently select the same product. This type of consistency could come from a positive childhood association. To examine brand loyalty among fans of the Chicago Cubs, 371 Cubs fans among patrons of a restaurant located in Wrigleyville were surveyed prior to a game at Wrigley Field, the Cubs' home field.[39] The respondents were classified as "die-hard fans" or "less loyal fans." Of the 134 die-hard fans, 90.3% reported that they had watched or listened to Cubs games when they were children. Among the 237 less loyal fans, 67.9% said that they had watched or listened as children.

(a) Find the numbers of die-hard Cubs fans who watched or listened to games when they were children. Do the same for the less loyal fans.

(b) Use a significance test to compare the die-hard fans with the less loyal fans with respect to their childhood experiences relative to the team.

(c) Express the results with a 95% confidence interval for the difference in proportions.

8.89 Brand loyalty in action. The study mentioned in the previous exercise found that two-thirds of the die-hard fans attended Cubs games at least once a month, but only 20% of the less loyal fans attended this often. Analyze these data using a significance test and a confidence interval. Write a short summary of your findings.

8.90 More on gender bias in textbooks. Refer to the study of gender bias and stereotyping described in Exercise 8.69 (page 504). Here are the counts of "girl," "woman," "boy," and "man" for all of the syntax texts studied. The one we analyzed in Exercise 8.69 was number 6. GENDERBIAS

	Text number									
	1	**2**	**3**	**4**	**5**	**6**	**7**	**8**	**9**	**10**
Girl	2	5	25	11	2	48	38	5	48	13
Woman	3	2	31	65	1	12	2	13	24	5
Boy	7	18	14	19	12	52	70	6	128	32
Man	27	45	51	138	31	80	2	27	48	95

For each text perform the significance test to compare the proportions of juvenile references for females and males. Summarize the results of the significance tests for the 10 texts studied. The researchers who conducted the study note that the authors of the last three texts are women, while the other seven texts were written by men. Do you see any pattern that suggests that the gender of the author is associated with the results?

8.91 ⚔ **Even more on gender bias in textbooks.** Refer to the previous exercise. Let us now combine the categories "girl" with "woman" and "boy" with "man." For each text calculate the proportion of male references and test the hypothesis that male and female references are equally likely (that is, the proportion of male references is equal to 0.5). Summarize the results of your 10 tests. Is there a pattern that suggests a relation with the gender of the author?

8.92 ⚔ **Changing majors during college.** In a random sample of 975 students from a large public university, it was found that 463 of the students changed majors during their college years.

(a) Give a 95% confidence interval for the proportion of students at this university who change majors.

(b) Express your results from (a) in terms of the *percent* of students who change majors.

(c) University officials concerned with counseling students are interested in the number of students who change majors rather than the proportion. The university has 37,500 undergraduate students. Convert the confidence interval you found in (a) to a confidence interval for the *number* of students who change majors during their college years.

8.93 Gallup Poll study. Go to the Gallup Poll Web site gallup.com and find a poll that has several questions of interest to you. Summarize the results of the poll giving margins of error and comparisons of interest. (For this exercise, you may assume that the data come from an SRS.)

8.94 Parental pressure and gender. The Pew Research Center poll in Exercise 8.87 (page 507) also reported that 62% of the men and 51% of the women thought parents

are placing too little pressure on their children to succeed in school. Assuming that the respondents were 52% women, compare the proportions with a significance test and give a 95% confidence interval for the difference. Write a summary of your results.

8.95 ⚔ **Sample size and the *P*-value.** In this exercise we examine the effect of the sample size on the significance test for comparing two proportions. In each case suppose that $\hat{p}_1 = 0.5$ and $\hat{p}_2 = 0.4$, and take n to be the common value of n_1 and n_2. Use the z statistic to test $H_0: p_1 = p_2$ versus the alternative $H_a: p_1 \neq p_2$. Compute the statistic and the associated *P*-value for the following values of n: 40, 50, 80, 100, 400, 500, and 1000. Summarize the results in a table. Explain what you observe about the effect of the sample size on statistical significance when the sample proportions $\hat{p}_1$ and $\hat{p}_2$ are unchanged.

8.96 ⚔ **Sample size and the margin of error.** In the first section of this chapter, we studied the effect of the sample size on the margin of error of the confidence interval for a single proportion. In this exercise we perform some calculations to observe this effect for the two-sample problem. Suppose that $\hat{p}_1 = 0.7$ and $\hat{p}_2 = 0.6$, and n represents the common value of n_1 and n_2. Compute the 95% margins of error for the difference in the two proportions for $n = 40, 50, 80, 100, 400, 500,$ and 1000. Present the results in a table and with a graph. Write a short summary of your findings.

8.97 ⚔ **Calculating sample sizes for the two-sample problem.** For a single proportion the margin of error of a confidence interval is largest for any given sample size n and confidence level C when $\hat{p} = 0.5$. This led us to use $p^* = 0.5$ for planning purposes. The same kind of result is true for the two-sample problem. The margin of error of the confidence interval for the difference between two proportions is largest when $\hat{p}_1 = \hat{p}_2 = 0.5$. You are planning a survey and will calculate a 95% confidence interval for the difference in two proportions when the data are collected. You would like the margin of error of the interval to be less than or equal to 0.075. You will use the same sample size n for both populations.

(a) How large a value of n is needed?

(b) Give a general formula for n in terms of the desired margin of error m and the critical value z^*.

8.98 A corporate liability trial. A major court case on the health effects of drinking contaminated water took place in the town of Woburn, Massachusetts. A town well in Woburn was contaminated by industrial chemicals. During the period that residents drank water from this well, there were 16 birth defects among 414 births. In years when the contaminated well was shut off and water was supplied from other wells, there were 3 birth defects among 228 births. The plaintiffs suing the firm responsible for the contamination claimed that these data show that the rate of birth defects was higher when the contaminated well was in use.[40] How statistically significant is the evidence? What assumptions does your analysis require? Do these assumptions seem reasonable in this case?

8.99 ⚖ **Statistics and the law.** *Castaneda v. Partida* is an important court case in which statistical methods were used as part of a legal argument.[41] When reviewing this case, the Supreme Court used the phrase "two or three standard deviations" as a criterion for statistical significance. This Supreme Court review has served as the basis for many subsequent applications of statistical methods in legal settings. (The two or three standard deviations referred to by the Court are values of the z statistic and correspond to P-values of approximately 0.05 and 0.0026.) In *Castaneda* the plaintiffs alleged that the method for selecting juries in a county in Texas was biased against Mexican Americans. For the period of time at issue, there were 181,535 persons eligible for jury duty, of whom 143,611 were Mexican Americans. Of the 870 people selected for jury duty, 339 were Mexican Americans.

(a) What proportion of eligible jurors were Mexican Americans? Let this value be p_0.

(b) Let p be the probability that a randomly selected juror is a Mexican American. The null hypothesis to be tested is $H_0: p = p_0$. Find the value of $\hat{p}$ for this problem, compute the z statistic, and find the P-value. What do you conclude? (A finding of statistical significance in this circumstance does not constitute proof of discrimination. It can be used, however, to establish a prima facie case. The burden of proof then shifts to the defense.)

(c) We can reformulate this exercise as a two-sample problem. Here we wish to compare the proportion of Mexican Americans among those selected as jurors with the proportion of Mexican Americans among those not selected as jurors. Let p_1 be the probability that a randomly selected juror is a Mexican American, and let p_2 be the probability that a randomly selected nonjuror is a Mexican American. Find the z statistic and its P-value. How do your answers compare with your results in part (b)?

8.100 ⚖ **Home court advantage.** In many sports there is a home field or home court advantage. This means that the home team is more likely to win when playing at home than they are to win when playing at an opponent's field or court, all other things being equal. Go to the Web site of your favorite sports team and find the proportion of wins for home games and the proportion of wins for away games. Now consider these games to be a random sample of the process that generates wins and losses. A complete analysis of data like these requires methods that are beyond what we have studied, but the methods discussed in this chapter will give us a reasonable approximation. Examine the home court advantage for your team and write a summary of your results. Be sure to comment on the effect of the sample size.

8.101 ⚖ **Attitudes toward student loan debt.** The National Student Loan Survey asked the student loan borrowers in their sample about attitudes toward debt.[42] Here are some of the questions they asked, with the percent who responded in a particular way.

(a) "To what extent do you feel burdened by your student loan payments?" 55.5% said they felt burdened.

(b) "If you could begin again, taking into account your current experience, what would you borrow?" 54.4% said they would borrow less.

(c) "Since leaving school, my education loans have not caused me more financial hardship than I had anticipated at the time I took out the loans." 34.3% disagreed.

(d) "Making loan payments is unpleasant, but I know that the benefits of education loans are worth it." 58.9% agreed.

(e) "I am satisfied that the education I invested in with my student loan(s) was worth the investment for career opportunities." 58.9% agreed.

(f) "I am satisfied that the education I invested in with my student loan(s) was worth the investment for personal growth." 71.5% agreed.

Assume that the sample size is 1280 for all of these questions. Compute a 95% confidence interval for each of the questions, and write a short report about what student loan borrowers think about their debt.

Analysis of Two-Way Tables

Introduction

We continue our study of methods for analyzing categorical data in this chapter. Inference about proportions in one-sample and two-sample settings was the focus of Chapter 8. We now study how to compare two or more populations when the response variable has two or more categories and how to test whether two categorical variables are independent. A single statistical test handles both of these cases.

The first section of this chapter gives the basics of statistical inference that are appropriate in this setting. An optional second section provides some technical details, and a goodness of fit test is presented in the last section. The methods in this chapter answer questions such as

- Are men and women equally likely to suffer lingering fear symptoms after watching scary movies like *Jaws* and *Poltergeist* at a young age?

- Is there an association between texting while driving and automobile accidents?

- Does political preference predict making contributions online?

9.1 Inference for Two-Way Tables

When we studied inference for two proportions in Chapter 8, we started summarizing the raw data by giving the number of observations in each population (n) and how many of these were classified as "successes" (X).

EXAMPLE

9.1 Gender and the proportion of frequent binge drinkers. In Example 8.8, we compared the proportions of male and female college students who engage in frequent binge drinking. The following table summarizes the data used in this comparison:

Population	n	X	$\hat{p} = X/n$
1 (men)	5,348	1,392	0.260
2 (women)	8,471	1,748	0.206
Total	13,819	3,140	0.227

These data suggest that the men are 5.4% more likely to be frequent binge drinkers, with a 95% margin of error of 1.5%.

LOOK BACK
two-way table p. 137
In this chapter we consider a different summary of the data. Rather than recording just the count of binge drinkers, we record counts of all the outcomes in a two-way table.

EXAMPLE

9.2 Two-way table of frequent binge drinking and gender. Here is the two-way table classifying students by gender and whether or not they are frequent binge drinkers:

Two-way table for frequent binge drinking and gender

Frequent binge drinker	Gender		Total
	Men	Women	
Yes	1,392	1,748	3,140
No	3,956	6,723	10,679
Total	5,348	8,471	13,819

$r \times c$ table We use the term $r \times c$ table to describe a two-way table of counts with r rows and c columns. The two categorical variables in the 2×2 table of Example 9.2 are "Frequent binge drinker" and "Gender." "Frequent binge drinker" is the row variable, with values "Yes" and "No," and "Gender" is the column variable, with values "Men" and "Women." Since the objective in this example is to compare the genders, we view "Gender" as an explanatory variable, and therefore, we make it the column variable. The row variable is a categorical response variable, "Frequent binge drinker." The next example presents another two-way table.

EXAMPLE

9.3 Lingering symptoms from frightening movies. There is a growing body of literature demonstrating that early exposure to frightening movies is associated with lingering fright symptoms. As part of a class on media effects, college students were asked to write narrative accounts of their exposure to frightening movies before the age of 13. More than one-fourth of the respondents said that some of the fright symptoms were still present in waking life.[1] The following table breaks down these results by gender:

Observed numbers of students			
	Gender		
Ongoing fright symptoms	Men	Women	Total
Yes	7	29	36
No	31	50	81
Total	38	79	117

The two categorical variables in this example are "Ongoing fright symptoms," with values "Yes" and "No," and "Gender," with values "Men" and "Women." Again we view "Gender" as an explanatory variable and "Ongoing fright symptoms" as a categorical response variable.

In Chapter 2 we discussed two-way tables and the basics about joint, marginal, and conditional distributions. There we learned that the key to examining the relationship between two categorical variables is to look at conditional distributions. Figure 9.1 shows the output from CrunchIt! for the data of Example 9.3. Check this figure carefully. Be sure that you can identify the joint distribution, the marginal distributions, and the conditional distributions.

LOOK BACK
conditional distributions
p. 141

EXAMPLE

9.4 Two-way table of ongoing fright symptoms and gender. To compare the frequency of lingering fright symptoms across genders, we examine column percents. Here they are, rounded from the output for clarity:

CrunchIt!

Contingency Table – Selected Rows Gender

	Male	Female	Total
Yes	7 (19.44%) (18.42%) (5.98%)	29 (80.56%) (38.71%) (24.79%)	36 (100.00%) (30.77%) (30.77%)
No	31 (38.27%) (81.58%) (26.50%)	50 (61.73%) (63.29%) (42.74%)	81 (100.00%) (69.23%) (69.24%)
Total	38 (32.48%) (100.00%) (32.48%)	79 (67.52%) (100.00%) (67.52%)	117 (100.00%) (100.00%) (100.00%)

Chi-Square

DF	1
Value	4.0284
P-value	0.0447

FIGURE 9.1 Computer output for Example 9.3.

Column percents for gender

	Gender	
Ongoing fright symptoms	**Male**	**Female**
Yes	18%	37%
No	82%	63%
Total	100%	100%

The "Total" row reminds us that 100% of the male and female students have been classified as having ongoing fright symptoms or not. (The sums sometimes differ slightly from 100% because of roundoff error.) The bar graph in Figure 9.2 compares the percents. The data reveal a clear relationship: 37% of the women have ongoing fright symptoms, as opposed to only 18% of the men.

The difference between the percents of students with lingering fears is reasonably large. A statistical test will tell us whether or not this difference can be plausibly attributed to chance. Specifically, if there is no association between gender and having ongoing fright symptoms, how likely is it that a

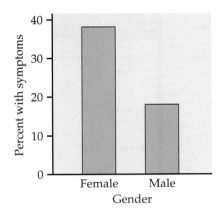

FIGURE 9.2 Bar graph of the percents of male and female students with lingering fright symptoms.

sample would show a difference as large or larger than that displayed in Figure 9.2? In the remainder of this section we discuss the significance test to examine this question.

The hypothesis: no association

The null hypothesis H_0 of interest in a two-way table is, There is *no association* between the row variable and the column variable. In Example 9.3, this null hypothesis says that gender and having ongoing fright symptoms are not related. The alternative hypothesis H_a is that there is an association between these two variables. The alternative H_a does not specify any particular direction for the association. For two-way tables in general, the alternative includes many different possibilities. Because it includes all sorts of possible associations, we cannot describe H_a as either one-sided or two-sided.

In our example, the hypothesis H_0 that there is no association between gender and having ongoing fright symptoms is equivalent to the statement that the distributions of the ongoing fright symptoms variable are the same across the genders. For other two-way tables, where the columns correspond to independent samples from c distinct populations, there are c distributions for the row variable, one for each population. The null hypothesis then says that the c distributions of the row variable are identical. The alternative hypothesis is that the distributions are not all the same.

Expected cell counts

expected cell counts

To test the null hypothesis in $r \times c$ tables, we compare the observed cell counts with **expected cell counts** calculated under the assumption that the null hypothesis is true. A numerical summary of the comparison will be our test statistic.

EXAMPLE

9.5 Expected counts from software. The observed and expected counts for the ongoing fright symptoms example appear in the Minitab computer output shown in Figure 9.3. The expected counts are given as the second entry in each cell. For example, in the first cell the observed count is 7 and the expected count is 11.69.

```
Minitab                                              [_][□][X]

Rows: Symptom    Columns: Gender

            1_Male   2_Female     All

1_Yes            7         29        36
             11.69      24.31     36.00

2_No            31         50        81
             26.31      54.69     81.00

All             38         79       117
             38.00      79.00    117.00

Cell Contents:        Count
                      Expected count


Pearson Chi-Square = 4.028, DF = 1, P-Value = 0.045
```

FIGURE 9.3   Minitab computer output for Example 9.5.

How is this expected count obtained? Look at the percents in the right margin of the table in Figure 9.1. We see that 30.77% of all students had ongoing fright symptoms. If the null hypothesis of no relation between gender and ongoing fright is true, we expect this overall percent to apply to both men and women. In particular, we expect 30.77% of the men to have lingering fright symptoms. Since there are 38 men, the expected count is 30.77% of 38, or 11.69. The other expected counts are calculated in the same way.

The reasoning of Example 9.5 leads to a simple formula for calculating expected cell counts. To compute the expected count of men with ongoing fright symptoms, we multiplied the proportion of students with fright symptoms (36/117) by the number of men (38). From Figures 9.1 and 9.3 we see that the numbers 36 and 38 are the row and column totals for the cell of interest and that 117 is n, the total number of observations for the table. The expected cell count is therefore the product of the row and column totals divided by the table total.

$$\text{expected cell count} = \frac{\text{row total} \times \text{column total}}{n}$$

The chi-square test

To test the H_0 that there is no association between the row and column classifications, we use a statistic that compares the entire set of observed counts with the set of expected counts. To compute this statistic,

- First, take the difference between each observed count and its corresponding expected count, and square these values so that they are all 0 or positive.

- Since a large difference means less if it comes from a cell that is expected to have a large count, divide each squared difference by the expected count. This is a type of standardization.

LOOK BACK
standardizing p. 58

- Finally, sum over all cells.

The result is called the *chi-square statistic* X^2. The chi-square statistic was proposed by the English statistician Karl Pearson (1857–1936) in 1900. It is the oldest inference procedure still used in its original form.

CHI-SQUARE STATISTIC

The **chi-square statistic** is a measure of how much the observed cell counts in a two-way table diverge from the expected cell counts. The formula for the statistic is

$$X^2 = \sum \frac{(\text{observed count} - \text{expected count})^2}{\text{expected count}}$$

where "observed" represents an observed cell count, "expected" represents the expected count for the same cell, and the sum is over all $r \times c$ cells in the table.

If the expected counts and the observed counts are very different, a large value of X^2 will result. Large values of X^2 provide evidence against the null hypothesis. To obtain a P-value for the test, we need the sampling distribution of X^2 under the assumption that H_0 (no association between the row and column variables) is true. We once again use an approximation, related to the Normal approximation for binomial distributions. The result is a new distribution, the **chi-square distribution,** which we denote by χ^2 (χ is the lowercase Greek letter chi).

LOOK BACK
Normal approximation for
binomial p. 324

chi-square distribution
χ^2

Like the t distributions, the χ^2 distributions form a family described by a single parameter, the degrees of freedom. We use $\chi^2(\text{df})$ to indicate a particular member of this family. Figure 9.4 displays the density curves of the $\chi^2(2)$ and $\chi^2(4)$ distributions. As you can see in the figure, χ^2 distributions take only positive values and are skewed to the right. Table F in the back of the book gives upper critical values for the χ^2 distributions.

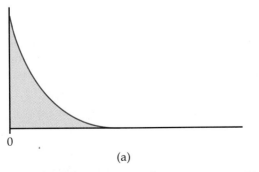
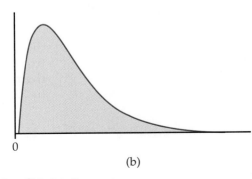

FIGURE 9.4   (a) The $\chi^2(2)$ density curve. (b) The $\chi^2(4)$ density curve.

CHI-SQUARE TEST FOR TWO-WAY TABLES

The null hypothesis H_0 is that there is no association between the row and column variables in a two-way table. The alternative is that these variables are related.

If H_0 is true, the chi-square statistic X^2 has approximately a χ^2 distribution with $(r-1)(c-1)$ degrees of freedom.

The P-value for the chi-square test is

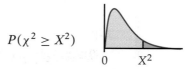

$$P(\chi^2 \geq X^2)$$

where χ^2 is a random variable having the $\chi^2(\mathrm{df})$ distribution with $\mathrm{df} = (r-1)(c-1)$.

The chi-square test always uses the upper tail of the χ^2 distribution, because any deviation from the null hypothesis makes the statistic larger. The approximation of the distribution of X^2 by χ^2 becomes more accurate as the cell counts increase. Moreover, it is more accurate for tables larger than 2×2 tables. For tables larger than 2×2, we will use this approximation whenever the average of the expected counts is 5 or more and the smallest expected count is 1 or more. For 2×2 tables, we require that all four expected cell counts be 5 or more.[2]

EXAMPLE

9.6 Chi-square significance test from software. The results of the chi-square significance test for the ongoing fright symptoms example appear in the computer outputs in Figures 9.1 and 9.3, labeled Chi-square and Pearson Chi-Square, respectively. Because all the expected cell counts are moderately large, the χ^2 distribution provides an accurate P-value. We see that $X^2 = 4.03$, $\mathrm{df} = 1$, and $P = 0.045$. As a check we verify that the degrees of freedom are correct for a 2×2 table:

$$\mathrm{df} = (r-1)(c-1) = (2-1)(2-1) = 1$$

The chi-square test confirms that the data contain clear evidence against the null hypothesis that there is no relationship between gender and ongoing fright symptoms. Under H_0, the chance of obtaining a value of X^2 greater than or equal to the calculated value of 4.03 is small—fewer than 5 times in 100.

The test does not provide insight into the nature of the relationship between the variables. It is up to us to see that the data show that women are more likely to have lingering fright symptoms. You should always accompany a chi-square test by percents such as those in Example 9.4 and Figure 9.2 and by a description of the nature of the relationship.

LOOK BACK
confounding p. 150

The observational study of Example 9.3 cannot tell us whether gender is a *cause* of lingering fright symptoms. The association may be explained by confounding with other variables. For example, other research has shown that there are gender differences in the social desirability of admitting fear.[3] *Our data don't allow us to investigate possible confounding variables.* Often a randomized comparative experiment can settle the issue of causation, but we cannot randomly assign gender to each student. The researcher who published the data of our example states merely that women are more likely to have lingering fright symptoms and that this conclusion is consistent with other studies.

The chi-square test and the *z* test

A comparison of the proportions of "successes" in two populations leads to a 2 × 2 table. We can compare two population proportions either by the chi-square test or by the two-sample *z* test from Section 8.2. In fact, *these tests always give exactly the same result,* because the X^2 statistic is equal to the square of the *z* statistic, and $\chi^2(1)$ critical values are equal to the squares of the corresponding $N(0, 1)$ critical values. The advantage of the *z* test is that we can test either one-sided or two-sided alternatives. The chi-square test always tests the two-sided alternative. Of course, the chi-square test can compare more than two populations, whereas the *z* test compares only two.

USE YOUR KNOWLEDGE

9.1 **Comparison of conditional distributions.** Consider the following 2 × 2 table.

	Observed counts		
	Explanatory variable		
Response variable	**1**	**2**	**Total**
Yes	70	90	160
No	130	110	240
Total	200	200	400

(a) Compute the conditional distribution of the response variable for each of the two explanatory-variable categories.

(b) Display the distributions graphically.

(c) Write a short paragraph describing the two distributions and how they differ.

9.2 **Expected cell counts and the chi-square test.** Refer to Exercise 9.1. You decide to use the chi-square test to compare these two conditional distributions.

(a) What is the expected count for the first cell (observed count is 70)?

(b) Computer software gives you $X^2 = 4.17$. What are the degrees of freedom for this statistic?

(c) Using Table F, give an appropriate bound on the *P*-value.

Meta-analysis

meta-analysis

Policymakers wanting to make decisions based on research are sometimes faced with the problem of summarizing the results of many studies. These studies may show effects of different magnitudes, some highly significant and some not significant. What *overall conclusion* can we draw? **Meta-analysis** is a collection of statistical techniques designed to combine information from different but similar studies. Each individual study must be examined with care to ensure that its design and data quality are adequate. The basic idea is to compute a measure of the effect of interest for each study. These are then combined, usually by taking some sort of weighted average, to produce a summary measure for all of the studies. Of course, a confidence interval for the summary is included in the results. Here is an example.

LOOK BACK
relative risk p. 500

EXAMPLE

9.7 Do we eat too much salt? Evidence from a variety of sources suggests that diets high in salt are associated with risks to human health. To investigate the relationship between salt intake and stroke, information from 14 studies were combined in a meta-analysis.[4] Subjects were classified based on the amount of salt in their normal diet. They were followed for several years and then classified according to whether or not they had developed cardiovascular disease (CVD). A total of 104,933 subjects were studied and 5161 of them developed CVD. Here are the data from one of the studies:[5]

	Low salt	High salt
CVD	88	112
No CVD	1081	1134
Total	1169	1246

Lets look at the relative risk for this study. We first find the proportion of subjects who develop CVD in each group. For the subjects with a low salt intake the proportion who develop CVD is

$$\frac{88}{1169} = 0.00753$$

or 7.5 per thousand; for the high salt group, the proportion is

$$\frac{112}{1246} = 0.00899$$

or 9.0 per thousand. We can now compute the relative risk as the ratio of these two proportions. We choose to put the high salt group in the numerator. The relative risk is

$$\frac{0.00899}{0.00753} = 1.19$$

Relative risk greater than 1 means that the high salt group developed more CVD than the low salt group.

When the data from all 14 studies were combined, the relative risk was reported as 1.17 with a 95% confidence interval of (1.02, 1.32). Since this interval does not include the value 1, corresponding to equal proportions in the two groups, we conclude that the higher CVD rate is statistically significant with $P < 0.05$. The high salt diet is associated with a 17% higher rate of CVD than the low salt diet.

USE YOUR KNOWLEDGE

9.3 In the previous example, we computed the relative risk using the high salt group in the numerator. Now, compute the relative risk using the low salt group in the numerator and then restate the last paragraph of the exercise with this change. (*Hint:* for the lower confidence limit, use 1 divided by the upper limit for the original ratio and do a similar calculation for the upper limit.)

SECTION 9.1 Summary

The **null hypothesis** for $r \times c$ tables of count data is that there is no relationship between the row variable and the column variable.

Expected cell counts under the null hypothesis are computed using the formula

$$\text{expected count} = \frac{\text{row total} \times \text{column total}}{n}$$

The null hypothesis is tested by the **chi-square statistic,** which compares the observed counts with the expected counts:

$$X^2 = \sum \frac{(\text{observed} - \text{expected})^2}{\text{expected}}$$

Under the null hypothesis, X^2 has approximately the χ^2 distribution with $(r - 1)(c - 1)$ degrees of freedom. The P-value for the test is

$$P(\chi^2 \geq X^2)$$

where χ^2 is a random variable having the $\chi^2(\text{df})$ distribution with df $= (r - 1)(c - 1)$.

The chi-square approximation is adequate for practical use when the average expected cell count is 5 or greater and all individual expected counts are 1 or greater, except in the case of 2×2 tables. All four expected counts in a 2×2 table should be 5 or greater.

The section we just completed assumed that you have access to software or a statistical calculator. If you do not, you now need to study the material on computations in the following optional section. All exercises appear at the end of the chapter.

9.2 Formulas and Models for Two-Way Tables*

Computations

The calculations required to analyze a two-way table are straightforward, but tedious. In practice, we recommend using software, but it is possible to do the work with a calculator, and some insight can be gained by examining the details. Here is an outline of the steps required.

> **COMPUTATIONS FOR TWO-WAY TABLES**
> 1. Calculate descriptive statistics that convey the important information in the table. Usually these will be column or row percents.
> 2. Find the expected counts and use these to compute the X^2 statistic.
> 3. Use chi-square critical values from Table F to find the approximate P-value.
> 4. Draw a conclusion about the association between the row and column variables.

The following examples illustrate these steps.

EXAMPLE

9.8 Health habits of college students. Physical activity generally declines when students leave high school and enroll in college. This suggests that college is an ideal setting to promote physical activity. One study examined the level of physical activity and other health-related behaviors in a sample of 1184 college students.[6] Let's look at the data for physical activity and consumption of fruits. We categorize physical activity as low, moderate, or vigorous and fruit consumption as low, medium, and high. Here is the two-way table that summarizes the data:

| | Physical Activity | | | |
Fruit consumption	Low	Moderate	Vigorous	Total
Low	69	206	294	569
Medium	25	126	170	321
High	14	111	169	294
Total	108	443	633	1184

This is a 3 × 3 table, to which we have added the marginal totals obtained by summing across rows and columns. For example, the first-row total is

---

*The analysis of two-way tables is based on computations that are a bit messy and on statistical models that require a fair amount of notation to describe. This section gives the details. By studying this material, you will deepen your understanding of the methods described in this chapter, but this section is optional.

$69 + 206 + 294 = 569$. The grand total, the number of students in the study, can be computed by summing the row totals, $569 + 321 + 294 = 1184$, or the column totals, $108 + 443 + 633 = 1184$. *It is easy to make an error in these calculations, so it is a good idea to do both as a check on your arithmetic.*

Computing conditional distributions

First, we summarize the observed relation between the physical activity and fruit consumption. We expect that there would be a positive association, but there is no clear distiction between an explanatory variable and a response variable in this setting. If we have such a distinction, then the clearest way to describe the relationship is to compare the conditional distributions of the response variable for each value of the explanatory variable. Otherwise, we can compute the condidional distribution each way and then decide which gives a better description of the data.

EXAMPLE

9.9 Health habits of college students: conditional distributions. Let's look at the data in the first column of the table in Example 9.8. There were 108 students with low physical activity. Of these, there were 69 with low fruit consumption. Therefore, the column proportion for this cell is

$$\frac{69}{108} = 0.639$$

That is, 63.9% of the low physical activity students had low fruit consumption. Similarly, 25 of the low physical activity students has moderate fruit consumption. This percent is 23.1%.

$$\frac{25}{108} = 0.231$$

In all, we calculate nine percents. Here are the results:

Column percents for fruit consumption and physical activity

| | Physical activity | | | |
Fruit consumption	Low	Moderate	Vigorous	Total
Low	63.9	46.5	46.4	48.1
Medium	23.1	28.4	26.9	27.1
High	13.0	25.1	26.7	24.8
Total	100.0	100.0	100.0	100.0

In addition to the conditional distributions of fruit consumption for each level of physical activity, the table also gives the marginal distribution of fruit consumption. These percents appear in the rightmost column, labeled "Total."

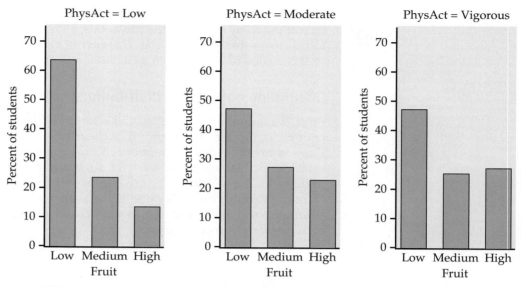

FIGURE 9.5 Comparison of the distribution of fruit consumption for different levels of physical activity, for Example 9.9.

The sum of the percents in each column should be 100, except for possible small roundoff errors. *It is good practice to calculate each percent separately and then sum each column as a check.* In this way we can find arithmetic errors that would not be uncovered if, for example, we calculated the column percent for the "High" row by subtracting the sum of the percents for "Low" and "Medium" from 100.

Figure 9.5 compares the distributions of fruit consumption for each of the three physical activity levels. For each activity level, the highest percent is for students who consume low amounts of fruit. For low physical activity, there is a clear decrease in the percent when moving from low to medium to high fruit consumption. The patterns for moderate physical activity and vigorous physical activity are similar. Low fruit consumption is still dominant, but the percents for medium and high fruit consumption are about the same. The percent of low fruit consumption is highest for the low physical activity students compared with those who have moderate or vigorous physical activity. These plots suggest that there is an association between these two variables.

Another way to look at these data is to examine the row percents. These compare the distribution of physical activity for each level of fruit consumption. Figure 9.6 gives these plots. The pattern in these plots is similar for the three levels of fruit consumption. The percents increase when we move from low physical activity to medium physical activity to high physical activity. The percent who exercise vigorously increases when we move from low fruit consumption to medium fruit consumption to high fruit consumption.

We observe a clear relationship between physical activity and fruit consumption in this study. The chi-square test assesses whether this observed association is statistically significant, that is, too strong to occur often just by chance. The test only confirms that there is some relationship. The percents we have compared describe the nature of the relationship. *The chi-square test does not in itself tell us what population our conclusion describes.* The subjects in this study were college students from four Midwestern universities. The researchers

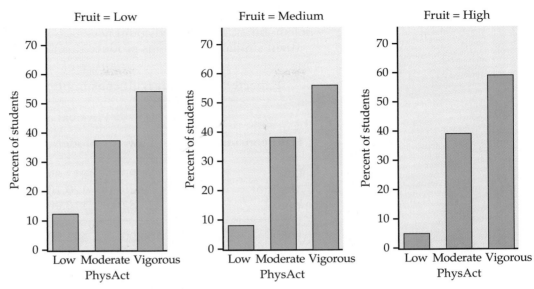

FIGURE 9.6 Comparison of the distribution of physical activity for different levels of fruit consumption, for Example 9.9.

could argue that these findings apply to college students in general. This type of inference is important but it is based on expert judgment and is beyond the scope of the statistical inference that we have been studying.

Computing expected cell counts

The null hypothesis is that there is no relationship between physical activity and fruit consumption. The alternative is that these two variables are related. Here is the formula for the expected cell counts under the hypothesis of "no relationship."

EXPECTED CELL COUNTS

$$\text{expected count} = \frac{\text{row total} \times \text{column total}}{n}$$

EXAMPLE

9.10 Health habits of college students: expected cell counts. What is the expected count in the upper-left cell in the table of Example 9.8, the number of students with low fruit consumption and low physical activity, under the null hypothesis that fruit consumption and physical activity are independent?

The column total, the number of students who have low physical activity is 108. The corresponding row total is 569, the number of students who have low fruit consumption. The total number of students is 1184. The expected cell count is therefore

$$\frac{(108)(569)}{1184} = 51.90$$

Note that although any observed count of the number of students must be a whole number, an expected count need not be.

Eight similar calculations produce this table of expected counts:

Expected counts for fruit consumption and physical activity

	Physical activity			
Fruit consumption	Low	Moderate	Vigorous	Total
Low	51.90	212.89	304.20	568.99
Medium	29.28	120.10	171.62	321.00
High	26.82	110.00	157.18	294.00
Total	108.00	442.99	633.00	1183.99

We can check our work by adding the expected counts to obtain the row and column totals, as in the table. These are the same as those in the table of observed counts except for small roundoff errors.

The X^2 statistic and its P-value

The expected counts are all large, so we proceed with the chi-square test. We compare the table of observed counts with the table of expected counts using the X^2 statistic.[7] We must calculate the term for each cell, then sum over all nine cells. For low fruit consumption and low physical activity, the observed number of students is 69 and the expected number of students is 51.90. The contribution to the X^2 statistic for this cell is therefore

$$\frac{(69 - 51.90)^2}{51.90} = 5.63$$

To find the X^2 statistic, we compute the contribution for each of the nine cells and add them. The result is

$$X^2 = 14.15$$

Because there are $r = 3$ levels of fruit consumption and $c = 3$ levels of physical activity, the degrees of freedom for this statistic are

$$\text{df} = (r - 1)(c - 1) = (3 - 1)(3 - 1) = 4$$

Under the null hypothesis that fruit consumption and physical activity are independent, the test statistic X^2 has a $\chi^2(4)$ distribution. To obtain the P-value, look at the df = 4 row in Table F.

The calculated value $X^2 = 14.15$ lies between the critical points for probabilities 0.01 and 0.005. The P-value is therefore between 0.01 and 0.005. (Software gives the value as 0.0068.) There is strong evidence ($X^2 = 14.15$, df = 4, $P < 0.01$) that there is a relationship between fruit consumption and physical activity.

The size and nature of the relationship between fruit consumption and physical activity are described by row and column percents. These are displayed in Figures 9.5 and 9.6.

df = 4

p	0.01	0.005
χ^2	13.28	14.86

USE YOUR KNOWLEDGE

9.4 Compute the chi-square statistic. Compute the nine terms that make up the chi-square statistic. Add them to verify that the statistic is 14.15. Are there any cells that have particularly large contributions? Explain how these cells influence the significance of the chi-square statistic.

Models for two-way tables

The chi-square test for the presence of a relationship between the two directions in a two-way table is valid for data produced from several different study designs. The precise statement of the null hypothesis "no relationship" in terms of population parameters is different for different designs. We now describe two of these settings in detail. *An essential requirement is that each experimental unit or subject is counted only once in the data table.*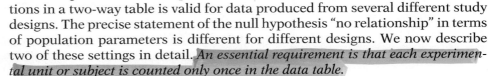

Comparing several populations: the first model   Let's think about the setting of Example 9.8 from a slightly different perspective. Suppose we are interested in the relationship between physical activity and year of study in college. We will assume that the design called for independent SRSs of students from each of the four years. Here we have an example of *separate and independent random samples* from each of c populations. The c columns of the two-way table represent the populations. There is a single categorical response variable, year of study. The r rows of the table correspond to the values of the response variable, physical activity.

We know that the z test for comparing the two proportions of successes and the chi-square test for the 2×2 table are equivalent. The $r \times c$ table allows us to compare more than two populations or more than two categories of response, or both. In this setting, the null hypothesis "no relationship between column variable and row variable" becomes

H_0: The distribution of the response variable is the same in all c populations.

Because the response variable is categorical, its distribution just consists of the probabilities of its r possible values. The null hypothesis says that these probabilities (or population proportions) are the same in all c populations.

EXAMPLE

9.11 Physical activity: comparing subpopulations based on year of study. In our scenario based on Example 9.8, we compare four populations:

Population 1: physical activity of first-year students
Population 2: physical activity of second-year students
Population 3: physical activity of third-year students
Population 4: physical activity of fourth-year students

The null hypothesis for the chi-square test is

H_0: The distribution of physical activity is the same in all four populations.

The parameters of the model are the proportions of low, moderate, and vigorous physical activity in each of the four years of study.

More generally, if we take an independent SRS from each of c populations and classify each outcome into one of r categories, we have an $r \times c$ table of population proportions. There are c different sets of proportions to be compared. There are c groups of subjects, and a single categorical variable with r possible values is measured for each individual.

MODEL FOR COMPARING SEVERAL POPULATIONS USING TWO-WAY TABLES

Select independent SRSs from each of c populations, of sizes $n_1, n_2, \ldots,$ n_c. Classify each individual in a sample according to a categorical response variable with r possible values. There are c different probability distributions, one for each population.

The null hypothesis is that the distributions of the response variable are the same in all c populations. The alternative hypothesis says that these c distributions are not all the same.

Testing independence: the second model A second model for which our analysis of $r \times c$ tables is valid is illustrated by the ongoing fright symptoms study, Example 9.3. There, a *single* sample from a *single* population was classified according to two categorical variables.

EXAMPLE

9.12 Ongoing fright symptoms and gender: testing independence. The single population studied is college students. Each college student was classified according to the following categorical variables: "Ongoing fright symptoms," with possible responses "Yes" and "No," and "Gender," with possible responses "Men" and "Women." The null hypothesis for the chi-square test is

H_0: "Ongoing fright symptoms" and "Gender" are independent.

The parameters of the model are the probabilities for each of the four possible combinations of values of the row and column variables. If the null hypothesis is true, the multiplication rule for independent events says that these can be found as the products of outcome probabilities for each variable alone.

LOOK BACK
multiplication rule p. 282

More generally, take an SRS from a single population and record the values of two categorical variables, one with r possible values and the other with c possible values. The data are summarized by recording the numbers of individuals for each possible combination of outcomes for the two random variables. This gives an $r \times c$ table of counts. Each of these $r \times c$ possible outcomes has its own probability. The probabilities give the joint distribution of the two categorical variables.

LOOK BACK
joint distribution p. 138

Each of the two categorical random variables has a distribution. These are the marginal distributions because they are the sums of the population proportions in the rows and columns.

LOOK BACK
marginal distributions p. 139

The null hypothesis "no relationship" now states that the row and column variables are independent. The multiplication rule for independent events tells us that the joint probabilities are the products of the marginal probabilities.

EXAMPLE

9.13 The joint distribution and the two marginal distributions. The joint probability distribution gives a probability for each of the four cells in our 2×2 table of "Ongoing fright symptoms" and "Gender." The marginal distribution for "Ongoing fright symptoms" gives probabilities for each of the two possible categories; the marginal distribution for "Gender" gives probabilities for each of the two possible gender categories.

Independence between "Ongoing fright symptoms" and "Gender" implies that the joint distribution can be obtained by multiplying the appropriate terms from the two marginal distributions. For example, the probability that a randomly chosen college student has ongoing fright symptoms *and* is male is equal to the probability that the student has ongoing symptoms *times* the probability that the student is male. The hypothesis that "Ongoing fright symptoms" and "Gender" are independent says that the multiplication rule applies to *all* outcomes.

MODEL FOR EXAMINING INDEPENDENCE IN TWO-WAY TABLES

Select an SRS of size n from a population. Measure two categorical variables for each individual.

The null hypothesis is that the row and column variables are independent. The alternative hypothesis is that the row and column variables are dependent.

Concluding remarks

You can distinguish between the two models by examining the design of the study. In the independence model, there is a single sample. The column totals and row totals are random variables. The total sample size n is set by the researcher; the column and row sums are known only after the data are collected. For the comparison-of-populations model, on the other hand, there is a sample from each of two or more populations. The column sums are the sample sizes selected at the design phase of the research. The null hypothesis in both models says that there is no relationship between the column variable and the row variable. The precise statement of the hypothesis differs, depending on the sampling design. Fortunately, *the test of the hypothesis of "no relationship" is the same for both models*; it is the chi-square test. There are yet other statistical models for two-way tables that justify the chi-square test of the null hypothesis "no relation," made precise in ways suitable for these models. Statistical methods related to the chi-square test also allow the analysis of three-way and higher-way tables of count data. You can find a discussion of these topics in advanced texts on categorical data.[8]

USE YOUR KNOWLEDGE

9.5 Find the P-value. For each of the following give the degrees of freedom and an appropriate bound on the P-value for the X^2 statistic.

(a) $X^2 = 5.32$ for a 2 by 2 table

(b) $X^2 = 2.7$ for a 2 by 2 table

(c) $X^2 = 25.2$ for a 4 by 5 table

(d) $X^2 = 25.2$ for a 5 by 4 table

9.6 **Frequent binge drinking and gender: the chi-square test.** Refer to Example 9.2 (page 512). Use the chi-square test to assess if frequent binge drinking is associated with gender. State your conclusion.

SECTION 9.2 Summary

For two-way tables we first compute percents or proportions that describe the relationship of interest. Then, we compute expected counts, the X^2 statistic, and the P-value.

Two different models for generating $r \times c$ tables lead to the chi-square test. In the first model, independent SRSs are drawn from each of c populations, and each observation is classified according to a categorical variable with r possible values. The null hypothesis is that the distributions of the row categorical variable are the same for all c populations. In the second model, a single SRS is drawn from a population, and observations are classified according to two categorical variables having r and c possible values. In this model, H_0 states that the row and column variables are independent.

9.3 Goodness of Fit*

In the last two sections, we discussed the use of the chi-square test to compare categorical-variable distributions of c populations. We now consider a slight variation on this scenario where we compare a sample from one population with a hypothesized distribution. Here is an example that illustrates the basic ideas.

EXAMPLE

9.14 Sampling in the Adequate Calcium Today (ACT) study. The ACT study was designed to examine relationships among bone growth patterns, bone development, and calcium intake. Participants were over 14,000 adolescents from six states, Arizona (AZ), California (CA), Hawaii (HI), Indiana, (IN), Nevada (NV), and Ohio (OH). After the major goals of the study were completed, the investigators decided to do an additional analysis of the written comments made by the participants during the study. Because the number of participants was so large, a sampling plan was devised to select sheets containing the written comments of approximately 10% of the participants. A systematic sample (see page 200) and every tenth comment sheet was retrieved from each storage container for analysis.[9] Here are the counts for each of the six states:

---

*This section is optional.

Number of study participants in the samples						
AZ	CA	HI	IN	NV	OH	Total
167	257	257	297	107	482	1567

There were 1567 study participants in the sample. We will use the proportions of students from each of the states in the original sample as the population values.[10] Here are the proportions:

Population proportions						
AZ	CA	HI	IN	NV	OH	Total
0.105	0.172	0.164	0.188	0.070	0.301	100.000

Let's see how well our sample reflects the state population proportions. We start by computing expected counts. Since 10.5% of the population is from Arizona, we expect the sample to have about 10.5% from Arizona. Therefore, since the sample has 1567 subjects, our expected count for Arizona is

$$\text{expected count for Arizona} = 0.105(1567) = 164.535$$

Here are the expected counts for all six states:

Expected counts						
AZ	CA	HI	IN	NV	OH	Total
164.54	269.52	256.99	294.60	109.69	471.67	1567.01

USE YOUR KNOWLEDGE

9.7 Why is the sum 1567.01? Refer to the table of expected counts in Example 9.14. Explain why the sum of the expected counts is 1567.01 and not 1567.

9.8 Calculate the expected counts. Refer to Example 9.14. Find the expected counts for the other five states. Report your results with three places after the decimal as we did for Arizona.

As we saw with the expected counts in the analysis of two-way tables in the previous section of this chapter, we do not really expect the observed counts to be *exactly* equal to the expected counts. Different samples under the same conditions would give different counts. We expect the average of these counts to be equal to the expected counts when the null hypothesis is true. How close do we think the counts and the expected counts should be?

We can think of our table of observed counts in Example 9.14 as a one-way table with six cells, each with a count of the number subjects sampled from a particular state. Our question of interest is translated into a null hypothesis that the observed proportions of students in the six states can be viewed as random samples from the subjects in the ACT study. The alternative is that the process generating the observed counts, a form of a systematic sampling in this case, does not provide samples that are compatible with this hypothesis. In other words, the alternative means that there is some bias in the way that we selected the subjects for which we will examine the written comments.

Our analysis of these data is very similar to the analyses of two-way tables that we studied in Section 9.1. We have already computed the expected counts. We now construct a chi-square statistic that measures how far the observed counts are from the expected counts. Here is a summary of the procedure:

THE CHI-SQUARE GOODNESS OF FIT TEST

Data for n observations of a categorical variable with k possible outcomes are summarized as observed counts, $n_1, n_2, \ldots, n_k$ in k cells. A null hypothesis specifies probabilities $p_1, p_2, \ldots, p_k$ for the possible outcomes.

For each cell, multiply the total number of observations n by the specified probability to determine the expected counts:

$$\text{expected count} = np_i$$

The **chi-square statistic** measures how much the observed cell counts differ from the expected cell counts. The formula for the statistic is

$$X^2 = \sum \frac{(\text{observed count} - \text{expected count})^2}{\text{expected count}}$$

The degrees of freedom are $k - 1$, and P-values are computed from the chi-square distribution.

EXAMPLE

9.15 The goodness of fit test for the ACT study. For Arizona, the observed count is 167. In Example 9.14, we calculated the expected count, 164.535. The contribution to the chi-square statistic for Arizona is

$$\frac{(\text{observed count} - \text{expected count})^2}{\text{expected count}} = \frac{(167 - 164.535)^2}{164.535} = 0.0369$$

We use the same approach to find the contributions to the chi-square for the other five states. The sum of these six values is the chi-square statistic,

$$X^2 = 0.93$$

The degrees of freedom are the number of cells minus 1, df $= 6 - 1 = 5$. We calculate the P-value using Table F or software. From Table F, we can determine $P > 0.25$. We conclude that the observed counts are compatible with the hypothesized proportions. The data do not provide any evidence that our systematic sample was biased with respect to selection of subjects from different states.

Software output from Minitab and SPSS for this problem is given in Figure 9.7. Both report the P-value as 0.968. Note that the SPSS output

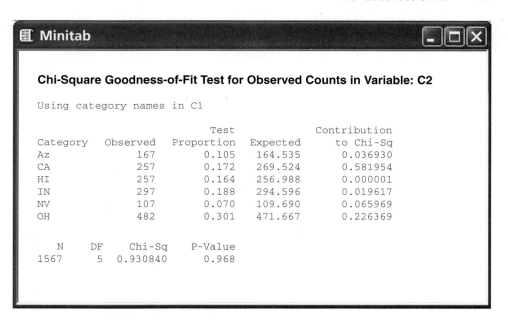

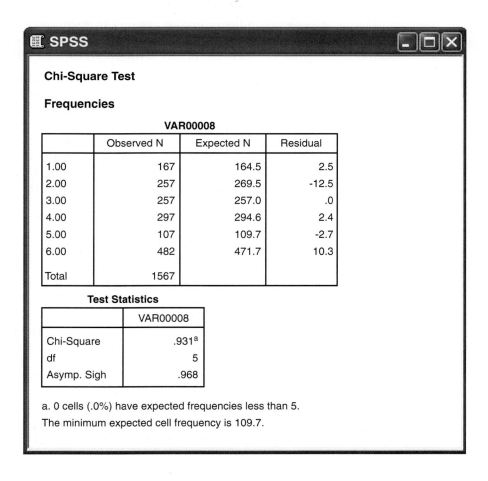

FIGURE 9.7 Minitab (a) and SPSS (b) output for Example 9.15.

includes a column titled "Residual." For tables of counts, a residual for a cell is defined as

$$\text{residual} = \frac{\text{observed count} - \text{expected count}}{\sqrt{\text{expected count}}}$$

Note that the chi-square statistic is the sum of the squares of these residuals.

Some software packages do not provide routines for computing the chi-square goodness of fit test. However, there is a very simple trick that can be used to produce the results from software that can analyze two-way tables. Make a two-way table where the first column contains k cells with the observed counts. Make a second column with counts that correspond *exactly* to the probabilities specified by the null hypothesis, with a very large number of observations.

USE YOUR KNOWLEDGE

9.9 **Compute the chi-square statistic.** For each of the other five states, compute the contribution to the chi-square statistic using the method illustrated for Arizona in Example 9.15. Use the expected counts that you calculated in Exercise 9.8 for these calculations. Show that the sum of these values is the chi-square statistic.

9.10 **Distribution of M&M colors.** M&M Mars Company has varied the mix of colors for M&M's Plain Chocolate Candies over the years. These changes in color blends are the result of consumer preference tests. Most recently, the color distribution is reported to be 13% brown, 14% yellow, 13% red, 20% orange, 24% blue, and 16% green.[11] You open up a 14-ounce bag of M&M's and find 61 brown, 59 yellow, 49 red, 77 orange, 141 blue, and 88 green. Use a goodness of fit test to examine how well this bag fits the percents stated by the M&M Mars Company.

SECTION 9.3 Summary

The **chi-square goodness of fit test** is used to compare the sample distribution of a categorical variable from a population with a hypothesized distribution. The data for n observations with k possible outcomes are summarized as observed counts, $n_1, n_2, \ldots, n_k$ in k cells. The **null hypothesis** specifies probabilities $p_1, p_2, \ldots, p_k$ for the possible outcomes.

The analysis of these data is similar to the analyses of two-way tables discussed in Section 9.1. For each cell, the **expected count** is determined by multiplying the total number of observations n by the specified probability p_i. The null hypothesis is tested by the usual **chi-square statistic**, which compares the observed counts, n_i, with the expected counts. Under the null hypothesis, X^2 has approximately the χ^2 distribution with df $= k - 1$.

CHAPTER 9   Exercises

For Exercises 9.1 and 9.2, see page 519; for Exercise 9.3, see page 521; for Exercise 9.4, see page 527; for Exercises 9.5 and 9.6, see pages 529–530; for Exercises 9.7 and 9.8, see page 531; and for Exercises 9.9 and 9.10, see page 534.

9.11 Trends in broadband use over time. The Pew Internet and American Life Project collects data about the impact of the Internet on various aspects of American life.[12] One set of surveys has tracked the use of broadband use in homes over a period of several years.[13] Here are some data on the percent of homes that address the Internet using broadband:

Date of survey	April 2001	April 2004	March 2007	April 2008
Homes with broadband	5%	24%	48%	55%

Assume a sample size of 2250 for each survey.

(a) Display the data in a two-way table of counts.

(b) Test the null hypothesis that the proportion of homes that access the Internet using broadband has not changed over this period of time. Report your test statistic with degrees of freedom and the *P*-value. What do you conclude?

(c) Now analyze the data from March 2007 and April 2008 only. Summarize your results and compare them with what you found in part (b).

9.12 What about dialup? Refer to the previous exercise. The same surveys provided data on access to the Internet using dialup. Here are the data:

Date of survey	April 2001	April 2004	March 2007	April 2008
Homes with dialup	41%	30%	16%	12%

(a) to (c) Answer the questions given in the previous exercise for these data.

(d) Write a short report summarizing the changes in broadband access that has occurred over this period of time using your analysis from this exercise and the previous one. Include a graph with information about both broadband and dialup access over time in your report.

9.13 How robust are the conclusions? Refer to Exercise 9.11 on the use of broadband to access the Internet. In that exercise, the percents were read from a graph and the sample size for the April 2008 survey was reported as 2251 in the report. We assumed that the sample size was 2250 for all surveys. Investigate the robustness of the conclusions that you drew in Exercise 9.11 to the use of 2250 as the sample size for all surveys and to roundoff and slight errors in reading the graphs. Assume that the actual sample sizes ranged from 2200 to 2600. Assume also that the percents reported are all accurate to within plus or minus 2%. In other words, if the reported percent is 16% then we can assume that the actual survey percent is between 14% and 18%. Reanalyze the data using at least five scenarios that vary the percents and the sample sizes within the assumed ranges. To what extent do the conclusions vary across your scenarios? Summarize your results in a report.

9.14 Switching from dialup to broadband. The Pew 2008 survey asked dialup users about their interest in switching to broadband.[14] Here is the summary given in the report.

Date of survey	October 2002	February 2004	December 2005	May 2008
Interested in switch	38%	40%	39%	36%

(a) The percents reported in these tables were computed by dividing the number of respondents who expressed an interest in switching from dialup to broadband by the number of respondents who said that they used dialup. Use the data given in Exercise 9.12 to estimate the numbers of dialup users in each of these surveys. Note that the dates do not match. Explain carefully how you obtain your estimates.

(b) Using the results from part (a), construct the 2 × 4 table of counts for this exercise.

(c) Analyze the table and write a summary of what you have found.

(d) The report states,

The roughly 60% of dial-up users consistently saying that they are not interested in broadband, in the face of the shrinking pool of dial-up users, suggests that the preferences of dial-up users change over time. That is, assuming that many of those interested in getting broadband switched over to it from December 2005 to May 2008, some of those who said that they were not interested in broadband replenished the supply of "yes, interested in broadband" responses in order to maintain the 40–60 ratio of those interested in broadband versus not interested in dial-up.

Approximately what percent of dialup users who said they were not interested in broadband in December 2005 would need to change their mind and become interested in broadband for this statement to be correct? Explain how you determined your answer.

9.15 Health care fraud. Most errors in billing insurance providers for health care services involve honest mistakes by patients, physicians, or others involved in the health care system. However, fraud is a serious problem. The National Health Care Anti-Fraud Association estimates that approximately $68 billion is lost to health care fraud each year.[15] When fraud is suspected, an audit of randomly selected billings is often conducted. The selected claims are then reviewed by experts and then each claim is classified as allowed or not allowed. The distributions of the amounts of claims are frequently highly skewed with a large number of small claims and a small number of large claims. Since simple random sampling would likely be overwhelmed by small claims and would tend to miss the large claims, stratification is often used. See the section on stratified sampling in Chapter 3 (page 193). Here are data from an audit that used three strata based on the sizes of the claims (small, medium, and large):[16]   🌀 BILLINGERRORS

Stratum	Sampled claims	Number not allowed
Small	57	6
Medium	17	5
Large	5	1

(a) Construct the 3 × 2 table of counts for these data and include the marginal totals.

(b) Find the percents of claims that were not allowed in each of the three strata.

(c) To calculate a significance test, combine the medium and large strata. Explain why we do this.

(d) State an appropriate null hypothesis to be tested for these data.

(e) Perform the significance test and report your test statistic with degrees of freedom, and the P-value. State your conclusion.

9.16 Population estimates. Refer to the previous exercise. One reason to do an audit such as this is to estimate the number of claims that would not be allowed if all claims in a population were examined by experts. We have estimates of the proportions of claims from each strata based on our sample. With our simple random sampling of claims from each stratum, we have unbiased estimates of the corresponding population proportions for each stratum. Therefore, if we take the sample proportions and multiply by the population sizes, we would have the estimates that we need. Here are the population sizes for the three strata:

Stratum	Claims in strata
Small	3342
Medium	246
Large	58

(a) For each strata, estimate the total number of claims that would be not allowed if all claims in the strata had been audited.

(b) (*Optional*) Give margins of error for your estimates. (*Hint:* you first need to find standard errors for your sample estimates using material in Chapter 8 (page 475). Then you need to use the rules for variances in Chapter 4 (page 271) to find the standard errors for the population estimates. Finally, you need to multiply by z^* to determine the margins of error.)

9.17 DFW rates. One measure of student success for colleges and universities is the percent of admitted students who graduate. Studies indicate that a key issue in retaining students is their performance in so-called gateway courses.[17] These are courses that serve as prerequisites for other key courses that are essential for student success. One measure of student performance in these courses is the DFW rate, the percent of students who receive grades of D, F, or W (withdraw). A major project was undertaken to improve the DFW rate in a gateway course at a large Midwestern university. The course curriculum was revised to make it more relevant to the majors of the students taking the course, a small group of excellent teachers taught the course, technology including clickers and online homework was introduced, and there was an increase in student support outside of the classroom. The following table gives data on the DFW rates for the course in three years.[18] In Year 1, the traditional course was given; in Year 2, a few changes were introduced; and in Year 3, the course was substantially revised.

Year	DFW rate	Number of students taking course
Year 1	42.3%	2408
Year 2	24.9%	2325
Year 3	19.9%	2126

Do you think that the changes in this gateway course had an impact on the DFW rate? Write a report giving your answer to this question. Support your answer by an analysis of the data.

9.18 Class attendance and DFW rates. One study that looked at DFW rates surveyed 719 students who were enrolled in one or more of seven gateway courses in business, mathematics, and science.[19] If a student was enrolled in more than one course, then a single course was randomly selected for analysis. In the survey, students were asked how often they attended the gateway class. Here are the data:

Class attendance	ABC percent	DFW percent
Less than 50%	2%	5%
51% to 74%	8%	14%
75% to 94%	24%	30%
95% or more	66%	51%
Total	100%	100%

In this table students are classified as earning an A, B, or C in the gateway course or earning a D. F, or withdrawing from the course, Notice that the data are given in terms of the marginal distributions of class attendance for each (ABC or DFW) group. In the survey, there were 539 students in the ABC group and 180 students in the DFW group.

(a) Use the data given to construct the 4 × 2 table of counts and add the marginal totals to the table. Sum the row totals and the marginal totals separately to verify that you have the correct total sample size, 719.

(b) Test the null hypothesis that there is no association between class attendance and a DFW grade. Give the test statistic with degrees of freedom, and the *P*-value.

(c) Summarize your conclusion in a short paragraph.

(d) Can you conclude that the data indicate that not going to class causes a student to get a bad grade? Can you conclude that the data are consistent with this scenario?

9.19 When do Canadian students enter private career colleges? A survey of 13,363 Canadian students who enrolled in private career colleges was conducted to understand student participation in the private postsecondary educational system.[20] In one part of the survey, students were asked about their field of study and about when they entered college. Here are the results:

Field of study	Number of students	Time of entry	
		Right after high school	Later
Trades	942	34%	66%
Design	584	47%	53%
Health	5085	40%	60%
Media/IT	3148	31%	69%
Service	1350	36%	64%
Other	2255	52%	48%

In this table, the second column gives the number of students in each field of study. The next two columns give the marginal distribution of time of entry for each field of study.

(a) Use the data provided to make the 6 × 2 table of counts for this problem.

(b) Analyze the data.

(c) Write a summary of your conclusions. Be sure to include the results of your significance testing as well as a graphical summary.

9.20 Government loans for Canadian students in private career colleges? Refer to the previous exercise. The survey also asked about how these college students paid for their education. A major source of funding was government loans. Here are the survey percents of Canadian private students who use government loans to finance their education by field of study:

Field of study	Number of students	Percent using government loans
Trades	942	45%
Design	599	53%
Health	5234	55%
Media/IT	3238	55%
Service	1378	60%
Other	2300	47%

(a) Construct the 6 × 2 table of counts for this exercise.

(b) Test the null hypothesis that the percent of students using government loans to finance their education does not vary with field of study. Be sure to provide all of details for your significance test.

(c) Summarize your analysis and conclusions. Be sure to include a graphical summary.

9.21 Other funding for Canadian students in private career colleges? Refer to the previous exercise. Another major source of funding was parents, family, or spouse. The following table gives the survey percents of Canadian private students who rely on these sources to finance their education by field of study.

Field of Study	Number of students	Percent using parents/family/spouse
Trades	942	20%
Design	599	37%
Health	5234	26%
Media/IT	3238	16%
Service	1378	18%
Other	2300	41%

Answer the questions in the previous exercise for these data.

9.22 Remote deposit capture. The Federal Reserve has called remote deposit capture (RDC) "the most important development the (U.S.) banking industry has seen in years." This service allows users to scan checks and to transmit the scanned images to a bank for posting.[21] In its annual survey of community banks, the American Bankers Association asked banks whether or not they offered this service.[22] Here are the results classified by the asset size (in millions of dollars) of the bank: 🔵 RDCASSET

Asset size	Offer RDC	
	Yes	No
Under $100	63	309
$101–200	59	132
$201 or more	112	85

(a) Summarize the results of this survey question numerically and graphically.

(b) Test the null hypothesis that there is no association between the size of a bank, measured by assets, and whether or not they offer RDC. Report the test statistic, the P-value, and your conclusion.

9.23 Exercise and adequate sleep. A survey of 656 boys and girls who were 13 to 18 years old asked about adequate sleep and other health-related behaviors. The recommended amount of sleep is six to eight hours per night.[23] In the survey 54% of the respondents reported that they got less than this amount of sleep on school nights. An exercise scale was developed and was used to classify the students as above or below the median in this domain. Here is the 2×2 table of counts with students classified as getting or not getting adequate sleep and by the exercise variable: 🔵 SLEEP

Enough sleep	Exercise	
	High	Low
Yes	151	115
No	148	242

Note that you answered parts (a) through (c) of this exercise if you completed Exercise 2.121 (page 146).

(a) Find the distribution of adequate sleep for the high exercisers.

(b) Do the same for the low exercisers.

(c) Summarize the relationship between adequate sleep and exercise using the results of parts (a) and (b).

(d) Perform the significance test for examining the relationship between exercise and getting enough sleep. Give the test statistic, the P-value (with a sketch similar to the one on page 518) and summarize your conclusion. Be sure to include numerical and graphical summaries.

9.24 Lying to a teacher. One of the questions in a survey of high school students asked about lying to teachers.[24] The following table gives the numbers of students who said that they lied to a teacher at least once during the past year, classified by gender. 🔵 LYING

Lied at least once	Gender	
	Male	Female
Yes	3,228	10,295
No	9,659	4,620

(a) Add the marginal totals to the table.

(b) Calculate appropriate percents to describe the results of this question.

(c) Summarize your findings in a short paragraph.

(d) Test the null hypothesis that there is no association between gender and lying to teachers. Give the test statistic, the P-value and summarize your conclusion. Be sure to include numerical and graphical summaries.

9.25 Trust and honesty in the workplace. The students surveyed in the study described in the previous exercise were also asked whether they thought trust and honesty were essential in business and the workplace. Here are the counts classified by gender: 🔵 TRUST

Trust and honesty are essential	Gender	
	Male	Female
Agree	11,724	14,169
Disagree	1,163	746

Answer questions given in the previous exercise for this question.

9.26 Why not use a chi-square test? As part of the study on ongoing fright symptoms due to exposure to horror movies at a young age, the following table was created based on the written responses from 119 students.[25] Explain why a chi-square test is not appropriate for this table.

	Percent of students who reported each problem			
	Type of Problem			
	Bedtime		Waking	
Movie or video	Short term	Enduring	Short term	Enduring
Poltergeist ($n = 29$)	68	7	64	32
Jaws ($n = 23$)	39	4	83	43
Nightmare on Elm Street ($n = 16$)	69	13	37	31
Thriller (music video) ($n = 16$)	40	0	27	7
It ($n = 24$)	64	0	64	50
The Wizard of Oz ($n = 12$)	75	17	50	8
E.T. ($n = 11$)	55	0	64	27

9.27 Age and time status of U.S. college students. The Census Bureau provides estimates of numbers of people in the United States classified in various ways.[26] Let's look at college students. The following table gives us data to examine the relation between age and full-time or part-time status. The numbers in the table are expressed as thousands of U.S. college students.   USCOLLEGE1

U.S. college students by age and status: October 2004

Age	Status	
	Full-time	**Part-time**
15–19	3553	329
20–24	5710	1215
25–34	1825	1864
35 and over	901	1983

(a) Give the joint distribution of age and status for this table.

(b) What is the marginal distribution of age? Display the results graphically.

(c) What is the marginal distribution of status? Display the results graphically.

(d) Compute the conditional distribution of age for each of the two status categories. Display the results graphically.

(e) Write a short paragraph describing the distributions and how they differ.

9.28 Time status versus gender for the 20–24 age category. Refer to Exercise 9.27. The table below breaks down the 20–24 age category by gender.   USCOLLEGE2

Status	Gender		Total
	Male	**Female**	
Full-time	2719	2991	5710
Part-time	535	680	1215
Total	3254	3671	6925

(a) Compute the marginal distribution for gender. Display the results graphically.

(b) Compute the conditional distribution of status for males and for females. Display the results graphically and comment on how these distributions differ.

(c) If you wanted to test the null hypothesis that there is no difference between these two conditional distributions, what would the expected cell counts be for the full-time status row of the table?

(d) Computer software gives $X^2 = 5.17$. Using Table F, give an appropriate bound for the P-value and state your conclusions at the 5% level.

9.29 Waking versus bedtime symptoms. As part of the study on ongoing fright symptoms due to exposure to horror movies at a young age, the following table was presented to describe the lasting impact these movies have had during bedtime and waking life: FRIGHTSYMPTOMS

Bedtime symptoms	Waking symptoms	
	Yes	**No**
Yes	36	33
No	33	17

(a) What percent of the students have lasting waking-life symptoms?

(b) What percent of the students have both waking-life and bedtime symptoms?

(c) Test whether there is an association between waking-life and bedtime symptoms. State the null and alternative hypotheses, the X^2 statistic, and the P-value.

9.30 Find the degrees of freedom and P-value. For each of the following situations give the degrees of freedom and an appropriate bound on the P-value (give the exact value if you have software available) for the X^2 statistic for testing the null hypothesis of no association between the row and column variables.

(a) A 2 by 2 table with $X^2 = 1.25$.

(b) A 4 by 4 table with $X^2 = 18.34$.

(c) A 2 by 8 table with $X^2 = 24.21$.

(d) A 5 by 3 table with $X^2 = 12.17$.

9.31 Can you construct the joint distribution from the marginal distributions? Here are the row and column totals for a two-way table with two rows and two columns:

a	b	50
c	d	150
100	100	200

Find *two different* sets of counts a, b, c, and d for the body of the table. This demonstrates that the relationship between two variables cannot be obtained solely from the two marginal distributions of the variables.

9.32 Construct a table with no association. Construct a 3×2 table of counts where there is no apparent association between the row and column variables.

9.33   Gender versus motivation for volunteer service. A study examined patterns and characteristics of volunteer service for young people from high school through early adulthood.[27] Here are some data that can be used to compare males and females on participation in unpaid volunteer service or community service and motivation for participation: VOLUNTEER

| | Participants | | | |
| | Motivation | | | |
Gender	Strictly voluntary	Court-ordered	Other	Non-participants
Men	31.9%	2.1%	6.3%	59.7%
Women	43.7%	1.1%	6.5%	48.7%

Note that the percents in each row sum to 100%.

(a) Graphically compare the volunteer service profiles for men and women. Describe any differences that are striking.

(b) Find the proportion of men who volunteer. Do the same for women. Refer to the section on relative risk in Chapter 8 (page 500) and the discussion on page 520 of this chapter. Compute the relative risk of being a volunteer for females versus males. Write a clear sentence contrasting females and males using relative risk as your numerical summary.

9.34 **Gender versus motivation for volunteer service, continued.** Refer to the previous exercise. Recompute the table for volunteers only. To do this, take the entries for each motivation and divide by the percent of volunteers. Do this separately for each gender. Verify that the percents sum to 100% for each gender. Give a graphical summary to compare the motivation of men and women who are volunteers. Compare this with your summary in part (a) of the previous exercise, and write a short paragraph describing similarities and differences in these two views of the data.

9.35 Dieting trends among male and female undergraduates. A recent study of undergraduates looked at gender differences in dieting trends.[28] There were 181 women and 105 men who participated in the survey. The following table summarizes whether a student tried a low-fat diet or not by gender:

| | Gender | |
Tried low-fat diet	Women	Men
Yes	35	8
No		

(a) Fill in the missing cells of the table.

(b) Summarize the data numerically and graphically.

(c) Test that there is no association between gender and the likelihood of trying a low-fat diet. Summarize the results.

9.36 Sexual imagery in magazine ads. In what ways do advertisers in magazines use sexual imagery to appeal to youth? One study classified each of 1509 full-page or larger ads as "not sexual" or "sexual," according to the amount and style of the dress of the male or female model in the ad. The ads were also classified according to the target readership of the magazine.[29] Here is the two-way table of counts: MAGAZINEADS

| | Magazine readership | | | |
Model dress	Women	Men	General interest	Total
Not sexual	351	514	248	1113
Sexual	225	105	66	396
Total	576	619	314	1509

(a) Summarize the data numerically and graphically.

(b) Perform the significance test that compares the model dress for the three categories of magazine readership. Summarize the results of your test and give your conclusion.

(c) All the ads were taken from the March, July, and November issues of six magazines in one year. Discuss this fact from the viewpoint of the validity of the significance test and the interpretation of the results.

9.37 Intended readership of ads with sexual imagery. The ads in the study described in the previous exercise were also classified according to the age group of the intended readership. Here is a summary of the data:

| | Magazine readership age group | |
Model dress	Young adult	Mature adult
Not sexual	72.3%	76.1%
Sexual	27.2%	23.9%
Number of ads	1006	503

Using parts (a) and (b) in the previous exercise as a guide, analyze these data and write a report summarizing your work.

9.38 Identity theft. A study of identity theft looked at how well consumers protect themselves from this increasingly prevalent crime. The behaviors of 61 college students were compared with the behaviors of 59 nonstudents.[30] One of the questions was "When asked to create a password, I have used either my mother's maiden name, or my pet's name, or my birth date, or the last four digits of my social security number, or a series of consecutive numbers." For the students, 22 agreed with this statement while 30 of the nonstudents disagreed.

(a) Display the data in a two-way table and perform the chi-square test. Summarize the results.

(b) Reanalyze the data using the methods for comparing two proportions that we studied in the previous chapter. Compare the results and verify that the chi-square statistic is the square of the z statistic.

(c) The students in this study were junior and senior college students from two sections of a course in Internet marketing at a large northeastern university. The nonstudents were a group of individuals who were recruited to attend commercial focus groups on the West Coast conducted by a lifestyle marketing organization. Discuss how the method of selecting the subjects in this study relates to the conclusions that can be drawn from it.

9.39 Student-athletes and gambling. A survey of student-athletes that asked questions about gambling behavior classified students according to the National Collegiate Athletic Association (NCAA) division.[31] For male student-athletes, the percents who reported wagering on collegiate sports are given here along with the numbers of respondents in each division:

Division	I	II	III
Percent	17.2%	21.0%	24.4%
Number	5619	2957	4089

(a) Use a significance test to compare the percents for the three NCAA divisions. Give details and a short summary of your conclusion.

(b) The percents in the preceding table are given in the NCAA report, but the numbers of male student-athletes in each division who responded to the survey question are estimated based on other information in the report. To what extent do you think this has an effect on the results? (*Hint:* Rerun your analysis a few times, with slightly different numbers of students but the same percents.)

(c) Some student-athletes may be reluctant to provide this kind of information, even in a survey where there is no possibility that they can be identified. Discuss how this fact may affect your conclusions.

(d) The chi-square test for this set of data assumes that the responses of the student-athletes are independent. However, some of the students are at the same school and even on the same team. Discuss how you think this might affect the results.

9.40 Which model? Refer to Exercises 9.15, 9.23, 9.24, and 9.39. For each, state whether you are comparing two or more populations (the first model for two-way tables) or testing independence between two categorical variables (the second model).

9.41 Hummingbirds of Saint Lucia. *E. jugularis* is a type of hummingbird that lives in the forest preserves of the Caribbean island of Saint Lucia. The males and the females of this species have bills that are shaped somewhat differently. Researchers who study these birds thought that the bill shape might be related to the shape of the flowers that they visit for food. The researchers observed 49 females and 21 males. Of the females, 20 visited the flowers of *H. bihai*, while none of the males visited these flowers.[32] Display the data in a two-way table and perform the chi-square test. Summarize the results and give a brief statement of your conclusion. Your two-way table has a count of zero in one cell. Does this invalidate your significance test? Explain why or why not.

9.42 Internet references in prominent journals. The World Wide Web (WWW) has led to an enormous increase in the amount of information that is easily available to anyone with Internet access. References to Internet pages are becoming quite common in the scientific literature. One study examined Internet references in articles in three prominent journals: the *New England Journal of Medicine* (*NEJM*), the *Journal of the American Medical Association* (*JAMA*), and *Science*.[33] In one part of the study, Internet references were classified according to the top-level domain. Here are the data: INTERNETREFS

	Journal		
Top-level domain	*NEJM*	*JAMA*	*Science*
.gov	41	103	111
.org	37	46	162
.com	6	17	14
.edu	4	8	47
Other	9	15	52

Analyze the data. Include numerical and graphical summaries as well as a significance test. Summarize your results and conclusions.

9.43 Pet ownership and education level. The Health, Aging, and Body Composition (Health ABC) study is a 10-year study of older adults. A research project based on this study examined the relationship between physical activity and pet ownership.[34] The data collected included information concerning pet owner characteristics and the type of pet owned. Here is a table of counts of subjects classified by pet ownership status and education level: PETSANDED

Education level	Pet ownership status		
	Non–pet owners	Dog owners	Cat owners
Less than high school	421	93	28
High school graduate	666	100	40
Postsecondary	845	135	99

Note that "Dog owners" and "Cat owners" designate individuals who own a dog only or a cat only, respectively. Individuals who own both a dog and a cat are not included in this table. Analyze the data. Include numerical and graphical summaries as well as a significance test. Summarize your results and conclusions.

9.44 Pet ownership and gender. Refer to the previous exercise. Here are similar data giving the relationship between pet ownership status and gender: PETSANDGENDER

Gender	Pet ownership status		
	Non–pet owners	Dog owners	Cat owners
Female	1024	157	85
Male	915	171	82

Analyze the data. Include numerical and graphical summaries as well as a significance test. Summarize your results and conclusions.

9.45 Changing majors. A task force set up to examine retention of students in the majors that they chose when starting college examined data on transfers to other majors.[35] Here are some data giving counts of students classified by initial major and the area that they transferred to:

Initial major	Area transferred to				
	Engineering	Management	Liberal arts	Other	Total
Biology	13	25	158		398
Chemistry	16	15	19		114
Mathematics	3	11	20		72
Physics	9	5	14		61

Complete the table by computing the values for the "Other" column. Write a short paragraph explaining what conclusions you can draw about the relationship between initial major and area transferred to. Be sure to include numerical and graphical summaries as well as the details of your significance test.

9.46 🛡 **Cracks in veneer.** Many furniture pieces are built with veneer, a thin layer of fine wood that is fastened to less expensive wood products underneath. Face checks are cracks that sometimes develop in the veneer. When face checks appear, the furniture needs to be reconstructed. Because this is a fairly expensive process, researchers seek ways to minimize the occurrence of face checks by controlling the manufacturing process. In one study, the type of adhesive used was one of the factors examined.[36] Because of the way that the veneer is cut, it has two different sides, called loose and tight, either of which can face out. Here is a table giving the numbers of veneer panels with and without face checks for two different adhesives, PVA and UF. Separate columns are given for the loose side and the tight side. VENEER

Adhesive	Loose side		Tight side	
	Face checks		Face checks	
	No	Yes	No	Yes
PVA	10	54	44	20
UF	21	43	37	27

Analyze the data. Write a summary of your results concerning the relationship between the adhesive and the occurrence of face checks. Be sure to include numerical and graphical summaries as well as the details of your significance tests.

9.47 Are Mexican Americans less likely to be selected as jurors? Refer to Exercise 8.99 (page 509) concerning *Castaneda v. Partida*, the case where the Supreme Court review used the phrase "two or three standard deviations" as a criterion for statistical significance. Recall that there were 181,535 persons eligible for jury duty, of whom 143,611 were Mexican Americans. Of the 870 people selected for jury duty, 339 were Mexican Americans. We are interested in finding out if there is an association between being a Mexican American and being selected as a juror. Formulate this problem using a two-way table of counts. Construct the 2×2 table using the variables Mexican American or not and juror or not. Find the X^2 statistic and its P-value. Square the z statistic that you obtained in Exercise 8.81 and verify that the result is equal to the X^2 statistic.

9.48 🛡 **Evaluation of an herbal remedy.** A study designed to evaluate the effects of the herbal remedy *Echinacea purpurea* randomly assigned healthy children who were 2 to 11 years old to receive either *echinacea* or a placebo.[37] Each time a child had an upper respiratory infection (URI), treatment with *echinacea* or the placebo was given for the duration of the URI. The dose for the *echinacea* was based on the age of the child according to the recommendation of the manufacturer. The *echinacea* children had 337 URIs, while the placebo children had 370 URIs. For each URI many variables were measured.

One of these was the parental assessment of the illness severity. Here are the data: 🐦 HERBALASSESS

	Group	
Parental assessment	*Echinacea*	**Placebo**
Mild	153	170
Moderate	128	157
Severe	48	40

They also recorded the presence or absence of various types of *adverse events*. Here is a summary: 🐦 HERBALEVENTS

	Group	
Adverse event	*Echinacea*	**Placebo**
Itchiness	13	7
Rash	24	10
"Hyper" behavior	30	23
Diarrhea	38	34
Vomiting	22	21
Headache	33	24
Stomachache	52	41
Drowsiness	63	48
Other	63	48
Any adverse event	152	146

(a) Analyze the parental assessment data. Write a summary of your analysis and conclusion. Be sure to include graphical and numerical summaries.

(b) Analyze each adverse event. Display the results graphically in a single graph. Make a table of the relevant descriptive statistics.

(c) Use a statistical significance test to compare the *echinacea* URIs with those of the placebo URIs for each type of adverse event. Summarize the results in a table and write a short report giving your conclusions about the effect of *echinacea* on URIs in healthy children who are 2 to 11 years old. Explain why you need to analyze each type of adverse event separately rather than performing a chi-square test on the preceding 10 × 2 table.

(d) One concern about analyzing several outcome variables in situations like this is that we may be able to find statistical significance by chance if we look at a sufficiently large number of outcome variables. Explain why this is a concern in general but is not a concern that is important for the interpretation of the results of your analysis here.

(e) The authors of the paper describing these results note that the unit of analysis for their computations is the URI and not the child. They state that similar results were found using more sophisticated statistical methods. Based on the descriptive statistics that you have computed, are you inclined to agree or to disagree with this statement of the authors? Explain your answer.

(f) This study was published in the *Journal of the American Medical Association* (*JAMA*) and was criticized in an article that appeared in *Alternative & Complementary Therapies*.[38] Three herbalists gave responses to the original article. Among their criticisms were (*i*) the dose of *echinacea* was too low; (*ii*) the treatment should have been given before the URI, not at the onset of symptoms; (*iii*) we should be skeptical of any positive trials on pharmaceuticals or negative trials on natural remedies that are published in *JAMA*; and (*iv*) *Echinacea angustifolia* (not *E. purpurea*) should have been used in combination with other herbs. Discuss these criticisms and write a summary of your opinions regarding *echinacea*.

The following exercises concern the optional material on goodness of fit discussed in Section 9.3.

9.49 Use of academic assistance services. The 2005 National Survey of Student Engagement reported on the use of campus services during the first year of college.[39] In terms of academic assistance (for example tutoring, writing lab), 43% never used the services, 35% sometimes used the services, 15% often used the services, and 7% very often used the services. You decide to see if your large university has this same distribution. You survey first-year students and obtain the counts 79, 83, 36, and 12, respectively. Use a goodness of fit test to examine how well your university reflects the national average.

9.50 Goodness of fit to a standard Normal distribution. Computer software generated 500 random numbers that should look as if they are from the standard Normal distribution. They are categorized into five groups: (1) less than or equal to −0.6; (2) greater than −0.6 and less than or equal to −0.1; (3) greater than −0.1 and less than or equal to 0.1; (4) greater than 0.1 and less than or equal to 0.6; and (5) greater than 0.6. The counts in the five groups are 139, 102, 41, 78, and 140, respectively. Find the probabilities for these five intervals using Table A. Then compute the expected number for each interval for a sample of 500. Finally, perform the goodness of fit test and summarize your results.

9.51 More on the goodness of fit to a standard Normal distribution. Refer to the previous exercise. Use software to generate your own sample of 500 standard Normal random variables, and perform the goodness of fit test. Choose a different set of intervals from the ones used in the previous exercise.

9.52 Goodness of fit to the uniform distribution. Computer software generated 500 random numbers that should look as if they are from the uniform distribution on the interval 0 to 1 (see page 70). They are categorized into five groups: (1) less than or equal to 0.2; (2) greater than 0.2 and less than or equal to 0.4; (3) greater than 0.4 and less than or equal to 0.6; (4) greater than 0.6 and less than or equal to 0.8; and (5) greater than 0.8. The counts in the five groups are 114, 92, 108, 101, and 85, respectively. The probabilities for these five intervals are all the same. What is this probability? Compute the expected number for each interval for a sample of 500. Finally, perform the goodness of fit test and summarize your results.

9.53 More on goodness of fit to the uniform distribution. Refer to the previous exercise. Use software to generate your own sample of 500 uniform random variables on the interval from 0 to 1, and perform the goodness of fit test. Choose a different set of intervals from the ones used in the previous exercise.

9.54 ⟁ **Suspicious results?** An instructor who assigned an exercise similar to the one described in the previous exercise received homework from a student who reported a *P*-value of 0.999. The instructor suspected that the student did not use the computer for the assignment but just made up some numbers for the homework. Why was the instructor suspicious? How would this scenario change if there were 1000 students in the class?

9.55 Is there a random distribution of trees? In Example 6.1 (page 342) we examined data concerning the longleaf pine trees in the Wade Tract and concluded that the distribution of trees in the tract was not random. Here is another way to examine the same question. First, we divide the tract into four equal parts, or quadrants, in the east-west direction. Call the four parts, Q_1 to Q_4. Then we take a random sample of 100 trees and count the number of trees in each quadrant. Here are the data:

Quadrant	Q_1	Q_2	Q_3	Q_4
Count	18	22	39	21

(a) If the trees are randomly distributed, we expect to find 25 trees in each quadrant. Why? Explain your answer.

(b) We do not really expect to get *exactly* 25 trees in each quadrant. Why? Explain your answer.

(c) Perform the goodness of fit test for these data to determine if these trees are randomly scattered. Write a short report giving the details of your analysis and your conclusion.

Inference for Regression

Introduction

In this chapter we describe methods for inference when there is a single quantitative response variable and a single quantitative explanatory variable. The descriptive tools we learned in Chapter 2—scatterplots, least-squares regression, and correlation—are essential preliminaries to inference and also provide a foundation for confidence intervals and significance tests.

We first met the sample mean $\bar{x}$ in Chapter 1 as a measure of the center of a collection of observations. Later we learned that when the data are a random sample from a population, the sample mean is an estimate of the population mean μ. In Chapters 6 and 7, we used $\bar{x}$ as the basis for confidence intervals and significance tests for inference about μ.

Now we will follow the same approach for the problem of fitting straight lines to data. In Chapter 2 we met the least-squares regression line $\hat{y} = b_0 + b_1 x$ as a description of a straight-line relationship between a response variable y

and an explanatory variable x. At that point we did not distinguish between sample and population. Now we will think of the least-squares line computed from a sample as an estimate of a *true* regression line for the population.

Following the common practice of using Greek letters for population parameters, we will write the population line as $\beta_0 + \beta_1 x$. This notation reminds us that the intercept b_0 of the fitted line estimates the intercept β_0 of the population line, and the slope b_1 estimates the slope β_1.

The methods detailed in this chapter will help us answers questions such as

- Is the trend in the annual number of tornadoes reported in the United States linear? If so, what is the average yearly increase in the number of tornadoes? How many are predicted for next year?

- What is the relationship between a college student's body mass index and physical activity level measured by a pedometer?

- Among North American universities, is there a strong correlation between the binge-drinking rate and the average price for a bottle of beer at establishments within a 2-mile radius of campus?

10.1 Simple Linear Regression

Statistical model for linear regression

Simple linear regression studies the relationship between a response variable y and a single explanatory variable x. We expect that different values of x will produce different mean responses. We encountered a similar but simpler situation in Chapter 7 when we discussed methods for comparing two population means. Figure 10.1 illustrates the statistical model for a comparison of blood pressure change in two groups of experimental subjects, one group taking a calcium supplement and the other a placebo. We can think of the treatment (placebo or calcium) as the explanatory variable in this example. This model has two important parts:

- The mean change in blood pressure may be different in the two populations. These means are labeled μ_1 and μ_2 in Figure 10.1.

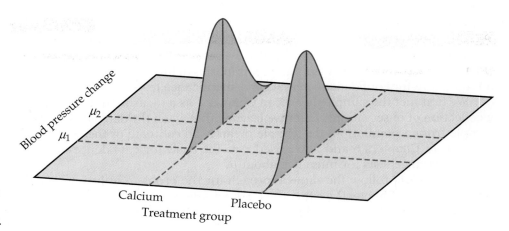

FIGURE 10.1 The statistical model for comparing responses to two treatments; the mean response varies with the treatment.

- Individual changes vary within each population according to a Normal distribution. The two Normal curves in Figure 10.1 describe these responses. These Normal distributions have the same spread, indicating that the population standard deviations are assumed to be equal.

In linear regression the explanatory variable x is quantitative and can have many different values. Imagine, for example, giving different amounts x of calcium to different groups of subjects. We can think of the values of x as defining different **subpopulations,** one for each possible value of x. Each subpopulation consists of all individuals in the population having the same value of x. If we conducted an experiment with five different amounts of calcium, we could view these values as defining five different subpopulations.

subpopulations

The statistical model for simple linear regression also assumes that for each value of x the observed values of the response variable y are Normally distributed with a mean that depends on x. We use μ_y to represent these means. In general, the means μ_y can change as x changes according to any sort of pattern. In **simple linear regression** we assume the means all lie on a line when plotted against x. To summarize, this model also has two important parts:

simple linear regression

- The mean of the response variable y changes as x changes. The means all lie on a straight line. That is, $\mu_y = \beta_0 + \beta_1 x$.

- Individual responses of y with the same x vary according to a Normal distribution. This variation, measured by the standard deviation σ, is the same for all values of x.

This statistical model is pictured in Figure 10.2. The line describes how the mean response μ_y changes with x. This is the **population regression line.** The three Normal curves show how the response y will vary for three different values of the explanatory variable x.

population regression line

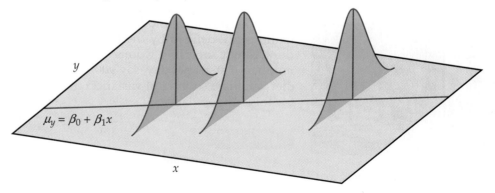

FIGURE 10.2 The statistical model for linear regression; the mean response is a straight-line function of the explanatory variable.

Data for simple linear regression

The data for a linear regression are observed values of y and x. The model takes each x to be a known quantity. In practice, x may not be exactly known. *If the error in measuring x is large, more advanced inference methods are needed.* The response y to a given x is a random variable. The linear regression model describes the mean and standard deviation of this random variable y. These unknown parameters must be estimated from the data.

We will use the following example to explain the fundamentals of simple linear regression. Because regression calculations in practice are always done by statistical software, we will rely on computer output for the arithmetic. In the next section, we give an example that illustrates how to do the work with a calculator if software is unavailable.

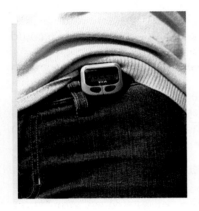

EXAMPLE

10.1 Relationship between physical activity and BMI. Decrease in physical activity is considered to be a major contributor to the increase in prevalence of overweight and obesity in the general adult population. Because the prevalence of physical inactivity among college students is similar to the adult population, many researchers feel a clearer understanding of college students' physical activity behaviors is needed to develop early interventions. As part of a recent study, researchers looked at the relationship between physical activity (PA) measured with a pedometer and body mass index (BMI).[1] Each participant wore the pedometer for a week and the average steps per day (in thousands) was recorded. Various body composition variables, including BMI (kg/m^2) were also measured. For this example, we focus on a sample of 100 female undergraduates.

Before starting our analysis, it is appropriate to consider the extent to which the results can reasonably be generalized. In the original study, undergraduate volunteers were obtained at a large public southeastern university through classroom announcements and campus flyers. *The potential for bias should always be considered when obtaining volunteers.* In this case, the participants were screened and those with severe health issues as well as varsity athletes were excluded. As a result, the researchers considered these volunteers as an SRS from the population of undergraduates at this university. They limited the extent of their scope, stating that similar investigations at universities of different sizes and in other climates of the United States are needed.

In the statistical model for predicting BMI from physical activity, subpopulations are defined by the explanatory variable, physical activity. We could think about sampling women from this university, each averaging the same number of steps per day, say 9000. Variation in genetic make-up, lifestyle, and diet would be sources of variation that would result in different values of BMI for this subpopulation.

LOOK BACK
scatterplot p. 84

EXAMPLE

10.2 Graphical display of the BMI and physical activity. We start our analysis with a graphical display of the data. Figure 10.3 is a plot of BMI versus physical activity for our sample of 100 observations. We use the variable names BMI and PA. The least-squares regression line is also shown in the plot. There is a negative association between BMI and PA that appears approximately linear. There is also a considerable amount of scatter about this least-squares regression line.

Always start with a graphical display of the data. There is no point in trying to do statistical inference if the data do not, at least approximately, meet the assumptions that are the foundation for our inference. Now that we have confirmed an approximate linear relationship, we return to predicting BMI for different subpopulations, defined by the explanatory variable physical activity.

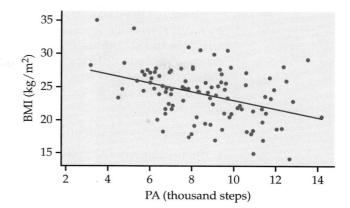

FIGURE 10.3 Scatterplot of BMI versus physical activity (PA) with the least-squares line, for Example 10.2.

Our statistical model assumes that the BMI values are Normally distributed with a mean μ_y that depends upon x in a linear way. Specifically,

$$\mu_y = \beta_0 + \beta_1 x$$

This population regression line gives the average BMI for all values of x. We cannot observe this line because the observed responses y vary about their means. The statistical model for linear regression consists of the population regression line and a description of the variation of y about the line. This was displayed in Figure 10.2 with the line and the three Normal curves. The following equation expresses this idea in an equation:

$$DATA = FIT + RESIDUAL$$

The FIT part of the model consists of the subpopulation means, given by the expression $\beta_0 + \beta_1 x$. The RESIDUAL part represents deviations of the data from the line of population means. We assume that these deviations are Normally distributed with standard deviation σ. We use ϵ (the lowercase Greek letter epsilon) to stand for the RESIDUAL part of the statistical model. A response y is the sum of its mean and a chance deviation ϵ from the mean. The deviations ϵ represent "noise," that is, variation in y due to other causes that prevent the observed (x, y)-values from forming a perfectly straight line on the scatterplot.

SIMPLE LINEAR REGRESSION MODEL

Given n observations of the explanatory variable x and the response variable y,

$$(x_1, y_1), \quad (x_2, y_2), \quad \ldots, \quad (x_n, y_n)$$

the **statistical model for simple linear regression** states that the observed response y_i when the explanatory variable takes the value x_i is

$$y_i = \beta_0 + \beta_1 x_i + \epsilon_i$$

Here $\beta_0 + \beta_1 x_i$ is the mean response when $x = x_i$. The deviations ϵ_i are assumed to be independent and Normally distributed with mean 0 and standard deviation σ.

The parameters of the model are β_0, β_1, and σ.

Because the means μ_y lie on the line $\mu_y = \beta_0 + \beta_1 x$, they are all determined by β_0 and β_1. Once we have estimates of β_0 and β_1, the linear relationship determines the estimates of μ_y for all values of x. Linear regression allows us to do inference not only for subpopulations for which we have data but also for those corresponding to x's not present in the data.

Given the simple linear regression model, we will now learn how to do inference about

- the slope β_1 and the intercept β_0 of the population regression line,

- the mean response μ_y for a given value of x, and

- an individual future response y for a given value of x.

Estimating the regression parameters

LOOK BACK
least-squares regression
p. 112

The method of least squares presented in Chapter 2 fits a line to summarize a relationship between the observed values of an explanatory variable and a response variable. Now we want to use the least-squares line as a basis for inference about a population from which our observations are a sample. We can do this only when the statistical model just presented holds. In that setting, the slope b_1 and intercept b_0 of the least-squares line

$$\hat{y} = b_0 + b_1 x$$

estimate the slope β_1 and the intercept β_0 of the population regression line.

Using the formulas from Chapter 2 (page 113), the slope of the least-squares line is

$$b_1 = r\frac{s_y}{s_x}$$

and the intercept is

$$b_0 = \bar{y} - b_1\bar{x}$$

LOOK BACK
correlation
p. 102

Here, r is the correlation between y and x, s_y is the standard deviation of y, and s_x is the standard deviation of x. Notice that if the slope is 0, so is the correlation, and vice versa. We will discuss this relationship more later in the chapter.

As a consequence of the model assumptions on the deviations ϵ_i, b_0 and b_1 are Normally distributed with means β_0 and β_1 and standard deviations that can be estimated from the data. In fact, even if the ϵ_i are not Normally distributed, a general form of the central limit theorem tells us that the distributions of b_0 and b_1 will be approximately Normal. Thus, our procedures for inference about the population regression line will be similar in flavor to those described in Chapter 7 for the population mean and difference in means. Moreover, we need to watch out for outliers and influential observations as they can invalidate the results of inference for regression.

LOOK BACK
central limit theorem
p. 303

LOOK BACK
outliers and influential
observations
p. 125

residual

The predicted value of y for a given value x^* of x is the point on the least-squares line $\hat{y} = b_0 + b_1 x^*$. This is an unbiased estimator of the mean response μ_y when $x = x^*$. The residual is

$$
\begin{aligned}
e_i &= \text{observed response} - \text{predicted response} \\
&= y_i - \hat{y}_i \\
&= y_i - b_0 - b_1 x_i
\end{aligned}
$$

The residuals e_i correspond to the model deviations ϵ_i. The e_i sum to 0, and the ϵ_i come from a population with mean 0.

The remaining parameter to be estimated is σ, which measures the variation of y about the population regression line. Because this parameter is the standard deviation of the model deviations, it should come as no surprise that we use the residuals to estimate it. As usual, we work first with the variance and take the square root to obtain the standard deviation. For simple linear regression, the estimate of σ^2 is the average squared residual

$$s^2 = \frac{\sum e_i^2}{n-2}$$

$$= \frac{\sum (y_i - \hat{y}_i)^2}{n-2}$$

LOOK BACK
sample variance
p. 39

degrees of freedom

We average by dividing the sum by $n-2$ in order to make s^2 an unbiased estimate of σ^2. The sample variance of n observations uses the divisor $n-1$ for this same reason. The quantity $n-2$ is called the **degrees of freedom** for s^2. The estimate of σ is given by

$$s = \sqrt{s^2}$$

We will use statistical software to calculate the regression for predicting BMI with physical activity for Example 10.1. In entering the data, we chose the names PA for the explanatory variable and BMI for the response. *It is good practice to use names, rather than just x and y, to remind yourself which data the output describes.*

> ### EXAMPLE
>
> **10.3 Statistical software output for BMI and physical activity.** Figure 10.4 gives the outputs for three commonly used statistical software packages and Excel. Other software will give similar information. The SPSS output reports estimates of our three parameters as $b_0 = 29.578$, $b_1 = -0.655$, and $s = 3.6549$. Be sure that you can find these entries in this output and the corresponding values in the other outputs.

The least-squares regression line is the straight line that is plotted in Figure 10.3. We would report it as

$$\widehat{\text{BMI}} = 29.578 - 0.655\text{PA}$$

with a model standard deviation of $s = 3.65$. Note that the number of digits provided varies with the software used and we have rounded off the values to three significant digits. *It is important to avoid cluttering up your report of the results of a statistical analysis with many digits that are not relevant.* Software often reports many more digits than are meaningful or useful.

The outputs contain other information that we will ignore for now. Computer outputs often give more information than we want or need. This is done to reduce user frustration when a software package does not print out the particular statistics wanted for an analysis. *The experienced user of statistical software learns to ignore the parts of the output that are not needed for the current problem.*

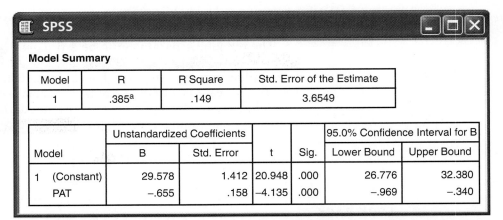

SPSS

Model Summary

Model	R	R Square	Std. Error of the Estimate
1	.385[a]	.149	3.6549

Model		Unstandardized Coefficients		t	Sig.	95.0% Confidence Interval for B	
		B	Std. Error			Lower Bound	Upper Bound
1	(Constant)	29.578	1.412	20.948	.000	26.776	32.380
	PAT	−.655	.158	−4.135	.000	−.969	−.340

Minitab

```
The regression equation is
BMI = 29.6 - 0.655 PA

Predictor        Coef      SE Coef         T         P
Constant       29.578        1.412     20.95     0.000
PA            -0.6547       0.1583     -4.13     0.000
```

Excel

	A	B	C	D	E	F	G
1	SUMMARY OUTPUT						
2							
3	*Regression Statistics*						
4	Multiple R	0.385409059					
5	R Square	0.148540143					
6	Adjusted R Square	0.139851777					
7	Standard Error	3.65488311					
8	Observations	100					
9							
10	ANOVA						
11		*df*	*SS*	*MS*	*F*	*Significance F*	
12	Regression	1	228.3771867	228.377	17.09644	7.50303E-05	
13	Residual	98	1309.100713	13.3582			
14	Total	99	1537.4779				
15							
16		*Coefficients*	*Standard Error*	*t Stat*	*P-value*	*Lower 95%*	*Upper 95%*
17	Intercept	29.57824714	1.411978287	20.9481	5.71E-38	28.77622225	32.38027203
18	X Variable 1	−0.65468576	0.158336132	−4.1347	7.5E-05	−0.96889865	−0.34047287

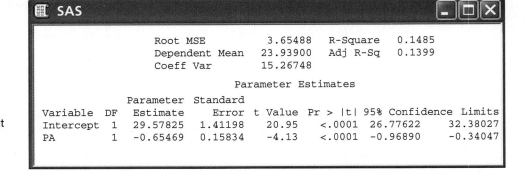

SAS

```
               Root MSE           3.65488   R-Square    0.1485
               Dependent Mean    23.93900   Adj R-Sq    0.1399
               Coeff Var         15.26748

                         Parameter Estimates

                  Parameter   Standard
Variable  DF      Estimate      Error   t Value  Pr > |t|  95% Confidence Limits
Intercept  1      29.57825    1.41198    20.95   <.0001    26.77622      32.38027
PA         1      -0.65469    0.15834    -4.13   <.0001    -0.96890      -0.34047
```

FIGURE 10.4 Regression output from SPSS, Minitab, Excel, and SAS for the physical activity example.

Now that we have fitted a line, we should examine the residuals for Normality and any remaining patterns in the data. We usually plot the residuals both against the case number (especially if this reflects the order in which the observations were collected) and against the explanatory variable. For this example, we will just look at the explanatory variable.

LOOK BACK
scatterplot smoothers
p. 92

Figure 10.5 gives a plot of the residuals versus physical activity with a smooth-function fit. The smooth function suggests that the residuals increase slightly at both low and high physical activity levels. This could mean that a curved relationship between BMI and physical activity would better fit the data. It also could just be chance variation. Notice that there is a large positive residual at each end of physical activity range. Since the effect does not appear to be particularly large, we will ignore this for the present analysis and investigate this further in Exercise 11.17 (page 614).

Finally, Figure 10.6 is a Normal quantile plot of the residuals. Because the plot looks fairly straight, we are confident that we do not have a serious violation of our assumption that the residuals are Normally distributed.

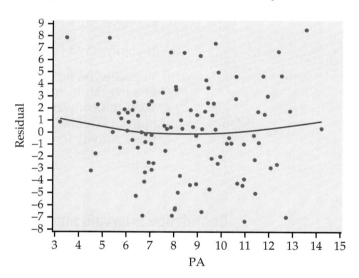

FIGURE 10.5 Plot of residuals versus physical activity (PA) with a smooth function, for the physical activity example.

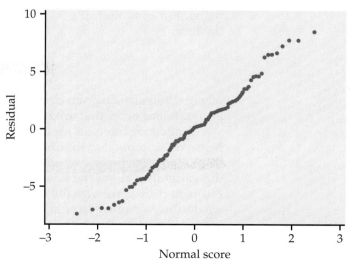

FIGURE 10.6 Normal quantile plot of the residuals for the physical activity example.

USE YOUR KNOWLEDGE

10.1 **Understanding a linear regression model.** Consider a linear regression model with $\mu_y = 43.4 + 2.8x$ and standard deviation $\sigma = 4.3$.

(a) What is the slope of the population regression line?

(b) Explain clearly what this slope says about the change in the mean of y for a change in x.

(c) What is the subpopulation mean when $x = 7$?

(d) Between what 2 values would approximately 95% of the observed responses, y, fall when $x = 7$?

10.2 **More on physical activity and BMI.** Refer to Example 10.3.

(a) What is the average BMI for a woman who averages 9500 steps per day?

(b) If an observed BMI at $x = 9.5$ were 22.8, what is the residual?

(c) Suppose you wanted to use the estimated population regression line to examine the average BMI for a woman who averages 2000, 7000, 12000, and 17000 steps per day. Discuss the appropriateness of using the equation to predict BMI for each of these activity levels.

Confidence intervals and significance tests

Chapter 7 presented confidence intervals and significance tests for means and differences in means. In each case, inference rested on the standard errors of estimates and on t distributions. Inference for the intercept and slope in a linear regression is similar in principle. For example, the confidence intervals have the form

$$\text{estimate} \pm t^* \text{SE}_{\text{estimate}}$$

where t^* is a critical point of a t distribution. It is the formulas for the estimate and standard error that are more complicated.

Confidence intervals and tests for the slope and intercept are based on the Normal sampling distributions of the estimates b_1 and b_0. The standard deviations of these estimates are multiples of σ, the model parameter that describes the variability about the true regression line. Because we do not know σ, we estimate it by s, the variability of the data about the least-squares line. When we do this, we move from the Normal distribution to t distributions with degrees of freedom $n - 2$, the degrees of freedom of s. We give formulas for the standard errors SE_{b_1} and SE_{b_0} in Section 10.2. For now we will concentrate on the basic ideas and let the computer do the computations.

CONFIDENCE INTERVALS AND SIGNIFICANCE TESTS FOR REGRESSION SLOPE AND INTERCEPT

A **level C confidence interval for the intercept β_0** is

$$b_0 \pm t^* \text{SE}_{b_0}$$

A **level C confidence interval for the slope β_1** is

$$b_1 \pm t^* \text{SE}_{b_1}$$

In these expressions t^* is the value for the $t(n-2)$ density curve with area C between $-t^*$ and t^*.

To test the hypothesis $H_0: \beta_1 = 0$, compute the **test statistic**

$$t = \frac{b_1}{\text{SE}_{b_1}}$$

The **degrees of freedom** are $n-2$. In terms of a random variable T having the $t(n-2)$ distribution, the P-value for a test of H_0 against

$H_a: \beta_1 > 0$ is $P(T \geq t)$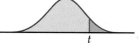

$H_a: \beta_1 < 0$ is $P(T \leq t)$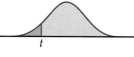

$H_a: \beta_1 \neq 0$ is $2P(T \geq |t|)$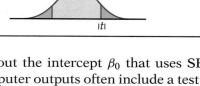

There is a similar significance test about the intercept β_0 that uses SE_{b_0} and the $t(n-2)$ distribution. Although computer outputs often include a test of $H_0: \beta_0 = 0$, this information usually has little practical value. From the equation for the population regression line, $\mu_y = \beta_0 + \beta_1 x$, we see that β_0 is the mean response corresponding to $x = 0$. In many practical situations, this subpopulation does not exist or is not interesting.

On the other hand, the test of $H_0: \beta_1 = 0$ is quite useful. When we substitute $\beta_1 = 0$ in the model, the x term drops out and we are left with

$$\mu_y = \beta_0$$

This model says that the mean of y does *not* vary with x. In other words, all the y's come from a single population with mean β_0, which we would estimate by $\bar{y}$. The hypothesis $H_0: \beta_1 = 0$ therefore says that there is *no straight-line relationship between y and x* and that linear regression of y on x is of no value for predicting y.

EXAMPLE

10.4 Statistical software output, continued. The computer outputs in Figure 10.4 for the BMI problem contain the information needed for inference about the regression slope and intercept. Let's look at the SPSS output.

The column labeled Std. Error gives the standard errors of the estimates. The value of SE_{b_1} appears on the line labeled with the variable name for the explanatory variable, PA. It is given as 0.158. In a summary we would report that the regression coefficient for the average steps per day is -0.655 with a standard error of 0.158.

The t statistic and P-value for the test of H_0: $\beta_1 = 0$ against the two-sided alternative H_a: $\beta_1 \neq 0$ appear in the columns labeled t and Sig. We can verify the t calculation from the formula for the standardized estimate:

$$t = \frac{b_1}{SE_{b_1}} = \frac{-0.655}{0.158} = -4.14$$

The P-value is given as 0.000. This is a rounded number and from that information we can conclude that $P < 0.0005$. The other outputs in Figure 10.5 also indicate that the P-value is very small. We will report the result as $P < 0.001$ because 1 chance in 1000 is sufficiently small for us to decisively reject the null hypothesis.

We have found a statistically significant linear relationship between physical activity and BMI. The estimated slope is more than 4 standard deviations away from zero. Because this is highly unlikely to happen if the true slope is zero, we have strong evidence for our claim. Note, however, that this is not the same as concluding that we have found a strong relationship between the response and explanatory variables in this example. We saw in Figure 10.3 that there was a lot of scatter about the regression line. *A very small P-value for the significance test for a zero slope does not necessarily imply that we have found a strong relationship.* A confidence interval will provide additional information about the relationship.

EXAMPLE

10.5 Confidence interval for the slope. A confidence interval for β_1 requires a critical value t^* from the $t(n - 2) = t(98)$ distribution. In Table D there are entries for 80 and 100 degrees of freedom. The values for these rows are very similar. To be conservative, we will use the larger critical value, for 80 degrees of freedom. Find the confidence level values at the bottom of the table. In the 95% confidence column the entry for 80 degrees of freedom is $t^* = 1.990$.

To compute the 95% confidence interval for β_1 we combine the estimate of the slope with the margin of error:

$$b_1 \pm t^* SE_{b_1} = -0.655 \pm (1.990)(0.158)$$
$$= -0.655 \pm 0.314$$

The interval is $(-0.969, -0.341)$. This agrees with the intervals given by the software outputs that provide this information in Figure 10.4. We estimate that an increase of 1000 steps per day is associated with a decrease in BMI between 0.341 and 0.969 kg/m^2.

Note that the intercept in this example is not of practical interest. It estimates average BMI when the activity level is 0, a value that isn't realistic. For this reason, we do not compute a confidence interval for β_0.

Confidence intervals for mean response

For any specific value of x, say x^*, the mean of the response y in this subpopulation is given by

$$\mu_y = \beta_0 + \beta_1 x^*$$

To estimate this mean from the sample, we substitute the estimates b_0 and b_1 for β_0 and β_1:

$$\hat{\mu}_y = b_0 + b_1 x^*$$

A confidence interval for μ_y adds to this estimate a margin of error based on the standard error $\mathrm{SE}_{\hat{\mu}}$. (The formula for the standard error is given in Section 10.2.)

CONFIDENCE INTERVAL FOR A MEAN RESPONSE

A **level C confidence interval for the mean response** μ_y when x takes the value x^* is

$$\hat{\mu}_y \pm t^* \mathrm{SE}_{\hat{\mu}}$$

where t^* is the value for the $t(n-2)$ density curve with area C between $-t^*$ and t^*.

Many computer programs calculate confidence intervals for the mean response corresponding to each of the x-values in the data. Some can calculate an interval for any value x^* of the explanatory variable. We will use a plot to illustrate these intervals.

EXAMPLE

10.6 Confidence intervals for the mean response. Figure 10.7 shows the upper and lower confidence limits on a graph with the data and the least-squares line. The 95% confidence limits appear as dashed curves. For any x^*, the confidence interval for the mean response extends from the lower dashed curve to the upper dashed curve. The intervals are narrowest for values of x^* near the mean of the observed x's and widen as x^* moves away from $\bar{x}$.

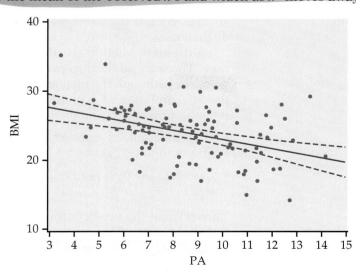

FIGURE 10.7 The 95% confidence limits (dashed curves) for the mean response for the physical activity example.

Some software will do these calculations directly if you input a value for the explanatory variable. Others will calculate the intervals for each value of x in the data set. Creating a new data set with an additional observation with x equal to the value of interest and y missing will often work.

EXAMPLE

10.7 Confidence interval for an average of 9000 steps per day. Let's find the confidence interval for the average BMI at $x = 9.0$. Our predicted BMI is

$$\widehat{BMI} = 29.578 - 0.655PA$$
$$= 29.578 - 0.655(9.0)$$
$$= 23.7$$

Software tells us that the 95% confidence interval for the mean response is 23.0 to 24.4 kg/m^2.

If we sampled many women that averaged 9000 steps per day, we would expect the average BMI to be between 23.0 and 24.4 kg/m^2. Note that many of the observations in Figure 10.7 lie outside the confidence bands. *These confidence intervals do not tell us what BMI to expect for a single observation at a particular average steps per day.* We need a different kind of interval for this purpose.

Prediction intervals

In the last example, we predicted the average BMI for an average of 9000 steps per day. Suppose we now want to predict a future observation of BMI for a woman averaging 9000 steps per day. Our best guess of BMI is what we obtained using the regression equation, that is, 23.7 kg/m^2. The margin of error, on the other hand, is larger because it is harder to predict an individual value than to predict the mean.

The predicted response y for an individual case with a specific value x^* of the explanatory variable x is

$$\hat{y} = b_0 + b_1 x^*$$

This is the same as the expression for $\hat{\mu}_y$. That is, the fitted line is used both to estimate the mean response when $x = x^*$ and to predict a single future response. We use the two notations $\hat{\mu}_y$ and $\hat{y}$ to remind ourselves of these two distinct uses.

A useful prediction should include a margin of error to indicate its accuracy. The interval used to predict a future observation is called a **prediction interval.** Although the response y that is being predicted is a random variable, the interpretation of a prediction interval is similar to that for a confidence interval. Consider doing the following many times:

- Draw a sample of n observations (x_i, y_i) and then one additional observation (x^*, y).

- Calculate the 95% prediction interval for y when $x = x^*$ using the sample of size n.

Then 95% of the prediction intervals will contain the value of y for the additional observation. In other words, the probability that this method produces an interval that contains the value of a future observation is 0.95.

The form of the prediction interval is very similar to that of the confidence interval for the mean response. The difference is that the standard error $SE_{\hat{y}}$ used in the prediction interval includes both the variability due to the fact that the least-squares line is not exactly equal to the true regression line *and* the variability of the future response variable y around the subpopulation mean. (The formula for $SE_{\hat{y}}$ appears in Section 10.2.)

PREDICTION INTERVAL FOR A FUTURE OBSERVATION

A **level C prediction interval for a future observation** on the response variable y from the subpopulation corresponding to x^* is

$$\hat{y} \pm t^* SE_{\hat{y}}$$

where t^* is the value for the $t(n - 2)$ density curve with area C between $-t^*$ and t^*.

Again, we use a graph to illustrate the results.

EXAMPLE

10.8 Prediction intervals for BMI. Figure 10.8 shows the upper and lower prediction limits, along with the data and the least-squares line. The 95% prediction limits are indicated by the dashed curves. Compare this figure with Figure 10.7, which shows the 95% confidence limits drawn to the same scale. The upper and lower limits of the prediction intervals are farther from the least-squares line than are the confidence limits. This results in most, but not all, of the observations in Figure 10.8 lying within the prediction bands.

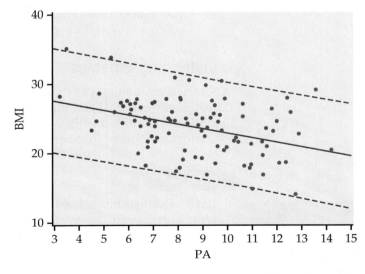

FIGURE 10.8 The 95% prediction limits (dashed curves) for individual responses for the physical activity example. Compare with Figure 10.7. The limits are wider because the margins of error incorporate the variability about the subpopulation means.

The comparison of Figures 10.7 and 10.8 reminds us that the interval for *a single future observation must be larger than an interval for the mean* of its subpopulation.

EXAMPLE

10.9   Prediction interval for an average of 9000 steps per day. Let's find the prediction interval for a future observation of BMI when a college-aged woman averages 9000 steps per day. The predicted value is the same as the estimate of the average BMI that we calculated in Example 10.7, that is, 23.7 kg/m². Software tells us that the 95% prediction interval is 16.4 to 31.0 kg/m². This interval is extremely wide, covering BMI values that are classified as underweight and obese. Because of the large amount of scatter about the regression line, prediction intervals here are relatively useless.

While a larger sample would better estimate the population regression line, it would not reduce the degree of scatter about the line. In other words, prediction intervals for BMI given activity level will always be wide. This example clearly demonstrates that a very small P-value for the significance test for a zero slope does not necessarily imply that we have found a strong predictive relationship.

USE YOUR KNOWLEDGE

10.3   Constructing confidence intervals for the mean response. Refer to Example 10.5. For the following three changes in average steps per day, construct a 95% confidence interval for the change in BMI.

(a) PA increases from 7.9 to 8.9.

(b) PA decreases from 10.7 to 9.7.

(c) PA increases from 11.8 to 12.3.

10.4   Standard error for the predicted response. Refer to Example 10.9. What is the standard error of $\hat{y}$ when $x = 9.0$ thousand steps? Would you expect the standard error of $\hat{y}$ to be larger, smaller, or the same when $x = 10.0$ thousand steps? Explain.

Transforming variables

We started our analysis of Example 10.1 with a scatterplot to check if the relationship between BMI and physical activity can be summarized with a straight line. When the relationship is not linear, there are times we can make it linear by a transformation of one or both of the variables. Here is an example.

MPH_MPG

EXAMPLE

10.10   Relationship between speed driven and fuel efficiency. Computers in some vehicles calculate various quantities related to the vehicle's performance. One of these is the fuel efficiency, or gas mileage, expressed as miles per gallon (mpg). Another is the average speed in miles per hour (mph). For one vehicle equipped in this way, mpg and mph were recorded each time the gas tank was filled, and the computer was then reset.[2] How does the speed at which the vehicle is driven affect the fuel efficiency? There are 234 observations available. We will work with a simple random sample of size 60.

Our statistical modeling for this data set is concerned about the process by which speed affects the fuel efficiency. Except for possibly the owner, no one cares about the particular vehicle. The results are interesting only if they can be applied to other similar vehicles that are driven under similar conditions. Although we would not expect the parameters that describe the relationship between speed and fuel efficiency to be *exactly* the same for similar vehicles, we would expect to find qualitatively similar results.

EXAMPLE

10.11 Graphical display of the fuel efficiency relationship. Figure 10.9 is a plot of fuel efficiency versus speed for our sample of 60 observations. We use the variable names MPG and MPH. The least-squares regression line and a smooth function are also shown in the plot. Although there is a positive association between MPG and MPH, the fit is not linear. The smooth function shows us that the relationship levels off somewhat with increasing speed.

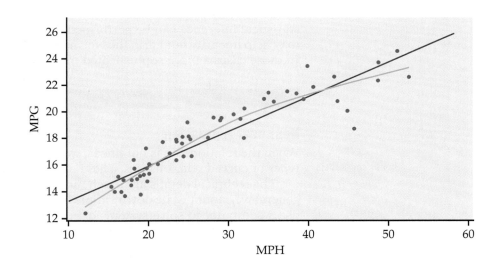

FIGURE 10.9 Scatterplot of MPG versus MPH with a smooth function and the least-squares line, for Example 10.11. The relationship between MPG and MPH does not appear linear.

Given this nonlinearity, we need to make a choice about how to proceed. One approach would be to confine our interest to speeds that are 30 mph or less, a region where it appears that a line would be a good fit to the data. Another possibility is to consider a transformation that will make the relationship approximately linear for the entire set of data.

LOOK BACK
log transformation
p. 90

EXAMPLE

10.12 Is this relationship linear? One type of function that looks similar to the smooth-function fit in Figure 10.9 is a logarithm. Therefore, we will examine the effect of transforming speed by taking the natural logarithm. The result is shown in Figure 10.10. In this plot the smooth function and the line are quite close. We are satisfied that the relationship between the log of speed and fuel efficiency is approximately linear for this set of data.

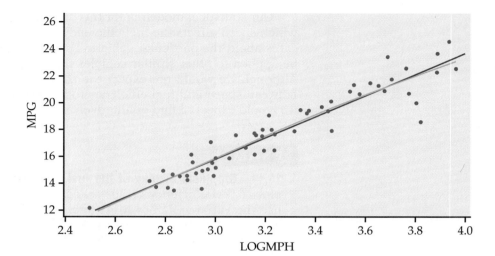

FIGURE 10.10 Scatterplot of MPG versus logarithm of MPH with a smooth function and the least-squares line, for Example 10.12. Here, the line and smooth function are very close.

Although this transformation has resulted in an approximate linear relationship, there are still other assumptions of the simple linear model that must be met. In this example, one can show these assumptions are also satisfied, so statistical inference can be performed. In other cases, transforming a variable may help linearity but harm the Normality and constant variance assumptions. In those cases a more sophisticated model is needed.

BEYOND THE BASICS

Nonlinear regression

nonlinear models

When the relationship is not linear, we often use models that allow for various types of curved relationships. These models are called **nonlinear models.**

The technical details are much more complicated for nonlinear models. In general we cannot write down simple formulas for the parameter estimates; we use a computer to solve systems of equations to find the estimates. However, the basic principles are those that we have already learned. For example,

$$DATA = FIT + RESIDUAL$$

still applies. The FIT is a nonlinear (curved) function, and the residuals are assumed to be an SRS from the $N(0, \sigma)$ distribution. The nonlinear function contains parameters that must be estimated from the data. Approximate standard errors for these estimates are part of the standard output provided by software. Here is an example.

EXAMPLE

10.13 Investing in one's bone health. As we age, our bones become weaker and are more likely to break. Osteoporosis (or weak bones) is the major cause of bone fractures in older women. Various researchers have studied this problem by looking at how and when bone mass is accumulated by young women. They've determined that up to 90 percent of a person's peak bone mass is acquired by age 18 in girls.[3] This makes youth the best time to invest in stronger bones.

Figure 10.11 displays data for a measure of bone strength, called "total body bone mineral density" (TBBMD), and age for a sample of 256 young women.[4] TBBMD is measured in grams per square centimeter (g/cm^2), and age is recorded in years. The solid curve is the nonlinear fit, and the dashed curves are 95% prediction limits. The fitted nonlinear equation is

$$\hat{y} = 1.162 \frac{e^{-1.162 + 0.28x}}{1 + e^{-1.162 + 0.28x}}$$

In this equation, $\hat{y}$ is the predicted value of TBBMD, the response variable; and x is age, the explanatory variable. A straight line would not do a very good job of summarizing the relationship between TBBMD and age. At first, TBBMD increases with age, but then it levels off as age increases. The value of the function where it is level is called "peak bone mass"; it is a parameter in the nonlinear model. The estimate is 1.162 and the standard error is 0.008. Software gives the 95% confidence interval as (1.146, 1.178).

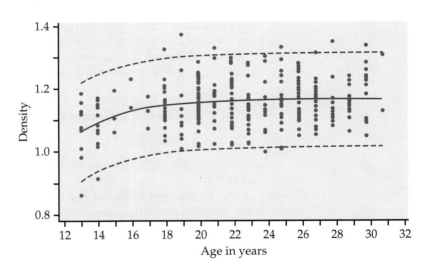

FIGURE 10.11 Plot of total body bone mineral density versus age.

The long-range goals of the researchers who conducted this study include developing intervention programs (exercise and increasing calcium intake have been shown to be effective) for young women that will increase their TBBMD.

SECTION 10.1 Summary

The statistical model for **simple linear regression** assumes the means of the response variable y fall on a line when plotted against x, with the observed y's varying Normally about these means. For n observations, this model can be written

$$y_i = \beta_0 + \beta_1 x_i + \epsilon_i$$

where $i = 1, 2, \ldots, n$, and the ϵ_i are assumed to be independent and Normally distributed with mean 0 and standard deviation σ. Here $\beta_0 + \beta_1 x_i$ is the mean response when $x = x_i$. The **parameters** of the model are β_0, β_1, and σ.

The **population regression line** intercept and slope, β_0 and β_1, are estimated by the intercept and slope of the **least-squares regression line,** b_0 and b_1. The parameter σ is estimated by

$$s = \sqrt{\frac{\sum e_i^2}{n-2}}$$

where the e_i are the **residuals**

$$e_i = y_i - \hat{y}_i$$

A **level C confidence interval for** β_1 is

$$b_1 \pm t^* \mathrm{SE}_{b_1}$$

where t^* is the value for the $t(n-2)$ density curve with area C between $-t^*$ and t^*.

The **test of the hypothesis** H_0: $\beta_1 = 0$ is based on the t **statistic**

$$t = \frac{b_1}{\mathrm{SE}_{b_1}}$$

and the $t(n-2)$ distribution. This tests whether there is a straight-line relationship between y and x. There are similar formulas for confidence intervals and tests for β_0, but these are meaningful only in special cases.

The **estimated mean response** for the subpopulation corresponding to the value x^* of the explanatory variable is

$$\hat{\mu}_y = b_0 + b_1 x^*$$

A **level C confidence interval for the mean response** is

$$\hat{\mu}_y \pm t^* \mathrm{SE}_{\hat{\mu}}$$

where t^* is the value for the $t(n-2)$ density curve with area C between $-t^*$ and t^*.

The **estimated value of the response variable** y for a future observation from the subpopulation corresponding to the value x^* of the explanatory variable is

$$\hat{y} = b_0 + b_1 x^*$$

A **level C prediction interval** for the estimated response is

$$\hat{y} \pm t^* \mathrm{SE}_{\hat{y}}$$

where t^* is the value for the $t(n-2)$ density curve with area C between $-t^*$ and t^*. The standard error for the prediction interval is larger than the confidence interval because it also includes the variability of the future observation around its subpopulation mean.

10.2 More Detail about Simple Linear Regression*

In this section we study three optional topics. The first is analysis of variance for regression. If you plan to read Chapter 11 on multiple regression, you should study this material. The second topic concerns computations for

---

*This material is optional.

regression inference. The section we just completed assumes that you have access to software or a statistical calculator. Here we present and illustrate the use of formulas for the inference procedures. Finally, we discuss inference for correlation.

Analysis of variance for regression

analysis of variance

The usual computer output for regression includes additional calculations called **analysis of variance.** Analysis of variance, often abbreviated ANOVA, is essential for multiple regression (Chapter 11) and for comparing several means (Chapters 12 and 13). Analysis of variance summarizes information about the sources of variation in the data. It is based on the

$$DATA = FIT + RESIDUAL$$

framework.

The total variation in the response y is expressed by the deviations $y_i - \bar{y}$. If these deviations were all 0, all observations would be equal and there would be no variation in the response. There are two reasons why the individual observations y_i are not all equal to their mean $\bar{y}$.

1. The responses y_i correspond to different values of the explanatory variable x and will differ because of that. The fitted value $\hat{y}_i$ estimates the mean response for the specific x_i. The differences $\hat{y}_i - \bar{y}$ reflect the variation in mean response due to differences in the x_i. This variation is accounted for by the regression line, because the $\hat{y}$'s lie exactly on the line.

2. Individual observations will vary about their mean because of variation within the subpopulation of responses to a fixed x_i. This variation is represented by the residuals $y_i - \hat{y}_i$ that record the scatter of the actual observations about the fitted line.

The overall deviation of any y observation from the mean of the y's is the sum of these two deviations:

$$(y_i - \bar{y}) = (\hat{y}_i - \bar{y}) + (y_i - \hat{y}_i)$$

In terms of deviations, this equation expresses the idea that DATA = FIT + RESIDUAL.

Several times we have measured variation by an average of squared deviations. If we square each of the preceding three deviations and then sum over all n observations, it is an algebraic fact that the sums of squares add:

$$\sum (y_i - \bar{y})^2 = \sum (\hat{y}_i - \bar{y})^2 + \sum (y_i - \hat{y}_i)^2$$

We rewrite this equation as

$$SST = SSM + SSE$$

where

$$SST = \sum(y_i - \bar{y})^2$$

$$SSM = \sum(\hat{y}_i - \bar{y})^2$$

$$SSE = \sum(y_i - \hat{y}_i)^2$$

sum of squares   The SS in each abbreviation stands for **sum of squares,** and the T, M, and E stand for total, model, and error, respectively. ("Error" here stands for deviations from the line, which might better be called "residual" or "unexplained variation.") The total variation, as expressed by SST, is the sum of the variation due to the straight-line model (SSM) and the variation due to deviations from this model (SSE). This partition of the variation in the data between two sources is the heart of analysis of variance.

If $H_0: \beta_1 = 0$ were true, there would be no subpopulations and all of the y's should be viewed as coming from a single population with mean μ_y. The variation of the y's would then be described by the sample variance

$$s_y^2 = \frac{\sum(y_i - \bar{y})^2}{n - 1}$$

degrees of freedom   The numerator in this expression is SST. The denominator is the total **degrees of freedom,** or simply DFT.

Just as the total sum of squares SST is the sum of SSM and SSE, the total degrees of freedom DFT is the sum of DFM and DFE, the degrees of freedom for the model and for the error:

$$DFT = DFM + DFE$$

The model has one explanatory variable x, so the degrees of freedom for this source are DFM $= 1$. Because DFT $= n - 1$, this leaves DFE $= n - 2$ as the degrees of freedom for error. For each source, the ratio of the sum of squares to **mean square**   the degrees of freedom is called the **mean square,** or simply MS. The general formula for a mean square is

$$MS = \frac{\text{sum of squares}}{\text{degrees of freedom}}$$

Each mean square is an average squared deviation. MST is just s_y^2, the sample variance that we would calculate if all of the data came from a single population. MSE is also familiar to us:

$$MSE = s^2 = \frac{\sum(y_i - \hat{y}_i)^2}{n - 2}$$

It is our estimate of σ^2, the variance about the population regression line.

interpretation of r^2   In Section 2.3 we noted that r^2 is the fraction of variation in the values of y that is explained by the least-squares regression of y on x. The sums of squares make this interpretation precise. Recall that SST $=$ SSM $+$ SSE. It is an algebraic fact that

$$r^2 = \frac{SSM}{SST} = \frac{\sum(\hat{y}_i - \bar{y})^2}{\sum(y_i - \bar{y})^2}$$

SUMS OF SQUARES, DEGREES OF FREEDOM, AND MEAN SQUARES

Sums of squares represent variation present in the responses. They are calculated by summing squared deviations. **Analysis of variance** partitions the total variation between two sources.

The sums of squares are related by the formula

$$SST = SSM + SSE$$

That is, the total variation is partitioned into two parts, one due to the model and one due to deviations from the model.

Degrees of freedom are associated with each sum of squares. They are related in the same way:

$$DFT = DFM + DFE$$

To calculate **mean squares,** use the formula

$$MS = \frac{\text{sum of squares}}{\text{degrees of freedom}}$$

Because SST is the total variation in y and SSM is the variation due to the regression of y on x, this equation is the precise statement of the fact that r^2 is the fraction of variation in y explained by x.

The ANOVA F test

F statistic

The null hypothesis H_0: $\beta_1 = 0$ that y is not linearly related to x can be tested by comparing MSM with MSE. The ANOVA test statistic is an **F statistic,**

$$F = \frac{MSM}{MSE}$$

LOOK BACK
F distribution
p. 460

When H_0 is true, this statistic has an F distribution with 1 degree of freedom in the numerator and $n-2$ degrees of freedom in the denominator. These degrees of freedom are those of MSM and MSE. Just as there are many t statistics, there are many F statistics. The ANOVA F statistic is not the same as the F statistic of equality of spread.

When $\beta_1 \neq 0$, MSM tends to be large relative to MSE. So large values of F are evidence against H_0 in favor of the two-sided alternative.

The F statistic tests the same null hypothesis as one of the t statistics that we encountered earlier in this chapter, so it is not surprising that the two are related. It is an algebraic fact that $t^2 = F$ in this case. For linear regression with one explanatory variable, we prefer the t form of the test because it more easily allows us to test one-sided alternatives and is closely related to the confidence interval for β_1.

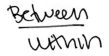

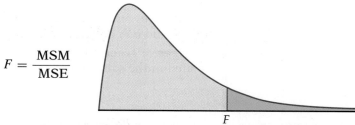

ANALYSIS OF VARIANCE F TEST

In the simple linear regression model, the hypotheses

$$H_0: \beta_1 = 0$$

$$H_a: \beta_1 \neq 0$$

are tested by the F **statistic**

$$F = \frac{\text{MSM}}{\text{MSE}}$$

The P-value is the probability that a random variable having the $F(1, n-2)$ distribution is greater than or equal to the calculated value of the F statistic.

The ANOVA calculations are displayed in an *analysis of variance table*, often abbreviated **ANOVA table.** Here is the format of the table for simple linear regression:

ANOVA table

Source	Degrees of freedom	Sum of squares	Mean square	F
Model	1	$\sum(\hat{y}_i - \bar{y})^2$	SSM/DFM	MSM/MSE
Error	$n - 2$	$\sum(y_i - \hat{y}_i)^2$	SSE/DFE	
Total	$n - 1$	$\sum(y_i - \bar{y})^2$	SST/DFT	

EXAMPLE

10.14   Interpreting SPSS output for BMI and physical activity. The entire output generated by SPSS for the physical activity study in Example 10.3 is given in Figure 10.12. Note that SPSS uses the labels Regression, Residual, and Total for the three sources of variation. We have called these Model, Error, and Total. Other statistical software packages may use slightly different labels. We round the calculated value of the F statistic to 17.10; the P-value is given as 0.000. This is a rounded value and we can conclude that $P < 0.0005$. There is strong evidence against the null hypothesis that there is no relationship between BMI and average steps per day (PA). Now look at the output for the regression coefficients. The t statistic for PA is given as -4.135. If we square this number, we obtain the F statistic (accurate up to roundoff error). The value of r^2 is also given in the output. Average steps per day explains only 14.9% of the variability in BMI. *Strong evidence against the null hypothesis that there is no relationship does not imply a large percentage of the total variability is explained by the model.*

CAUTION

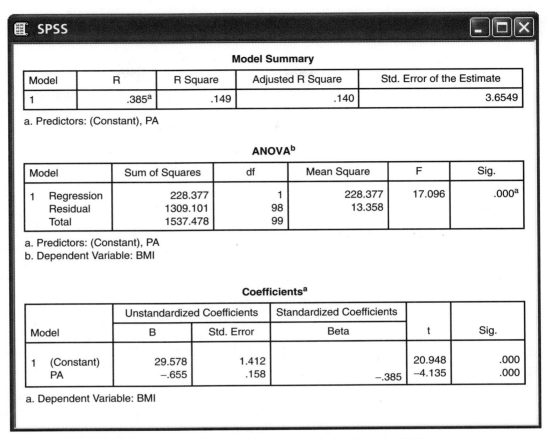

FIGURE 10.12   Regression output with ANOVA table for Example 10.14.

Calculations for regression inference

We recommend using statistical software for regression calculations. With time and care, however, the work is feasible with a calculator. We will use the following example to illustrate how to perform inference for regression analysis using a calculator.

> **EXAMPLE**
>
> **10.15 Protein requirements via nitrogen balance.** Nitrogen balance studies are used to determine protein requirements for people. Each subject is fed three different controlled diets during three separate experimental periods. The three diets are similar with regard to all nutrients except protein.
>
>   Nitrogen balance is the difference between the amount of nitrogen consumed and the amount lost in feces and urine and by other means. Since virtually all the nitrogen in a diet comes from protein, nitrogen balance is an indicator of the amount of protein retained by the body. The protein requirement for an individual is the intake corresponding to a balance of zero.
>
>   Linear regression is used to model the relationship between nitrogen balance, measured in milligrams of nitrogen per kilogram of body weight per day (mg/kg/d), and protein intake, measured in grams of protein per kilogram of body weight per day (g/kg/d). Here are the data for one subject:[5]

Protein intake (x)	0.543	0.797	1.030
Nitrogen balance (y)	−23.4	17.8	67.3

The data and the least-squares line are plotted in Figure 10.13. The strong straight-line pattern suggests that we can use linear regression to model the relationship between nitrogen balance and protein intake.

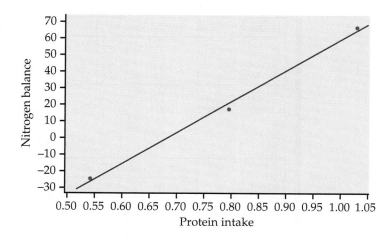

FIGURE 10.13 Data and regression line for Example 10.15.

We begin our regression calculations by fitting the least-squares line. Fitting the line gives estimates b_1 and b_0 of the model parameters β_1 and β_0. Next we examine the residuals from the fitted line and obtain an estimate s of the remaining parameter σ. These calculations are preliminary to inference. Finally, we use s to obtain the standard errors needed for the various interval estimates and significance tests. *Roundoff errors that accumulate during these calculations can ruin the final results. Be sure to carry many significant digits and check your work carefully.*

Preliminary calculations   After examining the scatterplot (Figure 10.13) to verify that the data show a straight-line pattern, we begin our calculations.

> ### EXAMPLE
>
> **10.16 Summary statistics for nitrogen balance study.**   We start by making a table with the mean and standard deviation for each of the variables, the correlation, and the sample size. These calculations should be familiar from Chapters 1 and 2. Here is the summary:
>
> | Variable | Mean | Standard deviation | Correlation | Sample size |
> |---|---|---|---|---|
> | Intake | $\bar{x} = 0.79000$ | $s_x = 0.24357545$ | $r = 0.99698478$ | $n = 3$ |
> | N balance | $\bar{y} = 20.56667$ | $s_y = 45.4132506$ | | |
>
> These quantities are the building blocks for our calculations.

We will need one additional quantity for the calculations to follow. It is the expression $\sum(x_i - \bar{x})^2$. We obtain this quantity as an intermediate step when we calculate s_x. You could also find it using the fact that $\sum(x_i - \bar{x})^2 = (n-1)s_x^2$. You should verify that the value for our example is

$$\sum(x_i - \bar{x})^2 = 0.118658$$

Our first task is to find the least-squares line. This is easy with the building blocks that we have assembled.

EXAMPLE

10.17 Computing the least-squares regression line.   The slope of the least-squares line is

$$b_1 = r\frac{s_y}{s_x}$$
$$= 0.99698478\frac{45.4132506}{0.24357545}$$
$$= 185.882$$

The intercept is

$$b_0 = \bar{y} - b_1\bar{x}$$
$$= 20.56667 - (185.882)(0.79000)$$
$$= -126.280$$

The equation of the least-squares regression line is therefore

$$\hat{y} = -126 + 186x$$

This is the line shown in Figure 10.13.

We now have estimates of the first two parameters, β_0 and β_1, of our linear regression model. We now find the estimate of the third parameter, σ: the standard deviation s about the fitted line. To do this we need to find the predicted values and then the residuals.

EXAMPLE

10.18 Computing the predicted values and residuals.   The first observation is an intake of $x = 0.543$. The corresponding predicted value of nitrogen balance is

$$\hat{y}_1 = b_0 + b_1x_1$$
$$= -126.280 + (185.882)(0.543)$$
$$= -25.346$$

and the residual is

$$e_1 = y_1 - \hat{y}_1$$
$$= -23.4 - (-25.346)$$
$$= 1.946$$

The residuals for the other intakes are calculated in the same way. You should verify that they are −4.068 and 2.122.

Notice that the sum of these three residuals is zero. When doing these calculations by hand, it is always helpful to check that the sum of the residuals is zero.

EXAMPLE

10.19 Computing s^2. The estimate of σ^2 is s^2, the sum of the squares of the residuals divided by $n - 2$. The estimated standard deviation about the line is the square root of this quantity.

$$s^2 = \frac{\sum e_i^2}{n - 2}$$
$$= \frac{(1.946)^2 + (-4.068)^2 + (2.122)^2}{1}$$
$$= 24.84$$

So the estimate of the standard deviation about the line is

$$s = \sqrt{24.84} = 4.984$$

Now that we have estimates of the three parameters of our model, we can proceed to the more detailed calculations needed for regression inference.

Inference for slope and intercept Confidence intervals and significance tests for the slope β_1 and intercept β_0 of the population regression line make use of the estimates b_1 and b_0 and their standard errors.

◄ **LOOK BACK**
rules for variances
p. 271

Some algebra using the rules for variances establishes that the standard deviation of b_1 is

$$\sigma_{b_1} = \frac{\sigma}{\sqrt{\sum(x_i - \bar{x})^2}}$$

Similarly, the standard deviation of b_0 is

$$\sigma_{b_0} = \sigma\sqrt{\frac{1}{n} + \frac{\bar{x}^2}{\sum(x_i - \bar{x})^2}}$$

To estimate these standard deviations, we need only replace σ by its estimate s.

STANDARD ERRORS FOR ESTIMATED REGRESSION COEFFICIENTS
The standard error of the slope b_1 of the least-squares regression line is

$$SE_{b_1} = \frac{s}{\sqrt{\sum(x_i - \bar{x})^2}}$$

The standard error of the intercept b_0 is

$$SE_{b_0} = s\sqrt{\frac{1}{n} + \frac{\bar{x}^2}{\sum(x_i - \bar{x})^2}}$$

The plot of the regression line with the data in Figure 10.13 *appears* to show a very strong relationship, but our sample size is very small. We assess the situation with a significance test for the slope.

EXAMPLE

10.20 Testing the slope. First we need the standard error of the estimated slope:

$$SE_{b_1} = \frac{s}{\sqrt{\sum(x_i - \bar{x})^2}}$$

$$= \frac{4.984}{\sqrt{0.118658}} = 14.47$$

To test

$$H_0: \beta_1 = 0$$

$$H_a: \beta_1 \neq 0$$

calculate the t statistic:

$$t = \frac{b_1}{SE_{b_1}}$$

$$= \frac{185.882}{14.47} = 12.85$$

Using Table D with $n - 2 = 1$ degree of freedom, we conclude that $P < 0.05$. (The exact value obtained from software is 0.0494.) The data provide evidence in favor of a relationship between nitrogen balance and protein intake ($t = 12.85$, df $= 1$, $P < 0.05$).

Three things are important to note about this example. First, the sample size is very small. Even though the estimated slope is more than 12 standard deviations away from zero, we have only barely attained the 0.05 standard for statistical significance. *It is important to remember that we need to have a very large effect if we expect to detect it with a small sample size.* Second, we would, of course, prefer to have more than three observations for this analysis. However, for each diet, data are collected for about a month. Because the requirement is assumed to be a distribution rather than a single number for everyone, it is important to measure several subjects. Because of the enormous expense involved, researchers typically use only three levels of intake. Third, because we expect balance to increase with increasing intake, a one-sided significance test is justified in this setting.

The significance test tells us that the data provide sufficient information to conclude that intake and balance are related. We use the estimate b_1 and its confidence interval to further describe the relationship.

EXAMPLE

10.21 Computing a 95% confidence interval for the slope. For the protein requirement problem, let's find a 95% confidence interval for the slope β_1. The degrees of freedom are $n - 2 = 1$, so t^* from Table D is 12.71. We compute

$$b_1 \pm t^* SE_{b_1} = 185.882 \pm (12.71)(14.47)$$

$$= 186 \pm 184$$

The interval is (2, 370).

Note the effect of the small sample size on the critical value t^*. With one additional observation, it would decrease to 4.303.

In this example, the intercept β_0 does not have a meaningful interpretation. A protein intake of zero is theoretically possible, but we would not expect our linear model to be reasonable when extended to such an extreme value. For problems where inference for β_0 is appropriate, the calculations are performed in the same way as those for β_1. Note that there is a different formula for the standard error, however.

Confidence intervals for the mean response and prediction intervals for a future observation When we substitute a particular value x^* of the explanatory variable into the regression equation and obtain a value of $\hat{y}$, we can view the result in two ways:

1. We have estimated the mean response μ_y.

2. We have predicted a future value of the response y.

The margins of error for these two uses are often quite different. Prediction intervals for an individual response are wider than confidence intervals for estimating a mean response. We now proceed with the details of these calculations. Once again, standard errors are the essential quantities. And once again, these standard errors are multiples of s, our basic measure of the variability of the responses about the fitted line.

STANDARD ERRORS FOR $\hat{\mu}$ AND $\hat{y}$

The standard error of $\hat{\mu}$ is

$$SE_{\hat{\mu}} = s\sqrt{\frac{1}{n} + \frac{(x^* - \overline{x})^2}{\sum(x_i - \overline{x})^2}}$$

The standard error for predicting an individual response $\hat{y}$ is[6]

$$SE_{\hat{y}} = s\sqrt{1 + \frac{1}{n} + \frac{(x^* - \overline{x})^2}{\sum(x_i - \overline{x})^2}}$$

Note that the only difference between the formulas for these two standard errors is the extra 1 under the square root sign in the standard error for prediction. This standard error is larger due to the additional variation of individual responses about the mean response. It produces prediction intervals that are wider than the confidence intervals for the mean response.

For the nitrogen balance example, we can think about the mean balance that would result if a particular protein intake was consumed many times. The confidence interval for the mean response would provide an interval estimate of this population value. On the other hand, we might want to predict a future observation under conditions similar to those used in the study, that is, for a one-month period, at a particular intake level. A prediction interval attempts to capture this future observation.

EXAMPLE

10.22 Computing the confidence intervals for $\hat{\mu}$. Let's find a 95% confidence interval for the mean balance corresponding to an intake of 0.7 g/kg/d. The estimated mean balance is

$$\begin{aligned} \hat{\mu} &= b_0 + b_1 x_1 \\ &= -126.280 + (185.882)(0.7) \\ &= 3.837 \end{aligned}$$

The standard error is

$$\begin{aligned} \text{SE}_{\hat{\mu}} &= s\sqrt{\frac{1}{n} + \frac{(x^* - \bar{x})^2}{\sum(x_i - \bar{x})^2}} \\ &= 4.984\sqrt{\frac{1}{3} + \frac{(0.70 - 0.79)^2}{0.118658}} \\ &= 3.158 \end{aligned}$$

To find the 95% confidence interval we compute

$$\begin{aligned} \hat{\mu} \pm t^*\text{SE}_{\hat{\mu}} &= 3.837 \pm (12.71)(3.158) \\ &= 3.837 \pm 40.138 \\ &= 4 \pm 40 \end{aligned}$$

The interval is -36 to 44 mg/kg/d of nitrogen.

Calculations for the prediction intervals are similar. The only difference is the use of the formula for $\text{SE}_{\hat{y}}$ in place of $\text{SE}_{\hat{\mu}}$. This results in a much wider interval.

Since the confidence interval for mean response includes the value 0, the corresponding intake 0.7 g/kg/d should be considered as a possible value for the intake requirement for this individual. Other intakes would also produce confidence intervals that would include the value of 0 for mean balance. Here is one method that is commonly used to determine a single value of the requirement for an individual.

EXAMPLE

10.23 Estimating the protein requirement. We define the estimated requirement for an individual to be the intake corresponding to zero balance using the fitted regression equation. To do this, we set the equation

$$\hat{\mu} = b_0 + b_1 x$$

equal to 0 and solve for the intake x. So,

$$\begin{aligned} x &= -b_0/b_1 \\ &= -(-126.280)/185.882 \\ &= 0.68 \end{aligned}$$

The estimated protein requirement for this individual is 0.68 g/kg/d.

If we repeat these calculations using data collected on a large number of individuals, we can estimate the requirement distribution for a population. There are many interesting statistical issues related to this problem.[7]

Inference for correlation

LOOK BACK
correlation
p. 102

The correlation coefficient is a measure of the strength and direction of the linear association between two variables. Correlation does not require an explanatory-response relationship between the variables. We can consider the sample correlation r as an estimate of the correlation in the population and base inference about the population correlation on r.

population correlation ρ

jointly Normal variables

The correlation between the variables x and y when they are measured for every member of a population is the **population correlation.** As usual, we use Greek letters to represent population parameters. In this case ρ (the Greek letter rho) is the population correlation. When $\rho = 0$, there is no linear association in the population. In the important case where the two variables x and y are both Normally distributed, the condition $\rho = 0$ is equivalent to the statement that x and y are independent. That is, there is no association of any kind between x and y. (Technically, the condition required is that x and y be **jointly Normal.** This means that the distribution of x is Normal and also that the conditional distribution of y, given any fixed value of x, is Normal.) We therefore may wish to test the null hypothesis that a population correlation is 0.

TEST FOR A ZERO POPULATION CORRELATION

To test the hypothesis H_0: $\rho = 0$, compute the t statistic:

$$t = \frac{r\sqrt{n-2}}{\sqrt{1-r^2}}$$

where n is the sample size and r is the sample correlation.

In terms of a random variable T having the $t(n-2)$ distribution, the P-value for a test of H_0 against

H_a: $\rho > 0$ is $P(T \geq t)$

H_a: $\rho < 0$ is $P(T \leq t)$

H_a: $\rho \neq 0$ is $2P(T \geq |t|)$

Most computer packages have routines for calculating correlations and some will provide the significance test for the null hypothesis that ρ is zero.

EXAMPLE

10.24 Correlation in the physical activity study. For the physical activity example, the SPSS output appears in Figure 10.14. The sample correlation between BMI and the average number of steps per day (PA) is $r = -0.385$. SPSS calls this a Pearson correlation to distinguish it from other kinds of correlations that it can calculate. The P-value for a two-sided test of $H_0: \rho = 0$ is given as 0.000. This means that the actual P-value is less than 0.0005. We conclude that there is a nonzero correlation between BMI and PA.

SPSS

Correlations

		PA	BMI
PA	Pearson Correlation	1	~.385**
	Sig. (2-tailed)		.000
	N	100	100
BMI	Pearson Correlation	~.385**	1
	Sig. (2-tailed)	.000	
	N	100	100

** . Correlation is significant at the 0.01 level (2-tailed).

FIGURE 10.14 Correlation output for Example 10.24.

If we wanted to test the one-sided alternative that the population correlation is negative, we divide the P-value in the output by 2, after checking that the sample coefficient is in fact negative.

If your software does not give the significance test, you can do the computations easily with a calculator.

EXAMPLE

10.25 Correlation test using a calculator. The correlation between BMI and PA is $r = -0.385$. Recall that $n = 100$. The t statistic for testing the null hypothesis that the population correlation is zero is

$$t = \frac{r\sqrt{n-2}}{\sqrt{1-r^2}}$$
$$= \frac{-0.385\sqrt{100-2}}{\sqrt{1-(-0.385)^2}}$$
$$= -4.13$$

The degrees of freedom are $n - 2 = 98$. From Table D we conclude that $P < 0.0001$. This agrees with the SPSS output in Figure 10.14, where the P-value is given as 0.000. The data provide clear evidence that BMI and PA are related.

There is a close connection between the significance test for a correlation and the test for the slope in a linear regression. Recall that

$$b_1 = r\frac{s_y}{s_x}$$

From this fact we see that if the slope is 0, so is the correlation, and vice versa. It should come as no surprise to learn that the procedures for testing H_0: $\beta_1 = 0$ and H_0: $\rho = 0$ are also closely related. In fact, the t statistics for testing these hypotheses are numerically equal. That is,

$$\frac{b_1}{s_{b_1}} = \frac{r\sqrt{n-2}}{\sqrt{1-r^2}}$$

Check that this holds in both of our examples.

In our examples, the conclusion that there is a statistically significant correlation between the two variables would not come as a surprise to anyone familiar with the meaning of these variables. The significance test simply tells us whether or not there is evidence in the data to conclude that the population correlation is different from 0. The actual size of the correlation is of considerably more interest. We would therefore like to give a confidence interval for the population correlation. Unfortunately, most software packages do not perform this calculation. Because hand calculation of the confidence interval is very tedious, we do not give the method here.[8]

USE YOUR KNOWLEDGE

10.5 Research and development spending. The National Science Foundation collects data on the research and development spending by universities and colleges in the United States.[9] Here are the data for the years 2003 to 2007:

Year	2003	2004	2005	2006	2007
Spending (billions of dollars)	40.1	43.3	45.8	47.7	49.4

NSF

Do the following by hand or with a calculator and verify your results with a software package.

(a) Make a scatterplot that shows the increase in research and development spending over time. Does the pattern suggest that the spending is increasing linearly over time?

(b) Find the equation of the least-squares regression line for predicting spending from year. Add this line to your scatterplot.

(c) For each of the five years, find the residual. Use these residuals to calculate the standard error s.

(d) Write the regression model for this setting. What are your estimates of the unknown parameters in this model?

(e) Compute a 95% confidence interval for the slope and summarize what this interval tells you about the increase in spending over time.

SECTION 10.2 Summary

The **ANOVA table** for a linear regression gives the degrees of freedom, sum of squares, and mean squares for the model, error, and total sources of variation. The **ANOVA F statistic** is the ratio MSM/MSE. Under $H_0: \beta_1 = 0$, this statistic has an $F(1, n-2)$ distribution and is used to test H_0 versus the two-sided alternative.

The **square of the sample correlation** can be expressed as

$$r^2 = \frac{\text{SSM}}{\text{SST}}$$

and is interpreted as the proportion of the variability in the response variable y that is explained by the explanatory variable x in the linear regression.

The **standard errors for b_0 and b_1** are

$$\text{SE}_{b_0} = s\sqrt{\frac{1}{n} + \frac{\bar{x}^2}{\sum(x_i - \bar{x})^2}}$$

$$\text{SE}_{b_1} = \frac{s}{\sqrt{\sum(x_i - \bar{x})^2}}$$

The **standard error that we use for a confidence interval** for the estimated mean response for the subpopulation corresponding to the value x^* of the explanatory variable is

$$\text{SE}_{\hat{\mu}} = s\sqrt{\frac{1}{n} + \frac{(x^* - \bar{x})^2}{\sum(x_i - \bar{x})^2}}$$

The **standard error that we use for a prediction interval** for a future observation from the subpopulation corresponding to the value x^* of the explanatory variable is

$$\text{SE}_{\hat{y}} = s\sqrt{1 + \frac{1}{n} + \frac{(x^* - \bar{x})^2}{\sum(x_i - \bar{x})^2}}$$

When the variables y and x are jointly Normal, the sample correlation is an estimate of the **population correlation** ρ. The test of $H_0: \rho = 0$ is based on the *t* **statistic**

$$t = \frac{r\sqrt{n-2}}{\sqrt{1-r^2}}$$

which has a $t(n-2)$ distribution under H_0. This test statistic is numerically identical to the t statistic used to test $H_0: \beta_1 = 0$.

CHAPTER 10 Exercises

For Exercises 10.1 and 10.2, see page 554; for Exercises 10.3 and 10.4, see page 560; and for Exercise 10.5, see page 578.

10.6 What's wrong? For each of the following, explain what is wrong and why.

(a) The slope describes the change in x for a change in y.

(b) The population regression line is $y = b_0 + b_1 x$.

(c) A 95% confidence interval for the mean response is the same width regardless of x.

10.7 What's wrong? For each of the following, explain what is wrong and why.

(a) The parameters of the simple linear regression model are b_0, b_1, and s.

(b) To test $H_0: b_1 = 0$, use a t test.

(c) For a particular value of the explanatory variable x, the confidence interval for the mean response will be wider than the prediction interval for a future observation.

10.8 95% confidence intervals for the slope. Find a 95% confidence interval for the slope in each of the following settings:

(a) $n = 25$, $\hat{y} = 5.3 + 1.10x$, and $SE_{b_1} = 0.58$

(b) $n = 25$, $\hat{y} = 5.3 + 2.10x$, and $SE_{b_1} = 0.58$

(c) $n = 100$, $\hat{y} = 5.3 + 1.10x$, and $SE_{b_1} = 0.58$

10.9 Significance test for the slope. For each of the settings in the previous exercise, test the null hypothesis that the slope is zero versus the two-sided alternative.

10.10 Public university tuition: 2000 versus 2008. Table 10.1 shows the instate undergraduate tuition and required fees for 33 public universities in 2000 and 2008.[10] TUITION

(a) Plot the data with the 2000 tuition on the x axis and describe the relationship. Are there any outliers or unusual values? Does a linear relationship between the tuition in 2000 and 2008 seem reasonable?

(b) Run the simple linear regression and state the least-squares regression line.

(c) Obtain the residuals and plot them versus the 2000 tuition amount. Is there anything unusual in the plot?

(d) Do the residuals appear to be approximately Normal? Explain.

(e) Give the null and alternative hypotheses for examining the relationship between 2000 and 2008 tuition amounts.

(f) Write down the test statistic and P-value for the hypotheses stated in part (e). State your conclusions.

10.11 More on public university tuition. Refer to the previous exercise.

(a) Construct a 95% confidence interval for the slope. What does this interval tell you about the increase in tuition between 2000 and 2008?

(b) What percent of the variability in 2008 tuition is explained by a linear regression model using the 2000 tuition?

(c) The tuition at Stat U was $5100 in 2000. What is the predicted tuition in 2008?

(d) The tuition at Moneypit U was $8700 in 2000. What is its predicted tuition in 2008?

(e) Discuss the appropriateness of using the fitted equation to predict tuition for each of these universities.

10.12 U.S. versus overseas stock returns. Returns on common stocks in the United States and overseas appear to be growing more closely correlated as economies become more interdependent. Suppose that the following population regression line connects the total annual returns (in percent) on two indexes of stock prices:

MEAN OVERSEAS RETURN $= 0.8 + 0.46 \times$ U.S. RETURN

(a) What is β_0 in this line? What does this number say about overseas returns when the U.S. market is flat (0% return)?

(b) What is β_1 in this line? What does this number say about the relationship between U.S. and overseas returns?

TABLE 10.1

In-state tuition and fees (in dollars) for 33 public universities

School	2000	2008	School	2000	2008	School	2000	2008
Penn State	7018	13,706	Virginia	4335	9300	Iowa State	3132	5524
Pittsburgh	7002	13,642	Indiana	4405	8231	Oregon	3819	6435
Michigan	6926	11,738	Texas A&M	3374	7844	Iowa	3204	6544
Rutgers	6333	11,540	Texas	3575	8532	Washington	3761	6802
Illinois	4994	12,106	Cal-Irvine	3970	8046	Nebraska	3450	6584
Minnesota	4877	10,634	Cal-San Diego	3848	8062	Kansas	2725	7042
Michigan State	5432	10,690	Cal-Berkeley	4047	7656	Colorado	3188	7278
Ohio State	4383	8679	UCLA	3698	8310	North Carolina	2768	5397
Maryland	5136	8005	Purdue	3872	7750	Arizona	2348	5542
Cal-Davis	4072	9497	Wisconsin	3791	7569	Florida	2256	3256
Missouri	4726	7386	Buffalo	4715	6385	Georgia Tech	3308	6040

(c) We know that overseas returns will vary in years having the same return on U.S. common stocks. Write the regression model based on the population regression line given above. What part of this model allows overseas returns to vary when U.S. returns remain the same?

10.13 Beer and blood alcohol. How well does the number of beers a student drinks predict his or her blood alcohol content? Sixteen student volunteers at Ohio State University drank a randomly assigned number of 12-ounce cans of beer. Thirty minutes later, a police officer measured their blood alcohol content (BAC). Here are the data:[11]

Student	1	2	3	4	5	6	7	8
Beers	5	2	9	8	3	7	3	5
BAC	0.10	0.03	0.19	0.12	0.04	0.095	0.07	0.06

Student	9	10	11	12	13	14	15	16
Beers	3	5	4	6	5	7	1	4
BAC	0.02	0.05	0.07	0.10	0.085	0.09	0.01	0.05

The students were equally divided between men and women and differed in weight and usual drinking habits. Because of this variation, many students don't believe that number of drinks predicts blood alcohol well. BAC

(a) Make a scatterplot of the data. Find the equation of the least-squares regression line for predicting blood alcohol from number of beers and add this line to your plot. What is r^2 for these data? Briefly summarize what your data analysis shows.

(b) Is there significant evidence that drinking more beers increases blood alcohol on the average in the population of all students? State hypotheses, give a test statistic and P-value, and state your conclusion.

(c) Steve thinks he can drive legally 30 minutes after he drinks 5 beers. The legal limit is BAC = 0.08. Give a 90% prediction interval for Steve's BAC. Can he be confident he won't be arrested if he drives and is stopped?

10.14 Correlations of an employment interview. To investigate the association between nonverbal clues and interview assessment, some researchers recruited 98 undergraduate students enrolled in an elective, one-credit career preparation course and had them participate in a mock interview. Participants were instructed to treat this experience just as they would a "real" interview (e.g., dress appropriately, research company and job characteristics). The following table of correlations summarizes the interviewers' impression of a series of characteristics associated with the interview. The characteristics were extraversion (XTRA), agreeableness (AGREE), overall handshake (HAND), eye contact (EYE), professional dress (DRESS), and interviewer assessment (ASSESS).

	XTRA	AGREE	HAND	EYE	DRESS
AGREE	0.42				
HAND	0.23	0.05			
EYE	0.18	0.14	0.82		
DRESS	0.17	0.11	0.42	0.43	
ASSESS	0.28	0.13	0.29	0.31	0.15

(a) These impressions were all rated on a 1 to 5 scale with 1 representing "weak" and 5 representing "strong." Do these sorts of responses pose a problem with correlation analysis? Explain.

(b) Perform a significance test for each of these correlations and summarize the results.

(c) You performed 15 significance tests in this exercise. Suppose that you wanted to do a Bonferroni correction (page 388) to take into account this fact and to keep the overall false rejection rate at 5%. In that case, you would declare a correlation to be statistically significant only if the P-value is 0.05/15 or less. Perform the significance tests using the Bonferroni correction. Summarize the results and compare them with the results you obtained in part (b).

(d) To what extent do you think that the results that you obtained apply to other interview situations? Give reasons for you answers.

10.15 Performance bonuses. In the National Football League (NFL), performance bonuses now account for roughly 25% of player compensation.[12] Does tying a player's salary into performance bonuses result in better individual or team success on the field? usatoday.com contains payroll data for most current NFL players. cbssports.com contains a player rating system that uses game statistics.[13] Focusing on linebackers, let's look at the relationship between a player's overall 2008 rating and the percent of his 2008 salary devoted to incentive payments. LINEBACKERS

(a) Use numerical and graphical methods to describe the two variables and summarize your results.

(b) Both variable distributions are non-Normal. Does this necessarily pose a problem with performing linear regression? Explain.

(c) Construct a scatterplot of the data and describe the relationship. Are there any outliers or unusual values? Does a linear relationship between the percent of salary and the player rating seem reasonable? Is it a very strong relationship? Explain.

(d) Run the simple linear regression and state the least-squares regression line.

(e) Obtain the residuals and assess whether the assumptions for the linear regression analysis are reasonable. Include all plots and numerical summaries used in doing this assessment.

10.16 🔺 **Performance bonuses, continued.** Refer to the previous problem. 📄 LINEBACKERS

(a) Now run the simple linear regression for the variables sqrt(rating) and percent of salary devoted to incentive payments.

(b) Obtain the residuals and assess whether the assumptions for the linear regression analysis are reasonable. Include all plots and numerical summaries used in doing this assessment.

(c) Construct a 95% confidence interval for the square root increase in rating given a 1% increase in the percent of salary devoted to incentive payments.

(d) Consider the values 0%, 20%, 40%, 60%, and 80% salary devoted to incentives. Compute the predicted rating for this model and the one in the previous exercise. For the model in this problem, you will need to square the predicted value to get back to the original units.

(e) Plot the predicted values versus the percent and connect those values from the same model. For which regions of percent do the predicted values from the two models vary the most?

(f) Based on the comparison of regression models (both predicted values and residuals), which model do you prefer? Explain.

10.17 Assessment value versus sales price. Real estate is typically reassessed annually for property tax purposes. This assessed value, however, is not necessarily the same as the fair market value of the property. Table 10.2 summarizes an SRS of 30 properties recently sold in a midwestern city.[14] Both variables are measured in thousands of dollars. 📄 HOMESALES

(a) Inspect the data. How many have a selling price greater than the assessed value? Do you think this trend would be true for the larger population of all homes recently sold? Explain your answer.

(b) Make a scatterplot with assessed value on the horizontal axis. Briefly describe the relationship between assessed value and selling price.

(c) Report the least-squares regression line for predicting selling price from assessed value.

(d) Obtain the residuals and plot them versus assessed value. Is there anything unusual to report? If so, explain.

(e) Do the residuals appear to be approximately Normal? Explain your answer.

(f) Based on your answers to parts, (b), (d), and (e), do the assumptions for the linear regression analysis appear reasonable? Explain your answer.

10.18 Assessment value versus sales price, continued. Refer to the previous exercise. 📄 HOMESALES

(a) Calculate the predicted selling prices for homes currently assessed at $155,000, $220,000, and $285,000.

(b) Suppose these houses sold for $142,900, $224,000, and $286,000 respectively. Calculate the residual for each of these sales.

TABLE 10.2

Assessed value and sales price (in $ thousands) of 30 homes in a midwestern city

Property	Sales Price	Assessed Value	Property	Sales Price	Assessed Value	Property	Sales Price	Assessed Value
1	167.9	152.7	11	230.0	225.4	21	283.0	303.9
2	168.0	163.8	12	230.0	170.4	22	269.0	233.7
3	155.0	167.6	13	222.5	200.4	23	255.0	233.6
4	158.5	127.3	14	225.5	209.6	24	285.0	234.2
5	159.9	155.7	15	220.0	205.2	25	146.0	145.1
6	162.0	169.0	16	216.0	220.9	26	128.0	108.3
7	165.0	187.1	17	215.0	194.9	27	126.5	136.2
8	174.5	153.6	18	228.0	231.4	28	129.9	113.3
9	175.0	167.1	19	209.0	224.2	29	150.0	121.4
10	159.0	148.9	20	267.0	235.1	30	195.0	184.0

(c) Construct a 95% confidence interval for both the slope and intercept.

(d) Using the result from part (c), compare the estimated regression line with $y = x$, which says, on average, the selling price is equal to the assessed value. Is there evidence that this model is not reasonable? In other words, is the selling price typically larger or smaller than the assessed value? Explain your answer.

10.19 Are the number of tornadoes increasing? The Storm Prediction Center of the National Oceanic and Atmospheric Administration contains listings of tornadoes, floods, and other weather phenomena. Table 10.3 summarizes the annual number of tornadoes in the United States between 1953 and 2008.[15] 🌪 TORNADOES

(a) Make a plot of the total number of tornadoes by year. Does a linear trend over years appear reasonable? Are there any outliers or unusual patterns? Explain your answer.

(b) Run the simple linear regression and summarize the results, making sure to construct a 95% confidence interval for the average annual increase in the number of tornadoes.

(c) Obtain the residuals and plot them versus year. Is there anything unusual in the plot?

(d) Are the residuals Normal? Justify your answer.

(e) The number of tornadoes in 2004 is much larger than expected under this linear model. Remove this observation and rerun the simple linear regression. Compare these results with the results in part (b).

10.20 Are the two fuel efficiency measurements similar? Refer to Exercise 7.26 (page 427). In addition to the computer calculating mpg, the driver also recorded the mpg by dividing the miles driven by the amount of gallons at fill-up. The driver wants to determine if these calculations are different.

Fill-up	1	2	3	4	5	6	7	8	9	10
Computer	41.5	50.7	36.6	37.3	34.2	45.0	48.0	43.2	47.7	42.2
Driver	36.5	44.2	37.2	35.6	30.5	40.5	40.0	41.0	42.8	39.2

Fill-up	11	12	13	14	15	16	17	18	19	20
Computer	43.2	44.6	48.4	46.4	46.8	39.2	37.3	43.5	44.3	43.3
Driver	38.8	44.5	45.4	45.3	45.7	34.2	35.2	39.8	44.9	47.5

(a) Consider the driver's mpg calculations as the explanatory variable. Plot the data and describe the relationship. Are there any outliers or unusual values? Does a linear relationship seem reasonable?

(b) Run the simple linear regression and state the least-squares regression line.

(c) Summarize the results. Does it appear that the computer and driver calculations are the same? Explain.

10.21 Gambling and alcohol use in first-year college students. Gambling and alcohol use are problematic behaviors for many college students. A recent study looked at 908 first-year students from a large northeastern university.[16] Each participant was asked to fill out the 10-item Alcohol Use Disorders Identification Test (AUDIT) and a 7-item inventory used in prior gambling research among college students. AUDIT assesses alcohol consumption and other alcohol-related risks and problems (a higher score means more risks). A correlation

TABLE 10.3

Annual number of tornadoes in the United States between 1953 and 2008

Year	Number of tornadoes	Year	Number of tornadoes	Year	Number of tornadoes	Year	Number of tornadoes
1953	421	1967	926	1981	783	1995	1235
1954	550	1968	660	1982	1046	1996	1173
1955	593	1969	608	1983	931	1997	1148
1956	504	1970	653	1984	907	1998	1449
1957	856	1971	888	1985	684	1999	1340
1958	564	1972	741	1986	764	2000	1076
1959	604	1973	1102	1987	656	2001	1213
1960	616	1974	947	1988	702	2002	934
1961	697	1975	920	1989	856	2003	1372
1962	657	1976	835	1990	1133	2004	1819
1963	464	1977	852	1991	1132	2005	1194
1964	704	1978	788	1992	1298	2006	1103
1965	906	1979	852	1993	1176	2007	1098
1966	585	1980	866	1994	1082	2008	1691

of 0.29 was reported between the frequency of gambling and the AUDIT score.

(a) What percent of the variability in AUDIT score is explained by frequency of gambling?

(b) Test the null hypothesis that the correlation between the gambling frequency and the AUDIT score is zero.

(c) The sample in this study represents 45% of the students contacted for the online study. To what extent do you think these results apply to all first-year students at this university? To what extent do you think those results apply to all first-year students? Give reasons for your answers.

10.22 ⚔️ **Predicting water quality.** The index of biotic integrity (IBI) is a measure of the water quality in streams. IBI and land use measures for a collection of streams in the Ozark Highland ecoregion of Arkansas were collected as part of a study.[17] Table 10.4 gives the data for IBI, the percent forest, and the area of the watershed in square kilometers for streams in the original sample with area less than or equal to 70 km^2. 💾 IBI

(a) Use numerical and graphical methods to describe the variable IBI. Do the same for area. Summarize your results.

(b) Plot the data and describe the relationship. Are there any outliers or unusual patterns?

(c) Give the statistical model for simple linear regression for this problem.

(d) State the null and alternative hypotheses for examining the relationship between IBI and area.

(e) Run the simple linear regression and summarize the results.

(f) Obtain the residuals and plot them versus area. Is there anything unusual in the plot?

(g) Do the residuals appear to be approximately Normal? Give reasons for your answer.

(h) Do the assumptions for the analysis of these data using the model you gave in part (c) appear to be reasonable? Explain your answer.

10.23 ⚔️ **More on predicting water quality.** The researchers who conducted the study described in the previous exercise also recorded the percent of the watershed area that was forest for each of the streams. The data are also given in Table 10.4. Analyze these data using the questions in the previous exercise as a guide. 💾 IBI

10.24 **Comparing the analyses.** In Exercises 10.22 and 10.23, you used two different explanatory variables to predict IBI. Summarize the two analyses and compare the results. If you had to choose between the two explanatory variables for predicting IBI, which one would you prefer? Give reasons for your answer. 💾 IBI

10.25 **How an outlier can affect statistical significance.** Consider the data in Table 10.4 and the relationship between IBI and the percent of watershed area that was forest. The relationship between these two variables is almost significant at the 0.05 level. In this exercise you will demonstrate the potential effect of an outlier on statistical significance. Investigate what happens when you decrease the IBI to 0.0 for (1) an observation with 0% forest and (2) an observation with 100% forest. Write a short summary of what you learn from this exercise. 💾 IBI

10.26 **Predicting water quality for an area of 40 km^2.** Refer to Exercise 10.22. 💾 IBI

(a) Find a 95% confidence interval for mean response corresponding to an area of 40 km^2.

(b) Find a 95% prediction interval for a future response corresponding to an area of 40 km^2.

TABLE 10.4

Watershed area, percent forest, and index of biotic integrity

Area	Forest	IBI	Area	Forest	IBI	Area	Forest	IBI	Area	Forest	IBI	Area	Forest	IBI
21	0	47	29	0	61	31	0	39	32	0	59	34	0	72
34	0	76	49	3	85	52	3	89	2	7	74	70	8	89
6	9	33	28	10	46	21	10	32	59	11	80	69	14	80
47	17	78	8	17	53	8	18	43	58	21	88	54	22	84
10	25	62	57	31	55	18	32	29	19	33	29	39	33	54
49	33	78	9	39	71	5	41	55	14	43	58	9	43	71
23	47	33	31	49	59	18	49	81	16	52	71	21	52	75
32	59	64	10	63	41	26	68	82	9	75	60	54	79	84
12	79	83	21	80	82	27	86	82	23	89	86	26	90	79
16	95	67	26	95	56	26	100	85	28	100	91			

(c) Write a short paragraph interpreting the meaning of the intervals in terms of Ozark Highland streams.

(d) Do you think that these results can be applied to other streams in Arkansas or in other states? Explain why or why not.

10.27 Compare the predictions. Consider case 37 in Table 10.4 (8th row, 2nd column). For this case the area is 10 km^2 and the percent forest is 63%. A predicted index of biotic integrity based on area was computed in Exercise 10.22, while one based on percent forest was computed in Exercise 10.23. Compare these two estimates and explain why they differ. Use the idea of a prediction interval to interpret these results. 🔘 IBI

10.28 Reading test scores and IQ. In Exercise 2.29 (page 97) you examined the relationship between a reading test score and an IQ score for a sample of 60 fifth-grade children. 🔘 READINGANDIQ

(a) Run the regression and summarize the results of the significance tests.

(b) Rerun the analysis with the four possible outliers removed. Summarize your findings, paying particular attention to the effects of removing the outliers.

10.29 🔘 **Neuron responses.** Exercise 2.162 (page 158) gives data on neuron responses to pure tones and to monkey calls. 🔘 MONKEYCALLS

(a) Describe each variable graphically and numerically.

(b) Plot the data with the pure tone response on the x-axis and the monkey call response on the y-axis. Describe the relationship and mark the point with the largest residual and the point with the extreme value of the tone response.

(c) Analyze the entire set of 37 observations and summarize the results.

(d) Perform additional analyses to assess the effects of the two marked points on the results. Summarize your findings.

10.30 School budget and number of students. Suppose that there is a linear relationship between the number of students x in an elementary school and the annual budget y. Write a population regression model to describe this relationship.

(a) Which parameter in your model is the fixed cost in the budget (for example, the salary of the principal and some administrative costs) that does not change as x increases?

(b) Which parameter in your model shows how total cost changes when there are more students in the school? Do you expect this number to be greater than 0 or less than 0?

(c) Actual data from schools will not fit a straight line exactly. What term in your model allows variation among schools of the same size x?

10.31 🔘 **Breaking strength of wood.** Exercise 2.163 (page 158) gives the modulus of elasticity (MOE) and the modulus of rupture (MOR) for 32 plywood specimens. Because measuring MOR involves breaking the wood but measuring MOE does not, we would like to predict the destructive test result, MOR, using the nondestructive test result, MOE.

(a) Describe the distribution of MOR using graphical and numerical summaries. Do the same for MOE.

(b) Make a plot of the two variables. Which should be plotted on the x axis? Give a reason for your answer.

(c) Give the statistical model for this analysis, run the analysis, summarize the results, and write a short summary of your conclusions.

(d) Examine the assumptions needed for the analysis. Are you satisfied that there are no serious violations that would cause you to question the validity of your conclusions?

10.32 Breaking strength of wood, continued. Refer to the previous exercise. Consider a MOE of 2,400,000.

(a) Interpret the confidence interval for mean response and the prediction interval for a future observation for this value of MOE.

(b) Which interval will include more values? Give a reason for your answer.

(c) (Optional) Calculate the two intervals.

10.33 Stocks and bonds. How is the flow of investors' money into stock mutual funds related to the flow of money into bond mutual funds? Table 10.5 shows the net new money flowing into stock and bond mutual funds in the years 1985 to 2008, in billions of dollars.[18] "Net" means that funds flowing out are subtracted from those flowing in. If more money leaves than arrives, the net flow will be negative. To eliminate the effect of inflation, all dollar amounts are in "real dollars" with constant buying power equal to that of a dollar in the year 2000. 🔘 MONEYFLOW

(a) Make a scatterplot with cash flow into stock funds as the explanatory variable. Find the least-squares line for predicting net bond investments from net stock investments. What do the data suggest?

(b) Is there statistically significant evidence that there is some straight-line relationship between the flows of cash into bond funds and stock funds? (State

TABLE 10.5

Net new money flowing into stock and bond mutual funds

Year	Stocks	Bonds	Year	Stocks	Bonds	Year	Stocks	Bonds
1985	12.8	100.8	1993	151.3	84.6	2001	31.1	85.0
1986	34.6	161.8	1994	133.6	−72.0	2002	−25.8	134.4
1987	28.8	10.6	1995	140.1	−6.8	2003	143.0	29.4
1988	−23.3	−5.8	1996	238.2	3.3	2004	161.8	−9.4
1989	8.3	−1.4	1997	243.5	30.0	2005	119.7	27.7
1990	17.1	9.2	1998	165.9	79.2	2006	134.9	51.9
1991	50.6	74.6	1999	194.3	−6.2	2007	76.8	90.1
1992	97.0	87.1	2000	309.0	−48.0	2008	−187.4	24.9

hypotheses, give a test statistic and its P-value, and state your conclusion.)

(c) Remove the data for 2008 and refit the remaining years. Is there now statistically significant evidence of a straight-line relationship?

(d) How would you report these results in a manuscript? In other words, how would you handle the change in statistical significance caused by this one observation?

10.34 Math pretest predicts success? Can a pretest on mathematics skills predict success in a statistics course? The 82 students in an introductory statistics class took a pretest at the beginning of the semester. The least-squares regression line for predicting the score y on the final exam from the pretest score x was $\hat{y} = 15.3 + 0.72x$. The standard error of b_1 was 0.38.

(a) Test the null hypothesis that there is no linear relationship between the pretest score and the score on the final exam against the two-sided alternative.

(b) Would you reject this null hypothesis versus the one-sided alternative that the slope is positive? Explain your answer.

10.35 Do wages rise with experience? We assume that our wages will increase as we gain experience and become more valuable to our employers. Wages also increase because of inflation. By examining a sample of employees at a given point in time, we can look at part of the picture. How does length of service (LOS) relate to wages? Consider a sample of 60 women who work in Indiana banks. LOS is measured in months and wages are yearly total income divided by the number of weeks worked. We have multiplied wages by a constant for reasons of confidentiality.[19] LOS

(a) Plot wages versus LOS. Describe the relationship. There is one woman with relatively high wages for her length of service. Circle this point and do not use it in the rest of this exercise.

(b) Find the least-squares line. Summarize the significance test for the slope. What do you conclude?

(c) State carefully what the slope tells you about the relationship between wages and length of service.

(d) Give a 95% confidence interval for the slope.

10.36 Do wages rise with experience? Refer to the previous exercise. Analyze the data with the outlier included. LOS

(a) How does this change the estimates of the parameters β_0, β_1, and σ?

(b) What effect does the outlier have on the results of the significance test for the slope?

(c) How has the width of the 95% confidence interval changed?

10.37 Leaning Tower of Pisa. The Leaning Tower of Pisa is an architectural wonder. Engineers concerned about the tower's stability have done extensive studies of its increasing tilt. Measurements of the lean of the tower over time provide much useful information. The following table gives measurements for the years 1975 to 1987. The variable "lean" represents the difference between where a point on the tower would be if the tower were straight and where it actually is. The data are coded as tenths of a millimeter in excess of 2.9 meters, so that the 1975 lean, which was 2.9642 meters, appears in the table as 642. Only the last two digits of the year were entered into the computer.[20] PISA

Year	75	76	77	78	79	80	81	82	83	84	85	86	87
Lean	642	644	656	667	673	688	696	698	713	717	725	742	757

(a) Plot the data. Does the trend in lean over time appear to be linear?

(b) What is the equation of the least-squares line? What percent of the variation in lean is explained by this line?

(c) Give a 99% confidence interval for the average rate of change (tenths of a millimeter per year) of the lean.

10.38 More on the Leaning Tower of Pisa. Refer to the previous exercise.

(a) In 1918 the lean was 2.9071 meters. (The coded value is 71.) Using the least-squares equation for the years 1975 to 1987, calculate a predicted value for the lean in 1918. (Note that you must use the coded value 18 for year.)

(b) Although the least-squares line gives an excellent fit to the data for 1975 to 1987, this pattern did not extend back to 1918. Write a short statement explaining why this conclusion follows from the information available. Use numerical and graphical summaries to support your explanation.

10.39 Predicting the lean in 2012. Refer to the previous two exercises.

(a) How would you code the explanatory variable for the year 2012?

(b) The engineers working on the Leaning Tower of Pisa were most interested in how much the tower would lean if no corrective action was taken. Use the least-squares equation to predict the tower's lean in the year 2012. (Note: The Tower was renovated in 2001 to make sure it does not fall down.)

(c) To give a margin of error for the lean in 2012, would you use a confidence interval for a mean response or a prediction interval? Explain your choice.

10.40 Correlation between binge drinking and the average price of beer. A recent study looked at 118 colleges to investigate the association between the binge-drinking rate and the average price for a bottle of beer at establishments within a 2-mile radius of campus.[21] A correlation of −0.36 was found. Explain this correlation.

10.41 Is this relationship significant? Refer to the previous exercise. Test the null hypothesis that the correlation between the binge-drinking rate and the average price for a bottle of beer within a 2-mile radius of campus is zero.

10.42 Capacity of DRAM. The capacity (bits) of the largest DRAM (dynamic random access memory) chips commonly available at retail has increased as follows:[22] 🔘 DRAM

Year	1971	1980	1987	1993	1998	2000	2002	2004
Kilobits	1	62.5	1000	16,000	125,000	250,000	500,000	976,526.5

(a) Make a scatterplot of the data. Growth is much faster than linear.

(b) Plot the logarithm of DRAM capacity against year. These points are close to a straight line.

(c) Regress the logarithm of DRAM capacity on year. Give a 95% confidence interval for the slope of the population regression line.

10.43 Net flow in stock and bond funds. Is there a nonzero correlation between net flow of money into stock mutual funds and into bond funds? Use the regression analysis you did in Exercise 10.33 (page 585) to answer this question with no additional calculations.
🔘 MONEYFLOW

10.44 Parental behavior and self-esteem. Chinese students from public schools in Hong Kong were the subjects of a study designed to investigate the relationship between various measures of parental behavior and other variables. The sample size was 713. The data were obtained from questionnaires filled in by the students. One of the variables examined was parental control, an indication of the amount of control that the parents exercised over the behavior of the students. Another was the self-esteem of the students.[23]

(a) The correlation between parental control and self-esteem was $r = -0.19$. Calculate the t statistic for testing the null hypothesis that the population correlation is 0.

(b) Find an approximate P-value for testing H_0 versus the two-sided alternative and report your conclusion.

10.45 Completing an ANOVA table. How are returns on common stocks in overseas markets related to returns in U.S. markets? Consider measuring U.S. returns by the annual rate of return on the Standard & Poor's 500 stock index and overseas returns by the annual rate of return on the Morgan Stanley Europe, Australasia, Far East (EAFE) index. Both are recorded in percents. Regressing the EAFE returns on the S&P 500 returns for the 20 years 1989 to 2008. Here is part of the Minitab output for this regression:

```
The regression equation is
EAFE = - 2.58 + 0.775 S&P

Analysis of Variance

Source          DF      SS       MS      F
Regression            4560.6
Residual Error
Total           19    8556.0
```

Complete the analysis of variance table by filling in the missing entries.

10.46 Interpreting statistical software output. Refer to the previous exercise. What are the values of the regression standard error s and the squared correlation r^2?

10.47 Standard error and confidence interval for the slope. Refer to the previous two exercises. The standard deviation of the S&P 500 returns for these years is 19.99%. From this and your work in the previous exercise, find the standard error for the least-squares slope b_1. Give a 95% confidence interval for the slope β_1 of the population regression line.

10.48 Quality of life in chronically ill patients. Concern about the quality of life for chronically ill patients is becoming as important as treating their physical symptoms. The SF-36, a questionnaire for measuring the health quality of life, was given to 50 patients with chronic obstructive lung disease.[24] A correlation of 0.68 was reported between the component of the questionnaire called general health perceptions (GHP) and a measure of lung function called forced vital capacity (FVC), expressed as a percent of normal. The mean and standard deviation of GHP are 43.5 and 20.3, and for FVC the values are 80.9 and 17.2.

(a) Find the equation of the least-squares line for predicting GHP from FVC.

(b) Give the results of the significance test for the null hypothesis that the slope is 0. (*Hint:* What is the relation between this test and the test for a zero correlation?)

10.49 Significance test of the correlation. A study reported a correlation $r = 0.6$ based on a sample size of $n = 20$; another reported the same correlation based on a sample size of $n = 10$. For each, perform the test of the null hypothesis that $\rho = 0$. Describe the results and explain why the conclusions are different.

10.50 Verifying the effect of bank size. Refer to the bank wages data described in Exercise 10.35 (page 586). The data also include a variable "Size," which classifies the bank as large or small. Obtain the residuals from the regression used to predict wages from LOS, and plot them versus LOS using different symbols for the large and small banks. Include on your plot a horizontal line at 0 (the mean of the residuals). Describe the important features of this plot. Explain what they indicate about wages in this set of data. 🌀 LOS

10.51 SAT versus ACT. The SAT and the ACT are the two major standardized tests that colleges use to evaluate candidates. Most students take just one of these tests. However, some students take both. Consider the scores of 60 students who did this. How can we relate the two tests? 🌀 SAT_ACT

(a) Plot the data with SAT on the x-axis and ACT on the y-axis. Describe the overall pattern and any unusual observations.

(b) Find the least-squares regression line and draw it on your plot. Give the results of the significance test for the slope.

(c) What is the correlation between the two tests?

10.52 🔺 **SAT versus ACT, continued.** Refer to the previous exercise. Find the predicted value of ACT for each observation in the data set. 🌀 SAT_ACT

(a) What is the mean of these predicted values? Compare it with the mean of the ACT scores.

(b) Compare the standard deviation of the predicted values with the standard deviation of the actual ACT scores. If least-squares regression is used to predict ACT scores for a large number of students such as these, the average predicted value will be accurate but the variability of the predicted scores will be too small.

(c) Find the SAT score for a student who is one standard deviation above the mean ($z = (x - \bar{x})/s = 1$). Find the predicted ACT score and standardize this score. (Use the means and standard deviations from this set of data for these calculations.)

(d) Repeat part (c) for a student whose SAT score is one standard deviation below the mean ($z = -1$).

(e) What do you conclude from parts (c) and (d)? Perform additional calculations for different z's if needed.

10.53 🔺 **Matching standardized scores.** Refer to the previous two exercises. An alternative to the least-squares method is based on matching standardized scores. Specifically, we set

$$\frac{(y - \bar{y})}{s_y} = \frac{(x - \bar{x})}{s_x}$$

and solve for y. Let's use the notation $y = a_0 + a_1 x$ for this line. The slope is $a_1 = s_y/s_x$ and the intercept is $a_0 = \bar{y} - a_1 \bar{x}$. Compare these expressions with the formulas for the least-squares slope and intercept (page 550). 🌀 SAT_ACT

(a) Using the data in the previous exercise, find the values of a_0 and a_1.

(b) Plot the data with the least-squares line and the new prediction line.

(c) Use the new line to find predicted ACT scores. Find the mean and the standard deviation of these scores. How do they compare with the mean and the standard deviation of the ACT scores?

10.54 Length, width, and weight of perch. Here are data for 12 perch caught in a lake in Finland:[25] 🐟 PERCH

Weight (grams)	Length (cm)	Width (cm)	Weight (grams)	Length (cm)	Width (cm)
5.9	8.8	1.4	300.0	28.7	5.1
100.0	19.2	3.3	300.0	30.1	4.6
110.0	22.5	3.6	685.0	39.0	6.9
120.0	23.5	3.5	650.0	41.4	6.0
150.0	24.0	3.6	820.0	42.5	6.6
145.0	25.5	3.8	1000.0	46.6	7.6

In this exercise we will examine different models for predicting weight.

(a) Run the regression using length to predict weight. Do the same using width as the explanatory variable. Summarize the results. Be sure to include the value of r^2.

(b) Plot weight versus length and weight versus width. Include the least-squares lines in these plots. Do these relationships appear to be linear? Explain your answer.

10.55 🔺 **Transforming the perch data.** Refer to the previous exercise.

(a) Try to find a better model using a transformation of length. One possibility is to use the square. Make a plot and perform the regression analysis. Summarize the results.

(b) Do the same for width.

10.56 🔺 **Creating a new explanatory variable.** Refer to the previous two exercises.

(a) Create a new variable that is the product of length and width. Make a plot and run the regression using this new variable. Summarize the results.

(b) Write a short report summarizing and comparing the different regression analyses that you performed in this exercise and the previous two exercises.

10.57 🔺 **Index of biotic integrity.** Refer to the data on the index of biotic integrity and area in Exercise 10.22 (page 584) and the additional data on percent watershed area that was forest in Exercise 10.23. Find the correlations among these three variables, perform the test of statistical significance, and summarize the results. Which of these test results could have been obtained from the analyses that you performed in Exercises 10.22 and 10.23? 🐟 IBI

10.58 Food neophobia. Food neophobia is a personality trait associated with avoiding unfamiliar foods. In one study of 564 children who were 2 to 6 years of age, food neophobia and the frequency of consumption of different types of food were measured.[26] Here is a summary of the correlations:

Type of food	Correlation
Vegetables	−0.27
Fruit	−0.16
Meat	−0.15
Eggs	−0.08
Sweet/fatty snacks	0.04
Starchy staples	−0.02

Perform the significance test for each correlation and write a summary about food neophobia and the consumption of different types of food.

10.59 A mechanistic explanation of popularity. Previous experimental work has suggested that the serotonin system plays an important and causal role in social status. In other words, genes may predispose individuals to be popular/likeable. As part of a recent study on adolescents, an experimenter looked at the relationship between the expression of a particular serotonin receptor gene, a person's "popularity," and the person's rule-breaking (RB) behaviors.[27] RB was measured both through a questionnaire and video observation. The composite score is an equal combination of these two assessments. Here is a table of the correlations:

Rule-Breaking Measure	Popularity	Gene Expression
Sample 1 (n = 123)		
RB.composite	0.28	0.26
RB.questionnaire	0.22	0.23
RB.video	0.24	0.20
Sample 1 Caucasians only (n = 96)		
RB.composite	0.22	0.23
RB.questionnaire	0.16	0.24
RB.video	0.19	0.16

For each correlation, test the null hypothesis that the corresponding true correlation is zero. Reproduce the table and mark the correlations that have $P < 0.001$ with ***, those that have $P < 0.01$ with **, and those that have $P < 0.05$ with *. Write a summary of the results of your significance tests.

10.60 Resting metabolic rate and exercise. Metabolic rate, the rate at which the body consumes energy, is important in studies of weight gain, dieting, and exercise. The following table gives data on the lean body mass and resting metabolic rate for 12 women and 7 men who are subjects in a study of dieting. Lean body mass, given in kilograms, is a person's weight leaving out all fat. Metabolic rate is measured in calories burned per 24 hours, the same calories used to describe the energy content of foods. The researchers believe that lean body mass is an important influence on metabolic rate.

METRATE

Subject	Sex	Mass	Rate	Subject	Sex	Mass	Rate
1	M	62.0	1792	11	F	40.3	1189
2	M	62.9	1666	12	F	33.1	913
3	F	36.1	995	13	M	51.9	1460
4	F	54.6	1425	14	F	42.4	1124
5	F	48.5	1396	15	F	34.5	1052
6	F	42.0	1418	16	F	51.1	1347
7	M	47.4	1362	17	F	41.2	1204
8	F	50.6	1502	18	M	51.9	1867
9	F	42.0	1256	19	M	46.9	1439
10	M	48.7	1614				

(a) Make a scatterplot of the data, using different symbols or colors for men and women. Summarize what you see in the plot.

(b) Run the regression to predict metabolic rate from lean body mass for the women in the sample and summarize the results. Do the same for the men.

10.61 🛡 **Resting metabolic rate and exercise, continued.** Refer to the previous exercise. It is tempting to conclude that there is a strong linear relationship for the women but no relationship for the men. Let's look at this issue a little more carefully.

(a) Find the confidence interval for the slope in the regression equation that you ran for the females. Do the same for the males. What do these suggest about the possibility that these two slopes are the same? (The formal method for making this comparison is a bit complicated and is beyond the scope of this chapter.)

(b) Examine the formula for the standard error of the regression slope given on page 572. The term in the denominator is $\sqrt{\Sigma(x_i - \bar{x})^2}$. Find this quantity for the females; do the same for the males. How do these calculations help to explain the results of the significance tests?

(c) Suppose you were able to collect additional data for males. How would you use lean body mass in deciding which subjects to choose?

10.62 🛡 **Inference over different ranges of X.** Think about what would happen if you analyzed a subset of a set of data by analyzing only data for a restricted range of values of the explanatory variable. What results would you expect to change? Examine your ideas by analyzing the fuel efficiency data described in Example 10.10 (page 560). First, run a regression of MPG versus MPH using all cases. This least squares regression line is shown in Figure 10.9. Next run a regression of MPG versus MPH for only those cases with speed less than or equal to 30 mph. Note that this corresponds to 3.4 in the log scale. Finally, do the same analysis with a restriction on the response variable. Run the analysis with only those cases with fuel efficiency less than or equal to 20 mpg. Write a summary comparing the effects of these two restrictions with each other and with the complete data set results. MPH_MPG

Multiple Regression

Introduction

In Chapter 10 we presented methods for inference in the setting of a linear relationship between a response variable y and a *single* explanatory variable x. In this chapter, we use *more than one* explanatory variable to explain or predict a single response variable. Many of the ideas that we encountered in our study of simple linear regression carry over to the multiple linear regression setting. For example, the descriptive tools we learned in Chapter 2—scatterplots, least-squares regression, and correlation—are still essential preliminaries to inference and also provide a foundation for confidence intervals and significance tests. However, the introduction of several explanatory variables leads to many additional considerations. In this short chapter we cannot explore all these issues. Rather, we will outline some basic facts about inference in the multiple regression setting and then illustrate the analysis with a case study whose purpose was to predict success in college based on several high school achievement scores.

11.1 Inference for Multiple Regression

Population multiple regression equation

The simple linear regression model assumes that the mean of the response variable y depends on the explanatory variable x according to a linear equation

$$\mu_y = \beta_0 + \beta_1 x$$

For any fixed value of x, the response y varies Normally around this mean and has a standard deviation σ that is the same for all values of x.

In the multiple regression setting, the response variable y depends on p explanatory variables, which we will denote by $x_1, x_2, \ldots, x_p$. The mean response depends on these explanatory variables according to a linear function

$$\mu_y = \beta_0 + \beta_1 x_1 + \beta_2 x_2 + \cdots + \beta_p x_p$$

population regression equation   Similar to simple linear regression, this expression is the **population regression equation**, and we do not observe these mean responses because the observed values of y vary about their means. We can also think of subpopulations of responses, just as we did in simple linear regression. Here, each subpopulation corresponds to a particular set of values for *all* of the explanatory variables $x_1, x_2, \ldots, x_p$. In each subpopulation, y varies Normally with a mean given by the population regression equation. The regression model assumes that the standard deviation σ of the responses is the same in all subpopulations.

EXAMPLE

11.1   Predicting early success in college.   Our case study uses data collected at a large university on all first-year computer science majors in a particular year.[1] The purpose of the study was to attempt to predict success in the early university years. One measure of success was the cumulative grade point average (GPA) after three semesters. Among the explanatory variables recorded at the time the students enrolled in the university were average high school grades in mathematics (HSM), science (HSS), and English (HSE).

We will use high school grades to predict the response variable GPA. There are $p = 3$ explanatory variables: $x_1 = $ HSM, $x_2 = $ HSS, and $x_3 = $ HSE. The high school grades are coded on a scale from 1 to 10, with 10 corresponding to A, 9 to A−, 8 to B+, and so on. These grades define the subpopulations. For example, the straight-C students are the subpopulation defined by HSM = 4, HSS = 4, and HSE = 4.

One possible multiple regression model for the subpopulation mean GPAs is

$$\mu_{\text{GPA}} = \beta_0 + \beta_1 \text{HSM} + \beta_2 \text{HSS} + \beta_3 \text{HSE}$$

For the straight-C subpopulation of students, the model gives the subpopulation mean as

$$\mu_{GPA} = \beta_0 + \beta_1 4 + \beta_2 4 + \beta_3 4$$

Data for multiple regression

The data for a simple linear regression problem consist of observations (x_i, y_i) of the two variables. Because there are several explanatory variables in multiple regression, the notation needed to describe the data is more elaborate. Each observation or case consists of a value for the response variable and for each of the explanatory variables. Call x_{ij} the value of the jth explanatory variable for the ith case. The data are then

$$\text{Case 1: } (x_{11}, x_{12}, \ldots, x_{1p}, y_1)$$
$$\text{Case 2: } (x_{21}, x_{22}, \ldots, x_{2p}, y_2)$$
$$\vdots$$
$$\text{Case } n: (x_{n1}, x_{n2}, \ldots, x_{np}, y_n)$$

Here, n is the number of cases and p is the number of explanatory variables. Data are often entered into computer regression programs in this format. Each row is a case and each column corresponds to a different variable. The data for Example 11.1, with several additional explanatory variables, appear in this format in the CSDATA data file. Figure 11.1 shows the first 5 rows. Grade point average (GPA) is the response variable, followed by $p = 6$ explanatory variables. There are a total of $n = 224$ students in this data set.

FIGURE 11.1 Format of data set for Example 11.1 in an Excel spreadsheet.

	A	B	C	D	E	F	G	H
1	obs	gpa	hsm	hss	hse	satm	satv	sex
2	1	3.32	10	10	10	670	600	1
3	2	2.26	6	8	5	700	640	1
4	3	2.35	8	6	8	640	530	1
5	4	2.08	9	10	7	670	600	1
6	5	3.38	8	9	8	540	580	1

USE YOUR KNOWLEDGE

11.1 Describing a multiple regression. As part of a recent study titled "Predicting Success for Actuarial Students in Undergraduate Mathematics Courses," data from 106 Bryant College actuarial graduates were obtained.[2] The researchers were interested in describing how students' overall math grade-point averages are explained by SAT Math and Verbal scores, class rank, and Bryant College's mathematics placement score.

(a) What is the response variable?

(b) What is n, the number of cases?

(c) What is p, the number of explanatory variables?

(d) What are the explanatory variables?

Multiple linear regression model

LOOK BACK
DATA = FIT + RESIDUAL
p. 549

We combine the population regression equation and assumptions about variation to construct the multiple linear regression model. The subpopulation means describe the FIT part of our statistical model. The RESIDUAL part represents the variation of observations about the means. We will use the same notation for the residual that we used in the simple linear regression model. The symbol ϵ represents the deviation of an individual observation from its subpopulation mean. We assume that these deviations are Normally distributed with mean 0 and an unknown standard deviation σ that does not depend on the values of the x variables. *These are assumptions that we can check by examining the residuals in the same way that we did for simple linear regression.*

MULTIPLE LINEAR REGRESSION MODEL

The **statistical model for multiple linear regression** is

$$y_i = \beta_0 + \beta_1 x_{i1} + \beta_2 x_{i2} + \cdots + \beta_p x_{ip} + \epsilon_i$$

for $i = 1, 2, \ldots, n$.

The **mean response** μ_y is a linear function of the explanatory variables:

$$\mu_y = \beta_0 + \beta_1 x_1 + \beta_2 x_2 + \cdots + \beta_p x_p$$

The **deviations** ϵ_i are independent and Normally distributed with mean 0 and standard deviation σ. In other words, they are an **SRS** from the $N(0, \sigma)$ distribution.

The parameters of the model are $\beta_0, \beta_1, \beta_2, \ldots, \beta_p$, and σ.

The assumption that the subpopulation means are related to the regression coefficients β by the equation

$$\mu_y = \beta_0 + \beta_1 x_1 + \beta_2 x_2 + \cdots + \beta_p x_p$$

implies that we can estimate all subpopulation means from estimates of the β's. To the extent that this equation is accurate, we have a useful tool for describing how the mean of y varies with the collection of x's.

USE YOUR KNOWLEDGE

11.2 Understanding the fitted regression line. The fitted regression equation for a multiple regression is

$$\hat{y} = -3.8 + 7.3x_1 - 2.1x_2$$

(a) If $x_1 = 3$ and $x_2 = 1$, what is the predicted value of y?

(b) For the answer to part (a) to be valid, is it necessary that the values $x_1 = 3$ and $x_2 = 1$ correspond to a case in the data set? Explain why or why not.

(c) If you hold x_2 at a fixed value, what is the effect of an increase of two units in x_1 on the predicted value of y?

Estimation of the multiple regression parameters

LOOK BACK
least squares p. 112

Similar to simple linear regression, we use the method of least squares to obtain estimators of the regression coefficients β. The details, however, are more complicated. Let

$$b_0, \ b_1, \ b_2, \ldots, \ b_p$$

denote the estimators of the parameters

$$\beta_0, \ \beta_1, \ \beta_2, \ldots, \ \beta_p$$

For the ith observation, the predicted response is

$$\hat{y}_i = b_0 + b_1 x_{i1} + b_2 x_{i2} + \cdots + b_p x_{ip}$$

LOOK BACK
residual p. 550

The ith residual, the difference between the observed and predicted response, is therefore

$$e_i = \text{observed response } - \text{ predicted response}$$
$$= y_i - \hat{y}_i$$
$$= y_i - b_0 - b_1 x_{i1} - b_2 x_{i2} - \cdots - b_p x_{ip}$$

The method of least squares chooses the values of the b's that make the sum of the squared residuals as small as possible. In other words, the parameter estimates $b_0, b_1, b_2, \ldots, b_p$ minimize the quantity

$$\sum (y_i - b_0 - b_1 x_{i1} - b_2 x_{i2} - \cdots - b_p x_{ip})^2$$

The formula for the least-squares estimates is complicated. We will be content to understand the principle on which it is based and to let software do the computations.

The parameter σ^2 measures the variability of the responses about the population regression equation. As in the case of simple linear regression, we estimate σ^2 by an average of the squared residuals. The estimator is

$$s^2 = \frac{\sum e_i^2}{n - p - 1}$$
$$= \frac{\sum (y_i - \hat{y}_i)^2}{n - p - 1}$$

LOOK BACK
degrees of freedom
p. 551

The quantity $n - p - 1$ is the degrees of freedom associated with s^2. The degrees of freedom equal the sample size, n, minus $p + 1$, the number of β's we must estimate to fit the model. In the simple linear regression case there is just one explanatory variable, so $p = 1$ and the degrees of freedom are $n - 2$. To estimate σ we use

$$s = \sqrt{s^2}$$

Confidence intervals and significance tests for regression coefficients

We can obtain confidence intervals and perform significance tests for each of the regression coefficients β_j as we did in simple linear regression. The standard errors of the b's have more complicated formulas, but all are multiples of s. We again rely on statistical software to do the calculations.

CONFIDENCE INTERVALS AND SIGNIFICANCE TESTS FOR β_j

A **level C confidence interval** for β_j is

$$b_j \pm t^* SE_{b_j}$$

where SE_{b_j} is the standard error of b_j and t^* is the value for the $t(n - p - 1)$ density curve with area C between $-t^*$ and t^*.

To test the hypothesis $H_0: \beta_j = 0$, compute the **t statistic**

$$t = \frac{b_j}{SE_{b_j}}$$

In terms of a random variable T having the $t(n - p - 1)$ distribution, the P-value for a test of H_0 against

$H_a: \beta_j > 0$ is $P(T \geq t)$

$H_a: \beta_j < 0$ is $P(T \leq t)$

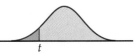

$H_a: \beta_j \neq 0$ is $2P(T \geq |t|)$

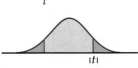

LOOK BACK
confidence intervals for mean response p. 557

LOOK BACK
prediction intervals p. 559

Because regression is often used for prediction, we may wish to use multiple regression models to construct confidence intervals for a mean response and prediction intervals for a future observation. The basic ideas are the same as in the simple linear regression case. In most software systems, the same commands that give confidence and prediction intervals for simple linear regression work for multiple regression. The only difference is that we specify a list of explanatory variables rather than a single variable. Modern software allows us to perform these rather complex calculations without an intimate knowledge of all the computational details. This frees us to concentrate on the meaning and appropriate use of the results.

ANOVA table for multiple regression

LOOK BACK
ANOVA F test p. 568

In simple linear regression the F test from the ANOVA table is equivalent to the two-sided t test of the hypothesis that the slope of the regression line is 0. For multiple regression there is a corresponding ANOVA F test, but it tests the hypothesis that *all* the regression coefficients (with the exception of the intercept) are 0. Here is the general form of the ANOVA table for multiple regression:

Source	Degrees of freedom	Sum of squares	Mean square	F
Model	p	$\sum(\hat{y}_i - \bar{y})^2$	SSM/DFM	MSM/MSE
Error	$n - p - 1$	$\sum(y_i - \hat{y}_i)^2$	SSE/DFE	
Total	$n - 1$	$\sum(y_i - \bar{y})^2$	SST/DFT	

The ANOVA table is similar to that for simple linear regression. The degrees of freedom for the model increase from 1 to p to reflect the fact that we now have p explanatory variables rather than just one. As a consequence, the degrees of freedom for error decrease by the same amount. *It is always a good idea to calculate the degrees of freedom by hand and then check that your software agrees with your calculations. In this way you can verify that your software is using the number of cases and number of explanatory variables that you intended.*

The sums of squares represent sources of variation. Once again, both the sums of squares and their degrees of freedom add:

$$SST = SSM + SSE$$
$$DFT = DFM + DFE$$

The estimate of the variance σ^2 for our model is again given by the MSE in the ANOVA table. That is, $s^2 = MSE$.

LOOK BACK
F statistic p. 568

The ratio MSM/MSE is an F statistic for testing the null hypothesis

$$H_0: \beta_1 = \beta_2 = \cdots = \beta_p = 0$$

against the alternative hypothesis

$$H_a: \text{at least one of the } \beta_j \text{ is not } 0$$

The null hypothesis says that none of the explanatory variables are predictors of the response variable when used in the form expressed by the multiple regression equation. The alternative states that at least one of them is a predictor of the response variable. As in simple linear regression, large values of F give evidence against H_0. When H_0 is true, F has the $F(p, n - p - 1)$ distribution. The degrees of freedom for the F distribution are those associated with the model and error in the ANOVA table. *A common error in the use of multiple regression is to assume that all the regression coefficients are statistically different from zero whenever the F statistic has a small P-value.* Be sure that you understand the difference between the F test and the t tests for individual coefficients.

ANALYSIS OF VARIANCE F TEST

In the multiple regression model, the hypothesis

$$H_0: \beta_1 = \beta_2 = \cdots = \beta_p = 0$$

is tested against the alternative hypothesis

$$H_a: \text{at least one of the } \beta_j \text{ is not } 0$$

by the analysis of variance F statistic

$$F = \frac{\text{MSM}}{\text{MSE}}$$

The P-value is the probability that a random variable having the $F(p, n - p - 1)$ distribution is greater than or equal to the calculated value of the F statistic.

Squared multiple correlation R^2

For simple linear regression we noted that the square of the sample correlation could be written as the ratio of SSM to SST and could be interpreted as the proportion of variation in y explained by x. A similar statistic is routinely calculated for multiple regression.

THE SQUARED MULTIPLE CORRELATION

The statistic

$$R^2 = \frac{\text{SSM}}{\text{SST}} = \frac{\sum(\hat{y}_i - \bar{y})^2}{\sum(y_i - \bar{y})^2}$$

is the proportion of the variation of the response variable y that is explained by the explanatory variables $x_1, x_2, \ldots, x_p$ in a multiple linear regression.

multiple correlation coefficient

Often, R^2 is multiplied by 100 and expressed as a percent. The square root of R^2, called the **multiple correlation coefficient,** is the correlation between the observations y_i and the predicted values $\hat{y}_i$.

USE YOUR KNOWLEDGE

11.3 Significance tests for regression coefficients. Recall Exercise 11.1 (page 593). Due to missing values for some students, only 86 students were used in the multiple regression analysis. The following table contains the estimated coefficients and standard errors:

Variable	Estimate	SE
Intercept	−0.764	0.651
SAT Math	0.00156	0.00074
SAT Verbal	0.00164	0.00076
High school rank	1.470	0.430
Bryant College placement	0.889	0.402

(a) All the estimated coefficients for the explanatory variables are positive. Is this what you would expect? Explain.

(b) What are the degrees of freedom for the model and error?

(c) Test the significance of each coefficient and state your conclusions.

11.4 ANOVA table for multiple regression. Use the following information to perform the ANOVA F test and compute R^2.

Source	Degrees of freedom	Sum of squares
Model		75
Error	50	
Total	53	515

11.2 A Case Study

Preliminary analysis

In this section we illustrate multiple regression by analyzing the data from the study described in Example 11.1. The response variable is the cumulative GPA, on a 4-point scale, after three semesters. The explanatory variables previously mentioned are average high school grades, represented by HSM, HSS, and HSE. We also examine the SAT Mathematics and SAT Verbal scores as explanatory variables. We have data for $n = 224$ students in the study. We use SAS, Excel, and Minitab to illustrate the outputs that are given by most software.

The first step in the analysis is to carefully examine each of the variables. Means, standard deviations, and minimum and maximum values appear in Figure 11.2. The minimum value for the SAT Mathematics (SATM) variable appears to be rather extreme; it is $(595 − 300)/86 = 3.43$ standard deviations below the mean. We do not discard this case at this time but will take care in our subsequent analyses to see if it has an excessive influence on our results. The mean for the SATM score is higher than the mean for the Verbal score (SATV), as we might expect for a group of computer science majors. The two standard deviations are about the same. The means of the three high school grade variables are similar, with the mathematics grades being a bit higher.

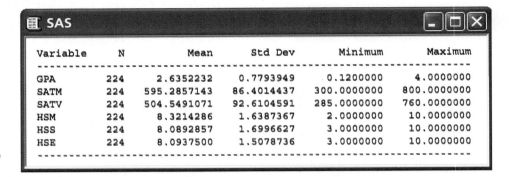

Variable	N	Mean	Std Dev	Minimum	Maximum
GPA	224	2.6352232	0.7793949	0.1200000	4.0000000
SATM	224	595.2857143	86.4014437	300.0000000	800.0000000
SATV	224	504.5491071	92.6104591	285.0000000	760.0000000
HSM	224	8.3214286	1.6387367	2.0000000	10.0000000
HSS	224	8.0892857	1.6996627	3.0000000	10.0000000
HSE	224	8.0937500	1.5078736	3.0000000	10.0000000

FIGURE 11.2 Descriptive statistics for the computer science student case study.

The standard deviations for the high school grade variables are very close to each other. The mean GPA is 2.635 on a 4-point scale, with standard deviation 0.779.

Because the variables GPA, SATM, and SATV have many possible values, we could use stemplots or histograms to examine the shapes of their distributions. Normal quantile plots indicate whether or not the distributions look Normal. *It is important to note that the multiple regression model does not require any of these distributions to be Normal.* Only the deviations of the responses y from their means are assumed to be Normal. The purpose of examining these plots is to understand something about each variable alone before attempting to use it in a complicated model. *Extreme values of any variable should be noted and checked for accuracy.* If found to be correct, the cases with these values should be carefully examined to see if they are truly exceptional and perhaps do not belong in the same analysis with the other cases. When our data on computer science majors are examined in this way, no obvious problems are evident.

The high school grade variables HSM, HSS, and HSE have relatively few values and are best summarized by giving the relative frequencies for each possible value. The output in Figure 11.3 provides these summaries. The distributions are all skewed, with a large proportion of high grades (10 = A and 9 = A−). Again we emphasize that these distributions need not be Normal.

Relationships between pairs of variables

LOOK BACK
correlation p. 102

The second step in our analysis is to examine the relationships between all pairs of variables. Scatterplots and correlations are our tools for studying two-variable relationships. The correlations appear in Figure 11.4. The output includes the P-value for the test of the null hypothesis that the population correlation is 0 versus the two-sided alternative for each pair. Thus, we see that the correlation between GPA and HSM is 0.44, with a P-value of 0.000 (i.e., $P < 0.0005$), whereas the correlation between GPA and SATV is 0.11, with a P-value of 0.087. The first is statistically significant by any reasonable standard, and the second is marginally significant.

The high school grades all have higher correlations with GPA than do the SAT scores. As we might expect, math grades have the highest correlation ($r = 0.44$), followed by science grades (0.33) and then English grades (0.29). The two SAT scores have a rather high correlation with each other (0.46), and the high school grades also correlate well with each other (0.45 to 0.58). SATM

SAS

hsm	Frequency	Percent	Cumulative Frequency	Cumulative Percent
2	1	0.45	1	0.45
3	1	0.45	2	0.89
4	4	1.79	6	2.68
5	6	2.68	12	5.36
6	23	10.27	35	15.63
7	28	12.50	63	28.13
8	36	16.07	99	44.20
9	59	26.34	158	70.54
10	66	29.46	224	100.00

hss	Frequency	Percent	Cumulative Frequency	Cumulative Percent
3	1	0.45	1	0.45
4	7	3.13	8	3.57
5	9	4.02	17	7.59
6	24	10.71	41	18.30
7	42	18.75	83	37.05
8	31	13.84	114	50.89
9	50	22.32	164	73.21
10	60	26.79	224	100.00

hse	Frequency	Percent	Cumulative Frequency	Cumulative Percent
3	1	0.45	1	0.45
4	4	1.79	5	2.23
5	5	2.23	10	4.46
6	23	10.27	33	14.73
7	43	19.20	76	33.93
8	49	21.88	125	55.80
9	52	23.21	177	79.02
10	47	20.98	224	100.00

FIGURE 11.3 The distributions of the high school grade variables.

Minitab

Correlations: gpa, satm, satv, hsm, hss, hse

	gpa	satm	satv	hsm	hss
satm	0.252 0.000				
satv	0.114 0.087	0.464 0.000			
hsm	0.436 0.000	0.454 0.000	0.221 0.001		
hss	0.329 0.000	0.240 0.000	0.262 0.000	0.576 0.000	
hse	0.289 0.000	0.108 0.106	0.244 0.000	0.447 0.000	0.579 0.000

Cell Contents: Pearson correlation
P-Value

FIGURE 11.4 Correlations among the case study variables.

correlates well with HSM (0.45), less well with HSS (0.24), and rather poorly with HSE (0.11). The correlations of SATV with the three high school grades are about equal, ranging from 0.22 to 0.26.

It is important to keep in mind that by examining pairs of variables we are seeking a better understanding of the data. *The fact that the correlation of a particular explanatory variable with the response variable does not achieve statistical significance does not necessarily imply that it will not be a useful (and statistically significant) predictor in a multiple regression.*

Numerical summaries such as correlations are useful, but plots are generally more informative when seeking to understand data. Plots tell us whether the numerical summary gives a fair representation of the data. For a multiple regression, each pair of variables should be plotted. For the six variables in our case study, this means that we should examine 15 plots. In general, there are $p+1$ variables in a multiple regression analysis with p explanatory variables, so that $p(p+1)/2$ plots are required. Multiple regression is a complicated procedure. If we do not do the necessary preliminary work, we are in serious danger of producing useless or misleading results. We leave the task of making these plots as an exercise.

USE YOUR KNOWLEDGE

11.5 Pairwise relationships among variables in the CSDATA data set. Using a statistical package, generate the pairwise correlations and scatterplots discussed previously. Comment on any unusual patterns or observations.

Regression on high school grades

To explore the relationship between the explanatory variables and our response variable GPA, we run several multiple regressions. The explanatory variables fall into two classes. High school grades are represented by HSM, HSS, and HSE, and standardized tests are represented by the two SAT scores. We begin our analysis by using the high school grades to predict GPA. Figure 11.5 gives the multiple regression output.

The output contains an ANOVA table, some additional descriptive statistics, and information about the parameter estimates. When examining any ANOVA table, it is a good idea to first verify the degrees of freedom. This ensures that we have not made some serious error in specifying the model for the software or in entering the data. Because there are $n = 224$ cases, we have $\text{DFT} = n - 1 = 223$. The three explanatory variables give $\text{DFM} = p = 3$ and $\text{DFE} = n - p - 1 = 223 - 3 = 220$.

The ANOVA F statistic is 18.86, with a P-value of 0.0001. Under the null hypothesis

$$H_0: \beta_1 = \beta_2 = \beta_3 = 0$$

the F statistic has an $F(3,220)$ distribution. According to this distribution, the chance of obtaining an F statistic of 18.86 or larger is 0.0001. We therefore conclude that at least one of the three regression coefficients for the high school grades is different from 0 in the population regression equation.

```
SAS

Dependent Variable: GPA

          Analysis of Variance

                             Sum of          Mean
Source            DF        Squares        Square        F Value       Prob>F

Model              3       27.71233       9.23744         18.861       <.0001
Error            220      107.75046       0.48977
C Total          223      135.46279

    Root MSE           0.69984      R-Square        0.2046
    Dep Mean           2.63522      Adj R-sq        0.1937
    C.V.              26.55711

                         Parameter Estimates

                    Parameter        Standard       T for H0:
Variable     DF      Estimate           Error     Parameter=0    Prob > |T|

INTERCEP      1      0.589877      0.29424324           2.005        0.0462
HSM           1      0.168567      0.03549214           4.749        0.0001
HSS           1      0.034316      0.03755888           0.914        0.3619
HSE           1      0.045102      0.03869585           1.166        0.2451
```

FIGURE 11.5 Multiple regression output for regression using high school grades to predict GPA.

In the descriptive statistics that follow the ANOVA table we find that Root MSE is 0.6998. This value is the square root of the MSE given in the ANOVA table and is s, the estimate of the parameter σ of our model. The value of R^2 is 0.20. That is, 20% of the observed variation in the GPA scores is explained by linear regression on high school grades. Although the P-value is very small, the model does not explain very much of the variation in GPA. Remember, a small P-value does not necessarily tell us that we have a large effect, particularly when the sample size is large.

From the Parameter Estimates section of the computer output we obtain the fitted regression equation

$$\widehat{\text{GPA}} = 0.590 + 0.169\text{HSM} + 0.034\text{HSS} + 0.045\text{HSE}$$

Let's find the predicted GPA for a student with an A– average in HSM, B+ in HSS, and B in HSE. The explanatory variables are HSM = 9, HSS = 8, and HSE = 7. The predicted GPA is

$$\begin{aligned}\widehat{\text{GPA}} &= 0.590 + 0.169(9) + 0.034(8) + 0.045(7)\\ &= 2.7\end{aligned}$$

Recall that the t statistics for testing the regression coefficients are obtained by dividing the estimates by their standard errors. Thus, for the coefficient of HSM we obtain the t-value given in the output by calculating

$$t = \frac{b}{\text{SE}_b} = \frac{0.168567}{0.03549214} = 4.749$$

The *P*-values appear in the last column. Note that these *P*-values are for the two-sided alternatives. HSM has a *P*-value of 0.0001, and we conclude that the regression coefficient for this explanatory variable is significantly different from 0. The *P*-values for the other explanatory variables (0.36 for HSS and 0.25 for HSE) do not achieve statistical significance.

Interpretation of results

The significance tests for the individual regression coefficients seem to contradict the impression obtained by examining the correlations in Figure 11.4. In that display we see that the correlation between GPA and HSS is 0.33 and the correlation between GPA and HSE is 0.29. The *P*-values for both of these correlations are <0.0005. In other words, if we used HSS alone in a regression to predict GPA, or if we used HSE alone, we would obtain statistically significant regression coefficients.

This phenomenon is not unusual in multiple regression analysis. Part of the explanation lies in the correlations between HSM and the other two explanatory variables. These are rather high (at least compared with the other correlations in Figure 11.4). The correlation between HSM and HSS is 0.58, and that between HSM and HSE is 0.45. Thus, when we have a regression model that contains all three high school grades as explanatory variables, there is considerable overlap of the predictive information contained in these variables. *The significance tests for individual regression coefficients assess the significance of each predictor variable assuming that all other predictors are included in the regression equation.* Given that we use a model with HSM and HSS as predictors, the coefficient of HSE is not statistically significant. Similarly, given that we have HSM and HSE in the model, HSS does not have a significant regression coefficient. HSM, however, adds significantly to our ability to predict GPA even after HSS and HSE are already in the model.

Unfortunately, we cannot conclude from this analysis that the *pair* of explanatory variables HSS and HSE contribute nothing significant to our model for predicting GPA once HSM is in the model. The impact of relations among the several explanatory variables on fitting models for the response is the most important new phenomenon encountered in moving from simple linear regression to multiple regression. We can only hint at the many complicated problems that arise.

Residuals

As in simple linear regression, we should always examine the residuals as an aid to determining whether the multiple regression model is appropriate for the data. Because there are several explanatory variables, we must examine several residual plots. It is usual to plot the residuals versus the predicted values $\hat{y}$ and also versus each of the explanatory variables. Look for outliers, influential observations, evidence of a curved (rather than linear) relation, and anything else unusual. Again, we leave the task of making these plots as an exercise. The plots all appear to show more or less random noise around the center value of 0.

If the deviations ϵ in the model are Normally distributed, the residuals should be Normally distributed. Figure 11.6 presents a Normal quantile plot of the residuals. The distribution appears to be approximately Normal. There

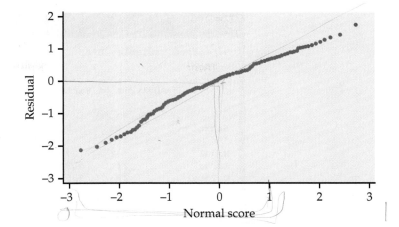

FIGURE 11.6 Normal quantile plot of the residuals from the high school grades model. There are no important deviations from Normality.

are many other specialized plots that help detect departures from the multiple regression model. Discussion of these, however, is more than we can undertake in this chapter.

USE YOUR KNOWLEDGE

11.6 Residual plots for the CSDATA analysis. Using a statistical package, fit the linear model with HSM and HSE as predictors and obtain the residuals and predicted values. Plot the residuals versus the predicted values, HSM, and HSE. Are the residuals more or less randomly dispersed around zero? Comment on any unusual patterns.

Refining the model

Because the variable HSS has the largest P-value of the three explanatory variables (see Figure 11.5) and therefore appears to contribute the least to our explanation of GPA, we rerun the regression using only HSM and HSE as explanatory variables. The SAS output appears in Figure 11.7. The F statistic indicates that we can reject the null hypothesis that the regression coefficients for the two explanatory variables are both 0. The P-value is still 0.0001. The value of R^2 has dropped very slightly compared with our previous run, from 0.2046 to 0.2016. Thus, dropping HSS from the model resulted in the loss of very little explanatory power. The measure s of variation about the fitted equation (Root MSE in the printout) is nearly identical for the two regressions, another indication that we lose very little dropping HSS. The t statistics for the individual regression coefficients indicate that HSM is still clearly significant ($P < 0.0001$), while the statistic for HSE is larger than before (1.747 versus 1.166) and approaches the traditional 0.05 level of significance ($P = 0.082$).

Comparison of the fitted equations for the two multiple regression analyses tells us something more about the intricacies of this procedure. For the first run we have

$$\widehat{\text{GPA}} = 0.590 + 0.169\text{HSM} + 0.034\text{HSS} + 0.045\text{HSE}$$

```
█ SAS                                                    _ □ X

Dependent Variable: GPA


            Analysis of Variance

                         Sum of         Mean
    Source        DF     Squares        Square      F Value      Prob>F

    Model          2     27.30349      13.65175      27.894      <.0001
    Error        221    108.15930       0.48941
    C Total      223    135.46279


        Root MSE           0.69958    R-Square       0.2016
        Dep Mean           2.63522    Adj R-sq       0.1943
        C.V.              26.54718


                    Parameter Estimates

                     Parameter     Standard      T for H0:
    Variable   DF     Estimate        Error     Parameter=0    Prob > |T|

    INTERCEP    1     0.624228     0.29172204       2.140        0.0335
    HSM         1     0.182654     0.03195581       5.716        0.0001
    HSM         1     0.060670     0.03472914       1.747        0.0820
```

FIGURE 11.7 Multiple regression output for regression using HSM and HSE to predict GPA.

whereas the second gives us

$$\widehat{\text{GPA}} = 0.624 + 0.183\text{HSM} + 0.061\text{HSE}$$

Eliminating HSS from the model changes the regression coefficients for all the remaining variables and the intercept. This phenomenon occurs quite generally in multiple regression. *Individual regression coefficients, their standard errors, and significance tests are meaningful only when interpreted in the context of the other explanatory variables in the model.*

Regression on SAT scores

We now turn to the problem of predicting GPA using the two SAT scores. Figure 11.8 gives the output. The fitted model is

$$\widehat{\text{GPA}} = 1.289 + 0.002283\text{SATM} - 0.000025\text{SATV}$$

The degrees of freedom are as expected: 2, 221, and 223. The F statistic is 7.476, with a P-value of 0.0007. We conclude that the regression coefficients for SATM and SATV are not both 0. Recall that we obtained the P-value 0.0001 when we used high school grades to predict GPA. Both multiple regression equations are highly significant, but this obscures the fact that the two models have quite different explanatory power. For the SAT regression, $R^2 = 0.0634$, whereas for the high school grades model even with only HSM and HSE (Figure 11.6), we have $R^2 = 0.2016$, a value more than three times as large. *Stating that we have a statistically significant result is quite different from saying that an effect is large or important.*

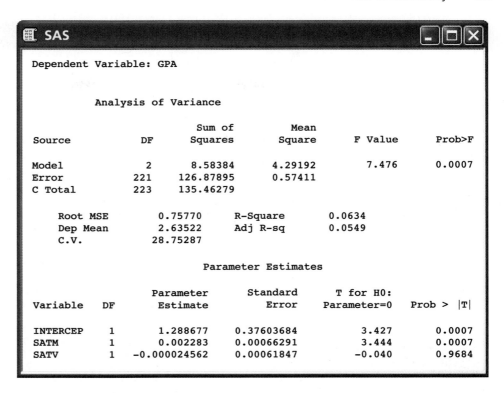

FIGURE 11.8 Multiple regression output for regression using SAT scores to predict GPA.

Dependent Variable: GPA

Analysis of Variance

Source	DF	Sum of Squares	Mean Square	F Value	Prob>F
Model	2	8.58384	4.29192	7.476	0.0007
Error	221	126.87895	0.57411		
C Total	223	135.46279			

Root MSE	0.75770	R-Square	0.0634
Dep Mean	2.63522	Adj R-sq	0.0549
C.V.	28.75287		

Parameter Estimates

Variable	DF	Parameter Estimate	Standard Error	T for H0: Parameter=0	Prob > |T|
INTERCEP	1	1.288677	0.37603684	3.427	0.0007
SATM	1	0.002283	0.00066291	3.444	0.0007
SATV	1	-0.000024562	0.00061847	-0.040	0.9684

Further examination of the output in Figure 11.8 reveals that the coefficient of SATM is significant ($t = 3.44$, $P = 0.0007$), and that for SATV is not ($t = -0.04$, $P = 0.9684$). For a complete analysis we should carefully examine the residuals. Also, we might want to run the analysis with SATM as the only explanatory variable.

Regression using all variables

We have seen that either the high school grades or the SAT scores give a highly significant regression equation. The mathematics component of each of these groups of explanatory variables appears to be the key predictor. Comparing the values of R^2 for the two models indicates that high school grades are better predictors than SAT scores. Can we get a better prediction equation using all the explanatory variables together in one multiple regression?

To address this question we run the regression with all five explanatory variables. The output appears in Figure 11.9. The F statistic is 11.69, with a P-value of 0.0001, so at least one of our explanatory variables has a nonzero regression coefficient. This result is not surprising, given that we have already seen that HSM and SATM are strong predictors of GPA. The value of R^2 is 0.2115, not much higher than the value of 0.2046 that we found for the high school grades regression.

Examination of the t statistics and the associated P-values for the individual regression coefficients reveals that HSM is the only one that is significant ($P = 0.0003$). That is, only HSM makes a significant contribution when it is added to a model that already has the other four explanatory variables. Once

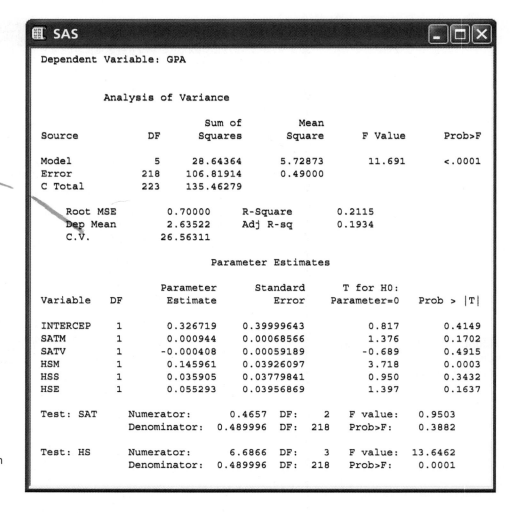

FIGURE 11.9 Multiple regression output for regression using all variables to predict GPA.

The SAS output shown:

SAS

Dependent Variable: GPA

Analysis of Variance

Source	DF	Sum of Squares	Mean Square	F Value	Prob>F
Model	5	28.64364	5.72873	11.691	<.0001
Error	218	106.81914	0.49000		
C Total	223	135.46279			

Root MSE	0.70000	R-Square	0.2115
Dep Mean	2.63522	Adj R-sq	0.1934
C.V.	26.56311		

Parameter Estimates

Variable	DF	Parameter Estimate	Standard Error	T for H0: Parameter=0	Prob > |T|
INTERCEP	1	0.326719	0.39999643	0.817	0.4149
SATM	1	0.000944	0.00068566	1.376	0.1702
SATV	1	-0.000408	0.00059189	-0.689	0.4915
HSM	1	0.145961	0.03926097	3.718	0.0003
HSS	1	0.035905	0.03779841	0.950	0.3432
HSE	1	0.055293	0.03956869	1.397	0.1637

Test: SAT	Numerator:	0.4657	DF:	2	F value:	0.9503
	Denominator:	0.489996	DF:	218	Prob>F:	0.3882

Test: HS	Numerator:	6.6866	DF:	3	F value:	13.6462
	Denominator:	0.489996	DF:	218	Prob>F:	0.0001

again it is important to understand that this result does not necessarily mean that the regression coefficients for the four other explanatory variables are *all* 0.

Figure 11.10 gives the Excel and Minitab multiple regression outputs for this problem. Although the format and organization of outputs differ among software packages, the basic results that we need are easy to find.

Many statistical software packages provide the capability for testing whether a collection of regression coefficients in a multiple regression model are *all* 0. We use this approach to address two interesting questions about this set of data. We did not discuss such tests in the outline that opened this section, but the basic idea is quite simple and discussed in Exercise 11.24 (page 616).

Test for a collection of regression coefficients

In the context of the multiple regression model with all five predictors, we ask first whether or not the coefficients for the two SAT scores are both 0. In other words, do the SAT scores add any significant predictive information to that already contained in the high school grades? To be fair, we also ask the

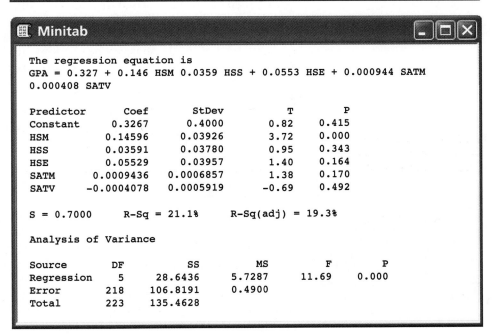

FIGURE 11.10 Excel and Minitab multiple regression outputs for regression using all variables to predict GPA.

complementary question—do the high school grades add any significant predictive information to that already contained in the SAT scores?

The answers are given in the last two parts of the output in Figure 11.9. For the first test we see that $F = 0.9503$. Under the null hypothesis that the two SAT coefficients are 0, this statistic has an $F(2,218)$ distribution and the P-value is 0.39. We conclude that the SAT scores are not significant predictors of GPA in a regression that already contains the high school scores as predictor variables. Recall that the model with just SAT scores has a highly significant F statistic. We now see that whatever predictive information is in the SAT scores can also be found in the high school grades. In this sense, the SAT scores are unnecessary.

The test statistic for the three high school grade variables is $F = 13.6462$. Under the null hypothesis that these three regression coefficients are 0, the statistic has an $F(3,218)$ distribution and the P-value is 0.0001. We conclude that high school grades contain useful information for predicting GPA that is not contained in SAT scores.

Of course, our statistical analysis of these data does not imply that SAT scores are less useful than high school grades for predicting college grades for all groups of students. We have studied a select group of students—computer science majors—from a specific university. Generalizations to other situations are beyond the scope of inference based on these data alone.

BEYOND THE BASICS

Multiple logistic regression

Many studies have yes/no or success/failure response variables. A surgery patient lives or dies; a consumer does or does not purchase a product after viewing an advertisement. Because the response variable in a multiple regression is assumed to have a Normal distribution, this methodology is not suitable for predicting such responses. However, there are models that apply the ideas of regression to response variables with only two possible outcomes.

logistic regression

One type of model that can be used is called **logistic regression.** We think in terms of a binomial model for the two possible values of the response variable and use one or more explanatory variables to explain the probability of success. Details are more complicated than those for multiple regression and are given in the supplemental Chapter 14 on this topic. However, the fundamental ideas are very much the same. Here is an example.

EXAMPLE

11.2 Tipping behavior in Canada. The Consumer Report on Eating Share Trends (CREST) contains data spanning all provinces of Canada detailing away-from-home food purchases by roughly 4000 households per quarter. Some researchers accessed these data but restricted their attention to restaurants at which tips would normally be given.[3] From a total of 73,822 observations, "high" and "low" tipping variables were created based on whether the observed tip rate was above 20% or below 10%, respectively. They then used logistic regression to identify explanatory variables associated with either "high" or "low" tips.

The model consisted of over 25 explanatory variables, grouped as "control" variables and "stereotype-related" variables. The stereotype-related explanatory variables were x_1, a variable having the value 1 if the age of the diner was greater than 65 years, and 0 otherwise; x_2, coded as 1 if the meal were on Sunday, and 0 otherwise; x_3, coded as 1 to indicate English was a second language; x_4, a variable coded 1 if the diner is a French-speaking Canadian; x_5, a variable coded 1 if alcoholic drinks were served with the meal; and x_6, a variable coded 1 if the meal involved a lone male.

Similar to the F test in multiple regression, there is a chi-square test for multiple logistic regression that tests the null hypothesis that *all* coefficients

LOOK BACK
chi-square distribution
p. 517

of the explanatory variables are zero. These results were not presented in the article because the focus was more on comparing the high- and low-tip models. In place of the t tests for individual coefficients in multiple regression, chi-square tests, each with 1 degree of freedom, are used to test whether individual coefficients are zero. They do report these tests. A majority of the variables considered in the models have P-values less than 0.01.

Interpretation of the coefficients is a little more difficult in multiple logistic regression because of the form of the model. For example, the high-tip model (using only the stereotype-related variables) is

$$\log\left(\frac{p}{1-p}\right) = \beta_0 + \beta_1 x_1 + \beta_2 x_2 + \cdots + \beta_6 x_6$$

odds

The expression $p/(1-p)$ is the **odds** that the tip was above 20%. Logistic regression models the "log odds" as a linear combination of the explanatory variables. Positive coefficients are associated with a higher probability that the tip is high. These coefficients are often transformed back (e^{β_j}) to the odds scale, giving us an **odds ratio**. An odds ratio greater than 1 is associated with a higher probability that the tip is high. Here is the table of odds ratios reported in the article for the high-tip model.

odds ratio

Explanatory variable	Odds ratio
Senior adult	0.7420*
Sunday	0.9970
English second language	0.7360*
French-speaking Canadian	0.7840*
Alcoholic drinks	1.1250*
Lone male	1.0220

Those starred were significant at the 0.01 level. We see the probability of a high tip is reduced (odds ratio less than 1) when the diner is over 65 years old, English is a second language, and the diner is a French-speaking Canadian. The probability of a high tip is increased (odds ratio greater than 1) if there is alcohol served with the meal.

CHAPTER 11 **Summary**

Data for multiple linear regression consist of the values of a response variable y and p explanatory variables $x_1, x_2, \ldots, x_p$ for n cases. We write the data and enter them into software in the form

	Variables				
Individual	x_1	x_2	...	x_p	y
1	x_{11}	x_{12}	...	x_{1p}	y_1
2	x_{21}	x_{22}	...	x_{2p}	y_2
$\vdots$					
n	x_{n1}	x_{n2}	...	x_{np}	y_n

The statistical model for **multiple linear regression** with response variable y and p explanatory variables $x_1, x_2, \ldots, x_p$ is

$$y_i = \beta_0 + \beta_1 x_{i1} + \beta_2 x_{i2} + \cdots + \beta_p x_{ip} + \epsilon_i$$

where $i = 1, 2, \ldots, n$. The ϵ_i are assumed to be independent and Normally distributed with mean 0 and standard deviation σ. The **parameters** of the model are $\beta_0, \beta_1, \beta_2, \ldots, \beta_p$, and σ.

The **multiple regression equation** predicts the response variable by a linear relationship with all the explanatory variables:

$$\hat{y} = b_0 + b_1 x_1 + b_2 x_2 + \cdots + b_p x_p$$

The β's are estimated by $b_0, b_1, b_2, \ldots, b_p$, which are obtained by the **method of least squares**. The parameter σ is estimated by

$$s = \sqrt{\text{MSE}} = \sqrt{\frac{\sum e_i^2}{n - p - 1}}$$

where the e_i are the **residuals,**

$$e_i = y_i - \hat{y}_i$$

Always examine the **distribution of the residuals** and plot them against the explanatory variables prior to inference.

A **level C confidence interval for β_j** is

$$b_j \pm t^* \text{SE}_{b_j}$$

where t^* is the value for the $t(n - p - 1)$ density curve with area C between $-t^*$ and t^*.

The test of the hypothesis $H_0\colon \beta_j = 0$ is based on the **t statistic**

$$t = \frac{b_j}{\text{SE}_{b_j}}$$

and the $t(n - p - 1)$ distribution.

The estimate b_j of β_j and the test and confidence interval for β_j are all based on a specific multiple linear regression model. The results of all of these procedures change if other explanatory variables are added to or deleted from the model.

The **ANOVA table** for a multiple linear regression gives the degrees of freedom, sum of squares, and mean squares for the model, error, and total sources of variation. The **ANOVA F statistic** is the ratio MSM/MSE and is used to test the null hypothesis

$$H_0\colon \beta_1 = \beta_2 = \cdots = \beta_p = 0$$

If H_0 is true, this statistic has an $F(p,\ n - p - 1)$ distribution.

The **squared multiple correlation** is given by the expression

$$R^2 = \frac{\text{SSM}}{\text{SST}}$$

and is interpreted as the proportion of the variability in the response variable y that is explained by the explanatory variables $x_1, x_2, \ldots, x_p$ in the multiple linear regression.

CHAPTER 11   Exercises

For Exercise 11.1, see page 593; for Exercise 11.2, see page 594; for Exercises 11.3 and 11.4, see pages 598–599; for Exercise 11.5, see page 602; and for Exercise 11.6, see page 605.

11.7   95% confidence intervals for regression coefficients. In each of the following settings, give a 95% confidence interval for the coefficient of x_1.

(a) $n = 27$, $\hat{y} = 10.6 + 6.4x_1 + 7.9x_2$, $SE_{b_1} = 3.1$.

(b) $n = 53$, $\hat{y} = 10.6 + 6.4x_1 + 7.9x_2$, $SE_{b_1} = 3.1$.

(c) $n = 27$, $\hat{y} = 10.6 + 6.4x_1 + 7.9x_2 + 5.2x_3$, $SE_{b_1} = 3.1$.

(d) $n = 124$, $\hat{y} = 10.6 + 6.4x_1 + 7.9x_2 + 5.2x_3$, $SE_{b_1} = 3.1$.

11.8   More on significance tests for regression coefficients. For each of the settings in the previous exercise, test the null hypotheses that the coefficient of x_1 is zero versus the two-sided alternative.

11.9   What's wrong? In each of the following situations, explain what is wrong and why.

(a) In a multiple regression with a sample size of 39 and 3 explanatory variables, the test statistic for the null hypothesis H_0: $b_2 = 0$ is a t statistic that follows the $t(35)$ distribution when the null hypothesis is true.

(b) The multiple correlation coefficient gives the proportion of the variation in the response variable that is explained by the explanatory variables.

(c) A small P-value for the ANOVA F test implies all explanatory variables are statistically different from zero.

11.10   What's wrong? In each of the following situations, explain what is wrong and why.

(a) One of the assumptions for multiple regression is that the distribution of each explanatory variable is Normal.

(b) The smaller the P-value for the ANOVA F test, the greater the explanatory power of the model.

(c) All explanatory variables that are significantly correlated with the response variable will have a statistically significant regression coefficient in the multiple regression model.

(d) The multiple correlation coefficient gives the average correlation between the response variable and each explanatory variable in the model.

11.11   Constructing the ANOVA table. Eight explanatory variables are used to predict a response variable using a multiple regression. There are 135 observations.

(a) Write the statistical model that is the foundation for this analysis. Also include a description of all assumptions.

(b) Outline the analysis of variance table giving the sources of variation and numerical values for the degrees of freedom.

11.12   More on constructing the ANOVA table A multiple regression analysis of 82 cases was performed with 6 explanatory variables. Suppose that SSM = 22.6 and SSE = 100.5.

(a) Find the value of the F statistic for testing the null hypothesis that the coefficients of all of the explanatory variables are zero.

(b) What are the degrees of freedom for this statistic?

(c) Find bounds on the P-value using Table E. Show your work.

(d) What proportion of the variation in the response variable is explained by the explanatory variables?

11.13   Predicting energy-drink consumption. Energy-drink advertising consistently emphasizes a physically active lifestyle and often features extreme sports and risk taking. Are these typical characteristics of an energy-drink consumer? A researcher decided to examine the links between energy-drink consumption, sport-related (jock) identity, and risk taking.[4] She invited over 1500 undergraduate students enrolled in large introductory-level courses at a public university to participate. Each participant had to complete a 45-minute anonymous questionnaire. From this questionnaire jock identity and risk-taking scores were obtained where the higher the score, the stronger the trait. She ended up with 795 respondents. The following table summarizes the results of a multiple regression analysis using the frequency of energy-drink consumption in the past 30 days as the response variable:

Explanatory Variable	b
Age	−0.02
Sex (1=female, 0=male)	−0.11**
Race (1=nonwhite, 0=white)	−0.02
Ethnicity (1=Hispanic, 0=non-Hispanic)	0.10**
Parental education	0.02
College GPA	−0.01
Jock Identity	0.05
Risk taking	0.19***

A superscript of ** means the individual coefficient t test had a P-value less than 0.01 and a superscript of *** had a P-value less than 0.001. All other P-values were greater than 0.05.

(a) The overall F statistic is reported to be 8.11. What are the degrees of freedom associated with this statistic?

(b) R is reported to be 0.28. What percent of the variation in energy-drink consumption is explained by the model? Is this a highly predictive model? Explain.

(c) Interpret each of the regression coefficients that are significant.

(d) The researcher states that "Controlling for gender, age, race, ethnicity, parental educational achievement, and college GPA, each of the predictors (risk taking and jock identity) was positively associated with energy-drink consumption frequency ..." Explain what is meant by "controlling for" these variables and how this helps strengthen her assertion that jock identity and risk taking are positively associated with energy-drink consumption.

11.14 Understanding the tests of significance. Using a new software package, you ran a multiple regression. The output reported an F statistic with $P < 0.05$, but none of the t tests for the individual coefficients were significant ($P > 0.05$). Does this mean that there is something wrong with the software? Explain your answer.

11.15 A mechanistic explanation of popularity. In Exercise 10.59 (page 589) correlations between an adolescent's "popularity," expression of a serotonin receptor gene, and rule-breaking behavior were assessed. An additional portion of the analysis looked at the relationship between the gene expression level and popularity, after adjusting for rule-breaking (RB) behavior. This adjustment was necessary because RB is positively associated with this gene expression and RB is positively associated with popularity in adolescents. The following summarizes these regression analyses using the composite RB score. A total of 202 individuals were included in this analysis.

	Estimate	Standard Error
Model 1		
Gene Expression	0.204	0.066
Model 2		
Gene Expression	0.161	0.066
RB.composite	0.100	0.030

For all analyses use the 0.05 significance level.

(a) What are the error degrees of freedom for Model 1 and Model 2?

(b) Test the null hypothesis that the serotonin gene receptor coefficient is equal to 0 in Model 1. State the test statistic and P-value.

(c) Perform both individual variable t tests for Model 2. Again state the test statistics and P-values.

(d) Is there still a positive relationship between the serotonin gene receptor expression level and popularity after adjusting for RB? If yes, compare the increase in popularity for a unit increase in gene expression (while RB remains unchanged) across the two models.

Results such as these suggest that adolescents with high serotonin receptor gene expression are not only predisposed to increased RB behavior, but also the advantageous social consequences of such behaviors.

11.16 Is the number of tornadoes increasing? In Exercise 10.19, data on the number of tornadoes in the United States between 1953 and 2008 were analyzed to see if there was a linear trend over time. Many argue that the probability of sighting a tornado has increased over time because there are more people living in the United States. Let's investigate this by including the U.S. census count as an additional explanatory variable. TORNADOES

(a) Using numerical and graphical summaries, describe the relationship between each pair of variables.

(b) Perform a multiple regression using both year and population count as explanatory variables. Write down the fitted model.

(c) Obtain the residuals from part (b). Plot them versus the two explanatory variables and generate a Normal probability plot. What do you conclude?

(d) Test the hypothesis that there is a linear increase over time. State the null and alternative hypotheses, test statistic, and P-value. What is your conclusion?

11.17 Checking for a polynomial relationship. When looking at the residuals from the simple linear model of BMI versus physical activity (PA), Figure 10.5 (page 553) suggested a possible curvilinear relationship. Let's investigate this further. Multiple regression can be used to fit the polynomial curve of degree q, $y = \beta_0 + \beta_1 x + \beta_2 x^2 + \cdots + \beta_q x^q$, through the creation of additional explanatory variables x^2, x^3, etc. Let's investigate a quadratic fit ($q = 2$) for the physical activity problem. PA_BMI

(a) It is often best to subtract off the sample mean, $\bar{x}$, before creating the necessary explanatory variables. In this case, the % average steps per day is 8.614. Create new explanatory variables $x_1 = (PA - 8.614)$ and

$x_2 = (PA - 8.614)^2$ and run a multiple regression for BMI using the explanatory variables % x_1 and x_2. Write down the fitted regression line.

(b) The regression model that just included PA had a $R^2 = 14.9$. What is the R^2 with the inclusion of this quadratic term?

(c) Obtain the residuals from part (b) and check the multiple regression assumptions. Are there any remaining patterns in the data? Are the residuals approximately Normal? Explain.

(d) Test the hypothesis that the coefficient of the variable $(PA - 8.614)^2$ is equal to 0. Report the t statistic, degrees of freedom, and P-value. Does the quadratic term contribute significantly to the fit? Explain your answer.

11.18 Architectural firm billings. A summary of firms engaged in commercial architecture in the Indianapolis, Indiana, area provides firm characteristics including total annual billing in millions of dollars and the number of architects, engineers, and staff employed in the firm.[5] Consider developing a model to predict total billing. ARCHITECT

(a) Using numerical and graphical summaries, describe the distribution of total billing and the number of architects, engineers, and staff.

(b) For each of the 6 pairs of variables, use graphical and numerical summaries to describe the relationship.

(c) Carry out a multiple regression. Report the fitted regression equation and the value of the regression standard error s.

(d) Analyze the residuals from the multiple regression. Are there any concerns?

(e) The firm HCO did not report its total billing but employs 3 architects, 1 engineer, and 17 staff members. What is the predicted total billing for this firm?

11.19 Comparing regression lines. In Exercise 10.50 (page 588), you plotted the residuals versus LOS and discovered a pattern in the residuals based on the size of the bank. This suggests the linear relationship between wages and LOS may be different for the two bank sizes. LOS

(a) Generate scatterplots of wages versus LOS for each bank size. Does there appear to be a linear relationship in each case? Describe any differences you see between the two plots.

(b) Generate a simple linear regression models for wages versus LOS using just the observations from small banks. Write down the fitted regression line, R^2, and the estimate of σ.

(c) Generate a simple linear regression model for wages versus LOS using just the observations from large banks. Write down the fitted regression line, R^2, and the estimate of σ.

(d) Compare the two fits. How do they differ?

11.20 Comparing regression lines, continued. Refer to the previous exercise. Provided one feels comfortable assuming the estimates of σ are equal, these two lines can be fit in the same multiple regression. This allows straightforward statistical comparisons of the regression lines. LOS

(a) Create a new explanatory variable LOSSIZE1 = LOS × SIZE1. This variable will take the value of LOS when SIZE1=1 and is 0 otherwise.

(b) Fit the regression model with LOS, SIZE1, and LOSSIZE1 as explanatory variables. Write down the fitted line.

(c) Compare the intercept and the coefficient of LOS with the coefficient estimates in part (b) of the previous exercise.

(d) Compare the sum of the intercept and coefficient of SIZE1 with the intercept estimate in part (c) of the previous problem. Similarly, compare the sum of the coefficients for LOS and LOSSIZE1 with the slope in part (c) of the previous problem.

(e) Use the relationships in part (d) to compare the two intercepts and slopes. How do the regression lines for small and large banks differ? Provide test statistics, degrees of freedom, and P-values.

The following six exercises use the MOVIES data set. This data set contains an SRS of 35 movies released in 2008. This sample was collected from the Internet Movie Database (IMDb) to see if information available soon after a movie's theatrical release can successfully predict total revenue.[6] All dollar amounts are measured in millions of U.S. dollars. MOVIES

11.21 Predicting movie revenue—preliminary analysis. The response variable is a movie's total U.S. revenue (USRevenue). Let's consider as explanatory variables the movie's budget (Budget); opening weekend revenue (Opening); the number of theaters (Theaters) the movie was in for the opening weekend; and the movie's IMDb rating (Opinion), which is on a 1 to 10 scale (10 being best). While this rating is updated continuously, we'll assume the rating in the data set is the rating at the end of the first week.

(a) Using numerical and graphical summaries, describe the distribution of each explanatory variable. Are there any unusual observations that should be monitored?

(b) Using numerical and graphical summaries, describe the relationship between each pair of explanatory variables.

11.22 Predicting movie revenue–simple linear regressions. Now let's look at the response variable and its relationship with each explanatory variable.

(a) Using numerical and graphical summaries, describe the distribution of the response variable, USRevenue.

(b) This variable is not Normally distributed. Does this violate one of key model assumptions? Explain.

(c) Generate scatterplots of each explanatory variable and USRevenue. Do all these relationships look linear? Explain what you see.

11.23 Predicting movie revenue–multiple linear regression. Now consider fitting a model using all the explanatory variables.

(a) Write out the statistical model for this analysis, making sure to specify all assumptions.

(b) Run the multiple regression model and specify the fitted regression equation.

(c) Obtain the residual from part (b) and check assumptions. Comment on any unusual residuals or patterns in the residuals.

(d) What percent of the variability in U.S. Revenue is explained by this model?

11.24 ⏃ **A simpler model.** In the multiple regression analysis using all of four variables, Theaters and Budget appear to be the least helpful (given that the other two explanatory variables are in the model).

(a) Perform a new analysis using only the movie's opening weekend revenue and IMDb rating. Give the estimated regression equation for this analysis.

(b) What percent of the variability in U.S. Revenue is explained by this model?

(c) In this chapter we discussed the F test for a collection of regression coefficients. In most cases, this capability is provided in the software. When it is not, the test can be performed using the R^2 values from the full and reduced models. The test statistic is

$$F = \left(\frac{n-p-1}{q}\right)\left(\frac{R_1^2 - R_2^2}{1 - R_1^2}\right)$$

with q and $n-p-1$ degrees of freedom. R_1^2 is the value for the full model and R_2^2 is the value for the reduced model.

Here $n = 35$ movies, $p = 4$ variables in the full model, and $q = 2$ variables were removed to form the reduced model. Plug in the values of R^2 from part (b) of this problem and part (d) of the previous problem and compute the test statistic and P-value. Do Theaters and Budget combined add any significant predictive information beyond what is already contained in Opening and Rating?

11.25 Predicting U.S. movie revenue. Refer to the previous two exercises. *Get Smart* was released in 2008, had a budget of $80.0 million dollars, was shown in 3911 theaters grossing $38.7 million dollars during the first weekend, and had an IMDb rating of 6.8. Use your software to construct

(a) A 95% prediction interval based on the model with all four predictors.

(b) A 95% prediction interval based on the model using only opening weekend revenue and IMDb rating.

(c) Compare the two intervals. Do the models give similar predictions?

11.26 Effect of potential outliers. Consider the simpler model of Exercise 11.24 for this analysis.

(a) There are two movies that have much larger U.S. revenues than those predicted. Which are they and how much more revenue did they obtain compared to that predicted?

(b) Remove these two movies and redo the multiple regression. Make a table giving the regression coefficients and their standard errors, t statistics, and P-values.

(c) Compare these results with those from Exercise 11.24. How does the removal of these outlying movies impact the estimated model?

(d) Obtain the residuals from this reduced data set and graphically examine their distribution. Do the residuals appear approximately Normal? Explain your answer.

The following three exercises use the RANKINGS data set. Since 2004, The Times Higher Education Supplement has provided an annual ranking of the world universities. A total score for each university is calculated based on the following scores: Peer Review (40%); Faculty-to-Student Ratio (20%); Citations-to-Faculty Ratio (20%); Recruiter Review (10%); Percent International Faculty (5%); and Percent International Student (5%). The percentages represent the contributions of each score to the total. For purposes here, we will assume these weights are unknown and will focus on the development of a model for the total score based on the first three explanatory variables. The report includes a table for the top 200 universities.[7] The RANKINGS data set is a random sample of 75 of these universities. This is not a random sample of all universities but for purposes here we will consider it to be. 🔵 RANKINGS

11.27 Annual ranking of world universities. Let's consider developing a model to predict total score based on the peer review score (PEER), faculty-to-student ratio (FtoS), and citations-to-faculty ratio (CtoF).

(a) Using numerical and graphical summaries, describe the distribution of each explanatory variable.

(b) Using numerical and graphical summaries, describe the relationship between each pair of explanatory variables.

11.28 Looking at the simple linear regressions. Now let's look at the relationship between each explanatory variable and the total score.

(a) Generate scatterplots for each explanatory variable and the total score. Do these relationships all look linear?

(b) Compute the correlation between each explanatory variable and the total score. Are certain explanatory variables more strongly associated with the total score?

11.29 Multiple linear regression model. Now consider a regression model using all three explanatory variables.

(a) Write out the statistical model for this analysis, making sure to specify all assumptions.

(b) Run the multiple regression model and specify the fitted regression equation.

(c) Generate a 95% confidence interval for each coefficient. Should any of these intervals contain 0? Explain.

(d) What percent of the variation in total score is explained by this model? What is the estimate for σ?

11.30 Predicting GPA of seventh-graders. Refer to the educational data for 78 seventh-grade students given in Table 1.4 (page 28). We view GPA as the response variable. IQ, gender, and self-concept are the explanatory variables. 🔵 SEVENTHGRADE

(a) Find the correlation between GPA and each of the explanatory variables. What percent of the total variation in student GPAs can be explained by the straight-line relationship for each of the explanatory variables?

(b) The importance of IQ in explaining GPA is not surprising. The purpose of the study is to assess the influence of self-concept on GPA. So we will include IQ in the regression model and ask, "How much does self-concept contribute to explaining GPA after the effect of IQ

on GPA is taken into account?" Give a model that can be used to answer this question.

(c) Run the model and report the fitted regression equation. What percent of the variation in GPA is explained by the explanatory variables in your model?

(d) Translate the question of interest into appropriate null and alternative hypotheses about the model parameters. Give the value of the test statistic and its P-value. Write a short summary of your analysis with an emphasis on your conclusion.

The following three exercises use the HAPPINESS data set. The World Database of Happiness *is an online registry of scientific research on the subjective appreciation of life. It is available at* worlddatabaseofhappiness.eur.nl *and is directed by Dr. Ruut Veenhoven, Erasmus University, Rotterdam. One inventory presents the "average happiness" score for various nations between 2007 and 2008. This average is based on individual responses from numerous general population surveys to a general life satisfaction (well-being) question. Scores ranged between 0 (dissatisfied) to 10 (satisfied). The NationMaster Web site,* www.nationmaster.com, *contains a collection of statistics associated with various nations. For this data set, the factors considered are the GINI Index: measures the degree of inequality in the distribution of income (higher score = greater inequality); the degree of corruption in government (higher score = less corruption); average life expectancy; and the degree of democracy (higher score = more political liberties).* 🔵 HAPPINESS

11.31 Predicting a nation's "average happiness" score. Consider the five statistics for each nation: LSI, the average life-satisfaction score; GINI, the GINI index; CORRUPT, the degree of corruption in government; LIFE, the average life expectancy; and DEMOCRACY, a measure of civil and political liberties.

(a) Using numerical and graphical summaries, describe the distribution of each variable.

(b) Using numerical and graphical summaries, describe the relationship between each pair of variables.

11.32 Building a multiple linear regression model. Let's now build a model to predict the life-satisfaction score, LSI.

(a) Consider a simple linear regression using GINI as the explanatory variable. Run the regression and summarize the results. Be sure to check assumptions.

(b) Now consider a model using GINI and LIFE. Run the multiple regression and summarize the results. Again be sure to check assumptions.

(c) Now consider a model using GINI, LIFE, and DEMOCRACY. Run the multiple regression and summarize the results. Again be sure to check assumptions.

(d) Now consider a model using all four explanatory variables. Again summarize the results and check assumptions.

11.33 Selecting from among several models. Refer to the results from the previous exercise.

(a) Make a table giving the estimated regression coefficients, standard errors, t statistics, and P-values.

(b) Describe how the coefficients and P-values change for the four models.

(c) Based on the table of coefficients, suggest another model. Run that model, summarize the results, and compare it with the other ones. Which model would you choose to explain LSI? Explain.

The following six exercises use the BIOMARKERS data set. Healthy bones are continually being renewed by two processes. Through bone formation, new bone is built; through bone resorption, old bone is removed. If one or both of these processes is disturbed, by disease, aging, or space travel, for example, bone loss can be the result. The variables VOPLUS and VOMINUS measure bone formation and bone resorption, respectively. Osteocalcin (OC) is a biochemical marker for bone formation: higher levels of bone formation are associated with higher levels of OC. A blood sample is used to measure OC, and it is much less expensive to obtain than direct measures of bone formation. The units are milligrams of OC per milliliter of blood (mg/ml). Similarly, tartrate resistant acid phosphatase (TRAP) is a biochemical marker for bone resorption that is also measured in blood. It is measured in units per liter (U/l). These variables were measured in a study of 31 healthy women aged 11 to 32 years.[8] Variables with the first letter "L" are the logarithms of the measured variables. ● BIOMARKERS

11.34 Bone formation and resorption. Consider the following four variables: VO^+, a measure of bone formation; VO^-, a measure of bone resorption; OC, a biomarker of bone formation; and TRAP, a biomarker of bone resorption.

(a) Using numerical and graphical summaries, describe the distribution of each of these variables.

(b) Using numerical and graphical summaries, describe the relationship between each pair of variables in this set.

11.35 Predicting bone formation. Let's use regression methods to predict VO^+, the measure of bone formation.

(a) Since OC is a biomarker of bone formation, we start with a simple linear regression using OC as the explanatory variable. Run the regression and summarize the results. Be sure to include an analysis of the residuals.

(b) Because the processes of bone formation and bone resorption are highly related, it is possible that there is some information in the bone resorption variables that can tell us something about bone formation. Use a model with both OC and TRAP, the biomarker of bone resorption, to predict VO^+. Summarize the results. In the context of this model, it appears that TRAP is a better predictor of bone formation, VO^+, than the biomarker of bone formation, OC. Is this view consistent with the pattern of relationships that you described in the previous exercise? One possible explanation is that, while all of these variables are highly related, TRAP is measured with more precision than OC.

11.36 More on predicting bone formation. Now consider a regression model for predicting VO^+ using OC, TRAP, and VO^-.

(a) Write out the statistical model for this analysis including all assumptions.

(b) Run the multiple regression to predict VO^+ using OC, TRAP, and VO^-. Summarize the results.

(c) Make a table giving the estimated regression coefficients, standard errors, and t statistics with P-values for this analysis and the two that you ran in the previous exercise. Describe how the coefficients and the P-values differ for the three analyses.

(d) Give the percent of variation in VO^+ explained by each of the three models and the estimate of σ. Give a short summary.

(e) The results you found in part (b) suggest another model. Run that model, summarize the results, and compare them with the results in part (b).

11.37 ⚠ **Predicting bone formation using transformed variables.** Because the distributions of VO^+, VO^-, OC, and TRAP tend to be skewed, it is common to work with logarithms rather than the measured values. Using the questions in the previous three exercises as a guide, analyze the log data.

11.38 ⚠ **Predicting bone resorption.** Refer to Exercises 11.34 to 11.36. Answer these questions with the roles of VO^+ and VO^- reversed; that is, run models to predict VO^-, with VO^+ as an explanatory variable.

11.39 ⚠ **Predicting bone resorption using transformed variables.** Refer to the previous exercise. Rerun using logs.

The following 11 exercises use the PCB data set. Polychlorinated biphenyls (PCBs) are a collection of synthetic compounds, called congeners, that are particularly toxic to fetuses and young children. Although PCBs are no longer produced in the United States, they are still found in the environment. Since human exposure to these PCBs is primarily through the consumption of fish, the Environmental Protection Agency (EPA) monitors the PCB levels in fish. Unfortunately, there are 209 different congeners and measuring all of them in a fish specimen is an expensive and time-consuming process. You've been asked to see if the total amount of PCBs in a specimen can be estimated with only a few, easily quantifiable congeners.[9] If this can be done, costs can be greatly reduced. PCB

11.40 Relationship among PCB congeners. Consider the following variables: PCB (the total amount of PCB) and four congeners: PCB52, PCB118, PCB138, and PCB180.

(a) Using numerical and graphical summaries, describe the distribution of each of these variables.

(b) Using numerical and graphical summaries, describe the relationship between each pair of variables in this set.

11.41 Predicting the total amount of PCB. Use the four congeners, PCB52, PCB118, PCB138, and PCB180, in a multiple regression to predict PCB.

(a) Write the statistical model for this analysis. Include all assumptions.

(b) Run the regression and summarize the results.

(c) Examine the residuals. Do they appear to be approximately Normal? When you plot them versus each of the explanatory variables, are any patterns evident?

11.42 Adjusting analysis for potential outliers. The examination of the residuals in part (c) of the previous exercise suggests that there may be two outliers, one with a high residual and one with a low residual.

(a) Because of safety issues, we are more concerned about underestimating PCB in a specimen than about overestimating. Give the specimen number for each of the two suspected outliers. Which one corresponds to an overestimate of PCB?

(b) Rerun the analysis with the two suspected outliers deleted, summarize these results, and compare them with those you obtained in the previous exercise.

11.43 More on predicting the total amount of PCB. Run a regression to predict PCB using the variables PCB52, PCB118, and PCB138. Note that this is similar to

the analysis that you did in Exercise 11.41, with the change that PCB180 is not included as an explanatory variable.

(a) Summarize the results.

(b) In this analysis, the regression coefficient for PCB118 is not statistically significant. Give the estimate of the coefficient and the associated *P*-value.

(c) Find the estimate of the coefficient for PCB118 and the associated *P*-value for the model analyzed in Exercise 11.41.

(d) Using the results in parts (b) and (c), write a short paragraph explaining how the inclusion of other variables in a multiple regression can have an effect on the estimate of a particular coefficient and the results of the associated significance test.

11.44 Multiple regression model for total TEQ. Dioxins and furans are other classes of chemicals that can cause undesirable health effects similar to those caused by PCB. The three types of chemicals are combined using toxic equivalent scores (TEQs), which attempt to measure the health effects on a common scale. The PCB data set contains TEQs for PCB, dioxins, and furans. The variables are called TEQPCB, TEQDIOXIN, and TEQFURAN. The data set also includes the total TEQ, defined to be the sum of these three variables.

(a) Consider using a multiple regression to predict TEQ using the three components TEQPCB, TEQDIOXIN, and TEQFURAN as explanatory variables. Write the multiple regression model in the form

$$\text{TEQ} = \beta_0 + \beta_1 \text{TEQPCB} + \beta_2 \text{TEQDIOXIN} + \beta_3 \text{TEQFURAN} + \epsilon$$

Give numerical values for the parameters β_0, β_1, β_2, and β_3.

(b) The multiple regression model assumes that the ϵ's are Normal with mean zero and standard deviation σ. What is the numerical value of σ?

(c) Use software to run this regression and summarize the results.

11.45 **Multiple regression model for total TEQ, continued.** The information summarized in TEQ is used to assess and manage risks from these chemicals. For example, the World Health Organization (WHO) has established the tolerable daily intake (TDI) as 1 to 4 TEQs per kilogram of body weight per day. Therefore, it would be very useful to have a procedure for estimating TEQ using just a few variables that can be measured cheaply. Use the four PCB congeners, PCB52, PCB118, PCB138, and PCB180, in a multiple regression to predict TEQ. Give a description of the model and assumptions, summarize the results, examine the residuals, and write a summary of what you have found.

11.46 ⚔ **Predicting total amount of PCB using transformed variables.** Because distributions of variables such as PCB, the PCB congeners, and TEQ tend to be skewed, researchers frequently analyze the logarithms of the measured variables. Create a data set that has the logs of each of the variables in the PCB data set. Note that zero is a possible value for PCB126; most software packages will eliminate these cases when you request a log transformation.

(a) If you do not do anything about the 16 zero values of PCB126, what does your software do with these cases? Is there an error message of some kind?

(b) If you attempt to run a regression to predict the log of PCB using the log of PCB126 and the log of PCB52, are the cases with the zero values of PCB126 eliminated? Do you think that is a good way to handle this situation?

(c) The smallest nonzero value of PCB126 is 0.0052. One common practice when taking logarithms of measured values is to replace the zeros by one-half of the smallest observed value. Create a logarithm data set using this procedure; that is, replace the 16 zero values of PCB126 by 0.0026 before taking logarithms. Use numerical and graphical summaries to describe the distributions of the log variables.

11.47 ⚔ **Predicting total amount of PCB using transformed variables, continued.** Refer to the previous exercise.

(a) Use numerical and graphical summaries to describe the relationships between each pair of log variables.

(b) Compare these summaries with the summaries that you produced in Exercise 11.40 for the measured variables.

11.48 ⚔ **Even more on predicting total amount of PCB using transformed variables.** Use the log data set that you created in Exercise 11.46 to find a good multiple regression model for predicting the log of PCB. Use only log PCB variables for this analysis. Write a report summarizing your results.

11.49 ⚔ **Predicting total TEQ using transformed variables.** Use the log data set that you created in Exercise 11.46 to find a good multiple regression model for predicting the log of TEQ. Use only log PCB variables for this analysis. Write a report summarizing your results and comparing them with the results that you obtained in the previous exercise.

11.50 Interpretation of coefficients in log PCB regressions. Use the results of your analysis of the log PCB data in Exercise 11.48 to write an explanation of how regression coefficients, standard errors of regression coefficients, and tests of significance for explanatory variables can change depending on what other explanatory variables are included in the multiple regression analysis.

The following nine exercises use the CHEESE data set. As cheddar cheese matures, a variety of chemical processes take place. The taste of matured cheese is related to the concentration of several chemicals in the final product. In a study of cheddar cheese from the LaTrobe Valley of Victoria, Australia, samples of cheese were analyzed for their chemical composition and were subjected to taste tests. The variable "Case" is used to number the observations from 1 to 30. "Taste" is the response variable of interest. The taste scores were obtained by combining the scores from several tasters. Three of the chemicals whose concentrations were measured were acetic acid, hydrogen sulfide, and lactic acid. For acetic acid and hydrogen sulfide (natural) log transformations were taken. Thus the explanatory variables are the transformed concentrations of acetic acid ("Acetic") and hydrogen sulfide ("H2S") and the untransformed concentration of lactic acid ("Lactic"). [10] 🧀 CHEESE*

11.51 Describing the explanatory variables. For each of the four variables in the CHEESE data set, find the mean, median, standard deviation, and interquartile range. Display each distribution by means of a stemplot and use a Normal quantile plot to assess Normality of the data. Summarize your findings. Note that when doing regressions with these data, we do not assume that these distributions are Normal. Only the residuals from our model need to be (approximately) Normal. The careful study of each variable to be analyzed is nonetheless an important first step in any statistical analysis.

11.52 Pairwise scatterplots of the explanatory variables. Make a scatterplot for each pair of variables in the CHEESE data set (you will have six plots). Describe the relationships. Calculate the correlation for each pair of variables and report the P-value for the test of zero population correlation in each case.

11.53 Simple linear regression model of Taste. Perform a simple linear regression analysis using Taste as the response variable and Acetic as the explanatory variable. Be sure to examine the residuals carefully. Summarize your results. Include a plot of the data with the least-squares regression line. Plot the residuals versus each of the other two chemicals. Are any patterns evident? (The concentrations of the other chemicals are lurking variables for the simple linear regression.)

11.54 Another simple linear regression model of Taste. Repeat the analysis of Exercise 11.53 using Taste as the response variable and H2S as the explanatory variable.

11.55 The final simple linear regression model of Taste. Repeat the analysis of Exercise 11.53 using Taste as the response variable and Lactic as the explanatory variable

11.56 Comparing the simple linear regression models. Compare the results of the regressions performed in the three previous exercises. Construct a table with values of the F statistic, its P-value, R^2, and the estimate s of the standard deviation for each model. Report the three regression equations. Why are the intercepts in these three equations different?

11.57 Multiple regression model of Taste. Carry out a multiple regression using Acetic and H2S to predict Taste. Summarize the results of your analysis. Compare the statistical significance of Acetic in this model with its significance in the model with Acetic alone as a predictor (Exercise 11.53). Which model do you prefer? Give a simple explanation for the fact that Acetic alone appears

to be a good predictor of Taste, but with H2S in the model, it is not.

11.58 Another multiple regression model of Taste. Carry out a multiple regression using H2S and Lactic to predict Taste. Comparing the results of this analysis with the simple linear regressions using each of these explanatory variables alone, it is evident that a better result is obtained by using both predictors in a model. Support this statement with explicit information obtained from your analysis.

11.59 The final multiple regression model of Taste. Use the three explanatory variables Acetic, H2S, and Lactic in a multiple regression to predict Taste. Write a short summary of your results, including an examination of the residuals. Based on all the regression analyses you have carried out on these data, which model do you prefer and why?

11.60 Finding a multiple regression model on the Internet. Search the Internet to find an example of the use of multiple regression. Give the setting of the example, describe the data, give the model, and summarize the results. Explain why the use of multiple regression in this setting was appropriate or inappropriate.

One-Way Analysis of Variance

Introduction

Many of the most effective statistical studies are comparative. For example, we may wish to compare customer satisfaction of men and women who use an online fantasy football site or compare the responses to various treatments in a clinical trial. We display these comparisons with back-to-back stemplots or side-by-side boxplots, and we measure them with five-number summaries or with means and standard deviations.

When only two groups are compared, Chapter 7 provides the tools we need to answer the question, "Is the difference between groups statistically significant?" Two-sample t procedures compare the means of two Normal populations, and we saw that these procedures, unlike comparisons of spread, are sufficiently robust to be widely useful.

In this chapter, we will compare any number of means by techniques that generalize the two-sample t and share its robustness and usefulness. These methods will allow us to address comparisons such as

• How does a user's number of Facebook friends affect his or her social attractiveness?

• Which of 10 brands of automobile tires wears longest?

• How long do cancer patients live under each of three therapies for their lung cancer?

12.1 Inference for One-Way Analysis of Variance

When comparing different populations or treatments, the data are subject to sampling variability. For example, we would not expect to observe the exact same sales data if we mailed an advertising offer to different random samples of households. We also wouldn't expect a new group of cancer patients to provide the same set of survival times. We therefore pose the question for inference in terms of the *mean* response.

LOOK BACK
comparing two means p. 432

In Chapter 7 we met procedures for comparing the means of two populations. We now extend those methods to problems involving more than two populations. The statistical methodology for comparing several means is called **analysis of variance,** or simply **ANOVA.** In this and the following section, we will examine the basic ideas and assumptions that are needed for ANOVA. Although the details differ, many of the concepts are similar to those discussed in the two-sample case.

ANOVA

We will consider two ANOVA techniques. When there is only one way to classify the populations of interest, we use **one-way ANOVA** to analyze the data. We call this categorical explanatory variable a **factor.** For example, to compare the tread lifetimes of 10 specific brands of tires we use one-way ANOVA with brand as our factor. This chapter presents the details for one-way ANOVA.

one-way ANOVA
factor

In many other comparison studies, there is more than one way to classify the populations. For the tire study, the researcher may also want to consider temperature. Are there brands that do relatively better in the heat? Analyzing the effect of two factors, brand and temperature, requires **two-way ANOVA.** This technique will be discussed in Chapter 13. While adding yet more factors necessitates even higher-way ANOVA techniques, most of the new ideas in ANOVA with more than one factor already appear in two-way ANOVA.

two-way ANOVA

Data for one-way ANOVA

One-way analysis of variance is a statistical method for comparing several population means. We draw a simple random sample (SRS) from each population and use the data to test the null hypothesis that the population means are all equal. Consider the following two examples:

EXAMPLE

12.1 Choosing the best video game box art. An upstart video game company wants to compare three different box art styles for a series of video games that will be offered for sale at electronics stores. The company is interested in whether there is a style that better captures shoppers' attention and results in more sales. To investigate this, they randomly assign each of 60 stores to one of the three box art styles and record the number of games that are sold in a one-week period.

EXAMPLE

12.2 Average age of coffee shop customers. How do five coffee houses around campus differ in the demographics of their customers? Are certain coffee houses more popular among graduate students? Do professors tend to go to one coffee house? A market researcher asks 50 customers of each coffee house to respond to a questionnaire. One variable of interest is the customer's age.

These two examples are similar in that

- There is a single quantitative response variable measured on many units; the units are stores in the first example and customers in the second.

- The goal is to compare several populations: stores displaying different box art styles in the first example and customers of five coffee houses in the second.

LOOK BACK
observation versus experiment
p. 167

There is, however, an important difference. Example 12.1 describes an experiment in which each store is randomly assigned to a box art style. Example 12.2 is an observational study in which customers are selected during a particular time period and not all agree to provide data. These samples of customers are not random samples, but we will treat them as such because we believe the selective sampling and nonresponse are ignorable sources of bias. This will not always be the case. *Always consider the various sources of bias in an observation study.*

In both examples, we will use ANOVA to compare the mean responses. The same ANOVA methods apply to data from random samples and to data from randomized experiments. *It is important to keep the data-production method in mind when interpreting the results. A strong case for causation is best made by a randomized experiment.*

Comparing means

The question we ask in ANOVA is, "Do all groups have the same population mean?" We will often use the term *groups* for the populations to be compared in a one-way ANOVA. To answer this question we compare the sample means. Figure 12.1 displays the sample means for Example 12.1. It appears that box-art

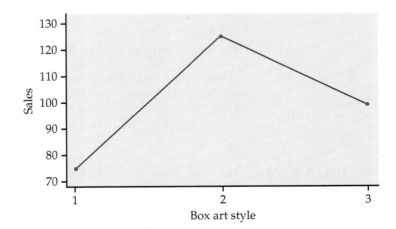

FIGURE 12.1 Mean sales of video games for three different box art styles.

style 2 has the highest average sales. But is the observed difference in sample means just the result of chance variation? We should not expect sample means to be equal, even if the population means are all identical.

The purpose of ANOVA is to assess whether the observed differences among sample means are *statistically significant*. Could a variation among the three sample means this large be plausibly due to chance, or is it good evidence for a difference among the population means? This question can't be answered from the sample means alone. Because the standard deviation of a sample mean $\bar{x}$ is the population standard deviation σ divided by $\sqrt{n}$, the answer also depends upon both the variation within the groups of observations and the sizes of the samples.

◄ LOOK BACK
standard deviation of $\bar{x}$ p. 301

Side-by-side boxplots help us see the within-group variation. Compare Figures 12.2(a) and 12.2(b). The sample medians are the same in both figures, but the large variation within the groups in Figure 12.2(a) suggests that the differences among the sample medians could be due simply to chance variation. The data in Figure 12.2(b) are much more convincing evidence that the populations differ. Even the boxplots omit essential information, however. To assess the observed differences, we must also know how large the samples are. Nonetheless, boxplots are a good preliminary display of the data. While ANOVA compares means and boxplots display medians, we expect the data to be approximately Normal and will consider a transformation if they are not. For distributions that are nearly symmetric, these two measures of center will be close together.

◄ LOOK BACK
transforming data p. 421

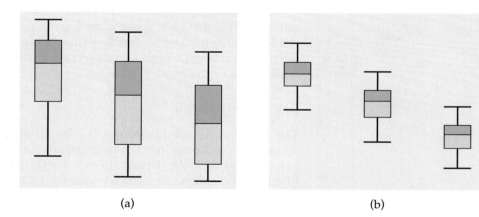

FIGURE 12.2 (a) Side-by-side boxplots for three groups with large within-group variation. The differences among centers may be just chance variation.
(b) Side-by-side boxplots for three groups with the same centers as in panel (a) but with small within-group variation. The differences among centers are more likely to be significant.

(a)   (b)

The two-sample *t* statistic

Two-sample t statistics compare the means of two populations. If the two populations are assumed to have equal but unknown standard deviations and the sample sizes are both equal to n, the t statistic is

◄ LOOK BACK
pooled two-sample t statistic
p. 447

$$t = \frac{\bar{x}_1 - \bar{x}_2}{s_p\sqrt{\dfrac{1}{n} + \dfrac{1}{n}}} = \frac{\sqrt{\dfrac{n}{2}}(\bar{x}_1 - \bar{x}_2)}{s_p}$$

The square of this t statistic is

$$t^2 = \frac{\frac{n}{2}(\overline{x}_1 - \overline{x}_2)^2}{s_p^2}$$

If we use ANOVA to compare two populations, the ANOVA F statistic is exactly equal to this t^2. We can therefore learn something about how ANOVA works by looking carefully at the statistic in this form.

between-group variation

The numerator in the t^2 statistic measures the variation **between** the groups in terms of the difference between their sample means $\overline{x}_1$ and $\overline{x}_2$. It includes a factor for the common sample size n. The numerator can be large because of a large difference between the sample means or because the sample sizes are large. The denominator measures the variation **within** groups by s_p^2, the pooled estimator of the common variance. If the within-group variation is small, the same variation between the groups produces a larger statistic and thus a more significant result.

within-group variation

Although the general form of the F statistic is more complicated, the idea is the same. To assess whether several populations all have the same mean, we compare the variation *among* the means of several groups with the variation *within* groups. Because we are comparing variation, the method is called *analysis of variance*.

An overview of ANOVA

ANOVA tests the null hypothesis that the population means are *all* equal. The alternative is that they are not all equal. This alternative could be true because all the means are different or simply because one of them differs from the rest. This is a more complex situation than comparing just two populations. If we reject the null hypothesis, we need to perform some further analysis to draw conclusions about which population means differ from which others and by how much.

The computations needed for an ANOVA are more lengthy than those for the t test. For this reason we generally use computer programs to perform the calculations. Automating the calculations frees us from the burden of arithmetic and allows us to concentrate on interpretation. *Complicated computations do not guarantee a valid statistical analysis. We should always start our ANOVA with a careful examination of the data using graphical and numerical summaries.*

EXAMPLE

12.3 Number of Facebook friends. A feature of each Facebook user's profile is the number of Facebook "friends," an indicator of the user's social network connectedness. Among college students on Facebook, the average number of Facebook friends has been estimated around 250.[1] Offline, having more friends is associated with higher ratings of positive attributes such as likability and trustworthiness. Is this also the case with Facebook friends? An experiment was run to examine the relationship between the number of Facebook friends and the user's perceived social attractiveness.[2] A total of 134 undergraduate participants were randomly assigned to observe one of five Facebook profiles. Everything about the profile was the same except the

FRIENDS

number of friends, which appeared on the profile as 102, 302, 502, 702, or 902. After viewing the profile, each participant was asked to fill out a questionnaire on the physical and social attractiveness of the profile user. Each attractiveness score is an average of several seven-point questionnaire items, ranging from 1 (strongly disagree) to 7 (strongly agree). Here is a summary of the data for the social attractiveness score:

Number of Friends	n	$\bar{x}$	s
102	24	3.82	1.00
302	33	4.88	0.85
502	26	4.56	1.01
702	30	4.41	1.43
902	21	3.99	1.02

Histograms for the five groups are given in Figure 12.3. Note that the heights of the bars in the histograms are percents rather than counts. This is

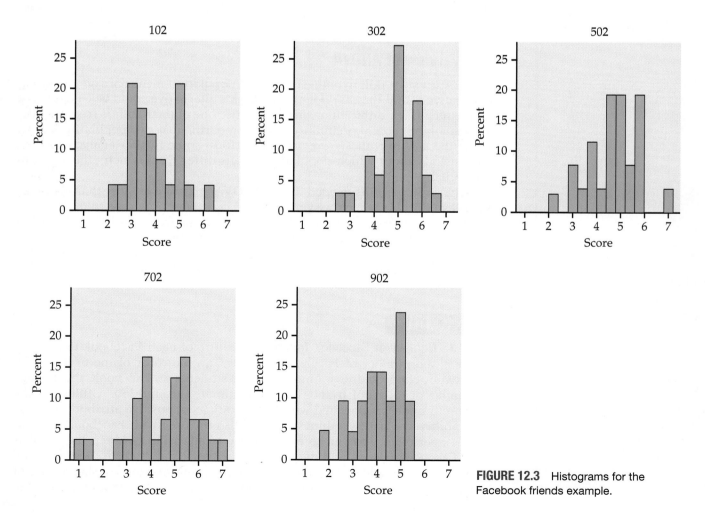

FIGURE 12.3 Histograms for the Facebook friends example.

commonly done when the group sample sizes vary. Figure 12.4 gives side-by-side boxplots for these data. We see that the scores covered the entire range of possible values, from 1.0 to 7.0. We also see a lot of overlap in scores across groups. The histograms are relatively symmetric and with the group sample sizes all more than 15, we can feel confident that the sample means are approximately Normal.

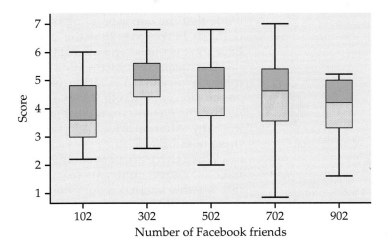

FIGURE 12.4 Side-by-side boxplots for the Facebook friends example.

The five sample means are plotted in Figure 12.5. It appears the mean scores rise and then fall as the number of friends increases. This would suggest having too many Facebook friends can harm a user's social attractiveness. However, given the variability in the data, this pattern could also just be the result of chance variation. We will use ANOVA to make this determination. In this setting, we have an experiment in which undergraduate Facebook users were randomly assigned to view one of five Facebook profiles. Each of these profile populations has a mean and our inference asks questions about these means.

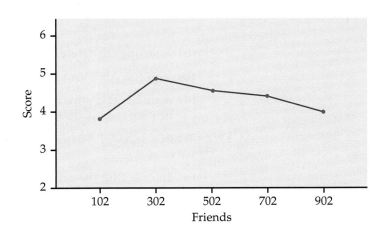

FIGURE 12.5 Social attractiveness means for the Facebook friends example.

Formulating a clear definition of the populations being compared with ANOVA can be difficult. Often some expert judgment is required, and different consumers of the results may have differing opinions. The undergraduates in this study were all from the same university. They also volunteered in exchange for course credit. Whether one can consider them as an SRS from the population of undergraduates at the university or from the population of all undergraduates is open for debate. Regardless, we are more confident in generalizing our conclusions to similar populations when the results are clearly significant than when the level of significance just barely passes the standard of $P = 0.05$.

We first ask whether or not there is sufficient evidence in the data to conclude that the corresponding population means are not all equal. Our null hypothesis here states that the population mean score is the same for all five Facebook profiles. The alternative is that they are not all the same.

Our inspection of the data for our example suggests that the means may follow a curvilinear relationship. *Rejecting the null hypothesis that the means are all the same using ANOVA is not the same as concluding that all of the means are different from one another.* The ANOVA null hypothesis can be false in many different ways. Additional analysis is required to distinguish among these possibilities.

contrasts   When there are particular versions of the alternative hypothesis that are of interest, we use **contrasts** to examine them. In our example, we might want to test whether there is a curvilinear relationship between the number of friends and acceptability score. *Note that, to use contrasts, it is necessary that the questions of interest be formulated before examining the data. It is cheating to make up these questions after analyzing the data.*

multiple comparisons   If we have no specific relations among the means in mind before looking at the data, we instead use a **multiple-comparisons** procedure to determine which pairs of population means differ significantly. In later sections we will explore both contrasts and multiple comparisons in detail.

USE YOUR KNOWLEDGE

12.1 What's wrong? For each of the following, explain what is wrong and why.

   (a) ANOVA tests the null hypothesis that the sample means are all equal.

   (b) A strong case for causation is best made in an observational study.

   (c) You use ANOVA to compare the variances of the populations.

   (d) A multiple comparisons procedure is used to compare a relation among means that was specified prior to looking at the data.

12.2 What's wrong? For each of the following, explain what is wrong and why.

   (a) In rejecting the null hypothesis, one can conclude all the means are different from one another.

   (b) A one-way ANOVA can be used only when there are two means to be compared.

   (c) A two-way ANOVA is used when the response variable has only two possible values.

The ANOVA model

When analyzing data, the following equation reminds us that we look for an overall pattern and deviations from it:

$$\text{DATA} = \text{FIT} + \text{RESIDUAL}$$

LOOK BACK
DATA = FIT + RESIDUAL
p. 565

In the regression model of Chapter 10, the FIT was the population regression line, and the RESIDUAL represented the deviations of the data from this line. We now apply this framework to describe the statistical models used in ANOVA. These models provide a convenient way to summarize the assumptions that are the foundation for our analysis. They also give us the necessary notation to describe the calculations needed.

First, recall the statistical model for a random sample of observations from a single Normal population with mean μ and standard deviation σ. If the observations are

$$x_1, x_2, \ldots, x_n$$

LOOK BACK
Normal distributions p. 54

we can describe this model by saying that the x_j are an SRS from the $N(\mu, \sigma)$ distribution. Another way to describe the same model is to think of the x's varying about their population mean. To do this, write each observation x_j as

$$x_j = \mu + \epsilon_j$$

The ϵ_j are then an SRS from the $N(0, \sigma)$ distribution. Because μ is unknown, the ϵ's cannot actually be observed. This form more closely corresponds to our

$$\text{DATA} = \text{FIT} + \text{RESIDUAL}$$

way of thinking. The FIT part of the model is represented by μ. It is the systematic part of the model, like the line in a regression. The RESIDUAL part is represented by ϵ_j. It represents the deviations of the data from the fit and is due to random, or chance, variation.

There are two unknown parameters in this statistical model: μ and σ. We estimate μ by $\bar{x}$, the sample mean, and σ by s, the sample standard deviation. The differences $e_j = x_j - \bar{x}$ are the sample residuals and correspond to the ϵ_j in the statistical model.

The model for one-way ANOVA is very similar. We take random samples from each of I different populations. The sample size is n_i for the ith population. Let x_{ij} represent the jth observation from the ith population. The I population means are the FIT part of the model and are represented by μ_i. The random variation, or RESIDUAL, part of the model is represented by the deviations ϵ_{ij} of the observations from the means.

THE ONE-WAY ANOVA MODEL

The **one-way ANOVA model** is

$$x_{ij} = \mu_i + \epsilon_{ij}$$

for $i = 1, \ldots, I$ and $j = 1, \ldots, n_i$. The ϵ_{ij} are assumed to be from an $N(0, \sigma)$ distribution. The **parameters of the model** are the population means $\mu_1, \mu_2, \ldots, \mu_I$ and the common standard deviation σ.

Note that the sample sizes n_i may differ, but the standard deviation σ is assumed to be the same in all of the populations. Figure 12.6 pictures this

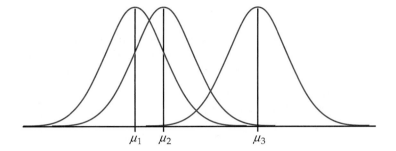

FIGURE 12.6 Model for one-way ANOVA with three groups. The three populations have Normal distributions with the same standard deviation.

model for $I = 3$. The three population means μ_i are different, but the shapes of the three Normal distributions are the same, reflecting the assumption that all three populations have the same standard deviation.

EXAMPLE

12.4 ANOVA model for the Facebook friends study. In the Facebook friends example there are five profiles that we want to compare, so $I = 5$. The population means $\mu_1, \mu_2, \ldots, \mu_5$ are the mean social attractiveness scores for the profiles with 102, 302, 502, 702, and 902 friends, respectively. The sample sizes n_i are 24, 33, 26, 30, and 21. It is common to use numerical subscripts to distinguish the different means, and some software requires that levels of factors in ANOVA be specified as numerical values. In this situation, it is very important to keep track of what each numerical value represents when drawing conclusions. In our example, we could use numerical values to suggest the actual groups by replacing μ_1 with μ_{102}, μ_2 with μ_{302}, and so on.

The observation $x_{1,1}$ is the social attractiveness score for the first participant that observed the profile with 102 friends. The data for the other participants assigned to this profile are denoted by $x_{1,2}, x_{1,3}, \ldots, x_{1,24}$. Similarly, the data for the other four groups have a first subscript indicating the profile and a second subscript indicating the participants assigned to that profile.

According to our model, the score for the first participant is $x_{1,1} = \mu_1 + \epsilon_{1,1}$, where μ_1 is the average score for *all* undergraduates after viewing profile 1 and $\epsilon_{1,1}$ is the chance variation due to this particular participant. The ANOVA model assumes that the ϵ_{ij} are independent and Normally distributed with mean 0 and standard deviation σ. We have clear evidence that the data are non-Normal in our example. The values are numbers ranging from 1.0 to 7.0 by 0.2. However, because our inference is based on the sample means, which will be approximately Normally distributed, we are not overly concerned about this violation of our assumptions.

Estimates of population parameters

The unknown parameters in the statistical model for ANOVA are the I population means μ_i and the common population standard deviation σ. To estimate μ_i we use the sample mean for the ith group:

$$\bar{x}_i = \frac{1}{n_i} \sum_{j=1}^{n_i} x_{ij}$$

The residuals $e_{ij} = x_{ij} - \bar{x}_i$ reflect the variation about the sample means that we see in the data.

The ANOVA model assumes that the population standard deviations are all equal. If we have unequal standard deviations, we generally try to transform the data so that they are approximately equal. We might, for example, work with $\sqrt{x_{ij}}$ or $\log x_{ij}$. Fortunately, we can often find a transformation that *both* makes the group standard deviations more nearly equal and also makes the distributions of observations in each group more nearly Normal. If the standard deviations are markedly different and cannot be made similar by a transformation, inference requires different methods that are beyond the scope of this book.

Unfortunately, formal tests for the equality of standard deviations in several groups share the lack of robustness against non-Normality that we noted in Chapter 7 for the case of two groups. Because ANOVA procedures are not extremely sensitive to unequal standard deviations, we do *not* recommend a formal test of equality of standard deviations as a preliminary to the ANOVA. Instead, we will use the following rule as a guideline.

◄ LOOK BACK
F test for equality of spread
p. 460

RULE FOR EXAMINING STANDARD DEVIATIONS IN ANOVA

If the largest standard deviation is less than twice the smallest standard deviation, we can use methods based on the assumption of equal standard deviations, and our results will still be approximately correct.[3]

When we assume that the population standard deviations are equal, each sample standard deviation is an estimate of σ. To combine these into a single estimate, we use a generalization of the pooling method introduced in Chapter 7.

POOLED ESTIMATOR OF σ

Suppose we have sample variances $s_1^2, s_2^2, \ldots, s_I^2$ from I independent SRSs of sizes $n_1, n_2, \ldots, n_I$ from populations with common variance σ^2. The **pooled sample variance**

$$s_p^2 = \frac{(n_1 - 1)s_1^2 + (n_2 - 1)s_2^2 + \cdots + (n_I - 1)s_I^2}{(n_1 - 1) + (n_2 - 1) + \cdots + (n_I - 1)}$$

is an unbiased estimator of σ^2. The **pooled standard deviation**

$$s_p = \sqrt{s_p^2}$$

is the estimate of σ.

Pooling gives more weight to groups with larger sample sizes. If the sample sizes are equal, s_p^2 is just the average of the I sample variances. *Note that s_p is not the average of the I sample standard deviations.*

EXAMPLE

12.5 Population estimates for the Facebook friends study. In the Facebook friends study there are $I = 5$ groups and the sample sizes are $n_1 = 24$, $n_2 = 33$, $n_3 = 26$, $n_4 = 30$, and $n_4 = 21$. The sample standard deviations are $s_1 = 1.00$, $s_2 = 0.85$, $s_3 = 1.07$, $s_4 = 1.43$, and $s_5 = 1.02$.

Because the largest standard deviation (1.43) is less than twice the smallest ($2 \times 0.85 = 1.70$), our rule indicates that we can use the assumption of equal population standard deviations.

The pooled variance estimate is

$$s_p^2 = \frac{(n_1 - 1)s_1^2 + (n_2 - 1)s_2^2 + (n_3 - 1)s_3^2 + (n_4 - 1)s_4^2 + (n_5 - 1)s_5^2}{(n_1 - 1) + (n_2 - 1) + (n_3 - 1) + (n_4 - 1) + (n_5 - 1)}$$

$$= \frac{(23)(1.00)^2 + (32)(0.85)^2 + (25)(1.07)^2 + (29)(1.43)^2 + (20)(1.02)^2}{23 + 32 + 25 + 29 + 20}$$

$$= \frac{154.85}{129} = 1.20$$

The pooled standard deviation is

$$s_p = \sqrt{1.20} = 1.10$$

This is our estimate of the common standard deviation σ of the social attractiveness scores for the five profiles.

USE YOUR KNOWLEDGE

12.3 Computing the pooled standard deviation. An experiment was run to compare three groups. The sample sizes were 23, 20, and 28, and the corresponding estimated standard deviations were 5, 5, and 6.

(a) Is it reasonable to use the assumption of equal standard deviations when we analyze these data? Give a reason for your answer.

(b) Give the values of the variances for the three groups.

(c) Find the pooled variance.

(d) What is the value of the pooled standard deviation?

12.4 Visualizing the ANOVA model. For each of the following situations, draw a picture of the ANOVA model similar to Figure 12.6 (page 632). Use the numerical values for the μ_i. To sketch the Normal curves, you may want to review the 68–95–99.7 rule on page 57.

(a) $\mu_1 = 18$, $\mu_2 = 12$, $\mu_3 = 13$, and $\sigma = 6$.

(b) $\mu_1 = 12$, $\mu_2 = 16$, $\mu_3 = 20$, $\mu_4 = 24$, and $\sigma = 1.5$.

(c) $\mu_1 = 18$, $\mu_2 = 12$, $\mu_3 = 13$, and $\sigma = 2$.

Testing hypotheses in one-way ANOVA

Comparison of several means is accomplished by using an F statistic to compare the variation among groups with the variation within groups. We now show how the F statistic expresses this comparison. Calculations are organized

LOOK BACK
ANOVA table p. 568

in an ANOVA table, which contains numerical measures of the variation among groups and within groups.

First we must specify our hypotheses for one-way ANOVA. As before, I represents the number of populations to be compared.

HYPOTHESES FOR ONE-WAY ANOVA

The **null and alternative hypotheses** for one-way ANOVA are

$$H_0: \mu_1 = \mu_2 = \ldots = \mu_I$$
$$H_a: \text{not all of the } \mu_i \text{ are equal}$$

We will now use the Facebook friends example to illustrate how to do a one-way ANOVA. Because the calculations are generally performed using statistical software, we focus on interpretation of the output.

EXAMPLE

12.6 Reading software output. Figure 12.7 gives descriptive statistics generated by SPSS for the ANOVA of the Facebook friends example. Summaries for each profile are given on the first five lines. In addition to the sample size, the mean, and the standard deviation, this output also gives the minimum and maximum observed value, standard error of the mean, and the 95% confidence interval for the mean of each profile. The five sample means $\bar{x}_i$ given in the output are estimates of the five unknown population means μ_i.

SPSS

Score   **Descriptives**

	N	Mean	Std. Deviation	Std. Error	95% Confidence Interval for Mean		Minimum	Maximum
					Lower Bound	Upper Bound		
102	24	3.817	.9990	.2039	3.395	4.239	2.2	6.0
302	33	4.879	.8514	.1482	4.577	5.181	2.6	6.4
502	26	4.562	1.0704	.2099	4.129	4.994	2.0	6.8
702	30	4.407	1.4283	.2608	3.873	4.940	1.0	7.0
902	21	3.990	1.0227	.2232	3.525	4.456	1.6	5.2
Total	134	4.382	1.1463	.0990	4.186	4.578	1.0	7.0

FIGURE 12.7 Software output with descriptive statistics for the Facebook friends example.

The output gives the estimates of the standard deviations, s_i, for each group but does not provide s_p, the pooled estimate of the model standard deviation, σ. We could perform the calculation using a calculator, as we did in Example 12.5. We will see an easier way to obtain this quantity from the ANOVA table in Figure 12.8. Some software packages report s_p as part of the standard ANOVA output. *Sometimes you are not sure whether or not a quantity given by software is what you think it is. A good way to resolve this dilemma is to do a sample calculation with a simple example to check the numerical results.*

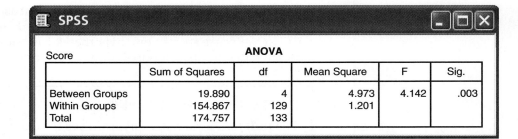

FIGURE 12.8 Software output giving the ANOVA table for the Facebook friends example.

Note that s_p is not the standard deviation given in the Total row of Figure 12.7. *This quantity is the standard deviation that we would obtain if we viewed the data as a single sample of 134 participants and ignored the possibility that the profile means could be different.* As we have mentioned many times before, it is important to use care when reading and interpreting software output.

EXAMPLE

12.7 Reading software output, continued. Additional output generated by SPSS for the ANOVA of the Facebook friends example is given in Figure 12.8. We will discuss some details in the next section. For now, we observe that the results of our significance test are given in the last two columns of the output. The null hypothesis that the five population means are the same is tested by the statistic $F = 4.142$, and the associated P-value is reported as $P = 0.003$. The data provide clear evidence to support the claim that there are some differences among the five profile population means.

The ANOVA table

The information in an analysis of variance is organized in an ANOVA table. To understand the table, it is helpful to think in terms of our

$$\text{DATA} = \text{FIT} + \text{RESIDUAL}$$

view of statistical models. For one-way ANOVA, this corresponds to

$$x_{ij} = \mu_i + \epsilon_{ij}$$

We can think of these three terms as sources of variation. The ANOVA table separates the variation in the data into two parts: the part due to the fit and the remainder, which we call residual.

EXAMPLE

12.8 ANOVA table for the Facebook friends study. The SPSS output in Figure 12.8 gives the sources of variation in the first column. Here, FIT is called Between Groups, RESIDUAL is called Within Groups, and DATA is the last entry, Total. Different software packages use different terms for these sources of variation but the basic concept is common to all. In place of FIT, some software packages use Between Groups, Model, or the name of the factor. Similarly, terms like Within Groups or Error are frequently used in place of RESIDUAL.

variation among groups

The Between Groups row in the table gives information related to the variation **among** group means. In writing ANOVA tables we will use the generic label "groups" or some other term that describes the factor being studied for this row.

variation within groups

The Within Groups row in the table gives information related to the variation **within** groups. We noted that the term "error" is frequently used for this source of variation, particularly for more general statistical models. This label is most appropriate for experiments in the physical sciences where the observations within a group differ because of measurement error. In business and the biological and social sciences, on the other hand, the within-group variation is often due to the fact that not all firms or plants or people are the same. This sort of variation is not due to errors and is better described as "residual" or "within-group" variation. Nevertheless, we will use the generic label "error" for this source of variation in writing ANOVA tables.

Finally, the Total row in the ANOVA table corresponds to the DATA term in our DATA = FIT + RESIDUAL framework. So, for analysis of variance,

$$DATA = FIT + RESIDUAL$$

translates into

$$Total = Between\ Groups + Within\ Groups$$

LOOK BACK
sum of squares p. 566

The second column in the software output given in Figure 12.8 is labeled Sum of Squares. As you might expect, each sum of squares is a sum of squared deviations. We use SSG, SSE, and SST for the entries in this column, corresponding to groups, error, and total. Each sum of squares measures a different type of variation. SST measures variation of the data around the overall mean, $x_{ij} - \bar{x}$. Variation of the group means around the overall mean $\bar{x}_i - \bar{x}$ is measured by SSG. Finally, SSE measures variation of each observation around its group mean, $x_{ij} - \bar{x}_i$.

EXAMPLE

12.9 ANOVA table for the Facebook friends study, continued. The Sum of Squares column in Figure 12.8 gives the values for the three sums of squares. They are:

$$SST = 174.757$$
$$SSG = 19.890$$
$$SSE = 154.867$$

Verify that SST = SSG + SSE for this example.

This fact is true in general. The total variation is always equal to the among-group variation plus the within-group variation. Note that software output frequently gives many more digits than we need, as in this case. In this example it appears that most of the variation is coming from within groups.

Associated with each sum of squares is a quantity called the degrees of freedom. Because SST measures the variation of all N observations around the overall mean, its degrees of freedom are DFT = $N - 1$. This is the same as the degrees of freedom for the ordinary sample variance with sample size N.

LOOK BACK
degrees of freedom p. 41

Similarly, because SSG measures the variation of the I sample means around the overall mean, its degrees of freedom are DFG $= I - 1$. Finally, SSE is the sum of squares of the deviations $x_{ij} - \bar{x}_i$. Here we have N observations being compared with I sample means, and DFE $= N - I$.

> ### EXAMPLE
>
> **12.10 Degrees of freedom for the Facebook friends study.** In the Facebook friends example, we have $I = 5$ and $N = 134$. Therefore,
>
> $$\text{DFT} = N - 1 = 134 - 1 = 133$$
> $$\text{DFG} = I - 1 = 5 - 1 = 4$$
> $$\text{DFE} = N - I = 134 - 5 = 129$$
>
> These are the entries in the df column of Figure 12.8.

Note that the degrees of freedom add in the same way that the sums of squares add. That is, DFT $=$ DFG $+$ DFE.

LOOK BACK
mean square p. 566

For each source of variation, the mean square is the sum of squares divided by the degrees of freedom. You can verify this by doing the divisions for the values given on the output in Figure 12.8.

SUMS OF SQUARES, DEGREES OF FREEDOM, AND MEAN SQUARES

Sums of squares represent variation present in the data. They are calculated by summing squared deviations. In one-way ANOVA there are three **sources of variation:** groups, error, and total. The sums of squares are related by the formula

$$\text{SST} = \text{SSG} + \text{SSE}$$

Thus, the total variation is composed of two parts, one due to groups and one due to error.

Degrees of freedom are related to the deviations that are used in the sums of squares. The degrees of freedom are related in the same way as the sums of squares are:

$$\text{DFT} = \text{DFG} + \text{DFE}$$

To calculate each **mean square,** divide the corresponding sum of squares by its degrees of freedom.

We can use the mean square for error to find s_p, the pooled estimate of the parameter σ of our model. It is true in general that

$$s_p^2 = \text{MSE} = \frac{\text{SSE}}{\text{DFE}}$$

In other words, the mean square for error is an estimate of the within-group variance, σ^2. The estimate of σ is therefore the square root of this quantity. So,

$$s_p = \sqrt{\text{MSE}}$$

EXAMPLE

12.11 MSE for the Facebook friends study. From the SPSS output in Figure 12.8 we see that the MSE is reported as 1.201. The pooled estimate of σ is therefore

$$s_p = \sqrt{\text{MSE}}$$
$$= \sqrt{1.201} = 1.10$$

The *F* test

If H_0 is true, there are no differences among the group means. The ratio MSG/MSE is a statistic that is approximately 1 if H_0 is true and tends to be larger if H_a is true. This is the ANOVA F statistic. In our example, MSG = 4.973 and MSE = 1.201, so the ANOVA F statistic is

$$F = \frac{\text{MSG}}{\text{MSE}} = \frac{4.973}{1.201} = 4.142$$

When H_0 is true, the F statistic has an F distribution that depends upon two numbers: the *degrees of freedom for the numerator* and the *degrees of freedom for the denominator*. These degrees of freedom are those associated with the mean squares in the numerator and denominator of the F statistic. For one-way ANOVA, the degrees of freedom for the numerator are DFG = $I - 1$, and the degrees of freedom for the denominator are DFE = $N - I$. We use the notation $F(I - 1, N - I)$ for this distribution.

APPLET

The *One-Way ANOVA* applet is an excellent way to see how the value of the F statistic and the P-value depend upon the variability of the data within the groups and the differences between the means. See Exercises 12.20 and 12.21 for use of this applet.

THE ANOVA *F* TEST

To test the null hypothesis in a one-way ANOVA, calculate the ***F* statistic**

$$F = \frac{\text{MSG}}{\text{MSE}}$$

F

When H_0 is true, the F statistic has the $F(I - 1, N - I)$ distribution. When H_a is true, the F statistic tends to be large. We reject H_0 in favor of H_a if the F statistic is sufficiently large.

The ***P*-value** of the F test is the probability that a random variable having the $F(I - 1, N - I)$ distribution is greater than or equal to the calculated value of the F statistic.

Tables of F critical values are available for use when software does not give the P-value. Table E in the back of the book contains the F critical values for probabilities $p = 0.100, 0.050, 0.025, 0.010,$ and 0.001. For one-way ANOVA we use critical values from the table corresponding to $I - 1$ degrees of freedom in the numerator and $N - I$ degrees of freedom in the denominator. *When determining the P-value, remember that the F test is always one-sided because any differences among the group means tend to make F large.*

EXAMPLE

12.12 The ANOVA *F* test for the Facebook friends study. In the Facebook friends study, we found $F = 4.14$. (Note that it is standard practice to round F statistics to two places after the decimal point.) There were five populations, so the degrees of freedom in the numerator are DFG $= I - 1 = 4$. For this example the degrees of freedom in the denominator are DFE $= N - I = 134 - 5 = 129$. In Table E we first find the column corresponding to 4 degrees of freedom in the numerator. For the degrees of freedom in the denominator, we see that there are entries for 100 and 200. These entries are very close. To be conservative we use critical values corresponding to 100 degrees of freedom in the denominator since these are slightly larger.

p	Critical value
0.100	2.00
0.050	2.46
0.025	2.92
0.010	3.51
0.001	5.02

We have $F = 4.14$. This is in between the critical value for $P = 0.010$ and $P = 0.001$. Using the table, we can conclude only that $0.001 < P < 0.010$. (Note that the more accurate calculations performed by software indicated that, in fact, $P = 0.003$.) For this example, we reject H_0 and conclude that the population means are not all the same.

The following display shows the general form of a one-way ANOVA table with the F statistic. The formulas in the sum of squares column can be used for calculations in small problems. There are other formulas that are more efficient for hand or calculator use, but ANOVA calculations are usually done by computer software.

Source	Degrees of freedom	Sum of squares	Mean square	F
Groups	$I - 1$	$\sum_{\text{groups}} n_i(\bar{x}_i - \bar{x})^2$	SSG/DFG	MSG/MSE
Error	$N - I$	$\sum_{\text{groups}} (n_i - 1)s_i^2$	SSE/DFE	
Total	$N - 1$	$\sum_{\text{obs}} (x_{ij} - \bar{x})^2$		

coefficient of determination

One other item given by some software for ANOVA is worth noting. For an analysis of variance, we define the **coefficient of determination** as

$$R^2 = \frac{\text{SSG}}{\text{SST}}$$

LOOK BACK
multiple correlation coefficient
p. 598

The coefficient of determination plays the same role as the squared multiple correlation R^2 in a multiple regression. We can easily calculate the value from the ANOVA table entries.

EXAMPLE

12.13 Coefficient of determination for the Facebook friends study. The software-generated ANOVA table for the Facebook friends study is given in Figure 12.8. From that display, we see that SSG = 19.890 and SST = 174.757. The coefficient of determination is

$$R^2 = \frac{\text{SSG}}{\text{SST}} = \frac{19.890}{174.757} = 0.11$$

About 11% of the variation in social attractiveness scores is explained by the different profiles. The other 89% of the variation is due to participant-to-participant variation within each of the profile groups. We can see this in the histograms of Figure 12.3. Each of the groups has a large amount of variation, and there is a substantial amount of overlap in the distributions. *The fact that we have strong evidence (P < 0.003) against the null hypothesis that the five population means are not all the same does not tell us that the distributions of values are far apart.*

USE YOUR KNOWLEDGE

12.5 What's wrong? For each of the following, explain what is wrong and why.

    (a) Within-group variation is the variation in the data due to the differences in the sample means.

    (b) The mean squares in an ANOVA table will add, that is, MST = MSG + MSE.

    (c) The pooled estimate s_p is a parameter of the ANOVA model.

    (d) A very small P-value implies the group distributions of responses are far apart.

12.6 Determining the critical value of F. For each of the following situations, state how large the F statistic needs to be for rejection of the null hypothesis at the 0.05 level.

    (a) Compare 4 groups with 4 observations per group.

    (b) Compare 4 groups with 6 observations per group.

    (c) Compare 4 groups with 8 observations per group.

    (d) Summarize what you have learned about F distributions from this exercise.

12.2 Comparing the Means

Contrasts

The ANOVA F test gives a general answer to a general question: are the differences among observed group means statistically significant? Unfortunately, a small P-value simply tells us that the group means are not all the same. It does not tell us specifically which means differ from each other. Plotting and inspecting the means give us some indication of where the differences lie, but we would like to supplement inspection with formal inference.

In the ideal situation, specific questions regarding comparisons among the means are posed before the data are collected. We can answer specific questions of this kind and attach a level of confidence to the answers we give. We now explore these ideas through the Facebook friends example.

EXAMPLE

12.14 Reporting the results. In the Facebook friends study we compared the social attractiveness scores for five profiles, which only varied in the number of friends. Let's use $\bar{x}_{102}, \bar{x}_{302}, \bar{x}_{502}, \bar{x}_{702}$, and $\bar{x}_{902}$ to represent the five sample means and a similar notation for the population means. From Figure 12.7 we see that the five sample means are

$$\bar{x}_{102} = 3.82, \bar{x}_{302} = 4.88, \bar{x}_{502} = 4.56, \bar{x}_{702} = 4.41, \text{ and } \bar{x}_{902} = 3.99$$

The null hypothesis we tested was

$$H_0: \mu_{102} = \mu_{302} = \mu_{502} = \mu_{702} = \mu_{902}$$

versus the alternative that the five population means are not all the same. We would report these results as $F(4, 129) = 4.14$ with $P = 0.003$. (Note that we have given the degrees of freedom for the F statistic in parentheses.) Because the P-value is very small, we conclude that the data provide clear evidence that the five population means are not all the same.

However, having evidence that the five population means are not the same does not tell us all we'd like to know. We would really like our analysis to provide us with more specific information. For example, the alternative hypothesis is true if

$$\mu_{102} < \mu_{302} = \mu_{502} = \mu_{702} = \mu_{902}$$

or if

$$\mu_{102} = \mu_{302} = \mu_{502} > \mu_{702} = \mu_{902}$$

or if

$$\mu_{102} \neq \mu_{302} \neq \mu_{502} \neq \mu_{702} \neq \mu_{902}$$

When you reject the ANOVA null hypothesis, additional analyses are required to clarify the nature of the differences between the means.

In terms of offline social networks, previous research has shown that the bigger one's social network, the higher one's social attractiveness. In fact, the relationship between the number of friends and social attractiveness appears linear. Therefore, a reasonable question to ask is whether or not this same sort of pattern exists within an online social network. We can take this question and translate it into a testable hypothesis.

EXAMPLE

12.15 An additional comparison of interest. The researchers hypothesize that, unlike an offline social network, the relationship between the number of friends and social attractiveness levels off as the number of friends increases. This can be assessed by comparing changes in the mean scores across various factor levels. These comparisons are simplified because the levels are equally spaced. To compare the change in mean between the lower friend levels to the change in mean between the upper friend levels we construct the following null hypothesis:

$$H_{01}: \mu_{502} - \mu_{102} = \mu_{902} - \mu_{502}$$

We could use the two-sided alternative

$$H_{a1}: \mu_{502} - \mu_{102} \neq \mu_{902} - \mu_{502}$$

but we could also argue that the one-sided alternative

$$H_{a1}: \mu_{502} - \mu_{102} > \mu_{902} - \mu_{502}$$

is appropriate for this problem because we expect there to be a leveling off.

In the example above we used H_{01} and H_{a1} to designate the null and alternative hypotheses. The reason for this is that there is an additional set of hypotheses to assesses linearity. We use H_{02} and H_{a2} for this set.

EXAMPLE

12.16 Another comparison of interest. This comparison tests the linearity across the factor levels. Here are the null and alternative hypotheses:

$$H_{02}: -2\mu_{102} - \mu_{302} + \mu_{702} + 2\mu_{902} = 0$$
$$H_{a2}: -2\mu_{102} - \mu_{302} + \mu_{702} + 2\mu_{902} \neq 0$$

Each of H_{01} and H_{02} says that a combination of population means is 0. These combinations of means are called contrasts because the coefficients sum to zero. We use ψ, the Greek letter psi, for contrasts among population means. For our first comparison, we have

$$\psi_1 = -\mu_{102} + 2\mu_{502} - \mu_{902}$$
$$= (-1)\mu_{102} + (2)\mu_{502} + (-1)\mu_{902}$$

and for comparing the second comparison

$$\psi_2 = (-2)\mu_{102} + (-1)\mu_{302} + (1)\mu_{702} + (2)\mu_{902}$$

In each case, the value of the contrast is 0 when H_0 is true. *Note that we have chosen to define the contrasts so that they will be positive when the alternative of interest (what we expect) is true. Whenever possible, this is a good idea because it makes some computations easier.*

sample contrast   A contrast expresses an effect in the population as a combination of population means. To estimate the contrast, form the corresponding **sample contrast**

by using sample means in place of population means. Under the ANOVA assumptions, a sample contrast is a linear combination of independent Normal variables and therefore has a Normal distribution. We can obtain the standard error of a contrast by using the rules for variances. Inference is based on t statistics. Here are the details.

LOOK BACK
rules for variances p. 274

CONTRASTS

A **contrast** is a combination of population means of the form

$$\psi = \sum a_i \mu_i$$

where the coefficients a_i sum to 0. The corresponding **sample contrast** is

$$c = \sum a_i \bar{x}_i$$

The **standard error of c** is

$$SE_c = s_p \sqrt{\sum \frac{a_i^2}{n_i}}$$

To test the null hypothesis

$$H_0: \psi = 0$$

use the t **statistic**

$$t = \frac{c}{SE_c}$$

with degrees of freedom DFE that are associated with s_p. The alternative hypothesis can be one-sided or two-sided.

A **level C confidence interval for** ψ is

$$c \pm t^* SE_c$$

where t^* is the value for the $t(DFE)$ density curve with area C between $-t^*$ and t^*.

LOOK BACK
addition rule for means p. 267

Because each $\bar{x}_i$ estimates the corresponding μ_i, the addition rule for means tells us that the mean μ_c of the sample contrast c is ψ. In other words, c is an unbiased estimator of ψ. Testing the hypothesis that a contrast is 0 assesses the significance of the effect measured by the contrast. It is often more informative to estimate the size of the effect using a confidence interval for the population contrast.

EXAMPLE

12.17 The contrast coefficients. In our example the coefficients in the contrasts are

$$a_1 = -1, a_2 = 0, a_3 = 2, a_4 = 0, a_5 = -1 \text{ for } \psi_1$$

and

$$a_1 = -2, a_2 = -1, a_3 = 0, a_4 = 1, a_5 = 2 \text{ for } \psi_2,$$

where the subscripts 1, 2, 3, 4, and 5 correspond to the profiles with 102, 302, 502, 702, and 902 friends, respectively. In each case the sum of the a_i is 0. We look at inference for each of these contrasts in turn.

EXAMPLE

12.18 Testing the first contrast of interest. The sample contrast that estimates ψ_1 is

$$c_1 = (-1)\bar{x}_{102} + (2)\bar{x}_{502} + (-1)\bar{x}_{902}$$
$$= -3.82 + (2)4.56 - 3.99 = 1.31$$

with standard error

$$SE_{c_1} = 1.10\sqrt{\frac{(-1)^2}{24} + \frac{(2)^2}{26} + \frac{(-1)^2}{21}}$$
$$= 0.54$$

The t statistic for testing $H_{01}: \psi_1 = 0$ versus $H_{a1}: \psi_1 > 0$ is

$$t = \frac{c_1}{SE_{c_1}}$$
$$= \frac{1.31}{0.54} = 2.43$$

Because s_p has 129 degrees of freedom, software using the $t(129)$ distribution gives the one-sided P-value as $P = 0.008$. If we used Table D, we would conclude that $0.005 < P < 0.01$. The P-value is small, so there is strong evidence against H_{01}.

We have evidence to conclude that the rate of change in the acceptability score at the lower levels is larger than the rate of change at the upper levels. This suggests either a leveling off or decrease in the acceptability score as the number of friends increases. The size of the difference can be described with a confidence interval.

EXAMPLE

12.19 Confidence interval for the first contrast. To find the 95% confidence interval for ψ_1, we combine the estimate with its margin of error:

$$c_1 \pm t^*SE_{c_1} = 1.31 \pm (1.984)(0.54)$$
$$= 1.31 \pm 1.07$$

The 1.984 is a conservative estimate of t^* using 100 degrees of freedom. The interval is $(0.24, \ 2.38)$. We are 95% confident that the difference is between 0.24 and 2.38 points.

We use the same method for the second contrast.

EXAMPLE

12.20 Testing the second contrast of interest. The second sample contrast, assesses the linear trend across the levels.

$$c_2 = (-2)\bar{x}_{102} + (-1)\bar{x}_{302} + (1)\bar{x}_{702} + (2)\bar{x}_{902}$$
$$= (-2)3.82 + (-1)4.88 + (1)4.41 + (2)3.99$$
$$= -7.64 - 4.88 + 4.41 + 7.98$$
$$= -0.13$$

with standard error

$$SE_{c_2} = 1.10\sqrt{\frac{(-2)^2}{24} + \frac{(-1)^2}{33} + \frac{(1)^2}{30} + \frac{(2)^2}{21}} = 0.71$$

The t statistic for assessing the significance of this contrast is

$$t = \frac{-0.13}{0.71} = -0.18$$

The P-value for the two-sided alternative is 0.861. The data do not provide much evidence in favor of a linear trend.

This contrast can be combined with others to assess the various polynomial contributions (e.g., linear, quadratic, cubic) to the relationship between acceptability score and the number of friends. As we saw in Figure 12.5, a quadratic trend appears most prominent. Further discussion of this contrast can be found in Exercise 12.46.

SPSS output for the contrasts is given in Figure 12.9. The results agree with the calculations that we performed in Examples 12.18 and 12.20 except for minor differences due to roundoff error in our calculations. Note that the output does not give the confidence interval that we calculated in Example 12.19. This is easily computed, however, from the contrast estimate and standard error provided in the output.

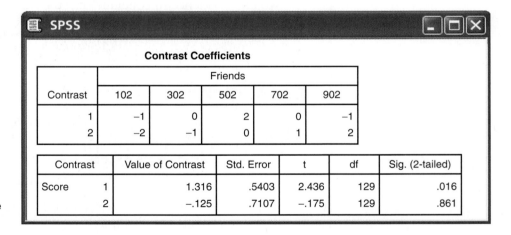

FIGURE 12.9 Software output giving the contrast analysis for the Facebook friends example.

Contrast Coefficients

Contrast	Friends				
	102	302	502	702	902
1	−1	0	2	0	−1
2	−2	−1	0	1	2

Contrast		Value of Contrast	Std. Error	t	df	Sig. (2-tailed)
Score	1	1.316	.5403	2.436	129	.016
	2	−.125	.7107	−.175	129	.861

Some statistical software packages report the test statistics associated with contrasts as F statistics rather than t statistics. These F statistics are the squares of the t statistics described previously. As with much statistical software output, P-values for significance tests are reported for the two-sided alternative. *If the software you are using gives P-values for the two-sided alternative, and you are using the appropriate one-sided alternative, divide the reported P-value by 2.* In our example, we argued that a one-sided alternative was appropriate for the first contrast. The software reported the P-value as 0.016, so we can conclude $P = 0.008$. Dividing this value by 2 has no effect on the conclusion.

Questions about population means are expressed as hypotheses about contrasts. A contrast should express a specific question that we have in mind when designing the study. *When contrasts are formulated before seeing the data,* *inference about contrasts is valid whether or not the ANOVA H_0 of equality of means is rejected.* Because the F test answers a very general question, it is less powerful than tests for contrasts designed to answer specific questions. Specifying the important questions before the analysis is undertaken enables us to use this powerful statistical technique.

Multiple comparisons

multiple-comparisons methods

In many studies, specific questions cannot be formulated in advance of the analysis. If H_0 is not rejected, we conclude that the population means are indistinguishable on the basis of the data given. On the other hand, if H_0 is rejected, we would like to know which pairs of means differ. **Multiple-comparisons methods** address this issue. It is important to keep in mind that multiple-comparisons methods are used only *after rejecting* the ANOVA H_0.

EXAMPLE

12.21 Comparing each pair of groups. Return once more to the Facebook friends data with five groups. We can make 10 comparisons between pairs of means. We can write a t statistic for each of these pairs. For example, the statistic

$$t_{12} = \frac{\bar{x}_1 - \bar{x}_2}{s_p \sqrt{\dfrac{1}{n_1} + \dfrac{1}{n_2}}}$$

$$= \frac{3.82 - 4.88}{1.10 \sqrt{\dfrac{1}{24} + \dfrac{1}{33}}} = -3.59$$

compares profiles with 102 and 302 friends. The subscripts on t specify which groups are compared.

The t statistics for two other pairs are

$$t_{23} = \frac{\bar{x}_2 - \bar{x}_3}{s_p \sqrt{\dfrac{1}{n_2} + \dfrac{1}{n_3}}}$$

$$= \frac{4.88 - 4.56}{1.10 \sqrt{\dfrac{1}{33} + \dfrac{1}{26}}} = 1.11$$

and

$$t_{25} = \frac{\bar{x}_2 - \bar{x}_5}{s_p \sqrt{\dfrac{1}{n_2} + \dfrac{1}{n_5}}}$$

$$= \frac{4.88 - 3.99}{1.10 \sqrt{\dfrac{1}{33} + \dfrac{1}{21}}} = 2.90$$

LOOK BACK
two-sample *t* procedures p. 447

These *t* statistics are very similar to the pooled two-sample *t* statistic for comparing two population means. The difference is that we now have more than two populations, so each statistic uses the pooled estimator s_p from all groups rather than the pooled estimator from just the two groups being compared. This additional information about the common σ increases the power of the tests. The degrees of freedom for all these statistics are DFE = 129, those associated with s_p.

Because we do not have any specific ordering of the means in mind as an alternative to equality, we must use a two-sided approach to the problem of deciding which pairs of means are significantly different.

MULTIPLE COMPARISONS

To perform a **multiple-comparisons procedure,** compute *t* **statistics** for all pairs of means using the formula

$$t_{ij} = \frac{\overline{x}_i - \overline{x}_j}{s_p\sqrt{\dfrac{1}{n_i} + \dfrac{1}{n_j}}}$$

If

$$|t_{ij}| \geq t^{**}$$

we declare that the population means μ_i and μ_j are different. Otherwise, we conclude that the data do not distinguish between them. The value of t^{**} depends upon which multiple-comparisons procedure we choose.

One obvious choice for t^{**} is the upper $\alpha/2$ critical value for the $t(\text{DFE})$ distribution. This choice simply carries out as many separate significance tests of fixed level α as there are pairs of means to be compared. The procedure based on this choice is called the **least-significant differences method,** or simply LSD.

LSD method

LSD has some undesirable properties, particularly if the number of means being compared is large. Suppose, for example, that there are $I = 20$ groups and we use LSD with $\alpha = 0.05$. There are 190 different pairs of means. If we perform 190 *t* tests, each with an error rate of 5%, our overall error rate will be unacceptably large. We expect about 5% of the 190 to be significant even if the corresponding population means are the same. Since 5% of 190 is 9.5, we expect 9 or 10 false rejections.

The LSD procedure fixes the probability of a false rejection for each single pair of means being compared. It does not control the overall probability of *some* false rejection among all pairs. Other choices of t^{**} control possible errors in other ways. The choice of t^{**} is therefore a complex problem, and a detailed discussion of it is beyond the scope of this text. Many choices for t^{**} are used in practice. One major statistical package allows selection from a list of over a dozen choices.

Bonferroni method

We will discuss only one of these, called the **Bonferroni method.** Use of this procedure with $\alpha = 0.05$, for example, guarantees that the probability of *any* false rejection among all comparisons made is no greater than 0.05. This is much stronger protection than controlling the probability of a false rejection at 0.05 for *each separate* comparison.

EXAMPLE

12.22 Applying the Bonferroni method. We apply the Bonferroni multiple-comparisons procedure with $\alpha = 0.05$ to the data from the Facebook friends study. The value of t^{**} for this procedure uses $\alpha = .05/10 = 0.005$ for each test. From Table D, this value is 2.63. Of the statistics $t_{12} = -3.59$, $t_{23} = 1.11$, and $t_{25} = 2.90$ calculated in the beginning of this section, only t_{12} and t_{25} are significant. These two statistics compare the profile of 302 friends with the two extreme levels.

Of course, we prefer to use software for the calculations.

EXAMPLE

12.23 Interpreting software output. The output generated by SPSS for Bonferroni comparisons appears in Figure 12.10. The software uses an asterisk

SPSS						
Multiple Comparisons						
Score						
Bonferroni						
(I) Friends	(J) Friends	Mean Difference (I − J)	Std. Error	Sig.	95% Confidence Interval	
					Lower Bound	Upper Bound
102	302	−1.0621*	.2939	.004	−1.902	−.223
	502	−.7449	.3102	.177	−1.631	.141
	702	−.5900	.3001	.514	−1.447	.267
	902	−.1738	.3274	1.000	−1.109	.761
302	102	1.0621*	.2939	.004	.223	1.902
	502	.3172	.2873	1.000	−.503	1.138
	702	.4721	.2764	.900	−.317	1.262
	902	.8883*	.3059	.043	.015	1.762
502	102	.7449	.3102	.177	−.141	1.631
	302	−.3172	.2873	1.000	−1.138	.503
	702	.1549	.2936	1.000	−.684	.993
	902	.5711	.3215	.780	−.347	1.489
702	102	.5900	.3001	.514	−.267	1.447
	302	−.4721	.2764	.900	−1.262	.317
	502	−.1549	.2936	1.000	−.993	.684
	902	.4162	.3117	1.000	−.474	1.307
902	102	.1738	.3274	1.000	−.761	1.109
	302	−.8883*	.3059	.043	−1.762	−.015
	502	−.5711	.3215	.780	−1.489	.347
	702	−.4162	.3117	1.000	−1.307	.474

*. The mean difference is significant at the 0.05 level.

FIGURE 12.10 Software output giving the multiple-comparisons analysis for the Facebook friends example.

to indicate that the difference in a pair of means is statistically significant. Here, all 10 comparisons are reported. These results agree with the calculations that we performed in Examples 12.21 and 12.22. There are no significant differences except those already mentioned. Note that each comparison is given twice in the output.

The data in the Facebook friends study provide a clear result: the social acceptability score increases as the number of friends increases to a point and then decreases. Unfortunately with these data, we cannot accurately describe this relationship in more detail. This lack of clarity is not unusual when performing a multiple-comparisons analysis. Here, the mean associated with 302 friends is different than the 102- and 902-friend profiles, but it is not found significantly different from the means for the profiles with 502 and 702 friends. To complicate things, the means for profiles with 502 and 702 friends were not found significantly different than the 102- and 902-friend profiles. *This kind of apparent contradiction points out dramatically the nature of the conclusions of statistical tests of significance.* The conclusion appears to be illogical. If μ_1 is the same as μ_3 and μ_3 is the same as μ_2, doesn't it follow that μ_1 is the same as μ_2? Logically, the answer must be Yes.

Some of the difficulty can be resolved by noting the choice of words used. In describing the inferences, we talk about failing to detect a difference or concluding that two groups are different. In making logical statements, we say things such as "is the same as." There is a big difference between the two modes of thought. Statistical tests ask, "Do we have adequate evidence to distinguish two means?" It is not illogical to conclude that we have sufficient evidence to distinguish μ_1 from μ_2, but not μ_1 from μ_3 or μ_2 from μ_3.

One way to deal with these difficulties of interpretation is to give confidence intervals for the differences. The intervals remind us that the differences are simultaneous confidence intervals not known exactly. We want to give **simultaneous confidence intervals,** that is, intervals for *all* differences among the population means at once. Again, we must face the problem that there are many competing procedures—in this case, many methods of obtaining simultaneous intervals.

SIMULTANEOUS CONFIDENCE INTERVALS FOR DIFFERENCES BETWEEN MEANS

Simultaneous confidence intervals for all differences $\mu_i - \mu_j$ between population means have the form

$$(\overline{x}_i - \overline{x}_j) \pm t^{**} s_p \sqrt{\frac{1}{n_i} + \frac{1}{n_j}}$$

The critical values t^{**} are the same as those used for the multiple-comparisons procedure chosen.

The confidence intervals generated by a particular choice of t^{**} are closely related to the multiple-comparisons results for that same method. If one of the confidence intervals includes the value 0, then that pair of means will not be declared significantly different, and vice versa.

EXAMPLE

12.24 Interpreting software output, continued. The SPSS output for the Bonferroni multiple-comparisons procedure given in Figure 12.10 includes the simultaneous 95% confidence intervals. We can see, for example, that the interval for $\mu_1 - \mu_3$ is -1.63 to 0.14. The fact that the interval includes 0 is consistent with the fact that we failed to detect a difference between these two means using this procedure. Note that the interval for $\mu_3 - \mu_1$ is also provided. This is not really a new piece of information, because it can be obtained from the other interval by reversing the signs and reversing the order, that is, -0.14 to 1.63. So, in fact, we really have only 10 intervals. Use of the Bonferroni procedure provides us with 95% confidence that *all 10* intervals simultaneously contain the true values of the population mean differences.

Software

We have used SPSS to illustrate the analysis of the worker safety data. Other statistical software gives similar output, and you should be able to read it without any difficulty.

 EYES

EXAMPLE

12.25 Do eyes affect ad response? Research from a variety of fields has found significant effects of eye gaze and eye color on emotions and perceptions such as arousal, attractiveness, and honesty. These findings suggest that a model's eyes may play a role in a viewer's response to an ad. In a recent study, students in marketing and management classes of a southern, predominantly Hispanic, university were each presented one of four portfolios.[4] Each portfolio contained a target ad for a fictional product, Sparkle Toothpaste. Students were asked to view the ad and then respond to questions concerning their attitudes and emotions about the ad and product. All questions were from advertising effects questionnaires previously used in the literature. Each response was on a seven-point scale. Although the researchers investigated nine attitudes and emotions, we will focus on the viewer's "attitudes toward the brand." This response was obtained by averaging 10 survey questions.

The target ads were created using two digital photographs of a model. In one picture the model is looking directly at the camera so the eyes can be seen. This picture was used in three target ads. The only difference was the model's eyes, which were made to be either brown, blue, or green. In the second picture, the model is in virtually the same pose but looking downward so the eyes are not visible. A total of 222 surveys were used for analysis. The following table summarizes the responses for the four portfolios. Outputs from SAS, Excel, and Minitab are given in Figure 12.11.

Group	n	Mean	Std. Dev
Blue	67	3.19	1.75
Brown	37	3.72	1.73
Down	41	3.11	1.53
Green	77	3.86	1.67

Excel

	A	B	C	D	E	F	G
1	Anova: Single Factor						
2							
3	SUMMARY						
4	*Groups*	*Count*	*Sum*	*Average*	*Variance*		
5	Blue	67	214	3.19403	3.079055		
6	Brown	37	137.8	3.724324	2.942447		
7	Down	41	127.4	3.107317	2.326695		
8	Green	77	297.2	3.85974	2.775332		
9							
10	ANOVA						
11	*Source of Variation*	*SS*	*df*	*MS*	*F*	*P-value*	*F crit*
12	Between Groups	24.41966	3	8.139886	2.894117	0.036184	2.646014
13	Within Groups	613.1387	218	2.812563			
14							
15	Total	637.5584	221				

SAS

The GLM Procedure

Dependent Variable: Score

Source	DF	Sum of Squares	Mean Square	F Value	Pr > F
Model	3	24.4196586	8.1398862	2.89	0.0362
Error	218	613.1387197	2.8125629		
Corrected Total	221	637.5583784			

R-Square	Coeff var	Root MSE	Score Mean
0.038302	47.95331	1.677070	3.497297

Level of Group	N	----------Score---------- Mean	Std Dev
Blue	67	3.19402985	1.75472355
Brown	37	3.72432432	1.71535636
Down	41	3.10731707	1.52535082
Green	77	3.85974026	1.66593262

FIGURE 12.11 SAS, Excel, and Minitab output for the advertising study in Example 12.25.

USE YOUR KNOWLEDGE

12.7 Why no multiple comparisons? Any pooled two-sample t problem can be run as a one-way ANOVA with $I = 2$. Explain why it is inappropriate to analyze the data using contrasts or multiple-comparisons procedures in this setting.

12.8 Growth of Douglas fir seedlings. An experiment was conducted to compare the growth of Douglas fir seedlings under three different levels of vegetation control (0%, 50%, and 100%). Forty seedlings were

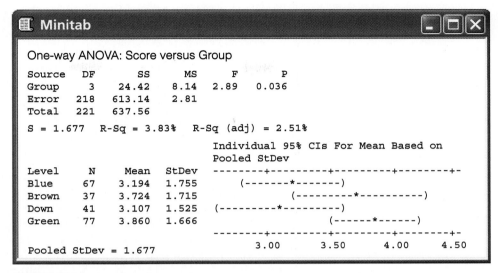

FIGURE 12.11   *(continued)*

randomized to each level of control. The resulting sample means for stem volume were 50, 75, and 120 cubic centimeters (cm^3) respectively with $s_p = 30 \ cm^3$. The researcher hypothesized that the average growth at 50% control would be less than the average of the 0% and 100% levels.

(a) What are the coefficients for testing this contrast?

(b) Perform the test. Do the data provide evidence to support this hypothesis?

Power*

Recall that the power of a test is the probability of rejecting H_0 when H_a is in fact true. Power measures how likely a test is to detect a specific alternative. When planning a study in which ANOVA will be used for the analysis, it is important to perform power calculations to check that the sample sizes are adequate to detect differences among means that are judged to be important. Power calculations also help evaluate and interpret the results of studies in which H_0 was not rejected. We sometimes find that the power of the test was so low against reasonable alternatives that there was little chance of obtaining a significant F.

LOOK BACK
power p. 463

In Chapter 7 we found the power for the two-sample t test. One-way ANOVA is a generalization of the two-sample t test, so it is not surprising that the procedure for calculating power is quite similar. Here are the steps that are needed:

1. Specify
   (a) an alternative (H_a) that you consider important; that is, values for the true population means $\mu_1, \mu_2, \ldots, \mu_I$;
   (b) sample sizes $n_1, n_2, \ldots, n_I$; usually these will all be equal to the common value n;

---

*This section is optional.

(c) a level of significance α, usually equal to 0.05; and

(d) a guess at the standard deviation σ.

2. Use the degrees of freedom DFG $= I - 1$ and DFE $= N - I$ to find the critical value that will lead to the rejection of H_0. This value, which we denote by F^*, is the upper α critical value for the $F(\text{DFG}, \text{DFE})$ distribution.

noncentrality parameter   3. Calculate the **noncentrality parameter**[5]

$$\lambda = \frac{\sum n_i (\mu_i - \overline{\mu})^2}{\sigma^2}$$

where $\overline{\mu}$ is a weighted average of the group means

$$\overline{\mu} = \sum \frac{n_i}{N} \mu_i$$

4. Find the power, which is the probability of rejecting H_0 when the alternative hypothesis is true, that is, the probability that the observed F is greater than F^*. Under H_a, the F statistic has a distribution known as the

noncentral *F* distribution   **noncentral F distribution.** SAS, for example, has a function for this distribution. Using this function, the power is

$$\text{Power} = 1 - \text{PROBF}(F^*, \text{DFG}, \text{DFE}, \lambda)$$

Note that, if the n_i are all equal to the common value n, then $\overline{\mu}$ is the ordinary average of the μ_i and

$$\lambda = \frac{n \sum (\mu_i - \overline{\mu})^2}{\sigma^2}$$

If the means are all equal (the ANOVA H_0), then $\lambda = 0$. The noncentrality parameter measures how unequal the given set of means is. Large λ points to an alternative far from H_0, and we expect the ANOVA F test to have high power. Software makes calculation of the power quite easy, but tables and charts are also available.

EXAMPLE

12.26 Power of a reading comprehension study. Suppose that a study on reading comprehension for three different teaching methods has 10 students in each group. How likely is this study to detect differences in the mean responses that would be viewed as important? A previous study performed in a different setting found sample means of 41, 47, and 44, and the pooled standard deviation was 7. Based on these results, we will use $\mu_1 = 41$, $\mu_2 = 47$, $\mu_3 = 44$, and $\sigma = 7$ in a calculation of power. The n_i are equal, so $\overline{\mu}$ is simply the average of the μ_i:

$$\overline{\mu} = \frac{41 + 47 + 44}{3} = 44$$

The noncentrality parameter is therefore

$$\begin{aligned}
\lambda &= \frac{n \sum (\mu_i - \overline{\mu})^2}{\sigma^2} \\
&= \frac{(10)[(41 - 44)^2 + (47 - 44)^2 + (44 - 44)^2]}{49} \\
&= \frac{(10)(18)}{49} = 3.67
\end{aligned}$$

Because there are three groups with 10 observations per group, DFG $= 2$ and DFE $= 27$. The critical value for $\alpha = 0.05$ is $F^* = 3.35$. The power is therefore

$$1 - \text{PROBF}(3.35, 2, 27, 3.67) = 0.3486$$

The chance that we reject the ANOVA H_0 at the 5% significance level is only about 35%.

If the assumed values of the μ_i in this example describe differences among the groups that the experimenter wants to detect, then we would want to use more than 10 subjects per group. Although H_0 is assumed to be false, the chance of rejecting it is only about 35%. This chance can be increased to acceptable levels by increasing the sample sizes.

> ### EXAMPLE
>
> **12.27 Changing the sample size.** To decide on an appropriate sample size for the experiment described in the previous example, we repeat the power calculation for different values of n, the number of subjects in each group. Here are the results:
>
> | n | DFG | DFE | F^* | λ | Power |
> |-----|-----|-----|-------|-----------|-------|
> | 20 | 2 | 57 | 3.16 | 7.35 | 0.65 |
> | 30 | 2 | 87 | 3.10 | 11.02 | 0.84 |
> | 40 | 2 | 117 | 3.07 | 14.69 | 0.93 |
> | 50 | 2 | 147 | 3.06 | 18.37 | 0.97 |
> | 100 | 2 | 297 | 3.03 | 36.73 | ≈ 1 |

With $n = 40$, the experimenters have a 93% chance of rejecting H_0 with $\alpha = 0.05$ and thereby demonstrating that the groups have different means. In the long run, 93 out of every 100 such experiments would reject H_0 at the $\alpha = 0.05$ level of significance. Using 50 subjects per group increases the chance of finding significance to 97%. With 100 subjects per group, the experimenters are virtually certain to reject H_0. The exact power for $n = 100$ is 0.99989. In most real-life situations the additional cost of increasing the sample size from 50 to 100 subjects per group would not be justified by the relatively small increase in the chance of obtaining statistically significant results.

CHAPTER 12 Summary

One-way analysis of variance (ANOVA) is used to compare several population means based on independent SRSs from each population. The populations are assumed to be Normal with possibly different means and the same standard deviation.

To do an analysis of variance, first compute sample means and standard deviations for all groups. Side-by-side boxplots give an overview of the data. Examine Normal quantile plots (either for each group separately or for the residuals) to detect outliers or extreme deviations from Normality. Compute the ratio of the largest to the smallest sample standard deviation. If this ratio is less than 2 and the Normal quantile plots are satisfactory, ANOVA can be performed.

The **null hypothesis** is that the population means are *all* equal. The **alternative hypothesis** is true if there are *any* differences among the population means.

ANOVA is based on separating the total variation observed in the data into two parts: variation **among group means** and variation **within groups.** If the variation among groups is large relative to the variation within groups, we have evidence against the null hypothesis.

An **analysis of variance table** organizes the ANOVA calculations. **Degrees of freedom, sums of squares, and mean squares** appear in the table. The *F* **statistic** and its *P***-value** are used to test the null hypothesis.

The ANOVA *F* test shares the **robustness** of the two-sample *t* test. It is relatively insensitive to moderate non-Normality and unequal variances, especially when the sample sizes are similar.

Specific questions formulated before examination of the data can be expressed as **contrasts.** Tests and confidence intervals for contrasts provide answers to these questions.

If no specific questions are formulated before examination of the data and the null hypothesis of equality of population means is rejected, **multiple-comparisons** methods are used to assess the statistical significance of the differences between pairs of means.

The **power** of the *F* test depends upon the sample sizes, the variation among population means, and the within-group standard deviation.

CHAPTER 12 Exercises

For Exercises 12.1 and 12.2, see page 630; for Exercises 12.3 and 12.4, see page 634; for Exercises 12.5 and 12.6, see page 641; and for Exercises 12.7 and 12.8, see page 652.

12.9 A one-way ANOVA example. A study compared 5 groups with 7 observations per group. An *F* statistic of 2.83 was reported.

(a) Give the degrees of freedom for this statistic and the entries from Table E that correspond to this distribution.

(b) Sketch a picture of this *F* distribution with the information from the table included.

(c) Based on the table information, how would you report the *P*-value?

(d) Can you conclude that all pairs of means are different? Explain your answer.

12.10 Calculating the ANOVA *F* test *P*-value. For each of the following situations, find the degrees of freedom for the *F* statistic and then use Table E to approximate the *P*-value.

(a) Seven groups are being compared with 5 observations per group. The value of the *F* statistic is 2.69.

(b) Five groups are being compared with 11 observations per group. The value of the *F* statistic is 2.43.

(c) Six groups are being compared using 34 total observations. The value of the *F* statistic is 3.06.

12.11 Calculating the ANOVA *F* test *P*-value, continued. For each of the following situations, find the *F* statistic and the degrees of freedom. Then draw a sketch of the distribution under the null hypothesis and shade in the portion corresponding to the *P*-value. State how you would report the *P*-value.

(a) Compare 3 groups with 11 observations per group, MSE = 50, and MSG = 127.

(b) Compare 4 groups with 8 observations per group, SSG = 58, and SSE = 182.

12.12 Calculating the pooled standard deviation. An experiment was run to compare three groups. The sample sizes were 29, 32, and 121, and the corresponding estimated standard deviations were 37, 28, and 42.

(a) Is it reasonable to use the assumption of equal standard deviations when we analyze these data? Give a reason for your answer.

(b) Give the values of the variances for the three groups.

(c) Find the pooled variance.

(d) What is the value of the pooled standard deviation?

(e) Explain why your answer in part (d) is much closer to the standard % deviation for the third group than to the other two standard deviations.

12.13 Describing the ANOVA model. For each of the following situations, identify the response variable and the populations to be compared, and give I, the n_i, and N.

(a) A poultry farmer is interested in reducing the cholesterol level in his marketable eggs. He wants to compare two different cholesterol-lowering drugs added to the hens' standard diet as well as an all-vegetarian diet. He assigns 25 of his hens to each of the three treatments.

(b) A researcher is interested in students' opinions regarding an additional annual fee to support non-income-producing varsity sports. Students were asked to rate their acceptance of this fee on a five-point scale. She received 94 responses, of which 31 were from students who attend varsity football or basketball games only, 18 were from students who also attend other varsity competitions, and 45 were from students who did not attend any varsity games.

(c) A professor wants to evaluate the effectiveness of his teaching assistants. In one class period, the 42 students were randomly divided into three equal-sized groups, and each group was taught power calculations from one of the assistants. At the beginning of the next class, each student took a quiz on power calculations, and these scores were compared.

12.14 Describing the ANOVA model, continued. For each of the following situations, identify the response variable and the populations to be compared, and give I, the n_i, and N.

(a) A developer of a virtual-reality (VR) teaching tool for the deaf wants to compare the effectiveness of different navigation methods. A total of 40 children were available for the experiment, of which equal numbers were randomly assigned to use a joystick, wand, dancemat, or gesture-based pinch gloves. The time (in seconds) to complete a designed VR path is recorded for each child.

(b) To study the effects of pesticides on birds, an experimenter randomly (and equally) allocated 65 chicks

to five diets (a control and four with a different pesticide included). After a month, the calcium content (milligrams) in a 1-centimeter length of bone from each chick was measured.

(c) A university sandwich shop wants to compare the effects of providing free food with a sandwich order on sales. The experiment will be conducted from 11:00 A.M. to 2:00 P.M. for the next 20 weekdays. On each day, customers will be offered one of the following: a free drink, free chips, a free cookie, or nothing. Each option will be offered five times.

12.15 Determining the degrees of freedom. Refer to Exercise 12.13. For each situation, give the following:

(a) Degrees of freedom for the model, for error, and for the total.

(b) Null and alternative hypotheses.

(c) Numerator and denominator degrees of freedom for the F statistic.

12.16 Determining the degrees of freedom, continued. Refer to Exercise 12.14. For each situation, give the following:

(a) Degrees of freedom for the model, for error, and for the total.

(b) Null and alternative hypotheses.

(c) Numerator and denominator degrees of freedom for the F statistic.

12.17 Data collection and the interpretation of results. Refer to Exercise 12.13. For each situation, discuss the method of obtaining the data and how this will affect the extent to which the results can be generalized.

12.18 Data collection, continued. Refer to Exercise 12.14. For each situation, discuss the method of obtaining the data and how this will affect the extent to which the results can be generalized.

12.19 Shopping and bargaining in Mexico. Price haggling and other bargaining behaviors among consumers have been observed for a long time. However, research addressing these behaviors, especially in a real-life setting, remains relatively sparse. A group of researchers recently performed a small study to determine whether gender or nationality of the bargainer has an effect in the final price obtained.[6] The study took place in Mexico because of the prevalence of price haggling in informal markets. Salespersons working at various informal shops were approached by one of three bargainers looking for a specific product. After an initial price was stated by the vendor, bargaining took place.

The response was the difference between the initial and final price of the product. The bargainers were a Spanish speaking Hispanic male, a Spanish speaking Hispanic female, and an Anglo non-Spanish speaking male. The following table summarizes the results:

Bargainer	n	Average reduction
Hispanic male	40	1.055
Anglo male	40	1.050
Hispanic female	40	2.310

(a) To compare the mean reductions in price, what are the degrees of freedom for the ANOVA F statistic?

(b) The reported test statistic is $F = 8.708$. Give an approximate (from a table) or exact (from software) P-value. What do you conclude?

(c) To what extent do you think the results of this study can be generalized? Give reasons for your answer.

12.20 The effect of increased variation within groups. The *One-Way ANOVA* applet lets you see how the F statistic and the P-value depend on the variability of the data within groups and the differences among the means.

(a) The black dots are at the means of the three groups. Move these up and down until you get a configuration that gives a P-value of about 0.01. What is the value of the F statistic?

(b) Now increase the variation within the groups by dragging the mark on the pooled standard error scale to the right. Describe what happens to the F statistic and the P-value. Explain why this happens.

12.21 The effect of increased variation between groups. Set the pooled standard error for the *One-Way ANOVA* applet at a middle value. Drag the black dots so that they are approximately equal.

(a) What is the F statistic? Give its P-value.

(b) Drag the mean of the second group up and the mean of the third group down. Describe the effect on the F statistic and its P-value. Explain why they change in this way.

12.22 Do labels matter? A study was performed to examine the self-identification of college students of Asian descent with various identity categories and assess whether there are attitudinal differences across these categories. Undergraduates at a large Midwestern university, who had identified themselves as being of Asian descent on their admission application, were asked to participate in the study. A total of 620 undergraduates filled out the survey. One question classified the participants into groups by asking them to indicate the

option they primarily identify: (a) Asian American, (b) specific ethnicity (e.g., Chinese), (c) ethnicity American (e.g., Chinese American), and (d) other. The responses to the remaining questions were then compared across these four groups. One question was, "The campus is supportive of Asian American students." Responses were on a 4-point scale (1 = strongly disagree, 4 = strongly agree. A summary of the results follows:

Label	n	$\bar{x}$
Asian American	130	2.93
Ethnicity	248	3.00
Ethnicity American	174	3.01
Other	68	3.39

(a) What are the numerator and denominator degrees of freedom for the F test?

(b) Using the formula on page 640 and the preceding results, calculate SSG.

(c) Given SSE = 797.25, use your result from part (b) to compute the F statistic.

(d) Compute the P-value and state your conclusions.

(e) Without doing any additional analysis, describe the pattern in the means that is likely responsible for your conclusions in part (d).

12.23 The multiple play strategy. Multiple play is a bundling strategy through which multiple services are provided over a single network. A common triple-play service these days is Internet, television, and telephone. The market for this service has become a key battleground among telecommunication, cable, and broadband service providers. A recent study compared the pricing (average monthly cost in U.S. dollars) among triple-play providers using DSL, cable, or fiber platforms.[7] The following table summarizes the results of 47 providers.

Group	n	$\bar{x}$	s
DSL	19	104.49	26.09
Cable	20	119.98	40.39
Fiber	8	83.87	31.78

(a) Plot the means versus the platform type. Does there appear to be a difference in pricing?

(b) Is it reasonable to assume that the variances are equal? Explain.

(c) The F statistic is 3.39. Give the degrees of freedom and either an approximate (from a table) or an exact (from software) P-value. What do you conclude?

12.24 A contrast. Refer to the previous exercise. Use a contrast to compare the fiber platform with the average of the other two. The hypothesis prior to collecting the data is that the fiber platform price would be smaller. Summarize your conclusion.

12.25 Sleep deprivation and reaction times. Sleep deprivation experienced by physicians during residency training and the possible negative consequences are of concern to many in the health care community. One study of 33 resident anesthesiologists compared their changes from baseline in reaction times on four tasks.[8] Under baseline conditions, the physicians reported getting an average of 7.04 hours of sleep. While on duty, however, the average was 1.66 hours. For each of the tasks the researchers reported a statistically significant increase in the reaction time when the residents were working in a state of sleep deprivation.

(a) If each task is analyzed separately as the researchers did in their report, what is the appropriate statistical method to use? Explain your answer.

(b) Is it appropriate to use a one-way ANOVA with $I = 4$ to analyze these data? Explain why or why not.

12.26 🏀 **The two-sample t test and one-way ANOVA.** Refer to the HDL level data in Exercise 7.69 (page xxx). Find the two-sample pooled t statistic for comparing men with women. Then formulate the problem as an ANOVA and report the results of this analysis. Verify that $F = t^2$.

12.27 The importance of recreational sports to college satisfaction. The National Intramural-Recreational Sports Association (NIRSA) performed a survey to look at the value of recreational sports on college campuses.[9] One of the questions asked each student to rate the importance of recreational sports to college satisfaction and success. Responses were on a 10-point scale with 1 indicating total lack of importance and 10 indicating very high importance. The following table summarizes these results:

Class	n	Mean score
Freshman	724	7.6
Sophomore	536	7.6
Junior	593	7.5
Senior	437	7.3

(a) To compare the mean scores across classes, what are the degrees of freedom for the ANOVA F statistic?

(b) The MSG $= 11.806$. If $s_p = 2.16$, what is the F statistic?

(c) Give an approximate (from a table) or exact (from software) P-value. What do you conclude?

12.28 Restaurant ambience and consumer behavior. There have been numerous studies investigating the effects of restaurant ambience on consumer behavior. A recent study investigated the effects of musical genre on consumer spending.[10] At a single high-end restaurant in England over a 3-week period, there were a total of 141 participants; 49 of them were subjected to background pop music (for example, Britney Spears, Culture Club, and Ricky Martin) while dining, 44 to background classical music (for example, Vivaldi, Handel, and Strauss), and 48 to no background music. For each participant, the total food bill, adjusted for time spent dining, was recorded. The following table summarizes the means and standard deviations (in British pounds):

Background music	Mean bill	n	s
Classical	24.130	44	2.243
Pop	21.912	49	2.627
None	21.697	48	3.332
Total	22.531	141	2.969

(a) Plot the means versus the type of background music. Does there appear to be a difference in spending?

(b) Is it reasonable to assume that the variances are equal? Explain.

(c) The F statistic is 10.62. Give the degrees of freedom and either an approximate (from a table) or an exact (from software) P-value. What do you conclude?

(d) Refer back to part (a). Without doing any formal analysis, describe the pattern in the means that is likely responsible for your conclusion in part (c).

(e) To what extent do you think the results of this study can be generalized to other settings? Give reasons for your answer.

12.29 The effects of two stimulant drugs. An experimenter was interested in investigating the effects of two stimulant drugs (labeled A and B). She divided 20 rats equally into 5 groups (placebo, Drug A low, Drug A high, Drug B low, and Drug B high) and, 20 minutes after injection of the drug, recorded each rat's activity level (higher score is more active). The following table summarizes the results:

Treatment	$\bar{x}$	s^2
Placebo	14.00	9.00
Low A	15.25	14.00
High A	18.25	12.25
Low B	15.75	6.75
High B	22.50	11.00

(a) Plot the means versus the type of treatment. Does there appear to be a difference in the activity level? Explain.

(b) Is it reasonable to assume that the variances are equal? Explain your answer, and if reasonable, compute s_p.

(c) Give the degrees of freedom for the F statistic.

(d) The F statistic is 4.28. Find the associated P-value and state your conclusions.

12.30 Exam accommodations and end-of-term grades. The Americans with Disabilities Act (ADA) requires that students with learning disabilities (LD) and/or attention deficit disorder (ADD) be given certain accommodations when taking examinations. One study designed to assess the effects of these accommodations examined the relationship between end-of-term grades and the number of accommodations given.[11] The researchers reported the mean grades with sample sizes and standard deviations versus the number of accommodations in a table similar to this:

Accommodations	Mean grade	n	s
0	2.7894	160	0.85035
1	2.8605	38	0.83068
2	2.5757	37	0.82745
3	2.6286	7	1.03072
4	2.4667	3	1.66233
Total	2.7596	245	0.85701

(a) Plot the means versus the number of accommodations. Is there a pattern evident?

(b) A large number of digits are reported for the means and the standard deviations. Do you think that all of these are necessary? Give reasons for your answer and describe how you would report these results.

(c) Should we pool to obtain an estimate of an assumed standard deviation for these data? Explain your answer and give the pooled estimate if your answer is Yes.

(d) The small numbers of observations with 3 or 4 accommodations lead to estimates that are highly variable in these groups compared with the other groups. Inclusion of groups with relatively few observations in an ANOVA can also lead to low power. We could eliminate these two levels from the analysis or we could combine them with the 37 observations in the group above to form a new group with 2 or more accommodations. Which of these options would you prefer? Give reasons for your answer.

(e) The 245 grades reported in the table were from a sample of 61 students who completed three, four, or five courses during a spring term at one college and who were qualified to receive accommodations. Students in the sample were self-identified, in the sense that they had to request qualification. Even when qualified, some students chose not to request accommodations for some or all of their courses. Based on these facts, would you advise that ANOVA methods be used for these data? Explain your answer. (The authors did not present the results of an ANOVA in their publication.)

(f) To what extent do you think the results of this study can be generalized to other settings? Give reasons for your answer.

(g) Most reasonable approaches to the analysis of these data would conclude that the data fail to provide evidence that the number of accommodations is related to the mean grades. Does this imply that the accommodations are not needed or does it suggest that they are effective? Discuss your answer.

12.31 Do we experience emotions differently? Do people from different cultures experience emotions differently? One study designed to examine this question collected data from 416 college students from five different cultures.[12] The participants were asked to record, on a 1 (never) to 7 (always) scale, how much of the time they typically felt eight specific emotions. These were averaged to produce the global emotion score for each participant. Here is a summary of this measure:

Culture	n	Mean (s)
European American	46	4.39 (1.03)
Asian American	33	4.35 (1.18)
Japanese	91	4.72 (1.13)
Indian	160	4.34 (1.26)
Hispanic American	80	5.04 (1.16)

Note that the convention of giving the standard deviations in parentheses after the means saves a great deal of space in a table such as this.

(a) From the information given, do you think that we need to be concerned that a possible lack of Normality in the data will invalidate the conclusions that we might draw using ANOVA to analyze the data? Give reasons for your answer.

(b) Is it reasonable to use a pooled standard deviation for these data? Why or why not?

(c) The ANOVA F statistic was reported as 5.69. Give the degrees of freedom and either an approximate (from a table) or an exact (from software) P-value. Sketch a picture of the F distribution that illustrates the P-value. What do you conclude?

(d) Without doing any additional formal analysis, describe the pattern in the means that appears to be responsible for your conclusion in part (c). Are there pairs of means that are quite similar?

12.32 ⚔️ **The emotion study, continued.** Refer to the previous exercise. The experimenters also measured emotions in some different ways. For a period of a week, each participant carried a device that sounded an alarm at random times during a 3-hour interval 5 times a day. When the alarm sounded, participants recorded several mood ratings indicating their emotions for the time immediately preceding the alarm. These responses were combined to form two variables: frequency, the number of emotions recorded, expressed as a percent; and intensity, an average of the intensity scores measured on a scale of 0 to 6. At the end of the 1-week experimental period, the subjects were asked to recall the percent of time that they experienced different emotions. This variable was called "recall." Here is a summary of the results:

Culture	n	Frequency mean (s)	Intensity mean (s)	Recall mean (s)
European American	46	82.87 (18.26)	2.79 (0.72)	49.12 (22.33)
Asian American	33	72.68 (25.15)	2.37 (0.60)	39.77 (23.24)
Japanese	91	73.36 (22.78)	2.53 (0.64)	43.98 (22.02)
Indian	160	82.71 (17.97)	2.87 (0.74)	49.86 (21.60)
Hispanic American	80	92.25 (8.85)	3.21 (0.64)	59.99 (24.64)
F statistic		11.89	13.10	7.06

(a) For each response variable state whether or not it is reasonable to use a pooled standard deviation to analyze these data. Give reasons for your answer.

(b) Give the degrees of freedom for the F statistics and find the associated P-values. Summarize what you can conclude from these ANOVA analyses.

(c) Summarize the means, paying particular attention to similarities and differences across cultures and across variables. Include the means from the previous exercise in your summary.

(d) The European American and Asian American subjects were from the University of Illinois, the Japanese subjects were from two universities in Tokyo, the Indian subjects were from eight universities in or near Calcutta, and the Hispanic American subjects were from California State University at Fresno. Participants were paid $25 or an equivalent monetary incentive for the Japanese and Indians. Ads were posted on or near the campuses to recruit volunteers for the study. Discuss how these facts influence your conclusions and the extent to which you would generalize the results.

(e) The percents of female students in the samples were as follows: European American, 83%; Asian American, 67%; Japanese, 63%; Indian, 64%; and Hispanic American, 79%. Use a chi-square test to compare these proportions (see Section 9.2, page 517) and discuss how this information influences your interpretation of the results that you have found in this exercise.

12.33 Writing contrasts. Return to the eye study described in Example 12.25 (page 659). Let μ_1, μ_2, μ_3, and μ_4 represent the mean scores for blue, brown, gaze down, and green eyes.

(a) Because a majority of the population is Hispanic (eye color predominantly brown), we want to compare the average score of the brown eyes with the average of the other two eye colors. Write a contrast that expresses this comparison.

(b) Write a contrast to compare the average score when the model is looking at you versus the score when looking down.

12.34 Analyzing contrasts. Answer the following questions for the two contrasts that you defined in Exercise 12.33. 📀 EYES

(a) For each contrast give H_0 and an appropriate H_a. In choosing the alternatives you should use information given in the description of the problem, but you may not consider any impressions obtained by inspection of the sample means.

(b) Find the values of the corresponding sample contrasts c_1 and c_2.

(c) Calculate the standard errors SE_{c_1} and SE_{c_2}.

(d) Give the test statistics and approximate P-values for the two significance tests. What do you conclude?

(e) Compute 95% confidence intervals for the two contrasts.

12.35 ⚔️ **A comparison of tropical flower varieties.** Different varieties of the tropical flower *Heliconia* are fertilized by different species of hummingbirds. Over time, the lengths of the flowers and the form of the hummingbirds' beaks have evolved to match each other. Here are data on the lengths in millimeters of three varieties of these flowers on the island of Dominica:[13] 📀 HUMMINGBIRD

H. bihai							
47.12	46.75	46.81	47.12	46.67	47.43	46.44	46.64
48.07	48.34	48.15	50.26	50.12	46.34	46.94	48.36

H. caribaea red							
41.90	42.01	41.93	43.09	41.47	41.69	39.78	40.57
39.63	42.18	40.66	37.87	39.16	37.40	38.20	38.07
38.10	37.97	38.79	38.23	38.87	37.78	38.01	

H. caribaea yellow							
36.78	37.02	36.52	36.11	36.03	35.45	38.13	37.1
35.17	36.82	36.66	35.68	36.03	34.57	34.63	

Do a complete analysis that includes description of the data and a significance test to compare the mean lengths of the flowers for the three species.

12.36 Animals on product labels? Recall Exercise 7.66 (page 453). This experiment actually involved comparing product preference for a group of consumers that were "primed" and two groups of consumers that served as controls. A bottle of MagicCoat pet shampoo was the product and participants indicated their attitude toward this product on a seven-point scale (from 1 = dislike very much to 7 = like very much). The bottle of shampoo either had a picture of a collie on the label or just the wording. Also, prior to giving this score, participants were asked to do a word find where four of the words were common across groups (pet, grooming, bottle, label) and four were either related to the image (dog, collie, puppy, woof) or image conflicting (cat, feline, kitten, meow). A summary of the groups follows: 🐾 BRANDPREF1

Group	Label w/dog	Dog "primed"	n
1	Y	Y	22
2	Y	N	20
3	N	Y	10

(a) Use graphical and numerical methods to describe the data.

(b) Run the ANOVA and report the results.

(c) Examine the assumptions necessary for inference using your results in part (a) and an examination of the residuals. Summarize your findings.

(d) Use a multiple-comparisons method to compare the three groups. State your conclusions.

12.37 🎯 **Taking the log of the response variable.** The distributions of the flower lengths in Exercise 12.35 are somewhat skewed. Take natural logs of the lengths

and reanalyze the data. Write a summary of your results and include a comparison with the results you found in Exercise 12.35. 🐦 HUMMINGBIRD

12.38 🎯 **Do poets die young?** According to William Butler Yeats, "She is the Gaelic muse, for she gives inspiration to those she persecutes. The Gaelic poets die young, for she is restless, and will not let them remain long on earth." One study designed to investigate this issue examined the age at death for writers from different cultures and genders.[14] Three categories of writers examined were novelists, poets, and nonfiction writers. The ages at death for female writers in these categories from North America are given in Table 12.1. Most of the writers are from the United States, but Canadian and Mexican writers are also included. 🎯 DEADPOETS

(a) Use graphical and numerical methods to describe the data.

(b) Examine the assumptions necessary for ANOVA. Summarize your findings.

(c) Run the ANOVA and report the results.

(d) Use a contrast to compare the poets with the two other types of writers. Do you think that the quotation from Yeats justifies the use of a one-sided alternative for examining this contrast? Explain your answer.

(e) Use another contrast to compare the novelists with the nonfiction writers. Explain your choice for an alternative hypothesis for this contrast.

(f) Use a multiple-comparisons procedure to compare the three means. How do the conclusions from this approach compare with those using the contrasts?

12.39 Do isoflavones increase bone mineral density? Kudzu is a plant that was imported to the United States from Japan and now covers over seven million acres in the South. The plant contains chemicals called isoflavones

TABLE 12.1

Age at death for women writers

Type	Age at death														
Novels	57	90	67	56	90	72	56	90	80	74	73	86	53	72	86
(n = 67)	82	74	60	79	80	79	77	64	72	88	75	79	74	85	71
	78	57	54	50	59	72	60	77	50	49	73	39	73	61	90
	77	57	72	82	54	62	74	65	83	86	73	79	63	72	85
	91	77	66	75	90	35	86								
Poems	88	69	78	68	72	60	50	47	74	36	87	55	68	75	78
(n = 32)	85	69	38	58	51	72	58	84	30	79	90	66	45	70	48
	31	43													
Nonfiction	74	86	87	68	76	73	63	78	83	86	40	75	90	47	91
(n = 24)	94	61	83	75	89	77	86	66	97						

that have been shown to have beneficial effects on bones. One study used three groups of rats to compare a control group with rats that were fed either a low dose or a high dose of isoflavones from kudzu.[15] One of the outcomes examined was the bone mineral density in the femur (in grams per square centimeter). Here are the data: 🔵 KUDZU

Treatment	Bone mineral density (g/cm^2)					
Control	0.228	0.207	0.234	0.220	0.217	0.228
	0.209	0.221	0.204	0.220	0.203	0.219
	0.218	0.245	0.210			
Low dose	0.211	0.220	0.211	0.233	0.219	0.233
	0.226	0.228	0.216	0.225	0.200	0.208
	0.198	0.208	0.203			
High dose	0.250	0.237	0.217	0.206	0.247	0.228
	0.245	0.232	0.267	0.261	0.221	0.219
	0.232	0.209	0.255			

(a) Use graphical and numerical methods to describe the data.

(b) Examine the assumptions necessary for ANOVA. Summarize your findings.

(c) Run the ANOVA and report the results.

(d) Use a multiple-comparisons method to compare the three groups.

(e) Write a short report explaining the effect of kudzu isoflavones on the femur of the rat.

12.40 Developing marketing strategies for travel to Hawaii. In 1997 approximately one-third of all tourists to Hawaii were from Japan. Since that time the percent has steadily decreased and is now around 20%.[16] To better understand the reasons for travel to Hawaii, a group of researchers surveyed 315 Japanese tourists who plan to visit to Hawaii. The tourists were divide into groups based on their purpose for travel. They were (1) honeymoon, (2) fraternal association, (3) sports, (4) leisure, and (5) business. Their responses to various survey questions were compared across these groups. The responses were on a 7-point Likert scale ranging from 1 (strongly disagree) to 7 (strongly agree). The following table summarizes the mean response and F test statistic for several questions.

Question	Grp 1 $n = 34$	Grp 2 $n = 56$	Grp 3 $n = 105$	Grp 4 $n = 26$	Grp 5 $n = 94$	F
I'd like to experience native Hawaiian culture	3.97	4.26	4.25	5.33	4.23	2.46
I'd prefer a group tour to a individual one	3.18	3.38	2.39	2.58	2.98	3.97
I'd like to experience ocean sports	4.71	4.59	4.58	5.33	4.02	2.46
I respect Hawaiian residents' customs	4.88	5.39	5.14	5.83	5.46	1.62

(a) What are the numerator and denominator degrees of freedom for these F tests?

(b) The response variable is not Normally distributed. Explain why this should not cause difficulties in using ANOVA.

(c) Using a significance level of $\alpha = 0.05$ for each question, assess whether there are differences in the group means.

(d) For those questions with a significant F statistic, plot the means and describe their pattern.

12.41 Multiple comparisons. Refer to the previous exercise.

(a) Explain why it is inappropriate to perform a multiple-comparisons analysis for the last question.

(b) For each of the other questions, use the Bonferroni or another multiple-comparisons procedure to determine which group means differ significantly. The following table gives the MSE for each question.

Question	MSE
I'd like to experience native Hawaiian culture	3.261
I'd prefer a group tour to a individual one	2.841
I'd like to experience ocean sports	4.285
I respect Hawaiian residents' customs	2.905

Summarize your results in a short report.

12.42 Do piano lessons improve the spatial-temporal reasoning of preschool children? The data in Table 12.2 contain the change in spatial-temporal reasoning (after treatment minus before treatment) of 34 children who took piano lessons, 10 who took singing lessons, 20 who had some computer instruction, and 14 who received no extra lessons. 🔵 PIANO

(a) Make a table giving the sample size, the mean, the standard deviation, and the standard error for each group.

(b) Analyze the data using one-way analysis of variance. State the null and alternative hypotheses, the test statistic with degrees of freedom, the P-value, and your conclusion.

TABLE 12.2

Piano lesson data

Lessons	Scores									
Piano	2	5	7	−2	2	7	4	1	0	7
	3	4	3	4	9	4	5	2	9	6
	0	3	6	−1	3	4	6	7	−2	7
	−3	3	4	4						
Singing	1	−1	0	1	−4	0	0	1	0	−1
Computer	0	1	1	−3	−2	4	−1	2	4	2
	2	2	−3	−3	0	2	0	−1	3	−1
None	5	−1	7	0	4	0	2	1	−6	0
	2	−1	0	−2						

12.43 The piano lessons study, continued. Refer to the previous exercise. Use the Bonferroni or another multiple-comparisons procedure to compare the group means. Summarize the results and support your conclusions with a graph of the means. PIANO

12.44 More on the piano lessons study. The researchers in Exercise 12.42 based their research on a biological argument for a causal link between music and spatial-temporal reasoning. Therefore, it is natural to test the contrast that compares the mean of the piano lesson group with the average of the three other means. Construct this contrast, perform the significance test, and summarize the results. PIANO

12.45 College dining facilities. University and college food service operations have been trying to keep up with the growing expectations of consumers in regards to the overall campus dining experience. Since customer satisfaction has been shown to be associated with repeat patronage and new customers through word-of-mouth, a public university in the Midwest took a sample of patrons from their eating establishments and asked them about their overall dining satisfaction.[17] The following table summarizes the results for three groups of patrons:

Category	$\bar{x}$	n	s
Student - meal plan	3.44	489	0.804
Faculty meal plan	4.04	69	0.824
Student - no meal plan	3.47	212	0.657

(a) Is it reasonable to use a pooled standard deviation for these data? Why or why not? If yes, compute it.

(b) The ANOVA F statistic was reported as 17.66. Give the degrees of freedom and either an approximate (from a table) or an exact (from software) P-value. Sketch a picture of the F distribution that illustrates the P-value. What do you conclude?

(c) Prior to performing this survey, food service operations thought that satisfaction among faculty would be higher than satisfaction among students. Use the results in the table to test this contrast. Make sure to specify the null and alternative hypotheses, test statistic, and P-value.

12.46 Orthogonal polynomial contrasts. Recall the Facebook friends study. In Example 12.16 (page 643) we used a contrast to assess the linear trend between the social acceptability score and number of Facebook friends. With orthogonal polynomial contrasts, we can assess the contributions of different polynomial trends to the overall pattern. Given the 5 equally-spaced levels of the factor in this problem, we can investigate up to a quartic (x^4) trend. The derivation of the coefficients is beyond the scope of this book so we will just investigate the trends here. The coefficients for the linear, quadratic, and cubic trends follow: FRIENDS

Trend	a_1	a_2	a_3	a_4	a_5
Linear	−2	−1	0	1	2
Quadratic	2	−1	−2	−1	2
Cubic	−1	2	0	−2	1

(a) Plot the a_i versus i for the linear trend. Describe the pattern. Suppose all the μ_i were constant. What would the value of ψ equal?

(b) Plot the a_i versus i for the quadratic trend. Describe the pattern. Suppose all the μ_i were constant. What would the value of ψ equal? Suppose that $\mu_i = 5i$ (i.e., a linear trend). What would the value of ψ equal?

(c) Construct the sample contrasts for the quadratic and cubic trends using the Facebook data.

(d) Test the hypotheses that there is a quadratic and cubic trend. Combine this with the earlier linear trend results. What do you conclude?

12.47 Exercise and healthy bones. Many studies have suggested that there is a link between exercise and healthy bones. Exercise stresses the bones and this causes them to get stronger. One study examined the effect of jumping on the bone density of growing rats.[18] There were three treatments: a control with no jumping, a low-jump condition (the jump height was 30 centimeters), and a high-jump condition (60 centimeters). After 8 weeks of 10 jumps per day, 5 days per week, the bone density of the rats (expressed in mg/cm^3) was measured. Here are the data: BONEDENSITY

Group	Bone density (mg/cm^3)									
Control	611	621	614	593	593	653	600	554	603	569
Low jump	635	605	638	594	599	632	631	588	607	596
High jump	650	622	626	626	631	622	643	674	643	650

(a) Make a table giving the sample size, mean, and standard deviation for each group of rats. Is it reasonable to pool the variances?

(b) Run the analysis of variance. Report the F statistic with its degrees of freedom and P-value. What do you conclude?

12.48 Exercise and healthy bones, continued. Refer to the previous exercise. BONEDENSITY

(a) Examine the residuals. Is the Normality assumption reasonable for these data?

(b) Use the Bonferroni or another multiple-comparisons procedure to determine which pairs of means differ significantly. Summarize your results in a short report. Be sure to include a graph.

12.49 Does the type of cooking pot affect iron content? Iron-deficiency anemia is the most common form of malnutrition in developing countries, affecting about 50% of children and women and 25% of men. Iron pots for cooking foods had traditionally been used in many of these countries, but they have been largely replaced by aluminum pots, which are cheaper and lighter. Some research has suggested that food cooked in iron pots will contain more iron than food cooked in other types of pots. One study designed to investigate this issue compared the iron content of some Ethiopian foods cooked in aluminum, clay, and iron pots.[19] One of the foods was *yesiga wet'*, beef cut into small pieces and prepared with several Ethiopian spices. The iron content of four samples of *yesiga wet'* cooked in each of the three types of pots follows. The units are milligrams of iron per 100 grams of cooked food. **COOKINGPOT1**

Type of pot	Iron (mg/100 g food)			
Aluminum	1.77	2.36	1.96	2.14
Clay	2.27	1.28	2.48	2.68
Iron	5.27	5.17	4.06	4.22

(a) Make a table giving the sample size, mean, and standard deviation for each type of pot. Is it reasonable to pool the variances? Note that with the small sample sizes in this experiment, we expect a large amount of variability in the sample standard deviations.

(b) Run the analysis of variance. Report the F statistic with its degrees of freedom and P-value. What do you conclude?

12.50 The cooking pot study, continued. Refer to the previous exercise.

(a) Examine the residuals. Is the Normality assumption reasonable for these data?

(b) Use the Bonferroni or another multiple-comparisons procedure to determine which pairs of means differ significantly. Summarize your results in a short report. Be sure to include a graph.

12.51 A comparison of different types of scaffold material. One way to repair serious wounds is to insert some material as a scaffold for the body's repair cells to use as a template for new tissue. Scaffolds made from extracellular material (ECM) are particularly promising for this purpose. Because they are made from biological material, they serve as an effective scaffold and are then resorbed. Unlike biological material that includes cells, however, they do not trigger tissue rejection reactions in the body. One study compared six types of scaffold material.[20] Three of these were ECMs and the other three were made of inert materials (MAT). There were three mice used per scaffold type. The response measure was the percent of glucose phosphated isomerase (Gpi) cells in the region of the wound. A large value is good, indicating that there are many bone marrow cells sent by the body to repair the tissue. **ECM**

Material	Gpi (%)		
ECM1	55	70	70
ECM2	60	65	65
ECM3	75	70	75
MAT1	20	25	25
MAT2	5	10	5
MAT3	10	15	10

(a) Make a table giving the sample size, mean, and standard deviation for each of the six types of material. Is it reasonable to pool the variances? Note that the sample sizes are small and the data are rounded.

(b) Run the analysis of variance. Report the F statistic with its degrees of freedom and P-value. What do you conclude?

12.52 A comparison of different types of scaffold material, continued. Refer to the previous exercise. **ECM**

(a) Examine the residuals. Is the Normality assumption reasonable for these data?

(b) Use the Bonferroni or another multiple-comparisons procedure to determine which pairs of means differ significantly. Summarize your results in a short report. Be sure to include a graph.

(c) Use a contrast to compare the three ECM materials with the three other materials. Summarize your conclusions. How do these results compare with those that you obtained from the multiple-comparisons procedure in part (b)?

12.53 Two contrasts of interest for the stimulant study. Refer to Exercise 12.29 (page 659). There are two comparisons of interest to the experimenter. They are (1) placebo versus the average of the 2 low-dose treatments; and (2) the difference between High A and Low A versus the difference between High B and Low B.

(a) Express each contrast in terms of the means (μ's) of the treatments.

(b) Give estimates with standard errors for each of the contrasts.

(c) Perform the significance tests for the contrasts. Summarize the results of your tests and your conclusions.

12.54 A dandruff study. Analysis of variance methods are often used in clinical trials where the goal is to assess the effectiveness of one or more treatments for a particular medical condition. One such study compared three treatments for dandruff and a placebo. The treatments were 1% pyrithione zinc shampoo (PyrI), the same shampoo but with instructions to shampoo two times (PyrII), 2% ketoconazole shampoo (Keto), and a placebo shampoo (Placebo). After six weeks of treatment, eight sections of the scalp were examined and given a score that measured the amount of scalp flaking on a 0 to 10 scale. The response variable was the sum of these eight scores. An analysis of the baseline flaking measure indicated that randomization of patients to treatments was successful in that no differences were found between the groups. At baseline there were 112 subjects in each of the three treatment groups and 28 subjects in the Placebo group. During the clinical trial, 3 dropped out from the PyrII group and 6 from the Keto group. No patients dropped out of the other two groups. DANDRUFF

(a) Find the mean, standard deviation, and standard error for the subjects in each group. Summarize these, along with the sample sizes, in a table and make a graph of the means.

(b) Run the analysis of variance on these data. Write a short summary of the results and your conclusion. Be sure to include the hypotheses tested, the test statistic with degrees of freedom, and the P-value.

12.55 The dandruff study, continued. Refer to the previous exercise. DANDRUFF

(a) Plot the residuals versus case number (the first variable in the data set). Describe the plot. Is there any pattern that would cause you to question the assumption that the data are independent?

(b) Examine the standard deviations for the four treatment groups. Is there a problem with the assumption of equal standard deviations for ANOVA in this data set? Explain your answer.

(c) Prepare Normal quantile plots for each treatment group. What do you conclude from these plots?

(d) Obtain the residuals from the analysis of variance and prepare a Normal quantile plot of these. What do you conclude?

12.56 Comparing each pair of dandruff treatments. Refer to Exercise 12.54. Use the Bonferroni or another multiple-comparisons procedure that your software provides to compare the individual group means in the

dandruff study. Write a short summary of your conclusions. DANDRUFF

12.57 Testing several contrasts from the dandruff study. Refer to Exercise 12.54. There are several natural contrasts in this experiment that describe comparisons of interest to the experimenters. They are (1) Placebo versus the average of the three treatments; (2) Keto versus the average of the two Pyr treatments; and (3) PyrI versus PyrII. DANDRUFF

(a) Express each of these three contrasts in terms of the means (μ's) of the treatments.

(b) Give estimates with standard errors for each of the contrasts.

(c) Perform the significance tests for the contrasts. Summarize the results of your tests and your conclusions.

12.58 Changing the response variable. Refer to Exercise 12.51 (page 665), where we compared six types of scaffold material to repair wounds. The data are given as percents ranging from 5–75. ECM

(a) Convert these percents into their decimal form by dividing by 100. Calculate the transformed means, standard deviations, and standard errors and summarize them with the sample sizes in a table.

(b) Explain how you could have calculated the table entries directly from the table you gave in part (a) of Exercise 12.51.

(c) Analyze the percents using analysis of variance. Compare the test statistic, degrees of freedom, P-value, and conclusion you obtain here with the corresponding values that you found in Exercise 12.51.

12.59 More on changing the response variable. Refer to the previous exercise and Exercise 12.51 (page 665). A calibration error was found with the device that measured Gpi, which resulted in a shifted response. Add 5% to each response and redo the calculations. Summarize the effects of transforming the data by adding a constant to all responses. ECM

12.60 Linear transformation of the response variable. Refer to the previous two exercises. Can you suggest a general conclusion regarding what happens to the test statistic, degrees of freedom, P-value, and conclusion when you perform analysis of variance on data that have been transformed by multiplying the raw data by a constant and then adding another constant? (That is, if y is the original data, we analyze y^*, where $y^* = a + by$ and a and $b \neq 0$ are constants.)

12.61 ⚔ **Comparing three levels of reading comprehension instruction.** A study of reading comprehension in children compared three methods of instruction.[21] The three methods of instruction are called Basal, DRTA, and Strategies. As is common in such studies, several pretest variables were measured before any instruction was given. One purpose of the pretest was to see if the three groups of children were similar in their comprehension skills. The READING data set gives two pretest measures that were used in this study. Use one-way ANOVA to analyze these data and write a summary of your results. 🔵 READING

12.62 ⚔ **More on the reading comprehension study.** In the study described in the previous exercise, Basal is the traditional method of teaching, while DRTA and Strategies are two innovative methods based on similar theoretical considerations. The READING data set includes three response variables that the new methods were designed to improve. Analyze these variables using ANOVA methods. Be sure to include multiple comparisons or contrasts as needed. Write a report summarizing your findings. 🔵 READING

12.63 ⚔ **More on the Facebook friends study.** Refer to the Facebook friends study that we examined in Example 12.3 (page 627). The explanatory variable in this study is the number of Facebook friends, with possible values of 102, 302, 502, 702, and 902. When using analysis of variance we treat the explanatory variable as categorical. An alternative analysis is to use simple linear regression. Perform this analysis and summarize the results. Plot the residuals from the regression model versus the number of Facebook friends. What do you conclude? 🔵 FRIENDS

12.64 Overall standard deviation versus the pooled standard deviation. The last line of the summary table given in Exercise 12.30 (page 660) gives the mean and the standard deviation for all the data combined. Compare this standard deviation with the pooled standard deviation that you would use as an estimate of the model standard deviation. Explain why you would expect this standard deviation to be larger than the pooled standard deviation.

12.65 ⚔ **Search the Internet.** Search the Internet or your library to find a study that is interesting to you and that used one-way ANOVA to analyze the data. First describe the question or questions of interest and then give the details of how ANOVA was used to provide answers. Be sure to include how the study authors examined the assumptions for the analysis. Evaluate how well the authors used ANOVA in this study. If your evaluation finds the analysis deficient, make suggestions for how it could be improved.

12.66 A power calculation exercise (optional). In Example 12.27 (page 659) the power calculation indicated that there was a fairly small chance of detecting the alternative given. Redo the calculations for the alternative $\mu_1 = 40$, $\mu_2 = 47$, and $\mu_3 = 43$. Do you think that the choice of 10 students per treatment is adequate for this alternative?

12.67 Planning another emotions study. Scores on an emotional scale were compared for five different cultures in Exercise 12.31 (page 660). Suppose that you are planning a new study using the same outcome variable. Your study will use European American, Asian American, and Hispanic American students from a large university.

(a) Explain how you would select the students to participate in your study.

(b) (*Optional*) Use the data from Exercise 12.31 to perform power calculations to determine sample sizes for your study.

(c) Write a report that could be understood by someone with limited background in statistics and that describes your proposed study and why you think it is likely that you will obtain interesting results.

12.68 ⚔ **Planning another isoflavone study.** Exercise 12.39 (page 662) gave data for a bone health study that examined the effect of isoflavones on rat bone mineral density. In this study there were three groups. Controls received a placebo, and the other two groups received either a low or a high dose of isoflavones from kudzu. You are planning a similar study of a new kind of isoflavone. Use the results of the study described in Exercise 12.39 to plan your study. Write a proposal explaining why your study should be funded.

12.69 ⚔ **Planning another restaurant ambience study.** Exercise 12.28 (page 659) gave data for a study that examined the effect of background music on total food spending at a high-end restaurant. You are planning a similar study but intend to look at total food spending at a more casual restaurant. Use the results of the study described in Exercise 12.28 to plan your study.

Two-Way Analysis of Variance

Introduction

The *t* procedures of Chapter 7 compare the means of two populations. We generalized these procedures in Chapter 12 so that we could compare the means of several populations. In this chapter, we move from one-way ANOVA to two-way ANOVA. Two-way ANOVA compares the means of populations that can be classified in two ways or the mean responses in two-factor experiments.

Many of the key concepts are similar to those of one-way ANOVA, but the presence of more than one classification factor also introduces some new ideas. We once more assume that the data are approximately Normal and that groups may have different means, but the same standard deviation; we again pool to estimate the variance; and we again use *F* statistics for significance tests. *The major difference between one-way and two-way ANOVA is in the FIT part of the model.* We will carefully study this term, and we will find much that is

669

both new and useful. This will allow us to address comparisons such as the following:

- Can zinc supplementation reduce the occurrence and severity of malaria in both nutrient-sufficient and nutrient-deficient African children?

- What effects do the floral morphologies of male and female jack-in-the-pulpit plants have on herbivory?

- Do calcium supplements prevent bone loss in elderly people? Does this depend on whether the person is receiving adequate vitamin D?

13.1 The Two-Way ANOVA Model

We begin with a discussion of the advantages of the two-way ANOVA design and illustrate these with some examples. Then we discuss the model and the assumptions.

Advantages of two-way ANOVA

In one-way ANOVA, we classify populations according to one categorical variable, or factor. In the two-way ANOVA model, there are two factors, each with its own number of levels. When we are interested in the effects of two factors, a two-way design offers great advantages over several single-factor studies. Several examples will illustrate these advantages.

EXAMPLE

13.1 Design 1: Choosing the best box art style and game description. In Example 12.1, a video game company wants to compare three different box art styles. To do this, they plan to randomly assign the three styles equally among 60 electronics stores. The number of games sold during a one-week period is the outcome variable.

Now suppose a second experiment is planned for the following week to compare four different game descriptions on the back of the box. A similar experimental design will be used, with the four descriptions randomly assigned among the same 60 stores.

Here is a picture of the design of the first experiment with the sample sizes:

Style	n
1	20
2	20
3	20
Total	60

And this represents the second experiment:

Description	n
1	15
2	15
3	15
4	15
Total	60

In the first experiment 20 stores were assigned to each level of the factor for a total of 60 stores. In the second experiment 15 stores were assigned to each level of the factor for a total of 60 stores. The total amount of time for the two experiments is two weeks. Each experiment will be analyzed using one-way ANOVA. The factor in the first experiment is box art style with three levels, and the factor in the second experiment is game description with four levels. Let's now consider combining the two experiments into one.

EXAMPLE

13.2 Design 2: Choosing the best box art style and game description. Suppose we use a two-way approach for the video game problem. There are two factors, style and description. Since style has three levels and description has four levels, this is a 3 × 4 design. This gives a total of 12 possible combinations of style and description. With a total of 60 stores, we could assign each combination of style and description to 5 stores. The number of video games sold during a one-week period is the outcome variable.

Here is a picture of the two-way design with the sample sizes:

	Description				
Style	1	2	3	4	Total
1	5	5	5	5	20
2	5	5	5	5	20
3	5	5	5	5	20
Total	15	15	15	15	60

cell   Each combination of the factors in a two-way design corresponds to a **cell.** The 3 × 4 ANOVA for the video game experiment has 12 cells, each corresponding to a particular combination of box art style and game description.

With the two-way design for style, notice that we have 20 stores assigned to each level, the same as what we had for the one-way experiment for style alone. Similarly, there are 15 stores assigned to each level of description. Thus, the two-way design gives us the same amount of information for estimating the sales for each level of each factor as we had with the two one-way designs. The

difference is that we can collect all the information in only one week. By combining the two factors into one experiment, we have increased our efficiency by reducing the amount of data to be collected by half.

EXAMPLE

13.3 Can dietary supplementation with zinc prevent malaria? Malaria is a serious health problem causing an estimated one million deaths per year, mostly among Africa children.[1] Several studies, run in Asia, Latin America, and developed countries, have shown zinc supplementation to be an effective control of common infections in children. Can this supplementation program also be effective in Africa, where the primary threat to a child's health is malaria? A group of researchers have set out to answer this question.[2]

To design a study to answer this question the researchers first need to determine an appropriate target group. Since malaria is a serious problem for young children, they will concentrate on children who are 6 months to 5 years of age. A supplement will be prepared that contains either no zinc, or 10 mg of zinc. Because the response to zinc may be different in children who lack other important nutrients, they decide to also take this factor into account. Specifically, their supplement will either be a placebo, or contain daily dosages of essential vitamins and minerals.

EXAMPLE

13.4 Implementing the two-way ANOVA. The factors for their two-way ANOVA are zinc supplementation with two levels and vitamin supplementation with two levels. There are $2 \times 2 = 4$ cells in their study. They plan to enroll 600 children and randomly assign 150 to each of the cells. One outcome variable will be a measure of the child's T cell immune response.

Here is a table that summarizes the design:

| | **Vitamins** | | |
Zinc	**No**	**Yes**	**Total**
No	150	150	300
Yes	150	150	300
Total	300	300	600

This example illustrates another advantage of two-way designs. Although they are primarily interested in the possible benefit of zinc supplementation, they also included vitamin supplementation in the design because they thought the zinc effect may be different in children who are nutritionally deficient. Consider an alternative one-way design where we assign 300 children to the two levels of zinc and ignore nutritional status. With this design we will have the same number of children at each of the zinc levels, so in this way it is similar to our two-way design. However, suppose that there are, in fact, differences due to nutritional status. In this case, the one-way ANOVA would assign this variation to the RESIDUAL (within groups) part of the model. In the two-way ANOVA, vitamin supplementation is included as a factor, and therefore this variation is included in the FIT part of the model. Whenever we can move variation from RESIDUAL to FIT, we reduce the σ of our model and increase the power of our tests.

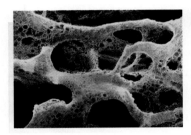

EXAMPLE

13.5 Vitamin D and osteoporosis. Osteoporosis is a disease primarily of the elderly. People with osteoporosis have low bone mass and an increased risk of bone fractures. Over 10 million people in the United States, 1.4 million Canadians, and many millions throughout the world have this disease. Adequate calcium in the diet is necessary for strong bones, but vitamin D is also needed for the body to efficiently use calcium. High doses of calcium in the diet will not prevent osteoporosis unless there is adequate vitamin D. Exposure of the skin to the ultraviolet rays in sunlight enables our bodies to make vitamin D. However, elderly people often avoid sunlight, and in northern areas such as Canada, there is not sufficient ultraviolet light to make vitamin D, particularly in the winter months.

Suppose we wanted to see if calcium supplements will increase bone mass (or prevent a decrease in bone mass) in an elderly Canadian population. Because of the vitamin D complication we will make this a factor in our design.

EXAMPLE

13.6 Designing the osteoporosis study. We will use a 2 × 2 design for our osteoporosis study. The two factors are calcium and vitamin D. The levels of each factor will be zero (placebo) and an amount that is expected to be adequate, 800 mg/day for calcium and 300 international units per day (IU/day) for vitamin D. Women between the ages of 70 and 80 will be recruited as subjects. Bone mineral density (BMD) will be measured at the beginning of the study, and supplements will be taken for one year. The change in BMD over the one-year period is the outcome variable. We expect a dropout rate of 20% and we would like to have about 20 subjects providing data in each group at the end of the study. We will therefore recruit 100 subjects and randomly assign 25 to each treatment combination.

Here is a table that summarizes the design with the sample sizes at baseline:

	Vitamin D		
Calcium	Placebo	300 IU/day	Total
Placebo	25	25	50
800 mg/day	25	25	50
Total	50	50	100

interaction

main effects

This example illustrates a third reason for using two-way designs. The effectiveness of the calcium supplement on BMD depends on having adequate vitamin D. We call this an **interaction.** In contrast, the average values for the calcium effect and the vitamin D effect are represented as **main effects.** The two-way model represents FIT as the sum of a main effect for each of the two factors *and* an interaction. One-way designs that vary a single factor and hold other factors fixed cannot discover interactions. We will discuss interactions more fully in a later section.

These examples illustrate several reasons why two-way designs are preferable to one-way designs.

ADVANTAGES OF TWO-WAY ANOVA

1. It is more efficient to study two factors simultaneously rather than separately.

2. We can reduce the residual variation in a model by including a second factor thought to influence the response.

3. We can investigate interactions between factors.

These considerations also apply to study designs with more than two factors. We will be content to explore only the two-way case. The choice of sampling or experimental design is fundamental to any statistical study. *Factors and levels must be carefully selected by an individual or team who understands both the statistical models and the issues that the study will address.*

The two-way ANOVA model

When discussing two-way models in general, we will use the labels A and B for the two factors. For particular examples and when using statistical software, it is better to use meaningful names for these categorical variables. Thus, in Example 13.2 we would say that the factors are box art style and description and in Example 13.4 we would say the factors are the zinc and vitamin supplementation.

The numbers of levels of the factors are often used to describe the model. Again using our earlier examples, we would say Example 13.2 represents a 3×4 ANOVA and Example 13.4 illustrates a 2×2 ANOVA. In general, Factor A will have I levels and Factor B will have J levels. Therefore, we call the general two-way problem an $I \times J$ ANOVA.

In a two-way design every level of A appears in combination with every level of B, so that $I \times J$ groups are compared. The sample size for level i of Factor A and level j of Factor B is n_{ij}.[3] The total number of observations is

$$N = \sum n_{ij}$$

ASSUMPTIONS FOR TWO-WAY ANOVA

We have independent SRSs of size n_{ij} from each of $I \times J$ Normal populations. The population means μ_{ij} may differ, but all populations have the same standard deviation σ. The μ_{ij} and σ are unknown parameters.

Let x_{ijk} represent the kth observation from the population having Factor A at level i and Factor B at level j. The statistical model is

$$x_{ijk} = \mu_{ij} + \epsilon_{ijk}$$

for $i = 1, \ldots, I$ and $j = 1, \ldots, J$ and $k = 1, \ldots, n_{ij}$, where the deviations ϵ_{ijk} are from an $N(0, \sigma)$ distribution.

LOOK BACK
one-way model p. 631

Similar to the one-way model, the FIT part is the group means μ_{ij}, and the RESIDUAL part is the deviations ϵ_{ijk} of the individual observations from their group means. To estimate a group mean μ_{ij} we use the sample mean of the observations in the samples from this group:

$$\overline{x}_{ij} = \frac{1}{n_{ij}} \sum_k x_{ijk}$$

The k below the $\sum$ means that we sum the n_{ij} observations that belong to the (i, j)th sample.

The RESIDUAL part of the model contains the unknown σ. We calculate the sample variances for each SRS and pool these to estimate σ^2:

$$s_p^2 = \frac{\sum(n_{ij} - 1)s_{ij}^2}{\sum(n_{ij} - 1)}$$

Just as in one-way ANOVA, the numerator in this fraction is SSE and the denominator is DFE. Also, DFE is the total number of observations minus the number of groups. That is, DFE = $N - IJ$. The estimator of σ is s_p.

Main effects and interactions

In this section we will further explore the FIT part of the two-way ANOVA, which is represented in the model by the population means μ_{ij}. The two-way design gives some structure to the set of means μ_{ij}.

So far, because we have independent samples from each of $I \times J$ groups, we have presented the problem as a one-way ANOVA with IJ groups. Each population mean μ_{ij} is estimated by the corresponding sample mean $\overline{x}_{ij}$, and we can calculate sums of squares and degrees of freedom as in one-way ANOVA. In accordance with the conventions used by many computer software packages, we use the term *model* when discussing the sums of squares and degrees of freedom calculated as in one-way ANOVA with IJ groups. Thus, SSM is a model sum of squares constructed from deviations of the form $\overline{x}_{ij} - \overline{x}$, where $\overline{x}$ is the average of all of the observations and $\overline{x}_{ij}$ is the mean of the (i, j)th group. Similarly, DFM is simply $IJ - 1$.

In two-way ANOVA, the terms SSM and DFM can be further broken down into terms corresponding to a main effect for A, a main effect for B, and an AB interaction. Each of SSM and DFM is then a sum of terms:

$$SSM = SSA + SSB + SSAB$$

and

$$DFM = DFA + DFB + DFAB$$

The term SSA represents variation among the means for the different levels of Factor A. Because there are I such means, DFA $= I - 1$ degrees of freedom. Similarly, SSB represents variation among the means for the different levels of Factor B, with DFB $= J - 1$.

Interactions are a bit more involved. We can see that SSAB, which is SSM $-$ SSA $-$ SSB, represents the variation in the model that is not accounted for by the main effects. By subtraction we see that its degrees of freedom are

$$DFAB = (IJ - 1) - (I - 1) - (J - 1) = (I - 1)(J - 1)$$

There are many kinds of interactions. The easiest way to study them is through examples.

EXAMPLE

13.7 Investigating differences in sugar-sweetened beverage consumption. Consumption of sugar-sweetened beverages (SSBs) has been linked to Type 2 diabetes and obesity. One study used data from the National Health and Nutrition Examination Survey (NHANES) to estimate SSB consumption among adults. More than 13,000 individuals provided data for this study. Adults were divided into 3 age categories: 20–44, 45–64, and ≥65 years old.[4] Here are the means for the number of calories per day in SSBs consumed during 1988–1994 and 1999–2004:

	Year		
Age (years)	**1994**	**2004**	**Mean**
20–44	231	289	260
45–64	124	160	142
≥65	68	83	76
Mean	141	177	159

The table includes averages of the means in the rows and columns. For example, in 1994 the mean of calories in SSBs consumed among adults is

$$\frac{231 + 124 + 68}{3} = 141$$

Similarly, the corresponding value for 2004 is

$$\frac{289 + 160 + 83}{3} = 177.3$$

marginal means

which is rounded to 177 in the table. These averages are called **marginal means** (because of their location at the *margins* of such tabulations). The grand mean can be obtained by averaging either set of marginal means. *It is always a good idea to do both as a check on your arithmetic.*

Figure 13.1 is a plot of the group means. From the plot we see that the calories in SSBs consumed by each group in 1994 are less than those consumed in 2004. In statistical language, there is a main effect for year. We also see that the means are different across age categories. This means there is a main effect of age. These main effects can be described by differences between the marginal means. For example, the mean for 1994 is 141 calories and then increases 36 calories to 177 calories in 2004. Similarly, the mean for adults 20–44 years old is 260, it drops by 118 calories to 142 for the adults 45–64 years old, and then drops an additional 66 calories to 76 adults older than 64.

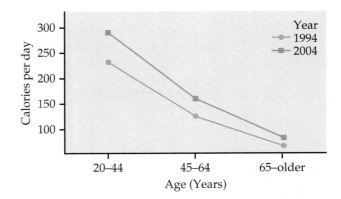

FIGURE 13.1 Plot of the mean calories of sugar-sweetened beverages consumed per day in 1988–1994 and 1999–2004, for Example 13.7.

To examine two-way ANOVA data for a possible interaction, always construct a plot similar to Figure 13.1. In this case, it is debatable whether the two profiles should be considered parallel. Profiles that are roughly parallel is another way of saying that there is *no* clear interaction between the two factors. When no interaction is present, the marginal means provide a reasonable description of the two-way table of means.

On the other hand, when there is an interaction, the marginal means do not tell the whole story. For example, with these data, the marginal mean difference between years is 36 calories. This equals the difference in calories for the 45–64 age class so it adequately describes the change for this age group. However, the mean difference between years for the 20–44 age group is 58 calories so this marginal mean difference underestimates the increase in calories by 18 calories per day. Likewise, this marginal mean difference overestimates the difference for the oldest age class by 21 calories per day. If differences of 20 calories per day are scientifically meaningful, then we would say there is evidence that there is an interaction present.

Interactions come in many shapes and forms. *When we find an interaction, a careful examination of the means is needed to properly interpret the data.* Simply stating that interactions are significant tells us very little. Plots of the group means are essential. Here is another example.

EXAMPLE

13.8 Eating in groups. Some research has shown that people eat more when they eat in groups. One possible mechanism for this phenomenon is that they may spend more time eating when in a larger group. A study designed to examine this idea measured the length of time spent (in minutes) eating lunch in different settings.[5] Here are some data from this study:

| | | | Number of people eating | | | |
Lunch setting	1	2	3	4	5 or more	Mean
Workplace	12.6	23.0	33.0	41.1	44.0	30.7
Fast-food restaurant	10.7	18.2	18.4	19.7	21.9	17.8
Mean		11.6	20.6	25.7	30.4	32.9	24.2

Figure 13.2 gives the plot of the means for this example. The patterns are not parallel, so it appears that we have an interaction. Meals take longer when there are more people present, but this phenomenon is much greater for the meals consumed at work. For fast-food eating, the meal durations are fairly similar when there is more than one person present.

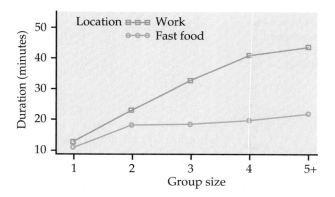

FIGURE 13.2 Plot of mean meal duration versus lunch setting and group size, for Example 13.8.

A different kind of interaction is present in the next example. Here, we must be very cautious in our interpretation of the main effects since one of them can lead to a distorted conclusion.

EXAMPLE

13.9 We got the beat? When we hear music that is familiar to us, we can quickly pick up the beat and our mind synchronizes with the music. However, if the music is unfamiliar, it takes us longer to synchronize. In a study that investigated the theoretical framework for this phenomenon, French and Tunisian nationals listened to French and Tunisian music.[6] Each subject was asked to tap in time with the music being played. A synchronization score, recorded in milliseconds, measured how well the subjects synchronized with the music. A higher score indicates better synchronization. Six songs of each music type were used. Here are the means:

Nationality	Music		Mean
	French	Tunisian	
French	950	750	850
Tunisian	760	1090	925
Mean	855	920	887

The means are plotted in Figure 13.3. In the study the researchers were not interested in main effects. Their theory predicted the interaction that we see in the figure. Subjects synchronize better with music from their own culture.

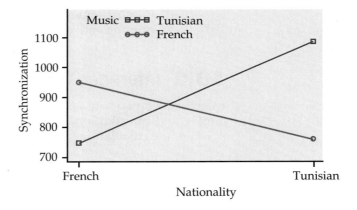

FIGURE 13.3 Plot of mean synchronization score versus type of music for French and Tunisian nationals, for Example 13.9.

The interaction in Figure 13.3 is very different from those that we saw in Figures 13.1 and 13.2. These examples illustrate the point that it is necessary to plot the means and carefully describe the patterns when interpreting an interaction.

The design of the study in Example 13.9 allows us to examine two main effects and an interaction. However, this setting does not meet all the assumptions needed for statistical inference using the two-way ANOVA framework of this chapter. *As with one-way ANOVA, we require that observations be independent.* In this study, we have a design that has each subject contributing data for two types of music, so these two scores will be dependent. The framework is similar to the matched pairs setting. The design is called a **repeated-measures design.** More advanced texts on statistical methods cover this important design.

LOOK BACK
matched pairs *t* test p. 414

repeated-measures design

USE YOUR KNOWLEDGE

13.1 What's wrong? For each of the following, explain what is wrong and why.

  (a) A two-way ANOVA is used when the outcome variable can only take two possible values.

(b) In a 2×3 ANOVA each level of Factor A appears with two levels of Factor B.

(c) The FIT part of the model in a two-way ANOVA represents the variation that is sometimes called error or residual.

(d) In an $I \times J$ ANOVA, DFAB $= IJ - 1$.

13.2 What's wrong? For each of the following, explain what is wrong and why.

(a) Parallel profiles of cell means implies a strong interaction is present.

(b) You can perform a two-way ANOVA only when the sample sizes are the same in all cells.

(c) The estimate s_p^2 is obtained by pooling the marginal sample variances.

(d) When interaction is present, the marginal means are always uninformative.

13.2 Inference for Two-Way ANOVA

Inference for two-way ANOVA involves F statistics for each of the two main effects and an additional F statistic for the interaction. As with one-way ANOVA, the calculations are organized in an ANOVA table.

The ANOVA table for two-way ANOVA

Two-way ANOVA is the statistical analysis for a two-way design with a quantitative response variable. The results of a two-way ANOVA are summarized in an ANOVA table based on splitting the total variation SST and the total degrees of freedom DFT among the two main effects and the interaction. Both the sums of squares (which measure variation) and the degrees of freedom add:

$$SST = SSA + SSB + SSAB + SSE$$
$$DFT = DFA + DFB + DFAB + DFE$$

The sums of squares are always calculated in practice by statistical software. *When the n_{ij} are not all equal, some methods of analysis can give sums of squares that do not add.* From each sum of squares and its degrees of freedom we find the mean square in the usual way:

$$\text{mean square} = \frac{\text{sum of squares}}{\text{degrees of freedom}}$$

The significance of each of the main effects and the interaction is assessed by an F statistic that compares the variation due to the effect of interest with the within-group variation. Each F statistic is the mean square for the source of interest divided by MSE. Here is the general form of the two-way ANOVA table:

Source	Degrees of freedom	Sum of squares	Mean square	F
A	$I - 1$	SSA	SSA/DFA	MSA/MSE
B	$J - 1$	SSB	SSB/DFB	MSB/MSE
AB	$(I - 1)(J - 1)$	SSAB	SSAB/DFAB	MSAB/MSE
Error	$N - IJ$	SSE	SSE/DFE	
Total	$N - 1$	SST		

There are three null hypotheses in two-way ANOVA, with an F test for each. We can test for significance of the main effect of A, the main effect of B, and the AB interaction. *It is generally good practice to examine the test for interaction first, since the presence of a strong interaction may influence the interpretation of the main effects.* Be sure to plot the means as an aid to interpreting the results of the significance tests.

SIGNIFICANCE TESTS IN TWO-WAY ANOVA

To test the main effect of A, use the F statistic

$$F_A = \frac{MSA}{MSE}$$

To test the main effect of B, use the F statistic

$$F_B = \frac{MSB}{MSE}$$

To test the interaction of A and B, use the F statistic

$$F_{AB} = \frac{MSAB}{MSE}$$

If the effect being tested is zero, the calculated F statistic has an F distribution with numerator degrees of freedom corresponding to the effect and denominator degrees of freedom equal to DFE. Large values of the F statistic lead to rejection of the null hypothesis. The P-value is the probability that a random variable having the corresponding F distribution is greater than or equal to the calculated value.

The following example illustrates how to do a two-way ANOVA. As with the one-way ANOVA, we focus our attention on interpretation of the computer output.

HEARTRATE

EXAMPLE

13.10 A study of cardiovascular risk factors. A study of cardiovascular risk factors compared runners who averaged at least 15 miles per week with a control group described as "generally sedentary." Both men and women were included in the study.[7] The design is a 2×2 ANOVA with the factors group and gender. There were 200 subjects in each of the four combinations. One of the variables measured was the heart rate after 6 minutes of exercise on a treadmill. SAS computer analysis produced the outputs in Figure 13.4 and Figure 13.5.

```
SAS                                                          _ □ X

  Analysis Variable : HR

    GROUP=Control GENDER=Female
    N           Mean        Std Dev     Std Error
  -------------------------------------------------
    200         148.00        16.27          1.15
  -------------------------------------------------


    GROUP=Control GENDER=Male
    N           Mean        Std Dev     Std Error
  -------------------------------------------------
    200         130.00        17.10          1.21
  -------------------------------------------------


    GROUP=Runners GENDER=Female
    N           Mean        Std Dev     Std Error
  -------------------------------------------------
    200         115.99        15.97          1.13
  -------------------------------------------------


    GROUP=Runners GENDER=Male
    N           Mean        Std Dev     Std Error
  -------------------------------------------------
    200         103.98        12.50          0.88
  -------------------------------------------------
```

FIGURE 13.4 Summary statistics for heart rates in the four groups of a 2×2 ANOVA, for Example 13.10.

```
SAS                                                          _ □ X

  General Linear Models Procedure
  Dependent Variable: HR

                              Sum of          Mean
  Source              DF      Squares         Square    F Value   Pr > F
  Model                3    215256.09       71752.03    296.35    0.0001
  Error              796    192729.83         242.12
  Corrected Total    799    407985.92

                   R-square          C.V.      Root MSE           HR Mean
                   0.527607      12.49924        15.560            124.49

  Source              DF     Type I SS   Mean Square   F Value    Pr > F
  GROUP                1    168432.08     168432.08     695.65    0.0001
  GENDER               1     45030.00      45030.00     185.98    0.0001
  GROUP*GENDER         1      1794.01       1794.01       7.41    0.0066
```

FIGURE 13.5 Two-way ANOVA output for heart rates, for Example 13.10.

We begin with the usual preliminary examination. From Figure 13.4 we see that the ratio of the largest to the smallest standard deviation is less than 2. Therefore, we are not concerned about violating the assumption of equal population standard deviations. Normal quantile plots (not shown) do not reveal any outliers, and the data appear to be reasonably Normal.

The ANOVA table at the top of the output in Figure 13.5 is in effect a one-way ANOVA with four groups: female control, female runner, male control, and male runner. In this analysis Model has 3 degrees of freedom, and Error has 796 degrees of freedom. The F test and its associated P-value for this analysis refer to the hypothesis that all four groups have the same population mean. We are interested in the main effects and interaction, so we ignore this test.

The sums of squares for the group and gender main effects and the group-by-gender interaction appear at the bottom of Figure 13.5 under the heading Type I SS. These sum to the sum of squares for Model. Similarly, the degrees of freedom for these sums of squares sum to the degrees of freedom for Model. Two-way ANOVA splits the variation among the means (expressed by the Model sum of squares) into three parts that reflect the two-way layout.

Because the degrees of freedom are all 1 for the main effects and the interaction, the mean squares are the same as the sums of squares. The F statistics for the three effects appear in the column labeled F Value, and the P-values are under the heading Pr > F. For the group main effect, we verify the calculation of F as follows:

$$F = \frac{\text{MSG}}{\text{MSE}} = \frac{168{,}432}{242.12} = 695.65$$

All three effects are statistically significant. The group effect has the largest F, followed by the gender effect and then the group-by-gender interaction. To interpret these results, we examine the plot of means with bars indicating one standard error in Figure 13.6. Note that the standard errors are quite small due to the large sample sizes. The significance of the main effect for group is due to the fact that the controls have higher average heart rates than the runners for both genders. This is the largest effect evident in the plot.

The significance of the main effect for gender is due to the fact that the females have higher heart rates than the men in both groups. The differences

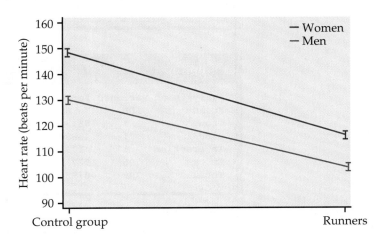

FIGURE 13.6 Plot of the group means with standard errors for heart rates in the 2 × 2 ANOVA, for Example 13.10.

are not as large as those for the group effect, and this is reflected in the smaller value of the F statistic.

The analysis indicates that a complete description of the average heart rates requires consideration of the interaction in addition to the main effects. The two lines in the plot are not parallel. This interaction can be described in two ways. The female-male difference in average heart rates is greater for the controls than for the runners. Alternatively, the difference in average heart rates between controls and runners is greater for women than for men. As the plot suggests, the interaction is not large. It is statistically significant because there were 800 subjects in the study.

Two-way ANOVA output for other software is similar to that given by SAS. Figure 13.7 gives the analysis of the heart rate data using Excel and Minitab.

Excel

	F	G	H	I	J	K	L	M
1	Anova Two Factor With Replication							
2								
3	SUMMARY	Control	Runners	Total				
4	*Female*							
5	Count	200	200	400				
6	Sum	29600	23197	52797				
7	Average	148	115.985	131.9925				
8	Variance	264.7437	255.0902	516.1478				
9								
10	*Male*							
11	Count	200	200	400				
12	Sum	26000	20795	46795				
13	Average	130	103.975	116.9875				
14	Variance	292.4221	156.2356	393.5161				
15								
16	*Total*							
17	Count	400	400					
18	Sum	55600	43992					
19	Average	139	109.98					
20	Variance	359.0877	241.2978					
21								
22								
23	ANOVA							
24	*Source of Variation*	SS	df	MS	F	P-value	F crit	
25	Sample	45030.01	1	45030.01	185.9799	3.29E-38	3.85316	
26	Columns	168432.1	1	168432.1	695.6471	1.10E-110	3.85316	
27	Interaction	1794.005	1	1794.005	7.409481	0.00663	3.85316	
28	Within	192729.8	796	242.1229				
29								
30	Total	407985.9	799					

Minitab

```
Analysis of Variance for HR
Source          DF        SS        MS
Group            1    168432    168432
Gender           1     45030     45030
Interaction      1      1794      1794
Error          796    192730       242
Total          799    407986
```

FIGURE 13.7 Excel and Minitab two-way ANOVA output for the heart rate study, for Example 13.10.

CHAPTER 13 Summary

Two-way analysis of variance is used to compare population means when populations are classified according to two factors.

ANOVA assumes that the populations are Normal with possibly different means and the same standard deviation and that independent SRSs are drawn from each population.

As with one-way ANOVA, preliminary analysis includes examination of means, standard deviations, and Normal quantile plots. **Marginal means** are calculated by taking averages of the cell means across rows and columns. Pooling is used to estimate the within-group variance.

ANOVA separates the total variation into parts for the **model** and **error.** The model variation is separated into parts for each of the **main effects** and the **interaction.**

The calculations are organized into an **ANOVA table.** F statistics and P-values are used to test hypotheses about the main effects and the interaction.

Careful inspection of the means is necessary to interpret significant main effects and interactions. Plots are a useful aid.

CHAPTER 13 Exercises

For Exercises 13.1 and 13.2, see pages 679–680.

13.3   What's wrong? For each of the following, explain what is wrong and why.

(a) You should reject the null hypothesis that there is no interaction in a two-way ANOVA when the AB F statistic is small.

(b) Sums of squares are equal to mean squares divided by degrees of freedom.

(c) The test statistics for the main effects in a two-way ANOVA have a chi-square distribution when the null hypothesis is true.

(d) The sums of squares always add in two-way ANOVA.

13.4   Is there an interaction? Each of the following tables gives means for a two-way ANOVA. Make a plot of the means with the levels of factor A on the x axis. State whether or not there is an interaction, and if there is, describe it.

(a)

	Factor A		
Factor B	1	2	3
1	11	16	21
2	6	16	26

(b)

	Factor A		
Factor B	1	2	3
1	10	15	20
2	40	45	50

(c)

	Factor A		
Factor B	1	2	3
1	10	15	20
2	50	45	40

(d)

	Factor A		
Factor B	1	2	3
1	10	15	20
2	50	55	52

13.5   Describing a two-way ANOVA model. A 3×2 ANOVA was run with 5 observations per cell.

(a) Give the degrees of freedom for the F statistic that is used to test for interaction in this analysis and the entries from Table E that correspond to this distribution.

(b) Sketch a picture of this distribution with the information from the table included.

(c) The calculated value of this F statistic is 3.72. How would you report the P-value?

(d) Would you expect a plot similar to Figure 13.1 to have mean profiles that look parallel? Explain your answer.

13.6 Determining the critical value of F. For each of the following situations, state how large the F statistic needs to be for rejection of the null hypothesis at the 5% level. Sketch each distribution and indicate the region where you would reject.

(a) The main effect for the first factor in a 2×4 ANOVA with 3 observations per cell.

(b) The interaction in a 4×4 ANOVA with 2 observations per cell.

(c) The interaction in a 2×2 ANOVA with 51 observations per cell.

13.7 Identifying the factors of a two-way ANOVA model. For each of the following situations, identify both factors and the response variable. Also, state the number of levels for each factor (I and J) and the total number of observations (N).

(a) A child psychologist is interested in studying how a child's percent of pretend play differs with gender and age (4, 8, and 12 months). There are 11 infants assigned to each cell of the experiment.

(b) Brewers malt is produced from germinating barley. A homebrewer wants to determine the best conditions to germinate the barley. A total of 30 lots of barley seed were equally and randomly assigned to 10 germination conditions. The conditions are combinations of the week after harvest (1, 3, 6, 9, or 12 weeks) and the amount of water used in the process (4 or 8 milliliters). The percent of seeds germinating is the outcome variable.

(c) The strength of concrete depends upon the formula used to prepare it. An experiment compares six different mixtures. Nine specimens of concrete are poured from each mixture. Three of these specimens are subjected to 0 cycles of freezing and thawing, three are subjected to 100 cycles, and three are subjected to 500 cycles. The strength of each specimen is then measured.

(d) A company wants to compare four different training programs for its new employees. Each of these programs takes 6 hours to complete. The training can be given for 6 hours on one day or for 3 hours on two consecutive days. The next 80 employees hired by the company will be the subjects for this study.

13.8 Determining the degrees of freedom. For each part in Exercise 13.7, outline the ANOVA table, giving the sources of variation and the degrees of freedom.

13.9 The influences of transaction history and a thank-you statement. A service failure is defined as any service-related problem (real or perceived) that transpires during a customer's experience with a firm. In the hotel industry, there is a high human component so these sorts of failures commonly occur regardless of extensive training and established policies. As a result, hotel firms must learn to effectively react to these failures. A recent study investigated the relationship between a consumer's transaction history (levels: long and short) and an employee thank-you statement (levels: yes and no) on a consumer's repurchase intent.[8] Each subject was randomly assigned to one of the four treatment groups and asked to read some service failure/resolution scenarios and respond accordingly. Repurchase intent was measured using a 9-point scale. Here is a summary of the means:

| | Thank-you ||
History	No	Yes
Short	5.69	6.80
Long	7.53	7.37

(a) Plot the means. Do you think there is an interaction? If yes, describe the interaction in terms of the two factors.

(b) Find the marginal means. Are they useful for understanding the results of this study? Explain your answer.

13.10 Transaction history and a thank-you statement, cont. Refer to the previous exercise. The number of subjects in each cell were not equal so the researchers used linear regression to analyze the data. This was done by creating an indicator variable for each factor and the interaction. Below is a partial ANOVA table. Complete it and state your conclusions regarding the main effects and interaction described in the previous exercise.

Source	DF	SS	MS	F	P-value
Transaction history		61.445			
Thank-you statement		21.810			
Interaction		15.404			
Error	160	759.904			

13.11 The effects of proximity and visibility on food intake. A recent study investigated the influence that proximity and visibility of food have on food intake.[9] A total of 40 secretaries from the University of Illinois participated in the study. A candy dish full of individually wrapped chocolates was placed either at the desk of the participant or at a location 2 meters from the participant. The candy dish was either a clear (candy visible) or opaque (candy not visible) covered bowl. After a week, the researchers noted not only the number of candies consumed per day but also the self-reported number of candies consumed by each participant. The following table summarizes the mean difference between these two values (reported minus actual):

Proximity	Visibility	
	Clear	**Opaque**
Proximate	−1.2	−0.8
Less proximate	0.5	0.4

(a) Make a plot of the means and describe the patterns that you see. Does the plot suggest an interaction between visibility and proximity?

(b) This study actually took four weeks, with each participant being observed at each treatment combination in a random order. Explain why a "repeated-measures" design like this may be beneficial.

13.12 Hypotension and endurance exercise. In sedentary individuals, low blood pressure (hypotension) often occurs after a single bout of aerobic exercise and lasts nearly two hours. This can cause dizziness, light-headedness, and possibly fainting upon standing. It is thought that endurance exercise training can reduce the degree of postexercise hypotension. To test this, researchers studied 16 endurance-trained and 16 sedentary men and women.[10] The following table summarizes the postexercise systolic arterial pressure (mmHg) after 60 minutes of upright cycling:

Group	n	$\bar{x}$	Std Err
Women, sedentary	8	100.7	3.4
Women, endurance	8	105.3	3.6
Men, sedentary	8	114.2	3.8
Men, endurance	8	110.2	2.3

(a) Make a plot similar to Figure 13.3 with the systolic blood pressure on the y axis and training level on the x axis. Describe the pattern you see.

(b) From the table, one can show that SSA = 677.12, SSB = 0.72, SSAB = 147.92, and SSE = 2478 where A is the gender effect and B is the training level. Construct the ANOVA table with F statistics and degrees of freedom, and state your conclusions regarding main effects and interaction.

(c) The researchers also measured the before-exercise systolic blood pressure of the participants and looked at a model that incorporated both the pre- and postexercise values. Explain why it is likely beneficial to incorporate both measurements in the study.

13.13 The effect of humor. In advertising, humor is often used to overcome sales resistance and stimulate

customer purchase behavior. A recent experiment looked at the use of humor as an approach to offset negative feelings often associated with web site encounters.[11] The setting of their experiment was an online travel agency and they used a three-factor design, each factor with two levels. They were humor (used, not used), process (favorable, unfavorable), and outcome (favorable, unfavorable). For the humor condition, cartoons and jokes of the day about skiing were used on the site. For the no humor condition, standard pictures of ski sites were used. Two hundred and forty-one business students from a large Dutch university participated in the experiment. Each was randomly assigned to one of the eight treatment conditions. The students were asked to book a skiing holiday and then rate their perceived enjoyment and satisfaction with the process. All responses were measured on a 7-point Likert scale. A summary of the results for satisfaction follows:

Treatment	n	$\bar{x}$	s
No humor - favorable process - unfavorable outcome	27	3.04	0.79
No humor - favorable process - favorable outcome	29	5.36	0.47
No humor - unfavorable process - unfavorable outcome	26	2.84	0.59
No humor - unfavorable process - favorable outcome	31	3.08	0.59
Humor - favorable process - unfavorable outcome	32	5.06	0.59
Humor - favorable process - favorable outcome	30	5.55	0.65
Humor - unfavorable process - unfavorable outcome	36	1.95	0.52
Humor - unfavorable process - favorable outcome	30	3.27	0.71

(a) Plot the means of the four treatments without humor. Do you think there is an interaction? If yes, describe the interaction in terms of the process and outcome factors.

(b) Plot the means of the four treatments that used humor. Do you think there is an interaction? If yes, describe the interaction in terms of the process and outcome factors.

(c) The three-factor interaction can be assessed by looking at the two interaction plots created in parts (a) and (b). If the relationship between process and outcome is different across the two humor conditions, there is evidence of an interaction among all three factors. Do you think there is a three-factor interaction? Explain your answer.

13.14 Pooling the standard deviations. Refer to the previous exercise. Find the pooled estimate of the standard deviation for these data. What are its degrees of freedom? Using the rule from Chapter 12 (page 633), is it reasonable to use a pooled standard deviation for the analysis? Explain your answer.

13.15 Describing the effects. Refer to Exercise 13.13. The *P*-values for all main effects and two-factor interactions are significant at the 0.05 level. Using the table, find the marginal means (i.e., the mean for the no humor treatment, the mean for the no humor and unfavorable process treatment combination, etc.) and use them to describe these effects.

13.16 Acceptance of functional foods. Functional foods are foods that are fortified with health-promoting supplements, like calcium-enriched orange juice or vitamin-enriched cereal. Although the number of functional foods is growing in the marketplace, very little is known about how the next generation of consumers views these foods. Because of this, a questionnaire was given to college students from the United States, Canada, and France.[12] This questionnaire measured both the students' general food and functional food attitudes and beliefs. One of the response variables collected was attitude towards cooking enjoyment. This variable was the average of numerous items, each measured on a 10-point scale, where 1 = most negative value and 10 = most positive value. Here are the means:

Gender	Culture		
	Canada	United States	France
Female	7.70	7.36	6.38
Male	6.39	6.43	5.69

(a) Make a plot of the means and describe the patterns that you see.

(b) Does the plot suggest that there is an interaction between culture and gender? If your answer is Yes, describe the interaction.

13.17 Estimating the within-group variance. Refer to the previous exercise. Here are the cell standard deviations and sample sizes for cooking enjoyment:

	Culture					
	Canada		United States		France	
Gender	*s*	*n*	*s*	*n*	*s*	*n*
Female	1.668	238	1.736	178	2.024	82
Male	1.909	125	1.601	101	1.875	87

Find the pooled estimate of the standard deviation for these data. Use the rule for examining standard deviations in ANOVA from Chapter 12 (page 633) to determine if it is reasonable to use a pooled standard deviation for the analysis of these data.

13.18 ⚔ Comparing the groups. The researchers presented a table of means with different superscripts indicating pairs of means that differed at the 0.05 significance level, using the Bonferroni method.

(a) What denominator degrees of freedom would be used here?

(b) How many pairwise comparisons are there for this problem?

(c) Perform these comparisons using $t^{**} = 2.94$ and summarize your results.

13.19 More on acceptance of functional foods. Refer to the Exercise 13.16. The means for four of the response variables associated with functional foods are as follows:

	General Attitude			Product Benefits		
	Culture			Culture		
Gender	Canada	United States	France	Canada	United States	France
Female	4.93	4.69	4.10	4.59	4.37	3.91
Male	4.50	4.43	4.02	4.20	4.09	3.87

	Credibility of Info			Purchase Intention		
	Culture			Culture		
Gender	Canada	United States	France	Canada	United States	France
Female	4.54	4.50	3.76	4.29	4.39	3.30
Male	4.23	3.99	3.83	4.11	3.86	3.41

For each of the four response variables, give a graphical summary of the means. Use this summary to discuss any interactions that are evident. Write a short report summarizing any differences in culture and gender on the response variables measured.

13.20 Interpreting the results. The goal of the study in the previous exercise was to understand cultural and gender differences in functional food attitudes and behaviors among young adults, the next generation of food consumers. The researchers used a sample of undergraduate students and had each participant fill out the survey during class time. How reasonable is it to generalize these results to the young adult population in these countries? Explain your answer.

13.21 Evaluation of an intervention program. The National Crime Victimization Survey estimates that there were over 400,000 violent crimes committed against women by their intimate partner that resulted in physical injury. An intervention study designed to increase safety behaviors of abused women compared the effectiveness of six telephone intervention sessions with a control group of abused women who received standard care. Fifteen different safety behaviors were examined.[13] One of the variables analyzed was the total number of behaviors (out of 15) that each woman performed. Here is a summary of the means of this variable at baseline (just before the first telephone call) and at follow-up three and six months later:

	Time		
Group	**Baseline**	**3 months**	**6 months**
Intervention	10.4	12.5	11.9
Control	9.6	9.9	10.4

(a) Find the marginal means. Are they useful for understanding the results of this study?

(b) Plot the means. Do you think there is an interaction? Describe the meaning of an interaction for this study. (*Note:* This exercise is from a repeated-measures design, and the data are not particularly Normal because they are counts with values from 1 to 15. Although we cannot use the methods in this chapter for statistical inference in this setting, the example does illustrate ideas about interactions.)

13.22 ⚠ **More on the assessment of an intervention program.** Refer to the previous exercise. Table 13.1 gives the percents of women who responded that they performed each of the 15 safety behaviors studied.

(a) Summarize these data graphically. Do you think that your graphical display is more effective than Table 13.1

for describing the results of this study? Explain why or why not.

(b) Note any particular patterns in the data that would be important to someone who would use these results to design future intervention programs for abused women.

(c) The study was conducted "at a family violence unit of a large urban District Attorney's Office that serves an ethnically diverse population of three million citizens." To what extent do you think that this fact limits the conclusions that can be drawn?

13.23 What can you conclude? Analysis of data for a 3×2 ANOVA with 5 observations per cell gave the F statistics in the following table:

Effect	F
A	1.87
B	3.49
AB	2.14

What can you conclude from the information given?

13.24 What can you conclude? A study reported the following results for data analyzed using the methods that we studied in this chapter:

Effect	F	P-value
A	1.50	0.236
B	13.66	0.001
AB	6.14	0.011

TABLE 13.1

Safety behaviors of abused women

	Intervention Group (%)			Control Group (%)		
Behavior	**Baseline**	**3 months**	**6 months**	**Baseline**	**3 months**	**6 months**
Hide money	68.0	60.0	62.7	60.0	37.8	35.1
Hide extra keys	52.7	76.0	68.9	53.3	33.8	39.2
Abuse code to alert family	30.7	74.7	60.0	22.7	27.0	43.2
Hide extra clothing	37.3	73.6	52.7	42.7	32.9	27.0
Asked neighbors to call police	49.3	73.0	66.2	32.0	45.9	40.5
Know Social Security number	93.2	93.2	100.0	89.3	93.2	98.6
Keep rent, utility receipts	75.3	95.5	89.4	70.3	84.7	80.9
Keep birth certificates	84.0	90.7	93.3	77.3	90.4	93.2
Keep driver's license	93.3	93.3	97.3	94.7	95.9	98.6
Keep telephone numbers	96.0	98.7	100.0	90.7	97.3	100.0
Removed weapons	50.0	70.6	38.5	40.7	23.8	5.9
Keep bank account numbers	81.0	94.3	96.2	76.2	85.5	94.4
Keep insurance policy number	70.9	90.4	89.7	68.3	84.2	94.8
Keep marriage license	71.1	92.3	84.6	63.3	73.2	80.0
Hide valuable jewelry	78.7	84.5	83.9	74.0	75.0	80.3

(a) What can you conclude from the information given?

(b) What additional information would you need to write a summary of the results for this study?

13.25 Conspicuous consumption and men's testosterone levels. It is argued that conspicuous consumption is a means by which men communicate their social status to prospective mates. One study looked at changes in a male's testosterone level in response to fluctuations in his status created by the consumption of a product.[14] The products considered were a new and luxurious sports car, and an old family sedan. Participants were asked to drive either on an isolated highway or a busy urban street. A table of cell means and standard deviations for the change (post-pre) in testosterone follows:

Car	Location Highway $\bar{x}$	s	City $\bar{x}$	s
Old sedan	0.03	0.12	−0.03	0.12
New sports car	0.15	0.14	0.13	0.13

(a) Make a plot of the means and describe the patterns that you see. Does the plot suggest an interaction between location and type of car?

(b) Compute the pooled standard error, s_p, assuming equal sample sizes.

(c) The researchers wanted to test the following hypotheses:
  (1) Testosterone levels will increase more in men who drive the new car
  (2) For men driving the new car, testosterone levels will increase more in men who drive in the city
  (3) For men driving the old car, testosterone levels will decrease less in men who drive the old car on the highway
Write out the contrasts for each of these hypotheses.

(d) This study actually involved each male participating in all four combinations. Half of them drove the sedan first and the other half drove the sports car first. Explain why a "repeated-measures" design like this may be beneficial.

13.26 🛡 **The effects of peer pressure on mathematics achievement.** Researchers were interested in comparing the relationship between high achievement in mathematics and peer pressure across several countries.[15] They hypothesized that in countries where high achievement is not valued highly, considerable peer pressure may exist. A questionnaire was distributed to 14-year-olds from three countries (Germany, Canada, and Israel). One of the questions asked students to rate how

often they fear being called a nerd or teacher's pet on a 4-point scale (1 = never, 4 = frequently). The following table summarizes the response:

Country	Gender	n	$\bar{x}$
Germany	Female	336	1.62
Germany	Male	305	1.39
Israel	Female	205	1.87
Israel	Male	214	1.63
Canada	Female	301	1.91
Canada	Male	304	1.88

(a) The P-values for the interaction and the main effects of country and gender are 0.016, 0.068, and 0.108, respectively. Using the table and P-values, summarize the results both graphically and numerically.

(b) The researchers contend that Germany does not value achievement as highly as Canada and Israel. Do the results from part (a) allow you to address their primary hypothesis? Explain.

(c) The students were also asked to indicate their current grade in mathematics on a 6-point scale (1 = excellent, 6 = insufficient). How might both responses be used to address the researchers' primary hypothesis?

13.27 The effect of chromium on insulin metabolism. The amount of chromium in the diet has an effect on the way the body processes insulin. In an experiment designed to study this phenomenon, four diets were fed to male rats. There were two factors. Chromium had two levels: low (L) and normal (N). The rats were allowed to eat as much as they wanted (M) or the total amount that they could eat was restricted (R). We call the second factor Eat. One of the variables measured was the amount of an enzyme called GITH.[16] The means for this response variable appear in the following table:

Chromium	Eat M	R
L	4.545	5.175
N	4.425	5.317

(a) Make a plot of the mean GITH for these diets, with the factor Chromium on the x axis and GITH on the y axis. For each Eat group, connect the points for the two Chromium means.

(b) Describe the patterns you see. Does the amount of chromium in the diet appear to affect the GITH mean? Does restricting the diet rather than letting the rats eat as much as they want appear to have an effect? Is there an interaction?

(c) Compute the marginal means. Compute the differences between the M and R diets for each level of Chromium. Use this information to summarize numerically the patterns in the plot.

13.28 Changing your major. A study of undergraduate computer science students examined changes in major after the first year.[17] The study examined the fates of 256 students who enrolled as first-year computer science students in the same fall semester. The students were classified according to gender and their declared major at the beginning of the second year. For convenience we use the labels CS for computer science majors, EO for engineering and other science majors, and O for other majors. The explanatory variables included several high school grade summaries coded as 10 = A, 9 = A−, etc. Here are the mean high school mathematics grades for these students:

	Major		
Gender	**CS**	**EO**	**O**
Males	8.68	8.35	7.65
Females	9.11	9.36	8.04

Describe the main effects and interaction using appropriate graphs and calculations.

13.29 More on changing your major. The mean SAT Mathematics scores for the students in the previous exercise are summarized in the following table:

	Major		
Gender	**CS**	**EO**	**O**
Males	628	618	589
Females	582	631	543

Summarize the results of this study using appropriate plots and calculations to describe the main effects and interaction.

13.30 Designing a study. The students studied in the previous two exercises were enrolled at a large Midwestern university more than two decades ago. Discuss how you would conduct a similar study at a college or university of your choice today. Include a description of all variables that you would collect for your study.

13.31 Trust of individuals and groups. Trust is an essential element in any exchange of goods or services. The following trust game is often used to study trust experimentally:

A *sender* starts with $X and can transfer any amount $x \leq X$ to a *responder*. The responder then gets $3x$ and can transfer any amount $y \leq 3x$ back to the sender. The game ends with final amounts $X - x + y$ and $3x - y$ for the sender and responder respectively.

The value x is taken as a measure of the sender's trust and the value $y/3x$ indicates the responder's trustworthiness. A recent study used this game to study the dynamics between individuals and groups of three.[18] The following table summarizes the average amount x sent by senders starting with $100.

Sender	Responder	n	$\bar{x}$	s
Individual	Individual	32	65.5	36.4
Individual	Group	25	76.3	31.2
Group	Individual	25	54.0	41.6
Group	Group	27	43.7	42.4

(a) Find the pooled estimate of the standard deviation for this study and its degrees of freedom.

(b) Is it reasonable to use a pooled standard deviation for the analysis? Explain your answer.

(c) Compute the marginal means.

(d) Plot the means. Do you think there is an interaction? If yes, then describe it.

(e) The F statistics for sender, responder, and interaction are 9.05, 0.001, and 2.08, respectively. Compute the P-values and state your conclusions.

13.32 🏅 **Does the type of cooking pot affect iron content?** Iron-deficiency anemia is the most common form of malnutrition in developing countries, affecting about 50% of children and women and 25% of men. Iron pots for cooking foods had traditionally been used in many of these countries, but they have been largely replaced by aluminum pots, which are cheaper and lighter. Some research has suggested that food cooked in iron pots will contain more iron than food cooked in other types of pots. One study designed to investigate this issue compared the iron content of some Ethiopian foods cooked in aluminum, clay, and iron pots.[19] In Exercise 12.49 (page 665), we analyzed the iron content of *yesiga wet'*, beef cut into small pieces and prepared with several Ethiopian spices. The researchers who conducted this study also examined the iron content of *shiro wet'*, a legume-based mixture of chickpea flour and Ethiopian spiced pepper, and *ye-atkilt allych'a*, a lightly spiced vegetable casserole. In the following table, these three foods are labeled meat, legumes, and vegetables. Four samples of each food were cooked in each type of pot. The iron in the food is measured in milligrams of iron per 100 grams of cooked food. The data are shown in Table 13.2 (page 692). 🍲 IRONCONTENT

(a) Make a table giving the sample size, mean, and standard deviation for each type of pot. Is it reasonable to pool the variances? Although the standard deviations

TABLE 13.2

Iron content

Type of pot	Meat				Legumes				Vegetables			
Aluminum	1.77	2.36	1.96	2.14	2.40	2.17	2.41	2.34	1.03	1.53	1.07	1.30
Clay	2.27	1.28	2.48	2.68	2.41	2.43	2.57	2.48	1.55	0.79	1.68	1.82
Iron	5.27	5.17	4.06	4.22	3.69	3.43	3.84	3.72	2.45	2.99	2.80	2.92

vary more than we would like, this is partially due to the small sample sizes and we will proceed with the analysis of variance.

(b) Plot the means. Give a short summary of how the iron content of foods depends upon the cooking pot.

(c) Run the analysis of variance. Give the ANOVA table, the F statistics with degrees of freedom and P-values, and your conclusions regarding the hypotheses about main effects and interactions.

13.33 Interpreting the results. Refer to the previous exercise. Although there is a statistically significant interaction, do you think that these data support the conclusion that foods cooked in iron pots contain more iron than foods cooked in aluminum or clay pots? Discuss.

13.34 Analysis using a one-way ANOVA. Refer to Exercise 13.32. Rerun the analysis as a one-way ANOVA with 9 groups and 4 observations per group. Report the results of the F test. Examine differences in means using a multiple-comparisons procedure. Summarize your results and compare them with those you obtained in Exercise 13.32.

13.35 Examination of a drilling process. One step in the manufacture of large engines requires that holes of very precise dimensions be drilled. The tools that do the drilling are regularly examined and are adjusted to ensure that the holes meet the required specifications. Part of the examination involves measurement of the diameter of the drilling tool. A team studying the variation in the sizes of the drilled holes selected this measurement procedure as a possible cause of variation in the drilled holes. They decided to use a designed experiment as one part of this examination. Some of the data are given in Table 13.3. The diameters in millimeters (mm) of five tools were measured by the same operator at three times (8:00 A.M., 11:00 A.M., and 3:00 P.M.). Three measurements were taken on each tool at each time. The person taking the measurements could not tell which tool was being measured, and the measurements were taken in random order.[20] 🛠 DRILLING

(a) Make a table of means and standard deviations for each of the 5 × 3 combinations of the two factors.

TABLE 13.3

Tool diameter data

Tool	Time	Diameter (mm)		
1	1	25.030	25.030	25.032
1	2	25.028	25.028	25.028
1	3	25.026	25.026	25.026
2	1	25.016	25.018	25.016
2	2	25.022	25.020	25.018
2	3	25.016	25.016	25.016
3	1	25.005	25.008	25.006
3	2	25.012	25.012	25.014
3	3	25.010	25.010	25.008
4	1	25.012	25.012	25.012
4	2	25.018	25.020	25.020
4	3	25.010	25.014	25.018
5	1	24.996	24.998	24.998
5	2	25.006	25.006	25.006
5	3	25.000	25.002	24.999

(b) Plot the means and describe how the means vary with tool and time. Note that we expect the tools to have slightly different diameters. These will be adjusted as needed. It is the process of measuring the diameters that is important.

(c) Use a two-way ANOVA to analyze these data. Report the test statistics, degrees of freedom, and P-values for the significance tests.

(d) Write a short report summarizing your results.

13.36 Examination of a drilling process, continued. Refer to the previous exercise. Multiply each measurement by 0.04 to convert from millimeters to inches. Redo the plots and rerun the ANOVA using the transformed measurements. Summarize what parts of the analysis have changed and what parts have remained the same. 🛠 DRILLING

13.37 Do left-handed people live shorter lives than right-handed people? A study of this question examined a sample of 949 death records and contacted next of kin to determine handedness.[21] Note that there are many possible definitions of "left-handed." The researchers examined the effects of different definitions on the results of their analysis and found that their conclusions were not sensitive to the exact definition used. For the results presented here, people were defined to be right-handed if they wrote, drew, and threw a ball with the right hand. All others were defined to be left-handed. People were classified by gender (female or male) and handedness (left or right), and a 2 × 2 ANOVA was run with the age at death as the response variable. The F statistics were 22.36 (handedness), 37.44 (gender), and 2.10 (interaction). The following marginal mean ages at death (in years) were reported: 77.39 (females), 71.32 (males), 75.00 (right-handed), and 66.03 (left-handed).

(a) For each of the F statistics given, find the degrees of freedom and an approximate P-value. Summarize the results of these tests.

(b) Using the information given, write a short summary of the results of the study.

13.38 A radon exposure study. Scientists believe that exposure to the radioactive gas radon is associated with some types of cancers in the respiratory system. Radon from natural sources is present in many homes in the United States. A group of researchers decided to study the problem in dogs because dogs get similar types of cancers and are exposed to environments similar to those of their owners. Radon detectors are available for home monitoring, but the researchers wanted to obtain actual measures of the exposure of a sample of dogs. To do this, they placed the detectors in holders and attached them to the collars of the dogs. One problem was that the holders might in some way affect the radon readings. The researchers therefore devised a laboratory experiment to study the effects of the holders. Detectors from four series of production were available, so they used a two-way ANOVA design (series with 4 levels and holder with 2, representing the presence or absence of a holder). All detectors were exposed to the same radon source and the radon measure in picocuries per liter was recorded.[22] The F statistic for the effect of series is 7.02, for holder it is 1.96, for the interaction it is 1.24, and $N = 69$.

(a) Using Table E or statistical software, find approximate P-values for the three test statistics. Summarize the results of these three significance tests.

(b) The mean radon readings for the four series were 330, 303, 302, and 295. The results of the significance test for series were of great concern to the researchers. Explain why.

13.39 A comparison of plant species under low water conditions. The PLANTS1 data set gives the percent of nitrogen in four different species of plants grown in a laboratory. The species are *Leucaena leucocephala*, *Acacia saligna*, *Prosopis juliflora*, and *Eucalyptus citriodora*. The researchers who collected these data were interested in commercially growing these plants in parts of the country of Jordan where there is very little rainfall. To examine the effect of water, they varied the amount per day from 50 millimeters (mm) to 650 mm in 100 mm increments. There were nine plants per species-by-water combination. Because the plants are to be used primarily for animal food, with some parts that can be consumed by people, a high nitrogen content is very desirable. PLANTS1

(a) Find the means for each species-by-water combination. Plot these means versus water for the four species, connecting the means for each species by lines. Describe the overall pattern.

(b) Find the standard deviations for each species-by-water combination. Is it reasonable to pool the standard deviations for this problem? Note that with sample sizes of size 9, we expect these standard deviations to be quite variable.

(c) Run the two-way analysis of variance. Give the results of the hypothesis tests for the main effects and the interaction.

13.40 Examination of the residuals. Refer to the previous exercise. Examine the residuals. Are there any unusual patterns or outliers? If you think that there are one or more points that are somewhat extreme, rerun the two-way analysis without these observations. Does this change the results in any substantial way? PLANTS1

13.41 Analysis using multiple one-way ANOVAs. Refer to Exercise 13.39. Run a separate one-way analysis of variance for each water level. If there is evidence that the species are not all the same, use a multiple-comparisons procedure to determine which pairs of species are significantly different. In what way, if any, do the differences appear to vary by water level? Write a short summary of your conclusions. PLANTS1

13.42 More on the analysis using multiple one-way ANOVAs. Refer to Exercise 13.39. Run a separate one-way analysis of variance for each species and summarize the results. Since the amount of water is a quantitative factor, we can also analyze these data using regression. Run simple linear regressions separately for each species to predict nitrogen percent from water. Use plots to determine whether or not a line is a good way to approximate this relationship. Summarize the regression results and compare them with the one-way ANOVA results. PLANTS1

13.43 Another comparison of plant species under low water conditions. Refer to Exercise 13.39. Additional data collected by the same researchers according to a similar design are given in the PLANTS2 data set. Here, there are two response variables. They are fresh biomass and dry biomass. High values for both these variables are desirable. The same four species and seven levels of water are used for this experiment. Here, however, there are four plants per species-by-water combination. Analyze each of the response variables in the PLANTS2 data set using the outline from Exercise 13.39. ![PLANTS2 icon] PLANTS2

13.44 Examination of the residuals. Perform the tasks described in Exercise 13.40 for the two response variables in the PLANTS2 data set. ![PLANTS2 icon] PLANTS2

13.45 Analysis using multiple one-way ANOVAs. Perform the tasks described in Exercise 13.41 for the two response variables in the PLANTS2 data set. ![PLANTS2 icon] PLANTS2

13.46 More on the analysis using multiple one-way ANOVAs. Perform the tasks described in Exercise 13.42 for the two response variables in the PLANTS2 data set. ![PLANTS2 icon] PLANTS2

13.47 Are insects more attracted to male plants? Some scientists wanted to determine if there are gender-related differences in the level of herbivory in the jack-in-the-pulpit, a spring-blooming perennial plant common in deciduous forests. A study was conducted in southern Maryland at forests associated with the Smithsonian Environmental Research Center (SERC).[23] To determine the effects of flowering and floral characteristics on herbivory, the researchers altered the floral morphology of male and female plants. The three levels of floral characteristics were (1) the spathes were completely removed; (2) in females, a gap was created in the base of the spathe, and in males, the gap was closed; (3) plants were not altered (control). The percent of leaf area damaged by thrips (an order of insects) between early May and mid-June was recorded for each of 30 plants per combination of sex and floral characteristic. A table of means and standard deviations (in parentheses) follows:

	Floral characteristic level		
Gender	1	2	3
Males	0.11 (0.081)	1.28 (0.088)	1.63 (0.382)
Females	0.02 (0.002)	0.58 (0.321)	0.20 (0.035)

(a) Give the degrees of freedom for the F statistics that are used to test for gender, floral characteristic, and the interaction.

(b) Describe the main effects and interaction using appropriate graphs.

(c) The researchers used the natural logarithm of percent area as the response in their analysis. Using the relationship between the means and standard deviations, explain why this was done.

13.48 Change-of-majors study: HSS. Refer to the data given for the change-of-majors study in the data set MAJORS. Analyze the data for HSS, the high school science grades. Your analysis should include a table of sample sizes, means, and standard deviations; Normal quantile plots; a plot of the means; and a two-way ANOVA using sex and major as the factors. Write a short summary of your conclusions. ![MAJORS icon] MAJORS

13.49 Change-of-majors study: HSE. Refer to the data given for the change-of-majors study in the data set MAJORS. Analyze the data for HSE, the high school English grades. Your analysis should include a table of sample sizes, means, and standard deviations; Normal quantile plots; a plot of the means; and a two-way ANOVA using sex and major as the factors. Write a short summary of your conclusions. ![MAJORS icon] MAJORS

13.50 Change-of-majors study: GPA. Refer to the data given for the change-of-majors study in the data set MAJORS. Analyze the data for GPA, the college grade point average. Your analysis should include a table of sample sizes, means, and standard deviations; Normal quantile plots; a plot of the means; and a two-way ANOVA using sex and major as the factors. Write a short summary of your conclusions. ![MAJORS icon] MAJORS

13.51 Change-of-majors study: SATV. Refer to the data given for the change-of-majors study in the data set MAJORS. Analyze the data for SATV, the SAT Verbal score. Your analysis should include a table of sample sizes, means, and standard deviations; Normal quantile plots; a plot of the means; and a two-way ANOVA using sex and major as the factors. Write a short summary of your conclusions. ![MAJORS icon] MAJORS

13.52 Search the Internet. Search the Internet or your library to find a study that is interesting to you and uses a two-way ANOVA to analyze the data. First describe the question or questions of interest and then give the details of how ANOVA was used to provide answers. Be sure to include how the study authors examined the assumptions for the analysis. Evaluate how well the authors used ANOVA in this study. If your evaluation finds the analysis deficient, make suggestions for how it could be improved.

TABLES

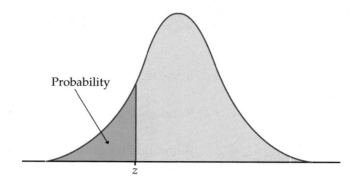

Table entry for z is
the area under the
standard Normal curve
to the left of z.

Probability

z

TABLE A

Standard Normal probabilities

z	.00	.01	.02	.03	.04	.05	.06	.07	.08	.09
−3.4	.0003	.0003	.0003	.0003	.0003	.0003	.0003	.0003	.0003	.0002
−3.3	.0005	.0005	.0005	.0004	.0004	.0004	.0004	.0004	.0004	.0003
−3.2	.0007	.0007	.0006	.0006	.0006	.0006	.0006	.0005	.0005	.0005
−3.1	.0010	.0009	.0009	.0009	.0008	.0008	.0008	.0008	.0007	.0007
−3.0	.0013	.0013	.0013	.0012	.0012	.0011	.0011	.0011	.0010	.0010
−2.9	.0019	.0018	.0018	.0017	.0016	.0016	.0015	.0015	.0014	.0014
−2.8	.0026	.0025	.0024	.0023	.0023	.0022	.0021	.0021	.0020	.0019
−2.7	.0035	.0034	.0033	.0032	.0031	.0030	.0029	.0028	.0027	.0026
−2.6	.0047	.0045	.0044	.0043	.0041	.0040	.0039	.0038	.0037	.0036
−2.5	.0062	.0060	.0059	.0057	.0055	.0054	.0052	.0051	.0049	.0048
−2.4	.0082	.0080	.0078	.0075	.0073	.0071	.0069	.0068	.0066	.0064
−2.3	.0107	.0104	.0102	.0099	.0096	.0094	.0091	.0089	.0087	.0084
−2.2	.0139	.0136	.0132	.0129	.0125	.0122	.0119	.0116	.0113	.0110
−2.1	.0179	.0174	.0170	.0166	.0162	.0158	.0154	.0150	.0146	.0143
−2.0	.0228	.0222	.0217	.0212	.0207	.0202	.0197	.0192	.0188	.0183
−1.9	.0287	.0281	.0274	.0268	.0262	.0256	.0250	.0244	.0239	.0233
−1.8	.0359	.0351	.0344	.0336	.0329	.0322	.0314	.0307	.0301	.0294
−1.7	.0446	.0436	.0427	.0418	.0409	.0401	.0392	.0384	.0375	.0367
−1.6	.0548	.0537	.0526	.0516	.0505	.0495	.0485	.0475	.0465	.0455
−1.5	.0668	.0655	.0643	.0630	.0618	.0606	.0594	.0582	.0571	.0559
−1.4	.0808	.0793	.0778	.0764	.0749	.0735	.0721	.0708	.0694	.0681
−1.3	.0968	.0951	.0934	.0918	.0901	.0885	.0869	.0853	.0838	.0823
−1.2	.1151	.1131	.1112	.1093	.1075	.1056	.1038	.1020	.1003	.0985
−1.1	.1357	.1335	.1314	.1292	.1271	.1251	.1230	.1210	.1190	.1170
−1.0	.1587	.1562	.1539	.1515	.1492	.1469	.1446	.1423	.1401	.1379
−0.9	.1841	.1814	.1788	.1762	.1736	.1711	.1685	.1660	.1635	.1611
−0.8	.2119	.2090	.2061	.2033	.2005	.1977	.1949	.1922	.1894	.1867
−0.7	.2420	.2389	.2358	.2327	.2296	.2266	.2236	.2206	.2177	.2148
−0.6	.2743	.2709	.2676	.2643	.2611	.2578	.2546	.2514	.2483	.2451
−0.5	.3085	.3050	.3015	.2981	.2946	.2912	.2877	.2843	.2810	.2776
−0.4	.3446	.3409	.3372	.3336	.3300	.3264	.3228	.3192	.3156	.3121
−0.3	.3821	.3783	.3745	.3707	.3669	.3632	.3594	.3557	.3520	.3483
−0.2	.4207	.4168	.4129	.4090	.4052	.4013	.3974	.3936	.3897	.3859
−0.1	.4602	.4562	.4522	.4483	.4443	.4404	.4364	.4325	.4286	.4247
−0.0	.5000	.4960	.4920	.4880	.4840	.4801	.4761	.4721	.4681	.4641

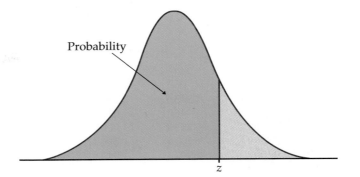

Table entry for z is the
area under the
standard Normal curve
to the left of z.

Probability

z

TABLE A

Standard Normal probabilities (continued)

z	.00	.01	.02	.03	.04	.05	.06	.07	.08	.09
0.0	.5000	.5040	.5080	.5120	.5160	.5199	.5239	.5279	.5319	.5359
0.1	.5398	.5438	.5478	.5517	.5557	.5596	.5636	.5675	.5714	.5753
0.2	.5793	.5832	.5871	.5910	.5948	.5987	.6026	.6064	.6103	.6141
0.3	.6179	.6217	.6255	.6293	.6331	.6368	.6406	.6443	.6480	.6517
0.4	.6554	.6591	.6628	.6664	.6700	.6736	.6772	.6808	.6844	.6879
0.5	.6915	.6950	.6985	.7019	.7054	.7088	.7123	.7157	.7190	.7224
0.6	.7257	.7291	.7324	.7357	.7389	.7422	.7454	.7486	.7517	.7549
0.7	.7580	.7611	.7642	.7673	.7704	.7734	.7764	.7794	.7823	.7852
0.8	.7881	.7910	.7939	.7967	.7995	.8023	.8051	.8078	.8106	.8133
0.9	.8159	.8186	.8212	.8238	.8264	.8289	.8315	.8340	.8365	.8389
1.0	.8413	.8438	.8461	.8485	.8508	.8531	.8554	.8577	.8599	.8621
1.1	.8643	.8665	.8686	.8708	.8729	.8749	.8770	.8790	.8810	.8830
1.2	.8849	.8869	.8888	.8907	.8925	.8944	.8962	.8980	.8997	.9015
1.3	.9032	.9049	.9066	.9082	.9099	.9115	.9131	.9147	.9162	.9177
1.4	.9192	.9207	.9222	.9236	.9251	.9265	.9279	.9292	.9306	.9319
1.5	.9332	.9345	.9357	.9370	.9382	.9394	.9406	.9418	.9429	.9441
1.6	.9452	.9463	.9474	.9484	.9495	.9505	.9515	.9525	.9535	.9545
1.7	.9554	.9564	.9573	.9582	.9591	.9599	.9608	.9616	.9625	.9633
1.8	.9641	.9649	.9656	.9664	.9671	.9678	.9686	.9693	.9699	.9706
1.9	.9713	.9719	.9726	.9732	.9738	.9744	.9750	.9756	.9761	.9767
2.0	.9772	.9778	.9783	.9788	.9793	.9798	.9803	.9808	.9812	.9817
2.1	.9821	.9826	.9830	.9834	.9838	.9842	.9846	.9850	.9854	.9857
2.2	.9861	.9864	.9868	.9871	.9875	.9878	.9881	.9884	.9887	.9890
2.3	.9893	.9896	.9898	.9901	.9904	.9906	.9909	.9911	.9913	.9916
2.4	.9918	.9920	.9922	.9925	.9927	.9929	.9931	.9932	.9934	.9936
2.5	.9938	.9940	.9941	.9943	.9945	.9946	.9948	.9949	.9951	.9952
2.6	.9953	.9955	.9956	.9957	.9959	.9960	.9961	.9962	.9963	.9964
2.7	.9965	.9966	.9967	.9968	.9969	.9970	.9971	.9972	.9973	.9974
2.8	.9974	.9975	.9976	.9977	.9977	.9978	.9979	.9979	.9980	.9981
2.9	.9981	.9982	.9982	.9983	.9984	.9984	.9985	.9985	.9986	.9986
3.0	.9987	.9987	.9987	.9988	.9988	.9989	.9989	.9989	.9990	.9990
3.1	.9990	.9991	.9991	.9991	.9992	.9992	.9992	.9992	.9993	.9993
3.2	.9993	.9993	.9994	.9994	.9994	.9994	.9994	.9995	.9995	.9995
3.3	.9995	.9995	.9995	.9996	.9996	.9996	.9996	.9996	.9996	.9997
3.4	.9997	.9997	.9997	.9997	.9997	.9997	.9997	.9997	.9997	.9998

TABLE B

Random digits

Line

101	19223	95034	05756	28713	96409	12531	42544	82853
102	73676	47150	99400	01927	27754	42648	82425	36290
103	45467	71709	77558	00095	32863	29485	82226	90056
104	52711	38889	93074	60227	40011	85848	48767	52573
105	95592	94007	69971	91481	60779	53791	17297	59335
106	68417	35013	15529	72765	85089	57067	50211	47487
107	82739	57890	20807	47511	81676	55300	94383	14893
108	60940	72024	17868	24943	61790	90656	87964	18883
109	36009	19365	15412	39638	85453	46816	83485	41979
110	38448	48789	18338	24697	39364	42006	76688	08708
111	81486	69487	60513	09297	00412	71238	27649	39950
112	59636	88804	04634	71197	19352	73089	84898	45785
113	62568	70206	40325	03699	71080	22553	11486	11776
114	45149	32992	75730	66280	03819	56202	02938	70915
115	61041	77684	94322	24709	73698	14526	31893	32592
116	14459	26056	31424	80371	65103	62253	50490	61181
117	38167	98532	62183	70632	23417	26185	41448	75532
118	73190	32533	04470	29669	84407	90785	65956	86382
119	95857	07118	87664	92099	58806	66979	98624	84826
120	35476	55972	39421	65850	04266	35435	43742	11937
121	71487	09984	29077	14863	61683	47052	62224	51025
122	13873	81598	95052	90908	73592	75186	87136	95761
123	54580	81507	27102	56027	55892	33063	41842	81868
124	71035	09001	43367	49497	72719	96758	27611	91596
125	96746	12149	37823	71868	18442	35119	62103	39244
126	96927	19931	36089	74192	77567	88741	48409	41903
127	43909	99477	25330	64359	40085	16925	85117	36071
128	15689	14227	06565	14374	13352	49367	81982	87209
129	36759	58984	68288	22913	18638	54303	00795	08727
130	69051	64817	87174	09517	84534	06489	87201	97245
131	05007	16632	81194	14873	04197	85576	45195	96565
132	68732	55259	84292	08796	43165	93739	31685	97150
133	45740	41807	65561	33302	07051	93623	18132	09547
134	27816	78416	18329	21337	35213	37741	04312	68508
135	66925	55658	39100	78458	11206	19876	87151	31260
136	08421	44753	77377	28744	75592	08563	79140	92454
137	53645	66812	61421	47836	12609	15373	98481	14592
138	66831	68908	40772	21558	47781	33586	79177	06928
139	55588	99404	70708	41098	43563	56934	48394	51719
140	12975	13258	13048	45144	72321	81940	00360	02428
141	96767	35964	23822	96012	94591	65194	50842	53372
142	72829	50232	97892	63408	77919	44575	24870	04178
143	88565	42628	17797	49376	61762	16953	88604	12724
144	62964	88145	83083	69453	46109	59505	69680	00900
145	19687	12633	57857	95806	09931	02150	43163	58636
146	37609	59057	66967	83401	60705	02384	90597	93600
147	54973	86278	88737	74351	47500	84552	19909	67181
148	00694	05977	19664	65441	20903	62371	22725	53340
149	71546	05233	53946	68743	72460	27601	45403	88692
150	07511	88915	41267	16853	84569	79367	32337	03316

TABLE B

Random digits (continued)

Line								
151	03802	29341	29264	80198	12371	13121	54969	43912
152	77320	35030	77519	41109	98296	18984	60869	12349
153	07886	56866	39648	69290	03600	05376	58958	22720
154	87065	74133	21117	70595	22791	67306	28420	52067
155	42090	09628	54035	93879	98441	04606	27381	82637
156	55494	67690	88131	81800	11188	28552	25752	21953
157	16698	30406	96587	65985	07165	50148	16201	86792
158	16297	07626	68683	45335	34377	72941	41764	77038
159	22897	17467	17638	70043	36243	13008	83993	22869
160	98163	45944	34210	64158	76971	27689	82926	75957
161	43400	25831	06283	22138	16043	15706	73345	26238
162	97341	46254	88153	62336	21112	35574	99271	45297
163	64578	67197	28310	90341	37531	63890	52630	76315
164	11022	79124	49525	63078	17229	32165	01343	21394
165	81232	43939	23840	05995	84589	06788	76358	26622
166	36843	84798	51167	44728	20554	55538	27647	32708
167	84329	80081	69516	78934	14293	92478	16479	26974
168	27788	85789	41592	74472	96773	27090	24954	41474
169	99224	00850	43737	75202	44753	63236	14260	73686
170	38075	73239	52555	46342	13365	02182	30443	53229
171	87368	49451	55771	48343	51236	18522	73670	23212
172	40512	00681	44282	47178	08139	78693	34715	75606
173	81636	57578	54286	27216	58758	80358	84115	84568
174	26411	94292	06340	97762	37033	85968	94165	46514
175	80011	09937	57195	33906	94831	10056	42211	65491
176	92813	87503	63494	71379	76550	45984	05481	50830
177	70348	72871	63419	57363	29685	43090	18763	31714
178	24005	52114	26224	39078	80798	15220	43186	00976
179	85063	55810	10470	08029	30025	29734	61181	72090
180	11532	73186	92541	06915	72954	10167	12142	26492
181	59618	03914	05208	84088	20426	39004	84582	87317
182	92965	50837	39921	84661	82514	81899	24565	60874
183	85116	27684	14597	85747	01596	25889	41998	15635
184	15106	10411	90221	49377	44369	28185	80959	76355
185	03638	31589	07871	25792	85823	55400	56026	12193
186	97971	48932	45792	63993	95635	28753	46069	84635
187	49345	18305	76213	82390	77412	97401	50650	71755
188	87370	88099	89695	87633	76987	85503	26257	51736
189	88296	95670	74932	65317	93848	43988	47597	83044
190	79485	92200	99401	54473	34336	82786	05457	60343
191	40830	24979	23333	37619	56227	95941	59494	86539
192	32006	76302	81221	00693	95197	75044	46596	11628
193	37569	85187	44692	50706	53161	69027	88389	60313
194	56680	79003	23361	67094	15019	63261	24543	52884
195	05172	08100	22316	54495	60005	29532	18433	18057
196	74782	27005	03894	98038	20627	40307	47317	92759
197	85288	93264	61409	03404	09649	55937	60843	66167
198	68309	12060	14762	58002	03716	81968	57934	32624
199	26461	88346	52430	60906	74216	96263	69296	90107
200	42672	67680	42376	95023	82744	03971	96560	55148

TABLE C

Binomial probabilities

Entry is $P(X = k) = \binom{n}{k} p^k (1-p)^{n-k}$

						p				
n	k	.01	.02	.03	.04	.05	.06	.07	.08	.09
2	0	.9801	.9604	.9409	.9216	.9025	.8836	.8649	.8464	.8281
	1	.0198	.0392	.0582	.0768	.0950	.1128	.1302	.1472	.1638
	2	.0001	.0004	.0009	.0016	.0025	.0036	.0049	.0064	.0081
3	0	.9703	.9412	.9127	.8847	.8574	.8306	.8044	.7787	.7536
	1	.0294	.0576	.0847	.1106	.1354	.1590	.1816	.2031	.2236
	2	.0003	.0012	.0026	.0046	.0071	.0102	.0137	.0177	.0221
	3				.0001	.0001	.0002	.0003	.0005	.0007
4	0	.9606	.9224	.8853	.8493	.8145	.7807	.7481	.7164	.6857
	1	.0388	.0753	.1095	.1416	.1715	.1993	.2252	.2492	.2713
	2	.0006	.0023	.0051	.0088	.0135	.0191	.0254	.0325	.0402
	3			.0001	.0002	.0005	.0008	.0013	.0019	.0027
	4									.0001
5	0	.9510	.9039	.8587	.8154	.7738	.7339	.6957	.6591	.6240
	1	.0480	.0922	.1328	.1699	.2036	.2342	.2618	.2866	.3086
	2	.0010	.0038	.0082	.0142	.0214	.0299	.0394	.0498	.0610
	3		.0001	.0003	.0006	.0011	.0019	.0030	.0043	.0060
	4						.0001	.0001	.0002	.0003
	5									
6	0	.9415	.8858	.8330	.7828	.7351	.6899	.6470	.6064	.5679
	1	.0571	.1085	.1546	.1957	.2321	.2642	.2922	.3164	.3370
	2	.0014	.0055	.0120	.0204	.0305	.0422	.0550	.0688	.0833
	3		.0002	.0005	.0011	.0021	.0036	.0055	.0080	.0110
	4					.0001	.0002	.0003	.0005	.0008
	5									
	6									
7	0	.9321	.8681	.8080	.7514	.6983	.6485	.6017	.5578	.5168
	1	.0659	.1240	.1749	.2192	.2573	.2897	.3170	.3396	.3578
	2	.0020	.0076	.0162	.0274	.0406	.0555	.0716	.0886	.1061
	3		.0003	.0008	.0019	.0036	.0059	.0090	.0128	.0175
	4				.0001	.0002	.0004	.0007	.0011	.0017
	5								.0001	.0001
	6									
	7									
8	0	.9227	.8508	.7837	.7214	.6634	.6096	.5596	.5132	.4703
	1	.0746	.1389	.1939	.2405	.2793	.3113	.3370	.3570	.3721
	2	.0026	.0099	.0210	.0351	.0515	.0695	.0888	.1087	.1288
	3	.0001	.0004	.0013	.0029	.0054	.0089	.0134	.0189	.0255
	4			.0001	.0002	.0004	.0007	.0013	.0021	.0031
	5							.0001	.0001	.0002
	6									
	7									
	8									

TABLE C

Binomial probabilities (continued)

Entry is $P(X = k) = \binom{n}{k} p^k (1-p)^{n-k}$

n	k	p .10	.15	.20	.25	.30	.35	.40	.45	.50
2	0	.8100	.7225	.6400	.5625	.4900	.4225	.3600	.3025	.2500
	1	.1800	.2550	.3200	.3750	.4200	.4550	.4800	.4950	.5000
	2	.0100	.0225	.0400	.0625	.0900	.1225	.1600	.2025	.2500
3	0	.7290	.6141	.5120	.4219	.3430	.2746	.2160	.1664	.1250
	1	.2430	.3251	.3840	.4219	.4410	.4436	.4320	.4084	.3750
	2	.0270	.0574	.0960	.1406	.1890	.2389	.2880	.3341	.3750
	3	.0010	.0034	.0080	.0156	.0270	.0429	.0640	.0911	.1250
4	0	.6561	.5220	.4096	.3164	.2401	.1785	.1296	.0915	.0625
	1	.2916	.3685	.4096	.4219	.4116	.3845	.3456	.2995	.2500
	2	.0486	.0975	.1536	.2109	.2646	.3105	.3456	.3675	.3750
	3	.0036	.0115	.0256	.0469	.0756	.1115	.1536	.2005	.2500
	4	.0001	.0005	.0016	.0039	.0081	.0150	.0256	.0410	.0625
5	0	.5905	.4437	.3277	.2373	.1681	.1160	.0778	.0503	.0313
	1	.3280	.3915	.4096	.3955	.3602	.3124	.2592	.2059	.1563
	2	.0729	.1382	.2048	.2637	.3087	.3364	.3456	.3369	.3125
	3	.0081	.0244	.0512	.0879	.1323	.1811	.2304	.2757	.3125
	4	.0004	.0022	.0064	.0146	.0284	.0488	.0768	.1128	.1562
	5		.0001	.0003	.0010	.0024	.0053	.0102	.0185	.0312
6	0	.5314	.3771	.2621	.1780	.1176	.0754	.0467	.0277	.0156
	1	.3543	.3993	.3932	.3560	.3025	.2437	.1866	.1359	.0938
	2	.0984	.1762	.2458	.2966	.3241	.3280	.3110	.2780	.2344
	3	.0146	.0415	.0819	.1318	.1852	.2355	.2765	.3032	.3125
	4	.0012	.0055	.0154	.0330	.0595	.0951	.1382	.1861	.2344
	5	.0001	.0004	.0015	.0044	.0102	.0205	.0369	.0609	.0937
	6			.0001	.0002	.0007	.0018	.0041	.0083	.0156
7	0	.4783	.3206	.2097	.1335	.0824	.0490	.0280	.0152	.0078
	1	.3720	.3960	.3670	.3115	.2471	.1848	.1306	.0872	.0547
	2	.1240	.2097	.2753	.3115	.3177	.2985	.2613	.2140	.1641
	3	.0230	.0617	.1147	.1730	.2269	.2679	.2903	.2918	.2734
	4	.0026	.0109	.0287	.0577	.0972	.1442	.1935	.2388	.2734
	5	.0002	.0012	.0043	.0115	.0250	.0466	.0774	.1172	.1641
	6		.0001	.0004	.0013	.0036	.0084	.0172	.0320	.0547
	7				.0001	.0002	.0006	.0016	.0037	.0078
8	0	.4305	.2725	.1678	.1001	.0576	.0319	.0168	.0084	.0039
	1	.3826	.3847	.3355	.2670	.1977	.1373	.0896	.0548	.0313
	2	.1488	.2376	.2936	.3115	.2965	.2587	.2090	.1569	.1094
	3	.0331	.0839	.1468	.2076	.2541	.2786	.2787	.2568	.2188
	4	.0046	.0185	.0459	.0865	.1361	.1875	.2322	.2627	.2734
	5	.0004	.0026	.0092	.0231	.0467	.0808	.1239	.1719	.2188
	6		.0002	.0011	.0038	.0100	.0217	.0413	.0703	.1094
	7			.0001	.0004	.0012	.0033	.0079	.0164	.0312
	8					.0001	.0002	.0007	.0017	.0039

(*Continued*)

TABLE C

Binomial probabilities (continued)

Entry is $P(X = k) = \binom{n}{k} p^k (1 - p)^{n-k}$

| | | | | | | p | | | | |
n	k	.01	.02	.03	.04	.05	.06	.07	.08	.09
9	0	.9135	.8337	.7602	.6925	.6302	.5730	.5204	.4722	.4279
	1	.0830	.1531	.2116	.2597	.2985	.3292	.3525	.3695	.3809
	2	.0034	.0125	.0262	.0433	.0629	.0840	.1061	.1285	.1507
	3	.0001	.0006	.0019	.0042	.0077	.0125	.0186	.0261	.0348
	4			.0001	.0003	.0006	.0012	.0021	.0034	.0052
	5						.0001	.0002	.0003	.0005
	6									
	7									
	8									
	9									
10	0	.9044	.8171	.7374	.6648	.5987	.5386	.4840	.4344	.3894
	1	.0914	.1667	.2281	.2770	.3151	.3438	.3643	.3777	.3851
	2	.0042	.0153	.0317	.0519	.0746	.0988	.1234	.1478	.1714
	3	.0001	.0008	.0026	.0058	.0105	.0168	.0248	.0343	.0452
	4			.0001	.0004	.0010	.0019	.0033	.0052	.0078
	5					.0001	.0001	.0003	.0005	.0009
	6									.0001
	7									
	8									
	9									
	10									
12	0	.8864	.7847	.6938	.6127	.5404	.4759	.4186	.3677	.3225
	1	.1074	.1922	.2575	.3064	.3413	.3645	.3781	.3837	.3827
	2	.0060	.0216	.0438	.0702	.0988	.1280	.1565	.1835	.2082
	3	.0002	.0015	.0045	.0098	.0173	.0272	.0393	.0532	.0686
	4		.0001	.0003	.0009	.0021	.0039	.0067	.0104	.0153
	5				.0001	.0002	.0004	.0008	.0014	.0024
	6							.0001	.0001	.0003
	7									
	8									
	9									
	10									
	11									
	12									
15	0	.8601	.7386	.6333	.5421	.4633	.3953	.3367	.2863	.2430
	1	.1303	.2261	.2938	.3388	.3658	.3785	.3801	.3734	.3605
	2	.0092	.0323	.0636	.0988	.1348	.1691	.2003	.2273	.2496
	3	.0004	.0029	.0085	.0178	.0307	.0468	.0653	.0857	.1070
	4		.0002	.0008	.0022	.0049	.0090	.0148	.0223	.0317
	5			.0001	.0002	.0006	.0013	.0024	.0043	.0069
	6						.0001	.0003	.0006	.0011
	7								.0001	.0001
	8									
	9									
	10									
	11									
	12									
	13									
	14									
	15									

TABLE C

Binomial probabilities (continued)

Entry is $P(X = k) = \binom{n}{k} p^k (1-p)^{n-k}$

n	k	.10	.15	.20	.25	.30	.35	.40	.45	.50
9	0	.3874	.2316	.1342	.0751	.0404	.0207	.0101	.0046	.0020
	1	.3874	.3679	.3020	.2253	.1556	.1004	.0605	.0339	.0176
	2	.1722	.2597	.3020	.3003	.2668	.2162	.1612	.1110	.0703
	3	.0446	.1069	.1762	.2336	.2668	.2716	.2508	.2119	.1641
	4	.0074	.0283	.0661	.1168	.1715	.2194	.2508	.2600	.2461
	5	.0008	.0050	.0165	.0389	.0735	.1181	.1672	.2128	.2461
	6	.0001	.0006	.0028	.0087	.0210	.0424	.0743	.1160	.1641
	7			.0003	.0012	.0039	.0098	.0212	.0407	.0703
	8				.0001	.0004	.0013	.0035	.0083	.0176
	9						.0001	.0003	.0008	.0020
10	0	.3487	.1969	.1074	.0563	.0282	.0135	.0060	.0025	.0010
	1	.3874	.3474	.2684	.1877	.1211	.0725	.0403	.0207	.0098
	2	.1937	.2759	.3020	.2816	.2335	.1757	.1209	.0763	.0439
	3	.0574	.1298	.2013	.2503	.2668	.2522	.2150	.1665	.1172
	4	.0112	.0401	.0881	.1460	.2001	.2377	.2508	.2384	.2051
	5	.0015	.0085	.0264	.0584	.1029	.1536	.2007	.2340	.2461
	6	.0001	.0012	.0055	.0162	.0368	.0689	.1115	.1596	.2051
	7		.0001	.0008	.0031	.0090	.0212	.0425	.0746	.1172
	8			.0001	.0004	.0014	.0043	.0106	.0229	.0439
	9					.0001	.0005	.0016	.0042	.0098
	10							.0001	.0003	.0010
12	0	.2824	.1422	.0687	.0317	.0138	.0057	.0022	.0008	.0002
	1	.3766	.3012	.2062	.1267	.0712	.0368	.0174	.0075	.0029
	2	.2301	.2924	.2835	.2323	.1678	.1088	.0639	.0339	.0161
	3	.0852	.1720	.2362	.2581	.2397	.1954	.1419	.0923	.0537
	4	.0213	.0683	.1329	.1936	.2311	.2367	.2128	.1700	.1208
	5	.0038	.0193	.0532	.1032	.1585	.2039	.2270	.2225	.1934
	6	.0005	.0040	.0155	.0401	.0792	.1281	.1766	.2124	.2256
	7		.0006	.0033	.0115	.0291	.0591	.1009	.1489	.1934
	8		.0001	.0005	.0024	.0078	.0199	.0420	.0762	.1208
	9			.0001	.0004	.0015	.0048	.0125	.0277	.0537
	10					.0002	.0008	.0025	.0068	.0161
	11						.0001	.0003	.0010	.0029
	12								.0001	.0002
15	0	.2059	.0874	.0352	.0134	.0047	.0016	.0005	.0001	.0000
	1	.3432	.2312	.1319	.0668	.0305	.0126	.0047	.0016	.0005
	2	.2669	.2856	.2309	.1559	.0916	.0476	.0219	.0090	.0032
	3	.1285	.2184	.2501	.2252	.1700	.1110	.0634	.0318	.0139
	4	.0428	.1156	.1876	.2252	.2186	.1792	.1268	.0780	.0417
	5	.0105	.0449	.1032	.1651	.2061	.2123	.1859	.1404	.0916
	6	.0019	.0132	.0430	.0917	.1472	.1906	.2066	.1914	.1527
	7	.0003	.0030	.0138	.0393	.0811	.1319	.1771	.2013	.1964
	8		.0005	.0035	.0131	.0348	.0710	.1181	.1647	.1964
	9		.0001	.0007	.0034	.0116	.0298	.0612	.1048	.1527
	10			.0001	.0007	.0030	.0096	.0245	.0515	.0916
	11				.0001	.0006	.0024	.0074	.0191	.0417
	12					.0001	.0004	.0016	.0052	.0139
	13						.0001	.0003	.0010	.0032
	14								.0001	.0005
	15									

(Continued)

TABLE C

Binomial probabilities (continued)

						p				
n	k	.01	.02	.03	.04	.05	.06	.07	.08	.09
20	0	.8179	.6676	.5438	.4420	.3585	.2901	.2342	.1887	.1516
	1	.1652	.2725	.3364	.3683	.3774	.3703	.3526	.3282	.3000
	2	.0159	.0528	.0988	.1458	.1887	.2246	.2521	.2711	.2818
	3	.0010	.0065	.0183	.0364	.0596	.0860	.1139	.1414	.1672
	4		.0006	.0024	.0065	.0133	.0233	.0364	.0523	.0703
	5			.0002	.0009	.0022	.0048	.0088	.0145	.0222
	6				.0001	.0003	.0008	.0017	.0032	.0055
	7						.0001	.0002	.0005	.0011
	8								.0001	.0002
	9									
	10									
	11									
	12									
	13									
	14									
	15									
	16									
	17									
	18									
	19									
	20									

						p				
n	k	.10	.15	.20	.25	.30	.35	.40	.45	.50
20	0	.1216	.0388	.0115	.0032	.0008	.0002	.0000	.0000	.0000
	1	.2702	.1368	.0576	.0211	.0068	.0020	.0005	.0001	.0000
	2	.2852	.2293	.1369	.0669	.0278	.0100	.0031	.0008	.0002
	3	.1901	.2428	.2054	.1339	.0716	.0323	.0123	.0040	.0011
	4	.0898	.1821	.2182	.1897	.1304	.0738	.0350	.0139	.0046
	5	.0319	.1028	.1746	.2023	.1789	.1272	.0746	.0365	.0148
	6	.0089	.0454	.1091	.1686	.1916	.1712	.1244	.0746	.0370
	7	.0020	.0160	.0545	.1124	.1643	.1844	.1659	.1221	.0739
	8	.0004	.0046	.0222	.0609	.1144	.1614	.1797	.1623	.1201
	9	.0001	.0011	.0074	.0271	.0654	.1158	.1597	.1771	.1602
	10		.0002	.0020	.0099	.0308	.0686	.1171	.1593	.1762
	11			.0005	.0030	.0120	.0336	.0710	.1185	.1602
	12			.0001	.0008	.0039	.0136	.0355	.0727	.1201
	13				.0002	.0010	.0045	.0146	.0366	.0739
	14					.0002	.0012	.0049	.0150	.0370
	15						.0003	.0013	.0049	.0148
	16							.0003	.0013	.0046
	17								.0002	.0011
	18									.0002
	19									
	20									

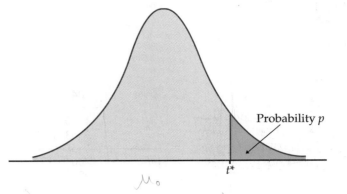

Table entry for p and C is the critical value t^* with probability p lying to its right and probability C lying between $-t^*$ and t^*.

Probability p

TABLE D

t distribution critical values

df	\multicolumn{11}{c}{Upper-tail probability p}
	.25	.20	.15	.10	.05	.025	.02	.01	.005	.0025	.001	.0005
1	1.000	1.376	1.963	3.078	6.314	12.71	15.89	31.82	63.66	127.3	318.3	636.6
2	0.816	1.061	1.386	1.886	2.920	4.303	4.849	6.965	9.925	14.09	22.33	31.60
3	0.765	0.978	1.250	1.638	2.353	3.182	3.482	4.541	5.841	7.453	10.21	12.92
4	0.741	0.941	1.190	1.533	2.132	2.776	2.999	3.747	4.604	5.598	7.173	8.610
5	0.727	0.920	1.156	1.476	2.015	2.571	2.757	3.365	4.032	4.773	5.893	6.869
6	0.718	0.906	1.134	1.440	1.943	2.447	2.612	3.143	3.707	4.317	5.208	5.959
7	0.711	0.896	1.119	1.415	1.895	2.365	2.517	2.998	3.499	4.029	4.785	5.408
8	0.706	0.889	1.108	1.397	1.860	2.306	2.449	2.896	3.355	3.833	4.501	5.041
9	0.703	0.883	1.100	1.383	1.833	2.262	2.398	2.821	3.250	3.690	4.297	4.781
10	0.700	0.879	1.093	1.372	1.812	2.228	2.359	2.764	3.169	3.581	4.144	4.587
11	0.697	0.876	1.088	1.363	1.796	2.201	2.328	2.718	3.106	3.497	4.025	4.437
12	0.695	0.873	1.083	1.356	1.782	2.179	2.303	2.681	3.055	3.428	3.930	4.318
13	0.694	0.870	1.079	1.350	1.771	2.160	2.282	2.650	3.012	3.372	3.852	4.221
14	0.692	0.868	1.076	1.345	1.761	2.145	2.264	2.624	2.977	3.326	3.787	4.140
15	0.691	0.866	1.074	1.341	1.753	2.131	2.249	2.602	2.947	3.286	3.733	4.073
16	0.690	0.865	1.071	1.337	1.746	2.120	2.235	2.583	2.921	3.252	3.686	4.015
17	0.689	0.863	1.069	1.333	1.740	2.110	2.224	2.567	2.898	3.222	3.646	3.965
18	0.688	0.862	1.067	1.330	1.734	2.101	2.214	2.552	2.878	3.197	3.611	3.922
19	0.688	0.861	1.066	1.328	1.729	2.093	2.205	2.539	2.861	3.174	3.579	3.883
20	0.687	0.860	1.064	1.325	1.725	2.086	2.197	2.528	2.845	3.153	3.552	3.850
21	0.686	0.859	1.063	1.323	1.721	2.080	2.189	2.518	2.831	3.135	3.527	3.819
22	0.686	0.858	1.061	1.321	1.717	2.074	2.183	2.508	2.819	3.119	3.505	3.792
23	0.685	0.858	1.060	1.319	1.714	2.069	2.177	2.500	2.807	3.104	3.485	3.768
24	0.685	0.857	1.059	1.318	1.711	2.064	2.172	2.492	2.797	3.091	3.467	3.745
25	0.684	0.856	1.058	1.316	1.708	2.060	2.167	2.485	2.787	3.078	3.450	3.725
26	0.684	0.856	1.058	1.315	1.706	2.056	2.162	2.479	2.779	3.067	3.435	3.707
27	0.684	0.855	1.057	1.314	1.703	2.052	2.158	2.473	2.771	3.057	3.421	3.690
28	0.683	0.855	1.056	1.313	1.701	2.048	2.154	2.467	2.763	3.047	3.408	3.674
29	0.683	0.854	1.055	1.311	1.699	2.045	2.150	2.462	2.756	3.038	3.396	3.659
30	0.683	0.854	1.055	1.310	1.697	2.042	2.147	2.457	2.750	3.030	3.385	3.646
40	0.681	0.851	1.050	1.303	1.684	2.021	2.123	2.423	2.704	2.971	3.307	3.551
50	0.679	0.849	1.047	1.299	1.676	2.009	2.109	2.403	2.678	2.937	3.261	3.496
60	0.679	0.848	1.045	1.296	1.671	2.000	2.099	2.390	2.660	2.915	3.232	3.460
80	0.678	0.846	1.043	1.292	1.664	1.990	2.088	2.374	2.639	2.887	3.195	3.416
100	0.677	0.845	1.042	1.290	1.660	1.984	2.081	2.364	2.626	2.871	3.174	3.390
1000	0.675	0.842	1.037	1.282	1.646	1.962	2.056	2.330	2.581	2.813	3.098	3.300
z^*	0.674	0.841	1.036	1.282	1.645	1.960	2.054	2.326	2.576	2.807	3.091	3.291
	50%	60%	70%	80%	90%	95%	96%	98%	99%	99.5%	99.8%	99.9%
	\multicolumn{11}{c}{Confidence level C}

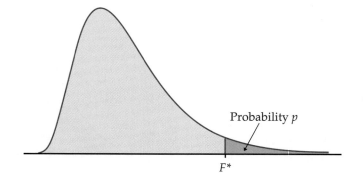

Table entry for *p* is the
critical value F^* with
probability *p* lying to
its right.

Probability *p*

F^*

TABLE E

F critical values

			Degrees of freedom in the numerator							
	p	1	2	3	4	5	6	7	8	9
	.100	39.86	49.50	53.59	55.83	57.24	58.20	58.91	59.44	59.86
	.050	161.45	199.50	215.71	224.58	230.16	233.99	236.77	238.88	240.54
1	.025	647.79	799.50	864.16	899.58	921.85	937.11	948.22	956.66	963.28
	.010	4052.2	4999.5	5403.4	5624.6	5763.6	5859.0	5928.4	5981.1	6022.5
	.001	405284	500000	540379	562500	576405	585937	592873	598144	602284
	.100	8.53	9.00	9.16	9.24	9.29	9.33	9.35	9.37	9.38
	.050	18.51	19.00	19.16	19.25	19.30	19.33	19.35	19.37	19.38
2	.025	38.51	39.00	39.17	39.25	39.30	39.33	39.36	39.37	39.39
	.010	98.50	99.00	99.17	99.25	99.30	99.33	99.36	99.37	99.39
	.001	998.50	999.00	999.17	999.25	999.30	999.33	999.36	999.37	999.39
	.100	5.54	5.46	5.39	5.34	5.31	5.28	5.27	5.25	5.24
	.050	10.13	9.55	9.28	9.12	9.01	8.94	8.89	8.85	8.81
3	.025	17.44	16.04	15.44	15.10	14.88	14.73	14.62	14.54	14.47
	.010	34.12	30.82	29.46	28.71	28.24	27.91	27.67	27.49	27.35
	.001	167.03	148.50	141.11	137.10	134.58	132.85	131.58	130.62	129.86
	.100	4.54	4.32	4.19	4.11	4.05	4.01	3.98	3.95	3.94
	.050	7.71	6.94	6.59	6.39	6.26	6.16	6.09	6.04	6.00
4	.025	12.22	10.65	9.98	9.60	9.36	9.20	9.07	8.98	8.90
	.010	21.20	18.00	16.69	15.98	15.52	15.21	14.98	14.80	14.66
	.001	74.14	61.25	56.18	53.44	51.71	50.53	49.66	49.00	48.47
	.100	4.06	3.78	3.62	3.52	3.45	3.40	3.37	3.34	3.32
	.050	6.61	5.79	5.41	5.19	5.05	4.95	4.88	4.82	4.77
5	.025	10.01	8.43	7.76	7.39	7.15	6.98	6.85	6.76	6.68
	.010	16.26	13.27	12.06	11.39	10.97	10.67	10.46	10.29	10.16
	.001	47.18	37.12	33.20	31.09	29.75	28.83	28.16	27.65	27.24
	.100	3.78	3.46	3.29	3.18	3.11	3.05	3.01	2.98	2.96
	.050	5.99	5.14	4.76	4.53	4.39	4.28	4.21	4.15	4.10
6	.025	8.81	7.26	6.60	6.23	5.99	5.82	5.70	5.60	5.52
	.010	13.75	10.92	9.78	9.15	8.75	8.47	8.26	8.10	7.98
	.001	35.51	27.00	23.70	21.92	20.80	20.03	19.46	19.03	18.69
	.100	3.59	3.26	3.07	2.96	2.88	2.83	2.78	2.75	2.72
	.050	5.59	4.74	4.35	4.12	3.97	3.87	3.79	3.73	3.68
7	.025	8.07	6.54	5.89	5.52	5.29	5.12	4.99	4.90	4.82
	.010	12.25	9.55	8.45	7.85	7.46	7.19	6.99	6.84	6.72
	.001	29.25	21.69	18.77	17.20	16.21	15.52	15.02	14.63	14.33

Degrees of freedom in the denominator

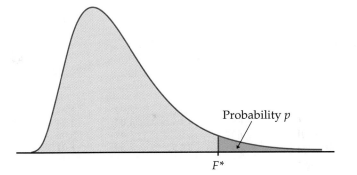

Table entry for *p* is the critical value *F** with probability *p* lying to its right.

Probability *p*

F*

TABLE E

F critical values (continued)

				Degrees of freedom in the numerator						
10	12	15	20	25	30	40	50	60	120	1000
60.19	60.71	61.22	61.74	62.05	62.26	62.53	62.69	62.79	63.06	63.30
241.88	243.91	245.95	248.01	249.26	250.10	251.14	251.77	252.20	253.25	254.19
968.63	976.71	984.87	993.10	998.08	1001.4	1005.6	1008.1	1009.8	1014.0	1017.7
6055.8	6106.3	6157.3	6208.7	6239.8	6260.6	6286.8	6302.5	6313.0	6339.4	6362.7
605621	610668	615764	620908	624017	626099	628712	630285	631337	633972	636301
9.39	9.41	9.42	9.44	9.45	9.46	9.47	9.47	9.47	9.48	9.49
19.40	19.41	19.43	19.45	19.46	19.46	19.47	19.48	19.48	19.49	19.49
39.40	39.41	39.43	39.45	39.46	39.46	39.47	39.48	39.48	39.49	39.50
99.40	99.42	99.43	99.45	99.46	99.47	99.47	99.48	99.48	99.49	99.50
999.40	999.42	999.43	999.45	999.46	999.47	999.47	999.48	999.48	999.49	999.50
5.23	5.22	5.20	5.18	5.17	5.17	5.16	5.15	5.15	5.14	5.13
8.79	8.74	8.70	8.66	8.63	8.62	8.59	8.58	8.57	8.55	8.53
14.42	14.34	14.25	14.17	14.12	14.08	14.04	14.01	13.99	13.95	13.91
27.23	27.05	26.87	26.69	26.58	26.50	26.41	26.35	26.32	26.22	26.14
129.25	128.32	127.37	126.42	125.84	125.45	124.96	124.66	124.47	123.97	123.53
3.92	3.90	3.87	3.84	3.83	3.82	3.80	3.80	3.79	3.78	3.76
5.96	5.91	5.86	5.80	5.77	5.75	5.72	5.70	5.69	5.66	5.63
8.84	8.75	8.66	8.56	8.50	8.46	8.41	8.38	8.36	8.31	8.26
14.55	14.37	14.20	14.02	13.91	13.84	13.75	13.69	13.65	13.56	13.47
48.05	47.41	46.76	46.10	45.70	45.43	45.09	44.88	44.75	44.40	44.09
3.30	3.27	3.24	3.21	3.19	3.17	3.16	3.15	3.14	3.12	3.11
4.74	4.68	4.62	4.56	4.52	4.50	4.46	4.44	4.43	4.40	4.37
6.62	6.52	6.43	6.33	6.27	6.23	6.18	6.14	6.12	6.07	6.02
10.05	9.89	9.72	9.55	9.45	9.38	9.29	9.24	9.20	9.11	9.03
26.92	26.42	25.91	25.39	25.08	24.87	24.60	24.44	24.33	24.06	23.82
2.94	2.90	2.87	2.84	2.81	2.80	2.78	2.77	2.76	2.74	2.72
4.06	4.00	3.94	3.87	3.83	3.81	3.77	3.75	3.74	3.70	3.67
5.46	5.37	5.27	5.17	5.11	5.07	5.01	4.98	4.96	4.90	4.86
7.87	7.72	7.56	7.40	7.30	7.23	7.14	7.09	7.06	6.97	6.89
18.41	17.99	17.56	17.12	16.85	16.67	16.44	16.31	16.21	15.98	15.77
2.70	2.67	2.63	2.59	2.57	2.56	2.54	2.52	2.51	2.49	2.47
3.64	3.57	3.51	3.44	3.40	3.38	3.34	3.32	3.30	3.27	3.23
4.76	4.67	4.57	4.47	4.40	4.36	4.31	4.28	4.25	4.20	4.15
6.62	6.47	6.31	6.16	6.06	5.99	5.91	5.86	5.82	5.74	5.66
14.08	13.71	13.32	12.93	12.69	12.53	12.33	12.20	12.12	11.91	11.72

(Continued)

TABLE E

F critical values (continued)

					Degrees of freedom in the numerator				
	p	1	2	3	4	5	6	7	8	9
8	.100	3.46	3.11	2.92	2.81	2.73	2.67	2.62	2.59	2.56
	.050	5.32	4.46	4.07	3.84	3.69	3.58	3.50	3.44	3.39
	.025	7.57	6.06	5.42	5.05	4.82	4.65	4.53	4.43	4.36
	.010	11.26	8.65	7.59	7.01	6.63	6.37	6.18	6.03	5.91
	.001	25.41	18.49	15.83	14.39	13.48	12.86	12.40	12.05	11.77
9	.100	3.36	3.01	2.81	2.69	2.61	2.55	2.51	2.47	2.44
	.050	5.12	4.26	3.86	3.63	3.48	3.37	3.29	3.23	3.18
	.025	7.21	5.71	5.08	4.72	4.48	4.32	4.20	4.10	4.03
	.010	10.56	8.02	6.99	6.42	6.06	5.80	5.61	5.47	5.35
	.001	22.86	16.39	13.90	12.56	11.71	11.13	10.70	10.37	10.11
10	.100	3.29	2.92	2.73	2.61	2.52	2.46	2.41	2.38	2.35
	.050	4.96	4.10	3.71	3.48	3.33	3.22	3.14	3.07	3.02
	.025	6.94	5.46	4.83	4.47	4.24	4.07	3.95	3.85	3.78
	.010	10.04	7.56	6.55	5.99	5.64	5.39	5.20	5.06	4.94
	.001	21.04	14.91	12.55	11.28	10.48	9.93	9.52	9.20	8.96
11	.100	3.23	2.86	2.66	2.54	2.45	2.39	2.34	2.30	2.27
	.050	4.84	3.98	3.59	3.36	3.20	3.09	3.01	2.95	2.90
	.025	6.72	5.26	4.63	4.28	4.04	3.88	3.76	3.66	3.59
	.010	9.65	7.21	6.22	5.67	5.32	5.07	4.89	4.74	4.63
	.001	19.69	13.81	11.56	10.35	9.58	9.05	8.66	8.35	8.12
12	.100	3.18	2.81	2.61	2.48	2.39	2.33	2.28	2.24	2.21
	.050	4.75	3.89	3.49	3.26	3.11	3.00	2.91	2.85	2.80
	.025	6.55	5.10	4.47	4.12	3.89	3.73	3.61	3.51	3.44
	.010	9.33	6.93	5.95	5.41	5.06	4.82	4.64	4.50	4.39
	.001	18.64	12.97	10.80	9.63	8.89	8.38	8.00	7.71	7.48
13	.100	3.14	2.76	2.56	2.43	2.35	2.28	2.23	2.20	2.16
	.050	4.67	3.81	3.41	3.18	3.03	2.92	2.83	2.77	2.71
	.025	6.41	4.97	4.35	4.00	3.77	3.60	3.48	3.39	3.31
	.010	9.07	6.70	5.74	5.21	4.86	4.62	4.44	4.30	4.19
	.001	17.82	12.31	10.21	9.07	8.35	7.86	7.49	7.21	6.98
14	.100	3.10	2.73	2.52	2.39	2.31	2.24	2.19	2.15	2.12
	.050	4.60	3.74	3.34	3.11	2.96	2.85	2.76	2.70	2.65
	.025	6.30	4.86	4.24	3.89	3.66	3.50	3.38	3.29	3.21
	.010	8.86	6.51	5.56	5.04	4.69	4.46	4.28	4.14	4.03
	.001	17.14	11.78	9.73	8.62	7.92	7.44	7.08	6.80	6.58
15	.100	3.07	2.70	2.49	2.36	2.27	2.21	2.16	2.12	2.09
	.050	4.54	3.68	3.29	3.06	2.90	2.79	2.71	2.64	2.59
	.025	6.20	4.77	4.15	3.80	3.58	3.41	3.29	3.20	3.12
	.010	8.68	6.36	5.42	4.89	4.56	4.32	4.14	4.00	3.89
	.001	16.59	11.34	9.34	8.25	7.57	7.09	6.74	6.47	6.26
16	.100	3.05	2.67	2.46	2.33	2.24	2.18	2.13	2.09	2.06
	.050	4.49	3.63	3.24	3.01	2.85	2.74	2.66	2.59	2.54
	.025	6.12	4.69	4.08	3.73	3.50	3.34	3.22	3.12	3.05
	.010	8.53	6.23	5.29	4.77	4.44	4.20	4.03	3.89	3.78
	.001	16.12	10.97	9.01	7.94	7.27	6.80	6.46	6.19	5.98
17	.100	3.03	2.64	2.44	2.31	2.22	2.15	2.10	2.06	2.03
	.050	4.45	3.59	3.20	2.96	2.81	2.70	2.61	2.55	2.49
	.025	6.04	4.62	4.01	3.66	3.44	3.28	3.16	3.06	2.98
	.010	8.40	6.11	5.19	4.67	4.34	4.10	3.93	3.79	3.68
	.001	15.72	10.66	8.73	7.68	7.02	6.56	6.22	5.96	5.75

Degrees of freedom in the denominator

TABLE E

F critical values (continued)

			Degrees of freedom in the numerator							
10	12	15	20	25	30	40	50	60	120	1000
2.54	2.50	2.46	2.42	2.40	2.38	2.36	2.35	2.34	2.32	2.30
3.35	3.28	3.22	3.15	3.11	3.08	3.04	3.02	3.01	2.97	2.93
4.30	4.20	4.10	4.00	3.94	3.89	3.84	3.81	3.78	3.73	3.68
5.81	5.67	5.52	5.36	5.26	5.20	5.12	5.07	5.03	4.95	4.87
11.54	11.19	10.84	10.48	10.26	10.11	9.92	9.80	9.73	9.53	9.36
2.42	2.38	2.34	2.30	2.27	2.25	2.23	2.22	2.21	2.18	2.16
3.14	3.07	3.01	2.94	2.89	2.86	2.83	2.80	2.79	2.75	2.71
3.96	3.87	3.77	3.67	3.60	3.56	3.51	3.47	3.45	3.39	3.34
5.26	5.11	4.96	4.81	4.71	4.65	4.57	4.52	4.48	4.40	4.32
9.89	9.57	9.24	8.90	8.69	8.55	8.37	8.26	8.19	8.00	7.84
2.32	2.28	2.24	2.20	2.17	2.16	2.13	2.12	2.11	2.08	2.06
2.98	2.91	2.85	2.77	2.73	2.70	2.66	2.64	2.62	2.58	2.54
3.72	3.62	3.52	3.42	3.35	3.31	3.26	3.22	3.20	3.14	3.09
4.85	4.71	4.56	4.41	4.31	4.25	4.17	4.12	4.08	4.00	3.92
8.75	8.45	8.13	7.80	7.60	7.47	7.30	7.19	7.12	6.94	6.78
2.25	2.21	2.17	2.12	2.10	2.08	2.05	2.04	2.03	2.00	1.98
2.85	2.79	2.72	2.65	2.60	2.57	2.53	2.51	2.49	2.45	2.41
3.53	3.43	3.33	3.23	3.16	3.12	3.06	3.03	3.00	2.94	2.89
4.54	4.40	4.25	4.10	4.01	3.94	3.86	3.81	3.78	3.69	3.61
7.92	7.63	7.32	7.01	6.81	6.68	6.52	6.42	6.35	6.18	6.02
2.19	2.15	2.10	2.06	2.03	2.01	1.99	1.97	1.96	1.93	1.91
2.75	2.69	2.62	2.54	2.50	2.47	2.43	2.40	2.38	2.34	2.30
3.37	3.28	3.18	3.07	3.01	2.96	2.91	2.87	2.85	2.79	2.73
4.30	4.16	4.01	3.86	3.76	3.70	3.62	3.57	3.54	3.45	3.37
7.29	7.00	6.71	6.40	6.22	6.09	5.93	5.83	5.76	5.59	5.44
2.14	2.10	2.05	2.01	1.98	1.96	1.93	1.92	1.90	1.88	1.85
2.67	2.60	2.53	2.46	2.41	2.38	2.34	2.31	2.30	2.25	2.21
3.25	3.15	3.05	2.95	2.88	2.84	2.78	2.74	2.72	2.66	2.60
4.10	3.96	3.82	3.66	3.57	3.51	3.43	3.38	3.34	3.25	3.18
6.80	6.52	6.23	5.93	5.75	5.63	5.47	5.37	5.30	5.14	4.99
2.10	2.05	2.01	1.96	1.93	1.91	1.89	1.87	1.86	1.83	1.80
2.60	2.53	2.46	2.39	2.34	2.31	2.27	2.24	2.22	2.18	2.14
3.15	3.05	2.95	2.84	2.78	2.73	2.67	2.64	2.61	2.55	2.50
3.94	3.80	3.66	3.51	3.41	3.35	3.27	3.22	3.18	3.09	3.02
6.40	6.13	5.85	5.56	5.38	5.25	5.10	5.00	4.94	4.77	4.62
2.06	2.02	1.97	1.92	1.89	1.87	1.85	1.83	1.82	1.79	1.76
2.54	2.48	2.40	2.33	2.28	2.25	2.20	2.18	2.16	2.11	2.07
3.06	2.96	2.86	2.76	2.69	2.64	2.59	2.55	2.52	2.46	2.40
3.80	3.67	3.52	3.37	3.28	3.21	3.13	3.08	3.05	2.96	2.88
6.08	5.81	5.54	5.25	5.07	4.95	4.80	4.70	4.64	4.47	4.33
2.03	1.99	1.94	1.89	1.86	1.84	1.81	1.79	1.78	1.75	1.72
2.49	2.42	2.35	2.28	2.23	2.19	2.15	2.12	2.11	2.06	2.02
2.99	2.89	2.79	2.68	2.61	2.57	2.51	2.47	2.45	2.38	2.32
3.69	3.55	3.41	3.26	3.16	3.10	3.02	2.97	2.93	2.84	2.76
5.81	5.55	5.27	4.99	4.82	4.70	4.54	4.45	4.39	4.23	4.08
2.00	1.96	1.91	1.86	1.83	1.81	1.78	1.76	1.75	1.72	1.69
2.45	2.38	2.31	2.23	2.18	2.15	2.10	2.08	2.06	2.01	1.97
2.92	2.82	2.72	2.62	2.55	2.50	2.44	2.41	2.38	2.32	2.26
3.59	3.46	3.31	3.16	3.07	3.00	2.92	2.87	2.83	2.75	2.66
5.58	5.32	5.05	4.78	4.60	4.48	4.33	4.24	4.18	4.02	3.87

(Continued)

TABLE E

F critical values (continued)

				Degrees of freedom in the numerator						
	p	1	2	3	4	5	6	7	8	9
18	.100	3.01	2.62	2.42	2.29	2.20	2.13	2.08	2.04	2.00
	.050	4.41	3.55	3.16	2.93	2.77	2.66	2.58	2.51	2.46
	.025	5.98	4.56	3.95	3.61	3.38	3.22	3.10	3.01	2.93
	.010	8.29	6.01	5.09	4.58	4.25	4.01	3.84	3.71	3.60
	.001	15.38	10.39	8.49	7.46	6.81	6.35	6.02	5.76	5.56
19	.100	2.99	2.61	2.40	2.27	2.18	2.11	2.06	2.02	1.98
	.050	4.38	3.52	3.13	2.90	2.74	2.63	2.54	2.48	2.42
	.025	5.92	4.51	3.90	3.56	3.33	3.17	3.05	2.96	2.88
	.010	8.18	5.93	5.01	4.50	4.17	3.94	3.77	3.63	3.52
	.001	15.08	10.16	8.28	7.27	6.62	6.18	5.85	5.59	5.39
20	.100	2.97	2.59	2.38	2.25	2.16	2.09	2.04	2.00	1.96
	.050	4.35	3.49	3.10	2.87	2.71	2.60	2.51	2.45	2.39
	.025	5.87	4.46	3.86	3.51	3.29	3.13	3.01	2.91	2.84
	.010	8.10	5.85	4.94	4.43	4.10	3.87	3.70	3.56	3.46
	.001	14.82	9.95	8.10	7.10	6.46	6.02	5.69	5.44	5.24
21	.100	2.96	2.57	2.36	2.23	2.14	2.08	2.02	1.98	1.95
	.050	4.32	3.47	3.07	2.84	2.68	2.57	2.49	2.42	2.37
	.025	5.83	4.42	3.82	3.48	3.25	3.09	2.97	2.87	2.80
	.010	8.02	5.78	4.87	4.37	4.04	3.81	3.64	3.51	3.40
	.001	14.59	9.77	7.94	6.95	6.32	5.88	5.56	5.31	5.11
22	.100	2.95	2.56	2.35	2.22	2.13	2.06	2.01	1.97	1.93
	.050	4.30	3.44	3.05	2.82	2.66	2.55	2.46	2.40	2.34
	.025	5.79	4.38	3.78	3.44	3.22	3.05	2.93	2.84	2.76
	.010	7.95	5.72	4.82	4.31	3.99	3.76	3.59	3.45	3.35
	.001	14.38	9.61	7.80	6.81	6.19	5.76	5.44	5.19	4.99
23	.100	2.94	2.55	2.34	2.21	2.11	2.05	1.99	1.95	1.92
	.050	4.28	3.42	3.03	2.80	2.64	2.53	2.44	2.37	2.32
	.025	5.75	4.35	3.75	3.41	3.18	3.02	2.90	2.81	2.73
	.010	7.88	5.66	4.76	4.26	3.94	3.71	3.54	3.41	3.30
	.001	14.20	9.47	7.67	6.70	6.08	5.65	5.33	5.09	4.89
24	.100	2.93	2.54	2.33	2.19	2.10	2.04	1.98	1.94	1.91
	.050	4.26	3.40	3.01	2.78	2.62	2.51	2.42	2.36	2.30
	.025	5.72	4.32	3.72	3.38	3.15	2.99	2.87	2.78	2.70
	.010	7.82	5.61	4.72	4.22	3.90	3.67	3.50	3.36	3.26
	.001	14.03	9.34	7.55	6.59	5.98	5.55	5.23	4.99	4.80
25	.100	2.92	2.53	2.32	2.18	2.09	2.02	1.97	1.93	1.89
	.050	4.24	3.39	2.99	2.76	2.60	2.49	2.40	2.34	2.28
	.025	5.69	4.29	3.69	3.35	3.13	2.97	2.85	2.75	2.68
	.010	7.77	5.57	4.68	4.18	3.85	3.63	3.46	3.32	3.22
	.001	13.88	9.22	7.45	6.49	5.89	5.46	5.15	4.91	4.71
26	.100	2.91	2.52	2.31	2.17	2.08	2.01	1.96	1.92	1.88
	.050	4.23	3.37	2.98	2.74	2.59	2.47	2.39	2.32	2.27
	.025	5.66	4.27	3.67	3.33	3.10	2.94	2.82	2.73	2.65
	.010	7.72	5.53	4.64	4.14	3.82	3.59	3.42	3.29	3.18
	.001	13.74	9.12	7.36	6.41	5.80	5.38	5.07	4.83	4.64
27	.100	2.90	2.51	2.30	2.17	2.07	2.00	1.95	1.91	1.87
	.050	4.21	3.35	2.96	2.73	2.57	2.46	2.37	2.31	2.25
	.025	5.63	4.24	3.65	3.31	3.08	2.92	2.80	2.71	2.63
	.010	7.68	5.49	4.60	4.11	3.78	3.56	3.39	3.26	3.15
	.001	13.61	9.02	7.27	6.33	5.73	5.31	5.00	4.76	4.57

Degrees of freedom in the denominator

TABLE E

F critical values (continued)

			Degrees of freedom in the numerator							
10	12	15	20	25	30	40	50	60	120	1000
1.98	1.93	1.89	1.84	1.80	1.78	1.75	1.74	1.72	1.69	1.66
2.41	2.34	2.27	2.19	2.14	2.11	2.06	2.04	2.02	1.97	1.92
2.87	2.77	2.67	2.56	2.49	2.44	2.38	2.35	2.32	2.26	2.20
3.51	3.37	3.23	3.08	2.98	2.92	2.84	2.78	2.75	2.66	2.58
5.39	5.13	4.87	4.59	4.42	4.30	4.15	4.06	4.00	3.84	3.69
1.96	1.91	1.86	1.81	1.78	1.76	1.73	1.71	1.70	1.67	1.64
2.38	2.31	2.23	2.16	2.11	2.07	2.03	2.00	1.98	1.93	1.88
2.82	2.72	2.62	2.51	2.44	2.39	2.33	2.30	2.27	2.20	2.14
3.43	3.30	3.15	3.00	2.91	2.84	2.76	2.71	2.67	2.58	2.50
5.22	4.97	4.70	4.43	4.26	4.14	3.99	3.90	3.84	3.68	3.53
1.94	1.89	1.84	1.79	1.76	1.74	1.71	1.69	1.68	1.64	1.61
2.35	2.28	2.20	2.12	2.07	2.04	1.99	1.97	1.95	1.90	1.85
2.77	2.68	2.57	2.46	2.40	2.35	2.29	2.25	2.22	2.16	2.09
3.37	3.23	3.09	2.94	2.84	2.78	2.69	2.64	2.61	2.52	2.43
5.08	4.82	4.56	4.29	4.12	4.00	3.86	3.77	3.70	3.54	3.40
1.92	1.87	1.83	1.78	1.74	1.72	1.69	1.67	1.66	1.62	1.59
2.32	2.25	2.18	2.10	2.05	2.01	1.96	1.94	1.92	1.87	1.82
2.73	2.64	2.53	2.42	2.36	2.31	2.25	2.21	2.18	2.11	2.05
3.31	3.17	3.03	2.88	2.79	2.72	2.64	2.58	2.55	2.46	2.37
4.95	4.70	4.44	4.17	4.00	3.88	3.74	3.64	3.58	3.42	3.28
1.90	1.86	1.81	1.76	1.73	1.70	1.67	1.65	1.64	1.60	1.57
2.30	2.23	2.15	2.07	2.02	1.98	1.94	1.91	1.89	1.84	1.79
2.70	2.60	2.50	2.39	2.32	2.27	2.21	2.17	2.14	2.08	2.01
3.26	3.12	2.98	2.83	2.73	2.67	2.58	2.53	2.50	2.40	2.32
4.83	4.58	4.33	4.06	3.89	3.78	3.63	3.54	3.48	3.32	3.17
1.89	1.84	1.80	1.74	1.71	1.69	1.66	1.64	1.62	1.59	1.55
2.27	2.20	2.13	2.05	2.00	1.96	1.91	1.88	1.86	1.81	1.76
2.67	2.57	2.47	2.36	2.29	2.24	2.18	2.14	2.11	2.04	1.98
3.21	3.07	2.93	2.78	2.69	2.62	2.54	2.48	2.45	2.35	2.27
4.73	4.48	4.23	3.96	3.79	3.68	3.53	3.44	3.38	3.22	3.08
1.88	1.83	1.78	1.73	1.70	1.67	1.64	1.62	1.61	1.57	1.54
2.25	2.18	2.11	2.03	1.97	1.94	1.89	1.86	1.84	1.79	1.74
2.64	2.54	2.44	2.33	2.26	2.21	2.15	2.11	2.08	2.01	1.94
3.17	3.03	2.89	2.74	2.64	2.58	2.49	2.44	2.40	2.31	2.22
4.64	4.39	4.14	3.87	3.71	3.59	3.45	3.36	3.29	3.14	2.99
1.87	1.82	1.77	1.72	1.68	1.66	1.63	1.61	1.59	1.56	1.52
2.24	2.16	2.09	2.01	1.96	1.92	1.87	1.84	1.82	1.77	1.72
2.61	2.51	2.41	2.30	2.23	2.18	2.12	2.08	2.05	1.98	1.91
3.13	2.99	2.85	2.70	2.60	2.54	2.45	2.40	2.36	2.27	2.18
4.56	4.31	4.06	3.79	3.63	3.52	3.37	3.28	3.22	3.06	2.91
1.86	1.81	1.76	1.71	1.67	1.65	1.61	1.59	1.58	1.54	1.51
2.22	2.15	2.07	1.99	1.94	1.90	1.85	1.82	1.80	1.75	1.70
2.59	2.49	2.39	2.28	2.21	2.16	2.09	2.05	2.03	1.95	1.89
3.09	2.96	2.81	2.66	2.57	2.50	2.42	2.36	2.33	2.23	2.14
4.48	4.24	3.99	3.72	3.56	3.44	3.30	3.21	3.15	2.99	2.84
1.85	1.80	1.75	1.70	1.66	1.64	1.60	1.58	1.57	1.53	1.50
2.20	2.13	2.06	1.97	1.92	1.88	1.84	1.81	1.79	1.73	1.68
2.57	2.47	2.36	2.25	2.18	2.13	2.07	2.03	2.00	1.93	1.86
3.06	2.93	2.78	2.63	2.54	2.47	2.38	2.33	2.29	2.20	2.11
4.41	4.17	3.92	3.66	3.49	3.38	3.23	3.14	3.08	2.92	2.78

(*Continued*)

TABLE E

F critical values (continued)

	p	\multicolumn{9}{c}{Degrees of freedom in the numerator}
		1	2	3	4	5	6	7	8	9
28	.100	2.89	2.50	2.29	2.16	2.06	2.00	1.94	1.90	1.87
	.050	4.20	3.34	2.95	2.71	2.56	2.45	2.36	2.29	2.24
	.025	5.61	4.22	3.63	3.29	3.06	2.90	2.78	2.69	2.61
	.010	7.64	5.45	4.57	4.07	3.75	3.53	3.36	3.23	3.12
	.001	13.50	8.93	7.19	6.25	5.66	5.24	4.93	4.69	4.50
29	.100	2.89	2.50	2.28	2.15	2.06	1.99	1.93	1.89	1.86
	.050	4.18	3.33	2.93	2.70	2.55	2.43	2.35	2.28	2.22
	.025	5.59	4.20	3.61	3.27	3.04	2.88	2.76	2.67	2.59
	.010	7.60	5.42	4.54	4.04	3.73	3.50	3.33	3.20	3.09
	.001	13.39	8.85	7.12	6.19	5.59	5.18	4.87	4.64	4.45
30	.100	2.88	2.49	2.28	2.14	2.05	1.98	1.93	1.88	1.85
	.050	4.17	3.32	2.92	2.69	2.53	2.42	2.33	2.27	2.21
	.025	5.57	4.18	3.59	3.25	3.03	2.87	2.75	2.65	2.57
	.010	7.56	5.39	4.51	4.02	3.70	3.47	3.30	3.17	3.07
	.001	13.29	8.77	7.05	6.12	5.53	5.12	4.82	4.58	4.39
40	.100	2.84	2.44	2.23	2.09	2.00	1.93	1.87	1.83	1.79
	.050	4.08	3.23	2.84	2.61	2.45	2.34	2.25	2.18	2.12
	.025	5.42	4.05	3.46	3.13	2.90	2.74	2.62	2.53	2.45
	.010	7.31	5.18	4.31	3.83	3.51	3.29	3.12	2.99	2.89
	.001	12.61	8.25	6.59	5.70	5.13	4.73	4.44	4.21	4.02
50	.100	2.81	2.41	2.20	2.06	1.97	1.90	1.84	1.80	1.76
	.050	4.03	3.18	2.79	2.56	2.40	2.29	2.20	2.13	2.07
	.025	5.34	3.97	3.39	3.05	2.83	2.67	2.55	2.46	2.38
	.010	7.17	5.06	4.20	3.72	3.41	3.19	3.02	2.89	2.78
	.001	12.22	7.96	6.34	5.46	4.90	4.51	4.22	4.00	3.82
60	.100	2.79	2.39	2.18	2.04	1.95	1.87	1.82	1.77	1.74
	.050	4.00	3.15	2.76	2.53	2.37	2.25	2.17	2.10	2.04
	.025	5.29	3.93	3.34	3.01	2.79	2.63	2.51	2.41	2.33
	.010	7.08	4.98	4.13	3.65	3.34	3.12	2.95	2.82	2.72
	.001	11.97	7.77	6.17	5.31	4.76	4.37	4.09	3.86	3.69
100	.100	2.76	2.36	2.14	2.00	1.91	1.83	1.78	1.73	1.69
	.050	3.94	3.09	2.70	2.46	2.31	2.19	2.10	2.03	1.97
	.025	5.18	3.83	3.25	2.92	2.70	2.54	2.42	2.32	2.24
	.010	6.90	4.82	3.98	3.51	3.21	2.99	2.82	2.69	2.59
	.001	11.50	7.41	5.86	5.02	4.48	4.11	3.83	3.61	3.44
200	.100	2.73	2.33	2.11	1.97	1.88	1.80	1.75	1.70	1.66
	.050	3.89	3.04	2.65	2.42	2.26	2.14	2.06	1.98	1.93
	.025	5.10	3.76	3.18	2.85	2.63	2.47	2.35	2.26	2.18
	.010	6.76	4.71	3.88	3.41	3.11	2.89	2.73	2.60	2.50
	.001	11.15	7.15	5.63	4.81	4.29	3.92	3.65	3.43	3.26
1000	.100	2.71	2.31	2.09	1.95	1.85	1.78	1.72	1.68	1.64
	.050	3.85	3.00	2.61	2.38	2.22	2.11	2.02	1.95	1.89
	.025	5.04	3.70	3.13	2.80	2.58	2.42	2.30	2.20	2.13
	.010	6.66	4.63	3.80	3.34	3.04	2.82	2.66	2.53	2.43
	.001	10.89	6.96	5.46	4.65	4.14	3.78	3.51	3.30	3.13

Degrees of freedom in the denominator

TABLE E

F critical values (continued)

Degrees of freedom in the numerator										
10	12	15	20	25	30	40	50	60	120	1000
1.84	1.79	1.74	1.69	1.65	1.63	1.59	1.57	1.56	1.52	1.48
2.19	2.12	2.04	1.96	1.91	1.87	1.82	1.79	1.77	1.71	1.66
2.55	2.45	2.34	2.23	2.16	2.11	2.05	2.01	1.98	1.91	1.84
3.03	2.90	2.75	2.60	2.51	2.44	2.35	2.30	2.26	2.17	2.08
4.35	4.11	3.86	3.60	3.43	3.32	3.18	3.09	3.02	2.86	2.72
1.83	1.78	1.73	1.68	1.64	1.62	1.58	1.56	1.55	1.51	1.47
2.18	2.10	2.03	1.94	1.89	1.85	1.81	1.77	1.75	1.70	1.65
2.53	2.43	2.32	2.21	2.14	2.09	2.03	1.99	1.96	1.89	1.82
3.00	2.87	2.73	2.57	2.48	2.41	2.33	2.27	2.23	2.14	2.05
4.29	4.05	3.80	3.54	3.38	3.27	3.12	3.03	2.97	2.81	2.66
1.82	1.77	1.72	1.67	1.63	1.61	1.57	1.55	1.54	1.50	1.46
2.16	2.09	2.01	1.93	1.88	1.84	1.79	1.76	1.74	1.68	1.63
2.51	2.41	2.31	2.20	2.12	2.07	2.01	1.97	1.94	1.87	1.80
2.98	2.84	2.70	2.55	2.45	2.39	2.30	2.25	2.21	2.11	2.02
4.24	4.00	3.75	3.49	3.33	3.22	3.07	2.98	2.92	2.76	2.61
1.76	1.71	1.66	1.61	1.57	1.54	1.51	1.48	1.47	1.42	1.38
2.08	2.00	1.92	1.84	1.78	1.74	1.69	1.66	1.64	1.58	1.52
2.39	2.29	2.18	2.07	1.99	1.94	1.88	1.83	1.80	1.72	1.65
2.80	2.66	2.52	2.37	2.27	2.20	2.11	2.06	2.02	1.92	1.82
3.87	3.64	3.40	3.14	2.98	2.87	2.73	2.64	2.57	2.41	2.25
1.73	1.68	1.63	1.57	1.53	1.50	1.46	1.44	1.42	1.38	1.33
2.03	1.95	1.87	1.78	1.73	1.69	1.63	1.60	1.58	1.51	1.45
2.32	2.22	2.11	1.99	1.92	1.87	1.80	1.75	1.72	1.64	1.56
2.70	2.56	2.42	2.27	2.17	2.10	2.01	1.95	1.91	1.80	1.70
3.67	3.44	3.20	2.95	2.79	2.68	2.53	2.44	2.38	2.21	2.05
1.71	1.66	1.60	1.54	1.50	1.48	1.44	1.41	1.40	1.35	1.30
1.99	1.92	1.84	1.75	1.69	1.65	1.59	1.56	1.53	1.47	1.40
2.27	2.17	2.06	1.94	1.87	1.82	1.74	1.70	1.67	1.58	1.49
2.63	2.50	2.35	2.20	2.10	2.03	1.94	1.88	1.84	1.73	1.62
3.54	3.32	3.08	2.83	2.67	2.55	2.41	2.32	2.25	2.08	1.92
1.66	1.61	1.56	1.49	1.45	1.42	1.38	1.35	1.34	1.28	1.22
1.93	1.85	1.77	1.68	1.62	1.57	1.52	1.48	1.45	1.38	1.30
2.18	2.08	1.97	1.85	1.77	1.71	1.64	1.59	1.56	1.46	1.36
2.50	2.37	2.22	2.07	1.97	1.89	1.80	1.74	1.69	1.57	1.45
3.30	3.07	2.84	2.59	2.43	2.32	2.17	2.08	2.01	1.83	1.64
1.63	1.58	1.52	1.46	1.41	1.38	1.34	1.31	1.29	1.23	1.16
1.88	1.80	1.72	1.62	1.56	1.52	1.46	1.41	1.39	1.30	1.21
2.11	2.01	1.90	1.78	1.70	1.64	1.56	1.51	1.47	1.37	1.25
2.41	2.27	2.13	1.97	1.87	1.79	1.69	1.63	1.58	1.45	1.30
3.12	2.90	2.67	2.42	2.26	2.15	2.00	1.90	1.83	1.64	1.43
1.61	1.55	1.49	1.43	1.38	1.35	1.30	1.27	1.25	1.18	1.08
1.84	1.76	1.68	1.58	1.52	1.47	1.41	1.36	1.33	1.24	1.11
2.06	1.96	1.85	1.72	1.64	1.58	1.50	1.45	1.41	1.29	1.13
2.34	2.20	2.06	1.90	1.79	1.72	1.61	1.54	1.50	1.35	1.16
2.99	2.77	2.54	2.30	2.14	2.02	1.87	1.77	1.69	1.49	1.22

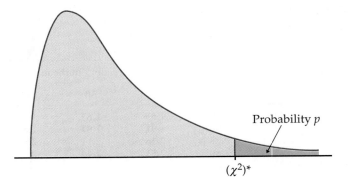

Table entry for *p* is the
critical value $(\chi^2)^*$ with
probability *p* lying to its
right.

Probability *p*

$(\chi^2)^*$

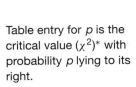

TABLE F

χ^2 distribution critical values

| df | \multicolumn{12}{c}{Tail probability p} |
	.25	.20	.15	.10	.05	.025	.02	.01	.005	.0025	.001	.0005
1	1.32	1.64	2.07	2.71	3.84	5.02	5.41	6.63	7.88	9.14	10.83	12.12
2	2.77	3.22	3.79	4.61	5.99	7.38	7.82	9.21	10.60	11.98	13.82	15.20
3	4.11	4.64	5.32	6.25	7.81	9.35	9.84	11.34	12.84	14.32	16.27	17.73
4	5.39	5.99	6.74	7.78	9.49	11.14	11.67	13.28	14.86	16.42	18.47	20.00
5	6.63	7.29	8.12	9.24	11.07	12.83	13.39	15.09	16.75	18.39	20.51	22.11
6	7.84	8.56	9.45	10.64	12.59	14.45	15.03	16.81	18.55	20.25	22.46	24.10
7	9.04	9.80	10.75	12.02	14.07	16.01	16.62	18.48	20.28	22.04	24.32	26.02
8	10.22	11.03	12.03	13.36	15.51	17.53	18.17	20.09	21.95	23.77	26.12	27.87
9	11.39	12.24	13.29	14.68	16.92	19.02	19.68	21.67	23.59	25.46	27.88	29.67
10	12.55	13.44	14.53	15.99	18.31	20.48	21.16	23.21	25.19	27.11	29.59	31.42
11	13.70	14.63	15.77	17.28	19.68	21.92	22.62	24.72	26.76	28.73	31.26	33.14
12	14.85	15.81	16.99	18.55	21.03	23.34	24.05	26.22	28.30	30.32	32.91	34.82
13	15.98	16.98	18.20	19.81	22.36	24.74	25.47	27.69	29.82	31.88	34.53	36.48
14	17.12	18.15	19.41	21.06	23.68	26.12	26.87	29.14	31.32	33.43	36.12	38.11
15	18.25	19.31	20.60	22.31	25.00	27.49	28.26	30.58	32.80	34.95	37.70	39.72
16	19.37	20.47	21.79	23.54	26.30	28.85	29.63	32.00	34.27	36.46	39.25	41.31
17	20.49	21.61	22.98	24.77	27.59	30.19	31.00	33.41	35.72	37.95	40.79	42.88
18	21.60	22.76	24.16	25.99	28.87	31.53	32.35	34.81	37.16	39.42	42.31	44.43
19	22.72	23.90	25.33	27.20	30.14	32.85	33.69	36.19	38.58	40.88	43.82	45.97
20	23.83	25.04	26.50	28.41	31.41	34.17	35.02	37.57	40.00	42.34	45.31	47.50
21	24.93	26.17	27.66	29.62	32.67	35.48	36.34	38.93	41.40	43.78	46.80	49.01
22	26.04	27.30	28.82	30.81	33.92	36.78	37.66	40.29	42.80	45.20	48.27	50.51
23	27.14	28.43	29.98	32.01	35.17	38.08	38.97	41.64	44.18	46.62	49.73	52.00
24	28.24	29.55	31.13	33.20	36.42	39.36	40.27	42.98	45.56	48.03	51.18	53.48
25	29.34	30.68	32.28	34.38	37.65	40.65	41.57	44.31	46.93	49.44	52.62	54.95
26	30.43	31.79	33.43	35.56	38.89	41.92	42.86	45.64	48.29	50.83	54.05	56.41
27	31.53	32.91	34.57	36.74	40.11	43.19	44.14	46.96	49.64	52.22	55.48	57.86
28	32.62	34.03	35.71	37.92	41.34	44.46	45.42	48.28	50.99	53.59	56.89	59.30
29	33.71	35.14	36.85	39.09	42.56	45.72	46.69	49.59	52.34	54.97	58.30	60.73
30	34.80	36.25	37.99	40.26	43.77	46.98	47.96	50.89	53.67	56.33	59.70	62.16
40	45.62	47.27	49.24	51.81	55.76	59.34	60.44	63.69	66.77	69.70	73.40	76.09
50	56.33	58.16	60.35	63.17	67.50	71.42	72.61	76.15	79.49	82.66	86.66	89.56
60	66.98	68.97	71.34	74.40	79.08	83.30	84.58	88.38	91.95	95.34	99.61	102.7
80	88.13	90.41	93.11	96.58	101.9	106.6	108.1	112.3	116.3	120.1	124.8	128.3
100	109.1	111.7	114.7	118.5	124.3	129.6	131.1	135.8	140.2	144.3	149.4	153.2

Chapter 1

1.1 Working in seconds means avoiding decimals and fractions.

1.3 Exam1 $= 79$, Exam2 $= 88$, Final $= 88$.

1.5 Cases: apartments. Five variables: rent (quantitative), cable (categorical), pets (categorical), bedrooms (quantitative), distance to campus (quantitative).

1.7 Scores are slightly left-skewed; most range from 70 to the low 90s.

1.9 (b) Use two stems, even though one is blank.

1.11 The larger classes hide a lot of detail.

1.13 A stemplot or histogram can be used; the distribution is left-skewed, centered near 80, and spread from 55 to 98.

1.15 For example, crime rates, income, cost of living, entertainment and cultural activities, or taxes.

1.19 For example, blue is by far the most popular choice; 70% of respondents chose 3 of the 10 options.

1.21 (a) 232 total respondents; 4.31%, 41.81%, 30.17%, 15.52%, 6.03%, 2.16%. **(c)** For example, 87.5% of the group were between 19 and 50. **(d)** The age-group classes do not have equal width.

1.23 Ordering bars by decreasing height shows the phones most affected by iPhone sales.

1.25 (c) When ordered by height, it may be easier to see in which categories recycling is common. **(d)** Each percent represents part of a different whole.

1.29 (a) Chile is an extreme outlier, as are (maybe) Venezuela and and Colombia. **(b)** List Chile and Venezuela separately, and split stems five ways. **(c)** For example, even without the outliers, the distribution is right-skewed. **(d)** The stemplot can show some of the detail of the low part of the distribution, if the outliers are omitted.

1.31 (c) For example, for each color, one could have two bars (one for each vehicle type) side-by-side.

1.33 359 mg/dl is an outlier; only four are in the desired range.

1.35 Roughly symmetric, centered near 7, spread from 2 to 13.

1.37 Right-skewed, centered near 5 or 6, spread from 0 to 18, no outliers.

1.39 Top-left histogram, 4; top right, 2; bottom left, 1; bottom right, 3.

1.41 (a) Most people will "round" their answers when asked to give an estimate like this, and some may

exaggerate. **(b)** The stemplots and midpoints (175 for women, 120 for men) suggest that women (claim to) study more than men.

1.43 (a) Four variables: GPA, IQ, and self-concept are quantitative; gender is categorical. **(c)** Skewed to the left, centered near 7.8, spread from 0.5 to 10.8. **(d)** There is more variability among the boys; in fact, there seem to be two groups of boys: those with GPAs below 5 and those with GPAs above 5.

1.45 Skewed to the left, centered near 59.5; most scores are between 35 and 73, with a few below that and one high score of 80 (probably not quite an outlier).

1.47 The new mean is 63.64 days.

1.49 The outlier increases the median, but the change is much less than the outlier's effect on the mean.

1.51 $M = 82.5$.

1.53 $\bar{x} = 196.575$ seconds (the value in the text was rounded). The quartiles and median are in positions 20.5, 40.5, and 60.5.

1.55 Use the five-number summary from Exercise 1.54 (55, 75, 82.5, 92, 98).

1.57 $s^2 \doteq 157.43$ and $s \doteq 12.55$.

1.59 Suriname changes the IQR from 23.5 to 30.5—much less drastic than the change in s.

1.61 (a) Use a stemplot or histogram. **(b)** In millions of dollars, the five-number summary is 3338, 4589, 7558.5, 13,416, and 66,667. **(c)** For example, the distribution is sharply right-skewed; the top nine could be considered outliers.

1.63 (a) $\bar{x}$ changes from 4.76% (with) to 4.81% (without); the median (4.7%) does not change. **(b)** s changes from 0.7523% to 0.5864%; Q_1 changes from 4.3% to 4.35%, while $Q_3 = 5\%$ does not change. **(c)** A low outlier decreases $\bar{x}$; any kind of outlier increases s. Outliers have little or no effect on the median and quartiles.

1.65 For example, 0, 1, 2, 998, 1000; the median changes from 2 to 500.

1.67 (a) The distribution is left-skewed; the five-number summary (in ounces) is 3.7, 4.95, 6.7, 7.85, and 8.2. **(b)** The numerical summary does not reveal the two weight clusters (visible in a stemplot or histogram). **(c)** For the small potatoes, $\bar{x} = 4.662$ and $s = 0.501$ ounces; for the large potatoes, $\bar{x} = 7.300$ and $s = 0.755$ ounces. Because there are clearly two groups, it seems appropriate to treat them separately.

1.69 (a) Min $= Q_1 = 0$, $M = 5.085$, $Q_3 = 9.47$, Max $= 73.2$. **(d)** The distribution is sharply right-skewed.

The histogram seems to convey the distribution better.

1.71 Min $= 0.24$, $Q_1 = 0.355$, $M = 0.76$, $Q_3 = 1.03$, Max $= 1.9$. The distribution is right-skewed. A histogram or stemplot reveals an important feature not evident from a boxplot: this distribution has two peaks.

1.73 This distribution would almost surely be strongly skewed to the right.

1.75 $\bar{x} = \$76,667$; six of the nine employees earn less than the mean. $M = \$35,000$.

1.77 The mean rises to $\$91,667$, while the median is unchanged.

1.79 Means are not the appropriate measure of center for skewed distributions.

1.81 (a) pH: $\bar{x} \doteq 5.4256$ and $s \doteq 0.5379$. Density: $\bar{x} \doteq 5.4479$ and $s \doteq 0.2209$. (b) Medians: 5.44 and 5.46.

1.83 The mean and median always agree for two observations.

1.85 (a) Place the new point at the current median.

1.87 (a) *Bihai:* $\bar{x} \doteq 47.5975$, $s \doteq 1.2129$. Red: $\bar{x} \doteq 39.7113$, $s \doteq 1.7988$. Yellow: $\bar{x} \doteq 36.1800$, $s \doteq 0.9753$ (all in mm). (b) *Bihai* and red appear to be right-skewed (although it is difficult to tell with such small samples).

1.89 The five-number summary is 1, 3, 4, 5, and 12 letters.

1.91 Take six or more numbers, with the largest number much larger than Q_3.

1.93 (a) Any set of four identical numbers works. (b) 0, 0, 20, 20 is the only possible answer.

1.95 Multiply each value by 0.03937.

1.97 $\bar{x} = 5.32$ pounds and $s = 2.60$ pounds.

1.99 Full data set: $\bar{x} = 196.575$ and $M = 103.5$ minutes. The 10% and 20% trimmed means are $\bar{x}^* \doteq 127.734$ and $\bar{x}^{**} \doteq 111.917$ minutes.

1.101 470 to 674.

1.103 $z = 0.94$.

1.105 Using Table A, the proportion below 620 is 0.8264, and the proportion at or above is 0.1736.

1.107 About 606.

1.109 (a) It is easiest to draw the curve first and then mark the scale on the axis. (b) Draw a copy of the first curve, with the peak over 20. (c) The curve has the same shape but is translated left or right.

1.113 (a) The 68–95–99.7 ranges for women are 8489 to 20919, 2274 to 27134, and -3941 to 33349. For men, they are 7158 to 22886, -706 to 30750, and -8570 to 38614. In both cases, some of the lower limits are negative, which does not make sense; this happens because the women's distribution is skewed, and the men's distribution has an outlier. Contrary to the conventional

wisdom, the men's mean is slightly higher. (b) The means suggest that Mexican men and women tend to speak more than those from the United States.

1.115 (a) -1.64, -1.04, 0.13, and 1.04. (b) 53.6, 59.6, 71.3, and 80.4.

1.117 (a) $1/4$. (b) 0.25. (c) 0.5.

1.119 (a) Mean C, median B. (b) Mean A, median A. (c) Mean A, median B.

1.121 (a) 0.6826; the 68–95–99.7 rule gives 0.68. (b) 0.9544 (compared to 0.95); 0.9974 (compared to 0.997).

1.123 (a) 327 to 345 days. (b) 16%.

1.125 $\bar{x} = 5.4256$ and $s = 0.5379$; 67.62% within $\bar{x} - s$ and $\bar{x} + s$, 95.24% within $\bar{x} \pm 2s$, and all within $\bar{x} \pm 3s$.

1.127 (a) 0.0359. (b) 0.9641. (c) 0.0548. (d) 0.9093.

1.129 (a) 0.3853. (b) 0.1257.

1.131 About 2.5%.

1.133 Jacob's score ($z = -1.02$) is higher than Emily's ($z = -1.52$).

1.135 About 2014.

1.137 32.3%.

1.139 1239 and below.

1.141 1239, 1428, 1590, and 1779.

1.143 (a) 33%. (b) 15%. (c) 52%.

1.145 (a) About 5.2%. (b) About 54.7%. (c) More than 279 days.

1.147 (a) About 1.35. (b) 1.35.

1.149 (a) ± 1.28. (b) 8.93 and 9.31 ounces.

1.151 (a) Roughly Normal, but with a high outlier. (b) Roughly Normal. (c) Skewed to the right.

1.153 (a) The yellow variety is the nearest to a straight line. (b) The other two distributions are both slightly right-skewed, and the *bihai* variety appears to have a couple of high outliers.

1.155 Histograms will suggest (but not exactly match) Figure 1.34. The uniform distribution does not extend as low or as high as a Normal distribution.

1.157 (a) The distribution appears to be roughly Normal. (b) One could justify either $\bar{x} = 15.27\%$ and $s = 3.118\%$, or the five-number summary (8.2%, 13%, 15.5%, 17.6%, and 22.8%). (c) For example, binge drinking rates are typically 10% to 20%. Which states are high, and which are low?

1.159 For example, white is considerably less popular in Europe, and gray is less common in China.

1.163 The distribution is somewhat right-skewed, with only one country (Bosnia and Herzegovina) in the 20s. The mean and standard deviation are 39.85 and 22.05 users per 100 people.

1.165 The given description is true on the average, but the curves (and a few calculations) give a more complete picture. For example, a score of about 675 is about the 97.5th percentile for both genders, so the top boys and girls have very similar scores.

1.167 Slightly right-skewed, with one (or more) high outliers. Five-number summary: 22, 23.735, 24.31, 24.845, and 28.55 hours.

1.169 The "Other" category includes 29.3 million subscribers.

1.171 **(a)** Makes: bar graph or pie chart. Age: histogram, stemplot, or boxplot. **(b)** Study time: histogram, stemplot, or boxplot. To show change over time, use a time plot. **(c)** Bar graph or pie chart. **(d)** Normal quantile plot.

1.173 No to both questions; no summary can exactly describe a distribution that can include any number of values.

1.175 Mean changes from 36.56 to 34.41 home runs; median from 35.5 to 34 home runs.

Chapter 2

2.1 Students.

2.3 Cases: cups of Mocha Frappuccino. Variables: size and price (both quantitative).

2.5 **(b)** 10 cases. **(c)** Choices will vary. **(d)** The second and third columns are quantitative.

2.7 **(b)** Bobax is the second point from the right.

2.9 Size should be explanatory. The scatterplot shows a positive association between size and price.

2.11 There is still a strong positive relationship, but the new points are far away from the others.

2.13 **(a)** A boxplot summarizes the distribution of one variable. **(b)** This is only correct if there is an explanatory/response relationship. **(c)** High values go with high values.

2.15 **(b)** There are 46,994 thousand uninsured; divide each number in the second column by this amount. **(d)** The plots differ only in the vertical scale. **(e)** The uninsured are found in similar numbers for the five lowest age groups (with slightly more in those aged 25–34 and 45–64), and fewer among those over 65.

2.17 The percents in Exercise 2.15 show what fraction of the uninsured fall in each age group. The percents in Exercise 2.16 show what fraction of each age group is uninsured.

2.19 **(b)** There is a moderate positive linear relationship.

2.21 There is a moderate positive linear relationship; the relationship for all countries is less linear because of the wide range in life expectancy among countries with low Internet use.

2.23 **(a)** "Month" (the passage of time) explains changes in temperature (not vice versa). **(b)** Temperature increases linearly with time (about 10 degrees per month); the relationship is strong.

2.25 **(a)** The second test happens before the final exam. **(b)** The plot shows a weak positive association. **(c)** Students' study habits are more established by the middle of the term.

2.27 **(a)** Explanatory: age. Response: weight. **(b)** Explore the relationship. **(c)** Explanatory: number of bedrooms. Response: price. **(d)** Explanatory: amount of sugar. Response: sweetness. **(e)** Explore the relationship.

2.29 **(a)** In general, we expect more intelligent children to be better readers, and less intelligent children to be weaker readers. The plot does show this positive association. **(b)** These four have moderate IQs but poor reading scores. **(c)** Roughly linear but weak (much scatter).

2.31 **(a)** Areas with many breeding pairs would correspondingly have more males that might potentially return. **(c)** The theory suggests a negative association; the scatterplot shows this.

2.33 A fairly strong, positive, linear association; social exclusion does appear to trigger a pain response.

2.35 **(b)** The association is linear and positive and is stronger for women. Males typically have larger values for both variables.

2.37 **(a)** Both show fairly steady improvement. Women have made more rapid progress but have not improved since 1993, while men's records may be dropping more rapidly in recent years. **(b)** The data support the first claim but do not seem to support the second.

2.39 **(a)** $r = 0.8839$. **(b)** They are equal. **(c)** Units do not affect correlation.

2.41 **(a)** $r = 1$. **(b)** $r = 1$.

2.43 $r = 0.2873$.

2.45 $r = 0.6701$.

2.47 **(a)** $r = 0.5194$. **(b)** The first-test/final-exam correlation will be lower.

2.49 The correlation increases.

2.51 Value and revenue: $r_1 = -0.3228$. Value and debt: $r_2 = 0.9858$. Value and income: $r_3 = 0.7177$.

2.53 **(a)** Positive, but not close to 1. **(b)** $r = 0.5653$. **(c)** r would not change; it does not tell us that the men were generally taller. **(d)** r would not change. **(e)** 1.

2.55 **(a)** $r = \pm 1$ for a line. **(c)** Leave some space above your vertical stack. **(d)** The curve must be higher at the right than at the left.

2.57 $r = 1$ for a positively sloped line.

2.59 There is little linear association between research and teaching—for example, knowing a professor is a good researcher gives little information about whether she is a good or bad teacher.

2.61 Both relationships are somewhat linear; GPA/IQ ($r = 0.6337$) is stronger than GPA/self-concept ($r = 0.5418$). The two students with the lowest GPAs stand out in both plots; a few others stand out in at least one plot. Generally speaking, removing these points raises r, except for the lower-left point in the self-concept plot.

2.63 1.441 kg.

2.65 Expressed as percentages, these fractions are 81%, 25%, 9%, 0%, 9%, 25%, and 81%.

2.67 Los Angeles (5995) is best, and Chicago (-7283) is worst.

2.69 For Baltimore, for example, this rate is $\frac{5091}{651} \doteq 7.82$ acres per thousand people. This scatterplot is much less linear; the regression equation $\hat{y} = 8.739 - 0.000424x$ explains only 8.7% of the variation in open space.

2.71 $\hat{y} = 3.379 + 1.6155x$.

2.73 **(a)** All correlations are approximately 0.816 or 0.817, and the regression lines are $\hat{y} = 3.000 + 0.500x$. We predict $\hat{y} \doteq 8$ when $x = 10$. **(c)** This regression formula is only appropriate for Set A.

2.75 **(a)** The relationship is linear, positive, and fairly strong. **(b)** $\hat{y} = 1.470 + 1.443x$. **(c)** The new point's residual is positive; the other residuals decrease as x increases. **(d)** 71.1%. **(e)** The new point makes the relationship stronger, but its location has a large impact on the regression equation.

2.77 **(a)** $y = 42$. **(b)** y increases by 6. **(c)** 12.

2.79 **(a)** $\hat{y} = 0.06078x - 0.1261$. **(b)** $\hat{y} \doteq 0$ (the formula gives -0.0045). **(c)** 77%.

2.81 $\bar{x} = 95$ min, $\bar{y} = 12.6611$ cm, $s_x = 53.3854$ min, $s_y = 8.4967$ cm, and $r = 0.9958$. For predicting length from time, $b_1 = r\,s_y/s_x \doteq 0.158$ cm/min and $a_1 = \bar{y} - b_1\bar{x} \doteq -2.39$ cm. For predicting time from length, $b_2 = r\,s_x/s_y \doteq 6.26$ min/cm and $a_2 = \bar{x} - b_2\bar{y} \doteq 15.79$ min.

2.83 IQ and GPA: $r_1 = 0.634$. Self-concept and GPA: $r_2 = 0.542$. IQ does a slightly better job.

2.85 When $x = \bar{x}$, $\hat{y} = a + b\bar{x} = (\bar{y} - b\bar{x}) + b\bar{x} = \bar{y}$.

2.87 $r = 0.40$.

2.89 **(a)** The regression equation is $\hat{y} = 10.27 - 0.5382x$. **(b)** The scatterplot looks more linear than Figure 2.5, but a line may not be appropriate for all values of log unemployment. **(c)** There are more negative residuals on the left and right, with more positive residuals in the middle.

2.91 Correlations were found in Exercise 2.51; of the three, the best regression line ($\hat{y} = 16.02 + 2.896x$) uses debt.

2.93 The sum is 0.01.

2.95 **(a)** A high correlation means strong association, not causation. **(b)** Outliers in the y direction (and some other data points) will have large residuals. **(c)** It is not extrapolation if $1 \leq x \leq 5$.

2.97 For example, it may be that doing well makes students or workers feel good about themselves, rather than vice versa.

2.99 The explanatory and response variables were "consumption of herbal tea" and "cheerfulness/health." One lurking variable is social interaction; many of the nursing home residents may have been lonely before the students started visiting.

2.101 **(a)** The plot is curved (low at the beginning and end of the year, high in the middle). **(b)** $\hat{y} = 39.392 + 1.4832x$; it does not fit well. **(c)** Residuals are negative for January through March and October through December, and positive from April to September. **(d)** A similar pattern would be expected in any city that is subject to seasonal temperature variation. **(e)** Seasons in the Southern Hemisphere are reversed.

2.103 The variation of individual data increases the scatter, thus decreasing the strength of the relationship.

2.105 For example, a student who in the past might have received a grade of B (and a lower SAT score) now receives an A (but has a lower SAT score than an A student in the past).

2.107 $r = 0.08795$ and $b = 0.000811$ kg/cal.

2.109 **(b)** The plot shows a strong positive linear relationship. **(c)** $\hat{y} = 20.40 + 0.7194x$. **(d)** Hernandez's point is in the lower left.

2.111 **(a)** Drawing the "best line" by eye is a very inaccurate process.

2.113 The plot should show a positive association when either group of points is viewed separately and should show a large number of bachelor's degree economists in business and graduate degree economists in academia.

2.115 1684 are binge drinkers; 8232 are not.

2.117 $8232/17{,}096 \doteq 0.482$.

2.119 $1630/7180 \doteq 0.227$.

2.121 **(a)** 50.5% get enough sleep; 49.5% do not. **(b)** 32.2% get enough sleep; 67.8% do not. **(c)** Those who exercise more than the median are more likely to get enough sleep.

2.123 3.0% of Hospital A's patients died, compared with 2.0% at Hospital B.

2.125 **(a)** About 3,388,000. **(b)** 0.207, 0.024; 0.320, 0.071; 0.104, 0.104; 0.046, 0.125. **(c)** 0.230, 0.391, 0.208, 0.171. **(d)** 0.677, 0.323.

2.127 Full-time: 0.305, 0.472, 0.154, 0.069. Part-time: 0.073, 0.220, 0.321, 0.386.

2.129 There are many possibilities, but the simplest such table would have all nine counts equal to one another.

2.131 **(a)** Females: 30.22% Accounting, 40.44% Administration, 2.22% Economics, and 27.11% Finance. Males: 34.78%, 24.84%, 3.7%, and 36.65%. Overall: 32.12%, 33.94%, 2.85%, 31.09%. **(b)** 46.5% did not respond.

2.133 This is a case of confounding.

2.135 For example, students who choose the online course might have more self-motivation or better computer skills.

2.137 No; self-confidence and improving fitness could be a common response to some other personality trait, or high self-confidence could make a person more likely to join the exercise program.

2.139 They may feel more confident after taking the course, or because they have taken the test before.

2.141 Patients suffering from more serious illnesses are more likely to go to larger hospitals and may require more time to recuperate afterwards.

2.143 People who are overweight are more likely to be on diets and so choose artificial sweeteners.

2.145 This is an observational study—students choose their "treatment" (to take or not take the refresher sessions).

2.149 **(a)** The residuals are positive at the beginning and end, and negative in the middle. **(b)** The behavior of the residuals agrees with the curved relationship seen in Figure 2.30.

2.151 **(a)** About $139,600. **(b)** About $160,040. **(c)** The second prediction is more reliable, although both involve extrapolation. **(d)** Interpreting relationships without a plot is risky.

2.153 **(a)** $\hat{y} = 5403 + 0.9816x$ **(b)** There are no clear outliers or influential observations.

2.155 A school that accepts weaker students but graduates a higher-than-expected number of them would have a positive residual, while a school with a stronger incoming class but a lower-than-expected graduation rate would have a negative residual. It seems reasonable to measure school quality by how much benefit students receive from attending the school.

2.157 **(a)** The scatterplot shows a moderate positive association. **(b)** The regression line ($y = 11.00 + 0.9344x$) fits the overall trend. **(c)** For example, a state whose point falls above the line has a higher percent of college graduates than we would expect based on the percent who eat five servings of fruits and vegetables. **(d)** No; association is not evidence of causation.

2.161 These results support the idea (the slope is negative), but the relationship is only moderately strong ($r^2 = 0.34$).

2.163 A scatterplot shows a moderate linear positive association. The regression equation $\hat{y} = 2653 + 0.004742x$ explains about 62% of the variation in MOR.

2.165 **(b)** $\hat{y} = 1.2027 + 0.3275x$; the relationship is not too strong ($r = 0.4916$, $r^2 = 0.2417$).

2.167 **(a)** $\hat{y} = 259.58 - 19.464x$; the relationship appears to be curved. **(b)** Either $\hat{y} = 5.9732 - 0.2184x$ or $\hat{y} = 2.5941 - 0.09486x$; the relationship appears to be linear.

2.169 Number of firefighters and amount of damage are common responses to the seriousness of the fire.

2.171 A notably higher percent of women are "strictly voluntary" participants.

2.173 **(a)** Males: 490 admitted, 310 not. Females: 400 admitted, 300 not. **(b)** Males: 61.25% admitted. Females: 57.14% admitted. **(c)** Business school: 66.67% of males, 66.67% of females. Law school: 45% of males, 50% of females. **(d)** Most male applicants apply to the business school, where admission is easier. More women apply to the law school, which is more selective.

2.175 First and second year: A has 8.3% small classes; B has 17.1% small. Upper-level: A has 77.5% small; B has 83.3% small.

2.177 **(a)** Wagering on collegiate sports appears to be more common in Division II, and even more in Division III. **(b)** Even with smaller sample sizes (1000 or more), the estimates should be fairly accurate (barring dishonest responses). **(c)** Our conclusion might not hold for the true percents.

Chapter 3

3.1 Any group of friends is unlikely to include a representative cross section of all students.

3.3 A hard-core runner (and her friends) are not representative of all young people.

3.5 For example, who owns the Web site? Do they have data to back up this statement, and if so, what was the source of those data?

3.7 An experiment: each subject is (presumably randomly) assigned to a treatment group. The explanatory variable is the drug received, and the response variables are adverse events, as well as some reaction (not specified in the exercise).

3.9 An experiment: each subject is (presumably randomly) assigned to a treatment group. Explanatory variable: teaching method; response variable: change in each student's test score.

3.11 Experimental units: food samples. Treatments: radiation exposure. Response variable: lipid oxidation. Factor: radiation exposure. Levels: nine different levels of radiation. It is likely that different lipids react to radiation differently.

3.13 Those who volunteer to use the software may be better students (or worse).

3.17 (a) Shopping patterns may differ on Friday and Saturday. **(b)** Responses may vary in different states. **(c)** A control is needed for comparison.

3.19 Students will know which method was used to teach them.

3.21 For example, new employees should be randomly assigned to either the current program or the new one.

3.23 (a) Factors: calcium dose, and vitamin D dose. There are nine treatments (each calcium/vitamin D combination). **(b)** Assign 20 students to each group, with 10 of each gender.

3.25 (a) For example, flip a coin for each customer to choose which variety (s)he will taste. **(b)** For example, flip a coin for each customer to choose which variety (s)he will taste *first*. **(c)** If each customer tastes both varieties, we only need to ask which was preferred.

3.27 Experimental units: pine tree seedlings. Factor: amount of light. Treatments: full light, or shaded to 5% of normal. Response variable: dry weight at end of study.

3.29 Subjects: adults from selected households. Factors: level of identification and offer of survey results. Six treatments: interviewer's name/university name/both names, with or without results. Response variable: whether or not the interview is completed.

3.31 Assign nine subjects to each treatment. The first three groups are 03, 22, 29, 26, 01, 12, 11, 31, 21; 32, 30, 09, 23, 07, 27, 20, 06, 33; 05, 16, 28, 10, 18, 13, 25, 19, 04.

3.33 For example, have some volunteers go through a training session, while others are given a written set of instructions, or watch a video. To measure training effectiveness, volunteers could analyze a sample of lake water and compare their results to some standard.

3.35 Assign six schools to each treatment group. Choose 16, 21, 06, 12, 02, 04 for Group 1; 14, 15, 23, 11, 09, 03 for Group 2; 07, 24, 17, 22, 01, 13 for Group 3; and the rest for Group 4.

3.37 (a) Population = 1 to 150, sample size 25, then click "Reset" and "Sample." **(b)** Without resetting, click "Sample" again.

3.39 Design (a) is an experiment, while (b) is an observational study; with the first, any difference in colon health between the two groups could be attributed to the treatment (bee pollen or not).

3.41 There are nine treatments. Choose the number of letters in each group and send them out at random times over several weeks.

3.43 Each player runs through the sequence (100 yards, four times) once with oxygen and once without (on different days to allow full recovery); observe the difference in times on the final run. We choose 12, 13, 04, 18, 19, 16, 02, 08, 17, 10 for the oxygen-first group.

3.45 (a) Randomly assign half the girls to get high-calcium punch; the other half will get low-calcium punch. Observe how each group processes the calcium. **(b)** Half receive high-calcium punch first; the rest get low-calcium punch first. For each subject, compute the difference in the response variable for each level. Matched pairs designs give more precise results. **(c)** The first five subjects are 38, 44, 18, 33, and 46.

3.47 (a) A block design. **(c)** Such results would rarely have occurred by chance if vitamin C were ineffective.

3.49 Population: area forest owners. Sample: the 772 forest owners contacted. Response rate: 348/772 = 45%. Additionally, we would like to know the sample design (among other things).

3.53 (a) This is a *census* rather than a sample. **(b)** Many people will probably not realize that dihydrogen monoxide is water. **(c)** In a public setting, few people will admit to cheating.

3.55 Population: local businesses. Sample: the 150 businesses surveyed (or the 73 businesses that responded). Nonresponse rate: 51.3%.

3.57 Note that the numbers add to 100% down the columns; that is, 39% is the percent of Fox viewers who are Republicans, *not* the percent of Republicans who watch Fox.

3.59 12, 14, 11, 16, and 08.

3.61 Population = 1 to 200, sample size 25, then click "Reset" and "Sample."

3.63 39 (block 3020), 10 (block 2003), 07 (block 2000), 11 (block 2004), and 20 (block 3001).

3.65 The simplest method is to assign the labels 0 to 5 to blocks in Group 1, then choose one of those blocks; use the last two digits of the blocks in Group 2, and choose two of those, etc.

3.67 Each student has chance 1/45 of being selected, but it is not an SRS, because the only possible samples have exactly one name from the first 45, one name from the second 45, and so on.

3.69 Assign labels 01 to 36 for the Climax 1 group, 01 to 72 for the Climax 2 group, and so on, then choose (from Table B) 12, 32, 13, 04; 51, 44, 72, 32, 18, 19, 40; 24, 28, 23; and 29, 12, 16, 25.

3.71 Each student has a 10% chance, but the only possible samples are those with 3 older and 2 younger students.

3.73 (a) Households without telephones or with unlisted numbers. Such households would likely be made up of poor individuals, those who choose not to have land lines, and those who do not wish to have their phone number published. **(b)** Those with unlisted numbers.

3.75 The first wording brought the higher numbers in favor of a tax cut.

3.79 Population: undergraduate college students. Sample: 2036 students.

3.81 The larger sample would have less sampling variability.

3.83 (a) Population: college students. Sample: 17,096 students. **(b)** Population: restaurant workers. Sample: 100 workers. **(c)** Population: longleaf pine trees. Sample: 40 trees.

3.87 (a) Multistage. **(b)** Attitudes in smaller countries (many of which were not surveyed) might be different. **(c)** An individual country's reported percent will typically differ from its true percent by no more than the stated margin of error.

3.89 The histograms should be centered at about 0.6 and 0.2.

3.93 (a) Mean GPA 2.6352, standard deviation 0.7794.

3.95 (a) If, for example, eight heads are observed, then $\hat{p} = \frac{8}{20} = 0.4 = 40\%$.

3.97 (a) A nonscientist might raise different viewpoints and concerns from those considered by scientists.

3.103 Alternate control and cancer samples.

3.111 Interviews conducted in person cannot be anonymous.

3.121 (a) You need information about a random selection of his games, not just the ones he chooses to talk about. **(b)** These students may have chosen to sit in the front; all students should be assigned to their seats.

3.123 This is an experiment, because treatments are assigned. Explanatory variable: price history (steady or fluctuating). Response variable: price the subject expects to pay.

3.127 Randomly choose the order in which the treatments (gear and steepness combination) are tried.

3.129 (a) One possibility: full-time undergraduate students in the fall term on a list provided by the registrar. **(b)** One possibility: a stratified sample with 125 students from each year. **(c)** Nonresponse might be higher with mailed (or emailed) questionnaires; telephone interviews exclude some students and may require repeated calling for those who are not home;

face-to-face interviews might be too costly. The topic might also be subject to response bias.

3.131 Use a block design: separate men and women, and randomly allocate each gender among the six treatments.

3.133 CASI will typically produce more honest responses to embarrassing questions.

Chapter 4

4.1 The proportion of heads is 0.3.

4.5 If you hear music (or talking) one time, you will almost certainly hear the same thing for several more checks after that.

4.7 Out of a very large number of patients taking this medication, the fraction who experience this side effect is about 0.00001.

4.11 One possibility: from 0 to _ hours (the largest number should be big enough to include all possible responses).

4.13 0.7 (add the probabilities of the other four colors and subtract from 1).

4.15 0.523.

4.17 1/4.

4.19 (a) The probability that both of two disjoint events occur is 0. **(b)** Probabilities must be no more than 1; $P(A \text{ and } B)$ will be no more than 0.5. **(c)** $P(A^c) = 0.65$.

4.21 There are 6 possible outcomes: {link1, link2, link3, link4, link5, leave}.

4.23 (a) 0.248. **(b)** 0.752.

4.25 (a) 0.03, so the sum equals 1. **(b)** 0.55.

4.27 (a) The probabilities sum to 2. **(b)** Legitimate (for a nonstandard deck). **(c)** Legitimate (for a nonstandard die).

4.29 (a) 0.28. **(b)** 0.88.

4.31 Take each blood type probability and multiply by 0.84 and by 0.16. For example, the probability for A-positive blood is $(0.42)(0.84) = 0.3528$.

4.33 (a) 0.006. **(b)** 0.001.

4.35 0.5160.

4.37 The two events (being 75 or older and being a woman) are probably not independent.

4.39 (a) 0.2746. **(b)** 0.35. **(c)** 0.545.

4.41 Observe that
$P(A \text{ and } B^c) = P(A) - P(A \text{ and } B) = P(A) - P(A)P(B).$

4.43 (a) Either B or O. **(b)** $P(B) = 0.75$, $P(O) = 0.25$.

4.45 (a) 0.25. **(b)** 0.015625; 0.140625.

4.47 Possible values: 0, 1, 2. Probabilities: 1/4, 1/2, 1/4.

4.49 (a) Discrete *random variable.* **(b)** Continuous random variables can take values from any interval. **(c)** Normal random variables are continuous.

4.51 (a) $P(T) = 0.19.$ **(b)** $P(TTT) = 0.0069,$ $P(TTT^c) = P(TT^cT) = P(T^cTT) = 0.0292,$ $P(TT^cT^c) = P(T^cT^cT^c) = P(T^cT^cT) = 0.1247,$ and $P(T^cT^cT^c) = 0.5314.$ **(c)** $P(X = 3) = 0.0069,$ $P(X = 2) = 0.0877, P(X = 1) = 0.3740,$ and $P(X = 0) = 0.5314.$

4.53 (a) Continuous. **(b)** Discrete. **(c)** Discrete.

4.55 (b) $P(X \geq 1) = 0.9.$ **(c)** "No more than two nonword errors." $P(X \leq 2) = 0.7; P(X < 2) = 0.4.$

4.57 (a) Note that, for example, "(1, 2)" and "(2,1)" are distinct outcomes. **(b)** 1/36. **(c)** For example, four pairs add to 5, so $P(X = 5) = 4/36 = 1/9.$ **(d)** 2/9. **(e)** 5/6.

4.59 $P(X = 2) = P(X = 14) = 1/48,$ $P(X = 3) = P(X = 13) = 2/48, \ldots,$ $P(X = 6) = P(X = 10) = 5/48,$ $P(X = 7) = P(X = 8) = P(X = 9) = 6/48,$

4.61 (a) 0.6. **(b)** 0.6. **(c)** "Equal to" has no effect on the probability.

4.63 (a) The height should be 1/2. **(b)** 0.8. **(c)** 0.6. **(d)** 0.525.

4.65 Very close to 1.

4.67 Possible values: $0 and $1. Probabilities: 0.5 and 0.5. Mean: $0.50.

4.69 $\mu_Y = 95.$

4.71 $\sigma_X^2 = 1$ and $\sigma_X = 1.$

4.73 Mean 2.3 servings; standard deviation 1.9 servings.

4.75 2.88.

4.77 0.3730 aces.

4.79 (a) $85.48. **(b)** This is larger; the negative correlation decreased the variance.

4.81 The exercise describes a positive correlation between calcium intake and compliance. Because of this, the variance of total calcium intake is greater than the variance we would see if there were no correlation.

4.83 (a) $\mu_1 = \sigma_1 = 0.5.$ **(b)** $\mu_4 = 2$ and $\sigma_4 = 1.$

4.85 Mean 16 cm and standard deviation 0.0042 mm.

4.87 (a) Not independent. **(b)** Independent.

4.89 Show that $\sigma_{X+Y}^2 = (\sigma_X + \sigma_Y)^2.$

4.91 If one of the 10 homes were lost, it would cost more than the collected premiums. For many policies, the average claim should be close to $300.

4.93 (a) 0.99749. **(b)** $623.22.

4.95 $2/6 = 1/3.$

4.97 $2/48 = 1/24.$

4.99 The addition rule for disjoint events.

4.101 The four probabilities are 5/39, 85/351, 85/351, and 136/351.

4.103 (a) 0.16. **(b)** 0.22. **(c)** 0.38. **(d)** For (a) and (b), use the addition rule for disjoint events; for (c), use the addition rule, and note that S^c and $E^c = (S \text{ or } E)^c.$

4.105 0.73; use the addition rule.

4.107 (a) The four probabilities sum to 1. **(b)** 0.77. **(c)** 0.7442. **(d)** The events are not independent.

4.109 (a) The four entries are 0.2684, 0.3416, 0.1599, 0.2301. **(b)** 0.5975.

4.111 For example, the probability of selecting a female student is 0.5717; the probability that she comes from a four-year institution is 0.5975.

4.113 $P(A|B) \doteq 0.3142.$ If A and B were independent, then $P(A|B)$ would equal $P(A).$

4.115 (a) $P(A^c) = 0.69.$ **(b)** $P(A \text{ and } B) = 0.08.$

4.117 (a) 0.6364. **(b)** Not independent.

4.119 1.

4.121 $P(B|C) = 1/3.$ $P(C|B) = 0.2.$

4.123 (a) $P(M) \doteq 0.4124.$ **(b)** $P(B|M) \doteq 0.6670.$ **(c)** $P(M \text{ and } B) \doteq 0.2751.$

4.125 (a) $P(L) \doteq 0.6722.$ **(b)** $P(L|C) \doteq 0.7837.$ **(c)** Not independent.

4.127 Retired persons are more likely than other adults to have not completed high school.

4.129 (a) Her brother has allele type aa, and he got one allele from each parent. **(b)** $P(aa) = 0.25, P(Aa) = 0.5,$ $P(AA) = 0.25.$ **(c)** $P(AA|\text{not } aa) = 1/3, P(Aa|\text{not } aa) = 2/3.$

4.131 0.9333.

4.133 Close to $\mu_X = 1.1.$

4.135 (a) Possible values 1 and 10, with probabilities 0.4 and 0.6. **(b)** $\mu_Y = 6.4$ and $\sigma_Y \doteq 4.4091.$ **(c)** Those rules are for transformations of the form $aX + b.$

4.137 (a) $P(A) = 5/36$ and $P(B) = 5/18.$ **(b)** $P(A) = 5/36$ and $P(B) = 7/12.$ **(c)** $P(A) = 5/12$ and $P(B) = 5/18.$ **(d)** $P(A) = 5/12$ and $P(B) = 5/18.$

4.139 For example, if the point is 4 or 10, the expected gain is $(1/3)(+\$20) + (2/3)(-\$10) = 0.$

4.141 (a) All probabilities are greater than or equal to 0, and their sum is 1. **(b)** 0.61. **(c)** Both probabilities are 0.39.

4.143 0.005352.

4.145 0.7356.

4.147 $P(\text{no point}) = 1/3.$ The probability of winning (losing) an odds bet is 1/36 (1/18) on 4 or 10, 2/45 (1/15) on 5 or 9, 25/396 (5/66) on 6 or 8.

4.149 0.1622.

4.151 $P(Y < 1/2|Y > X) = 1/4.$

Chapter 5

5.1 Population: iPhone users. Statistic: an average of 65 apps per device. Likely values will vary.

5.3 $\mu_{\bar{x}} = 240$, $\sigma_{\bar{x}} = 1.5$.

5.5 About 95% of the time, $\bar{x}$ is between 239 and 241.

5.7 **(a)** "Variance" should be "standard deviation." **(b)** Standard deviation decreases with increasing sample size. **(c)** $\mu_{\bar{x}}$ always equals μ.

5.9 **(a)** 338.8. **(b)** Answers will vary. **(c)** Answers will vary. **(d)** The center should get closer with increased repetition.

5.11 **(a)** We need a larger sample size to decrease the 95% range. **(b)** $\sigma_{\bar{x}} \le 0.4167$. **(c)** 762 students.

5.13 250 ml and 1.2247 ml.

5.15 **(b)** 0.7414. **(c)** 0.4122.

5.17 **(a)** $\bar{x}$ is not systematically higher than or lower than μ. **(b)** With large samples, $\bar{x}$ is more likely to be close to μ.

5.19 **(a)** 0.5 and 0.099 moths. **(b)** About 0.16.

5.21 **(a)** 0.0668. **(b)** 0.0047.

5.23 About 134.5 mg/dl.

5.25 About 0.0051.

5.27 **(a)** Nearly 1. **(b)** About 0.9641.

5.29 **(a)** $\bar{y}$ is $N(\mu_Y, \sigma_Y/\sqrt{m})$ and $\bar{x}$ is $N(\mu_X, \sigma_X/\sqrt{n})$. **(b)** $N\left(\mu_Y - \mu_X, \sqrt{\sigma_Y^2/m + \sigma_X^2/n}\right)$.

5.31 $N(32, 0.2)$; 0.0124.

5.33 **(a)** $n = 1500$. **(b)** The "Yes" count seems most reasonable, but either count is defensible. **(c)** $X = 825$ (or $X = 675$). **(d)** $\hat{p} = 0.55$ (or $\hat{p} = 0.45$).

5.35 $B(15, 0.5)$.

5.37 **(a)** $P(X = 0) = 0.0778$ and $P(X \ge 3) = 0.3174$. **(b)** $P(X = 5) = 0.0778$ and $P(X \le 2) = 0.3174$. **(c)** The number of "failures" in a $B(5, 0.4)$ distribution has a $B(5, 0.6)$ distribution.

5.39 **(a)** About 1. **(b)** About 0.9974.

5.41 **(a)** Separate flips are independent (coins have no "memory"). **(b)** Separate flips are independent (coins have no "memory"). **(c)** $\hat{p}$ can vary from one set of observed data to another; it is not a parameter.

5.43 **(a)** A $B(200, p)$ distribution seems reasonable (p not known). **(b)** Not binomial (no fixed n). **(c)** A $B(500, 1/12)$ distribution. **(d)** Not binomial (dependent trials).

5.45 **(a)** Caught: $B(10, 0.7)$. Missed: $B(10, 0.3)$. **(b)** 0.3503.

5.47 **(a)** 7 errors caught, 3 missed. **(b)** 1.4491 errors. **(c)** With $p = 0.9$, 0.9487 errors; with $p = 0.99$, 0.3146 errors. σ decreases toward 0 as p approaches 1.

5.49 $m = 6$.

5.51 **(a)** 0.4095. **(b)** 4.

5.53 **(a)** $n = 4$ and $p = 0.25$. **(b)** The probabilities are 0.3164, 0.4219, 0.2109, 0.0469, 0.0039. **(c)** $\mu = 1$ child.

5.55 **(a)** 0.8664. **(b)** 0.9990. **(c)** It increases.

5.57 **(a)** Mean 0.69, standard deviation 0.0008444. **(b)** 68.83% to 69.17%. **(c)** 67% and 70% is a much wider range than we would expect if the population proportion had not changed over time.

5.59 **(a)** 0.7. **(b)** 0.1841.

5.61 **(a)** $p = 0.25$. **(b)** 0.0139. **(c)** $\mu = 5$ and $\sigma = 1.9365$ successes. **(d)** No; the trials would not be independent.

5.63 **(a)** $\mu = 180$ and $\sigma = 12$ successes. **(b)** For $\hat{p}$, $\mu = 0.2$ and $\sigma = 0.01333$. **(c)** About 0.0013. **(d)** 0.2310 or higher.

5.65 **(a)** 0.1137. **(b)** 136.4 blacks. **(c)** About 0.0005.

5.67 **(a)** 0.1788 (or 0.2206). **(b)** 0.0594 (or 0.0721). **(c)** 400. **(d)** Yes.

5.69 Possible values: 1, 2, 3, ... $P(Y = k) = pq^{k-1}$.

5.71 **(a)** $N(123 \text{ mg}, 0.04619 \text{ mg})$. **(b)** 0.

5.73 **(a)** 0.9503. **(b)** 1.

5.75 0.0034.

5.77 0.9231.

5.79 4.0689 to 4.4311.

5.81 **(a)** $\hat{p}_F$: $N(0.82, 0.01921)$. $\hat{p}_M$: $N(0.88, 0.01625)$. **(b)** $N(0.06, 0.02516)$. **(c)** About 0.0087.

5.83 $P(Z \ge 4.56) \doteq 0$.

Chapter 6

6.1 $\sigma_{\bar{x}} = \$0.50$.

6.3 $1.00.

6.5 86.712 to 87.888.

6.7 $n = 554$.

6.9 The students who did not respond are (obviously) not represented in the results.

6.11 Margins of error: 4.3386, 3.0679, 2.1693, 1.5339; interval width decreases with increasing sample size.

6.13 **(a)** She did not divide the standard deviation by $\sqrt{n} = 20$. **(b)** Confidence interval concerns the *population* mean. **(c)** 0.95 is a confidence level, not a probability. **(d)** The large sample size affects the distribution of the sample mean, not the individual ratings.

6.15 **(a)** The margin of error is 0.1478; the interval is 7.3522 to 7.6478. **(b)** The margin of error is 0.1943; the interval is 7.3057 to 7.6943.

6.17 Margin of error, 2.29 U/l. Interval, 10.91 to 15.49 U/l.

6.19 Scenario B has a smaller margin of error; less variability in a single major.

6.21 **(a)** ± 14. **(b)** ± 14.

6.23 No; this is a range of values for the mean rent, not for individual rents.

6.25 (a) 18.6888 to 19.3112 hours. (b) No; this is a range of values for the mean time spent, not for individual times.

6.27 (a) Not certain (only 95% confident). (b) We obtained the interval 85% to 95% by a method that gives a correct result 95% of the time. (c) About 2.5%. (d) No (only random sampling error).

6.29 $\bar{x} = 18.3515$ kpl; the margin of error is 0.6521 kpl.

6.31 $n = 73$.

6.33 No; confidence interval methods can only be applied to an SRS.

6.35 (a) 0.7738. (b) 0.9774.

6.37 H_0: $\mu = 1.4 \, \text{g/cm}^2$; H_a: $\mu \neq 1.4 \, \text{g/cm}^2$.

6.39 $P = 0.1032$.

6.41 (a) 1.645. (b) $z > 1.645$.

6.43 (a) $z = 2.4$. (b) $P = 0.0082$. (c) $P = 0.0164$.

6.45 (a) No. (b) Yes.

6.47 (a) Yes. (b) No. (c) To reject, we need $P < \alpha$.

6.49 (a) $P = 0.022$ and $P = 0.978$. (b) We need to know whether the observed data (for example, $\bar{x}$) are consistent with H_a. (If so, use the smaller P-value.)

6.51 (a) Population mean, not sample mean. (b) H_0 should be that there is no change. (c) A small P-value is needed for significance. (d) Compare P, not z, to α.

6.53 (a) H_0: $\mu = 77$; H_a: $\mu \neq 77$. (b) H_0: $\mu = 20$ seconds; H_a: $\mu > 20$ seconds. (c) H_0: $\mu = 880 \, \text{ft}^2$; H_a: $\mu < 880 \, \text{ft}^2$.

6.55 (a) H_0: $\mu = \$42,800$; H_a: $\mu > \$42,800$. (b) H_0: $\mu = 0.4$ hr; H_a: $\mu \neq 0.4$ hr.

6.57 (a) $P = 0.9656$. (b) $P = 0.0344$. (c) $P = 0.0688$.

6.59 (b) Yes, because 24 is not in the interval. (a) No, because 30 is in the interval.

6.61 $P = 0.09$ means there is some evidence for the wage decrease, but it is not significant at the $\alpha = 0.05$ level.

6.63 The difference was large enough that it would rarely arise by chance. Health issues related to alcohol use are probably discussed in the health and safety class.

6.65 $z \doteq 3.33$, so $P = 0.0009$ (for a two-sided alternative).

6.67 H_0: $\mu = 100$; H_a: $\mu \neq 100$; $z = 5.75$; significant ($P < 0.0001$).

6.69 (a) $z = 2.13$, $P = 0.0166$. (b) The important assumption is that this is an SRS. We also assume a Normal distribution, but this is not crucial provided there are no outliers and little skewness.

6.71 (a) H_0: $\mu = 0$ mpg; H_a: $\mu \neq 0$ mpg, where μ is the mean difference. (b) $z \doteq 4.07$, which gives a very small P-value.

6.73 (a) H_0: $\mu = 0.9$ mg; H_a: $\mu > 0.9$ mg. (b) Yes. (c) No.

6.75 $\bar{x} = 0.8$ is significant, but 0.7 is not. Smaller α means that $\bar{x}$ must be farther away.

6.77 Something that occurs "fewer than 5 times in 100 repetitions" also occurs "fewer than 10 times in 100 repetitions."

6.79 Any z with $2.576 < |z| < 2.807$.

6.81 $P > 0.25$.

6.83 $0.05 < P < 0.10$; $P = 0.0548$.

6.85 In order to determine the effectiveness of alarm systems, we need to know the percent of all homes with alarm systems, and the percent of burglarized homes with alarm systems.

6.87 The first test was barely significant at $\alpha = 0.05$, while the second was significant at any reasonable α.

6.89 A significance test answers only Question b.

6.91 (a) The differences observed might occur by chance even if SES had no effect. (b) This tells us that the statistically insignificant test result did not occur merely because of a small sample size.

6.93 (a) $P = 0.2843$. (b) $P = 0.1020$. (c) $P = 0.0023$.

6.95 With a larger sample, we might have significant results.

6.101 n should be about 100,000.

6.103 Reject the fifth ($P = 0.002$) and eleventh ($P = 0.002$).

6.105 Larger samples give more power.

6.107 Higher; larger differences are easier to detect.

6.109 Power: about 0.99.

6.111 Power: 0.4641.

6.113 (a) Hypotheses: "subject should go to college" and "subject should join work force." Errors: recommending college for someone who is better suited for the work force, and recommending work for someone who should go to college.

6.115 (a) For example, if μ is the mean difference in scores, H_0: $\mu = 0$; H_a: $\mu \neq 0$. (b) No. (c) For example: Was this an experiment? What was the design? How big were the samples?

6.117 (a) 64.45 to 70.01; 67.59 to 73.15; 72.05 to 77.61. (c) 63.83 to 70.63; 66.97 to 73.77; 71.43 to 78.23. (d) With the larger margin of error, the intervals overlap more.

6.121 (a) 4.61 to 6.05 mg/dl. (b) $z = 1.45$, $P = 0.0735$; not significant.

6.123 (b) 26.0614 to 34.7386 μg/l. (c) $z = 2.44$, $P = 0.0073$.

6.125 (a) Under H_0, $\bar{x}$ has a $N(0\%, 5.3932\%)$ distribution. (b) $z = 1.28$, $P = 0.1003$. (c) Not significant.

6.127 It is essentially correct.

6.129 Find $\bar{x}$, then take $\bar{x} \pm 1.96(4/\sqrt{12}) = \bar{x} \pm 2.2632$.

6.131 Find $\bar{x}$, then compute $z = (\bar{x} - 23)/(4/\sqrt{12})$. Reject H_0 based on your chosen significance level.

Chapter 7

7.1 (a) $24. **(b)** 15.

7.3 $561.86 to $664.14.

7.5 (a) Yes. **(b)** No.

7.9 -1.33 to 10.13.

7.11 The sample size should be sufficient to overcome any non-Normality, but the mean μ might not be a useful summary of a bimodal distribution.

7.13 The power is about 0.9394.

7.15 (a) $t^* = 2.228$. **(b)** $t^* = 2.080$. **(c)** $t^* = 1.721$. **(d)** t^* decreases with increasing sample size and increases with increasing confidence.

7.17 $t^* = 1.740$ (or -1.740).

7.19 For the alternative $\mu < 0$, the answer would be the same ($P = 0.037$). For the alternative $\mu > 0$, the answer would be $P = 0.963$.

7.21 (a) df $= 27$. **(b)** $1.703 < t < 2.052$. **(c)** $0.05 < P < 0.10$. **(d)** Not significant at 5% or 1%. **(e)** $P = 0.0546$.

7.23 It depends on whether $\bar{x}$ is on the appropriate side of μ_0.

7.25 (a) $H_0: \mu = 10$; $H_a: \mu < 10$. **(b)** $t \doteq -5.26$, df $= 33$, $P < 0.0001$.

7.27 (a) Distribution is not Normal; it has two peaks and one large value. **(b)** Maybe; we have a large sample but a small population. **(c)** 27.29 ± 5.717, or 21.57 to 33.01 cm. **(d)** One could argue for either answer.

7.29 (a) Not Normal (lots of 1s and 10s), but no outliers. **(b)** 4.92 to 6.88. **(c)** Because this is not a random sample, it may not represent other children well.

7.31 90%: 2.119 to 2.321. 95%: 2.099 to 2.341. Width increases with confidence level.

7.33 (a) $t = 5.13$, df $= 15$, $P < 0.001$. **(b)** With 95% confidence, the mean NEAT increase is between 192 and 464 calories.

7.35 (a) $H_0: \mu_c = \mu_d$; $H_a: \mu_c \neq \mu_d$. **(b)** $t \doteq 4.358$, $P \doteq 0.0003$; we reject H_0.

7.37 (a) $H_0: \mu = 925$; $H_a: \mu > 925$. $t \doteq 3.27$ (df $= 35$), $P \doteq 0.0012$. **(b)** $H_0: \mu = 935$; $H_a: \mu > 935$. $t \doteq 0.80$, $P \doteq 0.2158$. **(c)** The confidence interval includes 935, but not 925.

7.39 (a) The differences are spread from -0.018 to 0.020 g. **(b)** $t = -0.347$, df $= 7$, $P = 0.7388$. **(c)** -0.0117 to 0.0087 g. **(d)** They may be representative of future

subjects, but the results are suspect because this is not a random sample.

7.41 (a) $H_0: \mu = 0$; $H_a: \mu > 0$. **(b)** Slightly left-skewed; $\bar{x} = 2.5$ and $s = 2.893$. **(c)** $t = 3.865$, df $= 19$, $P = 0.00052$. **(d)** 1.146 to 3.854.

7.43 $\bar{x} = 114.98$, $s = 14.80$; 111.14 to 118.82. This might adequately describe the mean IQ at this school, but the sample could not be considered representative of all fifth-graders.

7.45 For the sign test, $P = 0.0898$; not quite significant, unlike Exercise 7.34.

7.47 H_0: median $= 0$; H_a: median $\neq 0$; $P = 0.7266$. This is similar to the t test P-value.

7.49 H_0: median $= 0$; H_a: median > 0; $P = 0.0013$.

7.51 Reject H_0 if $|\bar{x}| \geq 0.00677$. The power is about 7%.

7.53 $n > 26$. (The power is about 0.7999 when $n = 26$.)

7.55 $0.02 < P < 0.04$.

7.57 -20.3163 to 0.3163; do not reject H_0.

7.59 SPSS and SAS give both results; the pooled t is 1.998, for which $P = 0.0808$.

7.61 (a) Hypotheses should involve μ_1 and μ_2. **(b)** The samples are not independent. **(c)** We need P to be small (for example, less than 0.10) to reject H_0. **(d)** t should be negative.

7.63 (a) No (in fact, $P \doteq 0.0542$). **(b)** Yes ($P \doteq 0.0271$).

7.65 (a) While the distributions do not look particularly Normal, they have no extreme outliers or skewness. **(b)** $\bar{x}_N \doteq 0.5714$, $s_N \doteq 0.7300$, $n_N = 14$; $\bar{x}_S \doteq 2.1176$, $s_S \doteq 1.2441$, $n_S = 17$. **(c)** $H_0: \mu_N = \mu_S$; $H_a: \mu_N < \mu_S$. **(d)** $t = -4.303$, so $P \doteq 0.0001$ (df $\doteq 26.5$) or $P < 0.0005$ (df $= 13$). **(e)** -2.2842 to -0.8082 (df $\doteq 26.5$) or -2.3225 to -0.7699 (df $= 13$).

7.67 (a) $\bar{x}_F \doteq 4.0791$, $s_F \doteq 0.9861$; $\bar{x}_M \doteq 3.8326$, $s_M \doteq 1.0677$. **(b)** Both distributions are somewhat skewed, but because the ratings range from 1 to 5, there are no outliers. **(c)** $t \doteq 2.898$, $P \doteq 0.0040$ (df $\doteq 402.2$) or $0.002 < P < 0.005$ (df $= 220$). **(d)** 0.0793 to 0.4137 (df $\doteq 402.2$) or 0.0777 to 0.4153 (df $= 220$). **(e)** The difference in means might not be as large as 0.25.

7.69 (a) Assuming we have SRSs from each population, this seems reasonable. **(b)** $H_0: \mu_f = \mu_m$; $H_a: \mu_f \neq \mu_m$. **(c)** $t \doteq 0.276$, $P \doteq 0.78$. **(d)** -11.75 to 15.53 (df $\doteq 76.1$) or -12.1 to 15.9 (df $= 30$) mg/dl. **(e)** It might not be appropriate to treat these students as SRSs from larger populations.

7.71 (a) Not Normal, because all numbers are integers. **(b)** Yes; we have two large samples, with no outliers. **(c)** $H_0: \mu_1 = \mu_2$; $H_a: \mu_1 >$ (or $\neq$) μ_2. **(d)** $t = 6.258$, df $= 354$ or 164, $P < 0.0001$. **(e)** 0.51 to 0.99 (regardless of df). **(f)** This may not generalize well to other areas of the country.

7.73 (a) This may be near enough to an SRS if this company's working conditions were similar to those of other workers. **(b)** 9.99 to 13.01 mg.y/m^3. **(c)** $t = 15.08$, $P < 0.0001$ with either df $= 137$ or 114. **(d)** The sample sizes are large enough that skewness should not matter.

7.75 You need either sample sizes and standard deviations or degrees of freedom and a more accurate value for the P-value. The confidence interval will give us useful information about the magnitude of the difference.

7.77 This is a matched pairs design.

7.79 The next 10 employees who need screens might not be an independent group—perhaps they all come from the same department, for example.

7.81 (a) The north distribution (five-number summary 2.2, 10.2, 17.05, 39.1, 58.8 cm) is right-skewed, while the south distribution (2.6, 26.1, 37.70, 44.6, 52.9) is left-skewed. **(b)** The methods of this section seem to be appropriate. **(c)** H_0: $\mu_n = \mu_s$; H_a: $\mu_n \neq \mu_s$. **(d)** $t = -2.63$ with df $= 55.7$ ($P = 0.011$) or df $= 29$ ($P = 0.014$). **(e)** Either -19.09 to -2.57 or -19.26 to -2.40 cm.

7.83 (a) Either -0.90 to 6.90 units (df $= 122.5$) or -0.95 to 6.95 units (df $= 54$). **(b)** Random fluctuation may account for the difference in the two averages.

7.85 (a) H_0: $\mu_B = \mu_F$; H_a: $\mu_B > \mu_F$; $t = 1.654$, $P = 0.053$ (df $= 37.6$) or $P = 0.058$ (df $= 18$). **(b)** -0.2 to 2.0. **(c)** We need two independent SRSs from Normal populations.

7.87 $s_p = 0.9347$; $t \doteq 3.636$, df $= 40$, $P \doteq 0.0008$; 0.4663 to 1.6337. Both results are similar to Exercise 7.66.

7.89 $s_p = 15.96$; $t = -2.629$, df $= 58$, $P = 0.0110$; -19.08 to -2.58 cm. All results are nearly the same as in Exercise 7.81.

7.91 df $= 55.725$.

7.93 (a) df $= 137.066$. **(b)** $s_p = 5.332$ (slightly closer to s_2, from the larger sample). **(c)** With no assumption, $SE_1 = 0.7626$; with the pooled method, $SE_2 = 0.6136$. **(d)** $t = 18.74$, df $= 333$, $P < 0.0001$. t and df are larger, so the evidence is stronger (although it was quite strong before). **(e)** df $= 121.503$; $s_p = 1.734$; $SE_1 = 0.2653$ and $SE_2 = 0.1995$. $t = 24.56$, df $= 333$, $P < 0.0001$.

7.95 (a) $F^* = 2.15$. **(b)** Significant at the 10% level but not at the 5% level.

7.97 A larger σ would yield less power.

7.99 $F = 1.742$ with df $= 60$ and 176; $P = 0.0057$. We do not know if the distributions are Normal, so this test may not be reliable.

7.101 $F = 1.017$ with df $= 211$ and 164; $P = 0.9114$. The distributions are not Normal, so this test may not be reliable (although the conclusion is reasonable). To reject at the 5% level, s_2 would need to be at least 1.39 (based on Table E) or 1.33 (software).

7.103 $F = 5.263$ with df $= 114$ and 219; $P < 0.0001$. The authors described the distributions as somewhat skewed, so the Normality assumption may be violated.

7.105 $F = 1.506$ with df $= 29$ and 29; $P = 0.2757$. The stemplots in Exercise 7.81 did not appear to be Normal.

7.107 (a) $F \doteq 1.12$; do not reject H_0. **(a)** The critical values are 9.60, 15.44, 39.00, and 647.79. With small samples, these are low-power tests.

7.109 Using a larger σ for planning the study is advisable because it provides a conservative (safe) estimate of the power.

7.111 $\bar{x} = 140.5$, $s \doteq 13.58$, $s_{\bar{x}} \doteq 6.79$. We cannot consider these four scores to be an SRS.

7.113 As df increases, t^* approaches 1.96.

7.115 Margins of error decrease with increasing sample size.

7.117 (a) Two independent samples. **(b)** Matched pairs. **(c)** Single sample.

7.119 (a) H_0: $\mu = 1.5$; H_a: $\mu < 1.5$. $t \doteq -2.344$, $P \doteq 0.01$. **(b)** 0.95 to 1.45 violations. **(d)** The sample size should be large enough to make t procedures safe.

7.121 (a) Body weight: mean -0.7 kg, SE 2.298 kg. Caloric intake: mean 14 cal, SE 56.125 cal. **(b)** $t_1 = -0.305$ (body weight) and $t_2 = 0.249$ (caloric intake), both with df $= 13$; both P-values are about 0.8. **(c)** -5.66 to 4.26 kg and -107.23 to 135.23 cal.

7.123 How much a person eats may depend on how many people he or she is sitting with.

7.125 For north/south differences: $t = 7.15$, df $= 575.4$ or 283, $P < 0.0001$; the confidence interval is 7.52 to 13.22 cm. For east/west differences: $t = 3.69$, df $= 472.7$ or 230, $P < 0.0005$; the confidence interval is 2.68 to 8.78 cm. With larger samples, t increases, P decreases, and the intervals shrink.

7.127 (a) At each nest, the same mockingbird responded on each day. **(b)** 6.9774 m. **(c)** $t \doteq 6.32$, $P < 0.0001$. **(d)** 5.5968 m; $t \doteq -1.05$, $P \doteq 0.3045$.

7.129 78.3% $\pm$ 13.8%, or 64.5% to 92.1%.

7.131 GPA: $t = -0.91$, df $= 74.9$ ($P = 0.1839$) or 30 ($0.15 < P < 0.20$). Confidence interval: -1.33 to 0.5. IQ: $t = 1.64$, df $= 56.9$ ($P = 0.0503$) or 30 ($0.05 < P < 0.10$). Confidence interval: -1.12 to 11.36.

7.133 $t = 3.65$, df $= 237.0$ or 115, $P < 0.0005$. 95% confidence interval for the difference: 0.78 to 2.60.

7.135 $t = -0.3533$, df $= 179$, $P = 0.3621$.

7.137 No; what we have is nothing like an SRS.

7.139 Basal: $\bar{x} = 41.0455$, $s = 5.6356$. DRTA: $\bar{x} = 46.7273$, $s = 7.3884$. Strat: $\bar{x} = 44.2727$, $s = 5.7668$. **(a)** $t \doteq 2.87$, $P < 0.005$. Confidence interval for difference: 1.7 to 9.7 points. **(b)** $t \doteq 1.88$, $P < 0.05$. Confidence interval for difference: -0.24 to 6.7 points.

7.141 (a) Three-bedroom homes are slightly right-skewed; four-bedroom homes are generally more expensive. The top two prices from the three-bedroom distribution qualify as outliers. **(b)** $t \doteq -4.475$ with either df $= 20.98$ ($P \doteq 0.0002$) or df $= 13$ ($P < 0.001$); we reject H_0. **(c)** It would be reasonable to guess that $\mu_3 < \mu_4$. **(d)** \$63,823 to \$174,642 (df $= 20.97$) or \$61,685 to \$176,779 (df $= 13$). **(e)** It seems that these houses should be a fair representation of three- and four-bedroom houses in West Lafayette.

Chapter 8

8.1 (a) $n = 760$ banks. **(b)** $X = 283$ banks expected to acquire another bank. **(c)** $\hat{p} \doteq 0.3724$.

8.3 (a) 0.01754. **(b)** 0.3724 ± 0.0344. **(c)** 33.8% to 40.7%.

8.5 Shade above 1.34 and below -1.34.

8.7 $\hat{p} = 0.75$, $z = 2.24$, $P = 0.0250$.

8.9 (a) $z = -1.34$, $P = 0.1802$. **(b)** 0.1410 to 0.5590—the complement of the interval shown in Figure 8.2.

8.11 The plot is symmetric about 0.5, where it has its maximum.

8.13 (a) Margin of error only accounts for random sampling error. **(b)** P-values measure the strength of the evidence against H_0, not the probability of it being true. **(c)** The confidence level cannot exceed 100%.

8.15 $\hat{p} = 0.6548$, 0.6416 to 0.6680.

8.17 (a) Without that number, we do not know the margin of error for the statistic. **(b)** About 2365. **(c)** 0.65 to 0.69. **(d)** We do not know the sampling methods used, which might make these methods unreliable.

8.19 (a) $\hat{p} \leq 0.2667$ and $\hat{p} \geq 0.5333$. **(b)** $\hat{p} \leq 0.30$ and $\hat{p} \geq 0.50$. **(c)** With a larger sample size, smaller values of $|\hat{p} - 0.4|$ lead to the rejection of H_0.

8.21 (a) About 67,179 students. **(b)** 0.4168 to 0.4232.

8.23 0.4043 to 0.4557.

8.25 (a) ± 0.001321. **(b)** Other sources of error are much more significant than sampling error.

8.27 (a) 0.3506 to 0.4094. **(b)** Yes; some respondents might not admit to such behavior.

8.29 (a) $\hat{p} = 0.3275$; 0.3008 to 0.3541. **(b)** Speakers and listeners probably perceive sermon length differently.

8.31 (a) No. **(b)** Yes. **(c)** Yes. **(d)** No.

8.33 0.6345 to 0.7455.

8.35 0.2180 to 0.2510.

8.37 0.8230 to 0.9370.

8.39 (a) $z = 1.34$, $P = 0.1802$. **(b)** 0.4969 to 0.5165.

8.41 $n = 171$ or 172.

8.43 The sample sizes are 35, 62, 81, 93, 97, 93, 81, 62, and 35; take $n = 97$.

8.45 Mean -0.1, standard deviation 0.1339.

8.47 (a) Means p_1 and p_2, standard deviations $\sqrt{p_1(1 - p_1)/n_1}$ and $\sqrt{p_2(1 - p_2)/n_2}$. **(b)** $p_1 - p_2$. **(c)** $p_1(1 - p_1)/n_1 + p_2(1 - p_2)/n_2$.

8.49 The interval for $q_w - q_m$ is -0.0030 to 0.2516.

8.51 The sample proportions support the alternative hypothesis $p_m > p_w$; $P = 0.0288$.

8.53 $z = 11.99$, for which P is tiny.

8.55 (a) $n_1 = 1063$, $\hat{p}_1 \doteq 0.54$, $n_2 = 1064$, $\hat{p}_2 \doteq 0.89$. (We can estimate $X_1 \doteq 574$ and $X_2 \doteq 947$.) **(b)** 0.35. **(c)** Yes; large, independent samples from two populations. **(d)** 0.3146 to 0.3854. **(e)** 35%; 31.5% to 38.5%. **(f)** A possible concern: adults were surveyed before Christmas.

8.57 (a) $n_1 = 1063$, $\hat{p}_1 \doteq 0.73$, $n_2 = 1064$, $\hat{p}_2 \doteq 0.76$. (We can estimate $X_1 \doteq 776$ and $X_2 \doteq 809$.) **(b)** 0.03. **(c)** Yes; large, independent samples from two populations. **(d)** -0.0070 to 0.0670. **(e)** 3%; -0.7% to 6.7%. **(f)** A possible concern: adults were surveyed before Christmas.

8.59 No; we need independent samples from different populations.

8.61 (a) H_0 should refer to p_1 and p_2. **(b)** Only if $n_1 = n_2$. **(c)** Confidence intervals account for only sampling error.

8.63 Pet owners: 0.4790. Others: 0.5281. Confidence interval: 0.0032 to 0.0950.

8.65 $z = 4.28$ and $P < 0.0001$. Confidence interval: 0.0381 to 0.1019.

8.67 (a) -0.0053 to 0.2335. **(b)** $z \doteq 1.83$, $P = 0.0336$. **(c)** We have fairly strong evidence that high-tech companies are more likely to offer stock options, but the difference in proportions could be very small or as large as 23%.

8.69 (a) $\hat{p}_f = 0.8$, SE $\doteq 0.05164$; $\hat{p}_m = 0.3939$, SE $\doteq 0.04253$. **(b)** 0.2960 to 0.5161.

8.71 $z = 2.10$, $P = 0.0360$.

8.73 (a) We have six chances to make an error. **(b)** Use $z^* = 2.65$. **(c)** 0.705 to 0.775, 0.684 to 0.756, 0.643 to 0.717, 0.632 to 0.708, 0.622 to 0.698, and 0.571 to 0.649.

8.75 (a) $\hat{p} = 0.164$ and $X = 2460$. **(b)** 0.1581 to 0.1699. **(c)** 16.4%; 15.8% to 17.0%. **(d)** 0.122. **(e)** 0.0067.

8.77 0.0747 to 0.1453.

8.79 (a) 0.0713 to 0.1487, $z = 5.48$. **(b)** 0.0772 to 0.1428, $z = 6.59$. **(c)** The confidence interval and z statistic change slightly, but the conclusions are roughly the same.

8.81 As sample size increases, the margin of error decreases.

8.83 −0.0298 to 0.0898.

8.85 **(a)** People have different symptoms; for example, not all who wheeze consult a doctor. **(b)** Sleep: 0.0864, 0.0280 to 0.1448. Number: 0.0307, −0.0361 to 0.0976. Speech: 0.0182, −0.0152 to 0.0515. Activities: 0.0137, −0.0395 to 0.0670. Doctor: −0.0112, −0.0796 to 0.0573. Phlegm: −0.0220, −0.0711 to 0.0271. Cough: −0.0323, −0.0853 to 0.0207. **(c)** It is reasonable to expect that the bypass proportions would be higher. **(d)** In the same order: $z = 2.64$, $P = 0.0042$; $z = 0.88$, $P = 0.1897$; $z = 0.99$, $P = 0.1600$; $z = 0.50$, $P = 0.3100$; $z = −0.32$, $P = 0.6267$; $z = −0.92$, $P = 0.8217$; $z = −1.25$, $P = 0.8950$. **(e)** 95% confidence interval for sleep improvement: 0.1168 to 0.2023. **(f)** Part (b) showed improvement relative to control group, which is a better measure of the effect of the bypass.

8.87 The margin of error is ±2.8%.

8.89 $z = 8.95$, $P < 0.0001$; 0.3720 to 0.5613.

8.91 All $\hat{p}$-values are greater than 0.5. Texts 3, 7, and 8 have (respectively) $z = 0.82$, $P = 0.4122$; $z = 3.02$, $P = 0.0025$; and $z = 2.10$, $P = 0.0357$. For the other texts, $z \geq 4.64$ and $P < 0.00005$.

8.95 z: 0.90, 1.01, 1.27, 1.42, 2.84, 3.18, 4.49. P: 0.3681, 0.3125, 0.2041, 0.1556, 0.0045, 0.0015, 0.0000.

8.97 **(a)** $n = 342$. **(b)** $n = (z^*/m)^2/2$.

8.99 **(a)** $p_0 = 0.7911$. **(b)** $\hat{p} = 0.3897$, $z = −29.1$, P is tiny. **(c)** $\hat{p}_1 = 0.3897$, $\hat{p}_2 = 0.7930$, $z = −29.2$, P is tiny.

8.101 **(a)** 0.5278 to 0.5822. **(b)** 0.5167 to 0.5713. **(c)** 0.3170 to 0.3690. **(d)** 0.5620 to 0.6160. **(e)** 0.5620 to 0.6160. **(f)** 0.6903 to 0.7397.

Chapter 9

9.1 **(a)** Given Explanatory = 1: 35% Yes, 65% No. Given Explanatory = 2: 45% Yes, 55% No. **(c)** When Explanatory = 2, "Yes" and "No" are closer to being evenly split.

9.3 The relative risk is 0.838.

9.5 **(a)** $0.02 < P < 0.025$. **(b)** $0.10 < P < 0.15$. **(c)** $0.01 < P < 0.02$. **(d)** $0.01 < P < 0.02$.

9.7 The expected counts were rounded.

9.9 The other five chi-square components are 0.5820, 0.0000, 0.0196, 0.0660, and 0.2264; the six values add up to 0.93.

9.11 **(a)** The broadband counts are 112.5, 540, 1080, and 1237.5. **(b)** $X^2 = 1601.8$, df = 3, $P < 0.0001$. **(c)** $X^2 = 22.07$, df = 1, $P < 0.0001$.

9.13 For any such scenario, the conclusion is the same.

9.15 **(a)** For example, in the "small" stratum, 51 claims were allowed, 6 not allowed, and 57 total. Altogether,

there were 79 claims; 67 were allowed and 12 were not. **(b)** 10.5% (small claims), 29.4% (medium), and 20.0% (large) were not allowed. **(c)** In the 3 × 2 table, the expected count for large/not allowed is too small. **(d)** There is no relationship between claim size and whether a claim is allowed. **(e)** $X^2 = 3.456$, df = 1, $P = 0.063$.

9.17 There is strong evidence of a change ($X^2 = 307.8$, df = 2, $P < 0.0001$).

9.19 **(a)** For example, among those students in trades, 320 enrolled right after high school, and 622 enrolled later. **(b)** For example, in addition to the given percents, 39.4% of these students enrolled right after high school. **(c)** $X^2 = 276.1$, df = 5, $P < 0.0001$.

9.21 **(a)** For example, among those students in trades, 188 relied on parents, family, or spouse, and 754 did not. **(b)** $X^2 = 544.8$, df = 5, $P < 0.0001$. **(c)** In addition to the given percents, 25.4% of all students relied on family support.

9.23 **(a)** 50.5% get enough sleep; 49.5% do not. **(b)** 32.2% get enough sleep; 67.8% do not. **(c)** Those who exercise more than the median are more likely to get enough sleep. **(d)** $X^2 = 22.577$, df = 1, $P < 0.0001$.

9.25 **(a)** There were 27,802 students total; 25,893 agree and 1909 disagree; 12,887 male and 14,915 female. **(b)** 91% of males and 95% of females agreed that trust and honesty are essential. **(d)** $X^2 = 175.0$, df = 1, $P < 0.0001$.

9.27 **(a)** 0.2044, 0.0189; 0.3285, 0.0699; 0.1050, 0.1072; 0.0518, 0.1141. **(b)** 0.2234, 0.3984, 0.2123, 0.1659. **(c)** 0.6898, 0.3102. **(d)** Full-time students: 0.2964, 0.4763, 0.1522, 0.0752. Part-time students: 0.0610, 0.2254, 0.3458, 0.3678.

9.29 **(a)** 57.98%. **(b)** 30.25%. **(c)** To test "There is no relationship between waking and bedtime symptoms" versus "There is a relationship," we find $X^2 \doteq 2.275$, df = 1, $P \doteq 0.132$.

9.31 Start by setting a equal to any number from 0 to 50.

9.33 **(a)** A notably higher percent of women are "strictly voluntary" participants. **(b)** 40.3% of men and 51.3% of women are participants; the relative risk is 1.27.

9.35 **(a)** 146 women/No, 97 men/No. **(b)** For example, 19.34% of women, versus 7.62% of men, have tried low-fat diets. **(c)** $X^2 = 7.143$, df = 1, $P = 0.008$.

9.37 **(b)** $X^2 = 2.591$, df = 1, $P = 0.108$.

9.39 **(a)** $X^2 = 76.7$, df = 2, $P < 0.0001$. **(b)** Even with much smaller numbers of students, P is still very small. **(c)** Our conclusion might not hold for the true percents. **(d)** Lack of independence could cause the estimated percents to be too large or too small.

9.41 $X^2 = 12.0$, df = 1, $P = 0.001$. The smallest expected count is 6, so the test is valid.

9.43 $X^2 = 23.1$, df $= 4$, $P < 0.0005$. Dog owners have less education, and cat owners more, than we would expect if there were no relationship between pet ownership and educational level.

9.45 The missing entries are 202, 64, 38, 33. $X^2 = 50.5$, df $= 9$, $P < 0.0005$. The largest contributions to X^2 come from chemistry/engineering, physics/engineering, and biology/liberal arts (more than expected), and biology/engineering and chemistry/liberal arts (less than expected).

9.47 $X^2 = 852.433$, df $= 1$, $P < 0.0005$.

9.49 $X^2 = 3.781$, df $= 3$, $P = 0.2861$.

9.55 (a) We expect each quadrant to contain one-fourth of the 100 trees. (b) *Some* random variation would not surprise us. (c) $X^2 = 10.8$, df $= 3$, $P = 0.0129$.

Chapter 10

10.1 (a) 2.8. (b) When x increases by 1, μ_y increases by 2.8. (c) 63. (d) 54.4 to 71.6.

10.3 (a) A decrease of 0.341 to 0.969 kg/m². (b) An increase of 0.341 to 0.969 kg/m². (c) A decrease of 0.1705 to 0.4845 kg/m².

10.5 (a) The plot suggests a linear increase. (b) $\hat{y} = -4566.24 + 2.3x$. (c) Residuals: -0.56, 0.34, 0.54, 0.14, -0.46. $s \doteq 0.5633$. (d) Given x (the year), spending comes from a $N(\mu_y, \sigma)$ distribution, where $\mu_y = \beta_0 + \beta_1 x$. Estimates: $b_0 = -4566.24$, $b_1 = 2.3$, and $s \doteq 0.5633$. (e) With 95% confidence, R&D spending increases between 1.733 and 2.867 billion dollars per year.

10.7 (a) β_0, β_1, and σ. (b) H_0 should refer to β_1. (c) The confidence interval will be narrower than the prediction interval.

10.9 (a) $t = 1.90$, $P = 0.0705$. (b) $t = 3.62$, $P = 0.0014$. (c) $t = 1.90$, $P = 0.0608$.

10.11 (a) 1.3649 to 2.0199; a $1 difference in tuition in 2000 changes 2008 tuition by between $1.36 and $2.02. (b) 78.2%. (c) $9764. (d) $15,856. (e) The Moneypit U prediction requires extrapolation.

10.13 (a) $\hat{y} = -0.0127 + 0.0180x$, $r^2 \doteq 80.0\%$. (b) H_0: $\beta_1 = 0$; H_a: $\beta_1 > 0$; $t = 7.48$, $P < 0.0001$. (c) The predicted mean is 0.07712; the interval is 0.040 to 0.114.

10.15 (a) Both distributions are sharply right-skewed; the five-number summaries are 0%, 0.306%, 1.43%, 17.65%, 85.01%, and 0, 2.25, 6.31, 12.69, and 27.88. (b) No; x and y do not need to be Normal. (c) There is a weak positive linear relationship. (d) $\hat{y} = 6.247 + 0.1063x$. (e) The residuals are slightly right-skewed.

10.17 (a) Twenty-two of these 30 homes sold for more than their assessed values. (b) A moderately strong linear association. (c) $\hat{y} = 21.50 + 0.9468x$. (d) There are no obvious unusual features. (e) A stemplot or histogram

looks reasonably Normal. (f) There are no clear violations of the assumptions.

10.19 (a) The plot is roughly linear and increasing. The 2004 and 2008 counts are noticeably high. (b) $\hat{y} \doteq -28{,}438 + 14.82x$; the confidence interval for the slope is 11.89 to 17.76 tornadoes per year. (c) The scatter might be greater in recent years, and the 2004 residual is particularly high. (d) The 2004 residual is an outlier; the other residuals appear to be roughly Normal. (e) $\hat{y} \doteq -26{,}584 + 13.88x$.

10.21 (a) 8.41%. (b) $t \doteq 9.12$, $P < 0.0001$. (c) The students who did not answer might have different characteristics.

10.23 (a) x (percent forested) is right-skewed; $\bar{x} = 39.3878\%$, $s_x = 32.2043\%$. y (IBI) is left-skewed; $\bar{y} = 65.9388$, $s_y = 18.2796$. (b) A weak positive association, with more scatter in y for small x. (c) $y_i = \beta_0 + \beta_1 x_i + \epsilon_i$, $i = 1, 2, \ldots, 49$; ϵ_i are independent $N(0, \sigma)$ variables. (d) H_0: $\beta_1 = 0$; H_a: $\beta_1 \neq 0$. (e) $\widehat{\text{IBI}} = 59.9 + 0.153\,\text{Area}$; $s = 17.79$. For testing the hypotheses in (d), $t = 1.92$ and $P = 0.061$. (f) Residual plot shows a slight curve. (g) Residuals are left-skewed.

10.25 The first change decreases P (that is, the relationship is more significant) because it accentuates the positive association. The second change weakens the association, so P increases (the relationship is less significant).

10.27 Using area $= 10$, $\hat{y} \doteq 57.52$; using forest $= 63$, $\hat{y} \doteq 69.55$. Both predictions have a lot of uncertainty (the prediction intervals are about 70 units wide).

10.29 (a) Both variables are right-skewed. Pure tones: $\bar{x} = 106.2$ and $s = 91.76$ spikes/second. Monkey calls: $\bar{y} = 176.6$ and $s_y = 111.85$ spikes/second. (b) A moderate positive association; the third point has the largest residual; the first point is an outlier for tone response. (c) $\widehat{\text{CALL}} = 93.9 + 0.778\,\text{TONE}$, $s = 87.30$, $t = 4.91$, $P < 0.0001$. (d) Without the first point, $\hat{y} = 101 + 0.693x$, $s = 88.14$, $t = 3.18$. Without the third point, $\hat{y} = 98.4 + 0.679x$, $s = 80.69$, $t = 4.49$. With neither, $\hat{y} = 116 + 0.466x$, $s = 79.46$, $t = 2.21$.

10.31 (a) Both distributions are fairly symmetric. For x (MOE), $\bar{x} = 1{,}799{,}180$ and $s_x = 329{,}253$; for y (MOR), $\bar{y} = 11{,}185$ and $s_y = 1980$. (b) Put MOE on the x axis. (c) $y_i = \beta_0 + \beta_1 x_i + \epsilon_i$, $i = 1, 2, \ldots, 32$; ϵ_i are independent $N(0, \sigma)$ variables. $\widehat{\text{MOR}} = 2653 + 0.00474\,\text{MOE}$, $s = 1238$, $t = 7.02$, $P < 0.0001$. (d) Assumptions appear to be met.

10.33 (a) Scatterplot shows a weak negative association. $\widehat{\text{Bonds}} = 55.58 - 0.1769\,\text{Stocks}$, $s = 54.55$. (b) H_0: $\beta_1 = 0$; H_a: $\beta_1 \neq 0$; $t = -1.66$, $P = 0.111$. (c) $\widehat{\text{Bonds}} = 69.46 - 0.2814\,\text{Stocks}$, $s = 53.12$. The slope is now significantly different from 0 ($t = -2.24$, $P = 0.036$). (d) We should explore whether something happened in 2008 that might explain why that point strayed from the line.

10.35 (a) Aside from the one high point, there is a moderate positive association. **(b)** $\widehat{\text{Wages}}$ = $43.4 + 0.0733\,\text{LOS}$, $t = 2.85$, $P = 0.006$. **(c)** Wages rise an average of 0.0733 wage units per week of service. **(d)** 0.0218 to 0.1247.

10.37 (a) It appears to be quite linear. **(b)** $\widehat{\text{Lean}}$ = $-61.12 + 9.3187\,\text{Year}$; $r^2 = 98.8\%$. **(c)** 8.36 to 10.28 tenths of a millimeter/year.

10.39 (a) $x = 112$. **(b)** 2.9983 m. **(c)** Use a prediction interval.

10.41 $t \doteq -4.16$, df = 116, $P < 0.0001$.

10.43 We cannot reject H_0: $\rho = 0$, because $t = -1.66$, $P = 0.111$.

10.45 DFM = 1, DFE = 18, SSE = 3995.4, MSM = 4560.6, MSE = 221.97, $F = 20.55$.

10.47 The standard error is 0.1710; the confidence interval is 0.4158 to 1.1342.

10.49 $n = 20$: $t = 3.18$, df = 18, $P = 0.0052$. $n = 10$: $t = 2.12$, df = 8, $P = 0.0667$.

10.51 (a) Strong positive linear association with one outlier (SAT 420, ACT 21). **(b)** $\widehat{\text{ACT}} = 1.63 + 0.0214\,\text{SAT}$, $t = 10.78$, $P < 0.0005$. **(c)** $r = 0.8167$.

10.53 (a) $a_1 = 0.02617$, $a_0 = -2.7522$. **(c)** Mean 21.1333 and standard deviation 4.7137—the same as for the ACT scores.

10.55 (a) For squared length: $\widehat{\text{Weight}} = -118 + 0.497\,\text{SQLEN}$, $r^2 = 0.977$. **(b)** For squared width: $\widehat{\text{Weight}} = -99.0 + 18.7\,\text{SQWID}$, $r^2 = 0.965$.

10.57 IBI and area: $r = 0.4459$, $t = 3.42$, $P = 0.001$ (from Exercise 10.14). IBI and percent forested: $r = 0.2698$, $t = 1.92$, $P = 0.061$ (Exercise 10.15). Area and percent forested: $r = -0.2571$, $t = -1.82$, $P = 0.074$.

10.59 The three smallest correlations (0.16 and 0.19) are the only ones that are not significant ($P = 0.1194$ and 0.0637). The three largest correlations for the whole group all had $P < 0.01$, and the remainder have $P < 0.05$.

10.61 (a) 95% confidence interval for women: 14.73 to 33.33. For men: -9.47 to 42.97. These intervals overlap quite a bit. **(b)** For women: 22.78. For men: 16.38. The women's slope standard error is smaller in part because it is divided by a large number. **(c)** Choose men with a wider variety of lean body masses.

Chapter 11

11.1 (a) Math GPA. **(b)** $n = 106$. **(c)** $p = 4$. **(d)** SAT Math, SAT Verbal, class rank, and mathematics placement score.

11.3 (a) Math GPA should increase when any explanatory variable increases. **(b)** DFM = 4, DFE = 81. **(c)** All four coefficients are significantly different from 0 (although the intercept is not).

11.5 The correlations are found in Figure 11.4. The scatterplots for the pairs with the largest correlations are easy to pick out. The whole-number scale for high school grades causes point clusters in those scatterplots.

11.7 (a) 0.0016 to 12.7984. **(b)** 0.1721 to 12.6279. **(c)** -0.0139 to 12.8139. **(d)** 0.2496 to 12.5504.

11.9 (a) H_0 should refer to β_2. **(b)** Squared multiple correlation. **(c)** Small P implies that at least one coefficient is different from 0.

11.11 (a) $y_i = \beta_0 + \beta_1 x_{i1} + \beta_2 x_{i2} + \cdots + \beta_8 x_{i8} + \epsilon_i$, where $i = 1, 2, \ldots, 135$; ϵ_i are independent $N(0, \sigma)$ random variables. **(b)** Model (df = 8), error (df = 126), and total (df = 134).

11.13 (a) 8 and 786. **(b)** 7.84%; this model is not very predictive. **(c)** Males and Hispanics consume energy drinks more frequently. Consumption increases with risk-taking scores. **(d)** Within a group of students with identical (or similar) values of those other variables, energy-drink consumption increases with increasing jock identity and increasing risk taking.

11.15 (a) Model 1: DFE = 200. Model 2: DFE = 199. **(b)** $t = 3.09$, $P = 0.0023$. **(c)** For gene expression: $t = 2.44$, $P = 0.0156$. For RB: $t = 3.33$, $P = 0.0010$. **(d)** The relationship is still positive. When gene expression increases by 1, popularity increases by 0.204 in Model 1, and by 0.161 in Model 2 (with RB fixed).

11.17 (a) $\widehat{\text{BMI}} = 23.4 - 0.682x_1 + 0.102x_2$. **(b)** $R^2 = 17.7\%$. **(c)** The residuals look roughly Normal and show no obvious remaining patterns. **(d)** $t = 1.83$, df = 97, $P = 0.070$.

11.19 (a) Small banks show a moderate positive linear relationship; for large banks, the relationship is very weak. **(b)** For small banks, $\widehat{\text{Wages}} = 35.9 + 0.1042\,\text{LOS}$. $R^2 = 46.6\%$, $s = 7.026$. **(c)** For large banks, $\widehat{\text{Wages}} = 49.5 + 0.0560\,\text{LOS}$. $R^2 = 3.5\%$, $s = 13.02$. **(d)** The large-bank regression is not significant (nor is it useful for prediction).

11.21 (a) Budget and Opening are right-skewed; Theaters and Opinion are roughly symmetric (slightly left-skewed). Five-number summaries for budget and opening are appropriate; mean and standard deviation could be used for the other two variables. **(b)** All relationships are positive. The Budget/Theaters and Opening/Theaters relationships appear to be curved; the others are reasonably linear. The correlations between Budget, Opening, and Theaters are all greater than 0.7. Opinion is less correlated with the other three variables—about 0.4 with Budget and Opening, and only 0.156 with Theaters.

11.23 (a) $USRevenue_i = \beta_0 + \beta_1 \, Budget_i + \beta_2 \, Opening_i + \beta_3 \, Theaters_i + \beta_4 \, Opinion_i + \epsilon_i$, where $i = 1, 2, \ldots, 35$; ϵ_i are independent $N(0, \sigma)$ random variables.
(b) $\widehat{USRevenue} = -67.72 + 0.1351 \, Budget + 3.0165 \, Opening - 0.00223 \, Theaters + 10.262 \, Opinion$.
(c) *The Dark Knight* may be influential. The spread of the residuals appears to increase with Theaters. **(d)** 98.1%.

11.25 (a) \$86.86 to \$154.91 million. **(b)** \$89.93 to \$155.00 million. **(c)** The intervals are very similar.

11.27 (a) PEER is left-skewed; the other two variables are irregular. **(b)** PEER and FtoS are negatively correlated $(r = -0.114)$; FtoS and CtoF are positively correlated $(r = 0.580)$; the other correlation is very small.

11.29 (a) $OVERALL_i = \beta_0 + \beta_1 \, PEER_i + \beta_2 \, FtoS_i + \beta_3 \, CtoS_i + \epsilon_i$, where ϵ_i are independent $N(0, \sigma)$ random variables. **(b)** $\widehat{OVERALL} = 18.85 + 0.5746 \, PEER + 0.0013 \, FtoS + 0.1369 \, CtoF$. **(c)** PEER: 0.4848 to 0.6644. FtoS: -0.0704 to 0.0730. CtoF: 0.0572 to 0.2166. The FtoS coefficient is not significantly different from 0. **(d)** $R^2 \doteq 72.2\%$, $s \doteq 7.043$.

11.31 (a) For example; all distributions are skewed to varying degrees—GINI and CORRUPT to the right, the other three to the left. CORRUPT and DEMOCRACY have the most skewness. **(b)** GINI is negatively correlated to the other four variables (ranging from -0.384 to -0.139), while all other correlations are positive and more substantial (0.533 or more).

11.33 (a) Refer to your regression output. **(b)** For example, the t statistic for the GINI coefficient grows from $t = -1.18$ ($P = 0.243$) to $t = 3.92$ ($P < 0.0005$). The DEMOCRACY t is 3.27 in the third model ($P = 0.002$) but drops to 0.60 ($P = 0.552$) in the fourth model. **(c)** A good choice is to use GINI, LIFE, and CORRUPT: all three coefficients are significant, and $R^2 = 77.3\%$ is nearly the same as the fourth model from Exercise 11.32.

11.35 (a) Plot suggests greater variation in VO+ for large OC. $\widehat{VO+} = 334 + 19.5 \, OC$, $t = 4.73$, $P < 0.0005$. Plot of residuals against OC is slightly curved. **(b)** $\widehat{VO+} = 57.7 + 6.41 \, OC + 53.9 \, TRAP$. Coefficient of OC is not significantly different from 0 ($t = 1.25$, $P = 0.221$), but coefficient of TRAP is ($t = 3.50$, $P = 0.002$). This is consistent with the correlations found in Exercise 11.34.

11.37 The correlations are 0.840 (LVO+ and LVO–), 0.774 (LVO+ and LOC), and 0.755 (LVO+ and LTRAP). Regression equations, t statistics, R^2, and s for each model: $\widehat{LVO+} = 4.38 + 0.706 \, LOC$; $t = 6.58$, $P < 0.0005$; $R^2 = 0.599$, $s = 0.3580$.
$\widehat{LVO+} = 4.26 + 0.430 \, LOC + 0.424 \, LTRAP$; $t = 2.56$, $P = 0.016$; $t = 2.06$, $P = 0.048$; $R^2 = 0.652$, $s = 0.3394$.
$\widehat{LVO+} = 0.872 + 0.392 \, LOC + 0.028 \, LTRAP + 0.672 \, LVO–$;

$t = 3.40$, $P = 0.002$; $t = 0.18$, $P = 0.862$; $t = 5.71$, $P < 0.0005$; $R^2 = 0.842$, $s = 0.2326$. As before, this suggests a model without LTRAP: $\widehat{LVO+} = 0.832 + 0.406 \, LOC + 0.682 \, LVO–$; $t = 4.93$, $P < 0.0005$; $t = 6.57$, $P < 0.0005$; $R^2 = 0.842$, $s = 0.2286$.

11.39 Regression equations, t statistics, R^2, and s for each model: $\widehat{LVO–} = 5.21 + 0.441 \, LOC$; $t = 3.59$, $P = 0.001$; $R^2 = 0.308$, $s = 0.4089$. $\widehat{LVO–} = 5.04 + 0.057 \, LOC + 0.590 \, LTRAP$; $t = 0.31$, $P = 0.761$; $t = 2.61$, $P = 0.014$; $R^2 = 0.443$, $s = 0.3732$.
$\widehat{LVO–} = 1.57 - 0.293 \, LOC + 0.245 \, LTRAP + 0.813 \, LVO+$; $t = -2.08$, $P = 0.047$; $t = 1.47$, $P = 0.152$; $t = 5.71$, $P < 0.0005$; $R^2 = 0.748$, $s = 0.2558$. $\widehat{LVO–} = 1.31 - 0.188 \, LOC + 0.890 \, LVO+$; $t = -1.52$, $P = 0.140$; $t = 6.57$, $P < 0.0005$; $R^2 = 0.728$, $s = 0.2611$.

11.41 (a) $y_i = \beta_0 + \beta_1 x_{i1} + \beta_2 x_{i2} + \beta_3 x_{i3} + \beta_4 x_{i4} + \epsilon_i$, where $i = 1, 2, \ldots, 69$; ϵ_i are independent $N(0, \sigma)$ random variables. **(b)** $\widehat{PCB} = 0.94 + 11.9 x_1 + 3.76 x_2 + 3.88 x_3 + 4.18 x_4$. All coefficients are significantly different from 0, although the constant 0.937 is not ($t = 0.76$, $P = 0.449$). $R^2 = 0.989$, $s = 6.382$. **(c)** The residuals appear to be roughly Normal, but with two outliers. There are no clear patterns when plotted against the explanatory variables.

11.43 (a) $\widehat{PCB} = -1.02 + 12.6 x_1 + 0.313 x_2 + 8.25 x_3$, $R^2 = 0.973$, $s = 9.945$. **(b)** $b_2 = 0.313$, $P = 0.708$. **(c)** In Exercise 11.41, $b_2 = 3.76$, $P < 0.0005$.

11.45 The model is $y_i = \beta_0 + \beta_1 x_{i1} + \beta_2 x_{i2} + \beta_3 x_{i3} + \beta_4 x_{i4} + \epsilon_i$, where $i = 1, 2, \ldots, 69$; ϵ_i are independent $N(0, \sigma)$ random variables. Regression gives $TEQ = 1.06 - 0.097 x_1 + 0.306 x_2 + 0.106 x_3 - 0.0039 x_4$, with $R^2 = 0.677$. Only the constant (1.06) and the PCB118 coefficient (0.306) are significantly different from 0. Residuals are slightly right-skewed and show no clear patterns when plotted with the explanatory variables.

11.47 (a) The correlations are all positive, ranging from 0.227 (LPCB28 and LPCB180) to 0.956 (LPCB and LPCB138). LPCB28 has one outlier (specimen 39) when plotted with the other variables; except for that point, all scatterplots appear fairly linear. **(b)** All correlations are higher with the transformed data.

11.49 It appears that a good model is LPCB126 and LPCB28 ($R^2 = 0.768$).

11.51 $\bar{x}$, M, s, and IQR for each variable: Taste: 24.53, 20.95, 16.26, 23.9. Acetic: 5.498, 5.425, 0.571, 0.656. H2S: 5.942, 5.329, 2.127, 3.689. Lactic: 1.442, 1.450, 0.3035, 0.430. None of the variables show striking deviations from Normality. Taste and H2S are slightly right-skewed, and Acetic has two peaks. There are no outliers.

11.53 $\widehat{Taste} = -61.5 + 15.6 \, Acetic$; $t = 3.48$, $P = 0.002$. The residuals seem to have a Normal distribution but are positively associated with both H2S and Lactic.

11.55 $\widehat{\text{Taste}} = -29.9 + 37.7\,\text{Lactic}$; $t = 5.25$, $P < 0.0005$. The residuals seem to have a Normal distribution; there are no striking patterns for residuals against the other variables.

11.57 $\widehat{\text{Taste}} = -26.9 + 3.80\,\text{Acetic} + 5.15\,\text{H2S}$. For the coefficient of Acetic, $t = 0.84$ and $P = 0.406$. This model is not much better than the model with H2S alone; Acetic and H2S are correlated ($r = 0.618$), so Acetic does not add significant information if H2S is included.

11.59 $\widehat{\text{Taste}} = -28.9 + 0.33\,\text{Acetic} + 3.91\,\text{H2S} + 19.7\,\text{Lactic}$. The coefficient of Acetic is not significantly different from 0 ($P = 0.942$). Residuals of this regression appear to be Normally distributed and show no patterns in scatterplots with the explanatory variables. It appears that the H2S/Lactic model is best.

Chapter 12

12.1 (a) H_0 says the population means are all equal. **(b)** Experiments are best for establishing causation. **(c)** ANOVA is used to compare means. **(d)** Multiple comparisons procedures are used when we wish to determine which means are significantly different but have no specific relations in mind before looking at the data.

12.3 (a) Yes: $6/5 \doteq 1.2 < 2$. **(b)** 25, 25, and 36. **(c)** 29.3676. **(d)** 5.4192.

12.5 (a) This is the description of *between*-group variation. **(b)** The *sums of* squares will add. **(c)** σ is a parameter. **(d)** A small P means the means are not all the same, but the distributions may still overlap.

12.7 Assuming the t (ANOVA) test establishes that the means are different, contrasts and multiple comparisons provide no further useful information.

12.9 (a) df 4 and 30; $2.69 < F < 3.25$. **(c)** $0.025 < P < 0.050$. **(d)** We can conclude that at least one mean is different.

12.11 (a) df 2 and 30; $F = 2.54$, $0.050 < P < 0.100$. **(b)** df 3 and 28; $F \doteq 2.97$, $0.025 < P < 0.050$.

12.13 (a) Response: egg cholesterol level. Populations: chickens with different diets or drugs. $I = 3$, $n_1 = n_2 = n_3 = 25$, $N = 75$. **(b)** Response: rating on five-point scale. Populations: the three groups of students. $I = 3$, $n_1 = 31$, $n_2 = 18$, $n_3 = 45$, $N = 94$. **(c)** Response: quiz score. Populations: students in each TA group. $I = 3$, $n_1 = n_2 = n_3 = 14$, $N = 42$.

12.15 For all three situations, we test $H_0: \mu_1 = \mu_2 = \mu_3$; H_a: at least one mean is different. **(a)** DFM 2, DFE 72, DFT 74. $F(2, 72)$. **(b)** DFM 2, DFE 91, DFT 93. $F(2, 91)$. **(c)** DFM 2, DFE 39, DFT 41. $F(2, 39)$.

12.17 (a) This sounds like a fairly well-designed

experiment, so the results should at least apply to this farmer's breed of chicken. **(b)** It would be good to know what proportion of the total student body falls in each of these groups—that is, is anyone overrepresented in this sample? **(c)** Effectiveness teaching one topic (power calculations) might not reflect overall effectiveness.

12.19 (a) 2 and 117. **(b)** $P < 0.001$, or $P = 0.0003$. **(c)** We should hesitate to generalize these results beyond similar informal shops in Mexico.

12.21 (a) F can be made very small (close to 0), and P close to 1. **(b)** F increases, and P decreases.

12.23 (a) Based on the sample means, fiber is cheapest and cable is most expensive. **(b)** Yes; the ratio is 1.55. **(c)** df 2 and 44; $0.025 < P < 0.050$, or $P = 0.0427$.

12.25 (a) Matched pairs t methods; we examine the change in reaction time for each subject. **(b)** No; we do not have four independent samples.

12.27 (a) df 3 and 2286. **(b)** $F \doteq 2.5304$. **(c)** $P = 0.0555$.

12.29 (a) Activity seems to increase with both drugs, and Drug B appears to have a greater effect. **(b)** Yes; the standard deviation ratio is 1.44. $s_p \doteq 3.256$. **(c)** df $= 4$ and 15. **(d)** $0.01 < P < 0.025$ ($P = 0.0165$).

12.31 (a) The variation in sample size is some cause for concern, but there can be no extreme outliers in a 1-to-7 scale, so ANOVA is probably reliable. **(b)** Yes: $1.26/1.03 = 1.22 < 2$. **(c)** $F(4, 405)$, $P = 0.0002$. **(d)** Hispanic Americans are highest, Japanese are in the middle, the other three are lowest.

12.33 (a) $\psi_1 = \mu_2 - (\mu_1 + \mu_4)/2$. **(b)** $\psi_2 = (\mu_1 + \mu_2 + \mu_4)/3 - \mu_3$.

12.35 *H. bihai* and *H. caribaea* red distributions are slightly skewed. *H. bihai*: $n = 16$, $\bar{x} = 47.597$, $s = 1.213$ mm. *H. caribaea* red: $n = 23$, $\bar{x} = 39.711$, $s = 1.799$ mm. *H. caribaea* yellow: $n = 15$, $\bar{x} = 36.180$, $s = 0.975$ mm. This just meets our rule for standard deviations. ANOVA gives $F = 259.12$, df 2 and 51, $P < 0.0005$, so we conclude the means are different.

12.37 *H. bihai*: $n = 16$, $\bar{x} = 3.8625$, $s = 0.0251$. *H. caribaea* red: $n = 23$, $\bar{x} = 3.6807$, $s = 0.0450$. *H. caribaea* yellow: $n = 15$, $\bar{x} = 3.5882$, $s = 0.0270$. ANOVA gives $F = 244.27$, df 2 and 51, $P < 0.0005$, so we conclude the means are different.

12.39 (a) All three distributions show no particular skewness. Control: $n = 15$, $\bar{x} = 0.21887$, $s = 0.01159\,\text{g/cm}^2$. Low dose: $n = 15$, $\bar{x} = 0.21593$, $s = 0.01151\,\text{g/cm}^2$. High dose: $n = 15$, $\bar{x} = 0.23507$, $s = 0.01877\,\text{g/cm}^2$. **(b)** All three distribution appear to be nearly Normal. **(c)** $F = 7.72$, df 2 and 42, $P = 0.001$. **(d)** For Bonferroni, $t^{**} = 2.49$ and MSD $= 0.0131$. The high-dose mean is significantly different from the other two. **(e)** High doses increase BMD.

12.41 (a) The mean responses were not significantly different. **(b)** For Bonferroni, $t^{**} = 2.827$. Only the largest difference within each set of means is significant: $t_{14} = -2.89$ (experience culture), $t_{23} = 3.55$ (group tour), and $t_{45} = 2.86$ (ocean sports).

12.43 For Bonferroni, $t^{**} = 2.71$ and $s_p \doteq 2.7348$. The Piano mean is significantly higher than the other three, but the other three means are not significantly different.

12.45 (a) Yes; the ratio is 1.25. $s_p = 0.7683$. **(b)** df 2 and 767; $P < 0.001$. **(c)** With $\psi = \mu_2 - (\mu_1 + \mu_3)/2$, we test $H_0: \psi = 0$; $H_a: \psi > 0$. We find $c = 0.585$, $t = 5.99$, and $P < 0.0001$.

12.47 (a) Control: $n = 10$, $\bar{x} = 601.10$, $s = 27.36$ mg/cm^3. Low jump: $n = 10$, $\bar{x} = 612.50$, $s = 19.33$ mg/cm^3. High jump: $n = 10$, $\bar{x} = 638.70$, $s = 16.59$ mg/cm^3. Pooling is reasonable. **(b)** $F = 7.98$, df 2 and 27, $P = 0.002$. We conclude that not all means are equal.

12.49 (a) Aluminum: $n = 4$, $\bar{x} = 2.0575$, $s = 0.2520$ mg/100 g. Clay: $n = 4$, $\bar{x} = 2.1775$, $s = 0.6213$ mg/100 g. Iron: $n = 4$, $\bar{x} = 4.6800$, $s = 0.6283$ mg/100 g. Pooling is risky because $0.6283/0.2520 = 2.49 > 2$. **(b)** $F = 31.16$, df 2 and 9, $P < 0.0005$. We cautiously conclude that the means are not the same.

12.51 (a) ECM1: $n = 3$, $\bar{x} = 65.0\%$, $s = 8.66\%$. ECM2: $n = 3$, $\bar{x} = 63.33\%$, $s = 2.89\%$. ECM3: $n = 3$, $\bar{x} = 73.33\%$, $s = 2.89\%$. MAT1: $n = 3$, $\bar{x} = 23.33\%$, $s = 2.89\%$. MAT2: $n = 3$, $\bar{x} = 6.67\%$, $s = 2.89\%$. MAT3: $n = 3$, $\bar{x} = 11.67\%$, $s = 2.89\%$. Pooling is risky because $8.66/2.89 > 2$. **(b)** $F = 137.94$, df 5 and 12, $P < 0.0005$. We conclude that the means are not the same.

12.53 (a) $\psi_1 = \mu_1 - (\mu_2 + \mu_4)/2$ and $\psi_2 = \mu_3 - \mu_2 - (\mu_5 - \mu_4)$. **(b)** $c_1 = -1.5$, SE$_{c_1} \doteq 1.9337$, $c_2 = -3.75$, and SE$_{c_2} \doteq 3.2558$. **(c)** Neither contrast is significant ($t_1 \doteq -0.752$ and $t_2 \doteq -1.152$).

12.55 (a) The plot shows granularity (which varies between groups), but that should not make us question independence; it is due to the fact that the scores are all integers. **(b)** The ratio of the largest to the smallest standard deviations is less than 2. **(c)** Apart from the granularity, the quantile plots are reasonably straight. **(d)** Again, apart from the granularity, the quantile plots look pretty good.

12.57 (a) $\psi_1 = (\mu_1 + \mu_2 + \mu_3)/3 - \mu_4$, $\psi_2 = (\mu_1 + \mu_2)/2 - \mu_3$, $\psi_3 = \mu_1 - \mu_2$. **(b)** The pooled standard deviation is $s_p = 1.1958$. SE$_{c_1} \doteq 0.2355$, SE$_{c_2} \doteq 0.1413$, SE$_{c_3} \doteq 0.1609$. **(c)** Testing $H_0: \psi_i = 0$; $H_a: \psi_i \neq 0$ for each contrast, we find $c_1 = -12.51$, $t_1 = -53.17$, $P_1 < 0.0005$; $c_2 = 1.269$, $t_2 = 8.98$, $P_2 < 0.0005$; $c_3 = 0.191$, $t_3 = 1.19$, $P_3 \doteq 0.2359$. The Placebo mean is significantly higher than the average of the other three, while the Keto mean is significantly lower than the average of the two Pyr means. The difference between the Pyr means is not

significant (meaning the second application of the shampoo is of little benefit).

12.59 The means all increase by 5%, but everything else (standard deviations, standard errors, and the ANOVA table) is unchanged.

12.61 All distributions are reasonably Normal, and standard deviations are close enough to justify pooling. For PRE1, $F = 1.13$, df 2 and 63, $P = 0.329$. For PRE2, $F = 0.11$, df 2 and 63, $P = 0.895$. Neither set of pretest scores suggests a difference in means.

12.63 $\widehat{\text{Score}} = 4.432 - 0.000102\,\text{Friends}$. The slope is not significantly different from 0 ($t = -0.28$, $P = 0.782$), and the regression only explains 0.1% of the variation in score. Residuals suggest a possible curved relationship.

12.67 (b) Answers will vary with choice of H_a and desired power. For example, with $\mu_1 = \mu_2 = 4.4$, $\mu_3 = 5$, $\sigma = 1.2$, three samples of size 75 will produce power 0.89.

12.69 The design can be similar, although the types of music might be different. Bear in mind that spending at a casual restaurant will likely be less than at the restaurants examined in Exercise 12.28; this might also mean that the standard deviations could be smaller. Decide how big a difference in mean spending you would want to detect and then do some power computations.

Chapter 13

13.1 (a) Two-way ANOVA is used when there are two factors. **(b)** Each level of A should occur with all three levels of B. **(c)** The RESIDUAL part of the model represents the error. **(d)** DFAB $= (I - 1)(J - 1)$.

13.3 (a) Reject H_0 when F is large. **(b)** Mean squares equal sum of squares divided by degrees of freedom. **(c)** The test statistics have an F distribution. **(d)** If the sample sizes are not the same, the sums of squares may not add.

13.5 (a) df 2 and 24. **(c)** $0.025 < P < 0.05$. **(d)** The interaction term is significantly different from 0, so the mean profiles should not be parallel.

13.7 (a) Factors: gender ($I = 2$) and age ($J = 3$). Response: percent of pretend play. $N = 66$. **(b)** Factors: time after harvest ($I = 5$) and amount of water ($J = 2$). Response: percent of seeds germinating. $N = 30$. **(c)** Factors: mixture ($I = 6$) and freezing/thawing cycles ($J = 3$). Response: Strength. $N = 54$. **(d)** Factors: training programs ($I = 4$) and number of days to give the training ($J = 2$). Response: some (unspecified) measure of the training's effectiveness. $N = 80$.

13.9 (a) There appears to be an interaction; a thank-you note increases repurchase intent for those with short history and decreases it for customers with long history.

(b) The marginal means for history (6.245 and 7.45) convey the fact that repurchase intent is higher for customers with long history. The thank-you note marginal means (6.61 and 7.085) are less useful because of the interaction.

13.11 (a) The plot suggests a possible interaction. **(b)** By subjecting the same individual to all four treatments, rather than four individuals to one treatment each, we reduce the within-groups variability.

13.13 (a) There may be an interaction; for a favorable process, a favorable outcome increases satisfaction quite a bit more than for an unfavorable process ($+2.32$ compared to $+0.24$). **(b)** This time, the increase in satisfaction from a favorable outcome is less for a favorable process ($+0.49$ compared to $+1.32$). **(c)** There seems to be a three-factor interaction, because the interactions in parts (a) and (b) are different.

13.15 Humor slightly increases satisfaction (3.58 with no humor, 3.96 with humor). The process and outcome effects are greater: favorable process 4.75, unfavorable process 2.79; favorable outcome 4.32, unfavorable outcome 3.22.

13.17 The largest-to-smallest ratio is 1.26, and the pooled standard deviation is 1.7746.

13.19 Except for female responses to purchase intention, means decreased from Canada to the United States to France. Females had higher means than men in almost every case, except for French responses to credibility and purchase intention (a modest interaction).

13.21 (a) Intervention, 11.6; control, 9.967; baseline, 10.0; 3 months, 11.2; 6 months, 11.15. Overall, 10.783. The row means suggest that the intervention group showed more improvement than the control group. **(b)** Interaction means that the mean number of actions changes differently over time for the two groups.

13.23 There are no significant effects (although B is close): F_A (df = 2 and 24) has $P = 0.1759$. F_B (df = 1 and 24) has $P = 0.0740$. F_{AB} (df = 2 and 24) has $P = 0.1396$.

13.25 (a) There is little evidence of an interaction. **(b)** $s_p \doteq 0.1278$. **(c)** $\psi_1 = (\mu_{new,city} + \mu_{new,highway})/2 - (\mu_{old,city} + \mu_{old,highway})/2$. $\psi_2 = \mu_{new,city} - \mu_{new,highway}$. $\psi_3 = \mu_{old,highway} - \mu_{old,city}$. **(d)** By subjecting the same individual to all four treatments, rather than four individuals to one treatment each, we reduce the within-groups variability.

13.27 (b) There seems to be a fairly large difference between the means based on how much the rats were allowed to eat, but not very much difference based on the chromium level. There may be an interaction: The NM mean is lower than the LM mean, while the NR mean is higher than the LR mean. **(c)** L mean: 4.86. N mean:

4.871. M mean: 4.485. R mean: 5.246. LR minus LM: 0.63. NR minus NM: 0.892. Mean GITH levels are lower for M than for R; there is not much difference between L and N. The difference between M and R is greater among rats that had normal chromium levels in their diets (N).

13.29 The "Other" category had the lowest mean SATM score for both genders; this is apparent from a graph of the means as well as from the marginal means (CS, 605; EO, 624.5; O, 566). Males had higher mean scores in CS and O, while females were slightly higher in EO; this seems to be an interaction. Overall, the marginal means are 611.7 (males) and 585.3 (females).

13.31 (a) $s_p \doteq \$38.14$, df = 105. **(b)** Yes; the largest-to-smallest ratio is 1.36. **(c)** Individual sender, $70.90; group sender, $48.85; individual responder, $59.75; group responder, $60.00. **(d)** There appears to be an interaction; individuals send more money to groups, while groups send more money to individuals. **(e)** $P = 0.0033$, $P = 0.9748$, and $P = 0.1522$. Only the main effect of sender is significant.

13.33 Yes; the iron-pot means are the highest, and F for testing the effect of the pot type is very large.

13.35 (a) In the order listed in the table: $\bar{x}_{11} = 25.0307$, $s_{11} = 0.0011541$; $\bar{x}_{12} = 25.0280$, $s_{12} = 0$; $\bar{x}_{13} = 25.0260$, $s_{13} = 0$; $\bar{x}_{21} = 25.0167$, $s_{21} = 0.0011541$; $\bar{x}_{22} = 25.0200$, $s_{22} = 0.0019999$; $\bar{x}_{23} = 25.0160$, $s_{23} = 0$; $\bar{x}_{31} = 25.0063$, $s_{31} = 0.0015275$; $\bar{x}_{32} = 25.0127$, $s_{32} = 0.0011552$; $\bar{x}_{33} = 25.0093$, $s_{33} = 0.0011552$; $\bar{x}_{41} = 25.0120$, $s_{41} = 0$; $\bar{x}_{42} = 25.0193$, $s_{42} = 0.0011552$; $\bar{x}_{43} = 25.0140$, $s_{43} = 0.0039997$; $\bar{x}_{51} = 24.9973$, $s_{51} = 0.0011541$; $\bar{x}_{52} = 25.0060$, $s_{52} = 0$; $\bar{x}_{53} = 25.0003$, $s_{53} = 0.0015277$. **(b)** Except for Tool 1, mean diameter is highest at Time 2. Tool 1 had the highest mean diameters, followed by Tool 2, Tool 4, Tool 3, and Tool 5. **(c)** $F_A = 412.98$, df 4 and 30, $P < 0.0005$. $F_B = 43.61$, df 2 and 30, $P < 0.0005$. $F_{AB} = 7.65$, df 8 and 30, $P < 0.0005$. **(d)** There is strong evidence of a difference in mean diameter among the tools (A) and among the times (B). There is also an interaction (specifically, Tool 1's mean diameters changed differently over time compared to the other tools).

13.37 (a) All three F-values have df 1 and 945, the P-values are < 0.001, < 0.001, and 0.1477. Gender and handedness both have significant effects on mean lifetime, but there is no interaction. **(b)** Women live about 6 years longer than men (on the average), while right-handed people average 9 more years of life than left-handed people. Handedness affects both genders in the same way, and vice versa.

13.39 (a) and **(b)** The first three means and standard deviations are $\bar{x}_{1,1} = 3.2543$, $s_{1,1} = 0.2287$; $\bar{x}_{1,2} = 2.7636$, $s_{1,2} = 0.0666$; $\bar{x}_{1,3} = 2.8429$, $s_{1,3} = 0.2333$. The standard deviations range from 0.0666 to 0.3437, for a ratio of

5.16—larger than we like. **(c)** For Plant, $F = 1301.32$, df 3 and 224, $P < 0.0005$. For Water, $F = 9.76$, df 6 and 224, $P < 0.0005$. For interaction, $F = 5.97$, df 18 and 224, $P < 0.0005$.

13.41 The seven F statistics are 184.05, 115.93, 208.87, 218.37, 220.01, 174.14, and 230.17, all with df 3 and 32 and $P < 0.0005$.

13.43 Fresh: Plant $F = 81.45$, df 3 and 84, $P < 0.0005$; Water $F = 43.71$, df 6 and 84, $P < 0.0005$; interaction $F = 1.79$, df 18 and 84, $P = 0.040$. Dry: Plant $F = 79.93$, df 3 and 84, $P < 0.0005$; Water $F = 44.79$, df 6 and 84, $P < 0.0005$; interaction $F = 2.22$, df 18 and 84, $P = 0.008$.

13.45 Fresh: The seven F statistics are 15.88, 11.81, 62.08, 10.83, 22.62, 8.20, and 10.81, all with df 3 and 12 and $P \leq 0.003$. Fresh: The seven F statistics are 8.14, 26.26, 22.58, 11.86, 21.38, 14.77, and 8.66, all with df 3 and 12 and $P \leq 0.003$.

13.47 (a) Gender: df 1 and 174. Floral characteristic: df 2 and 174. Interaction: df 2 and 174. **(b)** Damage to males was higher for all characteristics. For males, damage was highest under characteristic level 3, while for females, the highest damage occurred at level 2. **(c)** Three of the standard deviations are at least half as large as the means. Because the response variable (leaf damage) had to be nonnegative, this suggests that these distributions are right-skewed.

13.49 Men in CS: $n = 39$, $\bar{x} = 7.79487$, $s = 1.50752$. Men in EOS: $n = 39$, $\bar{x} = 7.48718$, $s = 2.15054$.

Men in Other: $n = 39$, $\bar{x} = 7.41026$, $s = 1.56807$. Women in CS: $n = 39$, $\bar{x} = 8.84615$, $s = 1.13644$. Women in EOS: $n = 39$, $\bar{x} = 9.25641$, $s = 0.75107$. Women in Other: $n = 39$, $\bar{x} = 8.61539$, $s = 1.16111$. The means suggest that females have higher HSE grades than males. For a given gender, there is not too much difference among majors. Normal quantile plots show no great deviations from Normality, apart from the granularity of the grades (most evident among women in EO). In the ANOVA, only the effect of gender is significant ($F = 50.32$, df 1 and 228, $P < 0.0005$).

13.51 Men in CS: $n = 39$, $\bar{x} = 526.949$, $s = 100.937$. Men in EOS: $n = 39$, $\bar{x} = 507.846$, $s = 57.213$. Men in Other: $n = 39$, $\bar{x} = 487.564$, $s = 108.779$. Women in CS: $n = 39$, $\bar{x} = 543.385$, $s = 77.654$. Women in EOS: $n = 39$, $\bar{x} = 538.205$, $s = 102.209$. Women in Other: $n = 39$, $\bar{x} = 465.026$, $s = 82.184$. The means suggest that students who stay in the sciences have higher mean SATV scores than those who end up in the "Other" group. Female CS and EO students have higher scores than males in those majors, but males have the higher mean in the Other group. Normal quantile plots suggests some right-skewness in the "Women in CS" group and also some non-Normality in the tails of the "Women in EO" group. Other groups look reasonably Normal. In the ANOVA, only the effect of major is significant ($F = 9.32$, df 2 and 228, $P < 0.0005$).

NOTES AND DATA SOURCES

Chapter 1

1. Data from the Bureau of Labor Statistics. See `www.bls.gov/iif/oshsum.htm`

2. Reoprted in the May 8, 2008 edition of *GPS Magazine*. See `www.gpsmagazine.com/print/000443.php`

3. Data collected in the lab of Connie Weaver, Department of Foods and Nutrition, Purdue University and provided by Linda McCabe.

4. Haipeng Shen, "Nonparametric regression for problems involving lognormal distributions," PhD thesis, University of Pennsylvania, 2003. Thanks to Haipeng Shen and Larry Brown for sharing the data.

5. From the Digest of Education Statistics at the Web site of the National Center for Education Statistics, `www.nces.ed.gov/programs/digest`

6. See Note 3.

7. Based on Barbara Ernst et al., "Seasonal variation in the deficiency of 25–hydroxyvitamin D_3 in mildly to extremely obese subjects," *Obesity Surgery*, 19 (2009), pp. 180–183.

8. From the Color Assignment Web site of Joe Hallock, `www.joehallock.com/edu/COM498/index.html`

9. From "The Apple iPhone: Successes and challenges for the mobile industry," March 31, 2008, Rubicon Win Markets. See `rubiconconsulting.com`

10. U.S. Environmental Protection Agency, *Municipal Solid Waste in the United States: 2007 Facts and Figures*.

11. September 2008 data from `marketshare.hitslink.com`

12. Robyn Greenspan, "The deadly duo: Spam and viruses, October 2003," found at `cyberatlas.internet.com`.

13. See `insidefacebook.com`

14. See previous note.

15. National Center for Education Statistics, NEDRC Table Library, `nces.ed.gov/surveys/npsas/table_library`

16. Color popularity for 2007 from the Dupont Automotive Color report; see `onlinepressroom.net/DuPont/MultimediaGallery`

17. Debora L. Arsenau, "Comparison of diet management instruction for patients with non-insulin dependent diabetes mellitus: Learning activity package vs. group instruction," MS thesis, Purdue University.

18. Data from Gary Community School Corporation, courtesy of Celeste Foster, Department of Education, Purdue University.

19. Found online at `earthtrends.wri.org`

20. We thank Heeseung Roh Ryu for supplying the data, from her dissertation in the Department of Health and Kinesiology, Purdue University,

21. This exercise was provided by Nicolas Fisher.

22. S. M. Stigler, "Do robust estimators work with real data?" *Annals of Statistics*, 5 (1977), pp. 1055–1078.

23. Data provided by Darlene Gordon, Purdue University.

24. Data from the World Bank, Quick Query option from Key Development Data & Statistics page, `worldbank.org`, 2008.

25. From the Interbrand Web site; see `interbrand.com/best_global_brands.aspx?langid=1000`

26. From `beer100.com/beercalories.htm`, on June 30, 2008.

27. See Noel Cressie, *Statistics for Spatial Data*, Wiley, 1993.

28. Data provided by Francisco Rosales of the Department of Nutritional Sciences, Pennsylvania State University.

29. Data provided by Betsy Hoza, Department of Psychological Sciences, University of Vermont.

30. From "Changes in U.S. family finances from 2004 to 2007: Evidence from the survey of consumer finances," Federal Reserve Bulletin, 2009.

31. *Extreme Weather Sourcebook 2001*, found online at `sciencepolicy.colorado.edu/sourcebook`

32. We thank Ethan J. Temeles of Amherst College for providing the data. His work is described in Ethan J. Temeles and W. John Kress, "Adaptation in a plant-hummingbird association," *Science*, 300 (2003), pp. 630–633.

33. Information about the Indiana Statewide Testing for Educational Progress program can be found at `doe.state.in.us/istep/`

34. See `stubhub.com`

35. From Matthias R. Mehl et al., "Are women really more talkative than men?" *Science*, 317 (5834), (2007), p. 82. The raw data were provided by Matthias Mehl.

36. From the American Heart Association Web site, `americanheart.org`

37. From `fueleconomy.gov/`

38. From `cdc.gov/brfss`

39. See Note 16.

40. See `worldbank.org`

41. Data are from the Open Accessible Space Information System for New York City. See `oasisnyc.net`

42. We thank C. Robertson McClung of Dartmouth College for supplying the data. The study is reported in Todd P. Michael et al., "Enhanced fitness conferred by naturally occurring variation in the circadian clock," *Science*, 302 (2003), pp. 1049–1053.

43. From `isp-planet.com`

Chapter 2

1. Hannah G. Lund et al, "Sleep patterns and predictors of disturbed sleep in a large population of college students," *Adolescent Health*, 46, No. 2 (2010), pp. 97–99.

2. See previous note.

3. See `cfs.purdue.edu/FN/campcalcium/public.htm` for information about the 2010 camp.

4. See "Happy spamiversary! Spam reaches 30," in *NewScientistTech*, April 25, 2008. See `NewScientist.com`

5. Thanks to Doug Crabill, Manager of Computer Systems for the Purdue University Department of Statistics, for providing this background information about spam botnets.

6. See `symantec.com/security_response/writeup.jsp?docid=2007-062007-0946-99`

7. OECD StatExtracts, Organization for Economic Co-operation Development, downloaded 6/29/08 from `stats.oecd.org/wbos`

8. See `forbes.com/lists/2008/6/biz_bizcountries08_Best-Countries-for-Business_Rank.html`

9. A sophisticated treatment of improvements and additions to scatterplots is W. S. Cleveland and R. McGill, "The many faces of a scatterplot," *Journal of the American Statistical Association*, 79 (1984), pp. 807–822.

10. From the Current Population Survey Web site, `census.gov/hhes/www/hlthins/hlthin07/p60no235_table6.pdf`

11. See `beer100.com/beercalories.htm`

12. See `worldbank.org`

13. James T. Fleming, "The measurement of children's perception of difficulty in reading materials," *Research in the Teaching of English*, 1 (1967), pp. 136–156.

14. Christer G. Wiklund, "Food as a mechanism of density-dependent regulation of breeding numbers in the merlin *Falco columbarius*," *Ecology*, 82 (2001), pp. 860–867.

15. We thank C. Robertson McClung of Dartmouth College for supplying the data. The study is reported in Todd P. Michael et al., "Enhanced fitness conferred by naturally occurring variation in the circadian clock," *Science*, 302 (2003), pp. 1049–1053.

16. Data from a plot in Naomi I. Eisenberger, Matthew D. Lieberman, and Kipling D. Williams, "Does rejection hurt? An fMRI study of social exclusion," *Science*, 302 (2003), pp. 290–292.

17. *Forbes*, September 19, 2009.

18. N. Maeno et al., "Growth rates of icicles," *Journal of Glaciology*, 40 (1994), pp. 319–326.

19. From `en.wikipedia.org/wiki/10000_metres`

20. A careful study of this phenomenon is W. S. Cleveland, P. Diaconis, and R. McGill, "Variables on scatterplots look more highly correlated when the scales are increased," *Science*, 216 (1982), pp. 1138–1141.

21. Data from a plot in James A. Levine, Norman L. Eberhardt, and Michael D. Jensen, "Role of nonexercise activity thermogenesis in resistance to fat gain in humans," *Science*, 283 (1999), pp. 212–214.

22. From the Web site `oasisnyc.net`

23. Frank J. Anscombe, "Graphs in statistical analysis," *The American Statistician*, 27 (1973), pp. 17–21.

24. From the Web site of the National Center for Education Statistics, `nces.ed.gov`

25. Debora L. Arsenau, "Comparison of diet management instruction for patients with non-insulin dependent diabetes mellitus: Learning activity package vs. group instruction," Master's thesis, Purdue University, 1993.

26. The facts in Example 2.22 come from Nancy W. Burton and Leonard Ramist, *Predicting Success in College: Classes Graduating Since 1980*, Research Report No. 2001-2, The College Board, 2001.

27. Zeinab E. M. Afifi, "Principal components analysis of growth of Nahya infants: size, velocity and two physique factors," *Human Biology*, 57 (1985), pp. 659–669.

28. D. A. Kurtz (ed.), *Trace Residue Analysis*, American Chemical Society Symposium Series No. 284, 1985, Appendix.

29. The scores are from the Purdue University 2008–2009 women's tennis team in the NCAA tournament.

30. Data from a plot in Feng Sheng Hu et al., "Cyclic variation and solar forcing of Holocene climate in the Alaskan subarctic," *Science*, 301 (2003), pp. 1890–1893.

31. Frank J. Anscombe, "Graphs in statistical analysis," *The American Statistician*, 27 (1973), pp. 17–21.

32. Results of this survey are reported in Henry Wechsler et al., "Health and behavioral consequences of binge drinking in college," *Journal of the American Medical Association*, 272 (1994), pp. 1672–1677.

33. You can find a clear and comprehensive discussion of numerical measures of association for categorical data in Chapter 2 of Alan Agresti, *Categorical Data Analysis*, 2nd ed., Wiley, 2002.

34. From M-Y Chen et al., "Adequate sleep among adolescents is positively associated with health status and health-related behaviors," *BMC Public Health*, 6 No. 59 (2006); available from `biomedicalcentral.com/1471-2458/6/59`

35. See the U.S. Bureau of the Census Web site at `census.gov/population/socdemo/school` for these and similar data.

36. From F. D. Blau and M. A. Ferber, "Career plans and expectations of young women and men," *Journal of Human Resources*, 26 (1991), pp. 581–607.

37. Data from D. M. Barnes, "Breaking the cycle of addiction," *Science*, 241 (1988), pp. 1029–1030.

38. M. S. Linet et al., "Residential exposure to magnetic fields and acute lymphoblastic leukemia in children," *New England Journal of Medicine*, 337 (1997), pp. 1–7.

39. *The Health Consequences of Smoking: 1983*, U.S. Public Health Service, 1983.

40. Contributed by Marigene Arnold of Kalamazoo College.

41. D. E. Powers and D. A. Rock, *Effects of Coaching on SAT I: Reasoning Test Scores*, Educational Testing Service Research Report 98-6, College Entrance Examination Board, 1998.

42. From the Social Security Web site, `ssa.gov/OACT/babynames`

43. Information about this procedure was provided by Samuel Flnigan of *U.S. News & World Report*. See `usnews.com/usnews/rankguide/rghome.htm` for a description of the variables used to construct the ranks and for the most recent ranks.

44. See `cdc.gov/brfss/` The data set BRFSS described in the data appendix contains several variables from this source.

45. Oskar Kindvall, "Habitat heterogeneity and survival in a bush cricket metapopulation," *Ecology*, 77 (1996), pp. 207–214.

46. Data from a plot in Josef P. Rauschecker, Biao Tian, and Marc Hauser, "Processing of complex sounds in the macaque nonprimary auditory cortex," *Science*, 268 (1995), pp. 111–114. The paper states that there are $n = 41$ observations, but only $n = 37$ can be read accurately from the plot.

47. We thank Zhiyong Cai of Texas A&M University for providing the data. The data are from work performed in connection with his PhD dissertation in the Department of Forestry and Natural Resources, Purdue University.

48. Data provided by Robert Dale, Purdue University.

49. S. Chatterjee and B. Price, *Regression Analysis by Example*, Wiley, 1977.

50. Gary Smith, "Do statistics test scores regress toward the mean?" *Chance*, 10, No. 4 (1997), pp. 42–45.

51. Data from National Science Foundation Science Resources Studies Division, *Data Brief*, 12 (1996), p. 1.

52. Based on data in Mike Planty et al., "Volunteer service by young people from high school through early adulthood," National Center for Educational Statistics Report, NCES 2004–365.

53. Although these data are fictitious, similar though less simple situations occur. See P. J. Bickel and J. W. O'Connell, "Is there a sex bias in graduate admissions?" *Science*, 187 (1975), pp. 398–404.

54. See George R. Milne, "How well do consumers protect themselves from identity theft?" *Journal of Consumer Affairs*, 37 (2003), pp. 388–402.

55. Based on information in "NCAA 2003 national study of collegiate sports wagering and associated health risks," which can be found at the NCAA Web site, `ncaa.org`

Chapter 3

1. See the NORC Web pages at `norc.uchicago.edu`

2. See `cdc.gov/mmwr/preview/mmwrhtml/mm5839a3.htm`

3. See Jeffrey G. Johnson et al., "Television viewing and aggressive behavior during adolescence and adulthood," *Science*, 295 (2002), pp. 2468–2471. The authors use statistical adjustments to control for the effects of a number of lurking variables. The association between TV viewing and aggression remains significant.

4. National Institute of Child Health and Human Development, Study of Early Child Care. The article appears in the July 2003 issue of *Child Development*.

5. The quotation is from the summary on the NICHD Web site, `nichd.nih.gov`

6. For a full description of the STAR program and its follow-up studies, go to `heros-inc.org/star.htm`

7. Simplified from Arno J. Rethans, John L. Swasy, and Lawrence J. Marks, "Effects of television commercial repetition, receiver knowledge, and commercial length: A test of the two-factor model," *Journal of Marketing Research*, 23 (February 1986), pp. 50–61.

8. Based on an experiment performed by Jake Gandolph under the direction of Professor Lisa Mauer in the Purdue University Department of Food Science.

9. Based on an experiment performed by Evan Whalen under the direction of Professor Patrick Connolly in the Purdue University Department of Computer Graphics Technology.

10. John H. Kagel, Raymond C. Battalio, and C. G. Miles, "Marijuana and work performance: results from an experiment," *Journal of Human Resources*, 15 (1980), pp. 373–395. A general discussion of failures of blinding is Dean Ferguson et al., "Turning a blind eye: The success of blinding reported in a random sample of randomised, placebo controlled trials," *British Medical Journal*, 328 (2004), p. 432.

11. Based on a study conducted by Sandra Simonis under the direction of Professor Jon Harbor from the Purdue University Earth and Atmospheric Sciences Department.

12. See the first citation in Note 14.

13. Joel Brockner et al., "Layoffs, equity theory, and work performance: Further evidence of the impact of survivor guilt," *Academy of Management Journal*, 29 (1986), pp. 373–384.

14. Simplified from Sanjay K. Dhar, Claudia González-Vallejo, and Dilip Soman, "Modeling the effects of advertised price claims: Tensile versus precise pricing," *Marketing Science*, 18 (1999), pp. 154–177.

15. Data from the source in the preceding note. The study also varied the percent of the price cut. The results in Figure 3.6 are for a 60% discount.

16. Based on Evan H. DeLucia et al., "Net primary production of a forest ecosystem with experimental CO_2 enhancement," *Science*, 284 (1999), pp. 1177–1179. The investigators used the block design.

17. E. M. Peters et al., "Vitamin C supplementation reduces the incidence of postrace symptoms of upper-respiratory tract infection in ultramarathon runners," *American Journal of Clinical Nutrition*, 57 (1993), pp. 170–174.

18. Based on a study conducted by Tammy Younts directed by Professor Deb Bennett of the Purdue University Department of Educational Studies. For more information about Reading Recovery, see readingrecovery.org/

19. Based on a study conducted by Rajendra Chaini under the direction of Professor Bill Hoover of the Purdue University Department of Forestry and Natural Resources.

20. From the Hot Ringtones list at billboard.com/ on November 9, 2009.

21. From the Rock Songs list at billboard.com/ on November 9, 2009.

22. From the online version of the Bureau of Labor Statistics, *Handbook of Methods*, modified April 17, 2003, at bls.gov. The details of the design are more complicated than the text describes.

23. For more detail on the material of this section and complete references, see P. E. Converse and M. W. Traugott, "Assessing the accuracy of polls and surveys," *Science*, 234 (1986), pp. 1094–1098.

24. The nonresponse rate for the CPS comes from the source cited in Note 22. The GSS reports its response rate on its Web site, norc.org/projects/gensoc.asp The Pew study is described in Gregory Flemming and Kimberly Parker, "Race and reluctant respondents: Possible consequences of non-response for pre-election surveys," Pew Research Center for the People and the Press, 1997, found at people-press.org

25. For more detail on the limits of memory in surveys, see N. M. Bradburn, L. J. Rips, and S. K. Shevell, "Answering autobiographical questions: The impact of memory and inference on surveys," *Science*, 236 (1987), pp. 157–161.

26. Robert F. Belli et al., "Reducing vote overreporting in surveys: Social desirability, memory failure, and source monitoring," *Public Opinion Quarterly*, 63 (1999), pp. 90–108.

27. Sex: Tom W. Smith, "The *JAMA* controversy and the meaning of sex," *Public Opinion Quarterly*, 63 (1999), pp. 385–400. Welfare: from a *New York Times*/CBS News poll reported in the *New York Times*, July 5, 1992. Scotland: †"All set for independence?" *Economist*, September 12, 1998. Many other examples appear in T. W. Smith, "That which we call welfare by any other name would smell sweeter," *Public Opinion Quarterly*, 51 (1987), pp. 75–83.

28. From gallup.com on November 10, 2009.

29. From pewresearch.org on November 10, 2009.

30. From *CIS Boletin 9, Spaniards' Economic Awareness*, found online at cis.es/ingles/opinion/economia.htm

31. From yankelovich.com on November 10, 2009.

32. From *Drawing the Line: Sexual Harassment on Campus*, a report from the American Association of University Women (AAUW) Educational Foundation published in 2006. See www.aauw.org

33. See gallup.com/poll/124028/700-Million-Worldwide-Desire-Migrate-Permanently.aspx

34. Warren McIsaac and Vivek Goel, "Is access to physician services in Ontario equitable?" Institute for Clinical Evaluative Sciences in Ontario, October 18, 1993.

35. John C. Bailar III, "The real threats to the integrity of science," *The Chronicle of Higher Education*, April 21, 1995, pp. B1–B2.

36. The difficulties of interpreting guidelines for informed consent and for the work of institutional review boards in medical research are a main theme of Beverly Woodward, "Challenges to human subject protections in U.S. medical research," *Journal of the American Medical Association*, 282 (1999), pp. 1947–1952. The references in this paper point to other discussions.

37. Quotation from the *Report of the Tuskegee Syphilis Study Legacy Committee*, May 20, 1996. A detailed history is James H. Jones, *Bad Blood: The Tuskegee Syphilis Experiment*, Free Press, 1993.

38. Dr. Hennekens's words are from an interview in the Annenberg/Corporation for Public Broadcasting video series *Against All Odds: Inside Statistics.*

39. See `ftc.gov/opa/2009/04/kellogg.shtm`

40. See `findarticles.com/p/articles/mi_m0CYD/is_8_40/ai_n13675065/`

41. R. D. Middlemist, E. S. Knowles, and C. F. Matter, "Personal space invasions in the lavatory: Suggestive evidence for arousal," *Journal of Personality and Social Psychology,* 33 (1976), pp. 541–546.

42. The report was issued in February 2009 and is available from `ftc.gov/os/2009/02/P085400behavadreport.pdf`

Chapter 4

1. An informative and entertaining account of the origins of probability theory is Florence N. David, *Games, Gods and Gambling,* Charles Griffin, London, 1962.

2. Based on infromation from the Color Assignment Web site of Joe Hallock, `joehallock.com/edu/COM498/index.html`

3. You can find a mathematical explanation of Benford's law in Ted Hill, "The first-digit phenomenon," *American Scientist,* 86 (1996), pp. 358–363; and Ted Hill, "The difficulty of faking data," *Chance,* 12, No. 3 (1999), pp. 27–31. Applications in fraud detection are discussed in the second paper by Hill and in Mark A. Nigrini, "I've got your number," *Journal of Accountancy,* May 1999, available online at `aicpa.org/pubs/jofa/joaiss.htm`

4. Royal Statistical Society news release, "Royal Statistical Society concerned by issues raised in Sally Clark case," October 23, 2001, at `www.rss.org.uk`. For background, see an editorial and article in the *Economist,* January 22, 2004. The editorial is entitled "The probability of injustice."

5. See `cdc.gov/mmwr/preview/mmwrhtml/mm57e618a1.htm`

6. See the previous note.

7. From `funtonia.com/top_ringtones_chart.asp/` / This Web site gives popularity scores based on download activity on the Internet. These scores were converted to probabilities for this exercise by dividing each popularity score by the sum of the scores for the top ten ringtones.

8. See `bloodbook.com/world-abo.html` for the distribution of blood types for various groups of people.

9. From Statistics Canada, `www.statcan.ca`

10. Robert P. Dellavalle et al., "Going, going, gone: lost Internet references," *Science,* 302 (2003), pp. 787–788.

11. A. Tversky and D. Kahneman, "Extensional versus intuitive reasoning: The conjunction fallacy in probability judgment," *Psychological Review,* 90 (1983), pp. 293–315.

12. We use $\bar{x}$ both for the random variable, which takes different values in repeated sampling, and for the numer-ical value of the random variable in a particular sample. Similarly, s and $\hat{p}$ stand both for random variables and for specific values. This notation is mathematically imprecise but statistically convenient.

13. We will consider only the case in which X takes a finite number of possible values. The same ideas, implemented with more advanced mathematics, apply to random variables with an infinite but still countable collection of values.

14. From the Web site of the North Carolina State University Department of Registration and Records, `ncsu.edu/class/grades`

15. Based on a Pew Internet report, "Teens and distracted driving," available from `pewinternet.org/Reports/2009/Teens-and-Distracted-Driving.aspx`

16. See `pewinternet.org/Reports/2009/17-Twitter-and-Status-Updating-Fall-2009.aspx`

17. The mean of a continuous random variable X with density function $f(x)$ can be found by integration:

$$\mu_X = \int xf(x)dx$$

This integral is a kind of weighted average, analogous to the discrete-case mean

$$\mu_X = \sum xP(X\bar{x})$$

The variance of a continuous random variable X is the average squared deviation of the values of X from their mean, found by the integral

$$\sigma_X^2 = \int (x - \mu)^2 f(x)dx$$

18. See A. Tversky and D. Kahneman, "Belief in the law of small numbers," *Psychological Bulletin,* 76 (1971), pp. 105–110, and other writings of these authors for a full account of our misperception of randomness.

19. Probabilities involving runs can be quite difficult to compute. That the probability of a run of three or more heads in 10 independent tosses of a fair coin is $(1/2) + (1/128) = 0.508$ can be found by clever counting. A general treatment using advanced methods appears in Section XIII.7 of William Feller, *An Introduction to Probability Theory and Its Applications,* Vol. 1, 3rd ed., Wiley, 1968.

20. R. Vallone and A. Tversky, "The hot hand in basketball: On the misperception of random sequences," *Cognitive Psychology,* 17 (1985), pp. 295–314. A later series of articles that debate the independence question is A. Tversky and T. Gilovich, "The cold facts about the `hot hand' in basketball," *Chance,* 2, No. 1 (1989), pp. 16–21; P. D. Larkey, R. A. Smith, and J. B. Kadane, "It's OK to believe in the `hot hand,'" *Chance,* 2, No. 4 (1989), pp. 22–30; and A. Tversky and T. Gilovich, "The `hot hand': statistical reality or cognitive illusion?" *Chance,* 2, No. 4 (1989), pp. 31–34.

21. Based on a study discussed in S. Atkinson, G. McCabe, C. Weaver, S. Abrams, and K O'Brien, "Are current calcium recommendations for adolescents higher than needed to achieve optimal peak bone mass? The controversy," *Journal of Nutrition* 138, No. 6 (2008) pp. 1182–1186.

22. Commissioners Standard Ordinary Task Force final report, 2002, American Academy of Actuaries, found online at `actuary.org/life/cso_0702.htm`

23. Means and standard deviations from Fidelity Investments, `fidelity.com`. Correlations from the *Fidelity Insight* newsletter, `fidelityinsight.com`. The correlations concern an unspecified period, and other online sources give different correlations, so these should be regarded as approximate at best.

24. Information about Internet users comes from sample surveys carried out by the Pew Internet and American Life Project, found online at `pewinternet.org/` The music downloading data were collected in 2003.

25. These probabilities come from studies by the sociologist Harry Edwards, reported in the *New York Times*, February 25, 1986.

26. From `irs.gov/taxstats/` The data are for taxes filed in 2008 for the tax year 2007.

27. Based on *The Ethics of American Youth---2008,* available from the Josephson Institute at `charactercounts.org/programs/reportcard/`

28. From `irs.gov/taxstats`

29. See `nces.ed.gov/programs/digest/` Data are from Table 311 of the 2007 *Digest of Education Statistics*.

30. From the 2006 *Statistical Abstract of the United States,* Table 474.

31. From the Bureau of Labor Statistics, found online at `bls.gov/data`

32. See Note 29.

Chapter 5

1. H.G. Lund et al., "Sleep patterns and predictors of disturbed sleep in a large population of college students," *Journal of Adolescent Health*, 46 (2010), pp. 124–132.

2. The description of the 2009 survey and results can be found at `blog.appsfire.com/tag/stats`

3. Haipeng Shen, "Nonparametric regression for problems involving lognormal distributions," PhD thesis, University of Pennsylvania, 2003. Thanks to Haipeng Shen and Larry Brown for sharing the data.

4. Statistics regarding Facebook usage can be found at `www.facebook.com/press/info.php?statistics`

5. From the Web site of the North Carolina State University Department of Registration and Records, `www-records.ncsu.edu/cgi-bin/grddist.pl`

6. Values based on Sherri A. Buzinski, "The effect of position of methylation on the performance properties of durable press treated fabrics," CSR490 honors paper, Purdue University, 1985.

7. Values based on David Hon-Kuen Chu, "A test of corporate advertising using the elaboration likelihood model," MS thesis, Purdue University, 1993.

8. S. A. Rahimtoola, "Outcomes 15 years after valve replacement with a mechanical vs. a prosthetic valve, final report of the Veterans Administration randomized trial," American College of Cardiology, found online at `www.acc.org/education/online/trials/acc2000/15yr.htm`

9. The full online clothing store ratings are featured in the December 2008 issue of Consumer Reports and online at `www.ConsumerReports.org`

10. Statistical methods for dealing with time-to-failure data, including the Weibull model, are presented in Wayne Nelson, *Applied Life Data Analysis*, Wiley, 1982.

11. A description and summary of this 2009 survey can be found at `www.musically.com/theleadingquestion/downloads/090713-filesharing.pdf`

12. Joseph Carroll, "Americans: 2.5 children is "ideal" family size," Gallup Poll press release, June 26, 2007, found online at `www.gallup.com`

13. Jeffrey M. Jones, "One in six Americans gamble on sports," Gallup Poll press release, February 1, 2008, found online at `www.gallup.com`

14. A summary of Larry Wright's study can be found at `www.nytimes.com/2009/03/04/sports/basketball/04freethrow.html`

15. Barbara Means et al., "Evaluation of evidence-based practices in online learning: A meta-analysis and review of online learning studies," U.S. Department of Education, Office of Planning, Evaluation, and Policy Development, 2009.

16. Henry Wechsler et al., "Binge drinking on America's college campuses," Harvard School of Public Health, 2001.

17. Statistics Canada, "Spending patterns in Canada," Income Statistics Division, 2008.

18. September 22, 2009 press release from `www.census.gov`

19. This information can be found at `www.census.gov/genealogy/names/dist.all.last`

Chapter 6

1. See Noel Cressie, *Statistics for Spatial Data*, Wiley, 1993. The significance test result that we report is one of several that could be used to address this question. See pp. 607–609 of the Cressie book for more details.

2. Based on information reported in "How undergraduate students use credit cards: Sallie Mae's National Study of Usage Rates and Trends 2009," found online at `www.salliemae.com/about/news_info/research/credit_card_study`

3. Average starting salary taken from a 2009 summer salary survey from the National Association of Colleges and Employers (NACE).

4. See Note 2.

5. The standard reference here is Bradley Efron and Robert J. Tibshirani, *An Introduction to the Bootstrap*, Chapman Hall, 1993. A less technical overview is in Bradley Efron and Robert J. Tibshirani, "Statistical data analysis in the computer age," *Science*, 253 (1991), pp. 390–395.

6. Site can be found at `ksolo.myspace.com`

7. "Value of Recreational Sports on College Campuses," National Intramural-Recreational Sports Association (2002).

8. C. M. Weaver et al., "Quantification of biochemical markers of bone turnover by kinetic measures of bone formation and resorption in young healthy females," *Journal of Bone and Mineral Research*, 12 (1997), pp. 1714–1720.

9. Press releases of these salary survey reports can be found at `www.naceweb.org/press`

10. Sara N. Bleich et al., "Increasing consumption of sugar-sweetened beverages among US adults: 1988ñ1994 to 1999ñ2004," *The American Journal of Clinical Nutrition*, 89 (2009), pp. 372–381.

11. 2009 press release from *The Student Monitor* located at `www.studentmonitor.com`

12. Lydia Saad, "U.S. workers' job satisfaction is relatively high," Gallup News Service, August 21, 2008. Found at `www.gallup.com/poll/`

13. The vehicle is a 2002 Toyota Prius owned by the third author.

14. Leading causes of death reports available `webapp.cdc.gov/sasweb/ncipc/leadcaus10.html`.

15. Phillip B. Sparling et al., "Serum cholesterol levels in college students: Opportunities for education and intervention," *Journal of American College Health*, 48 (1999), pp. 123–127.

16. Rodney G. Bowden et al., "Lipid levels in a cohort of sedentary university students," *The Internet Journal of Cardiovascular Research*, 2(2) (2005), pp. 1108–1115.

17. Some analyses of the data collected by the Gambian National Inpregnated Bednet Program are reported in M. C. Thompson et al., "A spatial model of malaria risk in the Gambia: Predicting the impact of insecticide treated and untreated bednets on malaria infection," *Proceedings of the Spatial Information Research Centre's 10th Colloquium*, University of Otago, New Zealand, 1998, pp. 321–328.

18. Giacomo DeGiorgi et al., "Be as careful of the company you keep as of the books you read: Peer effects in education and on the labor market," National Bureau of Economic Research, Working paper 14948 (2009).

19. Seung-Ok Kim, "Burials, pigs, and political prestige in neolithic China," *Current Anthropology*, 35 (1994), pp. 119–141.

20. These data were collected in connection with the Purdue Police Alcohol Student Awareness Program run by Police Officer D. A. Larson.

21. National Assessment of Educational Progress, *The Nations Report Card*, Mathematics 2009.

22. Sogol Javaheri et al., "Sleep quality and elevated blood pressure in adolescents," *Circulation*, 118 (2008), pp. 1034–1040.

23. Victor Lun et al., "Evaluation of nutritional intake in Canadian high-performance athletes," *Clinical Journal of Sports Medicine*, 19(5) (2009), pp. 405–411.

24. R. A. Fisher, "The arrangement of field experiments," *Journal of the Ministry of Agriculture of Great Britain*, 33 (1926), p. 504, quoted in Leonard J. Savage, "On rereading R. A. Fisher," *Annals of Statistics*, 4 (1976), p. 471. Fisher's work is described in a biography by his daughter: Joan Fisher Box, *R. A. Fisher: The Life of a Scientist*, Wiley, 1978.

25. The editorial was written by Phil Anderson. See *British Medical Journal*, 328 (2004), pp. 476–477. A letter to the editor on this topic by Doug Altman and J. Martin Bland appeared shortly after. See "Confidence intervals illuminate absence of evidence," *British Medical Journal*, 328 (2004), pp. 1016–1017.

26. A. Kamali et al., "Syndromic management of sexually-transmitted infections and behavior change interventions on transmission of HIV-1 in rural Uganda: A community randomised trial," *Lancet*, 361 (2003), pp. 645–652.

27. T. D. Sterling, "Publication decisions and their possible effects on inferences drawn from tests of significance—or vice versa," *Journal of the American Statistical Association*, 54 (1959), pp. 30–34. Related comments appear in J. K. Skipper, A. L. Guenther, and G. Nass, "The sacredness of 0.05: A note concerning the uses of statistical levels of significance in social science," *American Sociologist*, 1 (1967), pp. 16–18.

28. For a good overview of these issues see Bruce A. Craig, Michael A. Black, and Rebecca W. Doerge, "Gene expression data: The technology and statistical analysis," *Journal of Agricultural, Biological, and Environmental Statistics*, 8 (2003), pp. 1–28.

29. Erick H. Turner et al., "Selective publication of antidepressant trials and its influence on apparent efficacy," *The New England Journal of Medicine*, 358 (2008), pp. 252–260.

30. Robert J. Schiller, "The volatility of stock market prices," *Science*, 235 (1987), pp. 33–36.

31. Padmaja Ayyagari and Jody L. Sindelar, "The impact of job stress on smoking and quitting: evidence from the HRS," National Bureau of Economic Research, Working paper 15232 (2009).

32. Based on A. M. Garcia et al., "Why do workers behave unsafely at work? Determinants of safe work practices in industrial workers," *Occupational and Environmental Medicine*, 61 (2004), pp. 239–246.

33. Data from Joan M. Susic, "Dietary phosphorus intakes, urinary and peritoneal phosphate excretion and clearance in continuous ambulatory peritoneal dialysis patients," MS thesis, Purdue University, 1985.

34. Mugdha Gore and Joseph Thomas, "Store image as a predictor of store patronage for nonprescription medication purchases: A multiattribute model approach," *Journal of Pharmaceutical Marketing & Management*, 10 (1996), pp. 45–68.

35. Greg L. Stewart et al., "Exploring the handshake in employment interviews," *Journal of Applied Psychology*, 93 (2008), pp. 1139-1146.

Chapter 7

1. Average hours per month obtained from "A2/M2 Three Screen Report, 4th Quarter 2008," Nielsen Company (2009).

2. C. Don Wiggins, "The legal perils of `underdiversification'— a case study," *Personal Financial Planning*, 1, No. 6 (1999), pp. 16–18.

3. These data were collected as part of a larger study of dementia patients conducted by Nancy Edwards, School of Nursing, and Alan Beck, School of Veterinary Medicine, Purdue University.

4. These recommendations are based on extensive computer work. See, for example, Harry O. Posten, "The robustness of the one-sample *t*-test over the Pearson system," *Journal of Statistical Computation and Simulation*, 9 (1979), pp. 133–149; and E. S. Pearson and N. W. Please, "Relation between the shape of population distribution and the robustness of four simple test statistics," *Biometrika*, 62 (1975), pp. 223–241.

5. The data were obtained on August 24, 2006 from an iPod owned by George McCabe Jr.

6. Method described in the article by Naihua Duan titled "Smearing estimate: A nonparametric retransformation method," *Journal of the American Statistical Association*, 78 (1983), pp. 605–610.

7. You can find a practical discussion of distribution-free inference in Myles Hollander and Douglas A. Wolfe, *Nonparametric Statistical Methods*, 2nd ed., Wiley, 1999.

8. Statistics regarding Facebook usage can be found at www.facebook.com/press/info.php?statistics

9. Christine L. Porath and Amir Erez, "Overlooked but not untouched: How rudeness reduces onlookers' performance on routine and creative tasks," *Organizational Behavior and Human Decision Processes*, 109 (2009), pp. 29–44.

10. The vehicle is a 2002 Toyota Prius owned by the third author.

11. Data provided by Betsy Hoza of the Department of Psychological Sciences, Purdue University.

12. David R. Pillow et al., "Confirmatory factor analysis examining attention deficit hyperactivity disorder symptoms and other childhood disruptive behaviors," *Journal of Abnormal Child Psychology*, 26 (1998), pp. 293–309.

13. James A. Levine, Norman L. Eberhardt, and Michael D. Jensen, "Role of nonexercise activity thermogenesis in resistance to fat gain in humans," *Science*, 283 (1999), pp. 212–214. Data for this study are available from the *Science* Web site, www.sciencemag.org

14. These data were collected in connection with a bone health study at Purdue University and were provided by Linda McCabe.

15. Data provided by Joseph A. Wipf, Department of Foreign Languages and Literatures, Purdue University.

16. Data from Wayne Nelson, *Applied Life Data Analysis*, Wiley, 1982, p. 471.

17. Summary information can be found at the National Center for Health Statistics Web site www.cdc.gov/nchs/nhanes.htm

18. Detailed information about the conservative *t* procedures can be found in Paul Leaverton and John J. Birch, "Small sample power curves for the two sample location problem," *Technometrics*, 11 (1969), pp. 299–307; in Henry Scheffé, "Practical solutions of the Behrens-Fisher problem," *Journal of the American Statistical Association*, 65 (1970), pp. 1501–1508; and in D. J. Best and J. C. W. Rayner, "Welch's approximate solution for the Behrens-Fisher problem," *Technometrics*, 29 (1987), pp. 205–210.

19. This example is adapted from Maribeth C. Schmitt, "The effects of an elaborated directed reading activity on the metacomprehension skills of third graders," PhD dissertation, Purdue University, 1987.

20. See the extensive simulation studies in Harry O. Posten, "The robustness of the two-sample *t* test over the Pearson system," *Journal of Statistical Computation and Simulation*, 6 (1978), pp. 295–311.

21. Sogol Javaheri et al., "Sleep quality and elevated blood pressure in adolescents," *Circulation*, 118 (2008), pp. 1034–1040.

22. This study is reported in Roseann M. Lyle et al., "Blood pressure and metabolic effects of calcium supplementation in normotensive white and black men," *Journal of the American Medical Association*, 257 (1987), pp. 1772–1776. The individual measurements in Table 7.5 were provided by Dr. Lyle.

23. C. E. Cryfer et al., "Misery is not miserly: Sad and self-focused individuals spend more," *Psychological Science*, 19 (2008), pp. 525–530.

24. A. A. Labroo et al., "Of frog wines and frowning watches: Semantic priming, perceptual fluency, and brand evaluation," *Journal of Consumer Research*, 34 (2008), pp. 819–831.

25. The 2008 study can be found at http://www.qsrmagazine.com/reports/drive-thru_time_study/2008/consumers-1.phtml

26. Grant D. Brinkworth et al., "Long-term effects of a very low-carbohydrate diet and a low-fat diet on mood and cognitive function," *Archives of Internal Medicine*, 169 (2009), pp. 1873–1880.

27. M. K. Campbell et al., "A tailored multimedia nutrition education pilot program for low-income women receiving food assistance," *Health Education Research*, 14 (1999), pp. 257–267.

28. B. Bakke et al., "Cumulative exposure to dust and gases as determinants of lung function decline in tunnel construction workers," *Occupational Environmental Medicine*, 61 (2004), pp. 262–269.

29. Samara Joy Nielsen and Barry M. Popkin, "Patterns and trends in food portion sizes, 1977–1998," *Journal of the American Medical Association*, 289 (2003), pp. 450–453.

30. Gordana Mrdjenovic and David A. Levitsky, "Nutritional and energetic consequences of sweetened drink consumption in 6- to 13-year old children," *Journal of Pediatrics*, 142 (2003), pp. 604–610.

31. David Han-Kuen Chu, "A test of corporate advertising using the elaboration likelihood model," MS thesis, Purdue University, 1993.

32. M. F. Picciano and R. H. Deering, "The influence of feeding regimens on iron status during infancy," *American Journal of Clinical Nutrition*, 33 (1980), pp. 746–753.

33. The problem of comparing spreads is difficult even with advanced methods. Common distribution-free procedures do not offer a satisfactory alternative to the F test, because they are sensitive to unequal shapes when comparing two distributions. A good introduction to the available methods is W. J. Conover, M. E. Johnson, and M. M. Johnson, "A comparative study of tests for homogeneity of variances, with applications to outer continental shelf bidding data," *Technometrics*, 23 (1981), pp. 351–361. Modern resampling procedures often work well. See Dennis D. Boos and Colin Brownie, "Bootstrap methods for testing homogeneity of variances," *Technometrics*, 31 (1989), pp. 69–82.

34. G. E. P. Box, "Non-normality and tests on variances," *Biometrika*, 40 (1953), pp. 318–335. The quote appears on page 333.

35. This city's restaurant inspection data can be found at www.jsonline.com/watchdog/dataondemand/

36. G. E. Smith et al., "A cognitive training program based on principles of brain plasticity: Results from the improvement in memory with plasticity-based adaptive cognitive training (IMPACT) study," *Journal of the American Geriatrics Society*, epub (2009), pp. 1–10.

37. Based on Loren Cordain et al., "Influence of moderate daily wine consumption on body weight regulation and metabolism in healthy free-living males," *Journal of the American College of Nutrition*, 16 (1997), pp. 134–139.

38. B. Wansink et al., "Fine as North Dakota wine: Sensory expectations and the intake of companion foods," *Physiology & Behavior*, 90 (2007), pp. 712–716.

39. Douglas J. Levey et al., "Urban mockingbirds quickly learn to identify individual humans," *Proceedings of the National Academy of Sciences*, 106 (2009), pp. 8959–8962.

40. Anne Z. Hoch et al., "Prevalence of the female athlete triad in high school athletes and sedentary students," *Clinical Journal of Sports Medicine*, 19 (2009), pp. 421-428.

41. This exercise is based on events that are real. The data and details have been altered to protect the privacy of the individuals involved.

42. Based loosely on D. R. Black et al., "Minimal interventions for weight control: A cost-effective alternative," *Addictive Behaviors*, 9 (1984), pp. 279–285.

43. These data were provided by Professor Sebastian Heath, School of Veterinary Medicine, Purdue University.

44. J. W. Marr and J. A. Heady, "Within- and between-person variation in dietary surveys: number of days needed to classify individuals," *Human Nutrition: Applied Nutrition*, 40A (1986), pp. 347–364.

45. Data taken from the Web site www.realtor.com on December 19, 2009.

Chapter 8

1. The actual distribution of X based on an SRS from a finite population is the *hypergeometric distribution*. Details regarding this distribution can be found in Ross, Sheldon, *A First Course in Probability*, 8th ed., Prentice Hall, 2010.

2. From the PriceWaterhouseCooper Web site pwc.com

3. From a Pew Internet Project Memo, Amanda Lenhart et al., dated December 7, 2008. See pewinternet.org/pdfs/PIP_Adult_gaming_memo.pdf

4. From the "Community Bank Competitiveness Survey," 2008, *ABA Banking Journal*. The survey is available at nxtbook.com/nxtbooks/sb/ababj-compsurv08/index.php

5. Details of exact binomial procedures can be found in Myles Hollander and Douglas Wolfe, *Nonparametric Statistical Methods*, 2nd ed., Wiley, 1999.

6. See A. Agresti and B. A. Coull, "Approximate is better than `exact' for interval estimation of binomial proportions," *The American Statistician*, 52 (1998), pp. 119–126. A detailed theoretical study is Lawrence D. Brown, Tony Cai, and Anirban DasGupta, "Confidence intervals for a binomial proportion and asymptotic expansions," *Annals of Statistics*, 30 (2002), pp. 160–201.

7. See, for example, `pilatesmethodalliance.org`

8. Heather Tait, *Aboriginal Peoples Survey, 2006: Inuit Health and Social Conditions,* (2008) Social and Aboriginal Statistics Division, Statistics Canada. Available from `statcan.gc.ca/pub`

9. From a story posted on December 20, 2008 at `stuff.co.nz` A question based on this survey was used in Michael Feldman's "Whad' Ya Know Quiz" in December 2008. See `notmuch.com`

10. See `news.teamxbox.com/xbox/18254`

11. See the "National Survey of Student Engagement, The College Student Report 2009," available online at `nsse.iub.edu/index.cfm`

12. This survey and others that study issues related to college students can be found at `nelliemae.com`

13. See Note 11.

14. Results from the 2006 Pew Research Center report entitled "Americans and Their Cars: Is the Romance on the Skids?"

15. From the "Report Card 2004: The Ethics of American Youth," 2004, available online at `josephsoninstitute.org/reportcard`

16. Information about the survey can be found online at `saint-denis.library.arizona.edu/natcong`

17. From Heeseung Roh Ryu, Roseann M. Lyle, and George P. McCabe, "Factors associated with weight concerns and unhealthy eating patterns among young Korean females," *Eating Disorders*, 11 (2003), pp. 129–141.

18. Data from Roland J. Thorpe Jr. and based on his analysis of "Health ABC," a 10-year longitudinal study of older adults supported by the Laboratory of Epidemiology, Demography, and Biometry of the National Institute on Aging. Additional analyses are given in his PhD dissertation, "Relationship between pet ownership, physical activity, and human health in an elderly population," Purdue University, 2004.

19. From McCulloch et al., "Diagnostic accuracy of canine scent detection in early- and late-stage lung and breast cancers," *Integrative Cancer Therapies*, 5(1) (2006), pp. 30-39.

20. Data from Guohua Li and Susan P. Baker, "Alcohol in fatally injured bicyclists," *Accident Analysis and Prevention*, 26 (1994), pp. 543–548.

21. See, for example, `sciencedaily.com/releases/2009/11/091117094833.htm` for a collection of links concerning this problem.

22. See Alan Agresti and Brian Caffo, "Simple and effective confidence intervals for proportions and differences of proportions result from adding two successes and two failures," *The American Statistician*, 45 (2000), pp. 280–288. The Wilson interval is a bit conservative (true coverage probability is higher than the confidence level) when p_1 and p_2 are equal and close to 0 or 1, but the traditional interval is much less accurate and has the fatal flaw that the true coverage probability is *less* than the confidence level.

23. J. M Tanner, "Physical growth and development," in J. O. Forfar and G. C Arneil, *Textbook of Paediatrics*, 3rd ed., Churchill Livingston, 1984, pp. 1–292.

24. This information was posted on the Podcast Alley Web site, `podcastalley.com` on January 13, 2009.

25. From a Pew Internet Project Data Memo by Mary Madden, dated August 2008. Available at `pewinternet.org`

26. From a Pew Internet Project Data Memo by Amanda Lenhart et al., dated December 2008. Available at `pewinternet.org`

27. See Note 18.

28. From M. L. Burr et al., "Effects on respiratory health of a reduction in air pollution from vehicle exhaust emissions," *Occupational and Environmental Medicine*, 61 (2004), pp. 212–218.

29. Mary Madden and Lee Rainie, "Music and video downloading moves beyond P2P," Pew Internet Project Data Memo, 2005. This memo is available online at `pewinternet.org`

30. Based on Greg Clinch, "Employee compensation and firms' research and development activity," *Journal of Accounting Research*, 29 (1991), pp. 59–78.

31. From Monica Macaulay and Colleen Brice, "Don't touch my projectile: Gender bias and stereotyping in syntactic examples," *Language*, 73, No. 4 (1997), pp. 798–825. The first part of the title is a direct quote from one of the texts.

32. From a Pew Internet report "Teens, Video Games, and Civics," by Amanda Lehnart et al., September 16, 2008. Available at `pewinternet.org`

33. From the Entertainment Software Association Web site at `theesa.com/facts`

34. Data from the Neilsen Web site `nielsenmobile.com/documents/WirelessSubstitution.pdf`

35. See Note 12.

36. Lee Rainie et al., "The state of music downloading and file-sharing online," Pew Internet Project and Comscore Media Metrix Data Memo, 2004. This memo is available online at `pewinternet.org`

37. Based on Robert T. Driescher, "A quality swing with Ping," *Quality Progress*, August 2001, pp. 37–41.

38. Based on a Pew Research poll conducted in March–May, 2006. It can be found at `pewresearch.org`

39. From Dennis N. Bristow and Richard J. Sebastian, "Holy cow! Wait till next year! A closer look at the brand loyalty of Chicago Cubs baseball fans," *Journal of Consumer Marketing*, 18 (2001), pp. 256–275.

40. See S. W. Lagakos, B. J. Wessen, and M. Zelen, "An analysis of contaminated well water and health effects in Woburn, Massachusetts," *Journal of the American Statistical Association*, 81 (1986), pp. 583–596, and the following discussion. This case is the basis for the movie *A Civil Action*.

41. This case is discussed in D. H. Kaye and M. Aickin (eds.), *Statistical Methods in Discrimination Litigation*, Marcel Dekker, 1986; and D. C. Baldus and J. W. L. Cole, *Statistical Proof of Discrimination*, McGraw-Hill, 1980.

42. See Note 12.

Chapter 9

1. From J. Cantor, "Long-Term Memories of Frightening Media Often Include Lingering Trauma Symptoms," poster presented at the Association for Psychological Science Convention, New York, May 26, 2006.

2. When the expected cell counts are small, it is best to use a test based on the exact distribution rather than the chi-square approximation, particularly for 2×2 tables. Many statistical software systems offer an "exact" test as well as the chi-square test for 2×2 tables.

3. From E. Y. Peck, "Gender differences in film-induced fear as a function of type of emotion measure and stimulus content: A meta-analysis and laboratory study," unpublished doctoral dissertation, University of Wisconsin-Madison.

4. From P. Strazzullo et al., "Salt intake, stroke, and cardiovascular disease," *BMJ*, 339 (2009). The meta-analysis combined data from 14 study cohorts taken from 10 different studies.

5. N. R. Cook et al., "Long term effects of dietary sodium reduction on caradiovascular disease outcomes: Observational follow-up of the trials of the hypertension prevention (TOHP)," *BMJ*, (2007).

6. D-C Seo et al., "Relations between physical activity and behavioral and perceptual correlates among midwestern college students," *Journal of American College Health*, (2007) pp. 187–197.

7. An alternative formula that can be used for hand or calculator computations is

$$X^2 = \sum \frac{(\text{observed})^2}{\text{expected}} - n$$

8. See, for example, Alan Agresti, *Categorical Data Analysis*, Wiley, 2nd ed., 2007.

9. The sampling procedure was designed by George McCabe. It was carried out by Amy Conklin, an undergraduate Honors student in the Department of Foods and Nutrition at Purdue University.

10. The analysis could also be performed by using a two-way table to compare the states of the selected and not selected students. Since the selected students are a relatively small percentage of the total sample, the results will be approximately the same.

11. See the M&M Mars' Web site at `us.mms.com/us/about/products` for this and other information.

12. See `pewinternet.org/about.asp`

13. Data are from the report Home Broadband Adoption 2008 which was prepared by the Pew Internet American Life Project. See `pewinternet.org/pdfs/PIP_Broadband_2008.pdf`

14. See Note 13.

15. See `nhcaa.org`

16. These data are a composite based on several actual audits of this type.

17. See, for example, R. Benford and J. Gess-Newsome, "Factors affecting student academic success in gateway courses at Northern Arizona University," 2006; available from `eric.ed.gov`

18. Data provided by Professor Marcy Towns of the Purdue University Department of Chemistry.

19. See Note 17.

20. From the "Survey of Canadian Career College Students Phase II: In-School Student Survey," 2008. This report is available from `hrsdc.gc.ca/eng/publications_resources`

21. For an overview of remote deposit capture, see `remotedepositcapture.com/overview/rdc.overview.aspx`

22. From the "Community Bank Competitiveness Survey," 2008, *ABA Banking Journal*. The survey is available at `nxtbook.com/nxtbooks/sb/ababj-compsurv08/index.php`

23. From M-Y Chen et al., "Adequate sleep among adolescents is positively associated with health status and health-related behaviors," *BMC Public Health*, 6:59 (2006); available from `biomedicalcentral.com/1471-2458/6/59`

24. Based on *The Ethics of American Youth—2008*, available from the Josephson Institute at `charactercounts.org/programs/reportcard`

25. See Note 1.

26. See the U.S. Bureau of the Census Web site at `census.gov/population/socdemo/school` for these and similar data.

27. Based on data in Mike Planty et al., "Volunteer service by young people from high school through early adulthood," National Center for Educational Statistics Report NCES 2004–365.

28. From S.R. Davy, B.A. Benes, and J.A. Driskell, "Sex differences in dieting trends, eating habits, and nutrition beliefs of a group of Midwestern college students," *Journal of the American Dietetic Association*, 10 (2006), pp. 1673-1677.

29. Tom Reichert, "The prevalence of sexual imagery in ads targeted to young adults," *Journal of Consumer Affairs*, 37 (2003), pp. 403–412.

30. See George R. Milne, "How well do consumers protect themselves from identity theft?" *Journal of Consumer Affairs*, 37 (2003), pp. 388–402.

31. Based on information in "NCAA 2003 national study of collegiate sports wagering and associated health risks," which can be found at the NCAA Web site, `ncaa.org`

32. Ethan J. Temeles and W. John Kress, "Adaption in a plant-hummingbird association," *Science*, 300 (2003), pp. 630–633.

33. Robert P. Dellavalle et al., "Going, going, gone: Lost Internet references," *Science*, 302 (2003), pp. 787–788.

34. Data from Roland J. Thorpe Jr. and based on his analysis of "Health ABC," a 10-year longitudinal study of older adults supported by the Laboratory of Epidemiology, Demography, and Biometry of the National Institute on Aging. Additional analyses are given in his PhD dissertation "Relationship between pet ownership, physical activity, and human health in an elderly population," 2004, Purdue University.

35. These data are from an Undergraduate Task Force study of student retention in the Purdue University College of Science.

36. Daniel Cassens et al., "Face check development in veneered furniture panels," *Forest Products Journal*, 53 (2003), pp. 79–86.

37. See James A. Taylor et al., "Efficacy and safety of echinacea in treating upper respiratory tract infections in children," *Journal of the American Medical Association*, 290 (2003), pp. 2824–2830.

38. Karen L. Dean, "Herbalists respond to *JAMA* echinacea study," *Alternative & Complementary Therapies*, 10 (2004), pp. 11–12.

39. The report can be found at `nsse.iub.edu/pdf/NSSE2005\_annual\_report.pdf`

Chapter 10

1. Data based on Michael L. Mestek et al., "The relationship between pedometer-determined and self-reported physical activity and body composition variables in college-aged men and women," *Journal of American College Health*, 57 (2008), pp. 39–44.

2. The vehicle is a 1997 Pontiac transport van owned by the second author.

3. Information regarding bone health can be found in "Osteoporosis: Peak bone mass in women," last reviewed in May 2009 and available at `www.niams.nih.gov/bone`

4. The data were provided by Linda McCabe and were collected as part of a large study of women's bone health and another study of calcium kinetics, both directed by Professor Connie Weaver of the Department of Foods and Nutrition, Purdue University. Results related to this example are reported in Dorothy Teegarden et al., "Peak bone mass in young women," *Journal of Bone and Mineral Research*, 10 (1995), pp. 711–715.

5. These data were provided by Professor Wayne Campbell of the Purdue University Department of Foods and Nutrition.

6. This quantity is the estimated standard deviation of $\hat{y} - y$, not the estimated standard deviation of $\hat{y}$ alone.

7. For more information about nutrient requirements, see the Institute of Medicine publications on Dietary Reference Intakes available at `www.nap.edu`

8. The method is described in Chapter 2 of M. Kutner et al., *Applied Linear Statistical Models*, 5th ed., Irwin, 2005.

9. National Science Foundation, Division of Science Resources Statistics. 2008. *Academic Research and Development Expenditures: Fiscal Year 2007*. Detailed Statistical Tables NSF 09-303. Arlington, VA. Available at `www.nsf.gov/statistics/nsf09303/`

10. Tuition rates for 2000 from the "2000-2001 Tuition and Required Fees Report," University of Missouri. Tuition rates for 2008 are available at `colleges.collegetoolkit.com/college/main.aspx`

11. These are part of the data from the EESEE story "Blood alcohol content," found on the text Web site.

12. M. Mondello and J. Maxcy "The impact of salary dispersion and performance bonuses in NFL organizations'" *Management Decision*, 47 (2009), pp. 110-123.

13. These data were collected from `www.cbssports.com/nfl/playerrankings/regularseason/` and `content.usatoday.com/sports/football/nfl/salaries/`

14. Selling price and assessment value available at `php.jconline.com/propertysales/propertysales.php`

15. Data available at `www.ncdc.noaa.gov/oa/climate/sd`

16. Matthew P. Martens et al., "The co-occurrence of alcohol use and gambling activities in first-year college students," *Journal of American College Health,* 57 (2009), pp. 597–602.

17. Based on Dan Dauwalter's master's thesis in the Department of Forestry and Natural Resources at Purdue University. More information is available in Daniel C. Dauwalter et al., "An index of biotic integrity for fish assemblages in Ozark Highland streams of Arkansas," *Southeastern Naturalist,* 2 (2003), pp. 447–468. These data were provided by Emmanuel Frimpong.

18. Net cash flow data from Sean Collins, *Mutual Fund Assets and Flows in 2000,* Investment Company Institute, 2001 and "Trends in Mutual Fund Investing" at www.ici.org/stats/mf/arctrends/index.html The raw data were converted to real dollars using annual average values of the CPI.

19. The data were provided by Professor Shelly MacDermid, Department of Child Development and Family Studies, Purdue University, from a study reported in S. M. MacDermid et al., "Is small beautiful? Work-family tension, work conditions, and organizational size," *Family Relations,* 44 (1994), pp. 159–167.

20. These data are from G. Geri and B. Palla, "Considerazioni sulle piú recenti osservazioni ottiche alla Torre Pendente di Pisa," *Estratto dal Bollettino della Societá Italiana di Topografia e Fotogrammetria,* 2 (1988), pp. 121–135. Professor Julia Mortera of the University of Rome provided valuable assistance with the translation.

21. M. Kuo et al., "The marketing of alcohol to college students: The role of low prices and special promotions," *American Journal of Preventive Medicine,* 25(3) (2003), pp. 204–211.

22. Data for the first four years from Manuel Castells, *The Rise of the Network Society,* 2nd ed., Blackwell, 2000, p. 41. The remaining years come from Intel.

23. This study is reported in S. Lau and P. C. Cheung, "Relations between Chinese adolescents' perception of parental control and organization and their perception of parental warmth," *Developmental Psychology,* 23 (1987), pp. 726–729.

24. Donald A. Mahler and John I. Mackowiak, "Evaluation of the short-form 36-item questionnaire to measure health-related quality of life in patients with COLD," *Chest,* 107 (1995), pp. 1585–1589.

25. Data on a sample of 12 of 56 perch in a data set contributed to the *Journal of Statistics Education* data archive www.amstat.org/publications/jse/ by Juha Puranen of the University of Helsinki.

26. L. Cooke et al., "Relationship between parental report of food neophobia and everyday food consumption in 2–6-year-old children," *Appetite,* 41 (2003), pp. 205–206.

27. Alexandra Burt, "A mechanistic explanation of popularity: Genes, rule breaking, and evocative gene-environment correlations," *Journal of Personality and Social Psychology,* 96 (2009), pp. 783–794.

Chapter 11

1. Results of the study are reported in P. F. Campbell and G. P. McCabe, "Predicting the success of freshmen in a computer science major," *Communications of the ACM,* 27 (1984), pp. 1108-1113.

2. R.M. Smith and P.A. Schumacher "Predicting success for actuarial students in undergraduate mathematics courses," *College Student Journal,* 39(1) (2005), pp. 165-177.

3. Based on Leigh J. Maynard and Malvern Mupandawana, "Tipping behavior in Canadian restaurants," *International Journal of Hospitality Management,* 28 (2009), pp. 597-ñ603.

4. Kathleen E. Miller, "Wired: Energy drinks, jock identity, masculine norms, and risk taking," *Journal of American College Health,* 56 (2008), pp. 481-489.

5. From a table entitled "Largest Indianapolis-area architectural firms," *Indianapolis Business Journal,* December 16, 2003.

6. The data were obtained from the Internet Movie Database (IMDb), available at www.imdb.com on April 20, 2010.

7. 2009 table of 200 top universities can be found at www.timeshighereducation.co.uk

8. The results were published in C. M. Weaver et al., "Quantification of biochemical markers of bone turnover by kinetic measures of bone formation and resorption in young healthy females," *Journal of Bone and Mineral Research,* 12 (1997), pp. 1714-1720. The data were provided by Linda McCabe.

9. This data set was provided by Joanne Lasrado of the Purdue University Department of Foods and Nutrition.

10. These data are based on experiments performed by G. T. Lloyd and E. H. Ramshaw of the CSIRO Division of Food Research, Victoria, Australia. Some results of the statistical analyses of these data are given in G. P. McCabe, L. McCabe, and A. Miller, "Analysis of taste and chemical composition of cheddar cheese, 1982-83 experiments," CSIRO Division of Mathematics and Statistics Consulting Report VT85/6; and in I. Barlow et al., "Correlations and changes in flavour and chemical parameters of cheddar cheeses during maturation," *Australian Journal of Dairy Technology,* 44 (1989), pp. 7-18.

Chapter 12

1. Joseph B. Walther et al., "The role of friends' behavior on evaluations of individuals' Facebook profiles: Are we known by the company we keep?" *Human Communication Research*, 34 (2008), pp. 28–49.

2. Based on Stephanie T. Tong et al., "Too much of a good thing? The relationship between number of friends and interpersonal impressions on Facebook," *Journal of Computer-Mediated Communication*, 13 (2008), pp. 531–549.

3. This rule is intended to provide a general guideline for deciding when serious errors may result by applying ANOVA procedures. When the sample sizes in each group are very small, the sample variances will tend to vary much more than when the sample sizes are large. In this case, the rule may be a little too conservative. For unequal sample sizes, particular difficulties can arise when a relatively small sample size is associated with a population having a relatively large standard deviation. Careful judgment is needed in all cases. By considering P-values rather than fixed level α testing, judgments in ambiguous cases can more easily be made; for example, if the P-value is very small, say 0.001, then it is probably safe to reject H_0 even if there is a fair amount of variation in the sample standard deviations.

4. Penny M. Simpson et al., "The eyes have it, or do they? The effects of model eye color and eye gaze on consumer ad response," *The Journal of Applied Business and Economics*, 8 (2008), pp. 60–71.

5. Several different definitions for the noncentrality parameter of the noncentral F distribution are in use. When $I = 2$, the λ defined here is equal to the square of the noncentrality parameter δ that we used for the two-sample t test in Chapter 7. Many authors prefer $\phi = \sqrt{\lambda/I}$. We have chosen to use λ because it is the form needed for the SAS function PROBF.

6. Jesus Tanguma et al., "Shopping and bargaining in Mexico: The role of women," *The Journal of Applied Business and Economics*, 9 (2009), pp. 34–40.

7. Sangwon Lee and Seonmi Lee, "Multiple play strategy in global telecommunication markets: An empirical analysis," *International Journal of Mobile Marketing*, 3 (2008), pp. 44–53.

8. P. Bartel et al., "Attention and working memory in resident anaesthetists after night duty: Group and individual effects," *Occupational and Environmental Medicine*, 61 (2004), pp. 167–170.

9. "Value of recreational sports on college campuses," National Intramural-Recreational Sports Association (2002).

10. Adrian C. North et al., "The effect of musical style on restaurant consumers' spending," *Environment and Behavior*, 35 (2003), pp. 712–718.

11. Based on Jack K. Trammel, "The impact of academic accommodations on final grades in a post secondary setting," *Journal of College Reading and Learning*, 34 (2003), pp. 76–90.

12. Christie N. Scollon et al., "Emotions across cultures and methods," *Journal of Cross-cultural Psychology*, 35 (2004), pp. 304–326.

13. We thank Ethan J. Temeles of Amherst College for providing the data. His work is described in Ethan J. Temeles and W. John Kress, "Adaptation in a plant-hummingbird association," *Science*, 300 (2003), pp. 630–633.

14. The data were provided by James Kaufman. The study is described in James C. Kaufman, "The cost of the muse: poets die young," *Death Studies*, 27 (2003), pp. 813–821. The quote from Yeats appears in this article.

15. The experiment was performed in Connie Weaver's lab in the Purdue University Department of Foods and Nutrition. The data were provided by Berdine Martin and Yong Jiang.

16. Samuel S. Kim and Jerome Agrusa, "Segmenting Japanese tourists to Hawaii according to tour purpose," *Journal of Travel and Tourism Marketing*, 24 (2008), pp. 63–80.

17. Woo Gon Kim et al., "Influence of institutional DINESERV on customer satisfaction, return intention, and word-of-mouth," *International Journal of Hospitality Management*, 28 (2009), pp. 10–17.

18. Data provided by Jo Welch of the Purdue University Department of Foods and Nutrition.

19. Based on A. A. Adish et al., "Effect of consumption of food cooked in iron pots on iron status and growth of young children: A randomised trial," *The Lancet*, 353 (1999), pp. 712–716.

20. Steve Badylak et al., "Marrow-derived cells populate scaffolds composed of xenogeneic extracellular matrix," *Experimental Hematology*, 29 (2001), pp. 1310–1318.

21. This exercise is based on data provided from a study conducted by Jim Baumann and Leah Jones of the Purdue University School of Education.

Chapter 13

1. See `www.who.int/topics/malaria/en/` for more information about malaria.

2. This example is based on a 2009 study described at `clinicaltrials.gov/ct2/show/NCT00623857`

3. We present the two-way ANOVA model and analysis for the general case in which the sample sizes may be unequal. If the sample sizes vary a great deal, serious complications can arise. There is no longer a single standard ANOVA analysis. Most computer packages offer several options for the computation of the ANOVA table when cell counts are unequal. When the counts are approximately equal, all methods give essentially the same results.

4. Sara N. Bleich et al., "Increasing consumption of sugar-sweetened beverages among US adults: 1988-1994 to 1999-2004," *The American Journal of Clinical Nutrition,* 89 (2009), pp. 372–381.

5. Rick Bell and Patricia L. Pliner, "Time to eat: The relationship between the number of people eating and meal duration in three lunch settings," *Appetite,* 41 (2003), pp. 215–218.

6. Karolyn Drake and Jamel Ben El Hine, "Synchronizing with music: Intercultural differences," *Annals of the New York Academy of Sciences,* 99 (2003), pp. 429–437.

7. Example 13.10 is based on a study described in P. D. Wood et al., "Plasma lipoprotein distributions in male and female runners," in P. Milvey (ed.), *The Marathon: Physiological, Medical, Epidemiological, and Psychological Studies,* New York Academy of Sciences, 1977.

8. Vincent P. Magnini and Kiran Karande, "The influences of transaction history and thank you statements in service recovery," *International Journal of Hospitality Management,* 28 (2009), pp. 540–546.

9. Brian Wansink et al., "The office candy dish: proximity's influence on estimated and actual consumption," *International Journal of Obesity,* 30 (2006), pp. 871–875.

10. Annette N. Senitko et al., "Influence of endurance exercise training status and gender on postexercise hypotension," *Journal of Applied Physiology,* 92 (2002), pp. 2368-2374.

11. Willemijn M. van Dolen, Ko de Ruyter, and Sandra Streukens. "The effect of humor in electronic service encounters," *Journal of Economic Psychology,* 29 (2008), pp. 160–179.

12. Jane Kolodinsky et al., "Sex and cultural differences in the acceptance of functional foods: A comparison of American, Canadian, and French college students," *Journal of American College Health,* 57 (2008), pp. 143–149.

13. Judith McFarlane et al., "An intervention to increase safety behaviors of abused women," *Nursing Research,* 51 (2002), pp. 347–354.

14. Gad Saad and John G. Vongas, "The effect of conspicuous consumption on menís testosterone levels," *Organizational Behavior and Human Decision Processes,* 110 (2009), pp. 80–92.

15. Klaus Boehnke et al., "On the interrelation of peer climate and school performance in mathematics: A German-Canadian-Israeli comparison of 14-year-old school students," in B. N. Setiadi, A. Supratiknya, W. J. Lonner, & Y. H. Poortinga (Eds.) *Ongoing themes in psychology and culture* (Online Ed.), International Association for Cross-Cultural Psychology.

16. Data provided by Julie Hendricks and V. J. K. Liu of the Department of Foods and Nutrition, Purdue University.

17. Results of the study are reported in P. F. Campbell and G. P. McCabe, "Predicting the success of freshmen in a computer science major," *Communications of the ACM,* 27 (1984), pp. 1108–1113.

18. Tamar Kugler et al., "Trust between individuals and groups: Groups are less trusting than individuals but just as trustworthy," *Journal of Economic Psychology,* 28 (2007), pp. 646–657.

19. Based on A. A. Adish et al., "Effect of consumption of food cooked in iron pots on iron status and growth of young children: A randomised trial," *Lancet,* 353 (1999), pp. 712–716.

20. Based on a problem from Renée A. Jones and Regina P. Becker, Department of Statistics, Purdue University.

21. For a summary of this study and other research in this area, see Stanley Coren and Diane F. Halpern, "Left-handedness: A marker for decreased survival fitness," *Psychological Bulletin,* 109 (1991), pp. 90–106.

22. Data provided by Neil Zimmerman of the Purdue University School of Health Sciences.

23. See I. C. Feller et al. "Sex-biased herbivory in Jack-in-the-pulpit (*arisaema triphyllum*) by a specialist thrips (*Heterothrips arisaemae*)," in *Proceedings 7th International Thysanoptera Conference,* Reggio Callabrio, Italy, pp. 163–172.

PHOTO CREDITS

Chapter 1

UNP1.1 (An iPod Nano 5G) GLEN ARGOV/Landov
UNP1.2 (Students in classroom) Alamy
UNP1.3 (Garmin GPS) Garmin Ltd.
UNP1.4 (People answering phones at a call center): Alamy
UNP1.5 (Young women in a line with various heights): Alamy
UNP1.6 (Meghan Duggan (7) of the University of Wisconsin holds up the championship trophy): Adam Hunger/NCAA Photos
UNP1.7 (The STubHub! Logo): © 2000-2010 StubHub, Inc. All rights reserved.

Chapter 2

UNP2.1 (Student sleeping) iStockphoto
UNP2.2 (A Starbucks Mocha Frappuccino) Alamy
UNP2.3 (Filling out loan application) David Young-Wolff/Photo Edit
UNP2.4 (An Olympic figure skater) TORU YAMANAKA/AFP/Getty Images
UNP2.5 (Young-woman-with-diabetes-checking-her-blood-sugar) iStockphoto
UNP2.6 (Power lines) Alamy

Chapter 3

UNP3.1 (H1N1 vaccine) © FILIP SINGER/epa/Corbis
UNP3.2 (Apple juice) iStockphoto
UNP3.3 (Watching television) Alamy
UNP3.4 (Cellphone) iStockphoto
UNP3.5 (Young students reading) iStockphoto
UNP3.6 (People voting) Alamy
UNP3.7 (Least flycatcher birds migrating): © Brian E. Small/VIREO

Chapter 4

UNP4.1 (Diana Taurasi shooting a free throw) AP Photo/Al Goldis
UNP4.2 (College student(s) studying) iStockphoto
UNP4.3 (Hand with two aces) iStockphoto
UN4.4 (Peas) iStockphoto
UNP4.5 (Someone texting while driving) iStockphoto
UNP4.6 (Someone sleeping; someone exercising) © Randy Faris/Corbis

Chapter 5

UNP5CO (Picture of student pulling late nighter) © Macdougy | Dreamstime.com
UNP5.1 (Person answering a service call for a financial institution) Howard Grey/GettyImages
UNP5.2 (Text messaging) istockphoto
UNP5.3 (Pressure parents put on students) BananaStock/Picturequest
UNP5.4 (Tax audits of business records) Alamy
UNP5.5 (Buying clothes) © Michael Newman / PhotoEdit

Chapter 6

UNP6.1 (Credit card bill shock) Alamy
UNP6.2 (Forest from above) USDA Forest Service Archives
UNP6.3 (Student fretting over SAT exam) © Gary Conner/PhotoEdit
UNP6.4 (Various credit cards) Alamy
UNP6.5 (Water analysis.) Olivier Voisin/Photo Researchers
UNP6.6 (Ball bearings for skates) Photo by The Photo Works

Chapter 7

UNP7.1 (Moon) Steve Satushek/Getty
UNP7.2 (Young women using a cell phone) Alamy
UNP7.3 (Young person looking at alarm clock in the middle of the night) iStockphoto
UNP7.4 (iPod songs being shuffled.) Courtesy powerbooktrance
UNP7.5 (Group of students (10) standing in line) Robert Warren/Getty
UNP7.6 (Four girls and boys looking at the same textbook) Getty Images/Digital Vision

Chapter 8

UNP8.1 (Couple playing computer game) Alamy
UNP8.2 (Woman eating noodles with soy sauce) Photolibrary
UNP8.3 (Woman applying sunscreen) Alamy
UNP8.4 (Pilates class) iStockphoto

Chapter 9

UNP9.1 (Boy and girl with popcorn frightened at movie theater) Alamy

UNP9.2 (Close up of a girl putting salt on her fries) Alamy

UNP9.3 (Three happy young women working out on exercise bicycle at the gym) Alamy

UNP9.4 (Woman eating fruit salad) Alamy

Chapter 10

UNP10.1 (Tornado) © Chuck Doswell / Visuals Unlimited

UNP10.2 (Close up of pedometer) Alamy

UNP10.2B (Students walking) istockphoto

UNP10.3 (1997 Pontiac Transport Van) © General Motors Corp. Used with permission. GM Media Archives

UNP10.4 (Bone density scan legs) Custom Medical Stock Photo

Chapter 11

UNP11.1 (Students in graduation cap and gown) Bill Losh/Getty

UNP11.2 (Tipping in Canada) Chris Spavins

Chapter 12

UNP12.1 (Facebook) Michael Newman / Photo Edit

UNP12.2 (The video game Gears of War) Mario Tama/Getty Images

UNP12.3 (Cafe, Montreal, Quebec, Canada) Alamy

UNP12.5 (Loreal ad featuring Penelope Cruz) The Advertising Archives

UNP12.6 (Picture of Douglas fir seedlings) © Gary Braasch/CORBIS

Chapter 13

UNP13.1 (Different packed Xbox 360 games) Alamy

UNP13.2 (Zinc dietary supplement) Alamy

UNP13.3 (Osteoporosis. False-color scanning electron micrograph) Professor Pietro M. Motta / Photo Researchers

UNP13.5 (People eating together) © Banana Stock/ Agefotostock

UNP13.4 (Assorted drinks) Alamy

Chapter 14

UNP14.1 Photodisc

UNP14.2 (Playing video games) Alamy

UNP14.3 (Moviegoers) Alamy

UNP14.4 (Aphid adults and nymphs) Rod Planck / Photo Researchers

Chapter 15

UNP15.1 (Lambsquarters) Nigel Cattlin/Photo Researchers

UNP15.2 (Hotel and Casino The Bellagio) DWC/Alamy

UNP15.3 (Parents reading to a child) © Iofoto/Dreamstime

UNP15.4 (2010 NCAA Women's Golf Championship) Photo by Jason Barnette; courtesy of Purdue University

Chapter 16

UNP16.1 (Telephone technician) Alamy

UNP16.2 (Seattle Suburbs) Stuart Kelly/Alamy

UNP16.3 ($330 million Mega Millions jackpot) AP Photo/David Gard

UNP16.4 (Los Angeles Dodgers pitcher Hong-Chih Kuo, of Taiwan) AP Photo/Danny Moloshok

UNP16.5 (Full moon) istockphoto

UNP16.6 (Drawing of people) Dreamstime

Chapter 17

UNP17.1 © A. T. Willett/Alamy

UNP17.2 Michael Rosenfeld/Getty

UNP17.3 Alamy

UNP17.4 © Jeff Greenberg/The Image Works

UNP17.5 David Young-Wolff/Photo Edit

INDEX

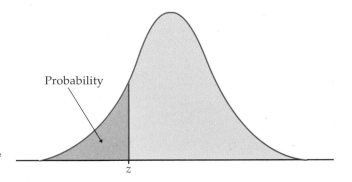

Table entry for z is the area under the standard Normal curve to the left of z.

Probability

z

TABLE A

Standard Normal probabilities

z	.00	.01	.02	.03	.04	.05	.06	.07	.08	.09
−3.4	.0003	.0003	.0003	.0003	.0003	.0003	.0003	.0003	.0003	.0002
−3.3	.0005	.0005	.0005	.0004	.0004	.0004	.0004	.0004	.0004	.0003
−3.2	.0007	.0007	.0006	.0006	.0006	.0006	.0006	.0005	.0005	.0005
−3.1	.0010	.0009	.0009	.0009	.0008	.0008	.0008	.0008	.0007	.0007
−3.0	.0013	.0013	.0013	.0012	.0012	.0011	.0011	.0011	.0010	.0010
−2.9	.0019	.0018	.0018	.0017	.0016	.0016	.0015	.0015	.0014	.0014
−2.8	.0026	.0025	.0024	.0023	.0023	.0022	.0021	.0021	.0020	.0019
−2.7	.0035	.0034	.0033	.0032	.0031	.0030	.0029	.0028	.0027	.0026
−2.6	.0047	.0045	.0044	.0043	.0041	.0040	.0039	.0038	.0037	.0036
−2.5	.0062	.0060	.0059	.0057	.0055	.0054	.0052	.0051	.0049	.0048
−2.4	.0082	.0080	.0078	.0075	.0073	.0071	.0069	.0068	.0066	.0064
−2.3	.0107	.0104	.0102	.0099	.0096	.0094	.0091	.0089	.0087	.0084
−2.2	.0139	.0136	.0132	.0129	.0125	.0122	.0119	.0116	.0113	.0110
−2.1	.0179	.0174	.0170	.0166	.0162	.0158	.0154	.0150	.0146	.0143
−2.0	.0228	.0222	.0217	.0212	.0207	.0202	.0197	.0192	.0188	.0183
−1.9	.0287	.0281	.0274	.0268	.0262	.0256	.0250	.0244	.0239	.0233
−1.8	.0359	.0351	.0344	.0336	.0329	.0322	.0314	.0307	.0301	.0294
−1.7	.0446	.0436	.0427	.0418	.0409	.0401	.0392	.0384	.0375	.0367
−1.6	.0548	.0537	.0526	.0516	.0505	.0495	.0485	.0475	.0465	.0455
−1.5	.0668	.0655	.0643	.0630	.0618	.0606	.0594	.0582	.0571	.0559
−1.4	.0808	.0793	.0778	.0764	.0749	.0735	.0721	.0708	.0694	.0681
−1.3	.0968	.0951	.0934	.0918	.0901	.0885	.0869	.0853	.0838	.0823
−1.2	.1151	.1131	.1112	.1093	.1075	.1056	.1038	.1020	.1003	.0985
−1.1	.1357	.1335	.1314	.1292	.1271	.1251	.1230	.1210	.1190	.1170
−1.0	.1587	.1562	.1539	.1515	.1492	.1469	.1446	.1423	.1401	.1379
−0.9	.1841	.1814	.1788	.1762	.1736	.1711	.1685	.1660	.1635	.1611
−0.8	.2119	.2090	.2061	.2033	.2005	.1977	.1949	.1922	.1894	.1867
−0.7	.2420	.2389	.2358	.2327	.2296	.2266	.2236	.2206	.2177	.2148
−0.6	.2743	.2709	.2676	.2643	.2611	.2578	.2546	.2514	.2483	.2451
−0.5	.3085	.3050	.3015	.2981	.2946	.2912	.2877	.2843	.2810	.2776
−0.4	.3446	.3409	.3372	.3336	.3300	.3264	.3228	.3192	.3156	.3121
−0.3	.3821	.3783	.3745	.3707	.3669	.3632	.3594	.3557	.3520	.3483
−0.2	.4207	.4168	.4129	.4090	.4052	.4013	.3974	.3936	.3897	.3859
−0.1	.4602	.4562	.4522	.4483	.4443	.4404	.4364	.4325	.4286	.4247
−0.0	.5000	.4960	.4920	.4880	.4840	.4801	.4761	.4721	.4681	.4641

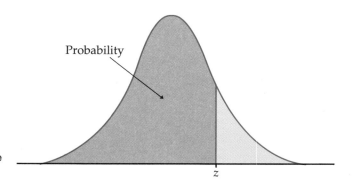

Table entry for z is the area under the standard Normal curve to the left of z.

Probability

z

Standard Normal probabilities (continued)

z	.00	.01	.02	.03	.04	.05	.06	.07	.08	.09
0.0	.5000	.5040	.5080	.5120	.5160	.5199	.5239	.5279	.5319	.5359
0.1	.5398	.5438	.5478	.5517	.5557	.5596	.5636	.5675	.5714	.5753
0.2	.5793	.5832	.5871	.5910	.5948	.5987	.6026	.6064	.6103	.6141
0.3	.6179	.6217	.6255	.6293	.6331	.6368	.6406	.6443	.6480	.6517
0.4	.6554	.6591	.6628	.6664	.6700	.6736	.6772	.6808	.6844	.6879
0.5	.6915	.6950	.6985	.7019	.7054	.7088	.7123	.7157	.7190	.7224
0.6	.7257	.7291	.7324	.7357	.7389	.7422	.7454	.7486	.7517	.7549
0.7	.7580	.7611	.7642	.7673	.7704	.7734	.7764	.7794	.7823	.7852
0.8	.7881	.7910	.7939	.7967	.7995	.8023	.8051	.8078	.8106	.8133
0.9	.8159	.8186	.8212	.8238	.8264	.8289	.8315	.8340	.8365	.8389
1.0	.8413	.8438	.8461	.8485	.8508	.8531	.8554	.8577	.8599	.8621
1.1	.8643	.8665	.8686	.8708	.8729	.8749	.8770	.8790	.8810	.8830
1.2	.8849	.8869	.8888	.8907	.8925	.8944	.8962	.8980	.8997	.9015
1.3	.9032	.9049	.9066	.9082	.9099	.9115	.9131	.9147	.9162	.9177
1.4	.9192	.9207	.9222	.9236	.9251	.9265	.9279	.9292	.9306	.9319
1.5	.9332	.9345	.9357	.9370	.9382	.9394	.9406	.9418	.9429	.9441
1.6	.9452	.9463	.9474	.9484	.9495	.9505	.9515	.9525	.9535	.9545
1.7	.9554	.9564	.9573	.9582	.9591	.9599	.9608	.9616	.9625	.9633
1.8	.9641	.9649	.9656	.9664	.9671	.9678	.9686	.9693	.9699	.9706
1.9	.9713	.9719	.9726	.9732	.9738	.9744	.9750	.9756	.9761	.9767
2.0	.9772	.9778	.9783	.9788	.9793	.9798	.9803	.9808	.9812	.9817
2.1	.9821	.9826	.9830	.9834	.9838	.9842	.9846	.9850	.9854	.9857
2.2	.9861	.9864	.9868	.9871	.9875	.9878	.9881	.9884	.9887	.9890
2.3	.9893	.9896	.9898	.9901	.9904	.9906	.9909	.9911	.9913	.9916
2.4	.9918	.9920	.9922	.9925	.9927	.9929	.9931	.9932	.9934	.9936
2.5	.9938	.9940	.9941	.9943	.9945	.9946	.9948	.9949	.9951	.9952
2.6	.9953	.9955	.9956	.9957	.9959	.9960	.9961	.9962	.9963	.9964
2.7	.9965	.9966	.9967	.9968	.9969	.9970	.9971	.9972	.9973	.9974
2.8	.9974	.9975	.9976	.9977	.9977	.9978	.9979	.9979	.9980	.9981
2.9	.9981	.9982	.9982	.9983	.9984	.9984	.9985	.9985	.9986	.9986
3.0	.9987	.9987	.9987	.9988	.9988	.9989	.9989	.9989	.9990	.9990
3.1	.9990	.9991	.9991	.9991	.9992	.9992	.9992	.9992	.9993	.9993
3.2	.9993	.9993	.9994	.9994	.9994	.9994	.9994	.9995	.9995	.9995
3.3	.9995	.9995	.9995	.9996	.9996	.9996	.9996	.9996	.9996	.9997
3.4	.9997	.9997	.9997	.9997	.9997	.9997	.9997	.9997	.9997	.9998